Ecology of
Insects

CONCEPTS AND
APPLICATIONS

Martin R. Speight
Mark D. Hunter
Allan D. Watt

WILEY-BLACKWELL

A John Wiley & Sons, Ltd., Publication

This edition first published 2008, © 2008 by Martin R. Speight, Mark D. Hunter and Allan D. Watt

Blackwell Publishing was acquired by John Wiley & Sons in February 2007. Blackwell's publishing program has been merged with Wiley's global Scientific, Technical and Medical business to form Wiley-Blackwell.

Registered office: John Wiley & Sons Ltd, The Atrium, Southern Gate, Chichester, West Sussex, PO19 8SQ, UK

Editorial offices: 9600 Garsington Road, Oxford, OX4 2DQ, UK
    The Atrium, Southern Gate, Chichester, West Sussex, PO19 8SQ, UK
    111 River Street, Hoboken, NJ 07030-5774, USA

For details of our global editorial offices, for customer services and for information about how to apply for permission to reuse the copyright material in this book please see our website at www.wiley.com/wiley-blackwell

Library of Congress Cataloguing-in-Publication Data
Data available

ISBN: 978-1-4051-3114-8

A catalogue record for this book is available from the British Library.

Set in 9/11 Photina
by SPi Publisher Services, Pondicherry, India
Printed in Singapore by COS Pte

1   2008

CONTENTS

Colour plate section appears between pp. 310 and 311

PREFACE TO SECOND EDITION

The first edition of this book was published in 1999. In the 9 years since then, many changes have occurred in the worlds of ecology and entomology. First, and very sadly, the author of the foreword of our first edition, Professor Sir Richard Southwood, died in 2005. Dick was an eminent taxonomist, a classical ecologist and population dynamisist, a great diplomat and leader, and a guiding light to many students, academics, and authors alike.

Second, major advances have been made in our understanding of the ecology and evolution of insects, especially in the fields of molecular evolution and systematics. We have completely rewritten much of this second edition to take account of recent changes in these fields. For example, who would have thought back in 1999 that the comfortable and to many, satisfying, place in the Arthropoda occupied by the insects would be considered very different less than a decade later? When the first edition was published, the Insecta sat neatly in the Hexapoda as part of the subphylum Uniramia, along with their clear close relatives the Myriapoda (surely, insects are just milli-pedes or centipedes with fewer legs?). However, tech-niques in molecular biology have revealed previously unrecognized relationships among taxa, details about evolutionary pathways, and resulting taxonomic associations that have overthrown our long-accepted comforts. As we explain in detail in Chapter 1, the Insecta now belong in the major group known as the Pan-Crustacea along with the true Crustacea as a sister group. The unfortunate Myriapods appear to be closely related to no-one.

The past decade has seen almost universal accept-ance by the scientific community of the role that humans are playing in global climate change and our second edition reflects that acceptance. For ecologists, climate change represents one of the most important topics for debate and research that has arisen this century. Entomologists, too, have a clear role to play in predicting changes in the distribution and abundance of plant pests and disease vectors. As air currents, precipitation, and temperature change, many insect species will be lost from current geo-graphic areas and gained in others. In Chapter 2, we describe how speckled wood butterflies are moving northwards in the UK while skipper butterflies are moving northward in the USA. What about insect diversity? Despite uncertainty about the future climate, there is concern among some scientists that biodiversity losses could be serious (see Chapter 10). Changing distributions of pest species are already apparent. The western corn rootworm and the Colorado potato beetle are moving further and further into northern Europe (see Chapter 12), while remote areas in Alaska are experiencing range expansions of agricultural and forest pests.

There have also been major advances in the applied entomological fields of disease epidemiology and pest management since the first edition of this book. Emerging diseases vectored by insects such as West Nile virus have gained prominence, whereas old ones, such as malaria and barley yellow dwarf virus, have become even more deleterious despite many attempts at control. Chapter 11 describes in detail how both plant and animal (including human) diseases have continued to proliferate because of the actions of insects as vectors or predispositioning agents, and how keen efforts are being made to reduce or even prevent disease incidence by insect popul-ation management. Newer techniques in vector management are based largely in molecular biology. An example is the manipulation of the mosquito genome to render it immune to the malarial protozoa *Plasmodium*, thus removing the ability of mosquitoes to vector the pathogenic agent. Similarly, modern crop pest management is becoming more and more reliant on genetic techniques. While continuing to refine and improve the efficiency of traditional methods in biological and chemical control, pest managers are now increasingly exploring ways to manipulate crop genomes to outwit insect pests. Chapter 12 discusses the pros and cons of genetic modification of crops to express genes of the highly efficient insecticidal bacterium, *Bacillus thuringiensis*. Like it or not, this technology has a vital role to play in the future of global food production.

The power of evolution by natural selection continues to structure much of what we see in the biological world, and we now know considerably more about its strengths and limitations than we did a decade ago. As we discuss in Chapter 6, it seems that that precise matching of insect herbivore and plant phylogenies might be rarer than we once thought. At the same time, students of insect evolution continue to provide us with wonderful new discoveries (or old discoveries recently brought to the notice of authors such as ourselves). One of the most exciting new inclusions in the second edition of this book is the story of the barnacle predator, *Oedoparena*, described in Chapter 6. This insect has managed to get its own back on what we are now told are its closest relatives, the Crustacea. While not being marine in the true sense of the term, the larvae of this insect are able to kill and eat a truly marine animal, when said animal has the audacity to try to invade a realm of the insects, in this case, the littoral zone of the North American intertidal.

We would like to acknowledge the assistance, advice, help, and support of our families, friends,

and colleagues in the production of this new edition. Two anonymous reviewers provided a wealth of comments. Generations of Oxford undergraduates read some of the chapters, in particular numerous (in fact annual) updates of Chapter 12. Graduate students at the University of Georgia helped to remove (unwanted) bugs from several chapters. Clive Hambler, Peter Henderson, Angela Speight, Mary-Carol Hunter, and Katy Watt were always there for us with advice, support, and in some cases, refreshments. Peter Holland, David Ferrier, Charlie Gibson, Dave Thompson, Charles Godfray, Angela Douglas, and George McGavin advised us on aspects of ecology and entomology. Jane Andrew, Rosie Hayden, and Ward Cooper from Blackwell Publishing got the product to the bookshelves.

Grateful acknowledgement is made to all publishers who granted us permission to use tables, graphs, and photos from journals, books, and websites. In particular, we thank Wiley-Blackwell Publishing, Elsevier, Springer Science and Business Media, Ecological Society of America, Entomological Society of America, Cambridge University Press, National Academy of Sciences of the United States of America, American Association for the Advancement of Science, Inter-Research Science Centre, Japanese Society of Applied Entomology and Zoology, Annual Reviews Organisation, American Medical Association, Ingenta Publishers, Company of Biologists Ltd., and Utah State University. We have tried very hard to seek permissions from all concerned. If we have failed on any occasion, we apologize, and will rectify the situation in future editions. All photos were taken by Martin Speight unless otherwise acknowledged.

Martin Speight, Oxford
Mark Hunter, Michigan
Allan Watt, Edinburgh
August 2007

FOREWORD TO FIRST EDITION

Today, it is not necessary to stress the significance of ecology or its place in the world. The authors of this volume have provided a succinct overview of the subject: both its fundamental principles and the application of these to matters affecting human welfare. They have limited their canvas to insects, but these provide excellent models. Without in any way decrying other branches of the subject, one can point out that in so many respects insects provide the mid-point optimum on the spectrum between the study of mammals and birds on the one hand and that of microbes on the other. They may be seen with the naked eye, yet their generation times are sufficiently short and the sizes of their populations sufficiently large to permit the quantitative study of population dynamics on time and spatial scales appropriate for students—undergraduate and postgraduate.

Insects have other claims on our attention. The vast number of species make them the major contributor to biodiversity in the great majority of habitats other than the sea. From a more anthropocentric view, one must also recognize that they are our major competitors taking about 10% of the food that we grow and infecting one in six of the world's population with a pathogen. All these facets are explored in this volume.

The authors bring a diversity of experience to the task of encapsulating the essence of the subject in a compact book. A glance at the colour plate illustrations shows the range of insects and habitats that have interested the principal author. From their hands-on experience in the field the authors bring a freshness to even the most mathematical of concepts. Not surprisingly the text emphasizes the truth that 'it all starts with good observations in the field'. As this book shows, as well as providing a good exemplar of a key subject for today's world, insect ecology is also fun.

T.R.E. Southwood
October 1998

Chapter 1

AN OVERVIEW OF INSECT ECOLOGY

1.1 INTRODUCTION

In this first chapter we provide a brief overview of the major concepts in insect ecology, and attempt to present a taste of what is to come in the 11 detailed chapters that follow. Unavoidably, a little repetition may therefore occur, in that topics briefly discussed in this chapter will appear again in more detail elsewhere in the book. This overlap is entirely intentional on our part, and we would encourage the random reader to explore this introductory chapter first, and then turn to the details in whichever subject and later chapter takes his or her fancy. Readers who know what they are looking for can, of course, proceed directly to the relevant chapter(s).

We have assumed that most readers will have some knowledge of insect taxonomy, and we have not attempted to provide an in-depth coverage of this topic. Some information is, however, provided to set the scene. Readers who wish to know more are recommended to obtain one of various excellent entomology textbooks such as that by Gullan and Cranston (2005).

1.2 HISTORY OF ECOLOGY AND ENTOMOLOGY

The science of ecology is broad ranging and difficult to define. Most of us think we know what it means, and indeed it can imply different things to different people. It is best considered as a description of interactions between organisms and their environment, and its basic philosophy is to account for the abundance and distribution of these organisms. In fact, ecology encompasses a whole variety of disciplines, both qualitative and quantitative, whole organism, cellular and molecular, from behavior and physiology, to evolution and interactions within and between populations. In 1933, Elton suggested that ecology represented, partly at least, the application of scientific method to natural history. As a science, Elton felt that ecology depended on three methods of approach: field observations, systematic techniques, and experimental work both in the laboratory and in the field. These three basic systems still form the framework of ecology today, and it is the appropriate integration of these systems, which provides our best estimates of the associations of living organisms with themselves and their environment, which we call ecology.

Insects have dominated the interests of zoologists for centuries. Those who could drag themselves away from the charismatic but species-poor vertebrates soon found the ecology of insects to be a complex and rich discipline, which has fascinated researchers for at least 200 years. At first, the early texts were mainly descriptive of the wonders of the insect world. Starting in 1822, Kirby and Spence published their four volumes of *An Introduction to Entomology*. This was a copious account of the lives of insects, written over three decades before Darwin first produced *On the Origin of Species* in 1859. The fact that Kirby and Spence did not have the benefit of explaining their observations on the myriad interactions of insects and their environment as results of evolution did not detract from the clear fascination that the world of insects provided. This fascination has withstood the test of time. Half a century or more after Kirby and Spence, Fabre published insightful and exciting accounts of the lives of insects, starting before the

Table 1.1 Summary of associations in ecology.

Type	Interaction level	Insect example
Competition		
Intraspecific	Individuals within a population of one species	Lepidoptera larvae on trees
Interspecific	Individuals within populations of two different species	Bark beetles in tree bark
Herbivory, or phytophagy	Between autotroph and primary consumer	Aphids on roses
Symbiosis		
Mutualism	Between individuals of two species	Ants and fungi
Sociality	Between individuals of one species	Termites
Predation	Between primary and secondary consumer trophic levels	Mantids and flies
Parasitism	Between primary and secondary consumer trophic levels	Ichneumonid wasps and sawflies

turn of the 20th century (Fabre 1882), wherein he presented in great detail observations made over many years of the ecologies of dung beetles, spider-hunting wasps, cicadas, and praying mantids. These books epitomized the first of Elton's three approaches to ecology, that of field observations. Later work began to enhance the third approach, that of experimental insect ecology, summarized in *Insect Population Ecology* by Varley et al. (1973). This later book was probably the first and certainly most influential, concise account of insect ecology derived from quantitative, scientific research.

1.3 ECOLOGICAL ASSOCIATIONS

The term "association" used by Elton covers a great many types of interaction, including those between individuals of the same species, between species in the same trophic level (a trophic level is a position in a food web occupied by organisms having the same functional way of gaining energy), and between different trophic levels. These associations may be mutually beneficial, or, alternatively, involve the advancement or enhancement of fitness of certain individuals at the expense of others. Of course, insects can live in association with other organisms without having any influence on them, or being influenced by them; such associations are "neutral".

Ecological associations among insects operate at one of three levels: at the level of individual organisms,

at a population level, or over an entire community. Table 1.1 summarizes some of these basic associations. With the exception of primary production, it should be possible to discover insect examples for all other kinds of ecological interactions. The fact that insects are so widespread and diverse in their ecological associations is not surprising when it is considered just how many of them there are, at least on land and in fresh water, how adaptable they are to changing and novel environmental conditions, and how long they have existed in geological time.

1.4 THE INSECTA

1.4.1 Structure

This is not a book about insect morphology or physiology in the main, but both are certainly worthy of some consideration, as the ability of an organism to succeed in its environment is dictated by form and function. Here, we merely provide a few basic details.

The basic structure of a typical adult insect is shown in Figure 1.1 (CSIRO 1979). Comparing the figure with a photograph of a grasshopper (Figure 1.2) will enable any new student of entomology to quickly realize how insects evolved a highly technologically efficient set of specialized body parts and appendages. The three basic sections (called tagmata) of an insect's body are admirably adapted for different purposes. The head specializes in sensory reception and food

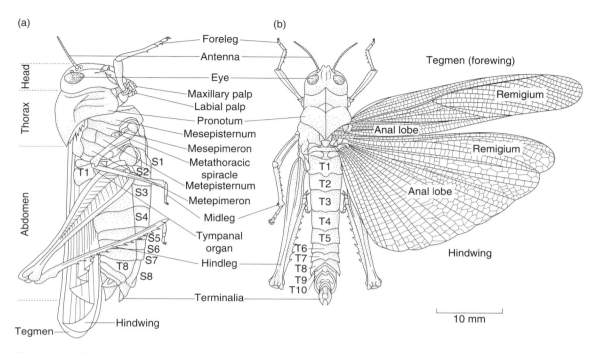

(a)   (b)

Head
Thorax
Abdomen

Foreleg
Antenna
Eye
Maxillary palp
Labial palp
Pronotum
Mesepisternum
Mesepimeron
Metathoracic spiracle
Metepisternum
Metepimeron
Midleg
Tympanal organ
Hindleg
Terminalia
Hindwing

T1
S1
S2
S3
S4
S5
S6
S7
S8
T8

Tegmen

Tegmen (forewing)
Remigium
Anal lobe
Remigium
Anal lobe
Hindwing

T1
T2
T3
T4
T5
T6
T7
T8
T9
T10

10 mm

Figure 1.1 The external structure of an adult insect, illustrating general features of a non-specialized species. (From CSIRO 1979.)

gathering, the thorax in locomotion, and the abdomen in digestion and reproduction. All but a minimum number of appendages have been lost when compared with ancestors, leaving a set of highly adapted mouthparts and a pair of immensely stable tripods, the legs. Throughout the book, we shall refer to these and

Figure 1.2 Side view of grasshopper showing general structure.

other body structures in terms of their evolution and ecology. This basic plan is of course highly variable, and the most specialized insects, for example, blowfly larvae (the fisherman's maggot), bear little or no superficial similarity to this plan.

1.4.2 Taxonomy

We hope that the reader does not get too bogged down with taxonomic nomenclature or classification. We are concerned with what insects do rather than what their names are, and although it is essential that detailed taxonomy is eventually carried out as part of an ecological investigation, in this book we have provided scientific names more as a reference system for comparison with other published work than as something people should automatically learn. However, some introduction to the taxonomic relationships of the insects will help to set the scene.

The class Insecta belongs within the superclass Hexapoda, within the phylum Arthropoda. These in turn belong to a huge group or clade of invertebrates known as the Ecdysozoa – animals that grow by shedding their cuticles (molting) (Mallatt & Giribet 2006). The Ecdysozoa bring together perhaps unlikely relatives, the nematodes and the arthropods for instance. The arthropod phylum itself is now satisfactorily monophyletic, having survived various attempts over the last few decades to invoke multiple origins for the admittedly hugely diverse group. The place within the arthropods for the insects, and their relationships to other major groups, has also had a rocky ride. For many years, the first edition of this book included, the hexapods containing the insects were placed with the Myriapoda in the Uniramia (otherwise known as the Atelocerata; Mallatt et al. 2004) (Figure 1.3a, b). Now, however, this convenient arrangement has been superseded by techniques involving nuclear and mitochondrial ribosomal RNAs and protein coding genes, with the most likely position for the Insecta shown in Figure 1.3c, d. The crucial point of the new phylogeny is that the Hexapods (most of which are insects) are relatively closely related to the Crustacea (in other words, Hexapods are terrestrial Crustacea), forming an all-encompassing group, the Pancrustacea (or Tetraconata; Richter 2002), whilst the Myriapods are separate, probably much more closely allied to the Chelicerates (Schultz & Regier 2000). The Pancrustacea makes a lot of sense ecologically, since insects and crustaceans have many similar lifestyles, niches, and so on; one just happens to be in the sea and the other on land (but see Chapter 6). However, controversy still exists; Bitsch and Bitsch (2004) for example found no support for the Pancrustacea concept using comparative morphological characters rather than molecular techniques.

Less controversial is the organization and relatedness of groups within the hexapods. Though the Insecta are by far the most numerous, they share the superclass with three other classes, each of which comprises just one order each, the Diplura, Protura, and Collembola (springtails) (Gullan & Cranston 2005). These latter three were until fairly recently included amongst the apterygote (wingless) insects, but their morphological and physiological features are more likely to mimic those of insects by virtue of convergent evolution than by true relatedness.

The taxonomic classification of the Insecta used in this book follows that described by Gullan and Cranston (2005). There are two orders within the Apterygota (wingless insects): the Zygentoma (silverfish;

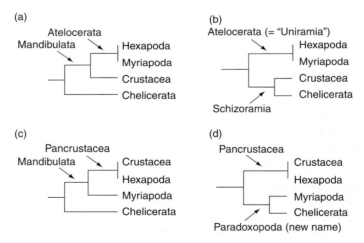

Figure 1.3 Major hypotheses of the relationships of arthropod groups: (a) classic Mandibulata (= Mandibulata + Chelicerata), in which the three groups possessing mandibles formed by the second post-oral appendage (hexapods, myriapods, crustaceans) are distinguished from chelicerates; (b) Atelocerata versus Schizoramia: clades with unbranched versus two-branched appendages; (c) Pancrustacea + Myriapoda within Mandibulata: that is, in the Mandibulata, hexapods group with crustaceans instead of with myriapods; (d) 'Chelicerata + Myriapoda' versus 'Crustacea + Hexapoda.' (From Mallatt et al. 2004.)

Figure 1.4 Silverfish (Thysanura).

Figure 1.4) and the Archaeognatha (bristletails). All other insects belong to the pterygote group (winged insects) and this is in turn divided into the Exopterygota (also known as the Hemimetabola), where wings develop gradually through several nymphal instars, and the Endopterygota (also known as the Holometabola), where there is usually a distinct larval stage separated from the adult by a pupa. Figure 1.5 summarizes the classification of insects, and indicates roughly the number of species so far described from each order. This great species richness within the Insecta is thought to have resulted from low extinction rates throughout their history (see below) (Labandeira & Sepkoski 1993), and it is important to realize that nearly 90% of insect species belong to the endopterygotes, indicating the overwhelming advantages for speciation provided by the specialized larval and pupal stages. As will be seen throughout the book, lay-people and entomologists alike tend to concentrate on adult insects, but the ecology of larval or nymphal stages may well have much more relevance to the success or otherwise of the species or order, as well as having much more direct impact on human life. After all, it is the larvae of peacock butterflies that devour their host plant, nettle, not the adults.

The success of a group of organisms is not just measured in terms of the number of species accrued in the group, but can also be discussed as the range of habitats or food types dealt with, the extremes of environments in which they are able to live, how long the group has been extant, and the relative abundance of individuals. Ecologically, it may be more useful to break down the insect orders into functional groups according to lifestyle or feeding strategies, rather than to merely count the number of species.

1.5 FOSSIL HISTORY AND INSECT EVOLUTION

The nature of insect bodies, at least the adults of both Exopterygota and Endopterygota and to some extent the nymphs of the Exopterygota, makes them very suitable for fossilization in a variety of preserving media from sediments to amber. The tough insect cuticle is composed of flexible chitin, and/or more rigid sclerotin. Chitin is certainly an ancient, widespread compound, known in animals from at least the Cambrian period, more than 550 million years ago (Ma) (Miller 1991), but estimating the origins of insects that use it to perfection is difficult and fraught with argument. These days, biologists employ molecular clocks to date events for which there is no physical evidence. With such rather imprecise tools, some remarkable claims of over 1 billion years ago have been made, for example, for when the groups of bilateral animals diversified (Graur & Martin 2004). Using more conservative and exacting molecular clocks, Peterson et al. (2004) have come up with a reasonable origin of the earliest arthropods, somewhere in the early to mid Cambrian, perhaps 540 Ma. Actual fossil hexapods, however, have not been found until more recently. *Rhyniella praecursor* is a fossil springtail (Collembola) found in Scottish sandstones of about 400 million years old (Fayers & Trewin 2005), and for many years this species has held the record for the oldest hexapod. Fragments of another fossil from the same rocks have now been identified as a "proper" insect (Engel & Grimaldi 2004). *Rhyniognatha hirsti* is not only the earliest true insect, but it may well also have had wings. This then places the origin of the Insecta somewhere in the Silurian period (417–443 Ma), arising from a common fairy shrimp-like ancestor (Gaunt & Miles 2002).

The fossil record of insects is therefore relatively complete when compared with that of most other animal groups, and fossil insects turn up extremely frequently in many types of material, from sandstones and shales to coal and amber. Amber is a particularly wonderful substance, a fossil tree resin that traps, entombs, and precisely preserves small particles such as seeds, spiders, and insects. Most amber is Cretaceous or Tertiary in origin, perhaps 30–90 million years old, and is particularly common in the Baltic and Siberia (see, for example, Arillo & Engel 2006), but parts of the Caribbean (Dominica) and Central and South America (Mexico to Peru) also have large

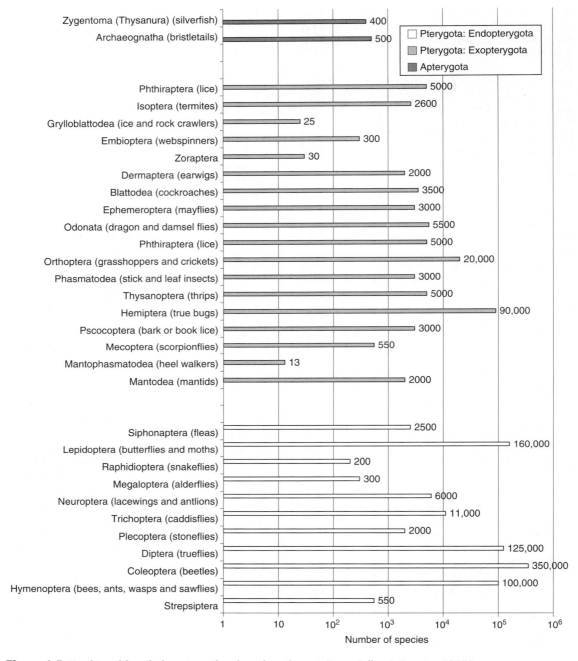

Figure 1.5 Numbers of described species within the orders of insect. (From Gullan & Cranston 2005.)

deposits (see, for example, Wichard et al. 2006). Amber preserves living things so well, in fact, that inferences can be made about whole fossil ecosystems. Antoine et al. (2006) examined amber from the Peruvian Amazon, and discovered diverse collections of fossil arthropods (13 insect families and three arachnid species), as well as large numbers of spores, pollen, and algae. These Middle Miocene deposits are indicative of lush, moist rainforests already in existence in the Amazonian region perhaps 10–18 Mya.

Table 1.2 shows the earliest known fossils from the major insect orders that are still extant today. Though wisdom might suggest that insects such as Odonata (dragonflies and damselflies) are primitive or early when compared with, for example, Hymenoptera (bees, ants, and wasps), there is little evidence from the fossil record of a clear progression or development from one group to the next. Even with modern molecular techniques such as nucleotide sequencing to investigate the relatedness of organisms, it is

Table 1.2 Fossil history of major insect orders alive today. (Data from Boudreaux 1987; Gullan & Cranston 1996.)

Order	Earliest fossils	Million years ago
Archaeognatha	Devonian	390
Thysanura	Carboniferous	300
Odonata	Permian	260
Ephemeroptera	Carboniferous	300
Plecoptera	Permian	280
Phasmatodea	Triassic	240
Dermaptera	Jurassic	160
Isoptera	Cretaceous	140
Mantodea	Eocene	50
Blattodea	Carboniferous	295
Thysanoptera	Permian	260
Hemiptera	Permian	275
Orthoptera	Carboniferous	300
Coleoptera	Permian	275
Strepsiptera	Cretaceous	125
Hymenoptera	Triassic	240
Neuroptera	Permian	270
Siphonaptera	Cretaceous	130
Diptera	Permian	260
Trichoptera	Triassic	240
Lepidoptera	Jurassic	200

difficult to explain the lineages of most modern insect orders (Whitfield & Kjer 2008). It is not even clear if the largest order, the Coleoptera, for example, is itself a single monophyletic entity (Caterino et al. 2002). In general, it is likely that the divergence of almost all of the orders is very ancient, and may have occurred too rapidly for easy resolution (Liu & Beckenbach 1992).

Most major orders were already distinguishable by 250 Ma, and only a few are known to be much more ancient. The most ancient winged insects probably included primitive cockroaches, the Palaeodictyoptera, whose fossils date back about 370 Ma to late Devonian time (Kambhampati 1995; Martinez-Delclos 1996, Prokop et al. 2006), illustrating the significance of a scavenging lifestyle in terms of early adaptability. Certain ways of life, however, do appear to be more recent than others. Specialist predators such as the praying mantids (Mantodea) are not found as fossils until around 150 Ma, and sociality might be expected to be an advanced feature of insect ecologies; termites, for example, do not appear in the fossil record until the Cretaceous period, around 130 Ma. In fact, one of the earliest records of termite activity comes from fossil wood found in the Isle of Wight, England, from where Francis and Harland (2006) describe pellet-filled, cylindrical borings in Early Cretaceous deposits (Table 1.3).

Various major orders or "cohorts" have become extinct in the last 250 million years, but relatively few when compared with many other major animal phyla, such as the molluscs and the tetrapods (Figure 1.6). The average time in which an insect species is in existence is conjectured to be an order of magnitude greater than that for, say, a bivalve or tetrapod (May et al. 1995), possibly well in excess of 10 million years. This is not to say that no insect orders have ever gone extinct – indeed they have. The first evidence of the extinction of an insect order comes from the Late Carboniferous (350 Ma), where the Cnemidolestodea vanished (Béthoux 2005). The Permian–Triassic border (around 250 Ma) saw the concurrent extinction of several insect orders (Béthoux et al. 2005), such as the Caloneurodea, an extinct order thought to be related to modern Orthoptera.

At the family level, it would appear that no insect families have become extinct over the past 100 million years or so. In some cases, it is thought that species we find alive today may extend much further back in time even than that. Dietrich and Vega (1995), by examining fossil leaf-hoppers (Hemiptera: Cicadellidae)

Table 1.3 A summary of evolutionary events in the ecology of insects, derived from fossil evidence (timescales very approximate).

Event	Period	Approx. time (Ma.)	Reference
Fossil weevils from rocks on the Beardmore Glacier, Antarctica	Pliocene to mid Miocene	15–60	Ashworth & Kuschel (2003)
Trace fossils of insect larvae from dessicating freshwater pools	Upper Oligocene	30	Uchman et al. (2004)
Oldest fossil of Strepsipteran larva	Eocene	45	Ge et al. (2003)
First fossil record of angiosperm leaf mimicry in phasmids	Middle Eocene	47	Wedmann et al. (2007)
Most recent aphid families present	Early Tertiary	50	Helie et al. (1996)
Radiation of higher (Cyclorrhaphan) flies (Dipera)	Tertiary (Paleocene)	55–65	Grimaldi (1997)
Radiation of sweat bees (Halictidae)	Late Cretaceous	80	Danforth et al. (2004)
Fossil bee nests forming oldest evidence of bees	Late Cretaceous	85	Genise et al. (2002)
Midges feeding on the blood of dinosaurs	Cretaceous	88–95	Borkent (1996)
Differentiation of various weevil families	Middle Cretaceous	100	Labandeira & Sepkoski (1993)
Fossils of aquatic hemiptera (Hydrometridae) in amber	Middle Cretaceous	100	Andersen & Grimaldi (2001)
Most recent common ancestor of modern ants groups	Early Cretaceous	115–135	Brady et al. (2006)
Earliest evidence of termite damage to wood	Early Cretaceous	130	Francis & Harland (2006)
Earliest fossil ant	Early Cretaceous	130	Brandao et al. (1989)
Trace fossil of basal Chalcidoidea (Hymenoptera: Parasitica)	Cretaceous/Jurassic	140	Rasnitsyn et al. (2004)
Bark beetles (Scolytidae) associated with gymnosperms	Late Jurassic	145	Sequeira & Farrell (2001)
Radiation of major lepidopteran lineages on gymnosperms	Late Jurassic	150	Labandeira et al. (1994)
Origin of praying mantids	Late Jurassic	150	Grimaldi (2003)
Establishment of intracellular symbionts in aphids	Jurassic/Permian	160–280	Fukatsu (1994)
Insect grazing damage on fern pinnules and gymnosperm leaves	Late Triassic	240	Ash (1996)
Assymetric wings (e.g. flies and beetles) in fossil record	Permian	250	Wootton (2002)
Evidence of first pollenivory	Early Permian	275	Krassilov & Rasnitsyn (1996)
Resting traces of primitive mayflies or stoneflies from aquatic muds	Early Permian	280	Braddy & Briggs (2002)
Evidence of leaf-mines and galls	Late Carboniferous	300	Scott et al. (1992)
Coprolites (fossil feces) and herbivore-induced plant galls from Holometabola	Late Carboniferous	300	Labandeira & Philips (2002)
Evidence from tree-ferns of insect feeding by piercing and sucking	Carboniferous	302	Labandeira & Phillips (1996)
Evidence of wood boring by insects	Early Carboniferous	330	Scott et al. (1992)

(*Continued*)

Table 1.3 (*Continued*)

Event	Period	Approx. time (Ma.)	Reference
Earliest fossil insect (bristletail) with significant structural detail	Early Devonian	400	Labandeira et al. (1988)
Earliest fossil insect fragment	Early Devonian	396–407	Engel & Grimaldi (2004)
Origin of insects	Silurian	417–440	Engel & Grimaldi (2004)

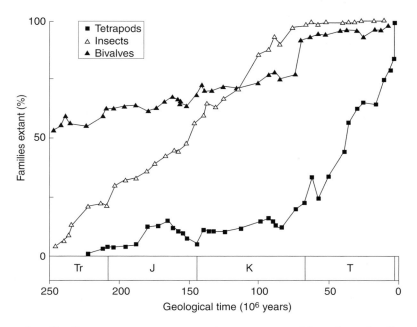

Figure 1.6 The number of families extant at various time periods as a percentage of those alive today. (From May et al. 1995.)

from the Tertiary period, suggested that modern genera of these sap-feeders existed as early as 55 Ma.

It would seem therefore that evolutionary pressures have not dramatically altered the fundamental interactions between insects and their environments for a very long time indeed, despite some major changes to terrestrial ecosystems. How the insects have managed to sustain such taxonomic constancy through large-scale climatic fluctuations is puzzling (Coope 1994), but this puzzle forms the very basis for investigations of insect ecology. Insects do show variations in the patterns of dominance of different species

and families as climates have changed. In Chile, for example, as the ice retreated from the south central lowlands about 18,000 years ago, the beetle fauna was characterized by species of moorland habitat. However, by 12,500 years ago, fossil beetle assemblages from the same region consisted entirely of rainforest species (Hoganson & Ashworth 1992). Thus, insects may not be prone to global extinctions, but are able to move from one region or another as conditions become more or less favorable for their ecologies. They can be thought of as ecologically "malleable" rather than "brittle".

Undoubtedly, the single most important development in the evolution of insects has been the development of wings, and indeed wings are usually the best-preserved structures in fossil insects (Béthoux et al. 2004). From fossil evidence, it is clear that the appearance of the Pterygota (winged insects) coincided with the tremendous diversification of insects that began in the Palaeozoic era (Kingsolver & Koehl 1994). Equally clear are the obvious advantages of flight to an animal, including avoiding predators, finding mates, and locating food, breeding sites, and new habitats. Any organism that has to walk or hop everywhere is bound to be at a disadvantage when compared with one that can fly. Maintaining steady flight in still air is a highly sophisticated process, which in a modern insect is achieved by flapping, twisting, and deforming the wings through a stroke cycle (Wootton 2002). It seems to be the case that insects only ever possessed wings that operated as flight organs on the second and third thoracic segments (the meso- and metathorax; see Figure 1.1) (Wootton 1992), and the fundamental problem concerns the evolutionary steps that led to this final, fully bi-winged state. Various theories exist to explain the origin of wings in insects, and their intermediate functions. One of the most popular suggests that so-called proto-winglets originated from unspecialized appendages on the basal segment of the legs (the pleural hypothesis, according to Kingsolver & Koehl 1994). These winglets were primitively articulated and hence movable, right from the start. Fossil mayfly nymphs (Ephemeroptera) from Lower Permian strata (of around 280 Ma) show both thoracic and even abdominal proto-wings, though this does not imply that that such appendages evolved initially in fresh water. However, other theories that utilize genetic examination of both crustacean and insect species have suggested that wings arose from gill-like appendages called epipodites, which were present on aquatic ancestors of the pterygote insects (Averof & Cohen 1997). Molecular phylogenetic studies have still to confirm this gill-to-wing hypothesis (Ogden & Whiting 2005).

Whatever the origin of proto-wings, there is still confusion as to what purpose or purposes they might have served before they became large enough to assist with aerodynamics. Sexual display and courtship, thermoregulation, camouflage, and aquatic respiration have all been suggested, but perhaps the most traditional idea involves some assistance in gliding after jumping. Whichever early route, once the winglets had increased in size sufficiently to be aerodynamically active, they would have been able to prolong airborne periods, as do for example various modern species of arboreal ant (Yanoviak et al. 2005). Stable gliding would then have become possible, especially with the assistance of long tails (cerci). If we assume that the proto-winglets were articulated from their earliest forms, then fairly easy evolutionary steps to fully powered and controlled flight can be imagined after the gliding habit was perfected. This ability to fly is likely to have been a great advantage in the colonization of new ecospace provided by early tree-like plants such as pteridophytes.

1.6 HABITS OF INSECTS

Not only have many species of insects been on Earth for many millions of years, but their various modern ecological habits also appeared at an early stage. Table 1.3 shows some major events in the development of insect ecology over the last 400 million years. The great radiation of insect species is thought to have begun about 245 Ma, in early Triassic times (Labandeira & Sepkoski 1993), and judging by various insect fossils, it is clear that insects were exploiting terrestrial habitats maybe 100 million years earlier. Note however that aquatic habitats, especially the sea (but see Chapter 6), have not been exploited – a "mere" 50,000 or so insect species are known to live in fresh water for at least part of their life cycles (Leveque et al. 2005). On land, conventional wisdom suggests that insect species richness has increased predominantly because of the appearance and subsequent radiation of plants, but examination of the numbers of new species in the fossil record suggests that the radiation of modern insects was not particularly accelerated by the expansion of the angiosperms in Cretaceous times. Instead, the basic "machinery" of insect trophic interactions was in place very much earlier. One example from Nishida and Hayashi (1996) describes how fossil beetle larvae have been found in the fruiting bodies of a now-extinct gymnosperm from the Late Cretaceous period in Japan.

In 1973, Southwood described the habits of the major insect orders in terms of their main food supplies, thus defining their general trophic roles (Figure 1.7). If it is assumed that each order has evolved but once, and that all species within the order are to some extent related so that they represent

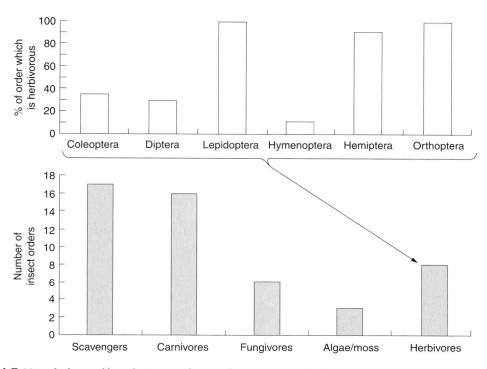

Figure 1.7 Major feeding guilds in the Insecta, showing the importance of herbivory. (From Southwood 1973.)

a common ancestral habit, then it can be seen that the major "trophic roles" at order level are scavenging (detritivory) followed by carnivory. Herbivory (or phytophagy), feeding on living plants (or the dead parts of living plants such as heart wood), is represented by only eight major orders, suggestive of what Southwood called an "evolutionary hurdle". It is very significant, however, to realize that over 50% of all insect species occur within the orders that contain herbivores, suggesting that once the "hurdle" was overcome, rapid and expansive species radiation was able to take place, and indeed it seems that herbivory has repeatedly led to high diversification rates in insects (Janz et al. 2006).

1.7 NUMBERS OF INSECTS: SPECIES RICHNESS

One problem that has at the same time excited and bewildered entomologists since the very earliest days has been the sheer number and variety of insects.

Linnaeus named a large number of insect species in the 18th century in the 10th edition of his *Systema Naturae* (1758), probably thinking that he had found most of them. At that time, only a small percentage of the world had been explored by entomologists (or indeed, anyone else), so it is not surprising that the earliest workers had no concept of the diversity and richness of insect communities, even less the complexities of their interactions. Two hundred and fifty years later we are not very much closer to estimating the total number of insect species on the planet with any degree of accuracy, though we do have a better feel for the subject than did Linnaeus. We have now described more than 1.5×10^6 organisms in total, of which more than 50% are insects. Figure 1.8 illustrates the dominance of the Insecta in the list of described species (May 1992). It is revealing to compare the number of species of insects with those of vertebrates (merely a subphylum of the Chordata, when all is said and done): around 47,000 insect species have been described, compared with a trivial 4000 or so for mammals. Furthermore, it is likely that the vast

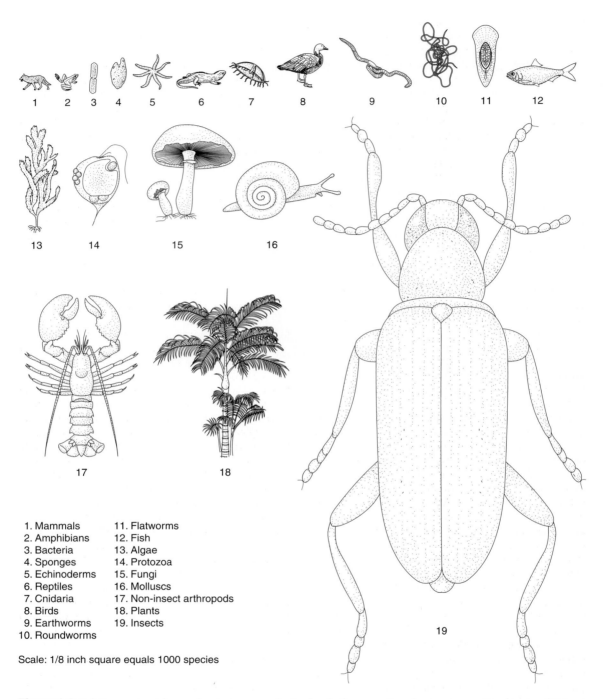

1. Mammals
2. Amphibians
3. Bacteria
4. Sponges
5. Echinoderms
6. Reptiles
7. Cnidaria
8. Birds
9. Earthworms
10. Roundworms
11. Flatworms
12. Fish
13. Algae
14. Protozoa
15. Fungi
16. Molluscs
17. Non-insect arthropods
18. Plants
19. Insects

Scale: 1/8 inch square equals 1000 species

Figure 1.8 Relative species richness of various groups of animal and other organisms shown as proportional to size of picture. (From May 1992.)

majority of vertebrate species have now been discovered. Insect species meanwhile continue to roll in apace. One problem with this has been the distribution and abundance of taxonomists, rather than that of the species of the other organisms they are dedicated to naming. Gaston and May (1992) have pointed out that there are gross discrepancies between both the number of taxonomists and the groups that they study, and also the regions of the world wherein they collect their animals. For every taxonomist devoted to tetrapods (amphibians, reptiles, birds, and mammals), there are only 0.3 for fish and around 0.02–0.04 for invertebrate species. To make matters worse, if published ecological papers are anything to go by, the study by Gaston and May suggested that 75% or so of authors come from North America, Europe, and Siberia together, whereas the areas of supposed highest species richness, the humid tropics, are seriously understudied.

As Gaston and Hudson (1994) have discussed, the total number of insect species in the world is an important but elusive figure. Exactly why it is important is also rather elusive, and different scientists would argue differently as to why we need to know just how many species there are in the world. First, by invoking Gause's axiom, which states that no two species may coexist in the same habitat if their niche requirements completely overlap, then by assessing the number of species in a given place, we can comment on the complexity of niche separation and thus the implied heterogeneity of the particular habitat. (Niche ecology is considered in Chapter 4.) Second, we know that an increase in environmental heterogeneity as associated, for example, with high densities of herbaceous vegetation reduces interspecific competition and this allows species coexistence (Corrêa et al. 2006). Thus a great deal of concern is being expressed worldwide about the decline of natural habitats through anthropogenic activities such as the logging of tropical rainforests and the urbanization of temperate regions (see Chapters 9 and 10). One consequence of habitat decline is species extinction, and with insects, as indeed with all other organisms, it is impossible to assess extinction rates effectively until we know how many species exist in the first place. Certainly, we cannot give a percentage extinction figure unless we know how many species were there originally. Third, in applied ecology, it is important to predict stabilities of forest or agroecosystems, in terms of pest outbreak potential (see Chapter 12). If we are to

rely more heavily in the future on population regulation by insect natural enemies, we need to inventory beneficial species in order to select targets for manipulation. One of the aims of evolutionary biologists is to explain the relationships between organisms, both extinct and extant, and the Insecta provide by far the most diverse group for this type of study. As all species have arisen through natural selection, a knowledge of the number of species must be an important baseline for discussing their origins. Finally, the intrinsic wonder of insect life wherever we turn in the nonmarine world must beg the question: "Just how many are there?"

Different researchers have used different methods to estimate the number of species in the world. Table 1.4 summarizes some of these conclusions. There is a surprising variation in opinion, spanning a whole order of magnitude from 2 or 3 million to a staggering 30 million or even more, though as we shall see, these higher figures seem relatively implausible (Gaston et al. 1996; Gering et al. 2007). The methods for arriving at these estimates vary considerably (May 1992), and only a brief résumé is presented here. In the case of well-known animal groups such as birds and mammals, the tropical species so far described (probably most of them) are twice as numerous as temperate ones. If it is assumed that this ratio also holds true for insects (though this has not been tested), and that we have described the majority of temperate insect species, then we reach an estimate of around 1.5–2 million species in total. As for all such estimates, some of the basic assumptions may be suspect. This theory assumes that insects, with an average size orders of magnitude smaller than those of birds and mammals, would show a proportional increase in species number as their tropical habitats diversify curvilinearly when compared with their temperate ones. Work in Borneo by Stork (1991) showed that the mean number of insect species in a rainforest canopy was 617 per tree, with one tree sample containing over 1000 species. This compares with the species numbers recorded from native British tree species, with an average of around 200 per tree (Southwood 1961) (see below), showing that tropical species might outnumber their temperate counterparts by a good deal more than 2:1.

Taxonomic biases are also to be found within the Insecta. Some orders of insects are better known than others, and have been collected heavily from only certain regions of the world. Butterflies, for example,

Table 1.4 Estimates of the numbers of insect species in the world (see also Chapter 9).

Estimated no. of species (x million)	Method or source of estimation	Reference
2–3	Hemiptera in Sulaweis	Hodkinson & Casson (1991)
3	Insects on plants	Gaston (1991)
3–5	Tropical vs temperate species	May (1992)
4.8	Tropical beetle communities	Ødegaard (2000a)
4.9–5.8	Specialist herbivores in rainforests	Novotny et al. (2002)
6	Butterflies in UK	Hammond (1992)
6.6	Host-plant specialization	Basset et al. (1996)
7–10	Insect host specificity	Stork (1988)
10	Biogeographic diversity patterns	Gaston & Hudson (1994)
12.2	Scale dependence and beetles	Gering et al. (2007)
30	Beetle–tree associations	Erwin (1982)

have fascinated amateur entomologists far longer than almost any other group of insects, and so we should by now know most of the temperate species within this order, and also have a fairly good notion of the tropical ones too. There are 67 species of butterfly in the UK, and between 18,000 and 20,000 world-wide. In total, we have 22,000 or so British insect species, encompassing all the orders, so by using the ratio of 67 : 22,000 and applying it to the 20,000 global butterfly species, we reach a grand total of around 6 million insect species worldwide. This estimate again, of course, relies on linear relationships of habitat diversity between temperate and tropical communities, though it seems reasonable to suggest that the latitudinal gradient of insect species richness increasing between these regions is a direct function of plant diversity also increasing in the same direction (Novotny et al. 2006).

Perhaps the most extravagant estimate of the richness of insect species comes from Erwin (1982), who predicted a figure of 30 million species for tropical forests on their own. Amongst other crude assumptions, this estimate relies heavily on the concept that tropical insect herbivores are plant-species specialists, so that a "one beetle, one tree" system has to operate. However, it is now suspected that host-generalist rainforest insects are more common than might have been thought (Basset 1992; Williams & Adam 1994; Novotny et al. 2002), so that rainforest trees may sustain herbivore faunas ranging from highly specialized to highly polyphagous. Note also

that it may be dangerous to make assumptions that cover all insects equally well. Very small insects, for example, have the ability to produce enormous numbers (see later in this chapter), and also to exploit microniches in their environment unavailable to much bigger species (Finlay et al. 2006). Hence, rules for speciation, diversification, and host specialization may not apply to the very small or the very large species.

Though the total number of species so far undescribed is certainly huge, some groups in some parts of the world are thought to be fairly complete. As mentioned earlier, British butterflies are certainly all described, as would befit a small group of easily recognizable species, as are the majority of other European insects. The degree of completion of species inventories of groups can sometimes be estimated using species accumulation curves and fitting asymptotic models to the data. Western Palaearctic dung beetle species, for example, are thought to be virtually complete (Cabrero-Sanudo & Lobo 2003), with only about 16% of the family Aphodiidae yet to be described (Figure 1.9).

1.8 VARIATIONS IN SPECIES NUMBER

Merely attempting to estimate the total number of insect species in the world, though a worthwhile exercise, conceals a myriad of ecological interactions that

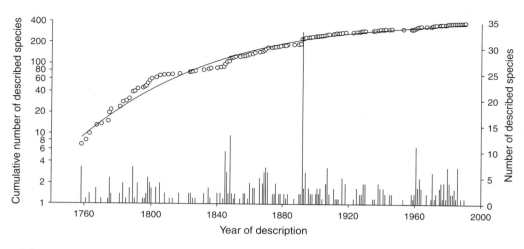

Figure 1.9 Number of described species (bars) and temporal variation in the logarithm of the cumulative number of described dung beetle species (○) from 1758 to 1990 in the western Palaearctic region, taking into account the exhaustive taxonomic information about the Palaearctic species of the three Scarabaeoidea families: Aphodiidae (395 species). The cumulative curves were fitted (continuous line) using the beta-p function & using the quasi-Newton method. (From Cabrero-Sanudo & Lobo 2003.)

influence the number of species found in any particular habitat. Many of these processes will be considered in detail in later chapters, but they are summarized here as an introduction.

1.8.1 Habitat heterogeneity

As stated above, if we invoke Gause's axiom, or the Competitive Exclusion Principle, it is clear that homogeneous habitats with relatively few "available niches" should support fewer species than heterogeneous ones. The latter are likely to allow for more coexistence of species by enabling them to partition resources within the habitat and hence avoid interspecific competition (see Chapter 4). From this, we may conclude that habitats that are more diverse in terms of greater plant species richness, for example, should exhibit greater insect species richness as a consequence. Undoubtedly, once Southwood's evolutionary hurdle of herbivory (see above) has been overcome, plants provide an enormous variety of new habitats and niches for insects (Fernandes 1994). Hutchinson's (1959) "environmental mosaic" concept describes the system well, and mosaics of mixed natural habitats such as small-sized crops, fields, and natural habitats in the

same landscape maximize the richness and diversity of insect species (Duelli et al. 1990; Yu et al. 2003). Such habitat mosaics are also thought to decrease the probability of the extinction of rare species (see Chapters 9 and 10).

Early seral stages of succession, for example, tend to possess higher plant species richness than those close to climaxes, so that it might be expected that early succession is also typified by higher insect species richness as well. A study on bees in set-aside fields in Western Europe illustrates this phenomenon well (Gathmann et al. 1994). Various types of crops and fallow fields in an agricultural landscape were assessed for bee species richness, and it was found that habitats with greater floral diversity offered better and richer food resources for flower visitors. Set-aside fields that were mown, and hence reduced to early successional stages, showed a greatly increased plant species richness, coupled with double the species richness of bees when compared with unmown, late-successional-stage fields.

So far, we have considered floral diversity at a phenotypic level, but it may be that even within a host species, plants with increasing genetic diversity may increase insect species richness (Wimp et al. 2005). A complex derivation of plant species richness involves hybrid zones, areas of habitats where two or more plant species hybridize to produce a third, F1, phenotype. These areas seem to provide a great diversity of resources for herbivorous insects (Drew & Roderick 2005). In Tasmania, Australia, for example, where two species of eucalyptus hybridize, this area was found to be a center for insect (and fungal) species richness (Whitham et al. 1994). In this study, out of 40 insect and fungal taxa, 53% more species were supported by hybrid trees than in pure (parent plant) zones. In a different study, Floate et al. (1997) found that not only were populations of leaf-galling aphids in the genus *Pemphigus* (Hemiptera: Aphididae) 28 times as abundant in hybrid zones of the host tree cottonwood (*Populus* spp.), but that, in the authors' opinion at least, preserving such small hybrid zones could have a disproportionately beneficial role in maintaining insect biodiversity. Figure 1.10 illustrates the use of multivariate statistics in detecting three distinct clusters of gall-forming and leaf-mining insects associated with two species of *Quercus* in Mexico, and a hybrid of the two (Tovar-Sánchez & Oyama 2006). Clearly, the hybrid supports an intermediate community, adding to both the genetic and species diversity of the forest ecosystem. Despite all this, the role of hybrid zones in determining insect numbers remains controversial (see Chapter 3).

It is not only herbivore communities that might be expected to show an increased species richness as their plant-derived habitat also becomes more species-rich. Consumer trophic levels such as specialist predators or parasites can also be influenced in this way. Parasitoid guilds constitute an important part of the biodiversity of terrestrial ecosystems (Mills 1994), and because many of them are host specialists, high species richness of their insect hosts results in a similarly higher parasitoid richness. This system parallels that of plant–herbivore interactions well. Insect herbivores feed on many different parts of the plant (see below), so that one plant species can often support many insect species. Similarly, it is possible to recognize a series of guilds, each composed of various species, depending on the life stage of the host insect (Figure 1.11). Habitat heterogeneity may also play an important part in promoting species coexistence, hence adding to regional species richness. Palmer (2003) found that termite mounds increased the complexity of habitats for African acacia-associated ant species, and allowed them to avoid competitive exclusion.

1.8.2 Plant architecture

The term "architecture" when applied to plants was coined by Lawton and Schroder (1977). It describes

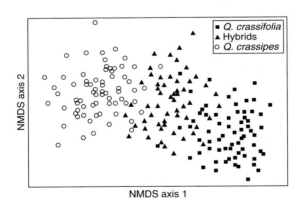

Figure 1.10 Differences in endophagous (gall-forming and leaf-mining insects) community composition among oak host *Quercus crassifolia*, hybrids, and *Q. crassipes*. Each point is a two-dimensional (axis 1 and axis 2) representation of endophagous species composition on an individual tree based on global, non-metric multidimensional scaling (NMDS). Distances between points reflect a dissimilarity matrix created using the Bray–Curtis dissimilarity coefficient. Points that are close together indicate endophagous communities that are more similar in composition. (From Tovar-Sánchez & Oyama et al. 2006.)

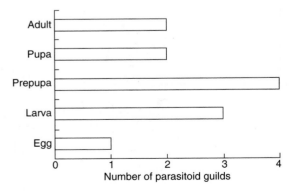

Figure 1.11 Number of guilds of parasitoid insect attacking different life stages of insects. (From Mills 1994.)

both the size, or spread, of plant tissues in space, and the variety of plant structures (Strong et al. 1984), from leaves to shoots to wood to roots; Lewinsohn et al. (2005) consider it to be a cornerstone of insect–plant relationships. Figure 1.12 shows the great diversity of niches thought to be available for just one, albeit the largest, order of insects, the Coleoptera (Evans 1977). All imaginable parts of the tree are utilized by one species of herbivore. Clearly, different plant types vary widely in their structures,

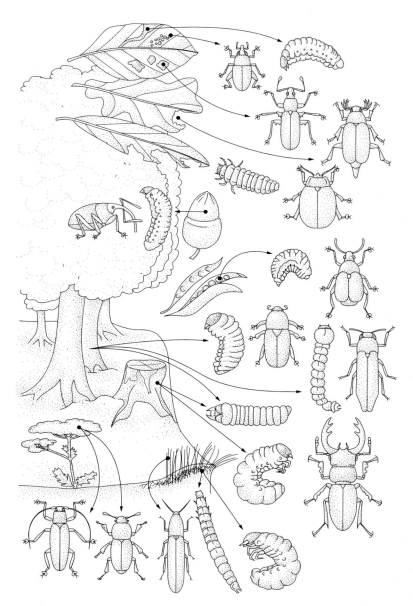

Figure 1.12 An illustration of the diversity of beetles found living on a single tree. (From Evans 1977.)

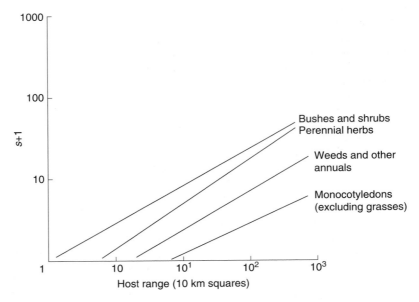

Figure 1.13 Number of insect species on plants with different architectures and varying commonness. (From Strong et al. 1984.)

and thus architectures. Ferns are very much less complex than trees, and hence in terms of habitat heterogeneity, a bush, shrub, or tree would be expected to support more insect species than would a fern, everything else being equal (Figure 1.13). If the ratios of species richness of herbivorous insects between more complex plant types or species and less complex ones are considered (Table 1.5) (Strong et al. 1984), it can be seen that, in all cases, the architecturally more complex plant has considerably more species associated with it. Summing over the ratios in the table, we can "guesstimate" that a tree should have something like 50 times as many species associated with it than a monocotyledon, although this is probably somewhat of an overestimate. A most striking case compares Lepidoptera on trees and shrubs with those on herbs and grasses (Niemalä et al. 1982), where there are over 10 times as many species on the former plant types than on the latter.

Table 1.5 Ratio of numbers of herbivore insect species on plants of different structures. (From Strong et al. 1984.)

More complex plant	Ratio	Less complex plant
Trees	27:1	Shrubs
Shrubs	21:1	Herbs
Bushes	1.3:1 to 25:1	Perennial herbs
Perennial herbs	1.5:1 to 27:1	Weeds
Weeds	1.3:1 to 29:1	Monocotyledons

1.8.3 Plant chemistry

Chemicals in plants that may have an influence on the ecology of insects fall into two basic categories: food and defense. Insect herbivores are particularly limited by the suboptimal levels of organic nitrogen provided by plants as food (White 1993). Though many insects are highly resistance to plant toxins (Torrie et al. 2004), plant defenses reduce the efficiencies of feeding processes even further. However, although plant chemistry influences the abundance of insects, it is difficult to detect a major influence on insect species richness. Jones and Lawton (1991) looked at the effects of plant chemistry in British umbellifers, and though there was thought to be some influence on species richness via changes in natural enemy responses to hosts or prey on biochemically diverse plants, they concluded that there was no evidence that plant species with complex or unusual biochemistries supported less species-rich assemblages of insects. Essentially then, though a "niche" on a toxic or low nutrient plant may be harder to utilize, the number of niches (i.e. habitat heterogeneity) remains the same. This result is difficult to reconcile with the fact that most insect herbivores (about 80%) are specialists, unable to feed on plants with different chemistries. Indeed, it seems that feeding specializations in herbivorous insects depend on recognition of stimulants specific to the host plant (del Campo et al. 2003).

1.8.4   Habitat abundance (time and space)

For insects to adapt to a new habitat or resource of any sort, there must be sufficient opportunity for natural selection to do its job, and for selection pressure to "push" a species into a new and stable form. This type of opportunity consists of the regional abundance of the resource and/or how long in evolutionary time the resource has been available. The best illustrations of this principle again come from insects feeding on plants.

In the UK, both abundance through time, measured as the number of Quaternary (last million years) remains, and commonness, measured as the number of kilometer squares of the UK wherein the species of tree was recorded, have a significant positive influence on the numbers of insects on the trees, as noted by Southwood (1961) and Claridge and Wilson (1978). Both studies showed that the *Quercus* genus (oaks), which is both the most common and has been in the landscape for the longest time, shows the highest number of insects associated with it. Newly introduced tree species such as the sycamore and horse chestnut have relatively few (see Chapter 9). In summary, there appears to be tremendous variation in numbers of insect species on British trees, all of which may be assumed to possess roughly the same architecture.

1.8.5   Habitat size and isolation

A fundamental concept in ecology is the equilibrium theory of island biogeography, first proposed by MacArthur and Wilson (1967). Simply put, the theory makes predictions about how island area and distance from a source of colonists affect immigration and extinction rates (Schoener 1988; see also Chapter 9). The term "island" can indicate an island in the true, geographic sense, but it can also mean any patch of relatively homogeneous habitat that is surrounded by a different one. The latter broad category could include woodlands in farmland, ponds in fields, or even one plant species surrounded by different ones. A nice example is provided by Krauss et al. (2003) who studied butterfly species in different-sized calcareous grasslands in Germany. As Figure 1.14 shows, there is a highly significant relationship (note the log scale on the *x*-axis) between the size of the grassland "patch" and the number of species of butterfly living in it. This association is robust enough not to differ between years of sampling.

Theoretically, there should be a balance between the migration of species into an island or habitat patch and extinction rate, such that for a given size of island there is an equilibrium number of species present (Durrett & Levin 1996). The species–area relationship is firmly established in ecology (Hanski

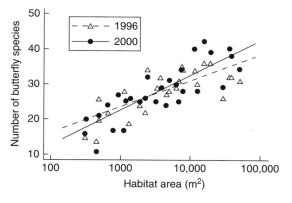

Figure 1.14 Relationship between the number of butterfly species and grassland area ($n = 31$ fragments) in 1996 and 2000. 1996, $y = 1.69 + 7.23 \log_{10} x$, $F = 40.37$, $r^2 = 0.582$, $P_R < 0.0001$; 2000, $y = -6.28 + 9.58 \log_{10} x$, $F = 62.48$, $r^2 = 0.683$, $P_R < 0.0001$. Comparison of regressions: slopes, $F = 2.00$, $P_S = 0.163$. (From Krauss et al. 2003.)

& Gyllenberg 1997), and it describes the rate at which the number of animal and/or plant species increases with the area of island available. The model is represented by the simple equation:

$$s = a^z$$

where s is the number of species, a is the island area, and z is a constant.

There are a host of assumptions to the theory, including those that insist that habitat type, age, and degree of isolation of the "islands" being compared are the same. There is some uncertainty about the existence of a maximum value for variability or heterogeneity of physical environments (Bell et al. 1993). It is possible that as islands become larger, their heterogeneity increases apace. However, field data fitted to the species–area model suggest that the curve will asymptote; in other words, the exponent value, z, is less than one. In fact, there are surprisingly few examples where z has been estimated with confidence for insects. Values range from 0.30 and 0.34 for ants and beetles, respectively, on oceanic islands (Begon et al. 1986) to 0.36 for ground beetles (Coleoptera: Carabidae) in Swedish wooded islands (Nilsson & Bengtsson 1988); z-values for specialist butterfly species in Germany were estimated to be 0.399 (Krauss et al. 2003), whereas generalist species only managed 0.096.

As mentioned above, the species–area model has some shortcomings (Williams 1995). There are no limits to the function that describes it, and the model is unable to handle zero values, so that if, by some quirk of fate, a particular island has no species representing the particular animal or plant group under investigation, then that island cannot be included in the model fitting. Despite these drawbacks, the island biogeography theory does fit observed data on occasion, but particularly few insect examples exist. One experimental illustration was carried out by Grez (1992), who set up patches of cabbages containing different numbers of plants, ranging from four to 225. Later sampling of the insect herbivores on and around the cabbages showed that species richness of insects was highest in the large host-plant patches, which also showed an enhanced presence of rare or infrequent species within the general area, when compared with the smaller patches. In a field study, Compton et al. (1989) were able to detect a significant but rather weak tendency for larger patches of bracken

(*Pteridium aquilinum*) to support more species of insect herbivores than smaller ones, in both Britain and South Africa. On the other side of the Atlantic Ocean, the Florida Keys have gained their fauna of longhorn beetles (Coleoptera: Cerambycidae) from both the islands of the West Indies, including the Bahamas and Cuba, and the mainland, represented by the peninsula of south Florida. Both species–area and species–distance relationships for cerambycids were found to conform to the island biogeographic theory (Browne & Peck 1996), even though the islands making up the Keys have been fragmented by rising sea levels for only the last 10,000 years or so. Such relationships may, though, be overshadowed by other biogeography factors, such as historical dispersal patterns. Beck et al. (2006) studied the number of hawkmoth (Lepidoptera: Sphingidae) species in the Malesian region of Southeast Asia (a region covering the Solomon Islands and New Guinea to the east and Peninsular Malaysia and Sumatra to the west). Though species–distance relationships were detectable for some species of hawkmoth, the main determinant of species richness on these islands was the amount and diversity of vegetation, especially rainforest.

The fragmentation of habitat islands may be a natural event over geological time, as evidenced by the Florida Keys above, but many aspects of human activity, such as logging and road building, over a very much shorter timescale, result in fragmentation as well. A particular case in point would be the impacts of landscape-scale modifications to natural habitats from agricultural practices such as the removal or reduction in woodland, the parceling up of land into fields, and the management of field boundaries and hedgerows as windbreaks or stock barriers (Pichancourt et al. 2006). Such practices can produce a further decrease in "island" size and an increase in isolation. Animal and plant populations remaining in these fragments constitute metapopulations, that is, local populations undergoing constant migration, extinction, and colonization on a regional rather than global scale (Husband & Barrett 1996). A British example of a metapopulation study is that by Hill et al. (1996) on the silver spotted skipper butterfly, *Hesperia comma* (Lepidoptera: Hesperiidae). The larvae of this butterfly are grass-feeders, preferring short swards where the conditions are warm. Patches of this type of habitat occur on chalk downland in the south of England, and studies of the metapopulation dynamics of the species showed that habitat patches were more likely to be colonized if they were relatively large and close to other large, occupied patches. Furthermore, adult butterflies were more likely to move between large patches close together, whereas local populations in small isolated patches were more likely to go extinct. A consideration of habitat "quality" is also required; two patches of grassland for instance may be of a similar size and equally isolated, but one may be of a higher quality than the other in terms of vegetation height and the abundance of flowers (Öckinger & Smith 2006).

1.8.6 Longitude and altitude

On a local scale, insect species richness may still vary even when habitat heterogeneity and abundance, or patch size, are constant. These variations would seem to be influenced by environmental factors such as latitude, longitude, and altitude, presumably via climatic interactions. There are two general predictions of how species richness and altitude are related. Either richness decreases fairly linearly with increasing altitude, or richness peaks at mid-elevations (Sanders 2002). Several examples describe how insects in a particular locale are most species-rich at a certain height above sea level. Libert (1994) found that many species of butterfly observed in Cameroon, West Africa, exhibited a preference for the tops of hills, whereas in Sulawesi, Indonesia, hemipteran communities showed highest species richness at elevations between 600 and 1000 m (Casson & Hodkinson 1991). The explanations of these observations are no doubt complex, and detailed microclimatic and vegetational data would be required to investigate causality. The roles of decreasing diversity with latitude or altitude of host-plant species, decreasing structural complexity, and increasingly unfavorable climatic conditions may all have an influence (Brehm et al. 2003). Indeed, weather conditions can influence insect species richness. In tropical Australia, leaf beetle (Coleoptera: Chrysomelidae) communities were found to be most species-rich during the hottest and wettest times of the year (December to March; Hawkeswood 1988). It can often be difficult in ecology to assign cause and effect with confidence, but various effects of weather and climate on insect ecology are considered in detail in Chapter 2. Suffice it to say for now that, in this example, leaf beetle richness was not

correlated with plant richness, though it might be expected that these climatic influences would operate at least in part via the host plants of these totally herbivorous species.

On a larger geographic scale, it is usual to expect that insect species richness will decrease towards the poles. In North America, of 3550 species studied, 71% occurred south of a line along state boundaries from the Arizona–California to Georgia–South Carolina borders (Danks 1994), whereas in Scandinavia, out of a wide range of insect families examined, the number of species was generally highest in the southern provinces, declining to the north and north-west (Vaisanen & Heliovaara 1994). As always in insect ecology, exceptions to neat rules crop up. Also in Finland, Kouki et al. (1994) showed that the species richness of sawflies (Hymenoptera: Symphyta) showed an opposite latitudinal trend, so that species richness is highest in the north, not the south. Kouki et al. explained this by the fact that the principal host-plant group, willows (*Salix*), are also most species-rich further north.

1.8.7 Human interference

Various of the foregoing factors that influence insect species richness can be altered, usually detrimentally, by human activities such as agriculture, forestry, urbanization, and so on. In general, the diversity of insect communities in habitats such as grassland is often negatively correlated with management intensity (Nickel & Hildebrandt 2003). Certainly, the intensive growing of monoculture crops all over the world must result in a reduction in biodiversity of plants and the animals that associate with them (see Chapter 12). Rainforests all over the world are being depleted rapidly, and this also has an undoubted effect on species richness within them. In Sabah (northeast Borneo as was), primary lowland rainforest has largely been reduced to small or medium-sized forest "islands", often isolated from each other and surrounded by artificial agricultural ecosystems (Bruhl et al. 2003). Communities of leaf litter ants in these fragments of forest showed a significant reduction in species number and diversity when compared with communities in contiguous forest, with a maximum of only 47.5% of the original richness. Connectivity between habitat islands is crucial for organisms to move between them, and even a forest plantation is better than no forest at all, as found with dung beetle communities in Chile (Bustamante-Sanchez et al. 2004). In this example, pine plantations maintained the structural and functional biodiversity of native fauna, at least partially, and were able to connect native forest remnants across the landscape. Habitat specialists in particular suffer from this type of habitat isolation (Steffan 2003), whilst generalist species do better in a diverse landscape "matrix'".

Reductions in habitat heterogeneity result from practices such as selective logging, but the knock-on effects on insect species richness are not always so predictable. Some species of butterfly actually do better in logged forest, the gap specialists in particular (Hamer et al. 2003), and it has in fact proved difficult to detect significant changes in species richness of everything from moths to beetles when primary rainforest is selectively logged (Speight et al. 2003).

1.9 THE NUMBER OF INSECTS: ABUNDANCE

Species richness is only one measure of the success of insects; their abundance is another equally important one. No-one could deny that a plague of locusts, although very species-poor, is the epitome of success, at least as judged by the number of individuals and their rapacious ability to devour plant material of many kinds. Unfortunately, it is rather more difficult to obtain reliable estimates of the number of individuals (abundance) of a species than it is merely to score the species present in or absent from a habitat (species richness). Common but immobile or concealed species such as aphids or soil- and wood-borers may be underestimated or even completely overlooked by most ecological sampling systems, and it is only when we look very closely at tiny and, in most people's perceptions, inconsequential, insects in novel habitats that we find just how common some groups are. An example of booklice (Psocoptera; Figure 1.15) from Norway spruce canopies in lowland England shows how immensely abundant they are in a seemingly sterile habitat (Ozanne et al. 1997). Using insecticidal mist blowing of tree canopies, an astonishing 6500 per square meter were collected. As a side issue, this example illustrates the importance and magnitude of detritivore pathways in some ecosystems (Foggo et al. 2001). Booklice feed mainly on fungi, lichens, and fragments of leaf and bark material in dark tree

Figure 1.15 Booklice (Pscoptera).

canopies, and are likely to form the staple diet of a myriad of predatory insects and other arboreal arthropods.

The densities of insects that commonly act as disease vectors, such as mosquitoes, tsetse flies, and aphids, are especially significant when it is considered that these can be low-density pests, where only a very few (as few as one) individuals are required to pass on diseases such as malaria, sleeping sickness, or potato leaf curl (see Chapter 11). Although it is one thing to wonder at the sheer numbers of insects in a locust swarm, it is quite another to appreciate the potential for harm inherent in one small individual.

1.9.1 Variations in insect numbers

Of much more fundamental interest to ecologists is the manner in which insect population densities vary within a population through time. Some of the most basic ecological processes have been explored in response to such variations, and the explanation of the patterns and processes in population density changes has taxed ecologists the world over, and still does. Large variations in the population densities of insects occur under two different headings: population cycles and population eruptions (Speight & Wainhouse 1989; Speight & Wylie 2001). Cycles are periodic, with some degree of predictability about the time between peak numbers; high densities are rapidly followed by large declines. Eruptive populations, on the other hand, often remain at low densities for long periods before outbreaks occur suddenly and often unexpectedly. Once such outbreaks have developed, they may be sustained for some time.

1.9.2 Cycles

Most cycles of insect populations have been observed in species that inhabit perennial, non-disturbed habitats such as forests. Figure 1.16 shows a typical population cycle of a forest insect, the larch budmoth, *Zeiraphera diniana* (Lepidoptera: Tortricidae) (Baltensweiler 1984). This species shows remarkably regular cycles of abundance in the Engadine Valley in Switzerland, where outbreaks have a periodicity of 8 or 9 years (Dormont et al. 2006). Mean population density may vary 20,000-fold within five generations, though 100,000-fold increases have been observed locally. These huge numbers of defoliating caterpillars can have an enormous impact on the growth of the host trees, and larch budmoth outbreaks can be "recon-structed" over many hundreds of years by examining tree ring widths. In years when large numbers of larvae were eating larch needles, the density of latewood laid down by the trees is much reduced. Figure 1.17 shows that budmoth outbreaks have come and gone in fairly recognizable cycles, averaging every 9.3 years to be exact, for 1200 years. It is only in the last 20 or so years that these predictable outbreaks seem to have been absent (Esper et al. 2007), a possible consequence of climate change (see Chapter 2).

Many other forest insects show similarly predictable cycles; the Douglas fir tussock moth, *Orgyia pseudotsugata* (Lepidoptera: Lymantriidae), has exhibited regular outbreaks at 7–10-year intervals in British Columbia since the first recorded observations in 1916 (Vezina & Peterman 1985). What is more, these cycles may be synchronized over relatively large areas. Larvae of the autumnal moth, *Epirrita autumnata* (Lepidoptera : Geometridae) feed on mountain birch, *Betula pubescens*, over much of Fennoscandinavia, cycling to outbreak densities every 10 years or so. In Figure 1.18 it can be seen that moth outbreaks from 24 different localities, some over 1000 km apart, show remarkably similar patters over the years (Klemola et al. 2006). This spatial synchrony weakens as site distances increase, and it is likely to be linked to climatic variations within and between regions.

In practice, population cycles are very difficult to explain (Turchin et al. 2003). Unless climatic patterns are themselves cyclic (Hunter & Price 1998), most insect population cycles can usually be attributed to biotic interactions, such as competition and predation,

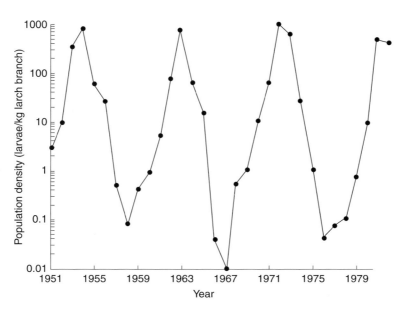

Figure 1.16 Population cycles for *Zeiraphera diniana* on larch in Switzerland. (From Baltensweiler 1984.)

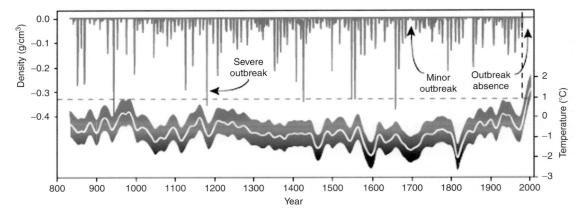

Figure 1.17 Long-term larch budmoth (LBM) and temperature reconstructions for the European Alps. (Top) Maximum latewood density (MXD)-based LBM outbreak reconstruction since AD 832. Time-series is the age-corrected difference series between gap-filled and original MXD data. Values less than $-0.005\,\mathrm{g/cm^3}$ are shown. (Below) The temperature model (white curve) is shown together with the standard error (gray band) derived from the fit with instrumental data. Dashed lines indicate the last LBM mass outbreak in 1981 (vertical) and the upperstandard error limit recorded in the late ninth century (horizontal). (From Esper et al. 2007.)

that have a delayed action on population growth rates. Delayed density dependence will be considered in detail in Chapter 5, and is introduced below. Population cycles are rarely as neat as in the case of *Zeiraphera*, and long-term examinations of abundance data for insect species, though revealing some degree of cycling, suggest that the levels of peaks and troughs are less regular or predictable.

1.9.3   Regulation

In ecological terms, we recognize that population cycles in insects are at least partially under the

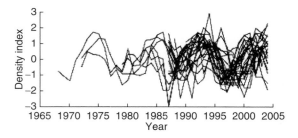

Figure 1.18 Autumnal moth population time series for each of 24 locations in Fennoscandia. (From Klemola et al. 2006.)

influence of a process known as regulation. Regulation describes the way in which a population's abundance varies through time as a decrease in population growth rate as population density increases (Agrawal et al. 2004). Declines in population growth rates with density can be manifested by: (i) increases in the rate (proportion) of mortality that the population suffers; (ii) decreases in birth rate; (iii) increases in emigration rates; or (iv) decreases in immigration rates. When birth, death, or movement rates vary proportionally with density (i.e. they are density dependent), they have the potential to maintain an insect population around some equilibrium density. If the density dependence occurs on a time delay, the population can overshoot this equilibrium, and exhibit cyclic behavior. In general, the higher the insect's fecundity, or the longer the time lag, the more dramatic the oscillation will be.

Various biotic factors are known to be potentially regulatory (i.e. may act in a density-dependent fashion), and each is considered in detail in later chapters. However, it is useful to separate such factors into those that act within a trophic level, such as competition, and those that act between trophic levels, either from below (so-called "bottom-up"), via food supply, or from above (so-called "top-down"), via the action of natural enemies such as predators, pathogens, or parasitoids (see Chapter 5).

In reality, these factors interact, so that, for example, competition often acts via the amount of food available for individual insects in what is known as resource-driven outbreaks (Steinbauer et al. 2004). Food limitation via intraspecific competition may result in decreased fecundity, or in increased migration (Azerefegne et al. 2001). Experimental studies frequently demonstrate this sort of competition. Cycles in populations of Indian meal moth, *Plodia interpunctella* (Lepidoptera: Pyralidae), for example, in laboratory containers appeared to be caused by density-dependent competition for food amongst the larvae (Sait et al. 1994), but field conditions are, of course, likely to be more complex. Though the cinnabar moth, *Tyria jacobaea* (Lepidoptera: Arctiidae), suffers periodic crashes in abundance because of competition for larval food (Van der Meijden et al. 1991), recovery after a crash is still delayed even with food available, presumably because of the

activities of natural enemies and reduced food quality. Clearly, competition for food might be something to avoid if possible. Various behavioral mechanisms have evolved that minimize competition, from eating siblings in tropical damselflies (Fincke 1994), to avoiding oviposition on host plants already bearing eggs in butterflies (Schoonhoven et al. 1990), or by selecting different types or size of food when competition becomes intense, as in stoneflies (Malmqvist et al. 1991).

The role of natural enemies in the population regulation of insects is discussed in full in Chapter 5. Classic work by Varley and Gradwell (1971) has shown that as herbivore populations vary with time, so do the abundances of various predators and parasitoids that feed on them (Figure 1.19). *Philonthus decorus* (Coleoptera: Staphylinidae) is a predatory rove beetle that eats the pupae of winter moth, *Operophtera brumata* (Lepidoptera: Geometridae), in the soil, whereas *Cratichneumon culex* (Hymenoptera:

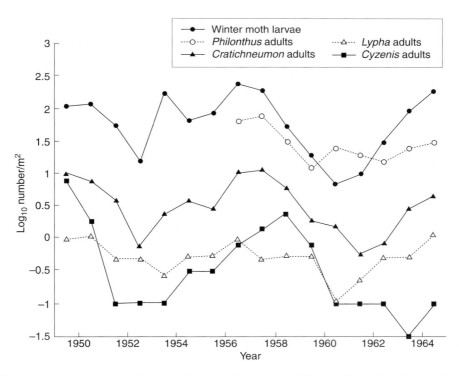

Figure 1.19 Densities of winter moth larva and its natural enemies in Wytham Wood, Oxfordshire. (From Varley & Gradwell 1971.)

Ichneumonidae) is a parasitic wasp, and *Lypha dubia* and *Cyzenis albicans* are both parasitic flies (Diptera: Tachinidae), all of which attack the larval stages whilst feeding in the canopy. Two points are worthy of note. First, the peaks of all the enemies seem to be 1 year (or in the case of *Cyzenis*, 2 years) later than that of the winter moth, illustrating the phenomenon of delayed density dependence described above. Second, it may be tempting to attribute the changes in winter moth abundance to regulation from its enemies. As the moth larvae population increases from year to year, so do the levels of predation and parasitism, thus knocking the herbivore numbers back again. In this instance, however, only pupal predation by *Philonthus* was shown to be density dependent and hence regulatory (Varley & Gradwell 1971). The other enemies, in fact, were merely tracking the variations in the host, with no regulatory impact. These host variations relate to nutrient and defense quality and quantity, and will be discussed in Chapter 3. Variations in predator–prey relationships

may also be detected on a much smaller timescale. Figure 1.20 shows how numbers of the aphid *Metopeurum fuscoviride* (Hemiptera: Aphididae), which feeds on tansy (*Tanacetum vulgare*), vary during one season in association with various natural enemies such as ladybirds, predatory bugs, and hoverflies (Stadler 2004). The increase in predator numbers at the end of July is associated with a sharp decline in aphid numbers, and it is of course tempting to attribute causality – i.e. the enemies are regulating the aphids. However, aphid numbers may decline for various other reasons, such as a decrease in host-plant quality, increased intraspecifc competition, or migration; such interactions therefore need further work to identify the factor or factors responsible for the observations.

Many pathogens are now known to act as regulators of insect populations, with fungi, bacteria, and especially viruses having a great impact on occasion (McVean et al. 2002). Forest insects known to be at least partially regulated by naturally

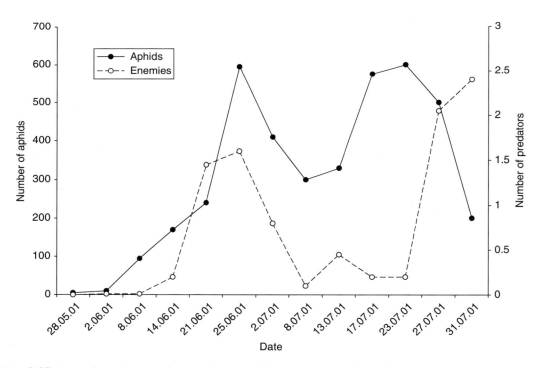

Figure 1.20 Seasonal trends in abundances of the aphid *Metopeurum fuscoviride* and its predators on tansy. (From Stadier 2004.)

occurring viruses include the nun moth *Lymantria monacha* (Lepidoptera: Lymantriidae) (Bakhvalov & Bakhvalova 1990), rhinoceros palm beetle *Oryctes rhinoceros* (Coleoptera: Scarabaeidae) (Hochberg & Waage 1991), and browntail moth *Euproctis chrysorrhoea* (Lepidoptera: Lymantriidae) (Speight et al. 1992). Chapter 12 provides more details.

The host plant may also influence the density of herbivorous insects via a factor or factors that cause feedbacks with time lags of suitable duration (Haukioja 1991). One example involves the autumnal moth, *Epirrita autumnata* (Lepidoptera: Geometridae) in northern Fennoscandia, mentioned earlier. Large-scale defoliation of the host tree, mountain birch, induces changes in the food quality of foliage produced subsequently, such that the reproductive potential of the moth is significantly reduced (see also Chapter 3). So for several years following outbreaks, insect populations are suppressed. However, *Epirrita* larvae also feed on apical buds of birch, which causes a change in plant hormone balance resulting in luxuriant growth of new leaves. This new foliage is particularly suitable for herbivores. Thus the insect population begins to build up again (Haukioja 1991). These complex reactions to herbivore density result in statistically significant cycles of the autumnal moth of 9–10 years' duration. Attempts to attribute *Epirrita* outbreaks to good birch mast years, when tree fruits are especially plentiful, have so far proved more difficult (Klemola et al. 2003).

Controversy has raged for years concerning the relative importance of natural enemies versus resource (e.g. food) limitation in the population ecology of insects and other animals. For now, it is clear that there are no easy answers. The science and practice of biological pest control relies on the ability of predators, parasitoids, and pathogens to regulate pest populations, whereas the ecology of pest outbreaks, and its links to crop husbandry of all types, emphasizes resource limitation as the driving force in insect epidemiology. Figure 1.21 summarizes the potential relationships between various possible mortality factors, and suggests that their relative importances vary as the population of the insect on which they act varies (Berryman 1987). In general terms, the figure suggests that regulation from natural enemies such as predators and, especially, parasitoids might be expected to be effective up to a medium host or prey density, but at high, epidemic levels, regulation via pathogens such as viruses, if present, is more

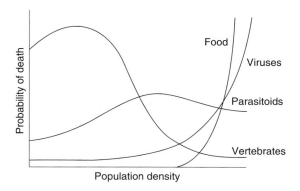

Figure 1.21 Relative importance of various mortality factors on insect populations at different densities. (From Berryman 1987.)

important. Food limitation is often the most important regulatory factor in high-density insect populations. These controversies are of fundamental relevance to insect pest management (see Chapter 12).

1.9.4 Eruptions

Eruptive outbreaks can develop when environmental changes, such as consecutive seasons of favorable weather acting directly on the insect or indirectly via its food supply, permit the rapid growth, dispersal, and/or reproduction of the insect population in question. In essence, regulation at low density is lost because of environmental conditions, and unpredictable outbreaks result. In many cases, this breakdown of natural regulation results from human intervention upsetting the coevolved semistability between insect and habitat, enabling the insect populations to exhibit much higher amplitudes of population oscillations, or indeed to remain at outbreak levels rather than declining again (Moreau et al. 2006).

Locusts are a classic example of this phenomenon. Globally, many species of locust are known to erupt into plagues from time to time, the most common being the desert locust, *Schistocerca gregaria*, migratory locust, *Locusta migratoria*, tree locust, *Anacridium melanorhodon*, Moroccan locust, *Dociostaurus maroccanus*, and Australian plague locust, *Chortoicetes terminifera* (all Orthoptera: Acrididae) (Wright et al. 1988; Showler 1995; Stride et al. 2003; Hunter 2004). Various species of *Schistocerca* are also serious pests in South America (Hunter & Cosenzo 1990). Many affected countries are some of the poorest in the world and are completely unable to afford control or prevention campaigns (Lecoq 2001). To make matters even worse, one of the major causes of desert locust outbreaks developing these days is armed conflict. Wars and revolutions prevent timely access to locust breeding sites so that the suppression of outbreaks before they occur is prevented (Showler 2003).

Locust plagues (Figure 1.22) have been ravaging crops for many centuries. Swarms of locusts invading southern Europe from Africa were described in Roman times by Pliny the Elder, and countries from Hungary to Spain were particularly badly attacked in the 14th, 16th, and 17th centuries (Camuffo & Enzi 1991). In 1693, for example, a swarm of *L. migratoria* built up on the northwestern shore of the Black Sea and between the rivers Danube and Theiss. Some of the swarm invaded the Tyrol, but most entered Austria via Budapest. Some ended up in Czechoslovakia and Poland, whereas others headed west into Germany. A few individuals actually reached the British Isles (Weidner 1986). Locust eruptions are, of course, still occurring. In the 1980s, for example, devastating plagues occurred in Algeria (Kellou et al. 1990), Argentina (Hunter & Cosenzo 1990), Australia (Bryceson 1989), Peru (Beingolea 1985), Chad (Ackonor & Vajime 1995), China (Kang & Chen 1989; Ma et al. 2005), the Arabian Peninsula (Showler & Potter 1991), and Sudan (Skaf et al. 1990). In the 1990s, successive generations of locusts gave rise to localized eruptions for 18 months as far west as Mauritania in West Africa and as far east as

Figure 1.22 *Locusta migratoria* nymphs in China. (Image courtesy of Steve Simpson.)

India (Showler 1995). The damage caused by these eruptions is enormous, as is the cost of control (Showler 2002). During one desert locust plague in Africa, 1.5×10^7 L of insecticide were used, at a cost of around US$200 million (Symmons 1992). Not only that, but locust plagues can adversely affect other animal populations. In central Saudi Arabia, many species of birds adapted to feeding on grasses and seeds were recorded inhabiting grassland and savanna areas (Newton & Newton 1997). A plague of desert locusts in spring and summer 1993 combined with poor spring rains to reduce the bird populations to the lowest numbers and species diversity recorded during the 28-month study.

The origins of locust swarms and their subsequent migrations are basically controlled by climatic factors (Camuffo & Enzi 1991; see also Chapter 2). Locust eggs may stay dormant in soil for months, waiting for random events to bring rain to their area. Depending on the type of soil, eggs survive best during rain (Showler 1995), and rain also promotes the growth of vegetation on which the hatching nymphs, or hoppers, can feed (Hunter 1989; Phelps & Gregg 1991; Ji et al. 2006). The El Niño phenomenon is a southward-flowing ocean current off the coasts of Peru and Ecuador. Apparently cyclical changes in the pattern of its flow are the cause of environmental and climatic disturbances that cause widespread damage every few years. In Peru in 1983, El Niño caused up to 3000 mm of rain in an otherwise arid region, resulting in enormous locust swarms (Beingolea 1985). In fact, the detection by satellite imagery of new areas of green vegetation caused by rain is one of the most important tools in international locust plague-monitoring programs (Bryceson 1990; Despland et al. 2004). Once the nymphs have depleted local resources, the migratory phase of a locust plague ensues, whose scale and direction is mainly dependent on winds (Symmons 1986; Camuffo & Enzi 1991). Under the right weather conditions, plagues can last for months or even years. Eventual declines occur because of extremely dry conditions (Wright et al. 1988) or drops in temperature (Camuffo & Enzi 1991). It can be seen, therefore, that population eruptions as typified by locust plagues are rarely if ever under tight regulation by density-dependent factors. Rather, density-independent factors such as climate are the most influential (see also Chapter 2).

Further information, updates, and international status of locust eruptions is provided by the Food and

Agriculture Organization (FAO) of the United Nations via its website.

1.10 INSECTS AND HUMANS

So far, we have briefly considered the success of insects as a function of their enormous species richness and their huge numbers of individuals. Finally, we need to introduce the all-important associations that humans have with insects, to see how the most successful group of animals interact, for better or worse, with ourselves. In this way, we move into the realm of applied ecology.

1.10.1 Pests

The term "pest" is entirely anthropocentric and subjective. It tries to attach a label to an organism that, via its ecological activities, causes some sort of detriment to humans, their crops, or livestock. Ecologically, an insect pest is merely a competitor with humans for another limited resource, such as a crop. A crop, after all, is merely a rather special type of host-plant community from the viewpoint of a herbivorous insect. It is only because humans planted the crop for their own uses that this herbivore then assumes pest status (see Plates 1.1 and 1.2, between pp. 310 and 311).

Insect pests first appeared in written records thousands of years ago, with lice, mosquitos, and other flies for example achieving mentions in Assyrian texts and the Christian Old Testament (Levinson & Levinson 1990). Plagues of locusts appear in the Book of Exodus, and the early Egyptians of the Third Dynasty (over 3500 years ago) had to fight to keep their mummified ancestors free of insect infestations. In modern times, crop losses on a global scale are rather difficult to assess, but clearly insects can have an enormous detrimental impact on humans and their activities. Crop losses averaged from many published studies reach almost 45% of total yield annually, a colossal loss in food and other products, with potential yield losses to pests (weeds and pathogens as well as insects) ranging from around 50% for wheat to a colossal 80% in cotton (Oerke 2006).

Let us take cotton as an example. This is one of the world's most important cash crops, with the capacity to earn international money for many countries, both developed and developing (Figure 1.23). It is grown on a colossal scale; the world total cotton lint production for the 2005 period exceeded 23×10^6 tonnes (FAO 2007). Despite long and bitter experiences of the development of high levels of resistance in insect pests to insecticides over the years, cotton growers still have to rely predominantly on this method of pest management (see Chapter 12). In the USA in the 1990s, for example, despite an average of six treatments with insecticide per growing season, nearly 7% of the entire crop was lost to insect damage at one stage of growth or another (Luttrell et al. 1994). Australia seems even worse. Here, an average of 10 applications of insecticide may be carried out per year; the average insecticide application per year in the 2002–3 and 2003–4 seasons, for example, for conventional cotton was 135 kg insecticide per hectare (Knox et al. 2006). The true benefits of new technologies such as transgenic crops, genetically engineered to express toxins lethal to feeding insects, can be appreciated. For cotton, crucial benefits of genetically modified varieties are a 70% reduction in insecticide applications in *Bt* cotton fields in India, resulting in a saving of up to US$30 per hectare in insecticide costs, and an 80–87% increase in harvested cotton yield (Christou et al. 2006).

The previous examples involved developed countries. Even more significant perhaps is the fact that developing countries may not have the infrastructure, knowledge, or technology to manage their pests effectively or safely. In Rwanda, as just one example, a country noted for its extreme socioeconomic and political difficulties, insect pests of various types attack common beans, a vital crop for subsistence farmers.

Figure 1.23 Cotton crop. (Courtesy of C. Hauxwell.)

Trutmann and Graf (1993) reported losses of between 158 and 233 kg/ha, which is equivalent to a national loss to Rwanda of dry beans worth somewhere in the region of an amazing US$32.7 million (£22 million) per year.

In fact, in the early 21st century, we are no closer to removing these threats. Instead, we are constantly attempting to develop new and often highly sophisticated techniques to combat insect pests – based these days on a sound knowledge of the target insect's ecology, how it interacts with its environment, and the influence of other organisms, natural enemies in particular. This is the subject of Chapter 12.

Insect pests also interact directly with humans via stings, rashes, and more serious medical conditions. The tiny hairs on the larvae, pupae, and egg masses of the browntail moth, *Euproctis chrysorrhoea* (Lepidoptera: Lymantriidae), cause very serious reactions in people in the UK, from serious rashes (urticaria; see Plate 1.3, between pp. 310 and 311) (Doutre 2006) to temporary blindness and even death via anaphylactic shock (Sterling & Speight 1989). This insect was also an extremely serious problem in the USA, though in recent years it has rather mysteriously reduced it's originally huge range to a mere handful of sites where it still occurs (Elkinton et al. 2006). The pine processionary caterpillar, *Thaumetopoea pityocampa* (Lepidoptera: Lymantriidae), causes similar dermatitic and conjunctivitic reactions in continental Europe (Lamy 1990). There was even a recent record of dogs in the eastern Mediterranean exhibiting severe tongue necrosis after eating processionary caterpillars (Bruchim et al. 2005), obviously a lapse in the evolved defense system of the moth. Contact with *Lonomia achelous* (Lepidoptera: Saturniidae) larvae is known to bring about hemorrhagic diathesis, clinical bleeding with reduced blood clotting, which, on rare occasions, can prove fatal (Arocha-Pinango et al. 1992). Bee and wasp stings are, of course, a regular occurrence, but with the spread of Africanized bees to warmer parts of the USA, for example, worries are increasing about the potential medical effects of multiple stings (Schumacher & Egen 1995).

1.10.2   Vectors

A particular type of insect pest is one that is able to carry diseases from one mammalian host to another, or from one plant to another. These hosts may both be human, or one may be another mammal such as another primate or a rodent. They may be wild plants that provide reservoirs of diseases for infection of crops, or they may both be crops. Both local and global epidemics are vectored in this way by insects; the rising importance of malaria, sleeping sickness, plague, encephalitis, and so on illustrate the vital need to explore the intimate ecology of insect–disease associations, in attempts to reduce the colossal and direct impact on human lives. The World Health Organization (WHO) estimates that malaria causes more than 300 million acute illnesses globally every year, of which at least 1 million die of the disease. Over 80% or so of cases occur in sub-Saharan Africa (Torre et al. 1997). Such problems may appear to those living in developed nations as irrelevant or remote. However, human suffering on such a large scale must impinge on all our lives. A steadily increasing number of Europeans, for instance, are treated in hospital after returning from trips to tropical countries where the disease is rife (see Chapter 11).

Though the Diptera (including mosquitoes, blackfly, and tsetse flies) dominate in terms of the number of human and livestock diseases with which they are associated, other orders such as the Hemiptera (true bugs) and the Siphonaptera (fleas) also have an enormous impact. Bubonic plague, the "Black Death", which is carried by fleas, is thought to have killed around a third of the entire population of Britain in a pandemic that first appeared in England in 1348 (Kettle 1995).

Insects also vector numerous pathogenic organisms that cause extremely serious diseases in annual and perennial plant species. The list of major problems includes Dutch elm disease, barley yellows virus, and potato leaf curl, and insects range from Hemiptera (such as aphids and hoppers) to Coleoptera (beetles). The details of the ecology of insect–pathogen–host associations are considered in Chapter 11.

1.10.3   Beneficials

Insects that in some way damage ourselves or our livelihoods tend to be uppermost in people's thoughts, and though we are also familiar with bees and their activities, a very large number of insect species that are important to us are largely ignored. Undoubtedly, insects have a variety of pivotal roles to play in human livelihoods and poverty alleviation. Chapter 12 describes the vital importance of predatory and parasitic insects in the ever-growing field of biological pest control, and the revenue from diverse systems such as pollination of crops and the silk industry is staggering. In Italy, for instance, the value of crop pollination by insects is estimated at around 2000 billion lire (US $1.2 billion); the profit from honey, wax, and other beekeeping products is about 30 billion lire ($1.8 million) (Longo 1994). In the USA in recent years, two parasitic mites have caused drastic declines in feral honeybee populations, and crop losses associated with a consequent reduction in pollination run into billions of dollars. Silk has been produced from silkworms (Lepidoptera: Saturnidae) commercially for many hundreds of years, and is now estimated to be worth US$1200 million globally.

On a much broader scale, the roles played by a myriad of insects in food webs are of supreme importance to the functioning of a very large number of terrestrial and non-marine aquatic ecosystems. The lives of many bird species, such as blue and great tits in an English oak wood, for example, are utterly dependent on a plentiful supply of lepidopteran larvae when the chicks are in the nest, and partridges in farmland have a similar reliance on insect food for their young.

1.10.4 Esthetics

Insects have been prized for decoration and ritual for thousands of years. Scarab beetles (Coleoptera: Scarabaeidae) were hugely significant in Egypt during the time of the Pharoahs, when their likenesses were used as seals, trinkets, and charms. This widespread decorative use was associated with a much more serious belief that the scarab beetle came into being spontaneously from balls of dung. The Egyptians associated this with their religious ideology of self-creation and resurrection, and the scarab was worshipped under the name "Khepri". Insects, in this case cicadas, were again associated with reincarnation by both the ancient Chinese and the Romans (Kysela 2002) and they were used as jewelry in Europe for many centuries after the Romans.

Most humans, if they have any interest at all in wildlife, will tend to think about higher vertebrates such as birds and especially mammals when it comes to considering the esthetics of the animals around them. The majority of conservation systems in the world are still heavily biased towards this minor subphylum, but the perceived value of insects for amenity and as natural and important components of habitats is increasing. Conservation projects are now directed on occasion at targets such as butterflies or dragonflies, which have really no economic value; they are encouraged for their own sake. The ecology of insect conservation and augmentation is considered further in Chapter 10. Insects can also take part in the educational process, being considered particularly appropriate to interest and excite pre-college students (Matthews et al. 1997). The handling, rearing, and simple admiration of moths and butterflies, beetles, stick insects and mantids, cockroaches, and locusts can play a very significant role within an educational framework, even beginning at primary school level (see Plates 1.4 and 1.5, between pp. 310 and 311).

1.10.5 Food

To a surprisingly large number of people in the world, insects provide a vital source of protein and other nutrients. Insects are undoubtedly very useful nutritional sources (see Plate 1.6, between pp. 310 and 311). Costa-Neto (2003) reports that over 1500 species of edible insects have been recorded from nearly 3000 ethnic groups from more than 120 countries. More than 50 species of edible insects from 11 orders are documented from just one country, Sabah (East Malaysia; Chung et al. 2002). According to Costa-Neto, the most important groups of insects eaten are beetles, bees and wasps, grasshoppers, and moths. Table 1.6 provides some details of the nutritional value of a variety of insects used by rural people in the state of Oaxaca in Mexico (Ramos-Elorduy et al. 1997). In some regions, insects are eaten every day as part of the staple diet. They may be roasted, fried, or stewed, usually as larvae. Ants, bees, and wasps are apparently the most popular, as the table suggests, though butterfly caterpillars (*Phasus triangularis*) are also full of fat and high in calories.

Chinese people in Yunnan eat the larvae and pupae of wasps, which are rich in protein and amino acids (Feng et al. 2001). Deep-fried grasshoppers can be bought on street corners in Bangkok, Thailand, and the Tukanoan Indians of the northwest Amazon eat over 20 species of insect, the most important being beetle larvae, ants, termites, and caterpillars (Dufour 1987). Insects provide up to 12% of the crude protein

Table 1.6 Nutritional value of insects used as food in rural Mexican communities. (From Ramos-Elorduy et al. 1997.)

Dietary component	Content or range (%)	Insect with highest dietary content
Dry protein	15–81	Wasp larvae
Fat content	4.2–77.2	Butterfly larvae
Carbohydrate	77.70	Ants
Essential amino acids	46–96*	–
Protein digestibility	76–98	–

* Percentage of total requirements.

derived from animal foods in men's diets, and 26% in those of women. In Irian Jaya (Indonesia), the Ekagi people regularly eat large species of cicada (Hemiptera: Cicadidae) (Duffels & van Mastrigt 1991). The mopane worm, the larva of the mopane emperor moth *Imbrasia belina* (Lepidoptera: Saturniidae), in southern Africa has become a cash "crop", with an annual production of nearly 2000 tonnes. Caterpillars contain high levels of crude protein, more than 50% according to some reports, as well as high concentrations of calcium and phosphorus (Madibela et al. 2007).

Ironically, it is possible that conservation projects attempting to boost populations of large grazing mammals in game reserves in the region may deprive locals of this food source in that, in Botswana at least, local absences of mopane worm may be caused by extensive herbivory on their host plant (Styles & Skinner 1996). Undoubtedly, the "mini-livestock" could be harvested more intensively but sustainably (van Huis 2005) with considerable nutritional and economic consequences for regions such as sub-Saharan Africa.

1.11 CONCLUSION

As we noted at the beginning, all of the concepts introduced in this chapter will appear again later in the book, where they are considered in detail. Although each chapter is necessarily separate, it is important to try to consider insect ecology as a series of interlocking systems. For instance, to fully understand how an insect population can be of great economic significance in a crop, we need to look at its relationships with its host plant, its enemies, its competitors, and with the climatic conditions in its environment, although these aspects are presented in separate chapters.

Chapter 2

INSECTS AND CLIMATE

2.1 INTRODUCTION

The *New Shorter Oxford English Dictionary* defines climate as "the prevailing atmospheric phenomena and conditions of temperature, humidity, wind, etc. (of a country or region)", whereas weather is "the condition of the atmosphere at a given place and time with respect to heat, cold, sunshine, rain, cloud, wind, etc." We shall use climate both to describe major patterns such as wind, rain, and heat (or lack of them), and local ones that operate on a smaller scale. The latter category can be labeled microclimate, which is "the climate of a very small or restricted area, or of the immediate surroundings of an object, especially where this differs from the climate generally". Whether large scale such as typhoons or droughts, or small scale such as relative humidity, these abiotic factors can have fundamental influences on the ecology of insects – ranging from reproductive success and dispersal to growth and interactions within and among species. This chapter considers a series of climatic factors, one by one, and illustrates the various ways in which they may impinge on the ecologies of insects. Note that more detail on the influence of some abiotic factors such as temperature and rainfall are described in Chapter 7 on physiological ecology. Also note that climatic factors are not independent from one another. Although we treat them separately here for simplicity, abiotic factors interact with one another to influence the ecology of insects, and are often highly correlated with one another. For example, Briers et al. (2003) studied the flight activity of stoneflies (Plecoptera) in Wales, and found that the numbers of insects caught per week in malaise (flight intercept) traps was positively related to air temperature, but negatively related to wind speed. In other words, warm, still, conditions caught most insects.

2.2 TEMPERATURE

2.2.1 Warmth

Temperature influences everything that an organisms does (Clarke 2003). All insects can be considered to be poikilotherms, that is, their body temperature varies with that of the surroundings. Although some, such as bees or moths, can elevate the temperature of their flight muscles by rapid contractions before take-off, their basic metabolism is a function of the temperature of their surroundings – such that within a certain range, the higher the temperature, the faster metabolic reactions are able to proceed (see Chapter 7). Development time is essentially the reciprocal of development rate. Figure 2.1a shows the mean development time of the pine false webworm, *Acantholyda erythrocephala* (Hymenoptera: Pamphiliidae), a sawfly whose ecology is basically similar to that of a defoliating lepidopteran caterpillar. The development rate of larvae increases as development time decreases. However, at least for the eggs of the webworm, development stops altogether at around 30°C, and the eggs fail to hatch (Figure 2.1b) (Lyons 1994). It is also important to note that no eggs hatch at temperatures below 6°C. Clearly there is a fairly restricted temperature range over which these (and other) insect eggs can survive and mature.

As we might expect, natural selection will generally act to match the physiology of insects to the

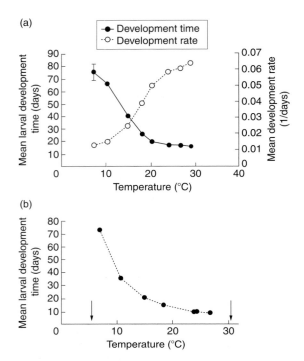

Figure 2.1 Mean development time for (a) larvae and (b) eggs of the pine false webworm reared at constant temperatures. Eggs failed to hatch at temperatures below 5.8°C and above 30.3°C (arrowed). (From Lyons 1994.)

environment in which they live. We might therefore expect warm-adapted species and cold-adapted species to vary in their temperature optima for vital processes such as growth and reproduction. For example, some species of butterfly in Europe belong to open, hot (and dry) habitats, such as the grayling, *Hipparchia semele*, and the small heath, *Coenonympha pamphilus* (Lepidoptera: Nymphalidae). On the other hand, species such as the ringlet, *Aphantopus hyperantus* (Lepidoptera: Satyridae), and the speckled wood, *Pararge aegeria* (Lepidoptera: Nymphalidae), are shade-dwelling (Karlsson & Wiklund 2005). Figure 2.2 shows that lifetime egg number (fecundity) peaks at near 30°C for open landscape species as compared with 25°C for shade species. Adaptations by insects to local temperatures can also include changes in phenology and morphology. Thus bumblebees in Western Europe are larger in northern climes such as Scotland when compared with southern England, and species from colder regions have longer setae (hairs) than those from warmer areas (Peat et al. 2005).

Of course, the ecologies of carnivorous insects such as predators and parasitoids will also be affected by temperature. Figure 2.3 illustrates how the development rate of a predatory ladybeetle (Coleoptera: Coccinellidae) increases linearly with temperature, while survival from egg to adult peaks at around 27°C (Omkar & Pervez 2004). Similarly, the development rate of the predatory bug *Joppeicus paradoxus* (Hemiptera: Joppeicidae) increases with temperature (Figure 2.4), but shows a decline, especially in the

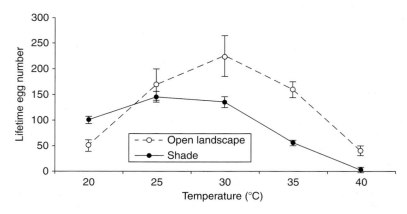

Figure 2.2 Total lifetime egg production in open landscape versus woodland species of butterfly at different temperatures. (From Karisson & Wiklund 2005.)

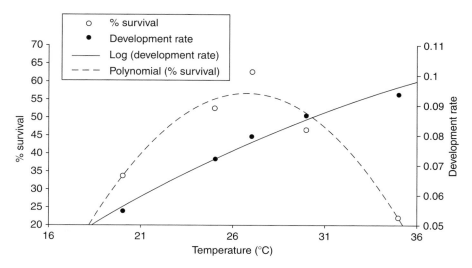

Figure 2.3 Development rate and percent survival from egg to adult emergence of the ladybird *Propylea dissecta* in relation to temperature. (From Omkar & Pervez 2004.)

nymph stage, after around 32°C (Morimoto et al. 2007). The activity and foraging success of predators or parasitoids can be temperature dependent. The probability that spruce budworm, *Choristoneura fumiferana* (Lepidoptera: Tortricidae), eggs will be parasitized by the parasitic wasp, *Trichogramma minutum* (Hymenoptera: Chalcidae), increases with temperature (Figure 2.5a) and declines with relative humidity (Figure 2.5b) (Bourchier & Smith 1996). The wasp is able to move more rapidly at higher temperatures, and hence to search out and lay eggs in more host eggs per unit time. Simultaneously, humid conditions, especially over 85% or so, are usually associated with rain, and it is well known that small parasitoids have difficulty moving around in wet conditions.

High temperatures may increase the movement rate of parasitoids, but not without cost. One of the drawbacks of a life at higher temperatures appears to be a reduction in longevity; insects live shorter lives at higher temperatures, everything else being equal. In the case of the parasitic wasp *Meteorus trachynotus* (Hymenoptera: Braconidae), which attacks larvae of the spruce budworm in North America, adult females live for a much shorter time when exposed to higher temperatures (Thireau & Regniere 1995) (Figure 2.6). A 40-day adult lifespan at 15°C is reduced to a mere 10 days or so at 30°C. Likewise, in the case of the European butterflies described above, adult lifespan

declines almost linearly as temperature increases, with shade species suffering greater declines in longevity than open landscape species (Figure 2.7). Life for insects under warm conditions may be active but short.

Understanding the effects of climatic variation on insects is critical for predicting the activity and density of insect crop pests and vectors of disease (see Chapters 11 and 12). For example, Figure 2.8 illustrates the development time (declining curves) and development rate (another hump-shaped curve) of the mosquito *Anopheles gambiae* (Diptera: Culicidae) (Bayoh & Lindsay 2003). This species is notorious as the principal vector of malaria in Africa (see Chapter 11). As temperature rises, there comes a point (in this case above 30°C) where mosquito development rate declines precipitously. These data can be combined with the development times of egg, larval, and pupal stages to create a three-dimensional surface that predicts the emergence of adult mosquitoes (Figure 2.9). This in turn may be useful for predicting malaria epidemics (but see Chapter 11). These sorts of prediction can be difficult on occasion, especially if the insect in question enters a period of quiescence known as diapause in response to certain climatic conditions. Diapause in the summer months (also called estivation) occurs to avoid one or more potentially damaging environmental conditions, such as drought, high temperatures, or starvation, and also to synchronize life

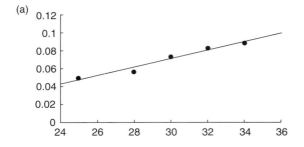

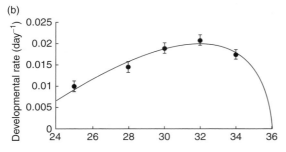

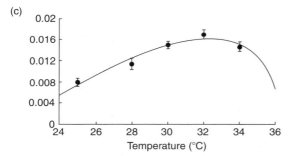

Figure 2.4 Relationship between temperature and the development rate of the (a) egg, (b) nymph, and (c) total immature stage of *Joppeicus paradoxus* as a function of temperature. Data points are mean observed rates ±95% CI. (From Morimoto *et al.* 2007.)

cycles such as the appearance of larval or adult stages with seasons (Liu et al. 2006b). We shall return to the topic of diapause again later in this chapter.

2.2.2 Cold

As temperatures decrease below some critical threshold, insect survival also declines, though frequently this relationship is not linear. Figure 2.10 provides an example from Crozier (2003) who studied the

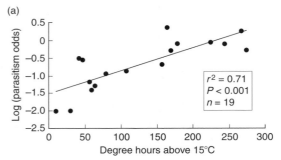

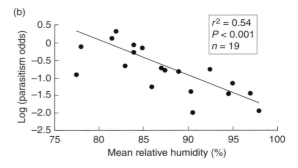

Figure 2.5 Relationship between (a) the number of hours above 15°C, and (b) the mean relative humidity, in a 3-day period following a release of the parasitoid *Trichogramma minutum*, and the probability of parasitism of spruce budworm egg masses. (From Bourchier & Smith 1996.)

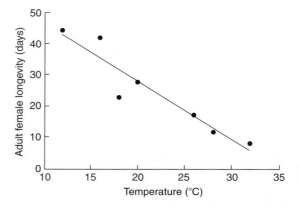

Figure 2.6 Longevity of adult female parasitoids, *Meteorus trachynotus*, at various temperatures. (From Thireau & Regniere 1995.)

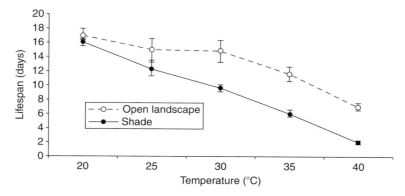

Figure 2.7 Lifespan of open landscape versus woodland species of butterfly at different temperatures. (From Karlsson & Wiklund 2005.)

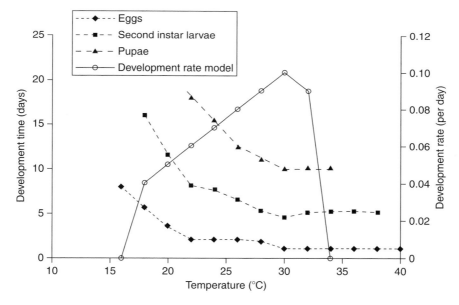

Figure 2.8 Comparison of a day-degree model of *Anopheles gambiae* with the development times of various aquatic stages. (From Bayoh & Lindsay 2003.)

skipper butterfly, *Atalopedes campestris* (Lepidoptera: Hesperiidae), in the USA. She found that minimum winter temperatures (T_{min}) constrained the range of the butterfly, and concluded that winter warming might thus allow range expansion of the species (see later). Like the skipper butterfly in Figure 2.10, all insect populations will exhibit a characteristic low-temperature threshold below which survival

declines. However, this does not mean that all ranges of cold temperature are necessarily bad. One advantage of living in the cold is that the energy demands made by metabolism are low, so that if resources such as food are in short supply, insects can survive longer without starving. The gypsy moth, *Lymantria dispar* (Lepidoptera: Lymantriidae), is an extremely serious defoliator of broad-leaved trees such as oak in

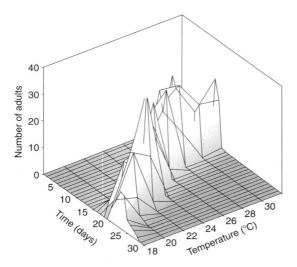

Figure 2.9 A three-dimensional view of the influence of temperature on adult *Anopheles gambiae* emergence times and numbers produced. (From Bayoh & Lindsay 2003.)

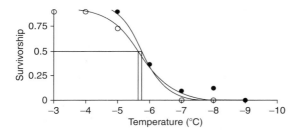

Figure 2.10 Lethal temperature for 50% of the sample (LT_{50}) for Californian (○) and Washingtonian (●) third instar larvae of *Atalopedes campestris* with a hyperbolic tangent curve fit. The estimated LT_{50} are $-5.6°C$ (California), and $-5.8°C$ (Washington). The difference between populations is not significant ($P > 0.05$). (From Crozier 2003.)

North America, and it is essential for its young larvae to find nutritious and palatable leaf material on which to feed as soon as possible after they hatch (see Plate 2.1, between pp. 310 and 311). The starvation rate of these insects increases fairly linearly with temperature above about 10°C, so that a warm spell during spring before most trees have broken bud can cause the larvae to starve to death. Below 10°C, however, the insects appear to remain in a kind of suspended animation until warmer weather returns (Hunter 1993).

Development in insects is often predicted and modeled by the accumulation of day-degrees. Put very simply, if an insect requires 100 day-degrees to develop from egg to adult, then in theory this could consist of 10 days at 10°C or 5 days at 20°C. Of course, it is not always that simple. Very low or very high temperatures may be deleterious or lethal. In addition, there is normally a base temperature below which the accumulation of day-degrees does not take place (Hunter 1993). For example, the scale insect *Hemiberlesia rapax* (Hemiptera: Diaspididae), a pest of kiwi-fruit in New Zealand, has a total development time of 1056 day-degrees above a base temperature of 9.3°C (Blank et al. 1995). One advantage of possessing such data for insect pests is that it becomes possible to predict fairly accurately when a certain life stage will be attained after a reference point such as egg laying, as long as temperature records are kept for the habitat in which the insect is active (see Chapter 12).

Insects that live in temperate or arctic regions must be able to survive extremes of cold, and in order to do so they may have to adopt a combination of tactics such as diapause or cold-hardiness to ensure that at least some of their population survives to the following spring (Andreadis et al. 2005). Rapid changes in insect physiology can accompany drops in temperature, and many insects become "cold hardened" after exposure to low temperatures for as little as a few hours (so-called rapid cold hardening or RCH) (Sinclair et al. 2003; Powell & Bale 2004). Most important, insects in cold regions must be able to prevent or withstand the freezing of their body fluids for extended periods without damage. In extreme cases, aquatic insects such as various dipteran larvae may have to diapause inside ice forming on their ponds in winter (Bouchard et al. 2006). This occurs by way of a process known as supercooling. The supercooling point (SCP) is considered to be the absolute lower limit for survival. Figure 2.11 illustrates changes in the body temperature of a cockroach, *Celatoblatte quinquemaculata* (Dictyoptera), as it cools. Notice that, as it reaches the SCP, the cockroach exhibits what is called a rebound temperature, whereby the body actually appears to resist freezing (where the cell contents solidify) for some time (Worland et al. 2004).

The elm bark beetle *Scolytus laevis* (Coleoptera: Scolytidae) is also capable of supercooling. This European species is a potential vector of Dutch elm

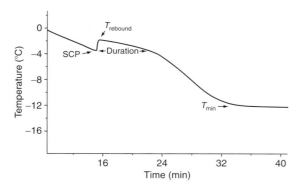

Figure 2.11 Typical cooling curve for a whole cockroach cooled at 0.5°C/min. It shows the supercooling point (SCP: −3.2°C), the rebound temperature ($T_{rebound}$: −1.5°C), the exotherm duration (7.2 min) and the minimum temperature (T_{min}: −12°C). (From Worland et al. 2004.)

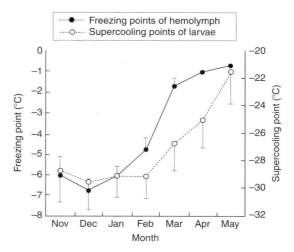

Figure 2.12 Mean freezing and supercooling points of overwintering larvae of *Scolytus laevis*. (From Hansen & Somme 1994.)

disease (see Chapter 11), and exhibits supercooling to withstand winter temperatures in Scandinavia (Hansen & Somme 1994). The mean SCP of larvae varies seasonally (Figure 2.12), reaching as low as −29°C in midwinter, and as high as −21°C in May. In experiments, the hemolymph of the insects actually froze at around −7°C in December and around −1°C in the spring. When frozen to temperatures corresponding to their minimum SCPs, larvae were able to recover muscular contractions for a while, but later died. In reality, larvae were able to survive temperatures as low as −19°C in winter, but were killed at that temperature in spring. Hansen and Somme (1994) conclude that the beetle (and therefore Dutch elm disease) has the potential to spread to habitats where winter temperatures remain above −19°C or so.

The mechanistic basis for insect supercooling seems to be enhanced concentrations of glycerol in the body fluids, which acts as an antifreeze. The reduction of supercooling ability in spring results from declining levels of this chemical. Many insects are capable of synthesizing glycerol. The spruce budworm, *Choristoneura fumiferana* (Lepidoptera: Tortricidae), produces higher levels of glycerol when exposed to lower temperatures, so that as winter sets in the insect is able to respond by setting up its supercooling abilities in advance of serious cold (Han & Bauce 1995). However, if these cold spells arrive too early, and the larvae are not in the appropriate

physiological state to synthesize glycerol, they may still be frozen to death. The biotic environment also has a role to play in freeze protection for insects. In the elm bark beetle described above, an insulating layer of bark will to some extent protect from, or at least delay, extremes of cold. Similarly, monarch butterflies, *Danaus plexippus* (Lepidoptera: Nymphalidae), need to overwinter in sites covered by forest canopy. This canopy layer insulates them from freezing winter storms at high altitude in Mexico (Anderson & Brower 1996). Another problem for monarchs is wetting. If it rains heavily before freezing, then their supercooling abilities are unable to prevent them from dying. The influence of rainfall on insect ecologies is discussed later in this chapter.

Finally, it has been suggested that prolonged periods of relatively extreme cold have shaped the life cycles of some modern-day insects. Periodical cicadas in the genus *Magicicada* (Hemiptera/Homoptera: Cicadidae) have unusually long life cycles for insects, with periodicities of either 13 or 17 years (Grant 2005). Nymphs develop very slowly in the soil, feeding on the xylem sap of tree roots, and emerge synchronously as adults to mate and return to obscurity in the soil for a further decade and a half. This life history may have evolved in response to perpetually low temperatures and hence extremely scarce resource availability during the last few ice ages.

2.3 DAYLENGTH (PHOTOPERIOD)

The length of daylight during a 24 h day–night cycle, known as photoperiod, may not exactly qualify as a climatic factor per se, but it does have a fundamental influence on the development and ecology of insects that live in seasonal climates. The daylength experienced by an insect provides information about the progression of the seasons, and the ability to vary growth and development rates enables the insect to achieve efficient timing relative to favorable conditions (Leimar 1996). *Kytorhinus sharpianus* (Coleoptera: Bruchidae) is a bean weevil from Japan that oviposits and develops as a larva inside both immature fresh and mature legume seeds (Ishihara & Shimada 1995). The durations of the various stages in the life cycle from egg to adult vary according to the photoperiod at constant temperature (Figure 2.13). The whole cycle can be accomplished between 50 and 80 days when the insect receives 15 or 16 h of daylight per day–night cycle, but this period increases dramatically as the hours of daylight diminish to 14 h and then 12 h. With only 12 h of daylight the pupal stage is never reached until longer hours of light return.

As in the bruchid example above, daylength can be used as a signal or trigger by insects to enter a quiescent period, diapause, within which to wait out potentially deleterious conditions such as summer heat or drought, or winter cold. The summer diapause may also be used by adult insects such as plant-feeding chrysomelid beetles (Coleoptera: Chrysomelidae) to lie quietly and become sexually mature before reproduction in the late summer or fall (Schops et al. 1996). Indeed, some insects show both types of diapause. The burnet moth, *Zygaena trifolii* (Lepidoptera: Zygaenidae), occurs over much of Europe but, in Mediterranean regions, it may undergo two types of diapause within 1 year (Wipking 1995). First, a facultative diapause may take place, lasting between 3 and 10 weeks in summer depending on weather conditions, followed by an obligate winter diapause, which may last several months if low temperatures persist. As with most climatic effects, it is difficult to separate the dual influences of temperature and daylength on both the onset and the cessation of diapause. However, daylength has been shown to have a fundamental effect on diapause induction and cessation in some cases. An example is shown in Figure 2.14. When larvae of the parasitic wasp, *Cotesia melanoscela* (Hymenoptera: Braconidae), were exposed to long daylengths (greater than 18 h), they developed virtually continuously all the way to adulthood. In contrast, larvae exposed to short daylengths (less than 16 h) entered diapause, which halted development in the cocooned prepupal stage (Nealis et al. 1996). In ecological terms, this system makes sure that adult parasitoids appear only during the long days of summer when their host insects, in this case gypsy moth larvae, are most likely to be abundant.

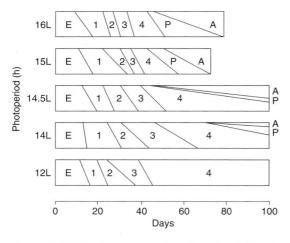

Figure 2.13 Development schedules for a bruchid beetle under five photoperiod cycles at 24°C. E, egg; 1–4, larvae instars; P, pupa; A, adult. (From Ishihara & Shimada 1995.)

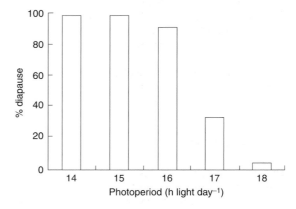

Figure 2.14 Percentage diapause in cohorts of *Cotesia melanoscela* reared at 21°C in relation to photoperiod. (From Nealis et al. 1996.)

The life cycles of aphids are complex (see Chapter 6), and because they are particularly influenced by intimate associations with the chemistry and phenology of sometimes several different species of specific host plant, they have to be closely controlled by environmental factors so as to stay "in tune" with their hosts. This they achieve in part by responding to changes in daylength and temperature. The damson-hop aphid, *Phorodon humuli* (Hemiptera: Aphididae), produces overwintering eggs that are laid on sloe (*Prunus spinosa*), damson (*P. insititia*), or plum (*P. domestica*), and when the eggs hatch in spring the resulting nymphs undergo several wingless generations on these host species before producing winged adults (the so-called emigrants) that fly to hops (*Humulus lupulus*), the only summer host (Campbell & Muir 2005). In Figure 2.15, it can be seen that the emigrants do not fly much at all at low temperatures, and only between sunrise and sunset. Low light seems to inhibit takeoff.

As we have mentioned before, no individual climatic effect is easily separable from another. For example, photoperiod combines with temperature to determine the proportion of adult female *Empoasca fabae* ovipositing (Taylor et al. 1995) (Figure 2.16) and female monarch butterflies entering diapause (Goehring & Oberhauser 2002) (Figure 2.17). Under both decreasing and long photoperiod regimes, constant rather than fluctuating temperatures reduce the incidence of diapause. Some interactions between temperature and photoperiod can be quite complex and subtle, as shown by diapause induction in the pine caterpillar, *Dendrolimus tabulaeformis* (Lepidoptera: Lasiocampidae), in China (Han et al. 2005) (Figure 2.18). After 9 or 10 h in the dark period of a light–dark cycle (scotophase), nearly 100% of larvae enter diapause. Understanding the independent and interactive effects of climatic factors on insect diapause is very important for predicting the activity of insect pests (see Chapters 11 and 12).

2.4 RAINFALL

The effects of rainfall on insects can be direct or indirect. Heavy rain can knock aphids off their host plant and both beetles and bugs may be killed by violent thunderstorms (T.R.E. Southwood, personal communication). Lack of rain can cause desiccation and death. Seasonal rains influence the ways in which host plants flush and grow and provide food, in particular, the quality of plant phloem (Hale et al. 2003) for herbivores (and hence, of course, for all those in higher trophic levels above them). Droughts can stress trees, rendering them susceptible to insect attack. Rainfall also affects humidity, which combines with temperature and wind to dictate local microclimatic conditions. Finally, of course, water itself is a vital habitat for around 3% of all insect species, the aquatic ones.

At a very simple level, lots of rain means lots of water that can flood ponds and lakes, or cause river spates and inundations. Such conditions may increase habitats for insects, or it may cause extra problems by creating fierce currents that may destroy insects or the places where they live (Wantzen 2006). Rainstorms regularly lead to extensive water runoff in water courses, causing erosion and high sediment levels which in turn can have significant effects on the distribution of aquatic insects such as midges (Diptera: Chironomidae) (Gresens et al. 2007).

The immature stages of orders such as dragonflies (Odonata), mayflies (Ephemeroptera), stoneflies (Plecoptera), and caddisflies (Trichoptera) are dependent on ponds and streams, as are myriads of species of flies (Diptera), beetles (Coleoptera), and bugs (Hemiptera) for at least part of their life cycles. Rainfall affects water levels and current strengths in these habitats, and also creates a number of more specialized and ephemeral aquatic habitats in some rather unlikely places. These include open-cast coal mine sites in Australia (Proctor & Grigg 2006), bromeliad "pools" high in rainforest canopies, and also bamboo internode communities illustrated in Figure 2.19. In this example from Amazonia, young bamboo shoots have slits cut in them by ovipositing katydid grasshoppers (Orthoptera: Tettigoniidae). As the shoot grows, reaching on average 45 mm in diameter, the oviposition scars deteriorate into a series of open slots and then to a single opening. Rain fills the cavity up to the level of the slot, and a succession of insects and other taxa colonize the little pool and its environs. These include mosquitoes (Diptera: Culicidae), rat-tailed maggots (Diptera: Syrphidae), and damselflies (Odonata: Zygoptera). Even amphibians take up residence to prey upon the emerging adult insects (Louton et al. 1996). Mosquito populations generally increase in wet weather, simply because rain fills all available containers in which the larvae can live (Vezzani et al. 2004) (see Chapter 11).

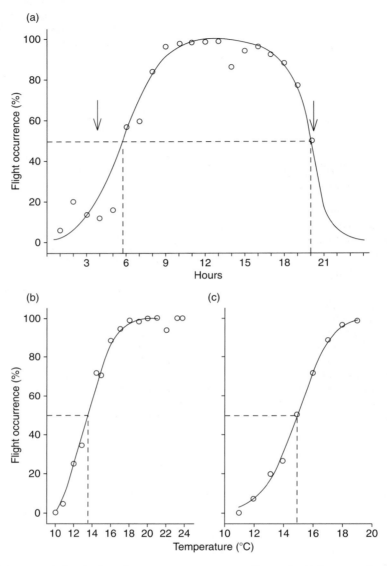

Figure 2.15 Flight activity of *Phorodon humuli* emigrants as percentage of hours that flight occurred against (a) time (mean sunrise and sunset, arrow), (b) in relation to temperature during hourly trapping periods, and (c) cumulative percentage takeoff in the laboratory. (From Campbell & Muir 2005.)

Rainfall patterns can influence the long-term abundance of insect populations. Figure 2.20 shows tree ring chronologies from 24 mixed conifer stands in northern New Mexico, from which large-scale outbreaks of the western spruce budworm, *Choristoneura occidentalis* (Lepidoptera: Tortricidae), have been detected over the last 300 years (Swetnam & Lynch 1993). The simple link is that heavy defoliation by the moth's larvae causes significant reductions in that period's ring widths. Superimposed on these are rainfall patterns, of which the older ones have been reconstructed by examining the growth rings of an

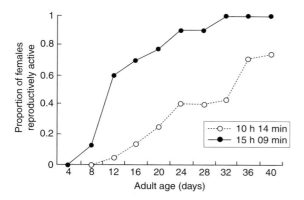

Figure 2.16 Proportion of female *Empoasca fabae* ovipositing by a given adult age at constant temperature (24°C) and two different daylengths. (From Taylor et al. 1995.)

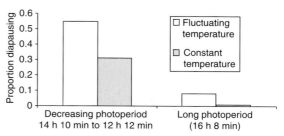

Figure 2.17 Proportions of female *Danaus plexippus* in diapause according to photoperiod and temperature regime. (From Goehring & Oberhauser 2002.)

insect-free tree species, limber pine. Although much variation in this type of data is to be expected, the figure shows fairly clearly that periods of increased and decreased budworm activity coincide with wetter and drier periods, respectively. Analysis of variance was carried out on these data, and Swetnam and Lynch felt that there was "compelling evidence" for a climate (rainfall)–budworm association that has existed for at least three centuries. The underlying mechanisms are rather harder to determine, however. The plant stress hypothesis (see Chapter 3), which

suggests that drought-struck trees will provide better nutrients for and possibly fewer defenses against herbivores, does not seem to apply here; wetter weather promotes the insects, not the reverse. Similar patterns have been found for gall-forming sawflies on willow trees in Arizona (Hunter & Price 1998; Price & Hunter 2005). At least in some systems, then, rainfall is associated with an increase in the abundance of insect herbivores, not a decrease.

Insect abundance is certainly linked to seasonal variations in rainfall, with some species being more abundant in the dry season, whereas others proliferate only during the rains. In the eucalyptus forests of northern Australia, Fensham (1994) found conflicting distributions of various insect taxa according

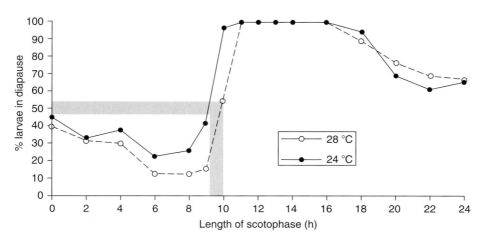

Figure 2.18 Photoperiodic response curves for the induction of diapause in *Dendrolimus tabulaeformis* under 24 h light–dark cycles at 24 and 28°C. (From Han et al. 2005.)

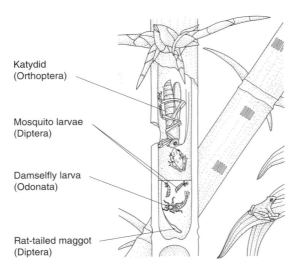

Katydid
(Orthoptera)

Mosquito larvae
(Diptera)

Damselfly larva
(Odonata)

Rat-tailed maggot
(Diptera)

Figure 2.19 Inhabitants of a bamboo internode community from Amazonia. (From Louton et al. 1996.)

to season. Two examples of herbivorous taxa from that work, one a sap-feeder (Hemiptera: Psylloidea) and the other a defoliator (Orthoptera) show interesting contrasts. The psyllids were very much more common in the late dry season than at any other time of year, whereas the grasshoppers, although present in the early and late dry seasons, were most abundant during the wet (rainy) season. Fensham suggests that the sap-feeders receive nutrients from sap produced by the regrowth of trees in response to fires that sweep through these forests in the late dry season. The defoliating Orthoptera, on the other hand, benefit from the relatively luxuriant production of new foliage during the rains. This suggests links between rainfall (or lack of it), host-plant physiology, and insect performance: a topic to which we shall return in various later chapters. For now, we simply point out that some insect species appear to benefit from wet conditions (Figure 2.21) while others benefit from dry conditions. An example of the latter is provided by the bird cherry-oat aphid, *Rhopalosiphum padi* (Hemiptera: Aphididae), that has a higher intrinsic rate of increase when feeding on host plants subjected to water stress (Hale et al. 2003). It may be that drought conditions increase the quality of phloem sap (in terms of nitrogen content; see also Chapter 3), but this is clearly not the whole story. On one of the four plant species, drought stress has no impact upon aphid growth rate.

Associations between insect distributions and rainfall patterns are frequently linked to the all-crucial host-plant availability, especially in semidesert conditions. The solitary phase of the infamous desert locust, *Schistocerca gregaria* (Orthoptera: Acrididae),

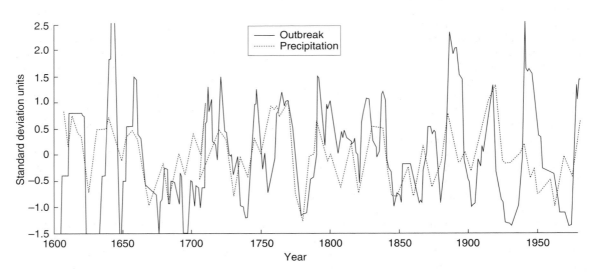

Figure 2.20 Tree ring chronologies from Donglasfir in New Mexico over nearly 300 years, depicting spruce budworm outbreaks, with reconstructed March–June rainfall patterns for the same period. (From Swetnam & Lynch 1993.)

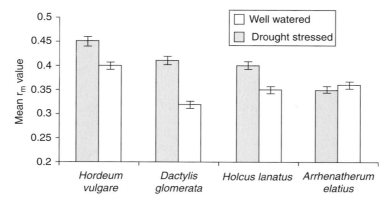

Figure 2.21 Intrinsic rate of increase (r_m) of *Rhopalosiphum padi* reared on four plant species grown under well-watered and drought-stressed conditions (all means are significantly different except for *A. elatius*). (From Hale et al. 2003.)

becomes concentrated into contracting areas of sufficient rainfall and humidity as dry conditions proliferate at the end of the wet season in Sudan (Woldewahid et al. 2004). Only in these areas will the insect's host plant *Heliotrophum arbainense* continue to be abundant (along with local crops of millet). Figure 2.22 illustrates the predicted distributions of locusts and host and crop plants in the dry midwinter. Notice that plants are distributed patchily, restricted to localities where the rain still falls, or where there is sufficient ground water in the form of wadis in the desert. The overcrowding of locusts within small patches is one of the factors that promotes the shift from the solitary to the migratory (outbreak) phase.

As well as direct effects of rain on insects, and indirect effects mediated by plant quality, the efficacy of natural enemies of insects may also be affected by rainfall. Insect-specific viruses, such as nuclear polyhedrosis viruses (NPVs; see Chapter 12), provide an example. These pathogens are able to remain infective outside the host-insect body, attached to foliage and bark, for considerable lengths of time. In an experiment with the gypsy moth, *Lymantria dispar* (Lepidoptera: Lymantriidae), in the USA, leaves of the host tree, red oak, were inoculated with gypsy moth NPV by using infected first instar caterpillars caged onto foliage in the upper canopy (D'Amico & Elkinton 1995). When these insects died, they released NPV onto the leaves, on which fresh, healthy third instar larvae were subsequently placed. Similar larvae were also placed on clean leaves on lower

branches, situated below the NPV-inoculated upper ones. Some trial branches were then protected from rain, and the mortalities of the larvae measured (Figure 2.23). Significantly fewer larvae died from NPV infection on the upper branches in wet conditions: the rain apparently washed some of the virus off the upper leaves and hence reduced the pathogen concentration. However, more larvae died of NPV on the lower branches exposed to rain, as NPV washed off the upper branches reached the lower ones, where some larvae picked up the infection.

The effects of rainfall on insects are often complex and difficult to separate from other allied climate patterns, wind in particular. Rain and wind can result in areas of low pressure and weather convergences that redistribute insect populations and at the same time provide fresh plant food for them to eat (see later). Rain may not even have an influence on insect performance at the time it falls, but instead may promote insect performance some months later. This phenomenon is well illustrated by the seasonal outbreaks of African armyworm, *Spodoptera exempta* (Lepidoptera: Noctuidae). Armyworms can reach enormous numbers, and have been serious pests of cereals and pasture in sub-Saharan Africa for over 100 years (Haggis 1996). In Kenya, long-term rainfall records were used to predict the number and size of armyworm outbreaks, and it was found that the number of outbreaks was negatively correlated with rainfall in the 6–8 months preceding the start of the armyworm season. In other words, severe outbreaks are very often preceded by periods of drought.

(a)

(b)

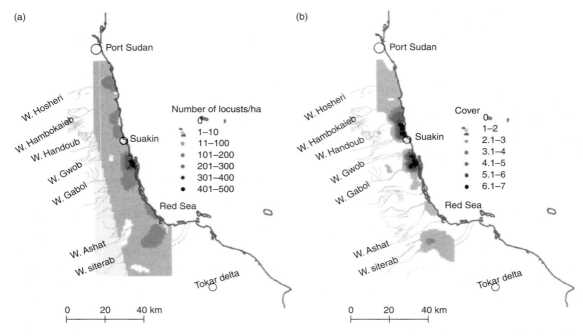

Figure 2.22 Kriged (interpolated) maps of (a) locust density estimates (number/ha) across the site: in 1999/2000 winter season for midseason, and (b) *Heliotropium*/millet cover abundance. (From Woldewahid et al. 2004.)

The explanation lies again with NPVs from which epidemic armyworm populations suffer. In wet, humid conditions, the larvae become stressed and succumb to viral infections, whereas in dry weather,

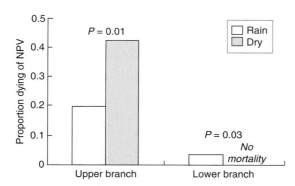

Figure 2.23 Mean proportion of third instar gypsy moth larvae killed by nuclear polyhedrosis virus (NPV) on upper and lower tree branches in wet and dry weather. (From D'Amico & Elkinton 1995.)

the sun rapidly kills the viruses. In these dry spells, as long as there is still enough rain for the host plants to grow, then low-density populations of armyworm can persist for some time, in fact until the next outbreak. So dependable are the links between rain and later armyworm outbreaks that Haggis (1996) has been able to construct a prediction system for Kenya, which is illustrated in Figure 2.24. In almost all cases, the forecasts produced by this system correctly predict the likelihood of armyworm outbreaks. It should be noticed that the system provides opportunities to modify decisions based on updated rainfall conditions. In a similar fashion, it is now possible to combine rainfall and temperature data to map existing populations of pests and to predict their future behavior (see also Chapter 11). Because insects such as mosquitoes are so dependent on particular rainfall and temperature conditions, it is possible to map the transmission of diseases such as malaria by plotting rainfall and temperature data in association with vegetation indices such as the normalized difference vegetation index (NDVI) (Gemperli et al. 2006).

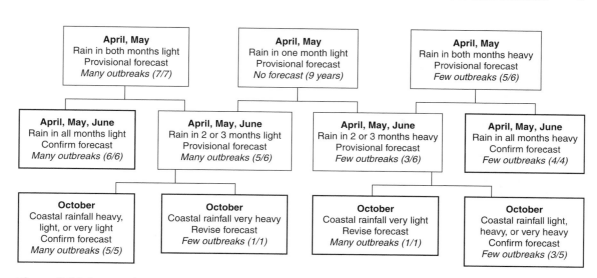

Figure 2.24 Stages in forecasting local or widespread armyworm outbreaks next season from current season's rainfall in southeast Kenya (based on 22 years data). Numbers in parentheses denote number of years when forecast was fulfilled/number of years with rainfall sequence. Provisional forecast with thin border; final forecast with bold border. (From Haggis 1996.)

Once these systems provide dependable predictions of malaria transmission, decisions can be made about where and when control such as insecticide-treated bednets may be necessary.

2.5 WIND

As with rain, wind can have a variety of influences on the ecology of insects. It can carry insects many miles to new habitats and regions (Harrison & Rasplus 2006), it can bring rain to produce new food (Ji et al. 2006), it can destroy trees to provide new breeding sites (Eriksson et al. 2005), and it can transmit volatile chemicals from one insect to another, or from a host plant to an insect (Wyatt 2003).

Many insects appear to undertake enormous migrations covering hundreds if not thousands of miles on occasion. The record for the sheer number of insects migrating seems to be held by the darner dragonfly, *Aeschna bonariensis* (Odonata: Aeshnidae), which is reported to migrate through Argentina in swarms containing 4–6 billion individuals, with a biomass of 4000 tons (Holland et al. 2006). Some species regularly move from continental Europe across the English Channel to southern and eastern England, and a few, such as the monarch butterfly, *Danaus*

plexippus (Lepidoptera: Nymphalidae), have been known to make it all the way from America across the Atlantic Ocean to Europe. Undoubtedly, they do not perform this feat unaided, and in fact it is air currents or wind that assists in both the distance and direction traveled. The long-distance migrations of monarchs were noted many years ago. Zalucki and Clarke (2004) quote J.J. Walker who in 1886 published a colleague's observation of monarch butterflies "flying at a great height above the ship, sometimes more than 200 miles from the nearest land. During a cruise between New Caledonia and the Solomon Islands, they were seen everyday, often in numbers ... the South East trade wind, which was blowing strongly at the time, was greatly in favour of the butterflies". By riding these trade winds, monarchs were able to spread across the Pacific Ocean in the 1880s, arriving in mainland Asia by 1900 (Zalucki & Clarke 2004) (Figure 2.25).

A butterfly related to the monarch, *Danaus chrysippus*, is a tropical species that occurs in a number of genetic races across Africa. Smith and Owen (1997) studied the seasonal cycles of abundance of several of these races in and around Dar es Salaam on the Tanzanian coast, and found that population oscillations were related to migratory activity of the butterflies. This activity in turn was related to the cyclical

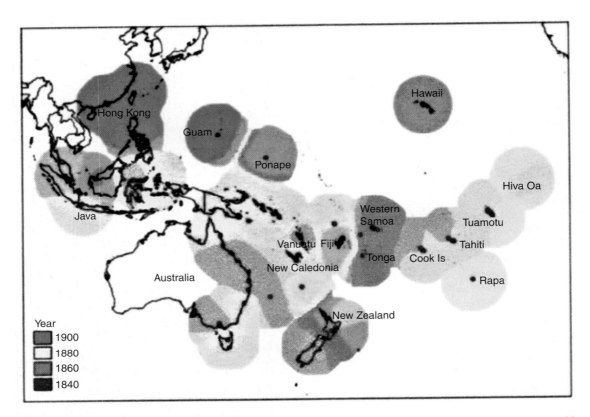

Figure 2.25 The spread of monarch butterflies, *Danaus plexippus*, across the Pacific in the 1800s. The map is generated by assuming that each new population was derived from the nearest neighboring population (in any direction) with a confirmed earlier arrival, unless an intervening island group was known to be free of the butterfly. Note that populations appear to stem from one or two incursion points in the South Pacific. (From Zalucki & Clarke 2004.)

north–south movement of wind, coupled with the allied rainfall and temperature regimes imposed by the monsoons (Figure 2.26). The problem, however, with the sorts of data shown in the figure is attributing cause and effect. Smith and Owen cited the Australian common name for *D. chrysippus* as "wanderer" – clearly an insect known to move about the place. Large butterflies such as this one can certainly fly against the wind when they have to, and so it is not yet clear in this example whether or not the monsoon winds are carrying the insects on their seasonal migrations, or whether it is simply the wind that creates the right environmental conditions (such as temperature and food plants) to which the butterflies return at certain times of the season.

Wind is a vital component of more general weather patterns, giving rise to fronts and convergence zones. Atmospheric convergence occurs on a number of scales (Drake & Farrow 1989). In the tropics and middle latitudes, belts of convergent and ascending air occur mainly in the Intertropical Convergence Zone (ITCZ) and produce heavy rain and storms. Two converging winds can concentrate airborne insects into a relatively localized region (Figure 2.27). The insects are assumed to rise to their flight ceiling and subsequently land to feed and reproduce. As the same convergence zone produces rain, semiarid tropical regions soon bloom with luxuriant vegetation on which the large densities of new-generation insect larvae feed voraciously. Serious African pests such as

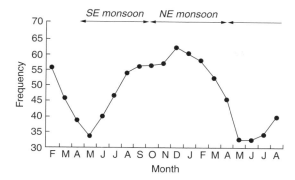

Figure 2.26 Frequencies (3-month moving average) of the orange phenotype of the butterfly *Danaus chrysippus* in Dar es Salaam, Tanzania, relative to monsoon patterns. (From Smith & Owen 1997.)

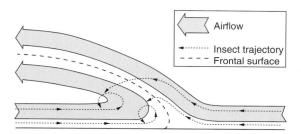

Figure 2.27 Vertical section through a frontal convergence, showing a likely mechanism of entrapment, and hence concentration, by the front. The insects are assumed to rise until they reach their flight ceiling, which will be higher in the warm air at the right than in the cooler undercutting air that is advancing from the left. (From Drake & Farrow 1989.)

Figure 2.28 Teak (*Tectona grandis*) defoliated by teak defoliator moth larvae (*Hyblaea Puera*) in Kerala, India.

the desert locust, *Schistocerca gregaria* (Orthoptera: Acrididae), and the armyworm, *Spodoptera exempta* (Lepidoptera: Noctuidae), provide classic examples of this phenomenon. Smaller scale wind convergences occur regularly as well, and may generate the unpredictable and sudden concentrations of insect pests in small patches of forest, such as happens in southern India with the teak defoliator moth, *Hyblaea puera* (Lepidoptera: Hyblaeidae) (Nair & Sudheendrakumar 1986) (Figure 2.28).

The diamondback moth, *Plutella xylostella* (Lepidoptera: Yponomeutidae) is a worldwide pest of *Brassica* crops, but feeds naturally on many plant species within the Cruciferae. In many parts of the world, where the winter climate is too extreme for overwinter survival, new diamondback moth outbreaks are entirely the result of regular, long-range immigrations of adult moths (Chapman et al. 2002). Some of these windborne immigrations can cover more than 2000 km, but they depend upon the right conditions for moth survival and dispersal coinciding with appropriate wind conditions above the ground surface. Figure 2.29 shows ideal conditions for moth dispersal, where high air temperatures coincide with the zone (height) of greatest insect density. The wind vectors (direction) at this height will dictate where *Plutella* ends up. Not only does *Plutella* routinely migrate from continental Europe into England every year on the wind, it is also possible, as long as the wind

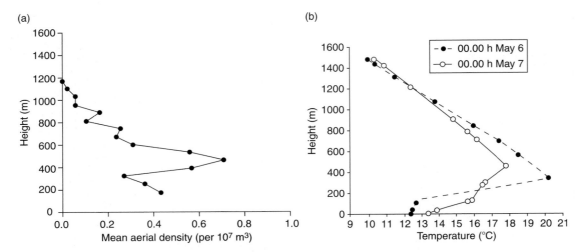

Figure 2.29 (a) Density–height profile of larger insects (body mass > 10 mg) detected by the vertical-looking radar in each of the 15 range gates (between 150 and 1166 m) during the first wave of *Plutella xylostella* immigrations (May 5–11, 2000). (b) Variation of air temperature with height at 00.00 hours on the nights of May 5/6 and 6/7, 2000 at Herstmonceux (East Sussex), UK. (From Chapman et al. 2002.)

speed and direction are suitable, that the moth can move thousands of kilometers north to the high Arctic islands such as Spitsbergen (Coulson et al. 2002). Quite what there is for *Plutella* larvae to eat in these northern climes is another matter.

The influence of wind patterns on the movement of insects in the Far East is particularly complex (Figure 2.30). In this region, trajectory analysis of winds at both 10 m and 1.5 km above ground has been used to determine the direction and extent of windborne movements of insect pests of rice (Mills et al. 1996). Plant-hoppers are thought to be transported by prevailing winds and, these days, the likely force and direction of wind currents can be predicted using weather forecasting models (Otuka et al. 2005). In the study by Mills et al. (1996), aerial traps caught two species predominantly, the brown rice plant-hopper, *Nilaparvata lugens*, and the white-backed plant-hopper, *Sogatella furcifera* (Hemiptera: Delphacidae). Northward migrations of these and other species were detected along a broad front in prevailing summer monsoon and trade winds. However, because most wind trajectories lasted no more than 40 h in any one direction, coupled with the fact that many of the associated weather systems were mobile and hence rather unpredictable in extent and direction, insect catches were very variable. This indicates that

although general insect movement occurs between the tropics and temperate areas during the spring and summer, it is difficult to predict where the highest populations will finally be. Undoubtedly, some movement within the tropical zone also occurs on a regular basis. The huge migrations of dragonflies in South America were mentioned above, but these large insects also perform spectacular aerial movements in East Asia. Feng et al. (2006) studied the nocturnal migrations of *Pantala flavescens* (Odonata: Libellulidae) in China, and found that these insects flew at heights up to and beyond 1000 m (Figure 2.31), traveling between 150 and 400 km in a single flight. Complex interactions of temperature, height above ground, wind direction, and, above all, wind speed, dictated the overall migration patterns.

Winds can also aid in the movement of insects over much smaller distances than discussed so far. *Bemisia tabaci* (Hemiptera: Aleyrodidae) is a worldwide whitefly pest of various agricultural crops, including sweet potato, tomato, tobacco, and cotton. To achieve integrated pest management (IPM; see Chapter 12) of *B. tabaci*, it is important to understand how far the insect is able to migrate. In this way, it should be possible to predict the likelihood of new crops becoming infested, especially if wind directions are known. Figure 2.32 illustrates the movement of whitefly

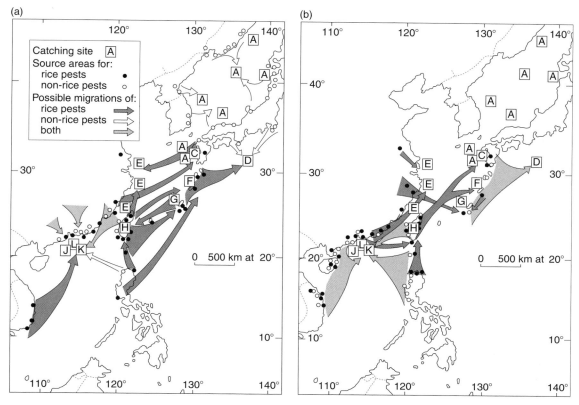

Figure 2.30 Schematic diagram of movements of rice pests and other insects and their likely source area at (a) 10 m and (b) 1.5 km above the ground in spring in southeastern and eastern Asia. (From Mills et al. 1996, © CABI Publishing.)

within fields of melons in the USA (Byrne et al. 1996). The figure shows that insects were caught up to 2.7 km from the source field but, because in this trial there were no traps further away, it is highly likely that the insects could travel much further. The figure also suggests that there were two distinct peaks of insect abundance. This can be explained by the fact that two types of whitefly exist within the population; one group are called "trivial" flying morphs, which move only in the immediate vicinity, whereas the other group contains strong migratory individuals who takeoff under the influence of cues from daylight. If the wind then picks up individuals from this latter group, they can be carried considerable distances. As we might expect, most whiteflies were caught approximately downwind, and in fact the published data from this trial suggest that the higher the wind

speed in any given direction, the greater the percentage of insects caught in traps in the corresponding downwind quadrant.

It is not only adult insects that can be dispersed by the wind, and some insect larvae can be carried considerable distances. Some larvae use wind-assisted dispersal in an attempt to escape from intraspecific competition when they occur in large densities on a limited food supply. The gypsy moth, *Lymantria dispar* (Lepidoptera: Lymantriidae), provides an example of larval dispersal by wind. In North America, as in Europe, female adult gypsy moths are flightless (except in the Asian strain), and in fact the pest population disperses mainly as first instar larvae, which use "ballooning" as a means of colonizing new areas (Diss et al. 1996). Recently hatched larvae produce fine silk threads that enable them to catch the wind

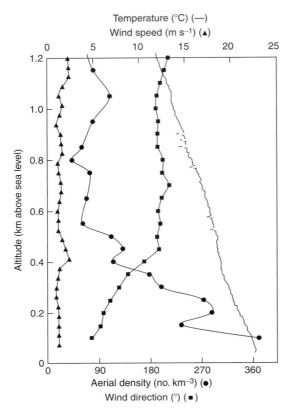

Figure 2.31 Profiles of insect density, temperature, wind speed, and direction during a balloon ascent that started at 20.48 hours on August 22, 2004. (From Feng et al. 2006.)

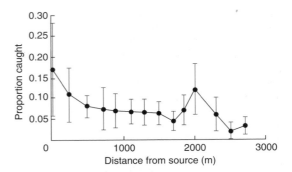

Figure 2.32 Proportion of whiteflies trapped in 200 m distance classes from the source field, averaged over eight trapping dates. (From Byrne et al. 1996.)

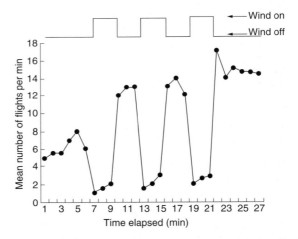

Figure 2.33 Flight response of 10 caged female parasitoids to 0.3 m/s pulses of wind. (From Messing et al. 1997.)

and sail away from their overcrowded host tree. Most of them settle again within 120 m or so of the takeoff point, but it is thought that some may be transported for many kilometers. Again, the strength and direction of the wind will be of paramount importance, although the final key to the success of the operation will be whether or not the little larvae can by chance reach a new, suitable host plant.

Another risk for insects that rely on wind to help them disperse is that of going too far. Strong winds may well carry small insects out of their habitat range altogether, and so some species will fly only in winds of a certain velocity, and refuse to take off if it becomes too strong. A small amount of wind helps movement and host finding; too much could kill by carrying individuals to unsuitable areas. *Diachasmimorpha longicaudata* (Hymenoptera: Braconidae)

parasitizes the larvae of fruit flies, and hence has biological control potential for these extremely serious pests (see Chapter 12). Messing et al. (1997) studied the flight behavior of this species in cages, and found that wind speeds of only 0.3 m/s inhibited flight activity (Figure 2.33). Adult parasitoids averaged 5–8 flights/min during the calm period, which later rose to 10 or more, whereas the number of flights decreased abruptly to less than 2 flights/min during wind pulses. In subsequent wind-tunnel studies, they found that wind speeds of up to 4 m/s stimulated flight, whereas when the speed approached 8 m/s, flight was suppressed (Messing et al. 1997). Clearly, the latter experiments produced somewhat different results from the former, but both suggest that a little wind may be beneficial to the insects; too much is to be avoided. In species such as this one, which have to find specific hosts in heterogeneous habitats, some air movement is advantageous.

Though no host odors were used in the trials described in the previous example, one vital property of wind is its ability to convey chemical messages to insects from point sources. These messages can include information about the distance and location of a specialized food plant, or of a suitable and receptive mate. The beauty of a wind-driven system like this is that as long as the wind flows in a laminar (non-turbulent) manner the chemical "plume" should provide an unbroken guide to the location of the source. Insects can then fly up a concentration gradient of chemical signal to locate their prize precisely.

Many herbivorous insects are host-plant specialists to one degree or another (see Chapter 3), and it is therefore of prime importance that they are able to locate suitable hosts without wasting too much energy. This may be a difficult task in a habitat that contains many species of plant, only one or two of which are in fact suitable, and it is common for insects to locate their host from a distance by olfaction, or odors carried downwind. This system works particularly well if the host plant itself has a distinctive chemical "signature". For example, in wind-tunnel experiments, adult females of the European corn borer, *Ostrinia nubilalis* (Lepidoptera, Pyralidae), flew upwind to three main host plants, corn (*Zea mays*), hemp (*Cannabis sativa*), and hop (*Humulus lupulus*) (Bengtsson et al. 2006). We now recognize that most, if not all, plant species can be distinguished by their chemical profiles, and that insects exploit those profiles in a variety of ways. In another example, ragwort (*Senecio jacobaea*) synthesizes a variety of chemical constituents including toxins such as cyanides. None the less, some insects have adapted to feed on this poisonous plant, including the ragwort flea beetle, *Longitarsus jacobaea* (Coleoptera: Chrysomelidae). Using wind tunnels, Zhang and McEvoy (1995) investigated the response by adult beetles to host plants placed 60 cm upwind (Figure 2.34). Both male and female adults orientated themselves in the upwind direction if they had been starved, though satiated insects showed no particular response to wind direction. Obviously, insects are able to turn off

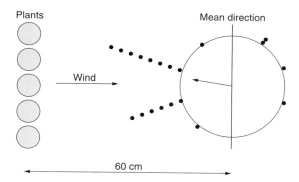

Figure 2.34 Movement directions of a group of starved male flea beetles with their host plant, ragwort, placed upwind of their release point. (From Zhang & McEvoy 1995.)

their food-finding systems if they are not hungry. Under field conditions, it is easy to imagine that specialist herbivores such as the flea beetles would be much more efficient at finding their host plant during a mildly windy day than on one where the air was completely still. However, a strong wind, even if did not break up the chemical plume, might well prevent insects flying upwind to the plant: they might be able to smell it, but would be unable to reach it.

A large number of insect species use chemical signals, pheromones, to locate mates. One sex, more often than not the female, sits in one place and releases complex chemical messages, which disperse around her. In still air, these volatile compounds remain in close proximity to the emitter, and can diffuse over only short distances. However, in a light wind, just as with host-plant odors, the pheromone forms a plume or concentration gradient up which males can fly once their immensely sensitive chemoreceptors have detected the scent. A very large body of expertise has been built up on the development of sex-attractant pheromone systems for the monitoring and management of many farm and forest pests (see Chapter 12). However, the interactions of insects, pheromones, wind, and habitat can be exceedingly complex and need to be understood fully before they can be harnessed successfully (Stelinski et al. 2004).

The codling moth, *Cydia pomonella* (Lepidoptera: Torticidae), is a very widespread pest of pome fruit such as apples and pears. Codling moth larvae (we know them as grubs in apples) are concealed within fruit and are therefore hard to control chemically or, to some extent, biologically too. Understanding the timing and duration of adult flight and egg-laying activities is key to the pest's management, for which pheromones are used in a mating disruption strategy (see Chapter 12). However, as field trials have shown, the nature of the habitat (orchard) and the wind speed and direction can have a fundamental influence on how moths respond to artificial pheromones (Milli et al. 1997). In field trials in Germany, codling moth pheromone could be detected up to 60 m downwind of a release point, and stronger winds lifted the compound higher into the air than more moderate ones. Winds blowing into an orchard from untreated (no pheromone) grassland actually produced a zone near the edge of the crop which became depleted in artificial pheromone, where pests might be able to proliferate. In addition, wind blowing along the edge of the orchard rather than into it tended to concentrate pheromone along the edge and into the grassland, where there was less wind resistance (no trees), thus wasting the treatment (Milli et al. 1997). Simply put, habitat architecture interacts with wind speed and direction to influence the spread and persistence of

pheromone treatments. Control strategies have to be designed with these variables in mind.

2.6 CLIMATE CHANGE

Climate as we have seen is a subtle and certainly complex integration of various abiotic effects, which include temperature, daylight, wind, and rain. These days, it is possible to construct computer simulations of climatic influences on insects, and to predict changes in their population density and distribution. For example, Hodges et al. (2003) have modeled attacks by the maize borer, *Prostephanus truncatus* (Coleoptera: Bostrichidae), on maize and cassava stored by smallholder farmers in Ghana. Their forecasting models use variables that include wind speed and maximum and minimum temperatures. The predictions of the climate-based models mimic very nicely the observed population fluctuations (Figure 2.35). In some cases, we may wish to predict the abundance and dynamics of a complex of species, rather than just one. This is illustrated in Figure 2.36, where climate, measured as a combination of temperature

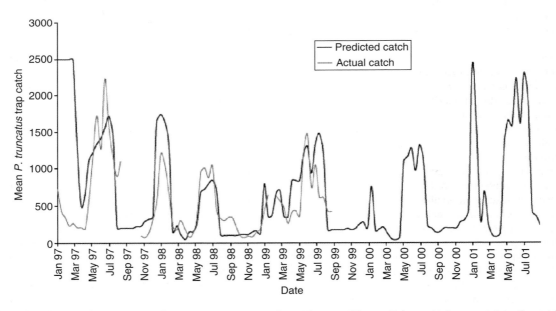

Figure 2.35 Actual mean *Prostephanus truncatus* trap catch in Nkwanta (Ghana, Volta region) compared to the catch predicted by the rule-based climate model. 1997–9 (*n* = 20). (From Hodges et al. 2003.)

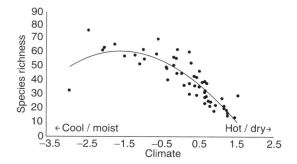

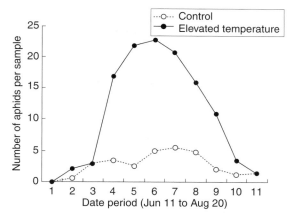

Figure 2.36 Relationships between butterfly species richness and environmental variables found to be significant in the multiple regression model Climate. (From Stefanescu et al. 2004.)

Figure 2.37 Population densities of aphids comparing control plants with plants placed under cloches and experiencing mean temperature elevations of 2.8°C at the leaf surface. Treatments were significantly different at $P < 0.05$. (From Strathdee et al. 1993.)

and rainfall conditions, predicts butterfly species richness in southern Europe (Stefanescu et al. 2004).

One of the most interesting and controversial topics in applied ecology today is that of climate change and global warming. In this chapter, we have shown how climate can influence the ecologies of insects in many varied ways, and it is of great significance to consider how actual and potential changes in the world's weather patterns have affected, and will affect, insects. Some work has already been carried out and given definitive results, whereas other work is based upon modeling and forecasts. The essential assumptions about global warming are centered on the increase in the concentration of atmospheric carbon dioxide (CO_2). General circulation models (GCMs) of the equilibrium response to the doubling of current levels of CO_2 indicate increases in global mean temperature in the range of $4 \pm 2°C$ (Jeffree & Jeffree 1996). Though the CO_2 levels themselves may influence phytophagous insects either directly or, more likely, via the changed physiology of their host plants (Guerenstein & Hildebrand 2008), there has been most speculation about the effects of temperature increases on insect development and distribution. If we couple increases in CO_2 with parallel elevations in ozone plus decreases in rainfall (i.e. increases in drought), then the effects on plants and the insects that feed on them may be very significant (Staley et al. 2006; Valkama et al. 2007).

A large number of experiments have been conducted on a variety of insects to investigate the potential effects of temperature increases of around 2 or

3°C. For example, in northern Norway, cloches were used to cover experimental host plants of the aphid *Acyrthosiphon svalbardicum* (Hemiptera: Aphididae) to study its reproductive ecology over a summer season (Strathdee et al. 1993). Figure 2.37 illustrates one set of results from these manipulations, and shows that reproductive output increased dramatically when the temperature was raised by 2.8°C. Higher reproduction also resulted in an 11-fold increase in the number of overwintering eggs produced by the end of the growing season. In other words, if global warming raises temperatures in this Arctic region, *A. svalbardicum* may be able to sustain much higher population densities. This will depend, of course, on concurrent responses of its hosts and natural enemies. Naturally, what concerns many entomologists, is that elevated temperatures may have very serious consequences for pest outbreaks. Another aphid, the green spruce aphid, *Elatobium abietinum* (Hemiptera: Aphididae), is probably the most serious pest of Sitka spruce in the UK. At present, the factor that limits population growth is mortality through freezing in late winter, as the adults begin to feed again. If winters in the west of the UK become on average warmer, and spring weather comes earlier in the year, then this occasionally serious pest could become much more harmful, perhaps jeopardizing the use of Sitka spruce as the UK's number one softwood species.

Once insects have survived the hazards of winter, the resumption of spring activity can be key to the success in the coming year. Early emergence from winter dormancy may enable exploitation of equally early leaf flushes, or may extend the available period for life cycle development such that more generations are possible. Gordo and Sanz (2006) studied the first appearances of adult bees and butterflies in Spain, and found that both the honeybee, *Apis mellifera* (Hymenoptera: Apidae), and the small white, *Pieris rapae* (Lepidoptera: Pieridae), appeared earlier in the spring as mean temperatures increased (Figure 2.38). These early appearances have been consistent since the 1970s, indicative, it is suggested, of the influences of global warming.

Insects may also extend their geographic ranges as the climate warms (Parmesan 2006), and indeed in the south of the UK, for instance, the number of species of migratory butterflies and moths reported every year has been rising steadily (Sparks et al. 2007). Other countries show similar patterns. Spittlebugs such as *Philaenus spumarius* (Hemiptera: Cercopidae) have moved their range northwards in California since 1988 (Karban & Strauss 2004), and simulation models abound that describe the likely expansion of insects through Europe as temperatures rise. One important example is illustrated in Figure 2.39. The Colorado potato beetle, *Leptinotarsa decemlineata* (Coleoptera: Chrysomelidae), is a notorious defoliator of potato crops, in both the larval and adult stages. At present, it is mainly restricted to southern parts of Europe, and its occasional forays into the UK are unlikely to result in establishment. However, using GCMs, Jeffree and Jeffree (1996) have developed predictions based upon increases in regional temperature equivalent to a doubling of CO_2 levels in the atmosphere. As shown in Figure 2.39, the potential geographic range of summer and winter temperatures suitable for the Colorado beetle could extend into Britain as far as the north of England. The consequences for agricultural production could be extremely serious.

It is not just insect pests that may be able to extend their ranges as global warming proceeds. In the UK, the speckled wood butterfly, *Pararge aegeria* (Lepidoptera: Satyridae), has already expanded its populations into new areas over the last decade or so (Pollard et al. 1996; Hill et al. 2001) (Figures 2.40

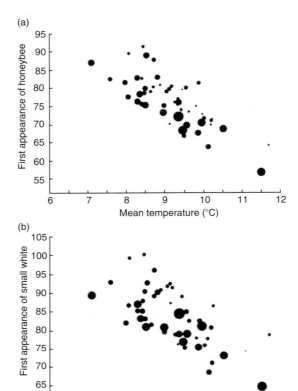

Figure 2.38 Illustrative scatterplots of the relationship between the appearance date (Julian day) of (a) the honeybee and (b) the small white and the mean temperature between February and April for Spain. Each dot represents a single year. Their sizes are weighted by the number of records. (From Gordo & Sanz 2006.)

and 2.41). The speckled wood feeds as a larva on various grass species, and is normally found at the edges of woodland rides or hedges where there is a mixture of dappled sunshine and shade. The new sites that have been colonized in recent years tend to be towards the north and east of England. It is not clear how significant the role of weather has been in these expansions; habitat change must also play a part, but it is clear that, ecologically, these new sites are now much more suitable for the butterfly than

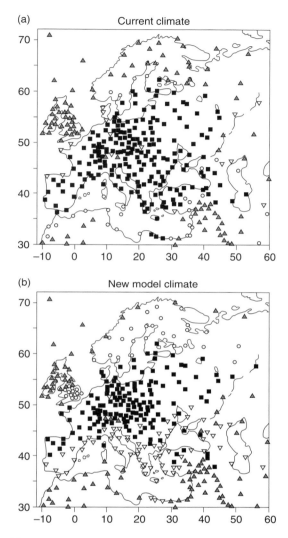

(a) Current climate

(b) New model climate

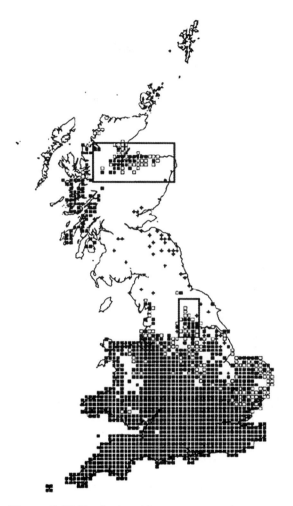

Figure 2.40 Distribution of *Pararge aegeria* in the UK at a 10 km grid resolution. Solid squares show records from 1940–89; hollow squares show 1990s range expansion. Crosses show 19th century records in areas that *P. aegeria* has not yet recolonized. Boxes show the outline of England (southern box; area = 50 × 100 km) and Scotland (northern box; area = 200 × 100 km) study areas. (From Hill et al. 2001.)

Figure 2.39 Changes in potential geographic distribution of Colorado beetle in Europe, resulting from a doubling of carbon dioxide levels and hence an increase in global mean temperature of 4 ± 2°C. Squares, locations within the present distribution that fall into the new model's prediction; circles, locations currently unoccupied but that are potentially suitable under the new model; inverted triangle, locations currently occupied but becoming marginal under new model; triangle, locations currently unoccupied and likely to remain so under new model. (From Jeffree & Jeffree 1996.)

they once were. In the USA, the skipper butterfly *Atalopedes campestris* (Lepidoptera: Hesperidae), has also expanded its range in recent years (Crozier 2003) (Figure 2.42) (see also Chapter 10).

Not all effects of climate warming are as predictable or as clear as the foregoing examples. As has already

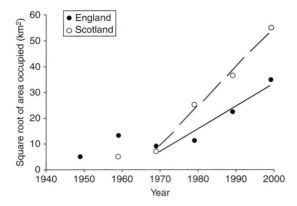

Figure 2.41 Range expansion of *Pararge aegeria* in the England and Scotland study areas plotted as square root of area occupied (5 km grid squares with records) against year. (From Hill et al. 2001.)

been mentioned, temperature increases will affect not just the insect; if the insect is a herbivore, the host plant may also be influenced. One of the best-known insect–plant interactions in ecology must be that of the winter moth, *Operophtera brumata* (Lepidoptera: Geometridae), and its host plant, oak. It has long been known that the synchrony between hatching winter moth larvae and the bud burst of oaks in the spring is critical for the moth's success (Wint 1983), and hence any climate-induced changes in this synchrony would have fundamental consequences for the insect's population dynamics. Buse and Good (1996) examined the effect of a 3°C temperature rise on this synchrony using experimental manipulations in solar domes. They found that the phenology of both eggs and trees were similarly advanced under higher temperature regimes. As the eggs hatched more rapidly, and the oak buds burst earlier, the synchrony was not upset. The 4–5-day delay between egg hatch and maximum leaf flush measured for larvae in field trials was also maintained at the higher temperatures.

These days, quite complex but realistic mathematical models can be used to simulate the effects of climate change on insect populations, and a variety of modeling systems have been applied. Most effort has been directed at serious pest species such as the southern pine beetle, *Dendroctonus frontalis* (Coleoptera: Scolytidae). The southern pine beetle is perhaps the most destructive pest of pine forests in southern

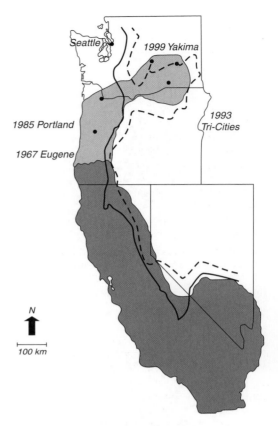

Figure 2.42 Overwintering range of *Atalopedes campestris* (darker shading) in Washington. Oregon, California, and Nevada, modified to include the western range expansion (lighter shading). Colonization dates of A. *campestris* by four cities in Oregon and Washington show the chronology of the range expansion. Contour lines represent the January average minimum −4°C isotherm from 1950–9 (solid line) and 1990–8 (dashed line). (From Crozier 2003.)

USA and, according to Gan (2004), trees killed by the beetle were valued at nearly US$1.5 billion between 1970 and 1996. In annual terms, this equates to somewhere in the region of US$366 to $630 million. Table 2.1 summarizes the predicted effects of global warming on bark beetle populations and the impact of their outbreaks. It is particularly telling that two distinctly different climate change models come up with remarkably similar results; both predict enormous increases in bark beetle abundance, and it appears very likely that pest damage will increase dramatically.

Table 2.1 Climate change scenario from two models and resultant impacts of southern pine beetle (SPB) outbreaks in USA. (From Gan 2004.)

Variable	Climate change scenario	
	GISS	UKMO
Change in winter temperature (°C)	4.6	6.7
Change in spring temperature (°C)	4.2	6.3
Change in summer temperature (°C)	4.1	6.5
Change in fall temperature (°C)	4.4	6.7
Change in winter rainfall (%)	−11	5
Change in spring rainfall (%)	5	2
Change in summer rainfall (%)	15	0
Change in fall rainfall (%)	−2	−4
Increase in pine forest area (%)	57	25
Increase in SPB outbreak risk (%)	396	508
Estimated annual value of trees killed by SPB, taking into account changes in pine forest area & productivity (US$ mil)	868.8	847.9

GISS, NASA Goddard Institute for Space Studies; UKMO, United Kingdom Meteorological office.

Finally, let us not forget that aquatic insects might also be influenced by a rise in water temperature, in ways similar to their terrestrial counterparts in air. Hogg and Williams (1996) manipulated the temperature of a Canadian stream to investigate the effects of a 2–3.5°C rise in water temperature. Various stream arthropods were studied, and Figure 2.43 presents the results for one species of insect, the stonefly, *Nemoura trispinosa* (Plecoptera). Stonefly nymphs are commonly found on the gravelly bottoms of flowing streams, where they feed mainly on algae, though some can also be predators. The warmer stream system saw earlier emergence of adult stoneflies, with new imagos appearing on the wing about 2 weeks in

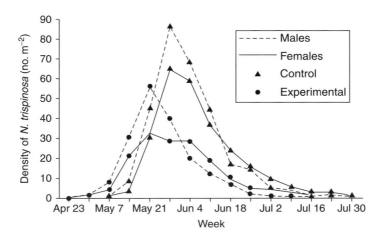

Figure 2.43 Weekly emergence for male and female *Nemoura trispinosa* from control and experimental channels of Valley Spring (April 23–July 30, 1993). (From Hogg & Williams 1996.)

advance of those from the cooler water. However, the density of new adults was somewhat reduced by warmer conditions, and the size of adult females was significantly lower from warm than from cool water. There are obvious consequences for the ecology of this species. Fewer and smaller adults would likely reduce population growth rates, simply as a function of lowered fecundities. There may also be genetic implications of stream warming. Hogg and Williams suggested that in areas where different streams are at varying temperatures, the earlier emergence of adult insects from the warmer ones might provide sufficient time for dispersal to the colder regions and hence a promotion of interbreeding between different habitats.

And what of disease vectors? In Chapter 11 we detail a variety of mammalian diseases such as encephalitis, plague, and malaria all of which are carried from host to host by insects such as kissing bugs, fleas, tsetse flies, and mosquitoes. According to Patz and Olson (2006), climate change since the mid-1970s may have caused an extra 150,000 human deaths per annum from various major infectious diseases. Whether or not global warming will bring at least some of these life-threatening problems such as malaria back to subtropical and warm temperate regions such as parts of the USA and Europe remains to be seen, but there seems to be no reason why it should not happen.

2.7 CONCLUSION

We hope you are convinced that climatic variables such as temperature, wind, and rain can have enormous consequences for the ecology of insects. Many aspects of host exploitation, growth, reproduction, and dispersal are influenced fundamentally by these abiotic factors. However, it must be remembered that, in general terms, these interactions cannot act in a density-dependent fashion (see Chapters 1 and 4). Instead, most climatic effects are density independent and, in isolation, cannot regulate insect populations around equilibrium densities. That being said, abiotic factors can cause large fluctuations in insect abundance, even if they do not regulate populations. Moreover, they can interact markedly with biotic factors such as competition and predation. The relative impacts of climate, competition, and predation on insect population change will be considered in some detail at the end of Chapter 5.

Chapter 3

INSECT HERBIVORES

3.1 THE TROUBLE WITH PLANTS AS FOOD

As was pointed out in Chapter 1, over 50% of extant insect species feed on plants. The fact that these insect species belong to only eight major orders suggests that, while the evolution of phytophagy may involve fundamental hurdles for insects, the radiation that results from "jumping those hurdles" is dramatic (Southwood 1961). What are the hurdles associated with eating plants? Difficulties include adequate nutrition, problems of attachment, maintaining water balance, and battling potential "evolutionary arms races" during which plants are hypothesized to protect themselves from herbivory by periodic development of novel physical and chemical defenses (see Chapter 6). Plant tissue contains lower concentrations of nitrogen than insect tissue, and gaining sufficient nitrogen appears to be a fundamental problem for insect herbivores (McNeill & Southwood 1978; Mattson 1980) (Figure 3.1). Moreover, the proportions of various amino acids differ between animal and plant proteins, and insect tissues have a higher energy content than those of land plants (Strong et al. 1984). Overall, these differences are reflected in the low assimilation and growth efficiencies of insect phytophages compared with those of predatory insects (2–38% for phytophages vs. 38–51% for predators; Southwood 1973).

Of course, before an insect can take a bite of plant tissue, it has to be able to attach itself to the plant. Insects exhibit a multitude of adaptations for holding on to plants, including modified tarsi, the ability to roll leaves, and the capacity to live within plant tissue (leaf-miners, gall-formers, stem-borers) (Juniper &

Southwood 1986). For insects that live on the exterior surfaces of plants, such as Orthoptera, many Lepidoptera, and Thysanoptera, there is the added problem of minimizing water loss when exposed to the drying influence of air currents. This problem is especially acute because most insects are small and hence have a high surface area to volume ratio (Telonis-Scott et al. 2006). Adaptations to avoid water loss include a waterproof exoskeleton (morphological), rolling leaves or producing webbing (behavioral, e.g. Hunter & Willmer 1989), and the reabsorption of water in the hindgut (physiological). Of course, maintaining water balance is necessary for flight, and therefore adult insects might be considered in part preadapted for phytophagy.

Finally, insect herbivores have to overcome the myriad of defenses exhibited by plants. Physical defenses include tough leaves, hairs and spines (trichomes), sticky secretions, and shiny surfaces. Chemical defenses include a dazzling array of both toxic substances (e.g. nicotene) and compounds that may reduce the digestibility of food (e.g. some tannins and lignin). Plant defenses will be discussed in detail later in this chapter.

3.1.1 Ecological stoichiometry

The low nutrient content of plant food in comparison with the nutrient content of insect tissues may impose a fundamental constraint on the development of insect herbivores. Nitrogen and phosphorus are particularly mismatched when herbivorous insects are compared with their host plants (Huberty & Denno 2006). Ecological stoichiometry is a conceptual

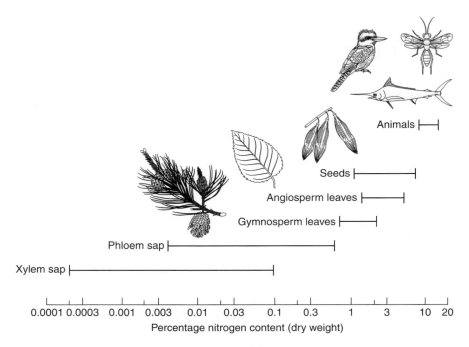

Figure 3.1 A comparison of the nitrogen content of plant and animal tissue. Low nitrogen levels may have represented a fundamental hurdle to the evolution of phytophagy in insects. (From Strong et al. 1984.)

framework that considers the relative balance of key elements (especially carbon, phosphorus, and nitrogen) in trophic interactions (Elser et al. 1996, Sterner & Elser 2002). High-quality food is defined by the ratios of these key elements and how well they match the ratios in the bodies of the insects consuming them. Plant material, for example, is generally much lower in concentrations of phosphorus and nitrogen than are insect bodies. Given the high C : P and C : N ratios of plants compared to insects, phytophagous insects must retain nitrogen and phosphorus from their food while excreting excess carbon. We might also expect insect herbivores to choose plants in which nutrient ratios best match those of their body tissues. According to stoichiometric theory, high insect growth rates also demand food sources with high concentrations of phosphorus and nitrogen (low C : P and N : P ratios) in part because fast growth is associated with many copies of ribosomal RNA which is very phosphorus-rich (Elser et al. 2003). Because most insects are thought not to be able to alter their own elemental ratios to any great degree

(they are relatively homeostatic), nutrient ratios in food may impose fundamental limitations on the growth and feeding of insect herbivores. According to Moe et al. (2005) stochiometric constraints can affect population stability and competitive interactions. Interestingly, insect predators have evolved body nitrogen concentrations that are, on average, 15% higher than those of herbivores, presumably because their food source (prey) is higher in nitrogen than is plant tissue (Fagan et al. 2002). Nonetheless, predators may still be limited by the nutrient availability of their prey and may include other predators as well as herbivores in their diet in an attempt to offset nutrient imbalance (Denno & Fagan 2003).

Stoichiometric theory implies that different food types do not have inherent "qualities" per se; food quality is relative, based upon the elemental requirements of individual consumers (Cross et al. 2003). Some elemental ratios from plants and different insect taxa are shown in Table 3.1. Cross et al. (2003) have shown that detritus-feeding insects in streams, which consume decomposing leaf litter, feed on a very

Table 3.1 Comparisons of C : P, N : P, and C : N ratios among invertebrate trophic groups and food resources from lake, stream, and terrestrial habitats. Values in bold are from the study by Cross et al. (2003) and compare a reference stream (C53) with a stream enriched with nitrogen and phosphorus (C54). Elemental imbalance is calculated as the arithmetic difference between a consumer and it food resource. (Data from Cross et al. 2003.)

Trophic group						Food resource		Elemental imbalance	
C53			C54			C53	C54	C53	C54
Mean	Median	Range	Mean	Median	Range	Mean	Mean		
Stream shredders						Leaf detritus			
C : P 498	493	136–877	252	221	123–610	4858	3063	4360	2565
C : N 6.7	6.4	5.4–8.9	6.4	6.3	5.0–7.7	73	82	66	75
N : P 73	76	17–125	39	30	19–97	67	39	–	–
Stream collectors						FPOM			
C : P 277	208	93–574	227	219	80–358	1015	673	738	396
C : N 6.4	6	5.2–9.0	6	5.8	5.3–7.2	34	29	28	23
N : P 43	38	14–78	37	37	14–59	28	23	–	–
Stream scraper herbivores						Stream epdirbon			
C : P 369	–	–	287	304	155–371	1741	845	1372	476
C : N 6.2	–	–	5.8	5.4	5.1–7.1	8.7	4.6	2.5	– 1.6
N : P 59	–	–	51	56	22–68	201	318	–	–
Stream predators						Stream prey			
C : P 223	215	102–351	227	215	78–430	324	236	101	13
C : N 5.1	5.2	4.9–5.4	5.6	5.5	5.0–6.8	6.1	5.9	1	0.8
N : P 43	42	20–65	40	37	15–75	52	40	–	–
Terrestrial herbivores						Terrestrial plants			
C : P			116			968	852		
C : N			6.5			36	29.5		
N : P			26			28	–		
Lake zooplankton						Lake phytoplankton			
C : P			124			307	183		
C : N			6.3			10.2	3.9		
N : P			22			30	–		
Lake benthic invertebrates						Lake benthic algae			
C : P			148			98–1496	– 50 to 1348		
C : N			5.5			–			
N : P			27			–			

FPOM, fine particulate organic matter.

nutrient-poor food source. The elemental ratios (C : P, C : N) in the bodies of these insects are therefore much higher than those of insects that feed on living plant material (Table 3.1). The idea that natural selection will act to minimize elemental imbalance between insects and their food sources is gaining popularity. For example, more recently derived orders of insect herbivores have lower body nitrogen concentrations than their more primitive ancestors (Fagan et al. 2002). The C : P and C : N ratios of terrestrial plants are much higher (poorer quality food) than those of aquatic plants and algae (Elser et al. 2000). However, terrestrial and aquatic herbivores do not differ significantly in their C : N : P stoichiometry. Combined, these observations suggest that terrestrial insect herbivores are faced with lower quality food than their aquatic counterparts, and this may be responsible in part for the high degree of specialization observed among insects that feed on terrestrial plants – herbivores may specialize on particular plant species that minimize their elemental imbalance. However, other characteristics of plants, including defensive chemistry, are thought to play a major role in the evolution of specialization (see Chapter 6). While ecological stoichiometry seems to play a significant role in the choices that insects make among sources of food, it is unlikely to represent the whole picture. For example, herbivory by insects on a wide diversity of tropical trees has been shown to be linked more to leaf digestibility (lignin and water content) than to nutritional value (Poorter et al. 2004). Likewise, high foliar nitrogen concentrations are not always associated with increases in herbivore performance. Females of the copper butterfly, *Lycaena tityrus*, do not distinguish among host plants based on nitrogen concentration (Fischer & Fiedler 2000). While larval growth rates do increase with foliar nitrogen concentrations, these effects are entirely offset by high pupal mortality and declines in adult weight. In many cases, plant chemical defenses (Section 3.3.3) may place greater constraints on insect herbivores than do stoichiometric ratios.

3.2 FEEDING STRATEGIES OF HERBIVOROUS INSECTS

Just as there are many ways to skin a cat, there are many different ways that phytophagous insects eat plant tissue. Most easily observed are those that live freely on the surface of plants, for example leaf-chewing insects that consume portions of leaves. Grasshoppers and locusts, stick insects, many Lepidoptera and Coleoptera, and some Hymenoptera are considered to be leaf-chewers, and they share similar cutting mandibles (see Plates 3.1 and 3.2, between pp. 310 and 311). When leaf-chewing insects become very abundant, they can defoliate their host plants. The gypsy moth, *Lymantria dispar* (Lepidoptera: Lymantriidae), and the winter moth, *Operophtera brumata* (Lepidoptera: Geometridae), are examples of important defoliators of deciduous forest in North America and Europe, respectively.

Not all free-living insects are leaf-chewers. Some (but far from all) insects that suck the phloem of plants live exposed on plant surfaces. Phloem-sucking insects are dominated by the Hemiptera/Homoptera. These include well-known insects such as aphids, plant-hoppers, leaf-hoppers, and shield bugs (see Plate 3.3, between pp. 310 and 311). The mouthparts of sucking insects are modified to form stylets to penetrate plant tissue. Many sap-sucking insects are important agricultural pests, not only because they reduce the growth of their host plants, but also because they can spread plant pathogens that can decimate crop production. Diseases transmitted by insects will be considered in detail in Chapter 11.

Other insect species with sucking mouthparts feed on xylem rather than phloem. They consume one of the most nutrient-poor parts of the plant (due especially to low nitrogen concentration) and yet are both diverse and abundant (Wu et al. 2006). Spittlebugs (Homoptera: Cercopidae) are the most polyphagous herbivores known. The spittle that surrounds these insects as nymphs is derived from a combination of fluid voided from the anus and a mucilagenous substance excreted from epidermal glands on the seventh and eighth abdominal segments. While we might question the personal hygiene habits of spittlebugs, anyone who has walked through a meadow in summer can testify to their abundance. Closely related to the spittlebugs are the sharpshooters (Hemiptera: Cicadellidae), a major group of xylem-feeding pests that transmit a large variety of plant pathogenic viruses. Finally, cicadas (Homoptera: Cicadidae) are also xylem-feeders and can be so abundant that their biomass over 1 ha is greater than the biomass of cows over 1 ha of rangeland. Some cicadas are "periodical" and take 13 or 17 years to develop.

Yet other insects feed between the upper and lower surfaces of leaves, and such species are called leaf-miners. Leaf-miners occur in four orders of insects (Coleoptera, Hymenoptera, Diptera, Lepidoptera) and the larvae leave characteristic trails or blotches on leaves as they feed (see Plate 3.4, between pp. 310 and 311). High densities of leaf-miners, such as often occur on white oak in eastern North America, can significantly reduce the photosynthetic area of plants. Other leaf-miners, particularly in the Dipteran family Agromyzidae, are important pests of agricultural crops. A fourth group of phytophagous insects form galls on the leaves, shoots, stems, roots, and reproductive parts of their host plants. Galls are made when the insects subvert the natural development of growing tissue of the plant so that it forms a chamber in which the nymphs or larvae live and feed (see Plate 3.5, between pp. 310 and 311). Most galling insects are Diptera, Hymenoptera, or Hemiptera/Homoptera, although some mites (also in the phylum Arthropoda, but not related to insects) form similar gall-like structures on plants. At least some gall-forming insects appear able to manipulate the chemistry of the plant tissue within the gall (Nyman & Julkunen-Tiitto 2000) so that it may be more palatable as a food source.

Another important strategy for feeding on plants is stem or shoot boring. Many of the most serious timber pests are wood-boring species such as the bark beetles (Coleoptera: Scolytidae), some long-horn beetles (Coleoptera: Cerambycidae), and jewel beetles (Coleoptera: Buprestidae). Bark beetles can form extensive galleries in the phloem of trees, and cut off nutrient supplies to and from the crown (see Plates 3.6 and 3.7, between pp. 310 and 311). Like sucking insects, wood-boring species are often associated with pathogens, particularly fungi, which can infect living trees and cause extensive mortality. Although many wood-boring species can only attack dead, dying, or stressed trees, the most serious timber pests will, if numbers are sufficiently high, attack and kill healthy trees (e.g. *Dendroctonus micans* in Europe and *D. pseudotsugae* in the United States; Pureswaran et al. 2006). Other strategies for feeding on plants include root-feeding, seed predation, flower predation, and nectar consumption (either during pollination, or as a "nectar robber"). Many feeding strategies employed by insects, particularly those relying on the digestion of cellulose, are enhanced by interactions with mutualistic organisms such as bacteria and fungi. Insect–mutualist interactions are considered in detail in Chapter 6.

3.2.1 Classifying insects into feeding guilds

One assumption that many insect ecologists have made is that insects that feed in similar ways should exhibit similar ecologies. This is the basis of the "feeding guild" as described by Root (1973). Insect herbivores can be divided among a series of guilds based on their feeding strategy, such as gall-former, sap-sucker, leaf-chewer, or leaf-miner. These guilds are not taxonomically based; leaf-chewing insects, for example, occur in at least five disparate orders. However, many ecologists have found it useful to group insects into guilds in order to study the ecological interactions between insects, their hosts, their natural enemies, and climate. Guilds are useful if, and only if, they provide us with generalities that would be missed by taxonomic studies alone. For example, that

wood-boring species appear to suffer more from the effects of intraspecific competition than from the effects of predation is a generality based upon their way of life, not their taxonomy (Denno et al. 1995; see Chapter 4 for details). Likewise, that phloem-feeding insects may be less vulnerable to phenolic-based defenses (e.g. tannins) in plant tissue than are leaf-chewers is another generality based on the study of guilds (Hunter 1997). In addition, phloem-feeders may belong to the only guild that consistently responds in a positive way to the addition of nitrogen to its host plant (Kyto et al. 1996).

However, some authors have found little evidence for common ecological factors operating on members of the same feeding guild (Fritz et al. 1994). They argue that different species within the same feeding guild respond in idiosyncratic ways to ecological pressures, and that generalities made at the guild level are suspect. Some insects, such as ants, often defy placement into a single feeding guild. Often considered as scavengers or predators, ants may use many different sources of food. In tropical forests, ants are much more abundant than would appear possible if they act only as predators and scavengers. Recent evidence suggests that many arboreal ants in the tropics may be cryptic "herbivores". By studying variation in the stable isotope ratios of nitrogen in insect tissue, which changes with trophic level, Davidson et al. (2003) have shown that many arboreal ants gets most of their nitrogen from plants and not from prey. The two dominant sources of plant nitrogen used by ants appear to be extrafloral nectar and the secretions of phloem-feeding insects. Evolutionary aspects of the interactions between ants and plants are considered in more detail in Chapter 6. Even when placed within a single feeding guild, insect herbivores may interact in different ways with their host plants. Agrawal (2000) has shown that four species of leaf-feeding Lepidoptera in the same feeding guild on wild radish plants induce different chemical responses when they feed on their host plant.

3.3 PLANT DEFENSES

Unless a species of insect herbivore benefits its host by pollination, seed dispersal, or confers some other advantage to the plant (Section 3.7), the impact of insects on plants should generally be deleterious. The removal of photosynthetic area, the loss of pollen, nectar, or seeds, the introduction of pathogenic organisms, or the interruption of nutrient, water, or hormone transport can all be associated with attack by phytophagous insects. If these plant parasites reduce plant fitness (proportional representation in the next generation), it should be expected that plants will evolve mechanisms by which they can reduce the impact of insect herbivores. The wide variety of traits expressed by plants that are associated with the avoidance, resistance, or tolerance of herbivory are generally called "plant defenses". All plant defenses are based on interference with one of the following steps of herbivory: (i) locating a host; (ii) accepting that host as suitable; (iii) attaching physically in some way to the host; (iv) avoiding natural enemies while on the host; (v) tolerating the microclimate during feeding; and (vi) gaining sufficient nutrition from the host during a period of time defined by the availability of suitable plant tissue and the insect's own life history. Any plant trait that interferes with one or more of the above steps can loosely be defined as a "defense".

Before describing the major classes of plant defenses, however, it is worth pointing out that not all traits expressed by plants that appear to ameliorate the effects of insect herbivores are necessarily adaptations resulting from depredation by specific insect herbivores (Janzen 1980; Hartley & Lawton 1990). Similarly, defensive traits that appear tightly linked to particular herbivore species may confer additional advantages to plants beyond those of protection (Schmitt et al. 1995). The origin and maintenance of defensive traits will be considered later in this chapter. For the present, we stress that many defense systems in plants are built upon cascading molecular interactions, wherein many individual plant genes participate in complex mechanisms to deter the feeding of not just insects but other potentially damaging herbivores such as nematodes (Smith & Boyko 2007).

3.3.1 Avoidance: escape in space and time

One obvious mechanism by which plants can reduce levels of herbivory is to be separated in space or time from their major herbivores. A plant might avoid phytophages in space if it were rare, or if its occurrence in a particular area was unpredictable. Rare or ephemeral plant species are, by definition, difficult to find. There is certainly some evidence that widely distributed plants support more species of insect herbivore than do rarer plants. In Chapter 1, data were presented that suggested that the number of insect species associated with British trees increased with the abundance of the tree species (Claridge & Wilson 1977). This pattern was firmly established during studies of agricultural systems where the number of species attacking a crop increased with the area planted. Examples include pests of cacao (Strong 1974), sugarcane (Strong et al. 1977), and agromyzid leaf-miners associated with crops in Kenya (Chemengich 1993). Rare plants are likely to accumulate lower levels of herbivory for both ecological and evolutionary reasons. In ecological time, rare plants might be difficult to find, even by herbivores that are able to feed on the plant should they manage to locate it. The probability of finding a suitable individual will decrease with decreasing plant density. In addition, rare or ephemeral plants may not provide a consistent resource on which insect populations can build over time: if the same plant species occurs in the same place at the same time each year, there is a greater probability that large populations of insect herbivores will develop.

It seems unlikely, however, that "rarity" is an adaptive strategy exhibited by plants to avoid herbivory. Natural selection should maximize an organism's lifetime reproductive output (or, strictly, its inclusive fitness; Hamilton 1964). Greatly sacrificing reproductive output to avoid insect herbivores is an unlikely evolutionary outcome. It is more likely that rare and ephemeral plants are responding to other limitations of the environment, such as lack of suitable habitat or competition with other species for limited resources. Most modern theories of plant succession, for example, emphasize limitations imposed by the abiotic environment or competitors to explain the distribution and abundance of plant species (Connell & Slatyer 1977; Tilman 1987). Low levels of herbivory are likely to be a consequence, rather than a cause, of rarity.

There is much better evidence that escaping herbivory in time rather than space is an adaptive strategy exhibited by plants. The phenological development of particular plant tissues (leaves, fruits, seeds, etc.) appears to have been modified in some plant species by the pressures of insect herbivores. Most oak trees in Britain, for example, support lower insect densities and suffer lower levels of defoliation if they burst bud late in the season rather than earlier (Hunter et al. 1997) (Figure 3.2). Because leaves generally decline in quality as food with age (Mattson 1980), insect larvae that feed on leaves that are even a few

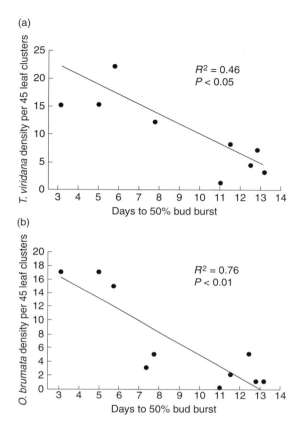

(a)

(b)

Figure 3.2 Densities of both (a) *Tortrix viridana* (Lepidop-tera: Tortricidae) and (b) *Operophtera brumata* (Lepidoptera: Geometridae) are higher on trees that burst bud early than on those that burst bud late. (From Hunter et al. 1997.)

days older than average exhibit slower growth rates and lower pupal weights than larvae feeding on younger leaves (Wint 1983). The selection pressure on plants to burst bud late must be balanced, however, by the needs of a long growing season to accumulate photosynthate (sugars accumulated by plants from carbon dioxide during the process of photosynthesis). None the less, oaks, which support a species-rich insect fauna because of their abundance and long evolution-ary history in Britain, are among the last woodland trees in Britain to burst bud. Similar phenological con-straints appear to operate for many tree species in tropical dry forests that burst bud before the start of the wet season. Since trees require relatively large amounts of water during leaf expansion, bursting

bud during the dry season would appear counterpro-ductive. However, insect densities are at their greatest during the wetter months, and several authors have suggested that early bud burst reduces the pressures of herbivory on valuable young, expanding leaves (Aide 1992; Borchert 1994). Overall, we can conclude that the degree of synchrony between tissue production by plants and the life history of insect herbivores has a major impact on the distribution and abundance of insects on plants.

3.3.2 Physical defenses

Plants exhibit a variety of physical traits that appear to reduce their susceptibility to insect herbivores. Hairs on the surfaces of leaves, called trichomes, have been associated with reduced levels of herbivory, as well as the maintenance of water balance, and the reflection of sunlight (Molina-Montenegro et al. 2006). Trichomes can be either purely structural, presenting simple barriers to attachment or consump-tion, or glandular, in which case the trichomes secrete sticky or noxious compounds (Figure 3.3). Variation among different genotypes of soybean and cowpea in the densities of non-glandular trichomes have been correlated with variation in damage by several species of phytophagous insects, including pyralid moths and coreid bugs (Turnipseed 1977; Jackai & Oghiakhe 1989). Likewise, whitefly density on 14 genotypes of soybean appears to be related to the erectness of the trichomes on the surface of the leaves (Lambert et al. 1995). In a manipulative experiment, leaves of the biennial plant mullein, *Verbascum thapsis*, were more readily colonized by the aphid *Aphis verbascae* (Hemiptera/Homoptera: Aphididae) after the trich-omes had been removed with an electric razor (Keenlyside 1989).

At least one species of aphid, *Myzocallis schreiberi* (Hemiptera/Homoptera: Calliphididae), appears to have modified both morphology and behavior to ameliorate the effects of non-glandular trichomes: rather than placing its tarsi flat on the surface of its oak host, *M. schreiberi* literally "tiptoes through the trichomes" on the bottom surface of *Quercus ilex* leaves by standing on the tips of its tarsi (Kennedy 1986) (Figure 3.4). Other aphid species have stylets that are much longer than is typical for their body size, apparently to reach the surface of leaves through the dense covering of trichomes on the leaves of their

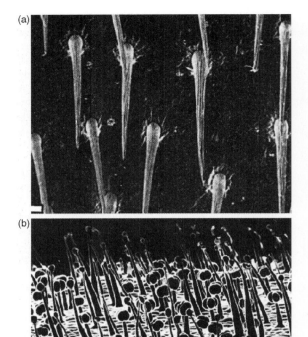

Figure 3.3 Trichomes on plant surfaces. (a) Non-glandular trichomes in the pitcher of *Sarracenia purpurea*. (From Jeffree 1986.) (b) Glandular trichomes on the foliage of *Solanum berthaultii*. (From Ryan et al. 1982.)

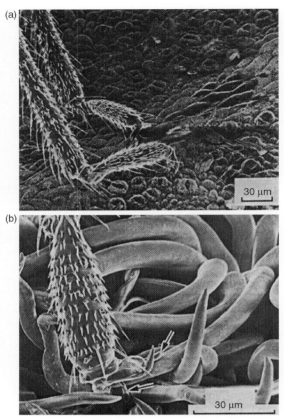

Figure 3.4 (a) Most aphids place their tarsi on the leaf surface while walking. (b) *Myzocallis schreiberi*, by contrast, walks on tip-toe through the trichomes on the under-side of *Quercus ilex* leaves. (From Kennedy 1986.)

hosts (Keenlyside 1989). Despite adaptations by certain insect species to overcome non-glandular trichomes, they are still effective barriers to insect colonization on some plants: differences in aphid density between two sympatric species of alder (*Alnus incana* and *A. glutinosa*) appear to be directly related to differences in trichome density (Gange 1995).

Glandular trichomes produce a variety of secretions that are either sticky enough to trap potential insect herbivores, or sufficiently noxious to deter herbivory. In a fascinating series of papers (Walters et al. 1989a, 1989b, 1990), Ralph Mumma and coworkers have demonstrated that there is genetic variation for resistance to herbivores on *Geranium* plants based on glandular trichomes. *Geranium* trichomes produce sticky secretions that can trap insects, particularly aphids, that attempt to feed on the plants. The "stickiness" of the secretions varies with genotype,

and is responsible for variation in resistance. However, the relative resistance of the genotypes varies with the temperature of the environment. Apparently, the stickiness of the secretions, based on the melting point of the anacardic acid-based compounds produced by the trichomes, varies with temperature. One genotype may be the most resistant at high temperatures and the least resistant at low temperatures. This is an example of a "genotype-by-environment" interaction, whereby the relative expression of a particular trait such as resistance among different genotypes varies with the environment (Figure 3.5). Genotype-by-environment interactions are probably a common feature of insect–plant interactions (Fritz & Simms 1992; Weis & Campbell 1992) and likely contribute

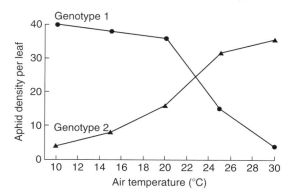

Figure 3.5 Hypothetical example of a "genotype-by-environment" interaction, where the relative susceptibility of two plant genotypes to an insect herbivore varies with temperature.

to the maintenance of genetic diversity for resistance traits. There is simply no "best" genotype for herbivore resistance. Rather, which genotype is best depends upon the current environment. Additionally, we must point out that trichome defenses may deter predatory or parasitic insects just as efficiently as they do herbivores. For example, Gassman and Hare (2005) report that the foraging efficiency of natural enemies is significantly reduced on some plants with glandular trichomes; putative plant defenses may not have uniformly positive effects upon plant fitness.

Work on the glandular trichomes of wild potato, *Solanum berthaultii* (see Figure 3.3), suggests that the secretions produced by the trichomes contain feeding deterrents that influence host choice by the Colorado potato beetle, *Leptinotarsa decemlineata* (Coleoptera: Chrysomelidae) (Yencho & Tingey 1994). Feeding deterrents produced by glandular trichomes combine a physical defense (the hairs) with a chemical defense (the feeding deterrent). Several other physical defenses will be familiar to many readers, such as the stinging hairs of nettles and the thorns of *Acacia* trees. In both these cases, the physical defense is "inducible", so that the density or length of the defensive structure increases following attack. Inducible defenses are considered in more detail in Section 3.3.5.

It is not only glandular trichomes that represent combined physical and chemical defenses. In a spectacular example, the "squirt-gun" defense of *Bursera schlechtendalii* combines the chemical defense of terpenes with the physical defense of resin fired

Figure 3.6 Antiherbivore defense in *Bursera*. In some species, terpene-containing resins are stored under pressure in networks of canals that run throughout the cortex of stems and in the leaves. When the resin canals are punctured or severed by an insect, a high-pressure squirt of resins can be propelled as far as 2 m and drench the attacker. (Courtesy Larry Venable.)

under pressure at attacking herbivores (Becerra 1994) (Figure 3.6). *Bursera* produces terpenes stored under pressure in networks of canals that run throughout the leaves. When a leaf is broken, it releases terpene resins that both bathe the leaf surface and can squirt up to 200 cm. Insect herbivores therefore face the dual challenge of terpene chemistry and physical attack. Inevitably, a specialist chrysomelid leaf beetle in the genus *Blepharida* has evolved a mechanism to cope, in part, with the squirt-gun defense of its host plant. Beetle larvae avoid the squirt response by cutting leaf veins before they consume the leaves. However, the time spent cutting veins on particularly active plants still imposes a cost on beetle development. The *Bursera–Blepharida* system is also interesting because it represents one of the oldest examples of apparent coevolution between insect herbivores and their host plants (Becerra 2003). There are many plant species in the genus *Bursera* and many beetle species in the genus *Blepharida*. During their 112 million years of evolutionary history together, speciation in both genera appears to be tightly linked, such that the appearance of new species of beetles is associated with the appearance of new species of plants. In both cases, speciation seems centered around the development of novel chemical defenses in the plants. Such reciprocal adaptation by plant and herbivore, mediated by changes in plant chemistry, may be an example of a coevolutionary "arms race" between the genera. Evidence for and against coevolution is presented in more detail in Chapter 6.

Another resin pressure defense of plants is worthy of note because of its importance in forest pest management. Trees that are attacked by wood-boring beetles often respond with copious flows of resin at the site of attack (Lombardero et al. 2006). The resin, released under pressure, is thought to physically exclude beetles from the tree by a process called "pitching out". Resin flow is considered by most forest entomologists to be an important component of the resistance of trees to attack by bark beetles. "Oleoresin pressure" is a measure of the flow rate of resin from wounds to the boles of trees, and is affected by levels of water deficit in trees. The reduction in resin pressure that can follow sustained drought conditions is thought to be responsible, in part, for the increased susceptibility of some trees to bark beetle attack following periods of low rainfall (Schwenke 1994; Lorio et al. 1995). However, even an apparently simple defense such as pitching out may be more complicated than it first appears. A variety of volatiles in resin are used by bark beetles to locate their hosts (Hobson et al. 1993) and are components of the pheromones used by beetles to attract conspecifics (Birch 1978). At the same time, other resin volatiles appear to act as deterrents (Nebeker et al. 1995). It seems likely that the resin flow caused by beetle attack has both physical and chemical components that influence the success of colonizing beetles in complex ways. The "bottom line" is that conifers are immensely long-lived organisms that in order to survive hundreds if not thousands of years have evolved complex anti-insect defense mechanisms, including repulsion and defense (Franceschi et al. 2005).

Some authors have suggested that waxy, shiny leaves act as physical defenses against insect herbivores. The assumption is that shiny leaves are difficult for insects to hold onto or colonize (Juniper & Southwood 1986; but see also Jayanthi et al. 1994). For example, leaf-mining by larvae of *Plutella xylostella* (Lepidoptera: Plutellidae) is apparently inhibited by the shiny surface of varieties of glossy-waxed cabbage, *Brassica oleracea* (Eigenbrode et al. 1995). However,

the important consequence of reduced mining activity is that larvae are more susceptible to a group of generalist natural enemies that are ultimately responsible for lower survival on shiny plants. This is an example of a "tri-trophic interaction" whereby variation in plant quality influences herbivorous insects indirectly by changing their susceptibility to natural enemies (Price et al. 1980). In the study by Eigenbrode et al. (1995), shiny leaves were shown to improve foraging efficiency of the natural enemies of *P. xylostella*. This illustrates the importance of experiments that examine the *mechanisms* underlying patterns observed by insect ecologists; without the experimental work, the natural conclusion might have been that glossy leaf surfaces acted directly to reduce colonization by the insect herbivore.

A second factor that can confound studies of the effects of shiny leaves on insect herbivores is that glossy leaves are generally the result of surface waxes, and waxes can contain a variety of chemical compounds that either stimulate or deter feeding by insects. For example, the surface waxes of the tomato cultivar *Lycopersicon hirsutum* f. *typicum* contain three sesquiterpenes, carbon-based compounds that are thought to act as defenses. Removal of the sesquiterpenes by washing the leaves with methanol increases the 10-day survival of beet armyworm *Spodoptera exigua* (Lepidoptera: Noctuidae) larvae from 0% to 65% (Eigenbrode et al. 1994). Indeed, surface extracts of plants such as wild tobacco, *Nicotiana gossei*, have been shown to exhibit antiherbivore activity against a number of commercially important insect pests, and are under investigation for the development of "biorational" pesticides (Neal et al. 1994). In contrast, surface waxes can contain compounds that stimulate feeding or oviposition by insect herbivores. The wild crucifer *Erysium cheiranthoides*, for example, contains both glucosinolates that stimulate oviposition by the cabbage butterfly, *Pieris rapae* (Lepidoptera: Pieridae), and cardenolides that deter oviposition (Hugentobler & Renwick 1995). One important conclusion from the study of plant-surface characteristics is that different defensive traits (trichomes, waxes, defensive chemicals, etc.) interact in complex ways to determine the ultimate susceptibility of plants to their insect herbivores.

Most insect ecologists agree, however, that tough plant tissue generally acts as an antiherbivore defense (Feeny 1970; Mattson 1980; but see also Feller 1995). Growing tissue is generally softer than mature tissue because growing tissue requires the flexibility to expand. After expansion, the leaves of most plants become tougher. The factors that cause leaves to become tough are still a source of debate (Casher 1996; Choong 1996). However, the thickness of the leaf per unit area, the number of structural bundle sheaths, the amount of lignin, and the volume of leaf occupied by cell wall appear to be the primary factors dictating toughness. Lignin and cellulose, neither of which are easily digested by insects, can be considered as very general plant defenses in that they are consistently associated with tough or indigestible tissues.

Variation in leaf toughness, within and among species, is often associated with variation in levels of herbivory. For example, authors have reported negative correlations between leaf toughness and insect development (Stevenson et al. 1993), insect preference (Bergivinson et al. 1995a, b), and levels of defoliation (Sagers & Coley 1995). In both temperate and tropical systems (Hunter 1987; Coley & Aide 1991), the vast majority of defoliation that an individual leaf receives occurs during the first few weeks of growth. The leaves of some tropical plants appear to be almost invulnerable to herbivores after leaf expansion is complete (Coley & Aide 1991). Developing grass leaves soon build up relatively high levels of silica, known as opaline phytoliths, which are thought to act as antiherbivore defenses by the rather simple mechanisms of increasing the abrasiveness of the food at the same time as reducing its digestibility (Massey et al. 2006). So rapid leaf expansion, in combination with increasing toughness, may be the most ubiquitous plant defense of all.

There are several potential mechanisms by which tough leaves could reduce insect densities and subsequent levels of defoliation on plants. Tough leaves may take longer to consume, and therefore result in slower growth rates of insect herbivores (Stevenson et al. 1993). Slow growth rates are thought to increase the probability of mortality from natural enemies or adverse climatic conditions (Clancy & Price 1987). Alternatively, larval and adult insects may avoid individuals or species with tough leaves, thereby leading to lower levels of herbivory (Roces & Hoelldobler 1994). Finally, one study has demonstrated that leaf toughness is associated with mandibular wear in leaf-feeding beetles on willow (Raupp 1985). Mandibular wear causes yet further declines in feeding rate and extends beetle development. In general, it seems clear that smaller, younger insects

feed predominantly on young leaves (Cizek 2005), such that early instars or juveniles are prevented from feeding in mature foliage simply because it is too big and tough.

As was noted for other defensive traits, leaf toughness is often correlated with other putative defenses. For example, within the leaves of several species of deciduous oak, increases in leaf toughness are correlated with declines in water content, declines in nitrogen content, increases in condensed tannin concentrations, and declines in hydrolysable tannin concentrations (tannins are a major class of carbon-based chemical "defenses" in some plant species, and will be discussed in some detail below) (Feeny 1970; Hunter & Schultz 1995) (Figure 3.7). In a similar example, lower levels of herbivory on the neotropical shrub *Psychotria horizontalis* are associated with

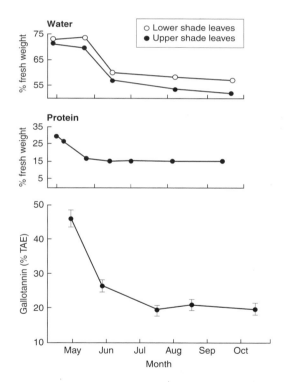

Figure 3.7 Seasonal changes in the quality of oak leaves for insect herbivores. When several measures of foliage quality change simultaneously, it becomes difficult to assess which has the greatest impact on insect populations. TAE, tannic acid equivalents. (Data collated from Strong et al. 1984 and the authors' unpublished data.)

both increased foliar toughness and higher levels of leaf tannins (Sagers & Coley 1995). When two or more plant traits correlate simultaneously with levels of herbivory, it is extremely difficult to determine which factor or factors are responsible for the observed trend. Even experimental work can be inconclusive: treatments that cause increases or decreases in leaf toughness often cause increases or decreases in other plant traits at the same time (Foggo et al. 1994; Sagers & Coley 1995). However, the frequency with which leaf toughness is correlated with low levels of herbivory in studies of plant–insect interactions makes it a strong candidate as an adaptive response by plants to reduce the pressures of herbivory.

3.3.3 Chemical defenses

The field of insect–plant interactions has been dominated by studies of the effects of plant chemistry on the preference (choices made by individuals) and performance (growth, survival, and reproduction) of insect herbivores. Since the classic paper by Ehrlich and Raven (1964) that considered apparent associations between host choice by butterflies and the occurrence of specific chemical compounds in the tissues of their hosts, there has been an explosion of studies that consider the effects of plant compounds on the distribution, abundance, and behavior of phytophagous insects. Put simply, plants protect themselves with a "diverse array of repellant and/or toxic secondary metabolites" (Wittstock et al. 2004). Thousands of different chemical compounds have been isolated from plant tissue and there are undoubtedly many more yet to be identified. Some common classes of plant-derived compounds are given in Table 3.2. Many of the chemical compounds isolated from plant tissue appear to have no primary metabolic function for either growth or reproduction. Consequently, these compounds are called "secondary plant metabolites" (or secondary compounds) and are often considered to serve defensive functions (Harborne 1982; Rosenthal & Berenbaum 1991). Berenbaum (1995) has pointed out, however, that primary metabolites also influence the preference and performance of insects on plants. In other words, the ratios of certain amino acids and proteins might be considered to be "defensive". In addition, certain enzymes, such as proteinase inhibitors (themselves protein based), influence the ability of insects to extract nutrition from plant material (Green & Ryan 1972) and are therefore "defensive". Clearly, it is difficult to make generalities about what kinds of compounds are defensive, and Duffey and Stout (1996) have suggested that terms such as "toxin", "digestability reducer", and "nutrient" signify ecological outcomes rather than specific properties of molecules. Generally, it has been suggested that because plant chemical defenses against herbivory are so complex, the term "defense syndromes" should be adopted, whereby suites of systems covary with insect adaptations (Agrawal & Fishbein 2006). In fact, it would be grossly naïve to assume that all secondary plant metabolites represent evolutionary adaptations directed against insect herbivores (Schultz 1992). Indeed, one single "secondary compound", such as the dihydrochalcone phloridzin from apple leaves, has been considered to be a metabolic end-product (Challice & Williams 1970), an auxin repressor (Growchowska & Ciurzynska 1979), a rooting cofactor (Bassuk et al. 1981), an antifungal defense (Raa 1968; Tschen & Fuchs 1969), an antibacterial defense (MacDonald & Bishop 1952), and an antiherbivore defense (Montgomery & Arn 1974; Hunter et al. 1994), depending upon the focus of the investigator. Moreover, insects are just one of many types of natural enemy that might influence the evolution of plant chemical defenses (Denno & McClure 1983; Hunter et al. 1992). Clearly, plant compounds can play a variety of primary and secondary roles, and searching for a single dominant selective agent for the production of specific chemicals in plant tissue may often be unrealistic.

None the less, there is overwhelming evidence that intra- and interspecific variation in the production of plant chemicals can influence the preference and performance of insect herbivores (Rosenthal & Berenbaum 1991). Until recently, about 80% of insect herbivore species were thought to be specialists (monophagous), feeding on plant species within a single genus. Conversely, 20% were thought to be oligophagous or polyphagous (feeding on several, or many, plant genera and families). These figures come from early studies of the British insect fauna such as aphids (Hemiptera: Aphididae) (Eastop 1973) and British Agromyzidae (Spencer 1972) that show high levels of specialization. Recent work in the tropics (Novotny et al. 2002) has questioned whether monophagy worldwide is really as prevalent as 80% of the fauna (see Chapter 9). Many insect herbivores feed on a significant number of species within the large genera

Table 3.2 Common classes of secondary metabolite in plant tissue; the numbers of structures of each are probably underestimates and new plant compounds are discovered daily by phytochemists. (After Harborne 1982.)

Class	Approx. no. of structures	Distribution	Physiological activity
Nitrogen compounds			
Alkaloids	5500	Widely in angiosperms, especially in root, leaf, and fruit	Many toxic and bitter tasting
Amines	100	Widely in angiosperms, often in flowers	Many repellent smelling; some hallucinogenic
Amino acids (non-proteins)	400	Especially in seeds of legumes but relatively widespread	Many toxic
Cyanogenic glycosides	30	Sporadic, especially in fruit and leaf	Poisonous (as HCN)
Glucosinolates	75	Cruciferae and 10 other families	Acrid and bitter (as isothiocynates)
Terpenoids			
Monoterpenes	1000	Widely, in essential oils	Pleasant smells
Sesquiterpene lactones	600	Mainly in Compositae, but increasingly in other angiosperms	Some bitter and toxic, also allergenic
Diterpenoids	1000	Widely, especially in latex and plant resins	Some toxic
Saponins	500	In over 70 plant families	Haemolyse blood cells
Limonoids	100	Mainly in Rutaceae, Meliaceae, and Simaroubaceae	Bitter tasting
Cucurbitacins	50	Mainly in Cucuribitacae	Bitter tasting and toxic
Cardenolides	150	Especially common in Apocynaceae, Asclepiadaceae, and Scrophulariaceae	Toxic and bitter
Carotenoids	350	Universal in leaf, often in flower and fruit	Colored
Phenolics			
Simple phenols	200	Universal in leaf, often in other tissues as well	Antimicrobial
Flavonoids	1000	Universal in angiosperms, gymnosperms, and ferns	Often colored
Quinones	500	Widely, especially Rhamnaceae	Colored
Other			
Polyacetylenes	650	Mainly in Compositae and Umbelliferae	Some toxic

of tropical plant species, and even among plant genera. This predominance of oligophagy in the tropics reduces the estimates of global insect species diversity from 31 million (Erwin 1982) to between 4 and 6 million, in broad agreement with other recent estimates (Dolphin & Quicke 2001; see also Chapter 9). As an aside, host specificity can actually be acquired during early instars. For example, naïve hatchlings of the tobacco hornworm, *Manduca sexta* (Lepidoptera: Sphingidae), can feed and grow successfully on many different plants and artificial diets. However, if they feed on their natural solanaceous hosts at an early age, they become specialist feeders on those hosts. This "induced specialization" appears to depend upon exposure to a plant-derived steroidal glycoside called indioside D (del Campo et al. 2001). In other words, specialization can actually be induced by the plants upon which insect herbivores feed.

However, it is clear that many insect species will feed on just a few plant species, and host chemistry is thought to be the dominant factor responsible for the tight linkage between most insect herbivores and specific plant taxa (Ehrlich & Raven 1964; Becerra 1997; but see also Bernays & Graham 1988). Effects of plant chemistry on insects can occur over evolutionary time and ecological time. The evolution of insect–plant associations is considered in detail in Chapter 6, and we concentrate on interactions in ecological time in this chapter. Suffice it to say that there is good evidence to support a dominant role for plant chemistry in the evolution of insect–plant associations (Ehrlich & Raven 1964; Becerra 2003). For example, the radiation and diversification of pierid butterflies such as the ubiquitous cabbage white, *Pieris rapae* (Lepidoptera: Pieridae), appear to be associated with using *Brassica* secondary compounds such as glucosinolates (Braby & Trueman 2006).

Effects of secondary metabolites on insects in ecological time can be negative (apparently defensive), neutral, or positive (acting as oviposition or feeding stimulants). The glucosinolates present in many crucifers, for example, appear to deter most generalist herbivores, but act as oviposition stimulants for the cabbage white butterfly, *P. rapae* (Hugentobler & Renwick 1995), and the cabbage seedpod weevil, *Ceutorhynchus obstrictus* (Coleoptera: Curculionidae) (Ulmer & Dosdall 2006). It seems logical that specialist insect herbivores would use a variety of cues, including plant chemistry, to locate and initiate consumption of their given hosts, and there are dozens of examples in the literature of oviposition stimulants (e.g. Huang et al. 1994; Nishida 1994; Honda 1995; Tebayashi et al. 1995) and feeding stimulants (Mehta & Sandhu 1992; Bartlet et al. 1994; Nakajima et al. 1995) isolated from plant tissue.

3.3.4 Endophytes

We now recognize that some chemical compounds with antiherbivore activity originate not from the plant itself, but from endophytic fungi that "infect" leaf tissue (Krauss et al. 2007). Most of the work on fungal-derived toxins has been conducted on grasses and sedges, many of which act as hosts to the fungi *Acremonium*, *Balansia*, and *Neotyphodium* (Clay et al. 1985). The fungi produce alkaloid mycotoxins (potent antiherbivore compounds) and/or lolitrem A and B (mammalian neurotoxins). The plants appear to benefit from the presence of the endophytes without significant costs to either growth or reproduction

(Keogh & Lawrence 1987). Infection by fungi confers resistance in grasses to a wide variety of insect herbivores, including two species of armyworm in the genus *Spodoptera* (Lepidoptera: Noctuidae) (Ahmad et al. 1987; Clay & Cheplick 1989), the Argentine stem weevil *Listronotus bonariensis* (Coleoptera: Curculionidae) (Prestidge & Gallagher 1988), house crickets *Acheta domesticus* (Ahmad et al. 1985), the hairy chinch bug *Blissus leucopterus* (Hemiptera: Lygaeidae) (Mathias et al. 1990), mealybugs, and aphids (Sabzalian et al. 2004). In a recent study, the deleterious effects of endophyte infection on the bird-cherry aphid, *Rhopalosiphum padi* (Hemiptera: Aphididae), were increased when the host grass was grown on high nutrient soils (Lehtonen et al. 2005). This suggests that natural variation in soil fertility under field conditions will influence the degree to which endophyte-derived toxins affect insect herbivores. Although not all insects are equally susceptible to the fungal toxins (Tibbets & Faeth 1999), they are under consideration as novel pesticides (Findlay et al. 1995). Studies of endophyte–insect interactions have moved beyond grasses to include oaks, firs, and cottonwood trees (Findlay et al. 1995; Wilson 1995; Gaylord et al. 1996; Wilson & Faeth 2001). Moreover, some studies suggest that endophytic fungi have stronger effects on natural enemies (Preszler et al. 1996) than they do on herbivores directly. Although it now appears that endophytes are present in the majority of plant species, at this time dramatic effects on insect herbivores have been primarily reported from grasses. Moreover, agronomic grasses seem to provide the most consistent evidence for deleterious effects on herbivores, whereas effects in natural plant communities appear to be much more variable (Faeth 2002; Faeth & Fagan 2002).

Although direct effects of endophyte toxins on insects may be of less general importance than effects of plant-derived toxins, endophytes may influence insect populations and communities in other ways. Long-term experiments suggest that changes in plant community structure, succession, and ecosystem function are associated with endophyte infection (Clay & Holah 1999; Rudgers et al. 2004). Changes in plant diversity and succession driven by endophytes will affect the insect herbivores that rely on plants for food and habitat (see Chapter 8), and we might expect to see endophytes playing a role in insect community ecology. In the remainder of this section, we focus upon plant-derived chemical defenses rather than those derived from endophyte infection.

3.3.5 Theories of plant chemical defense

When ecologists encounter overwhelming complexity in the environment, their first step is usually to search for repeated patterns that can be used to simplify or categorize the complexity. We saw an example of this earlier when Root (1973) suggested that we could use feeding guilds to categorize the way that insect herbivores feed on plants (Section 3.2.1). Given the many thousands of chemical compounds that occur in the tissues of plants, it should be no surprise that similar attempts have been made to simplify or categorize plant defenses into groups based upon their expression, mode of action, or evolutionary history. Unfortunately, at the current time, none of the proposed theories of plant defense are without flaws. In fact, the current status of plant defense theory has been described as a "quagmire" (Stamp 2003). Below, we present some history and some current views of plant defense theory, using the nomenclature of Stamp (2003) throughout. Interested readers are strongly encouraged to read Stamp's paper for a more detailed description of plant defense hypotheses. From what follows, we can conclude that an adequate theory of plant defense has yet to emerge. This may be because: (i) the hypotheses themselves are inadequate, typical of an immature science; or (ii) there is no single hypothesis of plant defense that is capable of explaining the diversity of defenses found in nature (Berenbaum 1995).

The primary goal of plant defense theory must be to explain and predict phenotypic, genetic, and geographic variation in plant defense in space and time. Remember, this process all crumbles to naught if the plant traits in question are not primarily defensive. As we have seen above, some plant traits that appear defensive may serve other functions and may have evolved under different or multiple selective forces. The idea that plant secondary metabolites act as defenses can be traced back to Dethier (1954) and Fraenkel (1959). However, the first comprehensive theory of plant defense was presented by Ehrlich and Raven (1964), who considered that coevolutionary processes (see Section 3.3.2 and Chapter 6) were responsible for generating most of the chemical diversity observed in plants through the action of an

evolutionary "arms race" between plants and insect herbivores. We now recognize that the arms race model cannot explain all of the phenotypic, genotypic, and spatial variation in plant defense that is observed in nature and that other hypotheses are required. The major ones are listed, then described, below:

(i) Optimal defense hypothesis (Feeny 1976; Rhoades & Cates 1976).

(ii) Carbon–nutrient balance hypothesis (Bryant et al. 1983).

(iii) Growth rate (resource availability) hypothesis (Coley et al. 1985).

(iv) Growth–differentiation balance hypothesis (Herms & Mattson 1992).

Note that all of these hypotheses are still used by entomologists and ecologists in their attempts to explain variation in plant defenses; none has emerged as the "winner".

Optimal defense hypothesis

The optimal defense hypothesis (Feeny 1976; Rhoades & Cates 1976) suggests that plants evolve to allocate defenses in ways that maximize fitness, but necessarily divert resources from other needs to do so. In other words, the hypothesis invokes the principle of trade-offs, where allocation to one function (say, defense), necessarily occurs at the expense of allocation to other functions (for example, growth). In the same way that we might have to choose between spending our money on music or clothes, plants make evolutionary and ecological "choices" between allocating limited resources to defense or other functions. One prediction that emerges from the theory is that defense allocation in evolutionary time should be in direct proportion to the risk of herbivory and inversely proportional to defense cost. When risk of herbivory is low, resources are wasted if they are allocated to defense. Likewise, if defense costs are very high, overall levels of defense should be low. These ideas are the cornerstone of "apparency theory", a part of the optimal defense hypothesis that has received a considerable amount of attention. Apparency theory (Feeny 1976; Rhoades & Cates 1976) suggests that long-lived and abundant plants are at greater risk of herbivory because they are more "apparent" to their herbivores. In contrast, rare and ephemeral plant species are at lower risk of herbivory (Section 3.3.1). As a result, we might expect different defensive strategies to have evolved in apparent and unapparent plants.

Apparency theory divides chemical defenses between two broad groups: digestibility reducers and toxins. Long-lived and abundant plant species should, over evolutionary time, be colonized by a large diversity of herbivore species (see Chapter 1). Because the reproductive rate of insects is so much higher than the reproductive rate of long-lived plants, insects would be likely to develop resistance to specific toxins in plant tissue in much the same way that pests develop resistance to insecticides (see Chapter 12). Long-lived or "apparent" plants should therefore invest in compounds that reduce the digestibility of their tissues rather than in narrow action toxins (Feeny 1976; Rhoades & Cates 1976). Tannins, large polyphenolic molecules (Figure 3.8), were considered to be examples of secondary plant metabolites that reduced the digestibility of plant tissue. Specifically, tannins have the property of precipitating protein from solution by hydrogen bonding (and, under certain conditions, by covalent bonding), and consequently can render plant proteins unavailable for digestion by insect herbivores. Conversely, rare or ephemeral plant species ("unapparent" plants) should be more likely to invest in toxins because their probability of frequent encounter by herbivores, and therefore the

selection pressure for resistance, should be lower. Broadly speaking, tannins do appear to be more common in the tissues of apparent plants than unapparent plants. However, there have been several criticisms leveled at apparency theory.

Firstly, tannins are a very diverse group of secondary metabolites, with a great variety of chemical forms (Zucker 1983; Waterman & Mole 1994). Based on an analysis of structure, it seems likely that some tannins act more like toxins than digestibility reducers (Zucker 1983). Secondly, not all apparent plants invest heavily in compounds that affect digestibility. Maples and aspens, for example, are common tree species that invest as much in toxin-like molecules as in tannins (Schultz et al. 1982; Lindroth & Hwang 1996). Thirdly, apparency theory requires that there are significant barriers to the evolution of adaptations by insects to combat digestibility-reducing compounds. Yet many insect species maintain conditions in their midgut – such as high pH (Berenbaum 1980), the presence of surfactants (Martin & Martin 1984), or a specific redox potential (Appel 1993; Appel & Maines 1995) – that ameliorate or negate the precipitation of tannin–protein complexes. None the less, there are strong indications that tannins can reduce the

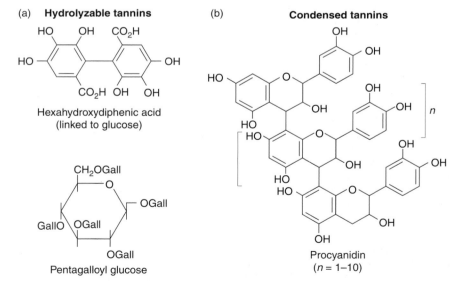

(a) **Hydrolyzable tannins**

Hexahydroxydiphenic acid
(linked to glucose)

Pentagalloyl glucose

(b) **Condensed tannins**

Procyanidin
(n = 1–10)

Figure 3.8 Two major classes of tannin in plant tissue. (a) Hydrolyzable tannins are derivatives of simple phenolic acids such as gallic acid and its dimeric form, hexahydroxydiphenic acid, combined with the sugar glucose. (b) Condensed tannins are oligomers formed by condensation of two or more hydroxyflavanol units. Gall, galloyl residue. (From Harborne 1982.)

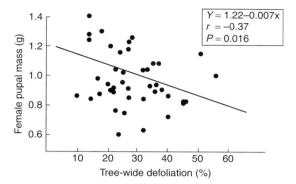

Figure 3.9 Defoliation of oak trees causes increases in foliar tannin concentrations, which, in turn, reduce female pupal mass of the gypsy moth *Lymantria dispar* (Lepidoptera: Lymantriidae). Because pupal mass is related to fecundity, oak tannin can reduce gypsy moth birth rates. (From Rossiter et al. 1988.)

performance of a wide variety of herbivores, including insects (Rossiter et al. 1988; Bryant et al. 1993; Hunter & Schultz 1995) (Figure 3.9).

While apparency theory may no longer be accepted as an adequate model for the evolution of plant defenses, it has served a vital role in stimulating research in plant–insect interactions. Moreover, apparency theory is just one part of the optimal defense hypothesis, which still has much to offer; allocation to defense based on the risks of herbivory and the costs of defense may still be valid. A further prediction of the theory is that defenses should decrease when herbivores are absent and increase when they are present. In other words, the optimal defense hypothesis predicts the phenomenon of "wound-induced defenses" by which the act of attack by insects stimulates a change or increase in defensive chemistry. The trick of wound-induced defenses is that many of the costs of defense are only paid when the risks are high (Rhoades 1985). In an early study, Green and Ryan (1972) reported that potato and tomato plants respond to herbivory by Colorado potato beetle (Coleoptera: Chrysomelidae) by producing proteinase inhibitors, compounds that can inhibit certain digestive enzymes in the midguts of insects. Subsequent work has demonstrated the existence of induced defenses in a wide variety of plant species including birch, oak, sedges, and alder, to name but a few. Authors have disagreed about whether induced defenses are the result of evolutionary pressures on

plants by insects, and even whether such defenses are powerful enough to influence insect populations (Haukioja & Niemelä 1977; Baldwin & Schultz 1983; Edwards & Wratten 1985; Hartley & Lawton 1987). However, this is one area where disagreement has slowly faded over time.

There is now little doubt that many, if not most, plant species undergo chemical changes following insect attack, and that some of these changes can deter subsequent herbivory. Examples abound. In Scandanavia, defoliation of birch hosts by the autumnal moth, *Epirrita autumnata* (Lepidoptera: Geometridae), causes a defense reaction in the host tree such that subsequent larval growth rate is retarded by about 10% (Kapari et al. 2006). On oak trees, defoliation induces increases in foliar condensed tannins (Figure 3.10). Induced defenses occur predominantly in young, developing leaves and may rely upon import of materials to support the induction response; induction of condensed tannins in poplar leaves relies upon carbohydrate that is assimilated in, and imported from, more mature leaves (Arnold & Schultz 2002). In addition, the type of damage caused by herbivores may influence the subsequent effect on others; for example, chewing lepidopteran larvae elicit a different set of plant responses compared

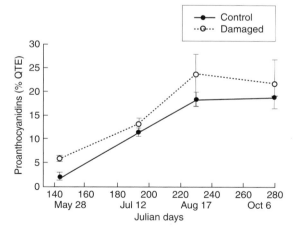

Figure 3.10 Wound-induced increases in the condensed tannin (proanthocyanidin) content of red oak, *Quercus rubra*, foliage following damage by the gypsy moth *Lymantria dispar* (Lepidoptera: Lymantriidae). Damaged trees suffered about 55% leaf area removed compared with 6% on controls. QTE, quebracho tannin equivalents. (From Hunter & Schultz 1995.)

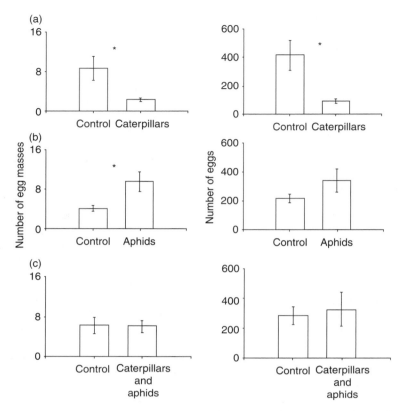

Figure 3.11 Number of egg masses and eggs laid by the herbivore *Spodoptera exigua* on tomato plants damaged by *S. exigua*, *Macrosiphum euphorbiae*, both, or neither (control). Three combinations were tested: (a) control versus caterpillar-damaged plants; (b) control versus aphid-damaged plants; and (c) control versus dual damaged plants. Each bar represents the mean ± 1SE An asterisk indicates significant differences between treatments (t-test, $P \leq 0.05$). (From Rodriguez-Saona et al. 2005.)

with phloem sap-feeders (Rodriguez-Saona et al. 2005) (Figure 3.11). Chemical changes that follow damage by herbivores, and the phenomenon of induction, are considered in more detail later in this chapter.

The optimal defense hypothesis also predicts that defenses should be compromised when plants are under stress (Barto & Cipollini 2005). If allocation to defense represents a tradeoff with competing functions such as growth, maintenance, and reproduction, we might expect that allocation to defense will decline when the environment becomes stressful for plants and overall resource budgets are low. Unfortunately, the links among environmental stress, plant defense, and attack by insects are far from clear, in part because there exist competing hypotheses that rely on

different mechanisms of operation. The optimal defense hypothesis predicts that attack by insect herbivores will increase when plants are under stress because defenses are lowered. Confusingly, the plant stress hypothesis of insect outbreak (White 1974, 1984) also predicts increases in defoliation under plant stress, but for a different reason. According to White, many plants that are under physiological stress respond to stress with increases in soluble nitrogen and free amino acids in their tissues. Recall that insect herbivores are generally limited by nitrogen availability (Section 3.1.1), so we might expect that stress-induced increases in soluble nitrogen might favor attack by insects. Two questions then emerge. First, are stressed plants really more susceptible to insect attack? Second, if so, is it because of reduced defenses or increased soluble nitrogen in their tissues? Unfortunately, many studies of plant stress and insect attack have not distinguished between potential mechanisms, and the questions are not easy to answer.

Most support for the plant stress hypothesis comes from studies of forest or range insects, particularly those of economic importance. Essentially it argues that when plants are stressed by environmental factors, they become more susceptible to insect herbivores via an alteration in foliar biochemistry which renders the leaves more palatable (Joern & Mole 2005). For example, drought stress increases the susceptibility of Norway spruce to the spruce needle miner, *Epinotia tedella* (Lepidoptera: Olethreutidae) (Muenster-Swendsen 1987). Susceptibility of balsam fir stands in Quebec to attack by spruce budworm, *Choristoneura fumiferana* (Lepidoptera: Tortricidae), is also associated with water availability (Dupont et al. 1991). Likewise, several studies have demonstrated that *Eucalyptus* species under moisture stress become more susceptible to wood-boring (Paine et al. 1990; Hanks et al. 1995) and leaf-chewing (Stone & Bacon 1994) insects. Where the mechanism of increased susceptibility under stress has been investigated, the conclusions have been varied. Certainly, there is some evidence that higher nitrogen availability increases the quality of stressed plants for insect herbivores (Gange & Brown 1989; Thomas & Hodkinson 1991). However, not all plants respond to stress in this way (Louda & Collinge 1992), and other mechanisms have been suggested. For example, stress-induced changes in leaf size (Stone & Bacon 1994), leaf toughness (Foggo et al. 1994), plant architecture (Waring & Price 1990), resin production (Schwenke 1994; Lorio et al. 1995), and plant physiology (Louda & Collinge 1992) have also been associated with increased susceptibility to insect attack. Table 3.3 lists some insect species that have been shown to increase in density following drought stress.

The links between plant stress and insect outbreak have been the subject of considerable controversy (Larsson 1989). Neither the optimal defense hypothesis nor White's stress hypothesis fare particularly well when a diversity of studies are considered. Certainly, water stress can increase the concentration of secondary plant metabolites (Zobayed et al. 2007), but at least some of the controversy arises from the fact that there are many different kinds of stress, many idiosyncratic responses by plants, and equally diverse responses by insect herbivores. Although water deficit is a common cause of stress in plants, other factors such as mammalian browsing and sun exposure (Linfield et al. 1993), root disturbance (Gange & Brown 1989; Foggo et al. 1994), and nutrient availability (Mopper & Whitham 1992) can alter susceptibility to insect herbivores. The timing, duration, location, and intensity of stress have also been shown to affect the responses of colonizing insects (Louda et al. 1987b; English-Loeb 1989; Mopper & Whitham 1992; McMillin & Wagner 1995). Even the genotype of individual plants within a population determines the magnitude of its stress responses (Mopper et al. 1991; McMillin & Wagner 1995). Perhaps most importantly, not all insects respond to plant stress in the same way. For example, although both leaf-chewing and leaf-mining insects occur at higher densities on stressed individuals of the crucifer *Cardimine cordifolia*, aphid densities are unaffected by stress (Louda & Collinge 1992). In a striking example, Waring and Price (1990) described the responses of eight species of leaf-galler in the genus *Asphondylia* (Diptera: Cecidomyiidae) to water stress of their host plant, creosote bush (*Larrea tridentata*). Five species were more abundant on stressed plants, two on unstressed plants, and one showed no preference (Table 3.4). Even within one genus, then, insects vary in their responses to plant stress. Finally, it has been suggested that plant stress may not be a continuous influence on the lives of herbivores (Huberty & Denno 2004). Instead, "pulses" of stress followed by recovery of turgor pressure may increase the performance of phloem-feeding insects.

Direct contradictions to the plant stress hypothesis are also common. Corn seedlings, for example, respond to water stress with increased concentrations of DIMBOA and DIBOA, two cyclic hydroxamic acids that deter both insect pests and pathogens (Richardson & Bacon 1993). Leaf-miners in the genus *Cameraria* occur in greater numbers on irrigated oak trees than on oaks under water stress (Bultman & Faeth 1987). The assumption that outbreaks of the pine beauty moth, *Panolis flammea* (Lepidoptera: Noctuidae), on lodgepole pine were caused by site-induced stress on deep peat soils was refuted by experimental manipulation: survival of larvae did not differ between stressed and unstressed trees (Watt 1988).

The most direct assault on the plant stress hypothesis has been the plant vigor hypothesis that contends that insect herbivores prefer, and perform better on,

Table 3.3 Some outbreaks of forest and range insects associated with drought. (From Mattson & Haack 1987.)

Species	Family	Genus of host	Location	Reference
Coleoptera				
Agrilus bilineatus	Buprestidae	Quercus	USA	Haack & Benjamin (1982)
Corthylus colambianus	Scolytidae	Acer	USA	McManus & Giese (1968)
Dendroctonus brevicomis	Scolytidae	Pinus	USA	Vite (1961)
Dendroctonus frontalis	Scolytidae	Pinus	USA	Craighead (1925); St George (1930)
Dendroctonus ponderosae	Scolytidae	Pinus	Canada	Thompson & Shrimptom (1984)
Ips calligraphus	Scolytidae	Pinus	USA	St George (1930)
Ips grandicollis	Scolytidae	Pinus	USA	St George (1930)
		Pinus	Australia	Witanachchi & Morgan (1981)
Ips spp.	Scolytidae	Pinus, Picea	Europe, Africa	Chararas (1979)
Scolytus quadrispinosa	Scolytidae	Carya	USA	Blackman (1924); St George (1930)
Scolytus ventralis	Scolytidae	Abies	USA	Berryman (1973); Ferrell & Hall (1975)
Tetropium abietis	Cerambycidae	Abies	USA	Ferrell & Hall (1975)
Homoptera				
Aphis pomi	Aphididae	Crataegus	Switzerland	Braun & Fluckiger (1984)
Cardiaspina densitexta	Psyllidae	Eucalyptus	Australia	White (1969)
Hymenoptera				
Neodiprion sertifer	Diprionidae	Pinus	Sweden	Larsson & Tenow (1984)
Lepidoptera				
Bupalus piniarius	Geometridae	Pinus	Europe	Schwenke (1968)
Choristoneura fumiferana	Tortricidae	Abies, Picea	Canada	Wellington (1950); Ives (1974)
Lambdina fiscellaria	Geometridae	Abies, Picea	Canada	Carroll (1956)
Lymantria dispar	Lymantriidae	Picea	Denmark	Bejer-Peterson (1972)
Orthoptera				
Several species	Acrididae	Grasses	World-wide	White (1976)

vigorous, not stressed, plants (Price 1991; Dhileepan 2004). Detailed studies by Peter Price and colleagues on gall-forming sawflies in the genus *Euura* and their willow hosts has provided a large body of evidence that more vigorous plants, and the most vigorous parts within plants, are preferred by some herbivores (Price 1989, 1990, 1991) (Figure 3.12). The plant vigor hypothesis has been supported by Price and others for an increasing number of insect–plant associations such as 13 of 15 phytophages on mountain birch trees (Hanhimaeki et al. 1995), five species of tropical insect herbivore on five different host plants

(Price et al. 1995), the rosette galls of *Rhopalomyia solidaginis* (Diptera: Cecidomyiidae) on *Solidago altissima* (Raman & Abrahamson 1995), and leaf-galling grape phylloxera on clones of wild grape (Kimberling et al. 1990).

Even insect herbivory on eucalypts, considered to be a classic case of the plant stress hypothesis (Paine et al. 1990; Stone & Bacon 1994; Hanks et al. 1995), appears to be controversial: Landsberg and Gillieson (1995), after a landscape-level analysis of eucalypt production, soil factors, and climate, found that insect herbivory was positively correlated with eucalypt

Table 3.4 Mean densities (per 10 g foliage) of eight galling insects in the genus *Asphondylia* (Diptera: Cecidomyiidae) on water-stressed and unstressed creosote bushes, *Larrea tridentata*. (Data from Waring & Price 1990)

Species	Density (unstressed)	Density (stressed)	Site favored
1	7.0	27.0	Stressed
2	6.6	6.6	None
3	10.4	18.9	Stressed
6	2.7	6.2	Stressed
7	30.7	144.1	Stressed
8	4.4	2.2	Unstressed
10	3.0	0.1	Unstressed
11	7.1	30.5	Stressed

production and unrelated to stress. It is difficult to reconcile the plant stress hypothesis, the optimal defense hypothesis, and the plant vigor hypothesis into a single theory to explain patterns of insect attack under stressful environmental conditions. Price (1991) points out that the plant vigor hypothesis may be better suited to galling and shoot-boring insects that rely on rapidly growing tissue, yet examples from other insect guilds exist as well. The authors of this book suggest that all three hypotheses fail primarily because they are not sufficiently mechanistic. They make simplifying assumptions about the ways that plants respond to environmental variation that are not well supported by physiological or molecular mechanisms. We will return to this point several times in our discussion of plant defenses and consider an alternative approach below.

Carbon–nutrient balance hypothesis

Note that the optimal defense hypothesis (above) makes predictions about plant defense in both evolutionary and ecological time. For example, it makes predictions about the kinds of defenses (apparency theory) that will evolve in evolutionary time and also predictions about how defenses will change under environmental stress in ecological time. In contrast, the carbon–nutrient balance (CNB) hypothesis (Bryant et al. 1983) concerns itself only with ecological time. It was developed to explain the influence of soil nutrients and shade on plant defenses during the lifetime of individual plants. It is concerned

primarily with phenotypic plasticity, the variation in defensive phenotype expressed by plants that is contingent upon their current growing conditions. For example, Figure 3.13 illustrates that foliar concentrations of total nitrogen and alkaloids (nitrogen-containing defenses) vary with the availability of soil nutrients to plants (Palumbo et al. 2007). As Stamp (2003) points out, the original statement of the hypothesis appears to contain some contradictions. The authors initially suggested that all allocation to defense by plants occurs after allocation to growth and that there is no predetermined (genetic) allocation to defense. This is not biologically likely (Karban & Baldwin 1997; Hamilton et al. 2001), and may explain, in part, why the hypothesis has been so severely criticized. However, the authors also allowed that plants may have some fixed allocation to defense which is then modified by flexible allocation. This is much more reasonable, and suggests that a plant's defensive phenotype reflects its underlying genotype (fixed), modified by resources that can be acquired from the environment (flexible). The resources in question are carbon and soil nutrients (primarily nitrogen), and it is the ratio of these resources acquired from the environment that will modify the defensive phenotype expressed by an individual plant. The CNB hypothesis predicts that carbon gained at levels greater than those required for growth and baseline (inflexible) defense will be allocated to high levels of carbon-based defense (flexible) such as tannins and terpenes (see Table 3.2). Similarly, if nitrogen is assimilated in concentrations higher than those demanded for growth and baseline (inflexible) defense, it will be allocated to nitrogen-based, flexible defense. What influences the relative accumulation of carbon and nutrients by plants? Carbon, of course, is assimilated during the process of photosynthesis, which generally increases with light availability. We might therefore expect that individual plants growing under high light might invest more in carbon-based defenses. If soil nutrient concentrations are very low, we might also expect that the ratio of carbon gain to nutrient gain will alter to favor allocation to carbon-based defenses. So, according to the CNB hypothesis, it is the relative availability of light and soil nutrients that should drive phenotypic variation in plant defenses in ecological time.

The CNB hypothesis has provided a framework for much valuable research on environmentally based variation in plant defense, yet it remains controversial

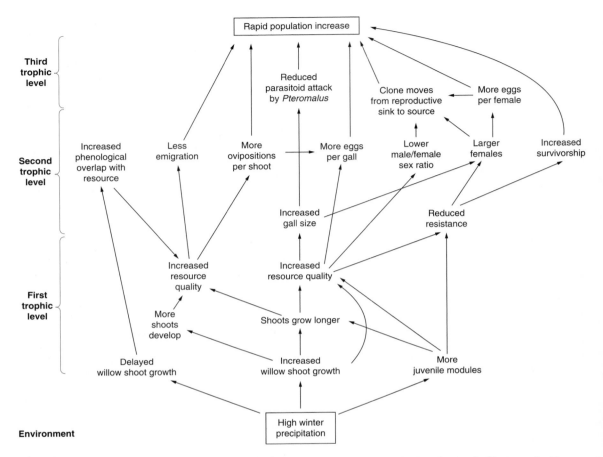

Figure 3.12 Price and coworkers have argued that more vigorous plants are more likely to be attacked by insect herbivores. In this diagram, populations of the gall-forming sawfly, *Euura lasiolepis*, increase on the vigorous willow plants that result from high levels of winter precipitation in Arizona. The effects of precipitation on plant vigor are felt throughout the trophic web. (From Price et al. 1998.)

(Gershenzon 1994; Karban & Baldwin 1997; Hamilton et al. 2001). The results of some experiments in which light or nutrient conditions are varied provide equivocal results. None the less, a significant number of studies support, at least partially, the CNB hypothesis (Muzika & Pregitzer 1992; Bryant et al. 1993; Holopainen et al. 1995; Hunter & Schultz 1995). Why should some experiments support the hypothesis while others do not? Of course, the simplest explanation might be that the hypothesis is not accurate or sufficient to explain phenotypic plasticity in plant defense. Alternatively, results might depend upon the relative importance of the "fixed" and

"flexible" allocation to defense. If the degree of flexibility, or phenotypic plasticity, is itself under genetic control, then some species or genotypes may simply not have the ability to respond to environmental variation in light and nutrients. If true, this will make clear tests of the CNB hypothesis rather difficult. Other problems with the CNB hypothesis remain. All chemical defenses rely upon enzymes and structures to make, store, and deliver the defenses. These all require nutrients such as nitrogen to make, precluding allocation of carbon to defense without some additional investment of nitrogen (Gershenzon 1994). Moreover, some plants may allocate their flexible

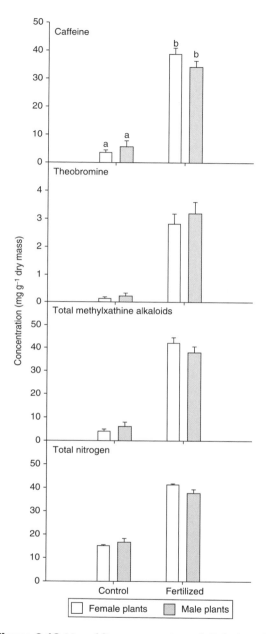

Figure 3.13 Mean foliar concentrations of alkaloids and total nitrogen by fertilizer treatment and gender in *Ilex vomitoria* (note difference in scales). White bars represent female plants, grey bars represent male plants. Lower case letters specify groupings as indicated by Tukey's test for multiple comparisons when interactions were indicated ($P < 0.05$). (From Palumbo et al. 2007.)

carbon gains to functions other than defense. For example, after meeting the demands of growth, we might expect plants to allocate some carbon to storage rather than to defense. Finally, the rapid induction of specific chemicals through signal cascades (active defense, see below) almost certainly does not fit the CNB model of more passive changes based on carbon to nutrient tissue ratios. In other words, the more we learn about the molecular mechanisms of chemical defense, the less appropriate the CNB hypothesis becomes. This does not mean, of course, that plant defenses do not vary with current growing environment. Rather, it means that the CNB hypothesis may be too simple to explain the patterns that occur in nature.

Growth rate hypothesis

While the CNB hypothesis is concerned with variable patterns of defense in ecological time, the growth rate hypothesis (also called the resource availability hypothesis) addresses the evolution of plant defense allocation in environments that differ in resource availability (Coley et al. 1985; McCall & Irwin 2006). As before, variation in the environment is key to the hypothesis, but it focuses on genotypic variation in defense over evolutionary timescales rather that phenotypic plasticity within a generation. The basic idea of the hypothesis is that plants that grow slowly defend their tissues with high levels of defense because those tissues are hard to replace. Plants that can grow rapidly allocate less to defense because losses to herbivores are easy to replace with new growth. Some environments (high light, high nutrients) favor the evolution of plants with low levels of defense because high growth rates in such environments allow the replacement of tissues lost to herbivores. Conversely, plants that evolve under conditions of low nutrients or low light cannot grow rapidly or replace damaged tissue, and protect themselves with higher levels of defense. Plant species native to resource-rich environments usually have short leaf lifetimes and should employ low concentrations of mobile defensive substances, such as alkaloids and cyanogenic glycosides (see Table 3.2), that can be recovered from leaves prior to senescence. In contrast, plant species typical of low-resource environments have long leaf lifetimes and should protect themselves with high concentrations of less mobile compounds such as tannins and lignin. In long-lived leaves, a one-time investment in immobile defenses should be less expensive than the task of continually replacing mobile defenses as they are rapidly turned over (Coley et al. 1985). In more recent work, Coley et al. (2006) have shown that leaf age and degree of host specialization both contribute to the relative growth rate of various species of caterpillar on tropical trees (Figure 3.14).

The growth rate hypothesis has some empirical support. For example, fast-growing provenances of *Eucalyptus globulus* are generally more susceptible to leaf-chewing herbivores than are slow-growing provenances of the same species (Floyd et al. 1994). However, some conifer species show patterns opposite to those predicted by the hypothesis. For example, pines exhibit both high constitutive defense and fast growth, whereas firs have low constitutive defense and slower growth (Cates 1996). In the authors' view, the growth rate hypothesis has not received sufficient empirical investigation as yet. Problems of phylogeny (relatedness among the plant species under comparison) make interpretations of data difficult because related species are likely to express more similar defensive strategies, even in different environments, than are unrelated species (Baldwin & Schultz 1988). We have yet to find any empirical studies that have controlled adequately for phylogeny while replicating sufficiently to test the hypothesis rigorously. Theoretical models of defense allocation suggest that defensive strategy should vary with herbivore pressure (Lundberg & Astrom 1990) and that reallocation

of mobile defenses may actually be more advantageous in slow-growing species (Van Dam et al. 1996). As with all modeling studies, however, the conclusions reached are only as valid as the assumptions built into the models, and we feel that there is much room for further empirical work on resource-based theories of defense allocation among plant species.

Growth–differentiation balance hypothesis

Another major hypothesis of plant defense allocation combines aspects of both the growth rate hypothesis and the CNB hypothesis. The growth–differentiation balance (GDB) hypothesis (Loomis 1932, 1953; Herms & Mattson 1992) argues that there exists a balance between allocation to growth (new stem, roots, leaves) and allocation to differentiation (everything else, including defensive chemicals and structures) (Barton 2007). Any ecological or evolutionary factor that slows plant growth faster than it slows photosynthesis will result in allocation favoring differentiation (including defense).

While the original hypothesis has a long history (Loomis 1932, 1953), Herms and Mattson (1992) first used it to consider plant allocation to defense under a variety of ecological conditions, including interactions with herbivores. These authors envisaged a growth–differentiation continuum, with growth favored in resource-rich environments typified by competition for resources, and differentiation favored in resource-poor environments where growth rates are slow. In environments that are very resource-poor, both growth and photosynthesis are likely to be low, favoring only moderate concentrations of secondary metabolites. At intermediate levels of resource availability, growth is limited more than is photosynthesis, favoring differentiation and high secondary metabolite production. Finally, when resource availability is high, neither growth nor photosynthesis are strongly limited. Growth is favored over differentiation, and defense levels are low (Figure 3.15). As you can see, one key prediction of the GDB hypothesis that separates it from the other hypotheses is that allocation to defense should be bell-shaped along a resource gradient. Is this prediction supported by data? Unfortunately, detecting such a bell-shaped curve requires experiments with many different resource levels and the majority of experiments to date have considered only a few resource levels (often just two,

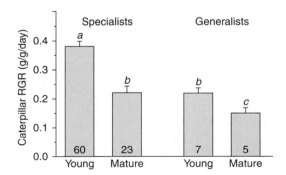

Figure 3.14 Relative growth rates (RGR) of specialist and generalist lepidopteran caterpillars feeding on young and mature leaves of 37 different plant species. The number of morphospecies in each category is presented; values with different letters are significantly different at $P < 0.05$ (ANOVA $F_{3,94} = 13.3$, $P < 0.001$). (From Coley et al. 2006.)

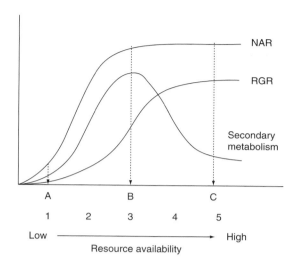

Figure 3.15 The growth–differentiation balance hypothesis. Both net assimilation rate (NAR) and relative growth rate (RGR) of plants vary across a resource gradient (values 1 through 5). The difference between NAR and RGR (e.g. points A, B, and C) determines the resources that are available for differentiation, in this case secondary metabolism (defense). In the example shown here, allocation to defense is highest at intermediate levels of resource availability (B), where the difference between NAR and RGR is greatest. Allocation to defense is lower when NAR and RGR differ less (A and C). (From Stamp 2003; after Herms & Mattson 1992.)

high and low). So one primary prediction of the GDB hypothesis has not been tested fully at this time.

Herms and Mattson (1992) also considered the evolutionary implications of a balance between growth and differentiation for allocation to defense. They argued that the dual forces of competition among plants and herbivory will combine to influence the evolution of defense. Given the tradeoff between growth and defense, environments in which there is intense competition should favor allocation to growth so that plants may outcompete their neighbors; defenses will therefore be low. In contrast, environments that support intense herbivory should favor allocation to defense over growth.

Slow growth, high mortality hypothesis

Although this is not a hypothesis developed specifically to explain differential patterns of plant allocation

to defense, the slow growth, high mortality hypothesis (Clancy & Price 1987) also deserves mention. In essence, the hypothesis suggests that poor-quality plant tissue, by reducing the growth rates of herbivorous insects, may increase the time-window during which herbivores are susceptible to natural enemies and other sources of mortality (Cornelissen & Stiling 2006). While it may be an appealing proposition, strong empirical support for the slow growth, high mortality hypothesis has been hard to come by. For every case in which slow growth of herbivore larvae appears to increase the risk of parasitism (e.g. Johnson & Gould 1992), there appears to be another case in which slow growth is unrelated to parasitoid attack, or actually reduces it (e.g. Clancy & Price 1987).

The best evidence in support of the hypothesis comes from studies that consider qualitative variation within a single species of plant. For example, during their studies of parasitism of *Pieris rapae* (Lepidoptera: Pieridae) by *Cotesia glomerata* (Hymenoptera: Braconidae), Benrey and Denno (1997) found that variation in growth rates of *P. rapae* among individuals of the same plant species was associated with variation in the proportion of larvae that were parasitized. However, variation in herbivore growth rates among different plant species was not a good predictor of parasitism. Indeed, the plant species upon which *P. rapae* grew most rapidly was also the species upon which the rates of parasitism were highest. The authors suggested that effects of plant species on parasitoid foraging behavior were more important to rates of parasitism than was the growth rate of the herbivore on a particular host species. In their literature review, Benrey and Denno (1997) found support for their suggestion that, while intraspecific variation in plant quality may influence rates of parasitism by altering herbivore growth rates, there is little evidence for the slow growth, high mortality hypothesis acting among plant species. Even within plant species, slow herbivore growth is not always linked to the risk of parasitism. For example, Lill and Marquis (2001) found that variation among herbivore families in development time on oak trees was unrelated to rates of parasitism. Low-quality oak foliage reduced caterpillar survival but had little impact upon development time and the authors concluded that low-quality foliage (high in total phenolics and hydrolysable tannins) acted primarily as a direct defense against herbivores rather than as an indirect defense mediated by predation or parasitism. At best, then, the slow growth, high mortality hypothesis has equivocal support. One reason may be that parasitoids may respond more to the quality of host herbivores than to their apparency. While herbivores that grow slowly may be more available for attack, they may be of reduced nutritional quality to foraging parasitoids (Harvey et al. 1995; Harvey 2000).

3.3.6 Molecular approaches to studying plant defense

None of the theories of plant defense described above has ever been firmly rejected (Berenbaum 1995), contributing to the current "quagmire" of plant defense theory. Over the past decade, however, an alternative approach to studying plant defense against insects has emerged and gained momentum. Rapid improvements in the techniques of molecular biology have provided entomologists and ecologists with a new toolbox with which to explore how plant defenses work and how they have evolved. In this new "school" of plant defense, scientists seek to understand the molecular mechanisms underlying physiological and chemical responses of plants to attack by insects (Kessler & Baldwin 2002). The search for pattern, then, has become the search for common molecular mechanisms underlying chemical defense. What genes are expressed when plants are attacked? What signals are sent through plant tissues? What chemical and physiological changes result from signal transduction?

The origin of this new school can be traced to studies of induced plant defenses (Green & Ryan 1972; Schultz & Baldwin 1982; see also Section 3.3.5). Plants are no longer seen as passive victims of attack, but rather are seen to respond in coordinated and predictable ways to herbivore damage. This new view of herbivory, emphasizing rapid and dynamic changes in the chemistry and physiology of both plants and insects, shifts our focus from the static chemical composition of plant tissue (see Table 3.2) to the factors that elicit change and the signals that coordinate it (Schultz 2002). Complex signaling pathways coordinate responses to attack in much the same ways as our own immune systems respond to infection. For example, Ian Baldwin and colleagues have studied the defensive responses of wild tobacco, *Nicotiana attenuata*, an annual plant native to the Great Basin desert of the western United States. Wild

tobacco responds to herbivore attack by the induction of the octadecanoid pathway and its associated metabolites, including jasmonic acid (JA) (Baldwin 1998). The octadecanoid pathway and JA are seen as key players in the signaling responses of plants to insect attack (see below). What do these signals do? Baldwin has used the techniques of mRNA differential display to investigate changes in gene expression that result from insect attack (Figure 3.16). By studying the mRNA molecules that are made by wild tobacco in the presence and absence of insect damage, Baldwin can tell how many genes are upregulated or downregulated following damage. By knowing what those genes do, the defensive responses can be accurately described. After scanning one-twentieth of the total mRNA pool, Baldwin found herbivore-induced changes in 27 transcripts, suggesting that herbivory affects the mRNA pools of over 500 genes. Unfortunately, about 40% of those 500 genes serve unknown functions so we still have a long way to go before we understand completely the defensive

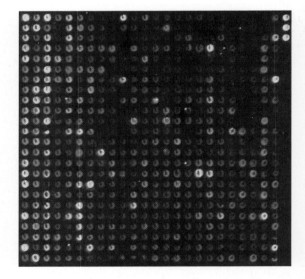

Figure 3.16 Part of a microscope slide (a microarray) showing differential display of mRNA. Each dot on the slide represents the level of expression of a gene, as measured by the amount of mRNA produced. Plants challenged by insect herbivores can be compared with herbivore-free controls to visualize changes in gene expression resulting from herbivory. Changes in color are used to infer whether gene expression increases, decreases, or remains unchanged following herbivory. (From Baldwin 1998.)

responses of wild tobacco. Of the mRNAs with known functions, transcripts involved in photosynthesis were strongly downregulated by insect elicitors whereas transcripts associated with stress, wounding, and pathogens were upregulated. Presumably, many of the mRNAs stimulated by insect damage represent so-called "civilian responses" to injury (Karban & Baldwin 1997), a range of non-defensive changes in plant physiology that result from enemy attack. These responses can include remobilization of reserves from storage tissues, increased partitioning of photosynthate to leaves over roots, and increased photosynthetic activity in undamaged tissues. The regrowth foliage produced by many oak trees following herbivory would be a typical example of a civilian response.

So the key to understanding this new school of plant defense is that molecular mechanisms and signaling pathways are the focus of the studies (Beckers & Spoel 2006). A necessary byproduct of this new approach is that insects are seen as only one of the enemies to which plants respond. Although this is a book about the ecology of insects, it would be naïve to assume that we could discuss plant defenses to insects in isolation from defenses to other forms of attack (see Chapter 11). Clearly, plants have a range of enemies including bacteria, fungi, viruses, insects, mammals, reptiles, molluscs, fish, and birds – we should not expect their responses to one enemy to be independent of all others. When we consider the signaling pathways by which plants respond to attack by insects, we must also consider their similarity to signaling pathways induced by other forms of attack. Specifically, we need to address how the pathways interact with each other – this is called "cross-talk". One possible outcome of cross-talk is interference among signaling pathways. For example, there is some evidence that defense pathways induced by insect damage can inhibit defense pathways induced by pathogen attack, and vice versa (Hunter 2000). In other words, plants may not be able to defend themselves efficiently against simultaneous attack by herbivores and pathogens.

For example, when pathogens attack plants, it usually leads to signal transduction by mitogen-activated protein (MAP) kinases, leading to what is called the "oxidative burst" (production of superoxide, hydrogen peroxide, and their pathway derivatives) associated with pathogen resistance, the induction of pathogenesis-related (PR) proteins, and the programmed cell death of the hypersensitive response

(Lamb & Dixon 1997; Delledonne et al. 1998). The oxidative burst and the induction of PR proteins are apparently ubiquitous responses of plants to attack by pathogens. The response includes the induction of phenylalanine ammonialyase, the products of which are required for both local induced resistance and systemic (whole plant) acquired resistance to pathogens via the salicylic acid (SA) defense pathway (Durner et al. 1998). As an aside, SA, produced by plants in response to pathogen attack, is the active ingredient in aspirin, and a constitutive defense in willow leaves. The critical point here is that the SA pathway appears to inhibit the JA pathway which is associated with plant resistance to insect herbivores (Cipollini et al. 2004). This is cross-talk. If the JA pathway is inhibited by the SA pathway, and vice versa, then insect attack can compromise defense against pathogens and pathogen attack can compromise defense against insects.

On the insect side of the signaling story, we know that the JA pathway in plants is induced following insect attack (Liechti & Farmer 2002) (Figure 3.17) and is responsible in part for the massive transcriptional reprogramming that occurs in response to wounding, and the induction of defenses that act against herbivores. Jasmonate synthesis begins with the formation of 12-oxo-phytodienoic acid (ODPA) from linolenic acid. ODPA can then be modified to form JA. ODPA and JA combine to help control the defensive responses of plants. Derivatives of JA, such as methyl jasmonate and cis-jasmone, are volatile and may act as signals that repel herbivores and attract enemies such as parasitic wasps. They may even allow communication among individual plants (Birkett et al. 2000). Volatile production, however, may be influenced by cross-talk because the upregulation of SA by pathogens interferes with steps in the JA pathway and therefore volatile synthesis. This suggests that pathogen infection may interfere with the production of volatiles that attract the natural enemies of insect herbivores. In other words, plant defense against insects that relies upon the third trophic level may be compromised by pathogen attack.

Given that plants respond to damage with predictable signal pathways, it should be no surprise that some herbivores may have developed the ability to "eavesdrop" on the signaling systems or to subvert the signaling process (Schultz 2002). Corn earworm, *Helicoverpa zea* (Lepidoptera: Noctuidae), larvae use both jasmonate and salicylate in their food as signals

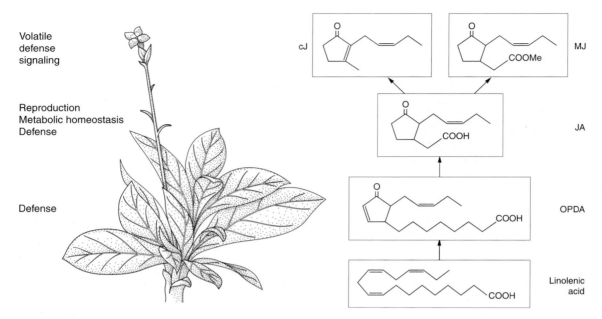

Figure 3.17 The jasmonate family of regulators involved in plant defense. OPDA is an octadecanoid derived from linolenic acid. It is one of the jasmonates which, together with jasmonic acid (JA), helps control defense responses. Two metabolites of JA, methyl jasmonate (MJ) and cis-jasmone (cJ) are volatile signals. (From Liechti & Farmer 2002.)

to induce the production of cytochrome P450s, enzymes that aid in the detoxification of plant defenses (Li et al. 2002). In other words, the insects are "stealing" information from the plant's signaling pathway to prepare for induced plant defenses. In at least one case, the salivary products of a herbivore actually inhibit defense production. Musser et al. (2002) performed an ingenious experiment in which they cauterized the spinnerets of *H. zea* larvae. Spinnerets are the site of secretion of enzymes, including glucose oxidase, from the salivary glands (Figure 3.18a). When larvae were reared on tobacco plants, nicotine levels declined if the spinnerets were intact, but remained unchanged when spinnerets were cauterized. A similar result was obtained using direct application of saliva and glucose oxidase (Figure 3.18b). In other words, larvae are capable of manipulating the defenses of their host plant with oral secretions.

Whether the cross-talk between signaling pathways demonstrated in the laboratory actually matters under field conditions is less clear-cut (Pauw & Memelink 2004). Certainly, some studies have shown that field plants stimulated to express SA-based resistance show a diminished JA response and a corresponding increase in vulnerability to certain insect herbivores (Bostock 1999; Thaler et al. 1999). However, Bostock has also reported that the induction of plant defenses by aphids behaves more like pathogen induction (SA-related) than insect induction (JA-related), suggesting that the mode of insect feeding (Section 3.2) may be related to the type of resistance mechanisms that are subsequently induced. For example, Grover (1995) demonstrated that when the hessian fly, *Mayetiola destructor* (Diptera: Cecidomyiidae), attacks wheat, the plant responses resemble those typical of pathogen attack, not herbivore attack. Indeed, the whole dichotomy of pathogen versus insect defense pathways seems to weaken in the face of field experiments. On *Rumex* plants, attack by a rust fungus, *Uromyces rumicis*, reduces the growth, survivorship, and fecundity of the chrysomelid beetle *Gastrophysa viridula* (Coleoptera: Chysomelidae). If there is any inhibition of herbivore-induced defense pathways (currently unknown in this system), the effect is weaker than the decline in nutritional quality of foliage for the beetles

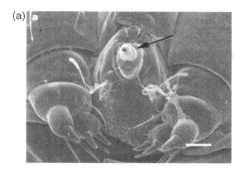

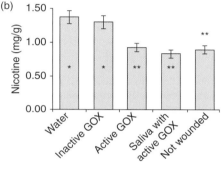

Figure 3.18 (a) The spinneret of *Helicoverpa zea* which produces saliva, and (b) the effects of salival glucose oxidase (GOX) on nicotine production by tobacco. (From Musser et al. 2002.)

(Hatcher et al. 1994a). Likewise, prior damage by *G. viridula* induces resistance to the rust fungus both locally and systemically (Hatcher et al. 1994b). Here, the result of any putative cross-talk appears to be positive rather than negative, with overall resistance levels to multiple attackers increasing following previous attack by any one. Field studies of *Rumex* have shown that pathogen resistance induced by herbivory is effective against additional pathogen species (Hatcher & Paul 2000). There also appears to be positive cross-talk between insect and pathogen resistance in common bean plants (Cipollini 1997) and zucchini (Moran & Schultz 1998), leading some authors to reject the notion of discrete herbivore and pathogen defense pathways (Schweizer et al. 1998; Terras et al. 1998). In part, separating positive and negative cross-talk between signaling pathways remains difficult because both pathways induce many civilian responses (physiological responses beyond defense) that are yet to be fully characterized (von Dahl & Baldwin 2004).

It is not only the chemical defenses of plants that can be induced by insect attack. Herbivore damage can cause increases in the production of spines or thorns (Bazely et al. 1991) and extrafloral nectar that attracts ants (Ness 2003). Attraction of natural enemies by plants can be seen as a form of indirect defense by which plants increase the mortality acting on herbivores. We consider these kinds of interaction below, and in Chapters 5 and 6.

3.3.7 Natural enemies as indirect defenses of plants

Volatiles induced in plants by insect attack may also attract natural enemies of the same herbivores (Carroll et al. 2006). Earlier in this chapter we learned that the jasmonic acid signaling pathways induced in plants by herbivore attack can lead to the production of volatile compounds that attract natural enemies of the herbivores. The list of crop plants that appear to use such volatile signals includes Lima bean, maize, cotton, tomato, cabbage, and Brussel sprouts (Hunter 2002). In addition, some species of trees are thought to produce volatile signals that influence the searching behavior of natural enemies (Drukker et al. 1995; Scutareanu et al. 1997) and the whole field of volatile signaling has become one of the fastest growing areas of entomological research. Volatile organic compounds released from herbivore-infested plants include the monoterpenes and sesquiterpenes of the isoprenoid pathway, green leaf volatiles of the fatty acid/lipoxygenase pathway, products of the octadecanoid pathway, and aromatic metabolites (e.g. indole and methyl salicylate) of the shikimate/tryptophan pathway (Paré & Tumlinson 1996).

So how exactly do plants produce volatiles in response to insect damage? The maize system has provided a number of important insights into herbivore-induced volatile emission and serves as a useful model. When beet armyworm, *Spodoptera exigua* (Lepidoptera: Nocuidae), caterpillars feed upon maize plants, they induce the systemic release of volatile chemicals that are attractive to at least two species of parasitoids, *Cotesia marginiventris* and *Microplitis croceipes* (Hymenoptera: Braconidae) (Turlings & Tumlinson 1992). Artificial damage alone does not induce significant volatile production. However, the addition of caterpillar regurgitant (spit) to sites of artificial damage does stimulate the emission of

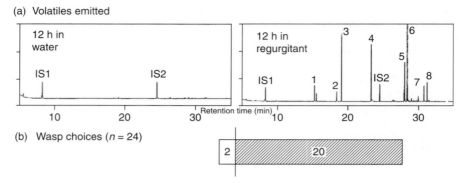

(a) Volatiles emitted

(b) Wasp choices (*n* = 24)

Figure 3.19 (a) Chromatographic profiles of volatiles collected from corn seedlings incubated with distilled water (left) and regurgitant of *Spodoptera exigua* larvae (right), and (b) choice of female parasitoids between the two types of corn seedlings. Volatiles induced by regurgitant are attractive to female parasitoids. (From Turlings et al. 2000.)

volatiles (Turlings et al. 1990) and has led to the isolation of *N*-(17-hydroxylinolenoyl)-L-glutamine (thankfully named volicitin) from the regurgitant of *S. exigua* larvae as a volatile-inducer (Alborn et al. 1997; Turlings et al. 2000) (Figure 3.19). What is the origin of volicitin? Chemical analyses of caterpillar oral secretions have shown that the fatty acid portion of volicitin is plant-derived whereas the 17-hydroxylation reaction and the conjugation with glutamine are carried out by the caterpillar with glutamine of insect origin (Paré et al. 1998). In other words, part of the volicitin molecule comes from the plant, and part from the insect. Volicitin appears to activate genes for indole emission (Frey et al. 2000) and terpenoid biosynthesis (Shen et al. 2000) in maize. In short, a substance in caterpillar spit induces the systemic production of volatile signals that attract enemies of the caterpillar (Yoshinaga et al. 2005).

In cotton, herbivore damage also induces the systemic release of volatiles that attract the natural enemies of cotton herbivores (Röse et al. 1998). In a fascinating twist, treatment of cotton plants with methyl jasmonate, a molecule that is known to participate in plant signaling (Section 3.3.6), results in volatile emissions that mimic those produced in response to herbivore damage (Rodriguez-Saona et al. 2001). This means that farmers might be able to attract enemies of herbivores by treating plants with signaling molecules rather than waiting for defoliation to occur. If volatiles are ever to be widely used in commercial pest management (see Chapter 12), such chemical methods for inducing volatiles will be essential.

Of course, volatiles will not be of much use unless they attract enemies that are capable of attacking the insect pest that is responsible for the damage. How specific are the volatiles produced in response to defoliation? Laboratory studies suggest that volatile blends from a single plant species can vary depending upon the herbivore species responsible for the damage, and that enemies can respond to herbivore-specific volatile emissions (Koch et al. 1999). For example, the specialist parasitoid *Cotesia rubecula* can distinguish among volatiles released from cabbage plants in response to its host insect, *Pieris rapae* (Lepidoptera: Pieridae), and those produced by non-host caterpillars, by snails, or by mechanical damage. Chemical analyses have shown that the volatiles released from cabbage plants in response to damage are herbivore-specific (Agelopoulos & Keller 1994a, b). Similarly, the specialist aphid parasitoid, *Aphidius ervi*, can distinguish between host (*Acyrthosiphon pisum*) and non-host (*Aphis fabae*) aphids (Hemiptera/Homoptera: Aphididae) on the host plant, *Vicia faba*, presumably as a result of variation in volatile emission (Du et al. 1996). However, many different environmental factors can act to interfere with herbivore-specific volatile signals. Emissions can vary with plant cultivar (Loughrin et al. 1995), plant species (Geervliet et al. 1997), and time of day (Loughrin et al. 1994), begging the question of how enemies can respond with any predictability to herbivore-induced volatiles. One solution may be associative learning. For example, field-collected anthocorid predators, *Anthocoris nemoralis* (Hemiptera/Heteroptera: Anthocoridae), are attracted to volatiles produced from psyllid-infested pear trees (Scutareanu et al. 1997). However, naïve predators do not distinguish between volatiles from psyllid-infested and control pear leaves. After experiencing volatiles in the presence of prey items, the predators prefer the odor from infested leaves. Associative learning may therefore help to counter the myriad of conflicting signals present under field conditions and such learning has been demonstrated for predatory mites on cotton as well as anthocorids on fruit trees (Drukker et al. 2000a, b).

If they are to be exploited in pest management, herbivore-induced volatiles must also operate reliably under field conditions as well as in the laboratory. In a field experiment, Thaler (1999) sprayed tomato plants with jasmonic acid to induce the octadecanoid pathway that is required for volatile release (Section 3.3.6). Densities of the pest *Spodoptera exigua* (Lepidoptera: Noctuidae) and its parasitoid *Hyposoter exiguae* (Hymenoptera: Ichneumonidae) were counted on control and induced plants 3 weeks after induction. Induced plants hosted twice as many parasitoid pupae as did control plants. In addition, parasitism rates of *S. exigua* larvae that were restrained beside induced plants increased by 37% (Figure 3.20). These results suggest that volatile release may attract or retain parasitoids under field conditions and that the system may be amenable to manipulation for biological control. However, in other field experiments, the placement of plants, the types of enemies, and the meteorological conditions have all been shown to influence the responses of predators and parasitoids to pest-induced volatiles (Ockroy et al. 2001).

It seems that predators and parasitoids distinguish among volatile cues from different insect herbivores under field conditions (Kost & Heil 2006). De Moraes et al. (1998) treated both tobacco and cotton plants with damage by *Heliothis virescens*, damage by *Helicoverpa zea* (Lepidoptera: Noctuidae), or no damage (controls). Plants were then placed within a field of cotton. Even after the removal of damaged leaves and herbivores, the specialist parasitoid *Cardiochiles nigriceps* (Hymenoptera: Braconidae) could distinguish among treatments and was attracted preferentially to plants that had been damaged by its host, *H. virescens*. In other words, the systemic cues released by plants were specific to the species of herbivore responsible for the damage, and could attract the appropriate enemy in the field. Gas chromatography confirmed that different chemicals were released from tobacco and cotton plants depending upon the species of herbivore responsible for the damage. These results suggest that specialist natural enemies may be able to distinguish among the diversity of volatile cues available in the environment in order to detect and attack the appropriate species of herbivore. However, generalist natural enemies may be less likely to distinguish among herbivore-specific volatile cues in nature. For example, Kessler and Baldwin (2001) studied *Nicotiana attenuata* plants in Utah that were attacked by three common herbivore species – a lepidopteran, a mirid, and a flea beetle. Volatiles were collected from attacked and unattacked plants and chemical analyses indicated that all herbivore species could induce the emission of volatiles. These volatiles were then applied to unattacked plants in a lanolin paste in association with eggs of *Manduca sexta* (Lepidoptera: Sphingidae) as a model herbivore. Predation by the geocorid bug, *Geocoris pallens* (Hemiptera/Heteroptera: Geocoridae), on *M. sexta* eggs was higher in the presence of volatiles than in the presence of lanolin-only controls. In this case, a generalist enemy is responding to volatiles from three

(a) 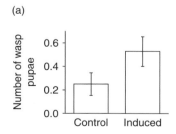 (b)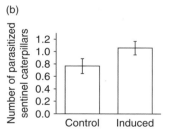

Figure 3.20 The number of parasitic wasp pupae (a) and the number of parasitized caterpillars (b) associated with tomato plants induced to produce volatiles by spraying with jasmonic acid. (From Thaler et al. 1999.)

different herbivores, and attacking the eggs of a fourth herbivore species. Clearly, specificity is not always observed in volatile defenses.

Another result from the Kessler and Baldwin (2001) study was that subsequent oviposition by herbivores also declined in the presence of volatiles. In other words, herbivore-induced volatile emission may serve a direct defensive function in addition to attracting enemies. Why would insect herbivores be deterred from plants emitting induced volatiles? The volatiles might act as a reliable indicator that the plant has already been colonized by herbivores. New herbivores might therefore be forced to compete with insects already on the plant (see Chapter 4), suffer the effects of wound-induced defenses (Section 3.3.5), or face imminent attack from natural enemies attracted to the same volatiles (see above). In any case, a number of studies have now shown that herbivore-induced volatile emission can sometimes act to deter herbivores as well as attract predators, doubling the impact of the defense system (Hunter 2002). The field application of *cis*-jasomone to winter wheat, for example, can reduce the densities of cereal aphids in plots (Birkett et al. 2000) by simultaneously repelling aphids while attracting predators and parasitoids. Likewise, tobacco plants appear to produce some volatiles only at night, which act to deter female moths from laying eggs (De Moraes et al. 2001).

Induced volatile production is only one way that plants attract natural enemies to act as indirect defenses. Plants can provide a variety of resources for predators and parasitoids, including shelter and food. One illustration of such an indirect defense is the extrafloral nectaries produced by some plant species (Figure 3.21a). Extrafloral nectaries usually play no role in pollination because they are distant from the sites of pollen transfer. Rather, they act as sources of food (sugars, amino acids) used by natural enemies. The assumption is that by attracting natural enemies such as ants, predatory wasps, or parasitoids, plants will gain an advantage by increased levels of predation on their insect herbivores. Extrafloral nectaries and ants have been collected from 35 million-year-old fossils of *Populus*, suggesting that this kind of ant–plant mutualism (where two organisms both benefit from an association; see Chapter 6) may be very ancient (Pemberton 1992). There is evidence, for example, that the extrafloral nectaries of bracken,

Pteridium aquilinum, attract wood ants, *Formica lugubris* (Hymenoptera: Formicidae), in Britain. Ants have been shown to lower the population densities of some native insect herbivores on bracken, although their greatest impact may be to deter colonization by naïve, or maladapted, herbivores (Strong et al. 1984). Extrafloral nectaries and ant–plant mutualisms are discussed in some detail in Chapter 6.

Plants can also provide shelter for natural enemies. Some ant–plant associations are obligate, called myrmecophytism, whereby a single species of trees is always inhabited by a single species of ant (Fiala & Maschwitz 1992). Ant–*Acacia* associations are classic examples of this phenomenon. Some *Acacia* trees have hollow thorns that are colonized by ants (Figure 3.21b); these so-called domatia become more numerous on a tree as it gets older (Trager & Bruna 2006). In return for shelter, the ants are supposed to reduce levels of herbivory on the *Acacia* trees. For example, ants living within *Acacia farnesiana* in Santa Rosa National Park in Costa Rica remove about 50% of eggs laid by bruchid beetles (seed predators) on seed pods (Traveset 1990). In Kenya, the symbiotic ants associated with *A. drepanolobium* are even thought to deter giraffe browsing (Madden & Young 1992). Similarly, the hollow stems (internodes) of *Cecropia* and *Macaranga* trees provide nest sites for ants which are then supposed to reduce levels of defoliation on these rapidly growing tropical trees. The evolution of ant–plant mutualisms will be considered in more detail in Chapter 6.

3.3.8 Interplant communication and "talking trees"

The current focus on volatile production by damaged plants (Section 3.3.7) has reawakened interest in an intriguing and controversial subject, communication among individual plants. In 1983, Baldwin and Schultz published a paper in the journal *Science* which appeared to describe interplant communication. They were studying the induced defenses of poplars and sugar maples, in which leaf damage causes an increase in polyphenolic concentrations (carbon-rich defenses) in their foliage. As well as recording increases in polyphenolics in damaged plants, Baldwin and Schultz (1983) observed increases in the same compounds in adjacent plants that had not been

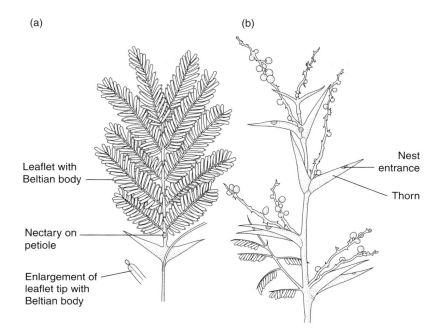

Figure 3.21 Features of a bull's horn acacia that assist in its mutualistic relationship with ants. (a) Extrafloral nectaries and Beltian bodies provide ants with sugars and proteins, respectively. (b) Hollow thorns provide shelter for *Pseudomyrmex* ants. (From Wheeler 1910; Price 1996.)

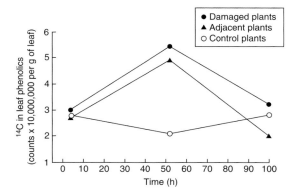

Figure 3.22 ^{14}C in foliar phenolics of poplar ramets that either received experimental damage, were adjacent to damaged ramets, or were controls. Increased phenolic synthesis in plants adjacent to those damaged experimentally provides evidence for interplant communication. (Data redrawn from Baldwin & Schultz 1983.)

damaged (Figure 3.22). They argued that undamaged plants must be able to detect volatile cues from damaged plants, and so "raise their own defenses" before attack. Quickly labeled the "talking tree" hypothesis, this study caught the public imagination, and was reported in the tabloids as well as in scientific journals. In the same year, an independent research study (Rhoades 1983) found a similar phenomenon in willow trees. The idea that plants might be warning each other of impending defoliation was met with disbelief by many. In one dismissive article, Fowler and Lawton (1985) described the results as statistically flawed, evolutionarily unreasonable, and explainable by other, more likely, factors. It is important to realize, however, that Baldwin and Schultz envisaged plants as "listening" for evidence of herbivory, rather than cooperating with one another by direct signals. None the less, their results proved difficult to reproduce, and interest in the subject declined.

In recent years, the evidence for interplant communication has increased, supporting the original conclusions of Baldwin and Schultz. The volatiles produced by damaged Lima bean plants, for example, are apparently detected by undamaged plants, which respond by producing volatiles of their own. Volatiles from attacked plants appear to activate five separate defense genes in unattacked plants, primarily through stimulation of the jasmonic acid pathway (Section 3.3.6) (Arimura et al. 2000). In similar experiments, uninfested cotton seedlings exposed to volatiles from infested seedlings become both a deterrent to herbivorous mites and attractive to predatory mites (Bruin et al. 1992). These studies have caused a reassessment of interplant communication (Bruin et al. 1995; Shonle & Bergelson 1995) and have stimulated new research in this fascinating area.

Two systems in particular appear to provide the best evidence of volatile-mediated interplant communication under field conditions. In the first, wild tobacco plants that are growing beside damaged sagebrush plants express induced chemical defenses and suffer lower levels of herbivory (Karban et al. 2004). Damaged sagebrush plants release methyl jasmonate (Section 3.3.6), a volatile signal that is known to induce defenses in wild tobacco. Tobacco plants adjacent to clipped sagebrush exhibit elevated concentrations of polyphenol oxidase, an enzyme used in the synthesis of chemical defenses. Moreover, tobacco plants near damaged sagebrush plants suffer greatly reduced levels of leaf damage by grasshoppers and cutworms. Although the precise mechanisms are unclear (Preston et al. 2004), the evidence for interplant communication appears quite strong. In a second system, artificial defoliation to alder trees appears to increase the resistance of nearby trees to leaf beetles (Dolch & Tscharntke 2000). When a single alder tree within a stand is defoliated, subsequent damage by beetles to neighboring alder trees increases with distance from the defoliated tree. In laboratory assays, beetles also avoid leaves from neighboring plants for both feeding and oviposition. The pattern of interplant resistance appears to hold for specialist insects, but not for generalist insect herbivores on alder (Tscharntke et al. 2001). Again, the precise mechanisms of information transfer among plants are not clear in this system. None the less, interplant communication has again become an active area of research by entomologists and ecologists and should provide some interesting results in the coming years.

3.4 PLANT HYBRID ZONES AND DEFENSE AGAINST HERBIVORES

A subset of the studies on plant genotype and insect herbivores has focused on plant hybrid zones, regions where two or more congeneric species of plant produce hybrid offspring. Insect ecologists became interested in plant hybrid zones when Whitham (1989) demonstrated that hybrid zones of *Populus* species acted as "sinks" for insect herbivores. In other words, hybrid poplars supported higher densities of insect herbivores and suffered higher levels of herbivory than either of the parental genotypes. The assumption was that the defenses of the parental species (chemistry, morphology, phenology, etc.) were somehow weakened or disrupted by hybridization. The concordance between insect densities and parental versus hybrid genotypes can be so strong that the distributions of insects can be used to segregate closely related plant taxa, their hybrids, and even complex backcrosses (Floate & Whitham 1995).

The "hybrid susceptibility" hypothesis has been controversial. Boecklen and Spellenberg (1990) reported that densities of leaf-mining Lepidoptera and gall-forming Hymenoptera were both lower in two different hybrid zones of oak (*Quercus* species), a direct contradiction of the hypothesis (Figure 3.23). In subsequent work, Aguilar and Boecklen (1992) found that leaf-miners and gall-formers in a third

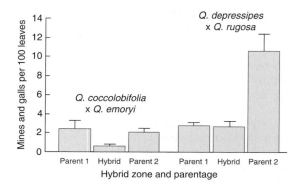

Figure 3.23 Combined densities of galling and mining insects in two oak (*Quercus*) hybrid zones. In neither hybrid zone are insects more common on hybrid trees than on parental trees. Data are the means of 100 trees per genotypic class. (From Boecklen & Spellenberg 1990.)

oak hybrid zone occurred at intermediate densities when compared with the parental oak genotypes. They argued that the responses of insect herbivores to hybridization were idiosyncratic, depending both upon the insect and hybrid zone in question. The efficacy of natural enemies of herbivores is also influenced by plant hybridization (Preszler & Boecklen 1994; Gange 1995), a so-called tri-trophic interaction (Price et al. 1980). Although several authors feel that there is no general response by insect herbivores to plant hybrid zones (Fritz et al. 1994; Gange 1995), it appears that at least some hybrid zones act as sinks for herbivores (Floate et al. 1993; Whitham et al. 1994). In some cases, the structure of insect communities and overall biodiversity vary widely between parental genotypes and hybrid zones (Whitham et al. 1999; Dungey et al. 2000). In other cases, the effects of plant ontogeny, leaf age, and keystone species can be more important than plant genotype (Martinsen et al. 2000; Lawrence et al. 2003). Consensus in this field will likely emerge when the relative impacts of hybridization and other ecological factors (soil type, plant age, climate, etc.) are measured concurrently with plant genetics in a wide range of systems.

3.5 WHEN IS A DEFENSE NOT A DEFENSE?

It is rarely possible to say unequivocally that a particular trait exhibited by a plant is a direct result of natural selection imposed by an insect herbivore. What we call "defenses" may have arisen for a number of different reasons and, serendipitously, may influence the preference or performance of insects on their hosts. It is all too easy for entomologists, fixated as they are with insects, to forget that plants are exposed to many other selection pressures from the abiotic (non-living) environment and from other members of the biotic environment (competitors, decomposers, pathogens, non-insect herbivores). It is likely that natural selection has molded most traits expressed by plants in response to multiple selection pressures. Trichomes, for example, have properties in addition to deterring herbivores. In Indian mustard, *Brassica juncea*, trichomes act as sites in which toxins from the soil such as cadmium are accumulated (Salt et al. 1995). Whether or not environmental toxin accumulation in trichomes led to the later development of trichomes as a defense against insects is

unknown. Trichomes can also improve the water status of leaves by reducing water loss and increasing water uptake. The hairs can trap and retain surface water, and assist in its final absorption into the mesophyll (Grammatikopoulous & Manetas 1994). There seems little doubt that trichomes can play a critical role in avoiding drought stress for some plant species.

Putative chemical defenses in leaf tissue may also play more than one role. For example, variation in phenolic compounds in different environments may have more to do with photoinhibition (reductions in photosynthesis under high light conditions) than with defense against insects (Close et al. 2003). If true, subsequent effects against insect herbivores would be consequences, rather than causes, of phytochemical variation. It might also explain why no single theory of plant chemical defense (Section 3.3.5) has been able to explain patterns of chemical variation found in nature. Multiple functions can be attributed to most plant "defenses" against insects. Is one function any more important than any other? What ecological factors were most important in their evolutionary development? It is unlikely that we will be able to answer these questions easily, and it may be pointless to try. It is much more important to understand the current role of putative defenses in the ecology of plants than their precise evolutionary origins. The controversy surrounding the tight evolutionary linkages between insects and plant defenses will be considered in the discussion of coevolution in Chapter 6.

3.6 COSTS OF RESISTANCE AND TOLERANCE TO HERBIVORES

Why are all plants not maximally and successfully defended against attack by insect herbivores? One possibility is that there is a dynamic, evolutionary "arms race" between insect herbivores and their hosts, with the development of novel plant defenses being followed by adaptations of herbivores to resist these defenses. This evolutionary arms race will be considered along with pollination in the chapter on evolutionary ecology (see Chapter 6). A second (not mutually exclusive) possibility is that the production of defenses is costly to plants. Indeed, this has been a common assumption by insect–plant ecologists for many years, especially those working on induced defenses (see the optimal defense hypothesis, Section 3.3.5). Presumably the advantage of an induced defense is that it is activated (and the costs paid) only when herbivores are known to be present. No costs are paid when the herbivore is absent (Rhoades 1985). If defenses truly are costly for plants, there ought to be tradeoffs between the production of defenses and other needs such as growth or reproduction. Ultimately, of course, costs should be reflected in reductions in lifetime reproductive success. However, since size and reproductive output are commonly linked in plants, growth is often substituted for reproduction in studies of defense costs.

For example, Coley (1986) studied the costs and benefits of tannin production in a neotropical tree, *Cecropia peltata*. She grew seedlings of equal age under uniform conditions, and measured their tannin concentrations, leaf production, and palatability to insect herbivores. The results demonstrated that plants with higher tannin content suffered lower levels of herbivory (benefit of defense) but produced fewer leaves (cost of defense) (Figure 3.24). Since *Cecropia* is a pioneer tree species that colonizes disturbed areas, rapid growth is presumably advantageous so that reproduction occurs before competitively superior, but slower growing, trees dominate the area. *Cecropia* appears to face a tradeoff between growth and defense (Coley 1986). Similarly, there is a negative correlation between growth and tannin production in the neotropical shrub *Psychotria horizontalis*.

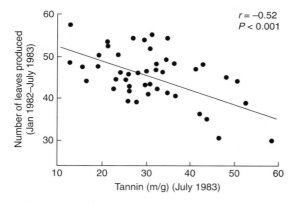

Figure 3.24 Relationship between tannin (quebracho) concentration in the follage of *Cecropia* and the number of leaves produced per plant. Investment in tannin appears to occur at the expense of leaf production, and may represent the cost of defense. (From Coley 1986, with permission from Springer-Verlag New York, Inc.)

The tannins in this shrub confer some protection to the plants against insect herbivores (Sagers & Coley 1995). Tradeoffs between growth and defense presumably help to explain natural variation in defense production among individual plants. Simply put, there is no perfect way to "be a plant" in the presence of herbivores. Some individuals are heavily defended but grow slowly. Others are less well defended, suffer insect attack, but grow more rapidly. Both strategies can work; both can be maintained in populations over time, and result in the natural variation we observe in plant populations (Coley et al. 2006).

However, substituting growth for reproduction when measuring the costs of defense may be misleading. This is illustrated by the perennial plant *Senecio jacobaea*. Although vegetative (growth) costs associated with the production of chemical defenses have been measured in *S. jacobaea*, there are no apparent reproductive costs (flowers and seeds) to pyrrolizidine alkaloid production (chemical defense) or the production of tough leaves (Vrieling 1991). Indeed, costs of resistance initially proved difficult to detect in many different studies, generating some anxiety as to the relevance of evolutionary theories based on optimal allocation to defense (Simms & Rausher 1989; Adler & Karban 1994). Defense costs may not be particularly high if defenses last a reasonable length of time without renewal. For example, there may not be as rapid a turnover of some chemical defenses, such as the monoterpenes of peppermint, as was previously thought (Gershenzon 1994) (Figure 3.25). Similarly, there are no measurable costs associated with the production of the thorns on bramble plants that deter herbivores (Bazely et al. 1991). Measurable costs of defense have also proven illusive in the desert shrub, *Gossypium thurberi* (Karban 1993), in *Plantago lanceolata* (Adler et al. 1995), and in the annual morning glory, *Ipomoea purpurea* (Simms & Rausher 1989). However, as techniques for estimating the costs of resistance have improved, the number of studies reporting significant costs has increased (Strauss et al. 2002). In his studies of wild tobacco defenses, Baldwin and colleagues have measured substantial costs to alkaloid production. You will recall from Section 3.3.6 (above), that wild tobacco plants respond to herbivore attack by the induction of octadecanoid metabolites, including jasmonic acid (Baldwin 1998). The induction occurs at a substantial fitness cost to plants, measured as a significant decline in seed production. Interestingly, the costs are associated

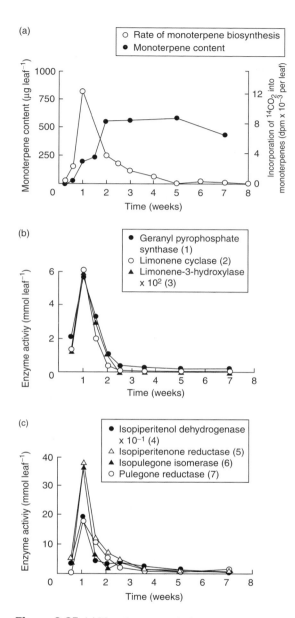

Figure 3.25 (a) Monoterpene content versus monoterpene synthesis in peppermint. Content is stable after synthesis declines, showing that monoterpenes are not being turned over. (b, c) Enzymes involved in monoterpene biosynthesis are virtually inactive after leaves are 2 weeks old. (From Gershenzon 1994.)

with impaired competitive ability for below-ground resources (Van Dam & Baldwin 1998). Additionally, it may be that tradeoffs between defenses and growth only occur under resource-limiting conditions (Osier & Lindroth 2006), so that the costs of resistance vary with the state of the plant. Some costs to resistance can be measured as "ecological costs", whereby the cost is paid through interactions with other species. For example, if chemical resistance comes at the expense of nectar production, resistance costs can be measured as reductions in pollinator visits (Strauss et al. 1999). This has been shown by artificial selection experiments with *Brassica rapa* plants. Pollinators spend less time per flower on *Brassica* strains selected for high chemical resistance than they do on plants selected for low chemical resistance (Figure 3.26).

Some insect ecologists have explored alternative costs imposed by herbivores. One of these is "tolerance". Tolerance is defined as the ability of plants to compensate, in part, for the effects of defoliation on plant fitness. In other words, a plant with high tolerance maintains higher levels of growth or reproduction for a given level of defoliation than a plant with low tolerance. As originally formulated (Van der Meijden et al. 1988), the hypothesis suggests that there is a tradeoff between tolerance and resistance, and that both strategies act to reduce the impact of insect herbivores on plant fitness. Certainly, some plants appear to be able to tolerate herbivory more than others, losing little in overall growth or reproduction despite defoliation (Rosenthal & Kotanen 1994). If tolerance is expensive to plants, we might expect to see lower reproductive rates from tolerant plants than less tolerant plants *in the absence of herbivory* (Simms & Triplett 1994) (Figure 3.27). Is there empirical evidence to support a cost for tolerance? Fineblum and Rausher (1995) have studied the effects of damage to the apical meristems of the annual morning glory, *Ipomoea purpurea*, by insects, and have shown that genotypes of *I. purpurea* with relatively high levels of tolerance to this type of damage pay the "cost" of lower resistance. On the same plant species, Simms and Triplett (1994) demonstrated that individuals that were more tolerant to disease had lower fitness in the absence of disease, suggesting a significant cost of tolerance. Like Fineblum and Rausher, Stowe (1998) found a negative relationship between resistance (glucosinolate content) and tolerance to herbivory

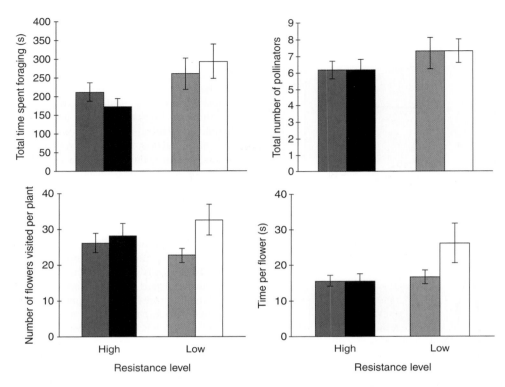

Figure 3.26 The tradeoff between resistance to herbivores by *Brassica rapa* and visits by pollinating insects. Plants selected to express low levels of myrosinase (low resistance) were visited more frequently, and for longer periods, than were plants selected for high resistance. The left (grey) bar in each pair represents plants that were damaged by herbivores, suggesting that induction of defense can also reduce pollinator visitation. (From Strauss et al. 1999.)

in *Brassica rapa*. However, other studies have failed to find a cost for tolerance. For example, Agrawal et al. (1999) found no negative relationship between tolerance to herbivory and five measures of fitness in wild radish, *Raphanus raphanistrum*. Interestingly, costs of induced resistance in wild radish were expressed as reductions in pollen grains per flower (male function) as opposed to seed production (female function). This suggests that the methods by which costs of resistance and tolerance are measured need to be carefully considered (Purrington 2000; Strauss et al. 2002).

3.7 OVERCOMPENSATION

Some studies of plant tolerance to herbivory have revealed that certain plant species appear to overcompensate for defoliation. In other words, the growth or

reproduction of the plant is actually higher in the presence of herbivores than it is in the absence of herbivores. This has led some workers to speculate that some levels of herbivory may be beneficial to plants, and increase their fitness (Dyer & Bokhari 1976; Owen & Wiegert 1976). It has further been suggested that tight coevolution between plants and herbivores may have resulted in "plant–herbivore mutualisms" by which some degree of defoliation is "expected" by plants, and rates of plant production rise following damage by herbivores (Paige & Whitham 1987). Given the limited resource budget of plants and the rigors of natural selection, it is difficult to understand why plants that are capable of increasing growth or reproduction following damage would grow or reproduce maximally in the absence of herbivores. This argument was forcibly made by Belsky et al. (1993) who concluded that,

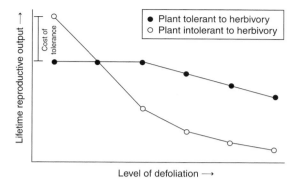

Figure 3.27 Hypothetical costs associated with a plant's tolerance to insect herbivory. A tolerant plant exhibits a slower decline in fitness than an intolerant plant as defoliation increases. There may be a cost for this tolerance in reduced fitness when herbivores are absent.

"There is no evolutionary justification and little evidence to support the idea that plant–herbivore mutualisms are likely to evolve. Neither life-history theory nor recent theoretical models provide plausible explanations for the benefits of herbivory."

Although convincing evolutionary explanations for overcompensation may be illusive, there is now little doubt that overcompensation occurs in some circumstances. For example, flower production by scarlet gilia, *Ipomopsis aggregata*, increases following herbivory: undefoliated plants produce only one inflorescence whereas defoliated plants produce multiple inflorescences (Figure 3.28). The result is a 2.4-fold increase in relative fitness following defoliation (Paige & Whitham 1987; Paige 1992). Similarly, growth of blue grama grass increases following herbivory by the big-headed grasshopper, *Aulocara elliotti* (Orthoptera: Acrididae) (Williamson et al. 1989), and the production of tallgrass prairie increases following defoliation (Turner et al. 1993). Some willow trees produce more and longer shoots after attack by spittlebugs, increasing resource availability for subsequent spittlebug generations (Nozawa & Ohgushi 2002). Some genetic lines of *Arabidopsis thaliana* exhibit overcompensation, producing more fruit in the damaged than the undamaged state (Weinig et al. 2003). Finally, the exotic plant spotted knapweed, *Centaurea maculosa*, which is invasive in North America, exudes much higher concentrations of allelopathic chemicals when attacked by root-feeding insects (Thelen et al. 2005) and is hence able to deter the growth of neighboring

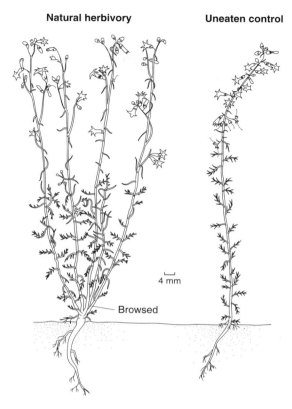

Figure 3.28 Illustration showing the positive effects of herbivory on scarlet gilia, *Ipomopsis aggregata*. Plants damaged by herbivores have multiple stems and multiple inflorescences. Control plants have a single stem and inflorescence. (Adapted from Paige & Whitham 1987, with permission from University of Chicago Press.)

native plants. So, with at least some empirical support for overcompensation, theoreticians have become more willing to explore the conditions under which natural selection might favor palatable plants. De Mazancourt and Loreau (2000) have modeled overcompensation, and concluded that palatable plants might indeed evolve so long as herbivores are small (like insects) and contribute to local nutrient cycling that benefits the defoliated plant (see Chapter 8).

Studies of grasshopper crop and midgut secretions suggest that there may be animal biochemical messengers that indirectly increase the growth rates of defoliated plants by up to 295% (Moon et al. 1994; Dyer et al. 1995). Termed "reward feedback" by Mel Dyer and colleagues, these biochemical messengers may exert a major control on ecosystem productivity (Dyer et al. 1995). In Section 3.3.7 above, we described how plants can recognize chemical messengers in insect herbivore saliva and, in response, produce volatiles to attract parasitoids. We should at least consider the possibility that insect herbivores can also produce salival messengers to manipulate the growth patterns of their host plants. However, several studies have found little or no evidence of overcompensation following defoliation (Bergelson & Crawley 1992; Edenius et al. 1993; Oba 1994) and it appears that many factors, including plant species (Alward & Joern 1993; Escarre et al. 1996), resource availability (Alward & Joern 1993), type of defoliation (Hjalten et al. 1993), and defoliation history (Turner et al. 1993) all influence overcompensation responses. It has been suggested that overcompensation is at one end of a continuum of plant responses to herbivory, from negative to positive, and that we should expect to observe considerable variation in plant responses to defoliation in natural systems (Maschinski & Whitham 1989).

3.8 PLANT DEFENSE UNDER ELEVATED CARBON DIOXIDE

Global environmental change is very likely to influence the interactions between insect herbivores and their host plants. Entomologists have recognized for some time that elevated concentrations of atmospheric carbon dioxide (CO_2) may influence the distribution, abundance, and performance of insects that feed on plants (Lincoln et al. 1984; Fajer 1989). Many ecological factors can change as CO_2 levels rise, including weather patterns, plant quality, predation pressure, and ecosystem processes. Effects of changing climate on insects were discussed in Chapter 2. Here, we focus upon potential changes in plant quality, and what the consequences might be for insects.

Atmospheric CO_2 concentrations have already risen by about 25% since the industrial revolution and are expected to increase from current ambient levels of 350–360 ppm (or μL/L) to around 770 ppm by the end of the century (Houghton et al. 1995). CO_2 is the principle source of carbon for photosynthesis, and changes in concentration have marked effects upon the phenotype of plants. For example, elevated concentrations of CO_2 generally result in increased rates of photosynthesis (Drake et al. 1997), increased rates of growth (Saxe et al. 1998),

and increased plant biomass (Owensby et al. 1999). These changes in physiology, growth, and biomass are sometimes referred to as the effects of carbon fertilization. However, when plants grow more without any concurrent changes in soil nutrient concentrations, the result appears to be a dilution of the nutrient content in plant tissue. For example, elevated CO_2 concentrations often dilute foliar concentrations of nitrogen by 15–25% (Lincoln et al. 1993) thereby increasing C : N ratios. Recall from Section 3.1.1 that the stoichiometric ratios of carbon to nutrients in plant tissue can have profound effects upon insect preference and performance. We might therefore expect that increases in foliar C : N ratios under elevated CO_2 should be deleterious for insect herbivores. Also, based on the still-cited CNB hypothesis (Section 3.3.5), high concentrations of CO_2 might also lead to the allocation of carbon to some carbon-rich secondary metabolites such as tannins (Lindroth et al. 1995; Agrell et al. 2000). Again, the results should be negative for insect herbivores.

To what extent do these generalities hold true? Unfortunately, not all plant species respond identically to elevated concentrations of CO_2. For example, elevated CO_2 results in reduced foliar nitrogen levels and increased condensed tannin levels in paper birch but not in white pine (Roth & Lindroth 1994). In further studies with paper birch, quaking aspen, and sugar maple (Agrell et al. 2000), all species show increases in foliar concentrations of condensed tannins under elevated CO_2. However, the foliage of quaking aspen also expresses higher concentrations of phenolic glycosides and sugar maple is the only species to show elevated foliar concentrations of hydrolysable tannins. In a study contrasting a C_3 sedge with a C_4 grass in marsh habitat, Thompson and Drake (1994) reported CO_2-mediated declines in foliar nitrogen only in the sedge. Foliar concentrations of the iridoid glycosides in *Plantago lanceolata* are unaffected under CO_2 enrichment (Fajer 1989). Likewise, the volatile terpenoids of peppermint (Lincoln & Couvet 1989), big sagebrush (Johnson & Lincoln 1990), and loblolly pine (Williams et al. 1997) do not vary with experimental increases in CO_2. It seems as if the CNB hypothesis is inadequate to explain the effects of carbon fertilization on all aspects of plant quality.

Although there are no perfect generalities for the way that plant phenotypes change under elevated CO_2, in nearly every case examined to date, foliar nitrogen concentrations decline (Hunter 2001c). This level of generality is somewhat heartening and

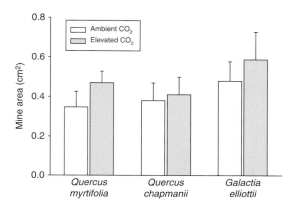

Figure 3.29 The area of leaf mines on three plant species growing under ambient and elevated levels of carbon dioxide (CO_2). To compensate for lower foliar nitrogen concentrations, leaf-miners consume more tissue on plants growing under elevated CO_2. (From Stiling et al. 2003.)

Figure 3.30 Open-top chambers at the Kennedy Space Center (Florida, USA) used to manipulate levels of atmospheric carbon dioxide. (Courtesy Peter Stiling.)

allows us to predict that overall decreases in foliar quality should induce at least some insect herbivores to eat more. And that prediction usually holds true. Lower levels of nitrogen and higher C : N ratios in plants under elevated CO_2 have generally been associated with compensatory feeding and subsequent increases in levels of damage or defoliation, at least under laboratory conditions. Leaf-chewing insects such as grasshoppers and caterpillar larvae generally consume more leaf area when fed plants that have been grown under elevated CO_2 (Lindroth et al. 1993). Likewise, the area damaged by leaf-mining insects may also increase. For example, the area of leaf mines on *Quercus myrtifolia* increases by over 25% under elevated CO_2 (Figure 3.29), apparently because nitrogen concentrations fall by over 11% (Stiling et al. 2003).

However, simply because the amount of foliage consumed by each individual insect increases under elevated CO_2 does not mean that plants suffer more damage overall. Two additional effects that mediate the ultimate level of damage that plants receive are CO_2-induced increases in plant biomass and changes in insect density. Increases in plant biomass under elevated CO_2 can more than compensate for increases in defoliation, and Knepp et al. (2005) believe that damage to hardwood trees by herbivorous insects will actually decrease under elevated levels of CO_2. For example, even though per capita rates of consumption by insects on *Q. myrtifolia* increase with CO_2 enrichment, the proportion of leaves damaged by mining

and chewing insects actually declines because the plants produce many more leaves and overall insect densities are lower (Stiling et al. 2003; Hall et al. 2005). Even if they eat more, most insects appear to be unable to compensate fully for CO_2-mediated reductions in plant quality. Buckeye butterflies on *Plantago lanceolata* exhibit both higher rates of mortality and increased development time when fed on plants grown under elevated CO_2 (Fajer et al. 1991). Higher rates of insect mortality have been associated with nutritional deficiency that results from reduced foliar nitrogen concentrations under elevated CO_2.

Perhaps more dramatic effects upon insect performance will be mediated by the third trophic level. Given that rates of insect growth also seem to decline under elevated CO_2, we might expect that the risk of mortality from natural enemies will increase. Field studies using open-top chamber technology (Figure 3.30) have demonstrated that rates of leaf-miner parasitism increase under elevated CO_2 (Stiling et al. 1999). Nitrogen concentrations in the foliage of two dominant oak species, *Quercus myrtifolia* and *Q. geminata*, decline under elevated CO_2 and densities of leaf-mining insects are lower because of the combined effects of reduced foliage quality and increased rates of attack by parasitoids. Overall, death from plant effects increases by 50% and death from parasitoids increases by over 50% under elevated CO_2.

We should point out that not all insects respond negatively to the changes in plant phenotype that are mediated by elevated concentrations of CO_2. At least some phloem-feeding insects exhibit increases in

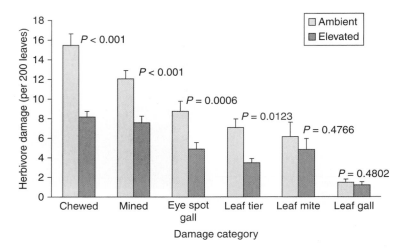

Figure 3.31 Damage by six groups of insect herbivores on plants growing under ambient and elevated concentrations of carbon dioxide. For four of the six groups, damage levels decline significantly under elevated concentrations of carbon dioxide. (From Hall et al. 2005.)

performance when provided with plants grown under elevated CO_2 (Bezemer & Jones 1998). However, there appears to be qualitative variation in the responses of phloem-feeding insects to changes in plant quality. For example, the same clone of the aphid *Aulacorthum solani* (Hemiptera/Homoptera: Aphididae) responds differently to elevated CO_2 on two different plant species (Awmack et al. 1997). On bean plants, the daily rate of nymph production increases by 16% whereas rates of development are unaffected. In contrast, aphids on tansy exhibit faster rates of development and no change in reproductive rate. Overall, aphid responses are positive under elevated CO_2 on both host plants, but the mechanisms differ between hosts. What this suggests is that we are a long way from being able to predict changes that may occur in the population dynamics of important crop pests under elevated concentrations of atmospheric CO_2.

In reality, the effects of elevated CO_2 on plant phenotype, and subsequent insect responses, will be modified by ecological complexity. Effects will be mediated by the availability of resources to plants such as water, light, and nutrients, and modified by ozone, climatic conditions, and biotic variability (Roth et al. 1997; Percy et al. 2002; Holton et al. 2003). For example, Valkama et al. (2007) have reported that elevated ozone levels tend to increase tree foliage quality for herbivores, an effect alleviated by elevated

CO_2 levels. Given this complexity, can we make any generalizations about the effects of elevated CO_2 on insects that feed on plants? Bezemer and Jones (1998) analyzed data from 61 plant–herbivore combinations and found some compelling patterns. First, they confirmed the general decreases in foliar nitrogen concentration (15%) and increases in carbohydrate (47%) and phenolic-based secondary metabolites (31%) reported in many individual studies. Second, consumption by herbivores was related primarily to changes in nitrogen and carbohydrate levels. Third, no differences were found between CO_2-mediated herbivore responses on woody and herbaceous plant species. Fourth, leaf-chewing insects generally increased their consumption of foliage (30%) under elevated CO_2 to compensate for reduced nutritional quality and suffered no adverse effects upon pupal weights. Fifth, leaf-mining insects could only partially compensate by increased consumption and their pupal weights did decline. Finally, phloem-feeding and whole-cell-feeding insects responded positively to elevated CO_2, with increases in population size and decreases in development time. The analysis by Bezemer and Jones agrees broadly with data from the longest running field study of CO_2 effects on insects to date (Stiling et al. 2003; Hall et al. 2005) in which populations of almost all herbivores in a scrub oak community declined under elevated CO_2 (Figure 3.31).

Chapter 4

RESOURCE LIMITATION

4.1 THE IMPORTANCE OF RESOURCE LIMITATION ON INSECT POPULATIONS

Like all organisms, insects have enormous potential for population growth. A fundamental principle of population biology is that, if left unchecked, populations will grow exponentially. What exactly does this mean? We illustrate the concept of exponential growth in Table 4.1. Imagine an insect population in which each individual is replaced, on average, by three individuals in the following year. If we start with a population size of two individuals, we will have six individuals in the following year. In subsequent years, we can see the population growing rapidly from six to 18, 54, and then 162 individuals. As you can see from Figure 4.1, the form of this growth is an accelerating curve, commonly referred to as exponential growth. You will also notice that the number of individuals in the population in any given year is three times the number of individuals in the population in the previous year. In other words:

$$\text{Number next year} = 3 \times \text{Number this year} \quad (4.1)$$

In this case, the yearly rate of growth of 3 is commonly referred to as the per capita rate of increase, denoted by λ. We can write a general expression for unrestrained population growth as:

$$N_{t+1} = \lambda N_t \quad (4.2)$$

where N_t is the population size in a given year, and λ is the per capita rate of increase. In some population models, particularly those in which population growth

is continuous and there are no discrete breeding seasons, we often use a second measure of per capita rate of increase, r. It is calculated as the natural logarithm of λ, or:

$$r = \ln \lambda \quad (4.3)$$

We call equation 4.2 a discrete time, or difference, equation. The continuous time analog of equation 4.2, for populations without discrete breeding seasons, is a differential equation. It also describes exponential growth.

$$dN/dt = rN \quad (4.4)$$

Returning to our example in Table 4.1, note that the population size in year four (162 individuals) can be calculated as:

$$\text{Starting density} \times 3 \times 3 \times 3 \times 3 \quad (4.5)$$

or,

$$\text{Starting density} \times 3^4 \quad (4.6)$$

We can write a general form of this expression as:

$$N_t = \lambda^t N_0 \quad (4.7)$$

where N_t is the number of individuals at time t and N_0 is the starting population size. In other words, if we know the starting size of the population, and its per capita rate of increase, we can calculate population size at any time in the future. Try it now. Given an initial population size of two, and a per capita rate of

Table 4.1 Unconstrained population growth of an imaginary insect population. The population grows by three-fold each year.

Year	Population size	Per capita growth rate
0	2	
1	6	× 3
2	18	× 3
3	54	× 3
4	162	× 3

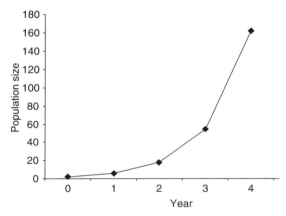

Figure 4.1 Unconstrained growth of an imaginary insect population. Data are from Table 4.1. The population approximates exponential growth.

increase of three, what will be the population size in year 6? By applying equation 4.7, you should get an answer of 1458 individuals. Our imaginary population is growing fast.

We can explore the importance of population limitation further by using data from the winter moth, *Operophtera brumata* (Lepidoptera: Geometridae), in Europe. Varley et al. (1973) have published long-term data on the winter moth that allow us to perform these calculations. First, we know that in any given year about 89% of all winter moth die from abiotic (non-living) sources of mortality. In Chapter 2 we learned that abiotic factors (floods, freezing temperatures, fire, etc.) can be very powerful sources of mortality on insects. We can calculate that, if 89% of

larvae die from abiotic sources of mortality and females lay approximately 150 eggs each, then the per capita rate of increase (λ) of an unconstrained winter moth population would be about eight (Table 4.2). If we start with a population size of 20 moths, and we know that λ of an unconstrained *O. brumata* population is approximately equal to eight, we could calculate the number of moths in 35 years time as follows:

$$\text{Population size in 35 years} = 8^{35} * 20$$
$$= 8.11 * 10^{32} \text{moths} \qquad (4.8)$$

If an average winter moth weighs 40 mg and the Earth weighs $5.9768 * 10^{24}$ kg, then after only 35 years of population growth, our population of winter moth would have a combined weight of over 5000 times the mass of the Earth!

It should be clear from this example that the growth of real insect populations must be limited in some way. We are not drowning in the bodies of insects and so ecological forces must impose constraints upon insect population growth. What have we failed to consider? In the calculations above, we made the assumption that the per capita rate of increase, λ, was a constant and did not change with population size. This is the source of our problem. In reality, population growth rates change with density, or are "density dependent". Density dependence acts to place limitations on population growth.

4.1.1   A brief introduction to density dependence

There are only four general factors, called "vital rates", which can change the size of any insect population. These are birth and immigration (which make populations increase) and death and emigration (which make populations decrease). In the simple equations above, we collapsed these four rates into a single measure which we called the per capita rate of increase, λ or *r*. In reality, the per capita growth rate is made up of all four vital rates and can be calculated as:

$$r = (\text{birth} + \text{immigration}) - (\text{death} + \text{emigration})$$
$$(4.9)$$

where birth, immigration, death, and emigration ("bide") are expressed as per capita rates (rates per

Table 4.2 Unconstrained population growth of the winter moth, *Operophtera brumata*.

Initial population size	= 20
Number of females	= 10
Fecundity (average)	= 150
Next generation	= 10 × 150
	= 1500
Abiotic mortality	= 89%
Survival	= 11%
	= 1500 × 0.11
	= 165
Per capita rate of increase, λ	= N_{t+1}/N_t
	= 165/20
	= 8.25
Population size in 35 years	= $\lambda^{35} \times N_0$
	= $8^{35} \times 20$
	= 811,000,000,000,000,000,000,000,000,000,000 moths

individual in the population). If any of these vital rates vary with population density, then they have the potential to impose limitations on population growth. Specifically, if rates of birth or immigration decline with density, or rates of death or emigration increase with density, population growth can be limited (Figure 4.2). What might cause birth rates to decline as density increases? Or emigration rates to increase with density? Competition for limited resources and the effects of natural enemies are the factors most commonly thought to introduce density dependence into insect populations. The circumstances under which natural enemies can limit insect populations will be considered in detail in Chapter 5. Here, we focus on competition.

One further comment about density dependence. If the effect of density on the four vital rates occurs after a delay, we call this "delayed density dependence". For example, if the population density of a given insect species is high in one generation, it might not result in a reduction in fecundity until the following generation; the density-dependent effect occurs on delay. Distinguishing between rapid and delayed density dependence is important because the latter can lead to regular oscillations, or cycles, in insect abundance. As we shall see in Chapter 5, population cycles are regularly observed in studies of insects such as the forest tent caterpillar, *Malacosoma disstria* (Lepidoptera: Lasiocampidae) (Cooke & Lorenzetti 2006).

(a)

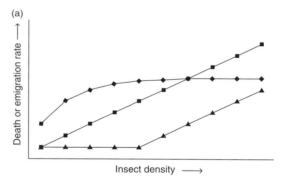

(b)

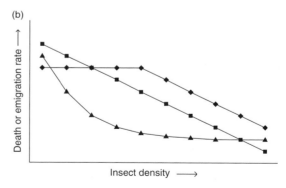

Figure 4.2 Hypothetical examples of density-dependent rates of (a) death or emigration, and (b) birth or immigration, acting on insect populations. While the exact form of the relationships can vary (i.e. the slopes of the lines can be different) death and emigration rates rise while birth and immigration rates fall with increasing insect density.

4.2 COMPETITION FOR LIMITED RESOURCES

Begon et al. (1990) defined competition in the following way: "Competition is an interaction between individuals, brought about by a shared requirement for a resource in limited supply, and leading to a reduction in the survivorship, growth, and/or reproduction of the competing individuals concerned." We would add to that definition that competition might also influence the movement (immigration or emigration) of individuals (Section 4.1.1). There are several important consequences that arise from defining competition in this way. First, organisms must overlap in their use of some limited resource; if the resource is unlimited, then competition does not occur. As an example, the readers of this book overlap completely in their use of oxygen with the authors of this book, but we do not compete with each other for the resource because it is not limiting. Similarly, if two species of herbivorous insect feed exclusively on leaves from the same plant species, they do not compete for the resource unless leaves, or at least leaves of high nutritional quality (see Chapter 3), are in limited supply.

A second consequence of this definition is that competition has the potential to limit the growth of insect populations by its influence on rates of birth, death, or movement. If the limited supply of some essential resource (space, food, shelter, etc.) causes a reduction in birth rates (or immigration) or an increase in death rates (or emigration) as population density increases, then competition for that resource is density dependent. We stress here that, just because competition has the potential to limit the growth of an insect population, it does not follow that competition regularly acts to limit natural insect populations; other factors such as predation can maintain populations below levels at which competition ever occurs (see Chapter 5). Indeed, the regularity with which competition influences insect populations has been the subject of considerable controversy. We will describe this controversy later in the chapter (Section 4.5.3).

One final consequence of the definition of competition is that the process can usually be envisaged as a negative–negative interaction (Figure 4.3). Unlike predation or mutualism, both parties engaged in competition suffer as a result of the interaction (we discuss asymmetric, or unequal, competition later in

Effect on species A	Effect on species B	Interaction
+	+	Mutualism
+	−	Predation
−	−	Competition
−	0	Amensalism
+	0	Comensalism
0	0	Neutralism

Figure 4.3 Potential types of interaction between insect species. Amensalism is a special form of asymmetric competition, where only one species suffers as a result of the interaction (Section 4.5.3).

the chapter). We might therefore expect that insects will attempt to avoid competition wherever possible. Avoidance can take several forms, including strategies that are expressed in ecological time and strategies expressed over evolutionary time. In other words, some are day-to-day events, whereas others have become characteristic ecological traits of populations, developed over the course of evolution. An example of a strategy to avoid competition expressed in ecological time might be for insects to leave a site already occupied by a potential competitor. Females of the codling moth, *Cydia pomonella* (Lepidoptera: Tortricidae), for example, avoid laying eggs beside other codling moth eggs in response to a chemical cue in egg tissue (Thiery et al. 1995). Similarly, larvae of the ladybird beetle, *Cryptolaemus montrouzieri* (Coleoptera: Coccinellidae), produce a chemical cue that deters oviposition by females of the same species (Merlin et al. 1996). Such "oviposition-deterring pheromones" are probably a common way for insects to avoid sites already occupied by potential competitors (McNeil & Quiring 1983; Quiring & McNeil 1984). In an applied context, competition among species of biological control agent can reduce the efficiency of pest suppression (Crowe & Bourchier 2006; see also Chapter 12).

Strategies expressed over evolutionary time to avoid competition might include fixed shifts in resource utilization that minimize similarity among potential competitors. One example is the late-season feeding habit of leaf-miners in the genus *Phyllonorycter* (Lepidoptera: Gracillariidae) on the pedunculate oak (West 1985). The miners begin development after early-season leaf-chewers have pupated, so avoiding the consumption of their mines by defoliating Lepidoptera. The late-season feeding of *Phyllonorycter* has become a fixed population trait over evolutionary time. In general, both behavioral and morphological changes can reduce overlap in resource use, and these will be discussed in the section on niche breadth (Section 4.3).

4.2.1 Types of competition

There are several more precise definitions of competition that are useful for understanding the struggle for limited resources. Intraspecific competition describes competition for limited resources by members of the same species, whereas interspecific competition refers to competition for limited resources among members of different species. There is considerable appeal to the notion that individuals of the same species will use resources in very similar ways and are therefore more likely to compete with one another than with individuals of different species. Indeed, there has been much support for the view that intraspecific competition plays a more dominant role in the population and community ecology of insects than does interspecific competition (Connell 1983; Strong et al. 1984). However, the relative dominance of competitive interactions within and among species is a complex issue that remains topical (Faeth 1987; Hunter 1990; Denno et al. 1995; see also Section 4.5.1).

When discussing intraspecific competition, we can also categorize the form of competition among individuals. Scramble competition refers to an interaction whereby limited resources are utilized evenly (or approximately so) by individuals. For example, if dragonfly nymphs (Odonata) in a pond are competing for the limited resource of prey items, and each nymph removes approximately the same biomass of prey from the stream, we might say that dragonfly larvae are "scrambling" for resources: when their food supply is depleted, all nymphs suffer equally. This is also sometimes referred to as exploitation competition. In contrast, if some individuals obtain an unequal share of resources by limiting the access of weaker individuals to that resource, or injure those individuals as they forage, we refer to this as contest or interference competition. Dragonfly adults that maintain territories around the edges of ponds, and exclude weaker individuals from their territories, may obtain a disproportionate share of limited resources (e.g. food, access to mates, etc.) by contest competition (Tsubaki et al. 1994). Contest competition is thought to lead to more

stable population dynamics than scramble competition because, as resources decline, some individuals will still have access to sufficient resources for growth, survival, and reproduction (i.e. fewer individuals, but with all the necessary resources). Scramble competition, in contrast, can lead to severe overexploitation of the habitat whereby few if any individuals receive sufficient resources. In such circumstances, reproduction may decline swiftly, emigration will be favored, and population crashes can occur (Figure 4.4).

In reality, most insect populations that compete for limited resources probably exhibit a blend of contest and scramble competition, with some individuals obtaining a greater proportion of resources than others, yet suffering some consequences of declining resource availability. Some insects switch from scramble to contest competition as they get older (Cameron et al. 2007) while others exhibit contest competition from a young age. For example, some solitary wasp larvae eliminate their competitors (including siblings) via contest competition (Pexton & Mayhew 2005). We should remember, however, that neither contest nor scramble (exploitation) competition will occur unless resources are limiting.

One final note on terminology. When the interaction between two organisms is mediated by the abundance or behavior of a third organism, we can refer to the interaction as an indirect effect. For example, if leaf-mining and phloem-feeding insects

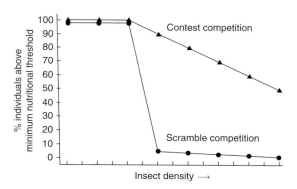

Figure 4.4 A comparison of scramble and contest competition. Because resources are divided approximately evenly under scramble competition, most individuals fall below the minimum requirements for survival at the same point. Contest competition is more stable because some individuals maintain sufficient resources while others have none.

are competing for the same plant species, and the competition results from depletion of the host plant, this is an example of an indirect effect. The interaction between the herbivore species is mediated by the abundance of the plant species. In contrast, if the herbivores were to fight over the plant, it would be an example of a direct effect. Indirect effects can become quite complex, occurring between species at different trophic levels, and will be considered further in Chapter 5.

4.3   THE NICHE CONCEPT

The niche concept dates back to the early 1900s. Grinnell (1904) equated an organism's niche with its distribution in space: a species would occur everywhere that suitable conditions existed and nowhere else. Elton (1927) expanded the concept to incorporate the behavior of the organism within the definition of the niche: he envisaged the niche as the "functional role" of the organism in the community. This was the first time that the niche was considered to be interaction based. Hutchinson (1957) provided the first quantitative view of the niche. He considered an organism's "fundamental niche" to be an n-dimensional hypervolume, "every point in which corresponds to a state of the environment that would permit a species to exist indefinitely". In other words, Hutchinson characterized the environment as complex and multidimensional, with an axis for each environmental factor that might impinge on the performance of a species. Hutchinson also considered that a species would be absent from those portions of the fundamental niche occupied by a dominant competitor through the action of competitive displacement. A species would therefore have a "realized niche" which was the subset of the environment from which it could not be displaced by competition. Put slightly differently, a fundamental niche is the full range of environments where a species could survive and increase in numbers, whereas a realized niche is the range actually occupied in nature (Traynor & Mayhew 2005). From Hutchinson onwards, the concept of the niche has been fundamental to discussions of the process of competition.

For our purposes, an insect's niche is the pattern of resource utilization that results from its interactions with the abiotic environment and other organisms (biota) in that environment. Inherent in this definition of the niche is the concept of constraint. Insects are constrained in their use of resources by the abiotic environment (suitable temperature, humidity, etc. to support life) and the biotic environment (interactions with food, mutualists, and natural enemies). For example, the niche of the monarch butterfly, *Danaeus plexippus* (Lepidoptera: Danaeidae) is constrained by the abiotic conditions to which it is adapted, the availability of its host plants in the genus *Asclepias*, and its interactions with natural enemies such as orioles and grosbeaks.

Niche theory and interspecific competition have been inextricably linked over the years by the simple principle that two species should not be able to coexist if they have identical niches (Gause's axiom): any slight variation in competitive advantage would allow the superior competitor to exclude the weaker competitor. This assertion spawned the study of "maximum tolerable niche overlap". Just how similar can two insect species be to one another before competitive exclusion will occur? Unfortunately, there seems to be no easy answer to this question. Although some ecologists have suggested on theoretical and empirical grounds that niche overlap has some maximum value (Hutchinson 1957; MacArthur & Levins 1967; May 1973), there has been considerable controversy surrounding both the validity of the concept of maximum tolerable overlap and appropriate ways to measure overlap in nature (Abrams 1983). To give just one example, Alvarez et al. (2006) studied the ecology of two closely related (sympatric) species of bean weevil, *Acanthoscelides obtectus* and *A. obvelatus* (Coleoptera : Bruchidae). These species have only recently been identified as separate from one another using molecular techniques. According to niche theory, the two species, so hard to tell apart, must differ ecologically. In fact, it turns out that they do differ slightly in their (realized) niches, in terms of preferred altitude of habitat, and the types of beans that they prefer.

In general, although the study of maximum tolerable niche overlap is still prevalent in some other fields of ecology, we detect a loss of interest in this topic among insect ecologists, and refer interested readers to broader ecological textbooks (Morin 1999; Stiling 2002) for further discussion. That being said, niche overlap may still be a useful concept when applied to biological control systems where multiple species of parasitoid or predator with some degree of niche separation may be required to control a broad-ranging pest species (Pedersen & Mills 2004).

4.4 THEORETICAL APPROACHES TO THE STUDY OF COMPETITION

We begin our description of theoretical approaches to the study of competition with the following caveat: we do not believe that most competition models, particularly the simple models presented here, accurately reflect competition among insects in natural populations. Algebraic models, by their very nature, behave according to the assumptions built into them, and many of the assumptions that underlie competition models do not hold true in natural populations. Yet there are two main reasons for presenting some simple models of competition here. First, the structure of the simple models can be used to develop more complex models with a greater degree of realism. Although complex models are beyond the scope of this text, students who are interested in theoretical approaches to studying insect populations should be familiar with the basic structure upon which more realistic models can be developed. Second, models often have their greatest value when they fail to reflect nature. Models are really hypotheses about the way that we believe natural populations operate. When we test these hypotheses experimentally (i.e. try to make them match our observations of real insects), and find them lacking, we learn that there is some basic assumption in our model that is incorrect. In other words, the failure of a model to match up to empirical and experimental observation forces us to reconsider our views of how competition operates, and to investigate the phenomenon further. We begin with a description of some simple competition models, then describe the predictions that we can derive from them, and conclude with experimental evidence that supports or refutes the models' predictions.

4.4.1 Competition models

Competition models for insects have developed in parallel with competition models for other taxa. We can distinguish two fundamental types of population model. The first type, continuous time models, are suitable for insect species that have overlapping generations (i.e. daughters begin to reproduce before mothers stop reproducing) and that exhibit continuous reproduction. The second type, discrete time models, are more suitable for insect populations without overlapping generations and with seasonal

reproduction. Neither type of model matches the real world perfectly. For example, many species of aphid (Hemiptera: Homoptera) exhibit overlapping generations. Indeed, aphids have "telescoping generations" in which the aphid's granddaughters are already formed in the body of the daughters before the mother gives birth! Yet these same species of aphid often pass the winter in the egg stage. Clearly, many aphids fulfill both the criteria of overlapping generations and seasonal reproduction. Should we use continuous time or discrete time models to study their population dynamics? The answer tends to be a compromise, depending on the insect species in question and the purpose of the model. We present only continuous time models for competition here because they illustrate the same general principles as discrete time models and provide a simple introduction. We encourage interested readers to explore discrete time models of competition in other texts (e.g. May 1978).

4.4.2 A continuous time model for competition

The simplest model of competition is based upon the continuous time model for exponential growth that we saw earlier. In equation 4.4, the left-hand side of the equation (dN/dt) represents the instantaneous rate of population growth. If we wish to impose some kind of limitation on our population, then we wish it to stop growing. In other words, we want dN/dt to become zero. A quick glance at equation 4.4 should tell you that dN/dt will be zero either when $r = 0$ or when $N = 0$. The latter is not very helpful (our population does not exist), and so we concentrate on the former; a population will stop growing when the per capita rate of change, r, is equal to zero. Specifically, a useful model will be one in which r gets closer and closer to zero as the population grows. In other words, r is density dependent and growth rate declines as the population increases.

A simple continuous time model which fits these criteria is the logistic growth equation of Verhulst (1838):

$$dN/dt = rN\left[\frac{(K - N)}{K}\right] \qquad (4.10)$$

where N is population size, r is the per capita exponential growth rate of the population, and K is the "carrying capacity" of the environment for species N. The carrying capacity represents the maximum average population size that the environment can support.

In English as opposed to "math-speak", the above model states that "the rate of population change depends upon population growth rate, the number of individuals in the population, and how far away current population size is from the carrying capacity". The term $(K - N)/K$ is designed to modify population growth rate, r, such that it is positive when $N < K$, negative when $N > K$, and zero when $N = K$ (choose some numbers for N and K, and try this yourself). In other words, Verhulst's model does exactly what we want. As the population grows towards carrying capacity (resource limitation), r becomes closer and closer to zero, and the population stops growing.

This model has a very simple equilibrium solution: independent of initial density, N will eventually reach K and stay there, unless some external force moves N from K (Figure 4.5). If some environmental perturbation causes an increase or decrease in N, it will again return to K. From where does this solution arise? By definition, the population will reach equilibrium when $dN/dt = 0$ (the rate of population change is zero at equilibrium). Readers should try rearranging equation 4.10 with the left-hand side (dN/dt) equal to zero. There are three possible solutions: $N = 0$, $r = 0$, or $[(K - N)/K] = 0$. If $N = 0$, then the population does not exist, so of course it cannot change size! When $r = 0$, we already know that the population is not changing: the per capita rate of change is zero. If $[(K - N)/K] = 0$, then $N = K$. This is the solution of interest. Simply,

when the population reaches equilibrium (stops changing), it does so at carrying capacity (Figure 4.5).

This is the simplest model that describes intraspecific competition – individuals within one species are competing for the limited resources that determine carrying capacity. We can expand this model to consider competition among different species. To do that, we need an additional equation for the second species and some way to influence the population growth rate of each species according to the density of the other. The simplest form of the two-species competition equations, developed by Lotka (1932) and Volterra (1931), are:

$$dN_1/dt = r_1 N_1 \left[\frac{K_1 - N_1 - \alpha N_2}{K_1} \right] \quad (4.11)$$

and

$$dN_2/dt = r_2 N_2 \left[\frac{K_2 - N_2 - \beta N_1}{K_2} \right] \quad (4.12)$$

We now have two equations, one for each of the competing species, and each species has its own specific r, N, and K (denoted by the subscripts 1 and 2). We have also modified equation 4.10 by adding the term αN_2 (or βN_1). The terms α and β are called the competition coefficients and are designed to transform individuals of the competing species into "equivalent units" of the original species: α in equation 4.11 reflects the effect of each individual of species 2 on species 1, whereas β in equation 4.12 reflects the effect of each individual of species 1 on species 2. In other words, if α in equation 4.11 is 0.75, it means that, as far as species 1 is concerned, each individual of species 2 uses about 0.75 of the resources of each individual of species 1. More precisely, if the population size of species 2 increases by 100, the population size of species 1 will drop by only 75. This suggests that individuals of species 1 suffer more from competition with each other than they do from competition with species 2: for species 1, intraspecific competition is more intense than interspecific competition.

In contrast, if β in equation 4.12 is 1.25, an increase of 100 individuals of species 1 would cause the population size of species 2 to fall by 125. If this were the case, we would consider interspecific competition stronger than intraspecific competition for species 2.

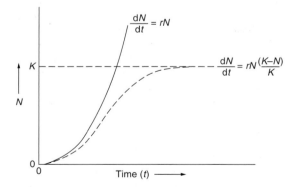

Figure 4.5 Unlike exponential growth, logistic population growth brings a population to carrying capacity.

We can determine equilibrium densities for both competing species as before by setting dN/dt equal to zero in equations 4.11 and 4.12. By the magic of mathematical manipulation, the equations reduce to:

$$N_1 = K_1 - \alpha N_2 \qquad (4.13)$$

and

$$N_2 = K_2 - \beta N_1 \qquad (4.14)$$

In other words, the equilibrium density for species 1 is no longer a single density, but varies depending on the density of species 2 and its competition coefficient. The same is true for species 2. We notice that equations 4.13 and 4.14 represent equations of straight lines of the general form $y = bx + c$. The competition coefficients represent the slopes of these lines. Ecologists have come up with the expression "zero-growth isocline" to describe the range of equilibrium densities (the straight lines) for each of the species. They are called zero-growth isoclines because they describe the range of densities for each species in which the population rate of change (dN/dt) equals zero. The zero-growth isocline for species 1 is shown in Figure 4.6. The line must cross the x-axis where $N_2 = 0$. When we set N_2 to zero in equation 4.13, we obtain the value $N_1 = K_1$. Similarly, the line must cross the y-axis when $N_1 = 0$. Substituting $N_1 = 0$ into equation 4.13, the line must cross the y-axis

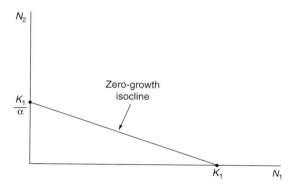

Figure 4.6 The zero-growth isocline for species 1. Each point along the line represents an equilibrium density for species 1. It should be noted that the equilibrium density declines for increasing densities of species 2.

when $N_2 = K_1/\alpha$. We can then calculate the zero-growth isocline for species 2 in the same way using equation 4.14. The zero-growth isocline for species 2 must cross the y-axis at K_2, and the x-axis at K_2/β. Here's the fun part: although we have established the equilibrium conditions (zero growth) for the two species, we have yet to establish whether or not they have a combined equilibrium where both species can coexist together. We determine this by drawing the isoclines for both species on the same graph, and there are four possible combinations (Figure 4.7). A common equilibrium exists only if the lines cross each other (Figure 4.7c, d) and that equilibrium is stable only if the lines cross as shown in Figure 4.7d.

Spend a few moments looking at Figure 4.7. Starting with Figure 4.7a, whatever the initial conditions, species 1 will eventually reach carrying capacity (K_1) and species 2 will become extinct. This occurs because the isocline for species 2 falls completely within that of species 1. When the density of either species is above its isocline (above equilibrium), its population must decline. Because N_2 is on the y-axis, we can illustrate population decline as a vertical line moving downwards. Because N_1 is on the x-axis, we illustrate population decline as a horizontal line moving from right to left. These lines represent vectors that we can add together to get the population trajectory (lines with arrows). At any combination of densities between the two isoclines, the population trajectory leads towards K_1 so that the density of species 1 is increasing while the density of species 2 declines. The reverse is true in Figure 4.7b: trajectories will lead the population of species 2 to K_2 and the population of species 1 to extinction because the species 2 isocline encloses the species 1 isocline. In Figure 4.7c, the winner of the competitive interaction depends on the starting densities of the two species and their relative growth rates. If conditions are such that population growth results in the population trajectory entering the stippled area of the graph (where the species 2 isocline covers the species 1 isocline) then species 2 will drive species 1 to extinction. Conversely, if the starting conditions cause the population trajectory to enter the striped area of the graph, species 1 will drive species 2 to extinction. Where the lines exactly overlap in Figure 4.7c, both species are at equilibrium. However, the equilibrium is unstable, and any perturbation that causes a change in density of either species will result in the extinction of one or the other.

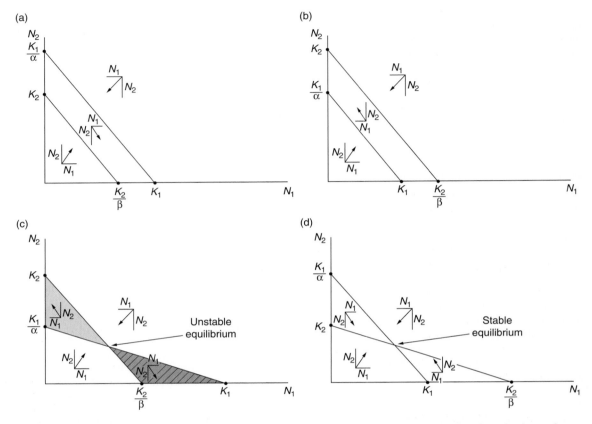

Figure 4.7 Zero-growth isoclines for two insect species in competition. Stable coexistence is possible only under the conditions shown in (d), where the isoclines cross such that the carrying capacities of each species fall within the isocline of the other species. For further details, see text.

Only Figure 4.7d represents a stable equilibrium with coexistence of both species. The carrying capacities of both species (K_1 and K_2) are enclosed below the isoclines of the other species. Simply put, this means that species 1 will inhibit its own population growth before it could ever drive species 2 to extinction and vice versa. This is another way of saying that, overall, intraspecific competition is greater than interspecific competition. From the x-axis, we can see directly that coexistence is possible if $K_1 < K_2/\beta$. From the y-axis, we see that coexistence is possible if $K_2 < K_1/\alpha$. By substituting and rearranging these equations (go on, give it a try), we obtain an additional condition for coexistence: $\alpha\beta < 1$. Earlier in this section we saw

that, if a competition coefficient is greater than one, interspecific competition is greater than intraspecific competition. Here, we have refined the concept to suggest that, as long as the product of the two competition coefficients is less than one, coexistence of competing species is possible.

4.4.3 Assumptions and predictions of Lotka–Volterra competition models

There are some important assumptions that underlie the simple competition models described above. We recall that the assumptions that are built into models can be considered hypotheses about the way the natural world operates. Assumptions of the Lotka–Volterra models include the following:

1 Competition coefficients are constant. The coefficients α and β in equations 4.11 and 4.12 are not permitted to vary with any changes in the abiotic or biotic environment.

2 The only density-dependent factors operating on populations N_1 and N_2 are limitations imposed by their own density (intraspecific competition) and limitations imposed by the density of the second species (interspecific competition). No other potentially density-dependent factors, such as predation or disease, act to limit either population.

3 There are no abiotic sources of mortality acting on the populations, and there is no stochastic (random) variation in the power of biotic factors.

In addition, a major prediction of the Lotka–Volterra competition equations is that coexistence among competing species is possible only if intraspecific competition is stronger than interspecific competition (or, more precisely, if $\alpha\beta < 1$). Such theoretical predictions have been applied to problems in the natural world, including potential effects of global change on insect ecology (Emmerson et al. 2004, 2005). Let us see how well these assumptions and predictions reflect reality.

4.5 COMPETITION AMONG INSECTS IN EXPERIMENTAL AND NATURAL POPULATIONS

Early laboratory studies with insects demonstrated that several assumptions of Lotka–Volterra theory were unrealistic. Chief among these is the assumption that competition coefficients are constant. For example, using laboratory populations of stored-product beetles, Birch (1953) demonstrated that competitive dominance among two species, *Rhizopertha dominica* (Coleoptera: Bostrichidae) and *Calandra oryzae* (Coleoptera: Curculionidae), varied with temperature. At 29.1°C, *C. oryzae* is the dominant competitor, and its abundance increases at the expense of that of *R. dominica*. This reverses at 32.3°C, where *R. dominica* becomes dominant (Figure 4.8). This simple study suggests that competition coefficients are temperature dependent. In a similar study, Ayala (1970) found that the coexistence of two fruit fly species, *Drosophila pseudoobscura* and *D. serrata*

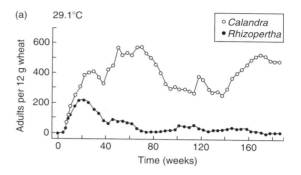

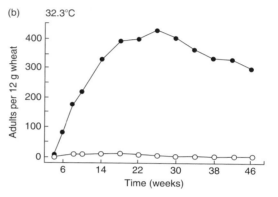

Figure 4.8 The competitive dominance of two beetle species switches in response to changing temperature. It seems probable that fluctuations in abiotic forces in nature help to maintain coexistence of competing species. (From Birch 1953.)

(Diptera: Drosophilidae), was temperature dependent. Below 23°C, *D. pseudoobscura* excluded *D. serrata* from laboratory populations, whereas above 23°C, *D. serrata* excluded *D. pseudoobscura*. When temperatures were maintained at 23°C, both species could coexist indefinitely.

We can draw two very important conclusions from Ayala's study. First, fluctuating environmental conditions are likely to mediate coexistence among species of insect through reversals of competitive advantage. We know that temperatures in nature fluctuate daily and seasonally, and this suggests that neither *D. pseudoobscura* nor *D. serrata* would maintain a long-term advantage over the other species. Without maintaining a long-term advantage, neither species should exclude the other, and therefore coexistence should occur. The role of fluctuating environmental conditions in mediating coexistence of species has been formalized mathematically by several researchers (Chesson & Warner 1981; Commins & Noble 1985) and environmentally mediated reversals in competitive advantage that promote coexistence have been shown to occur in natural insect populations. For example, coexistence of the winter moth, *Operophtera brumata* (Lepidoptera: Geometridae), and the green oak tortrix, *Tortrix viridana* (Lepidoptera: Tortricidae), on the English oak depends upon climatic variation. The winter moth is competitively superior when spring temperatures are relatively high, whereas the green oak tortrix is competitively superior when spring temperatures are low (Hunter 1990).

A second important conclusion that we can draw from Ayala's study is illustrated in Table 4.3 and Figure 4.9. We notice that, at 23°C, where both species could coexist, their population densities were much lower than their densities when reared alone. We can use the data in Table 4.3 to calculate the competition coefficients between the species at 23°C. Simply, the effect of *D. pseudoobscura* on *D. serrata* can be calculated as:

$$(1251 - 278)/252 = 3.86 \qquad (4.15)$$

and the effect of *D. serrata* on *D. pseudoobscura* can be calculated as:

$$(664 - 252)/278 = 1.48 \qquad (4.16)$$

We notice that both competition coefficients are greater than one, and that their product is necessarily also greater than one. According to Lotka–Volterra

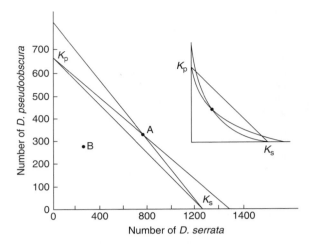

Figure 4.9 Competition between *Drosophila pseudoobscura* and *D. serrata*. Calculations based on the Lotka–Volterra theory suggest that coexistence should occur at the densities marked at point A. Experimental work by Ayala (1970) demonstrated, however, that coexistence occurs at point B, and the zero-growth isoclines must curve.

theory (Section 4.4.3), coexistence between species is not possible when the product of their competition coefficients exceeds unity. Clearly, the theory does not reflect competition among insects realistically, even in simple laboratory populations. Gilpin and Ayala's (1973) solution to this disparity was to curve the zero-growth isoclines of the two species (Figure 4.9) so that the lines intersected at a lower combined density. It should be recalled that the slopes of the zero-growth isoclines represent the competition coefficients, so by curving the lines, Ayala is suggesting that competition coefficients are frequency (or density) dependent (Ayala 1971). In essence, the curved isoclines suggest that the relative importance of intraspecific competition over interspecific competition increases with population density. This might occur if, as populations grow, scramble competition is replaced by interference (contest) competition and, in concert, species are more likely to interfere with members of their own species than with members of another species. The message to take away from these two simple laboratory experiments is that competition coefficients – that is, the strengths of interactions between competing species – are likely to vary in response to biotic and abiotic ecological factors.

Table 4.3 Densities of *Drosophila serrata* and *D. pseudoobscura* in laboratory experiments of competition. (Data from Ayala 1970.)

	Population size
Species raised separately	
D. pseudoobscura	664
D. serrata	1251
Species raised together	
D. pseudoobscura	252
D. serrata	278

4.5.1 Relative strengths of intra- and interspecific competition

It turns out that estimating the strength of inter-specific competition, and by extension comparing its strength to that of intraspecific competition, can pose significant challenges. Observational data are rarely sufficient. Take for example Forup and Memmott's (2005) work on potential competitive relationships between bumblebees and honeybees. Because both insects forage for pollen and nectar on similar flowers in the same localities, we might assume that competition would reduce resource availability. Indeed, such competition could explain the inverse relationship between honeybee and bumblebee abundance illustrated in Figure 4.10 and cannot be ruled out. However, detailed research has shown that pollinator declines could be due to other factors such as agricultural intensification that affect bumblebees independently of honeybees.

Experimental approaches are generally superior in estimating the power of competitive interactions. Experiments have shown, for example, that interspecific competition is more likely to be fatal for the loser than is intraspecific competition (Batchelor et al. 2005). However, this does not mean that interspecific competition is more prevalent in nature – indeed, the coexistence of multiple species in similar niches suggests just the opposite. If intraspecific competition is generally a more powerful force than interspecific competition, then coexistence among competing species will occur simply because populations will limit their own population growth before reaching

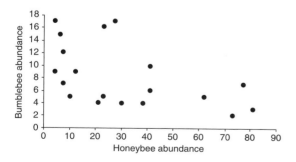

Figure 4.10 The abundance of bumblebees and honeybees on 19 heathlands sampled in 2002. (From Forup & Memmott 2005.)

levels at which they exclude their competitors. In literature reviews of manipulative experiments on competition among animals, both Connell (1983) and Schoener (1983) concluded that intraspecific competition was indeed often stronger than interspecific competition. However, the debate on the relative importance of interspecific competition for insect populations in general and phytophagous insects in particular, has been quite intense. We will describe studies of insect herbivores first, and then follow with a discussion of interactions among insects at other trophic levels.

In their classic paper, Lawton and Strong (1981) suggested that "interspecific competition is too rare or impuissant to regularly structure communities of insects on plants". They were in general agreement with the views of Hairston et al. (1960), who regarded the overall abundance of green plants as evidence that insect herbivores rarely deplete their food resources, and should not be limited by competition. Rather, those workers considered natural enemies the most likely factors regulating insect herbivore populations. Although most insect ecologists agree that competition among herbivorous insects can occur during periodic outbreaks (Varley 1949; Bylund & Tenow 1994; Carroll & Quiring 1994), the general view during the 1980s was that competition among insects was probably less important than previously conceived. For example, in a review of 31 insect life table studies (tables that categorize mortality factors operating during each generation), Strong et al. (1984) found convincing evidence of intraspecific competition in only six studies. It should be noted that, in the early to mid-1980s, there were not many rigorous experimental studies of competition among insects from which to draw generalizations, and that the reviews of Lawton and Strong (1981) and Strong et al. (1984) did much to stimulate a wave of manipulative experiments to study competition among insects.

4.5.2 Intraspecific competition among insect herbivores

Some studies have reported significant intraspecific competition among phytophagous insects. In a detailed series of experiments, Ohgushi has shown that populations of the herbivorous lady beetle, *Epilachna japonica* (Coleoptera: Coccinellidae), compete

Table 4.4 Females of the herbivorous lady beetle, *Epilachna japonica*, resorb eggs if defoliation of their host plant is too high. By resorbing eggs, females are more likely to survive the winter to reproduce the following year. (Data from Ohgushi 1992.)

Cage	No. of adults		% Leaf damage	% Females with egg resorption
	Male	Female		
A	1	1	20	0
B	1	1	20	0
C	1	1	20	0
D	1	2	50	100
E	4	5	80	100
F	4	7	90	100

strongly for their thistle host plants (Ohgushi & Sawada 1985; Ohgushi 1992). In this system, female beetles resorb eggs rather than lay them on defoliated thistle hosts (Table 4.4) and density-dependent birth rates regulate populations with remarkable stability over time. Density-dependent colonization of hosts may also be common in bark beetles, some of which switch from producing aggregation pheromones to producing dispersion pheromones when the density of adults on a tree surpasses some threshold (Borden 1984).

In Section 4.5.1, we described the views of Hairston et al. (1960) and Lawton and Strong (1981), who suggested that insect herbivores should not be limited by competition because they rarely deplete their host plants. A more modern view is one in which qualitative variation in plant traits is seen to shape the composition of herbivore communities (Hall et al. 2005; Rudgers & Whitney 2006). We learned in Chapter 3 that there is significant variation within and among individual plants in their quality as food for insects, and that not all green tissue is necessarily edible and therefore is not equally suitable for herbivore oviposition (Digweed 2006). As a consequence, some insect herbivores may be "trapped" in fierce competition with conspecifics for the limited high-quality plants (or plant parts) within a sea of unpalatable plant tissue. For example, Whitham (1978, 1986) has demonstrated that galling aphids in the genus *Pemphigus* (Homoptera: Aphididae) are clumped on high-quality individual leaves within their cottonwood (*Populus*) host plants. Most leaves on cottonwood trees cannot support high rates of aphid population growth, and intense intraspecific

competition for suitable leaves results. Even within leaves, some positions are better than others. When "stem mothers" are experimentally removed from the most suitable positions on individual leaves, there is an increase in the reproductive output of other aphids on those leaves. In other words, resource quality rather than quantity can limit insect herbivore populations long before all plant biomass is exploited. The chestnut weevil, *Curculio elephas*, uses only about one-quarter of all the chestnut fruits that are available for oviposition because of temporal variation in fruit quality (Debouzie et al. 2002). Similarly, the rare plant-hopper, *Delphacodes penedecta* (Hemiptera/Homoptera: Delphacidae), on salt marsh cordgrass, *Spartina alterniflora*, is kept at very low density as a result of strong intraspecific competition (Ferrenberg & Denno 2003). Even though there is plenty of *Spartina* biomass in the salt marsh environment in which these species live, a high requirement for plant nitrogen and intrinsically low lifetime fecundity combine to keep *D. penedecta* rare. The point is that simply observing the amount of green biomass on a plant may not inform us about the level of resource limitation and the intensity of competition.

Unfortunately, simple models of intraspecific competition (e.g. equation 4.10) generally ignore the fact that plant quality can vary in both space and time. Yet variation in plant quality is pervasive (see Chapter 3) and can have important effects upon the dynamics of insect herbivore populations (Hunter et al. 1996, 2000; Hunter 1997). In the past decade, insect ecologists have started to make some progress in understanding how we might incorporate variation in plant quality into a more realistic understanding of herbivore population dynamics. First, we learned in Chapter 3 (Section 3.3.5) that plants respond to herbivore attack with a suite of signals and induced defenses. As a result, plant quality can decline as a result of herbivory. Does the reduction in plant quality that is caused by insect attack affect the population dynamics of the herbivore? In part, the answer will depend upon the speed of the induced response relative to the lifespan of the herbivore, and the mobility of the herbivore (Underwood 1999). If the insect is relatively immobile and the induced defense is rapid, we might expect that herbivore performance will be reduced within the same herbivore generation. Moreover, if the strength of the induction response is dependent upon the density of herbivores attacking the plant, then the reductions in herbivore performance will be density dependent (Section 4.1.1) and capable in theory of regulating the herbivore population, keeping population levels stable over time. However, if the induction response is slow, such that the effects of declining plant quality are not felt by the herbivore population until the next generation, there exists the potential for delayed density dependence (Section 4.1.1). Delayed density dependence tends to cause populations to cycle, and some models suggest that cyclic fluctuations in insect populations might indeed be caused by induced changes in plant quality (Underwood 1999).

It has been hard to demonstrate, unequivocally, that induced responses influence the population dynamics of insect herbivores. One of the best examples to date comes from work on soybean plants. Underwood (2000) has shown that defense induction is density dependent on soybean, and that there is genetic variation among soybean plants for levels of induction. We might therefore expect that herbivore dynamics should differ among genetic lines of soybean and this also appears to be the case (Underwood & Rausher 2000). Populations of Mexican bean beetles reared upon different soybean genotypes vary in both their equilibrium densities and the periodicity of their population fluctuations (Figure 4.11). Moreover, soybean lines that vary in their levels of constitutive and induced resistance also vary in their effects upon Mexican bean beetle population dynamics (Underwood & Rausher 2002). In another example, Marchand and McNeil (2004) report that changes in cranberry chemistry after attack by the cranberry fruitworm, *Acrobasis vaccinii* (Lepidoptera: Pyralidae), reduce the general attractiveness of the plant to the

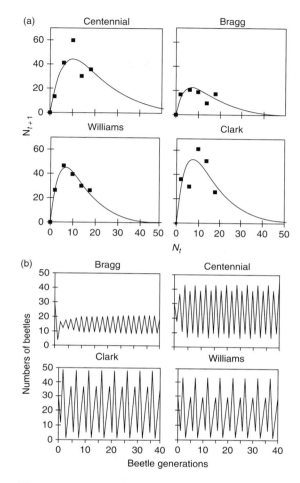

Figure 4.11 Population dynamics of Mexican bean beetles on four genotypes of soybean. Recruitment curves (a) from field experiments are used to estimate parameters of population growth. The parameters are then used to predict future population dynamics (b). (From Underwood & Rausher 2000.)

herbivore. As a result, population densities and subsequent intraspecific competition decline.

Defense induction is not the only mechanism by which variable plant quality can affect insect population dynamics. A second recent line of investigation has focused upon understanding the effects of spatial variation in plant quality on herbivore dynamics. The simple population models presented above (Section 4.4.2) kept parameters such as per capita growth rate, r, and carrying capacity, K, constant. In doing so, it is assumed either that these parameters do not vary among individual plants or, if they do, that average values are sufficient to capture population dynamics. Both assumptions appear to be false. For example, per capita growth rates of the milkweed-oleander aphid, *Aphis nerii* (Hemiptera/Homoptera: Aphididae), vary among species of their *Asclepias* host plants. Variation in per capita growth rates is related both to indices of plant quality (Agrawal 2004) and interactions with natural enemies (Helms et al. 2004) and results in different rates of aphid population growth among plants under field conditions (Figure 4.12). The strength of density dependence acting on *Aphis nerii* populations also varies among *Asclepias* species (Agrawal et al. 2004), confirming that variation in host-plant quality has the potential to influence long-term population dynamics. In addition, recent models suggest that using average values of insect growth rates and carrying capacities to predict population change can be grossly inaccurate (Underwood 2004; Helms & Hunter 2005). Rather, variation around average values of per capita growth rate and carrying capacity, caused by variation in plant quality, can cause substantial deviation from predictions made by simple models. The take-home message is simple. Insect population models that ignore spatial and temporal variation in plant quality are not likely to represent accurately the dynamics of natural populations.

In some cases, statistical analyses of long-term population data can be used to infer the strength of ecological forces such as competition. For example, the analysis of long-term population data of the green oak tortrix, *Tortrix viridana* (Lepidoptera: Tortricidae), has indicated that densities may be regulated by intraspecific competition (Hunter et al. 1997). By studying 16 years of sampling data, it was found that the per capita rate of increase of a European *T. viridana* population was strongly and negatively related to its density in the previous year (Figure 4.13). Simply put, if densities are high in one year, the population declines in the following year. Conversely, years of low population density are followed by years of increased population growth. In this system, parasitoids and predators appear to have little impact on *T. viridana* populations and, although there is occasional interspecific competition with the winter moth (see above), long-term dynamics are best explained by intraspecific competition among foraging larvae (Hunter 1998). One word of caution here – detailed genetic analyses of a single species such as *T. viridana* suggest that within-species genotypes may show different responses to environmental factors (Simchuk & Ivashov 2006), thus adding complexity to population dynamics.

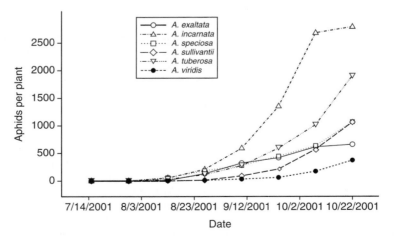

Figure 4.12 Population growth of the milkweed-oleander aphid, *Aphis nerii*, varies under field conditions among six species of host plant in the genus *Asclepias*. (From Helms et al. 2004.)

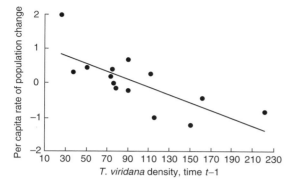

Figure 4.13 The per capita rate of change of *Tortrix viridana* is negatively associated with its population density in the previous year (time *t*–1). Years of high density are followed by low population growth rates, and vice versa. (From Hunter et al. 1997.)

Statistical analyses of long-term population data have also shown that different populations of insect herbivore may exhibit different population dynamics. The dynamics of the interaction between the cinnabar moth, *Tyria jacobaeae* (Lepidoptera: Arctiidae), and its host plant ragwort, *Senecio jacobeae*, differ between coastal dunes in the Netherlands and grasslands in southeast England (Bonsall et al. 2003). In the Netherlands, both moth and ragwort populations cycle over time. As is typical, the cycles are generated by the action of delayed density dependence (Section 4.1.1). This delayed density dependence arises because defoliation by cinnabar moths in one year influences ragwort abundance the following year. In turn, changes in plant abundance drive changes in cinnabar moth abundance (Figure 4.14). In southeast England, however, the cinnabar moth has little impact upon ragwort density in the next generation. Ragwort density is more a function of competition among plants than it is a function of herbivory. As a result, there is no delayed density dependence and no cycles. This example illustrates the fact that the effects of resource limitation on insect herbivores are contingent upon other ecological factors that influence plant demography.

4.5.3 Interspecific competition among insect herbivores

As we mentioned previously, there has been much debate on the frequency and strength of competitive interactions among different species of insect herbivore. Certainly, there is compelling evidence that interspecific competition is rare in some systems. Bracken ferns in the genus *Pteridium* are well-known for their high concentrations of hydrogen cyanide, especially when young, which act as a potent defense against herbivory (Alonso-Amelot & Oliveros-Bastidas 2005). In a comparison of the insect faunas feeding on bracken, *Pteridium aquilinum*, in Britain, Papua

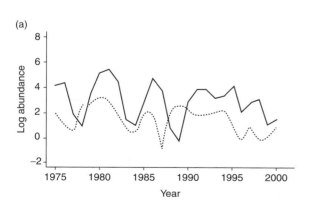

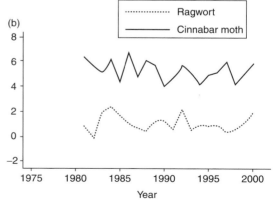

Figure 4.14 Population dynamics of ragwort and cinnabar moth in (a) the Netherlands and (b) southern England. The dynamics are cyclic in the Netherlands because of a delayed density-dependent interaction between the moth and host plant. (From Bonsall et al. 2003.)

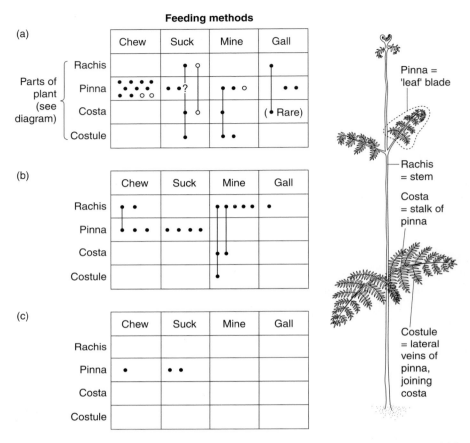

Figure 4.15 The use of bracken resources by insect herbivores in (a) the UK, (b) Papua New Guinea, and (c) New Mexico. "Gaps" in resource use suggest that the insect herbivore communities are not "saturated" by competing species. Open circles refer to species recorded in less than 50% of studies. (Data from Lawton 1982, 1984; from Strong et al. 1984.)

New Guinea, and New Mexico, Lawton (1982, 1984) observed very different numbers of herbivore species using the same resources in different regions, and idiosyncratic gaps in resource utilization (Figure 4.15). Lawton concluded that there was evidence of "empty niches" on bracken in some regions and no corresponding increase in the densities of resident species. In other words, the lack of certain taxa in species-poor bracken communities did not provide any apparent increase in resources for the remaining species. These results are consistent with the view that the pool of colonizing species is exhausted, and that the community is not "saturated" with competing species of herbivore.

In a study of leaf-miners on oak, Bultman and Faeth (1985) found that leaf-miner mortality was in general higher when individuals shared an oak leaf with members of the same species than with members of a different species, suggesting that intraspecific effects were more important than interspecific effects (Table 4.5). Likewise, Rathcke (1976) demonstrated that members of a stem-boring guild of insects in tall grass prairie avoid conspecifics (members of the same species) more often than heterospecifics (members of different species). Finally, Strong (1982) studied eight species of hispine beetle (Coleoptera: Chrysomelidae) that live in leaf rolls on *Heliconia* at one site in Costa Rica. Despite considerable dietary overlap, including

Table 4.5 The mortality of five species of leaf-miners on three species of oak tree. In general, rates of mortality are higher in the presence of conspecifics than in the presence of heterospecifics. (Data from Bultman & Faeth 1985.)

Species	Cohabitants			d.f.	χ^2
	Alone	Conspecifics	Heterospecifics		
Bucculatrix cerina					
Survival	926	326	116[a]	2	20.75†
Mortality	306	101	8[b]		
Cameraria sp.nov.					
Survival	53	0	8	2	4.00
Mortality	531	12	40		
Tischeria sp.nov.					
Survival	87	18[b]	19	2	8.47*
Mortality	347	165[a]	74		
Stigmella sp.					
Survival	422	21[b]	47[a]	2	13.11†
Mortality	630	48	34[b]		
Stilbosis quadricustatella					
Survival	60	3[b]	5	2	5.80*
Mortality	168	33	10		

Numbers followed by superscript letter make major contributions to overall χ^2: [a] Observed > expected. [b] Observed < expected.
* $P < 0.05$, † $P < 0.01$.

many different species found living in the same leaf roll, there was no evidence of significant interspecific interactions among the beetle species.

However, the concept that phytophagous insects can become trapped into intense competition by variation in host-plant quality, described above for *Pemphigus* aphids (Section 4.5.2), can be applied to interspecific competitive interactions also. For example, Gibson (1980) observed that two species of grassbug (*Notostira elongata* and *Megalocerea recticornis*) (Hemiptera: Miridae) that feed in limestone grasslands in England are forced to shift between host-plant species during the season to track nitrogen availability. Because nitrogen limitation affects both herbivore species, however, they are forced into interspecific competition on the same plant species (see Section 3.1 for a discussion of insect herbivores and nitrogen limitation). Similarly, Fritz et al. (1986) suggested that a stem-galling sawfly on willow had negative effects on three other willow sawfly species because they were constrained to feed on a subset of available willow genotypes. In fact, willow sawflies provide

intriguing topics of study for a variety of related reasons. The common name refers to the fact that female adults cut slots in plant stems in which to lay eggs with their saw-like ovipositors. As a result, females are restricted to laying eggs in only a subset of the shoots and stems of their host plants, the young soft ones (Price & Hunter 2005). Females also prefer longer shoots over shorter shoots because more eggs can be laid in the former (Ferrier & Price 2004). Overall, the number of suitable shoots and stems is low, generating intense competition for oviposition sites. Competition is relaxed in year of high precipitation, when willow plants can produce many new long shoots (Price & Hunter 2005).

Some of the confusion that exists over the relative importance of interspecific competition for insect herbivores may have arisen because of just such forces that constrain the foraging of phytophagous insects. As originally described by Lawton and Strong (1981) and Strong et al. (1984), interspecific competition did not regularly structure insect communities on plants. Unfortunately, a subset of the studies that those researchers used to illustrate their point was population-level studies, and this was taken by many ecologists to mean that competition did not influence the population dynamics of insect herbivores. There is a profound difference between population dynamics and community structure. Using Gibson's (1980) study of grassbugs above, it is clear that nitrogen limitation determines host utilization patterns by grassbug species: it is the major determinant of community structure. None the less, the herbivore species are forced to compete with one another, and this influences their population dynamics. It is ironic that competition probably has its greatest impact on insect herbivores when other forces structure the community. It should be remembered that interspecific competition is a negative–negative interaction where both species suffer, and we should expect insects to develop strategies, over ecological or evolutionary time, to avoid competition. Only where some more powerful force constrains the foraging of insects should we expect to see the effects of competition still operating. In other words, when competition structures communities, interactions should decline over time to trivial levels, leaving only the "ghost of competition past" as a reminder of once fierce interactions (Connell 1980). Ghosts of competition past might include morphological or behavioral adaptations that reduce overlap in resource use. An extreme ghost of competition past might be the competitive exclusion of one insect species by a second species. Exclusion is likely to be a transient phenomenon and therefore difficult to observe. In contrast, communities structured by forces other than competition may retain significant competitive interactions that influence population dynamics.

In any case, the tide has turned once more, and a significant number of insect ecologists are re-evaluating the role of interspecific competition in the population ecology of insect herbivores. This has arisen in part because techniques such as life table analysis, previously used to infer the strength of competitive effects (Strong et al. 1984), can miss critical aspects of life history, particularly oviposition preference, during which resource limitation can be expressed (Price 1990). In addition, life tables do not reliably detect density dependence when insects are patchily distributed (Hassell 1985b; Hassell et al. 1987). Patchy distributions can greatly enhance the regulatory power of competitive interactions (De Jong 1981; Chesson 1985). In addition, a plethora of

experimental studies during the late 1980s and 1990s have suggested that interspecific competitive interactions can occur through subtle and unexpected pathways. For example, the recognition that many plants respond to damage by herbivores with changes in their chemistry (see Chapter 3) led several ecologists to search for temporally separated interactions among phytophagous insect species. The possibility existed that damage by one insect species could influence populations of a second insect species through wound-induced changes in plant quality, even if the insects lived at different times of the year.

Across-season competitive effects were reported by West (1985) in studies of the English oak, *Quercus robur*. He demonstrated that spring defoliation by two caterpillars, *Operophtera brumata* (Lepidoptera: Geometridae) and *Tortrix viridana* (Lepidoptera: Tortricidae), on oak leaves resulted in reduced leaf nitrogen content. Low nitrogen levels adversely affected the survival of *Phyllonorycter* (Lepidoptera: Gracillariidae) species of leaf-miner that attacked oak leaves late in the season (Figure 4.16). In the same study system, Silva-Bohorquez (1987) showed that late-season aphids were also negatively affected by spring defoliation. Similarly, Faeth (1985, 1986) demonstrated that spring defoliation on *Quercus emoryi* caused declines in late-season leaf-miner success although, in this case, effects seemed to be caused by the attraction of parasitoids to previously damaged foliage. In a more recent reanalysis, Faeth (1992) has concluded that such cross-season effects may be weaker than intraspecific competition among leaf-miners in this system.

Plant-mediated competitive effects can occur on much shorter timescales too. High densities of phloem-feeding insects can cause rapid declines in the nutritional quality of their hosts for other insect species (McClure 1980; Denno & Roderick 1992). Aphids, in particular, have been shown to exhibit reduced survival (Itô 1960) and increased emigration (Edson 1985; Lamb & Mackay 1987) following competition-induced declines in foliage quality. As the number of studies reporting interspecific competition mediated by wound-induced changes in plant quality has increased, two additional facts have emerged. First, the strength and magnitude of the effects can depend upon the identity of the initial herbivores attacking the plants. For example, on the milkweed *Asclepias syriaca*, the duration of the competitive effects, and the identity of the herbivores

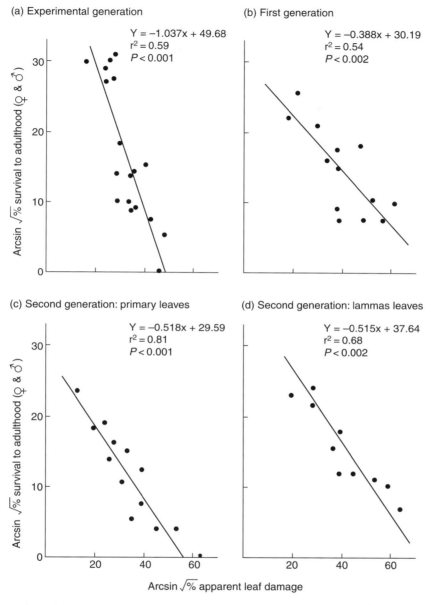

Figure 4.16 Spring defoliation by oak caterpillars reduces the survival of late-season leaf-miners. This across-season interaction results from defoliation-induced declines in foliage quality. (From West 1985.)

that respond, depend upon whether beetles or caterpillars are the first to damage the plants (Van Zandt & Agrawal 2004). Second, competition among herbivore species mediated by changes in plant quality is not

necessarily reciprocal or symmetric. For example, two species of plant-hopper, *Prokelisia dolus* and *P. marginata* (Hemiptera/Homoptera: Delphacidae), appear to compete for their common host plant,

Spartina alterniflora, via induced resistance (Denno et al. 2000). In this case, the effects of *P. dolus* on *P. marginata* growth rates, body size, and survival appear to be stronger than the reverse effects of *P. marginata* on *P. dolus* (Figure 4.17). In a second example, interspecific competition between a leaf-miner and a phloem-feeding whitefly, mediated by wound-induced resistance, is strongly asymmetric. Tomato plants are colonized by both a leaf-miner, *Liriomyza trifolii* (Diptera: Agromyzidae), and a whitefly, *Bemisia argentifolii* (Hemiptera/Homoptera: Aleyrodidae), and both induce substantial increases in defensive proteins in tomato leaves (Inbar et al. 1999). Previous feeding by the whitefly has significant negative effects upon leaf-miner performance, reducing feeding, oviposition, and survival by 47.7%, 30.7%, and 26.5%, respectively. The results of similar experiments are shown

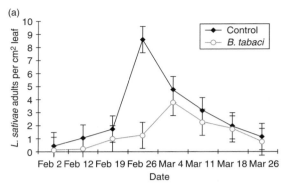

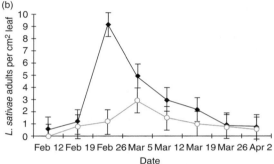

Figure 4.18 (a) Effect of preinfestation by pumpkin *Bemisia tabaci* on *Liriomyza sativae* population dynamics in the greenhouse. (b) Effect of preinfestation by cucumber *B. tabaci* on *L. sativae* population dynamics in the greenhouse. Values are means ± SE. (From Zhang et al. 2005.)

in Figure 4.18 where "preinfestations" of both pumpkin and cucumber plants by whitefly have a very clear negative effect on leaf-miner densities (Zhang et al. 2005). In contrast, there are no reciprocal effects of feeding by the leaf-miners on whitefly performance, even though the same chemical defenses are induced. In this case, it appears that the phloem-feeding whitefly has some mechanism of ameliorating the effects of chemical induction. The overall result is that the competitive interaction between the two species of insect herbivore is strongly asymmetric.

How frequently is interspecific competition important for insect herbivore populations? A review of 193 pairwise interactions among insect herbivore species (Denno et al. 1995) has provided a detailed analysis. We will review some of the conclusions of Denno et al. (1995) and we encourage interested readers to refer to the original paper. Of the 193 pairwise interactions

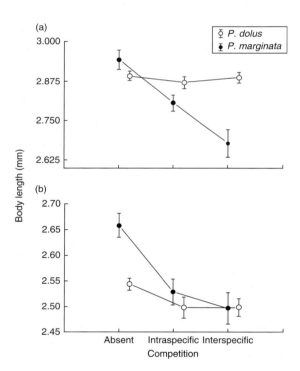

Figure 4.17 Effects of intra- and interspecific competition on the body length of (a) female and (b) male plant-hoppers. The effects of interspecific competition are much more pronounced for *Prokelisia marginata* than for *P. dolus*. (From Denno et al. 2000.)

analyzed, 76% provided evidence for interspecific competition, 6% demonstrated facilitation, and 18% recorded no interactions among insect herbivore species. The frequency of competition was therefore much greater than that described in previous reviews (Lawton & Strong 1981; Strong et al. 1984; but see also Damman 1993).

Insect species with haustellate (piercing and sucking) mouthparts provided more evidence of competition than species with mandibulate (chewing) mouthparts (93% vs. 78% of species tested, respectively). In short, interspecific competition appeared particularly prevalent among phytophagous Hemiptera. Within the mandibulate herbivores, interactions were more common among enclosed feeders such as stem-borers, wood-borers, and seed-feeders than among external feeders such as folivorous Orthoptera, Coleoptera, Hymenoptera, and Lepidoptera (89% vs. 59% of species tested). Moreover, competitive interactions among pairs of enclosed feeders occurred with greater frequency than interactions among free-living folivores. In fact, folivores appear to exhibit the lowest levels of interspecific competition of all herbivore guilds. In contrast, enclosed feeders, such as nut-feeders (Harris et al. 1996), galling insects (Craig et al. 1990; Akimoto & Yamaguchi 1994), and leaf-miners (Bultman & Faeth 1985), appear to compete frequently.

Most of the competitive interactions reviewed by Denno et al. (1995) were asymmetric. Asymmetric competition occurs when one species suffers to a greater extent from the interaction than does a second species (above). The extreme example of asymmetric competition, where one of the species is entirely unaffected by the presence of the other, is known as amensalism. A theoretical depiction of amensalism is shown in Figure 4.19. It should be noted that the zero-growth isocline for species 1 is vertical, illustrating that the density of species 1 remains constant over any range of density of species 2. Qureshi and Michaud (2005) provide an example of amenalism among three species of aphids feeding simultaneously on wheat. Greenbug, *Schizaphis graminum*, Russian wheat aphid, *Diuraphis noxia*, and bird cherry-oat aphid, *Rhopalosiphum padi* (all Hemiptera: Aphidae) were released onto wheat plants and population trends monitored according to species combinations. *D. noxia* experienced delayed development when associated with *S. graminum*, and reduced fecundity when combined with both *R. padi* and *S. graminum*.

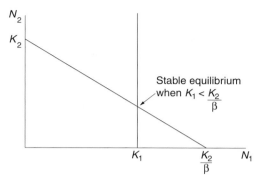

Figure 4.19 Asymmetric competition (amensalism) between two insect species. Species 1 is unaffected by species 2, whereas densities of species 2 decline with increases in species 1.

In addition, *S. graminum* nymphs showed reduced survival when their mothers had grown up on the same plants as *R. padi*. Asymmetric competitive interactions may be more stable than symmetric interactions because a decline in one competitor species does not result in any increase in other competitors. In Figure 4.19, there is no positive feedback on the population of species 1 that would tend towards exclusion of species 2.

The prevalence of asymmetric competition (84% of cases analyzed by Denno et al. 1995) agrees with the paper by Lawton and Hassell (1981), which suggested that amensalism was a common feature of interactions among insects. This, at least, appears to be a point of consensus. Strong asymmetry has been noted in insect phytophages, carrion-feeders, aquatic detritivores, predators, and nectarivores (Lawton & Hassell 1981). For example, the carrion flies (Diptera: Calliphoridae) *Calliphora auger* and *Chrysomya rufifacies* both dominate *Lucilla cuprina* on sheep carrion, and are essentially unaffected by *L. cuprina* (Andrewartha & Birch 1954). Similarly, *Danaus plexippus* (Lepidoptera: Danaidae) is a dominant competitor of *Oncopeltus* (Heteroptera: Lygaeidae) species on milkweed (Blakely & Dingle 1978). None the less, symmetric competition appears relatively common among sap-feeding insects, and may also be common among bark beetle species that respond negatively and reciprocally to the pheromones of other species (Birch et al. 1980; Coulson et al. 1980; Flamm et al. 1987). For example, the two bark beetle species, *Ips typographus* and *Pityogenes chalcographus*

(Coleoptera: Scolytidae), respond negatively to each other's aggregation pheromone during colonization of Norway spruce (Byers 1993). Since Denno et al.'s review in 1995, ecologists and entomologists have continued to investigate the importance of competition in the population dynamics and community structure of insects. However, the controversy seems to have died down somewhat. Rather than arguing about whether or not competition is important, more studies are exploring techniques by which we might estimate the relative importance of all the ecological factors that impinge upon the dynamics of insects (Hunter 2001a). We consider this in detail in Chapter 5.

4.5.4 Competition among different guilds of insect herbivore

In Section 4.3, we described the underlying assumption of niche theory that insects should compete most strongly with individuals that are very similar, occupying similar niches. However, insect species can still compete with one another if they occupy rather different niches. In 1973, Janzen pointed out that species of root-feeder, leaf-chewer, stem-borer, etc., might still compete for shared host plants because they are linked by the common resource budget of the plant (Janzen 1973). With the additional caveat that plant resources must be limiting for insect herbivores if competition is to occur, we might expect to see competition among insects in different feeding guilds (see Chapter 3) if they share a host-plant species. However, studies of competition among different guilds of insect herbivore are not as common as they should be. At this time, it is not possible to say whether herbivores in different guilds suffer less or more from interspecific competition than do herbivores in the same guild.

One area in which progress has been made is in the study of interactions between root-feeding and above-ground insect herbivores. Whilst much less apparent for obvious reasons, root feeding is thought to be highly significant in the lives of plants. The number of studies of insect herbivores above ground exceeds the number of studies of below-ground insect herbivores by about two orders of magnitude (Hunter 2001b). However, feeding by below-ground insect herbivores is known to influence the diversity of plant communities, the rate and direction of succession,

competitive interactions among plants, their allocation of resources to different functions, the susceptibility of plants to other herbivores and pathogens, and, ultimately, the yield of agricultural and forest systems. With this variety of effects, it seems likely that there exist interactions between insect herbivores above and below ground. What is the form of these interactions? Are they generally competitive? New techniques are now available to investigate these questions without disturbing the insects or the roots on which they feed. Such high-tech approaches include X-ray tomography and acoustic field detection (Johnson et al. 2007).

The effects of root feeding on above-ground herbivores will likely depend upon the effects of root damage on plant physiology, growth, and defense. Obviously, if root-feeders actually kill the plants, this will have a negative impact on above-ground herbivores. Root-feeding insects are associated with pine seedling mortality in the southeastern United States (Mitchell et al. 1991) and *Eucalyptus* seedling mortality in plantations in India (Nair & Varma 1985). In California, ghost moth caterpillars can cause up to 41% mortality in bush lupine stands (Strong et al. 1995) and appear to contribute to the long-term dynamics of lupine populations. However, root damage is far from always lethal. More subtle effects include reductions in foliar biomass, changes in tissue nitrogen concentrations, and loss of productivity. For example, root herbivory on white clover by *Sitona flavescens* (Cdeoptera: Curculionidae) reduces foliar biomass, total nitrogen and carbon contents, and impairs nitrogen fixation (Murray et al. 1996). Likewise, feeding by *Cerotomia arcuata* (Cdeoptera: Chrysomelidae) larvae on bean roots and nodules reduces plant growth and subsequent yield (Teixeira et al. 1996). When the effects of root herbivores on plants are non-lethal, how do above-ground herbivores respond? The answer is somewhat surprising.

On dicotyledonous plants, damage to roots by below-ground feeders almost always causes an increase in the performance of above-ground feeders. In contrast, above-ground feeding by insect herbivores almost always causes a reduction in the performance of below-ground herbivores (Van Dam et al. 2003) (Table 4.6). For example, Moran and Whitham (1990) demonstrated that leaf galling by aphids on *Chenopodium* can lead to a large reduction in the populations of root-feeding aphids. Likewise, root-feeding aphid populations on *Cardamine* are

Table 4.6 Published studies of interactions between above-(AG) and below-ground (BG) herbivores. The symbols +, −, and 0 represent positive, negative, and neutral effects, respectively. (Data from Van Dam et al. 2003.)

Plant species	AG phytophage	BG phytophage	Phytophage density Manipulation	Design	Effect of BG treatment on AG phytophage	Effect of AG treatment on BG phytophage	Refence
Capsella bursa-pastoris (Cruciferae)	*Aphis fabae* (sap-sucker)	*Phyllopertha horticola* (root chewer)	Addition	Controlled environment	Development time 0 Adult weight + RGR + Longevity + Fecundity +		Gange & Brown 1989
Chenopodium album (Chenopodiaceae)	*Hayhurstia atriplicus* (gall aphid)	*Pemphigus betae* (sap-sucker)	Addition Natural density	Field experiment Field census		Abundance − Adult size − Abundance −	Moran & Whitham 1990
Sonchus oleraceus (Compositae)	*Chromatomyia syngenesiae* (leaf-miner)	*Phyllopertha horticola* (root chewer)	Addition/soil insecticide	Controlled environment	Food consumption 0 Pupal weight +	RGR −	Masters & Brown 1992
Sonchus oleraceus (Compositae)	*Uroleucon sonchi* (sap-sucker) + other aphid sp.	Root chewers	Addition/soil insecticide	Field	Abundance +		Masters 1995
Sonchus oleraceus (Compositae)	*Myzus persicae* (sap-sucker)	*Phyllopertha horticola* (root chewer)	Addition	Controlled environment	Adult weight + RGR + Fecundity 0		Masters 1995

(Continued)

Table 4.6 (*Continued*)

Plant species	AG phytophage	BG phytophage	Phytophage density Manipulation	Design	Effect of BG treatment on AG phytophage	Effect of AG treatment on BG phytophage	Refence
Tripleurospermum perforatum (Compositae)	Seed eaters and foliar feeders	Root chewers	Soil/foliar insecticide	Field	Abundance	0	Müller-Schärer & Brown 1995
Cardamine pratensis (Cruciferae)	*Aphis fabae fabae* (sap-sucker)	*Pemphigus populitransversus* (sap-sucker)	Addition	Controlled environment	Abundance	0 Abundance −	Salt et al. 1996
Zea mays (Gramineae)	*Romalea guttata* (leaf-chewer)	*Phytophagous nematodes*	Natural density	Field	Abundance	0 Abundance −	Fu et al. 2001
Fragana × *ananassa* (Rosaceae)	*Otiorhynchus sulcatus* (adult, leaf-chewer)	*Otiorhynchus sulcatus* (larva, root chewer)	Addition	Controlled environment	Food consumption	+	Gange 2001
Cirsium palustre (Compositae)	*Terellia ruficauda* (seed eater)	Root chewers	Soil insecticide Field		Damaged flowers Parasitoid abundance Parasitism (%)	+ + 0	Masters et al. 2001

RGR, relative growth rate.

significantly reduced in the presence of shoot-feeding aphids (Salt et al. 1996). The growth rates of chafer larvae feeding upon the roots of *Sonchus oleraceus* are reduced when leaves are concurrently mined by Agromyzid larvae (Masters & Brown 1992). In contrast, the pupal weights of leaf-miners are higher in the presence of chafers (Figure 4.20). Other guilds of above-ground herbivores (leaf-chewers, phloem-feeders)

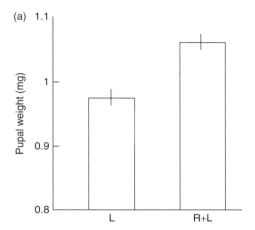

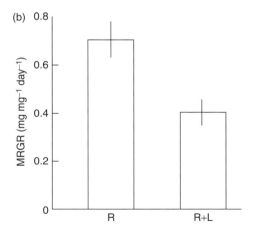

Figure 4.20 Effects of above- and below-ground damage on (a) leaf-miner pupal weight and (b) chafer mean relative growth rate (MRGR). (a) Leaf-miner pupal weight is higher in the presence of the root-feeding chafer (R+L) than it is when leaf-miners are alone (L). (b) Chafer growth rate is lower in the presence of the aerial-feeding leaf-miner (R+L) than it is when chafers are alone (R). (From Masters & Brown 1992.)

also respond positively to chafer damage in this system (Masters 1999) while simultaneously depressing chafer growth rates. Field experiments employing soil insecticides (Masters 1995b; Masters & Brown 1997) and artificial root damage (Foggo & Speight 1993; Foggo et al. 1994) support the view that below-ground herbivores facilitate population growth of insect herbivores above ground. However, the negative effects of above-ground herbivores on root-feeding insects may decline as succession proceeds (Blossey & Hunt-Joshi 2003).

Opposing (contramensal) interactions between above- and below-ground herbivores appear to be mediated primarily by increases in soluble nitrogen and sugar contents in above-ground tissues (root-feeders increase foliage quality) and decreases in root biomass below ground (aerial herbivores depress root growth) (Holland & Detling 1990; Masters 1995a). More detailed investigations of mechanism are, however, clearly needed in such systems, including the potential induction or suppression of defenses and the signaling pathways responsible for both defensive and "civilian" plant responses to damage (Van Dam et al. 2003). You will recall from Chapter 3 that many defenses induced in plants by insect damage are systemic, affecting the phytochemistry of the whole plant. In a systemic response, a signal compound is transported through the plant to elicit a response in plant parts that have not been directly attacked by the herbivore (Karban & Baldwin 1997). It is therefore possible that damage by herbivores above ground might induce defenses that could affect feeding by herbivores below ground and vice versa (Van Dam et al. 2003). So induced defenses provide one avenue by which above- and below-ground herbivores might interact and compete with one another. Data suggest that above-ground induction does indeed increase concentrations of defense compounds in the roots of some plants such as wild tobacco and *Brassica campestris*. Of course, just to complicate matters, some studies appear to show trends opposite to those of the majority. Figure 4.21 illustrates the response of small white butterfly, *Pieris rapae* (Lepidoptera: Pieridae), larvae feeding on the shoots and leaves of black mustard *Brassica nigra* (Van Dam et al. 2005). Some of the mustard roots were damaged by larvae of the cabbage root fly, *Delia brassicae* (Diptera: Anthomyiidae), and/or the root lesion nematode *Pratylenchus penetrans* (do not forget that not all herbivores are necessarily insects). As the figure illustrates,

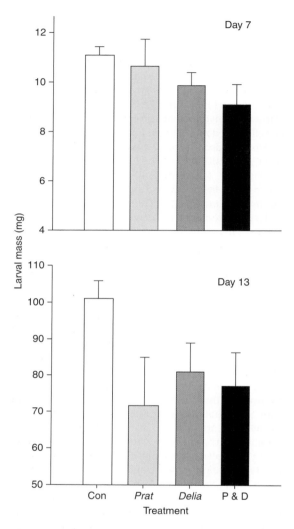

Figure 4.21 Larval body masses (+SEM) at 7 days and 13 days of *Pieris rapae* larvae feeding on shoots of *Brassica nigra* plants, whose roots were not infested with root feeders (control; Con) or whose roots were infested with either *Pratylenchus penetrans* nematodes (Prat), *Delia radicum* root fly larvae (Delia), or both (P & D). (From Van Dam et al. 2005.)

caterpillars weighed less when growing on plants damaged below ground. It seems that root feeding alters the nutritional quality of shoots by changing their secondary metabolite concentrations. In turn, lower shoot quality reduces the performance of the

specialist herbivore. In general, however, it is too early to say whether systemic induction represents a major pathway of interaction between above- and below-ground herbivores.

As an added thought, the complex interactions between below- and above-ground herbivores are potentially very important in agriculture and forestry. When we observe high densities of insect herbivores feeding upon the leaves and shoots of our agricultural and timber crops, we rarely consider the possibility that the severity of their depredations may result, in part, from high densities of root-feeding insects. Likewise, yield increases following the successful control of above-ground insect pests may be compromised by subsequent increases in the densities of root-feeding herbivores.

Interactions among other herbivore guilds are not yet well known. However, studies to date suggest that leaf-mining insects may often be the "losers" in competitive interactions with leaf-chewers (Faeth 1985; West 1985), if only because they are trapped by their mines and can rarely move among leaves. One fruitful area of current research is the interactions that occur between defoliating and pollinating insects. Plants have the dilemma of trying to attract beneficial visitors (pollinators) while deterring flower-feeding insects, nectar robbers (insects that take nectar without pollinating), and other herbivores. For example, flower-feeding thrips can deter pollinators from plants (Karban & Strauss 1993). Moreover, plants may face tradeoffs in their allocation of resources to resistance and allocation of resources to flowers. As we mentioned in Chapter 3, *Brassica* plants that are heavily defended chemically against foliar-feeding insects may suffer reductions in the length of time that pollinators spend on flowers (Strauss et al. 1999) (see Figure 3.26). In a second example, the reduction in fruit yield on cantaloupe that is caused by insect defoliators can be completely compensated for by supplemental pollination, suggesting that yield losses stem from poor pollination rather than a lack of resources for fruit (Strauss & Murch 2004). So what is the likely form of the interaction between defoliating insects and pollinating insects? To date, most studies have focused on the costs and benefits to plants rather than potential effects on the population dynamics of either defoliators or pollinators (interguild interactions). We might imagine, however, that pollinators have a generally positive influence on foliar-feeding insects by increasing plant reproductive output and the number of plants in the population. Conversely, we might expect that defoliators will reduce the allocation of resources to flowers, and therefore have a detrimental effect on pollinators. The extent to which this "contramensal" interaction plays out in nature is currently unclear.

4.5.5 Variation in the direction of interactions among insect herbivores

As we have seen, it is an oversimplification to assume that species of insect herbivore that share a common host plant must necessarily compete with one another. In Section 4.5.4, we saw that root-feeding herbivores may actually increase the quality of plant resources for foliage-feeders, a positive rather than negative effect. In reality, feeding by one species of insect herbivore on a host plant can have a range of effects on other herbivores in the community, from negative, through neutral, to positive. This suggests that herbivores sometimes increase rather than decrease resources for other consumers. The key here is to understand the type of resource that actually limits the herbivore population. For example, when root-feeding insects benefit foliar-feeding insects, it appears to result from increases in the limiting resource of available nitrogen (Section 4.5.4), even though root-feeders decrease the overall biomass of the plants.

In the late 1980s, entomologists recognized that defoliation could sometimes change the morphology of leaves, improving resources for subsequent insect herbivores. For example, caterpillars that roll or tie leaves together provide shelters for other insect species that later colonize the plants (Damman 1987; Hunter 1987). Such modifications of plant architecture can influence a wide variety of insect species, suggesting that suitable architecture can be a limiting resource. Herbivores that modify the physical, chemical, and nutritional environment for other insects can sometimes be considered as "keystone herbivores" (Hunter 1992a; Gonzales-Megias & Gomez 2003) because they have an impact on community structure that is disproportionate to their abundance. Subsequent work has confirmed that leaf-mining, leaf-rolling, and leaf-tying insects influence arthropod density, diversity, and community structure on plants (Johnson et al. 2002; Lill & Marquis 2003; Nakamura & Ohgushi 2003).

It is important to note that the effects of defoliation by one insect species on the densities of other insect species need not be linear. For example, damage at low levels may have a beneficial effect on a second species whereas damage at higher levels may have a deleterious effect. Nowhere is this more aptly demonstrated than on oak trees, where spring defoliation by caterpillars has non-linear effects on insect herbivores that feed late in the season (Hunter 1992b) (Figure 4.22). On the pedunculate oak, *Quercus robur*, spring defoliation by *Tortrix viridana* (Lepidoptera: Tortricidae) and *Operophtera brumata* (Lepidoptera: Geometridae) influences densities of leaf-chewing, phloem-feeding, and leaf-mining insects. Interestingly, the effects of spring defoliation differ among the three insect guilds, and are non-linear for two of the three (Figure 4.22). *T. viridana* is a spring leaf-roller, and moderate levels of damage improve the quality of the habitat for late-season, leaf-chewing insects. But spring damage does more than alter habitat – it also causes a decline in the nutritional quality of foliage. At high levels of defoliation, then, we see decreases in leaf-chewer abundance as the effects of poor nutritional quality outweigh the positive effects of habitat quality. In contrast, leaf-mining insects respond only to changes in leaf quality, and so their response is negative and linear across all levels of defoliation. Finally, phloem-feeding insects respond initially to the damage-induced declines in foliar nutritional quality. However, at high levels of defoliation, oaks produce new regrowth foliage which is of very high quality for aphids, and densities of phloem-feeding insects start to increase. Overall, then, spring defoliators affect the structure of late-season insect communities in complex and non-linear ways. A further example of non-linear interactions among insects comes from cotton fields. Aphids, which are normally thought of as serious pests, can be beneficial to cotton plants overall, but only when serious defoliating insects are present. Aphids attract predaceous fire-ants that forage in cotton canopies, killing more seriously destructive caterpillars (Kaplan & Eubanks 2005).

4.6 COMPETITION AMONG INSECTS OTHER THAN HERBIVORES

If the study of competition among insect herbivores has been controversial, the study of competition among other groups of insects can best be characterized as underdeveloped. Thanks to reviews by Lawton and Strong (1981), Strong et al. (1984), and Denno et al.

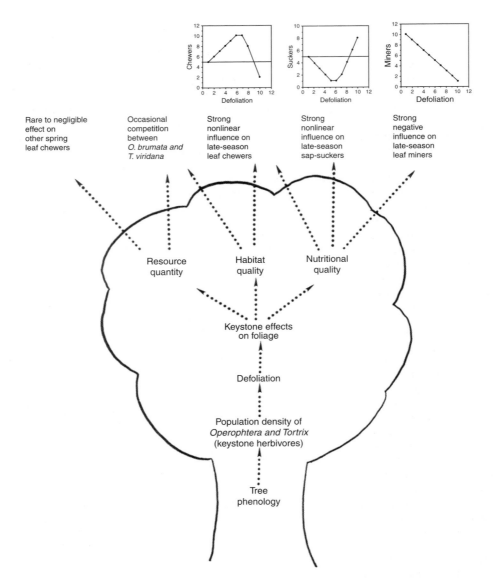

Figure 4.22 Spring defoliation by two species of Lepidoptera on the pedunculate oak has significant and non-linear effects on late-season insect herbivores. (From Hunter 1992b.)

(1995), there has been considerable interest in experimental investigations of competition among insect herbivores. Experiments designed to test the prevalence of competition among other groups of insects have not been as common in recent years. Some general patterns appear to be emerging, but we advocate caution until a substantial body of literature is available from which to draw firm conclusions.

One fairly reliable conclusion is that competition among Hymenoptera, particularly ants, may be widespread (Lawton & Hassell 1981). For example, ants in the genus *Camponotus* frequently kill, maim, and displace *Aphaenogaster* species in confrontations at feeding sites. The introduction of fire-ants (*Solenopsis invicta*; Hymenoptera: Formicidae) into the southern USA has been characterized by intense intra- and interspecific competition. Intraspecific competition among fire-ants occurs in several stages, including competition among queens during the founding of colonies (Balas & Adams 1996), density-dependent brood raids between young colonies that result in the "stealing" of broods and workers (Adams & Tschinkel 1995), and group fighting of individuals from older colonies at feeding sites. Overall, intra-specific competition among fire-ants results in the even spacing of colonies on the landscape (Adams & Tschinkel 1995). Aggression by fire-ants is not limited to conspecifics: laboratory studies have shown that *S. invicta* will kill species of termite and may be responsible for declines in native *Solenopsis* species. We can see from these studies of ants that, as competitive interactions become increasingly asymmetric and contest based, there is a very fine line between competition and predation.

Studies also suggest that hymenopteran parasitoids may suffer regularly from intra- and interspecific competition. Parasitoids lay eggs in or on other insects, and their larvae then kill the host (they are described in some detail in Chapter 5). Superparasitism occurs when individual parasitoids lay eggs in hosts that have been previously attacked by other individuals of the same or different species. Competition may then occur for the limited resource of the host insect (Kaneko 1995; Marris & Caspard 1996). The "winner" of competition that results from superparasitism is generally thought to be whichever individual or species attacks first (Bokononganta et al. 1996), although the first attacker is not always competitively dominant. In some cases, at least, female parasitoid larvae outcompete males in superparasitized hosts (Keasar et al. 2006).

In general, empirical studies of intraspecific competition among hymenopteran parasitoids support the predictions of combined host-quality (Charnov et al. 1981) and local mate competition (Hamilton 1967; Werren 1983) models. Specifically, parasitoids tend to produce a greater proportion of female offspring in large hosts than in small hosts (Pandey & Singh 1999; King 2002). The idea is that, as the number of emerging parasitoids increases, a greater proportion of females will reduce potential competition among males for mates. Moreover, large herbivore hosts generally produce more parasitoids (Mayhew & Godfray 1997)

of larger size, with greater egg loads, and greater longevity (Bernal et al. 1999). One striking example of intraspecific competition comes from the gregarious ectoparasitoid *Goniozus nephantidid* (Hymenoptera: Bethylidae) (Hardy et al. 1992). Female wasps lay clutches of up to 20 eggs on the caterpillars of micro-lepidopterans. Clutch size increases as larval host size increases, suggesting that host body size is limiting (Figure 4.23a). In addition, when the number of wasp larvae per host is increased experimentally, there is a concomitant decrease in female body size at maturation (Figure 4.23b). Larger female wasps both live longer and lay more eggs (Hardy et al. 1992). Incidentally, a preference by parasitoids for larger host

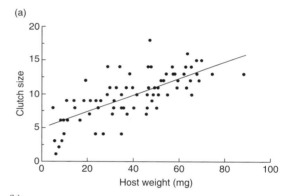

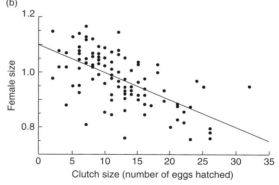

Figure 4.23 Evidence of intraspecific competition among individuals of the hymenoptera parasitoid, *Goniozus nephantidid*. (a) The size of clutch that a female parasitoid will lay on her caterpillar host increases with host size, suggesting that host body size is limiting. (b) The larger the clutch laid on a host of a given size, the lower the body size of females that emerge. (From Hardy et al. 1992.)

herbivores can skew attack ratios towards female herbivores when sexual dimorphism for herbivore size exists (Teder et al. 1999).

Interspecific competition among hymenopteran parasitoids also appears to be common (Kato 1994; Monge et al. 1995; Reitz 1996). One study has reported the facultative production of an "extraserosal envelope" around the developing embryo of the parasitoid wasp *Praon pequodorum* (Hymenoptera: Aphidiidae). The envelope, which separates the chorion and trophamnion of the developing embryo, is produced only when eggs are laid in hosts previously attacked by the heterospecific parasitoids *Aphidius ervi* and *A. smithi*. The envelope is not produced in singly or conspecifically superparasitized aphid hosts (Danyk & Mackauer 1996). This suggests that the extraserosal envelope has a defensive function and protects developing embryos from physical attack by mandibulate larvae of potential (interspecific) competitors. It is not yet clear whether limitations of host availability and competitive interactions support the suggestion of Hawkins (1992) that the dynamics of parasitoids mirror donor control (available food influences parasitoid densities, but parasitoids do not influence food availability).

Available data do suggest, however, qualitative similarities between competition among parasitoids and competition among insect herbivores. For example, interspecific competition among parasitoids is often asymmetric (Kato 1994; Monge et al. 1995; Reitz 1996), and can occur with unrelated taxa. Chilcutt and Tabashnik (1997) have reported competition between a parasitoid wasp, *Cotesia plutellae* (Hymenoptera: Braconidae), and a bacterial pathogen for the lepidopteran host, *Plutella xylostella* (Lepidoptera: Plutellidae). The outcome of the interaction depends upon the degree of susceptibility of *P. xylostella* to the pathogen. In susceptible hosts, the parasitoid does not affect performance of the pathogen, but the pathogen has a significant negative effect on the parasitoid. In moderately resistant hosts, the interaction between the parasitoid and pathogen is symmetric and competitive. Highly resistant hosts are not susceptible to infection by the pathogen, and this creates a refugium from competition for the parasitoid. In a similar study, Nakai and Kunimi (1997) demonstrated that larvae of the endoparasitoid *Ascogster reticulates* (Hymenoptera: Braconidae) are negatively affected by infection of their lepidopteran host with a granulosis virus. Virus capsules accumulate in the guts of developing parasitoids, and this lowers the rates of pupation and eclosion. Interactions between pathogens, insect hosts, and parasitoids provide fascinating opportunities for future research as well as potential difficulties for integrated pest management (see Chapter 12). Parasitoids have also been shown to compete with beetles (Evans & England 1996; Heinz & Nelson 1996) and ants (Itioka & Inoue 1996). Intraguild predation, a complex interaction among predators and their prey, is considered in Chapter 5 (Section 5.5.7).

Competition among insects is not restricted to terrestrial habitats. Larvae of insects such as mosquitoes that live in temporary pools must cope with rapidly dwindling resources (Schafer & Lundstrom 2006), with concomitant increases in competition for what remains (food, space, oxygen, etc). Intraspecific competition among individual larvae of the predatory mosquito *Topomyia tipuliformis* (Diptera: Culicidae) has been demonstrated within tiny pools of water (phytotelmata) held in plant tissue. In this case, the interaction is an extreme form of interference (contest) competition based on cannibalism (Mogi & Sembel 1996). Similarly, individual midge larvae in the species *Cricotopus bicintus* (Diptera: Chironomidae) compete for available periphyton (Wiley & Warren 1992). Although it is not yet clear how general the phenomenon is, competition among aquatic insect larvae for algal biomass has certainly been documented (Dudgeon & Chan 1992). We will return, in Chapter 8, to the role of aquatic insects in the regulation of primary production.

We do not wish to present an exhaustive list of all the various insect feeding groups, and the prevalence of competition in each. Indeed, we doubt that there are sufficient data to accomplish the task with any rigor. Suffice it to say that competition has been observed among most groups studied, including insect pollinators (Inoue & Kato 1992; Rathke 1992; Roubik 1992), predators (Vanbuskirk 1993; Moran et al. 1996), and decomposers (Trumbo & Fernandez 1995). In contrast, competition has not been observed in other studies of insects in these groups (Rosemond et al. 1992; Minckley et al. 1994). Like most ecological phenomena, the strength of competition is likely to vary in space, in time, and with community composition. Specifically, abiotic forces (Chapter 2), natural enemies (Chapter 5), and mutualism (Chapter 6) will interact with competition to determine the variation in insect populations and communities that we observe in space and time.

NATURAL ENEMIES AND INSECT POPULATION DYNAMICS

5.1 INTRODUCTION

Ecologists have long sought explanations for fluctuations in the abundance of insects and other animals. In previous chapters, we have considered the roles of climate, plant quality, and resource limitation in determining the abundance and distribution of insects. Here, we turn our attention to the role of natural enemies – predators, parasites, parasitoids, and pathogens. Historically, there has been considerable emphasis placed upon the role of natural enemies in the ecology of insects. One reason for this emphasis is, as Price (1975) wrote, "predation ... is certainly one of the most visible aspects of mortality". This visibility stems first from the variety of natural enemies (particularly the many insect species of predator and parasitoid) and their obvious effects on mortality. It is therefore hardly surprising that the action of parasitic wasps, predatory ladybirds, and other natural enemies has been observed by naturalists for over a hundred years: "very frequently it is not the obtaining of food, but the serving as prey to other animals, which determines the average numbers of a species" (Darwin 1859). In addition, the many adaptations of insects against attack by natural enemies, the successful cases of biological control, and population models have all given weight to the view that natural enemies have a major role in the population dynamics of insect herbivores.

In this chapter, we aim first to describe briefly the variety of insect and other natural enemies. We will then examine the role of natural enemies in insect population dynamics, and finally we will discuss the ways in which the effects of natural enemies combine with other factors, such as climate and host-plant quality, to determine insect abundance.

5.2 THE VARIETY OF NATURAL ENEMIES

5.2.1 Insect predators

A predator may be defined as an insect or other animal that kills its prey immediately or perhaps soon after attacking it. Predators also tend to kill many prey individuals. There are exceptions, of course, and many insect species may be both predatory and obtain their food resources by other means. For example, many predatory ant species also consume the honeydew produced by aphids (see Plate 5.1, between pp. 310 and 311), psyllids, or coccids as well as extrafloral nectar of plants (Davidson et al. 2004) (see Chapter 3). Some earwigs are both phytophagous (i.e. plant feeding) and predatory. Other insects have different feeding strategies at different stages in their life cycles, e.g. some staphylinid beetles are parasitoids when juvenile and predatory as adults.

Figure 5.1 Praying mantis (Mantodea). (Courtesy of P. Embden.)

Figure 5.2 Diving beetle larva (Coleoptera: Dytiscidae.) (Courtesy of P. Embden.)

Figure 5.3 Rove beetle (Coleoptera: Staphylinidae).

We consider the ecological consequences of omnivory and intraguild predation (predators that also feed on other predators) in more detail later in this chapter.

Many different taxonomic groups of insects contain predatory species, or species that are partly predatory. The major wholly or largely predatory groups are the dragonflies and damselflies (Odonata), mantids (Mantodea; Figure 5.1), ant-lions, lacewings, scorpion flies, and other Neuroptera. Within the Hymenoptera, the ants (Formicidae), although they adopt a diversity of lifestyles, are the major predatory family. Within the Coleoptera there are several predatory families such as the tiger beetles and ground beetles (Carabidae), the soldier beetles (Cantharidae), and the diving beetles (Dytiscidae; Figure 5.2). The Staphylinidae (Figure 5.3) includes many predators, parasitoids, scavengers, and species that show more than one type of feeding strategy. Perhaps the best-known predatory beetles are the ladybirds (Coccinellidae; Figure 5.4), but there are some notable plant-feeding coccinellids. The "true bugs" (Hemiptera, suborder Heteroptera) contain both predators and plant-feeding species. Some families (such as the largest family, the mirid or capsid bugs (Miridae)) contain examples of both lifestyles, but the assassin bugs (Reduviidae), damsel bugs (Nabidae), pond skaters (Gerridae), water scorpions (Nepidae), and other families contain mostly predatory species. Some Diptera, notably the robber flies (Asilidae) and the larvae of many hoverflies (Syrphidae; Figure 5.5), are predators. Although thrips (Thysanoptera), grasshoppers, and crickets (Orthoptera) are largely plant feeding, and include many serious crop pests, some are predatory, including some bush-cricket species (Tettigoniidae). Clearly, there is no such thing as a typical insect predator.

Figure 5.4 Ladybird larva and eggs (Coleoptera: Coccinellidae).

5.2.2 Other predators

Insects are subjected to predation from a wide range of insect and other species. The latter include other arthropods, such as centipedes (Chilopoda), scorpions, false scorpions (Pseudoscorpiones), harvestmen (Opiliones), and spiders (Araneae). In addition, there are many predatory as well as plant-feeding and scavenging mite (Acari) species. Many vertebrates are partly or wholly insectivorous and, as we will see below, bird and small mammal predators can play a significant role in the population dynamics of many insects such as several important forest pests.

5.2.3 Parasitic insects

A parasite may be defined as a species that, like a predator, obtains its nutritional requirements from another species, but, unlike a predator, usually does not kill its prey (Askew 1971). Parasites may live on or in the body of their hosts (ectoparasites and endoparasites, respectively). Ectoparasites include lice (Phthiraptera-Mallophaga and Anoplura) and fleas (Siphonaptera). These parasites, particularly lice, spend much of their lives on their hosts. Other ectoparasitic insects, in contrast, feed only briefly on their hosts. This type of parasite, which includes many vectors of serious diseases of humans and their livestock, is the subject of Chapter 11.

Most parasitic insects feed parasitically as larvae but not as adults, feed within or sometimes on their hosts, and eventually kill them. They are known as parasitoids (Askew 1971; Hassell 2000; Hochberg & Ives 2000). Approximately 10% of all insect species are parasitoids (Eggleton & Belshaw 1992). Most, about 75%, of these are Hymenoptera (Figures 5.6 and 5.7); the remaining 25% are Diptera (Figure 5.8) or Coleoptera. Within the Hymenoptera, the so-called Parasitica are almost all parasitic – the Ichneumonidae, Chalcidae, and

Figure 5.5 Hoverfly larva (Diptera: Syrphidae).

Figure 5.6 Adult hymenopteran parasitoid ovipositing in lepidopteran larval host.

Figure 5.7 Hymenopteran parasitoid larva and pupae emerged from the cadaver of lepidopteran larval host.

Figure 5.8 Adult tachinid parasitoid (Diptera).

Braconidae are the major parasitoid families in the Parasitica.

In most species of parasitoids, the adult female locates hosts but sometimes these hosts are located

by first instar larvae of the parasitoid (Eggleton & Belshaw 1992). The least common method of host location is for the host to ingest the parasitoid's egg. One example of the latter is *Cyzenis albicans* (Diptera: Tachinidae), a fly parasitoid that can develop only in the winter moth, *Operophtera brumata* (Section 5.3.4). Note that some ecologists classify many plant-feeding insects as parasites. For simplicity, we consider parasites only as natural enemies of other insects in this chapter.

5.2.4 Pathogens

Insects are attacked by a range of pathogenic bacteria, viruses, fungi, protozoans, and nematodes. Examples include the fungi *Entomophthora* and *Beauvaria* species, and the nuclear polyhedrosis viruses (NPVs). The increasing role of pathogens in biological control of insect pests is discussed in Chapter 12.

5.3 THE IMPACT OF NATURAL ENEMIES ON INSECT POPULATIONS

5.3.1 Observations, experiments, or models?

Ecologists have used different approaches to study the impact of natural enemies: observations, experiments, and models (Hunter 2001a). In this context, we define observations as studies of insects and their natural enemies in their natural environment without manipulation of either; experiments as studies where either or both are manipulated in some way (e.g. through the artificial introduction or exclusion of natural enemies); and models as mathematical representations of natural enemy–insect interactions.

Each of these approaches has its strengths and weaknesses, and each has contributed in many ways to our understanding of the impact of natural enemies on insects. Although we will present them as separate approaches below, each owes much to the other two (Hunter 2001a). For example, observations and experiments are used to test the predictions of models. It should be noted that we avoid using the term "theoretical models": each approach can be theoretical, in the sense that it tests a theory (or hypothesis) to explain the impact of natural enemies on their prey.

5.3.2   Observing natural enemies and their prey

Many ecologists have used observations of natural enemies and their prey to assess the impact of natural enemies on prey populations, particularly the populations of pests. Ecologists observe changes in the abundance of either the prey species or the natural enemy species, or the mortality caused by natural enemies.

Perhaps the best observations of enemy effects on prey populations come from "classical" biological control. The many forms of biological control are discussed in Chapter 12: classical biological control is the term often used to describe the introduction of non-native (or "exotic") natural enemies to control insect pests, which are, themselves, also exotic species (Gillespie et al. 2006). Two examples are shown in Figure 5.9. In each case the introduction (and therefore increase in the abundance) of the natural enemy results in a decline in the abundance of the pest species. The implication is clear: the natural enemy is responsible for the decline in the pest species. Generally, however, there are no "control" plots to assess whether the natural enemy introduction is responsible for the decline in prey numbers (Kidd & Jervis 1996). Moreover, the successes of biological control may not serve as a suitable indication of the power of natural communities of enemies to control the densities of insect herbivores (Hawkins et al. 1999). Biological control may overestimate the power of predation resulting, as it does, from the formation of a single strong link in simplified food webs. None the less, depending on the spatial scale over which the relationships are observed (Matsumoto et al. 2004), classical biological control can demonstrate a clear potential for natural enemies to greatly reduce prey populations.

Observations of the abundance of insects and their potential natural enemies can also be used to discover which natural enemy species are important in reducing the numbers of particular insect species. One situation where field observations, augmented by laboratory studies, have played a particularly important role is in the study of polyphagous predators (i.e. predators that consume a range of prey species). Agricultural pests, such as cereal aphids, are preyed upon by a wide range of arthropod predators. Some of these predators are aphid-specific, such as syrphid larvae and coccinelids; others, such as many ground-dwelling beetles, are polyphagous (consuming both aphids and other invertebrates). Often, polyphagous

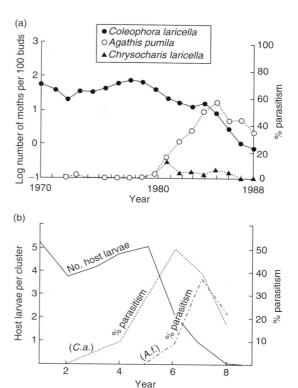

Figure 5.9 Two examples of successful biological control of insect pests by natural enemies. (a) The introduction of two parasitoids, *Agathis pumila* and *Chrysocharis laricella*, to control the larch casebearer, *Coleophora laricella*. (From Ryan 1990.) (b) The introduction of *Cyzenis albicans* and *Agrypon flaveolatum* to control the winter moth in Canada (where 'year' is the number of years from which data were recorded, starting 1 year before the appearance of *C. albicans*. (From Embree 1966.)

predators are much more abundant in ecosystems than aphid-specific ones (Pons et al. 2005). Although aphid-specific predators are conspicuous and consume large numbers of prey during aphid outbreaks, simple regressions between the numbers of cereal aphids and aphid-specific predators suggest that these predators take advantage of aphid outbreaks but are incapable of preventing them (Edwards et al. 1979) (Figure 5.10a). In contrast, regressions between the numbers of cereal aphids and the diversity of predatory and other arthropods in cereal fields suggest that polyphagous predators can prevent aphid outbreaks (Figure 5.10b). In more recent

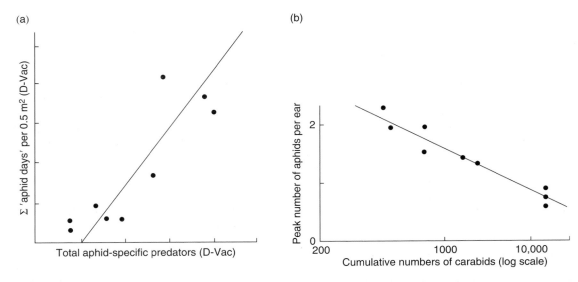

Figure 5.10 The relationship between the abundance of (a) aphid-specific predators, and (b) polyphagous predators and the abundance of cereal aphids. (From Edwards et al. 1979.)

work, Elliott et al. (2006) were unable to find any statistically significant relationships between predatory rove beetles (Coleoptera: Staphylinidae) in Oklahoma wheat fields and cereal aphid density on the crop. Unfortunately, these types of observation fail to quantify the importance of polyphagous predators on prey dynamics (Potts & Vickerman 1974). In fact, regressions between insects and their natural enemies can occur either because the natural enemies cause changes in prey numbers, or because they simply respond to changes in prey numbers (Kidd & Jervis 1996). In insect ecology, as in science in general, correlations do not demonstrate cause and effect. Nevertheless, this type of study may usefully identify which types of predator merit further (experimental) research.

A more useful form of observation than merely measuring the abundance of natural enemies and their prey is the estimation of mortality caused by natural enemies. Such observations are, however, much more difficult to make. For example, studies on the impact of parasitoids usually include observations on percentage parasitism. However, the technique used to quantify percentage parasitism can dramatically affect the estimate (Van Driesche 1983). Parasitism in natural populations can be measured by first taking a sample of the prey population and

then either: (i) dissecting each individual to detect parasitoid eggs or larvae within them; or (ii) rearing the sample to see how many parasitoids emerge. The most common technique used is rearing, but this typically underestimates parasitism in comparison with dissection by 12–44% (Day 1994). These underestimates are due to the relatively higher mortality of parasitized individuals because of disease and other factors during the rearing process. However, dissection can also lead to underestimates of parasitism because a proportion of parasitized hosts die of oviposition trauma and, in some species, after being fed upon by female parasitoids (Jervis et al. 1992). Day (1994) argued that to give the most comprehensive results both methods should be used concurrently. Kidd and Jervis (1996) have reviewed some of the more complex ways of assessing parasitism.

Despite the problems associated with measuring parasitism accurately, measuring the impact of predators is even more challenging. Vertebrate and many invertebrate predators completely consume their prey and leave little or no evidence that they have done so. In some cases, measuring mortality caused by predators can be accomplished by subtraction – accounting for all other potential sources of loss, and assuming that predators were responsible for the rest. For example, the number of pupae of a univoltine (i.e. single generation per year) forest pest (such as the winter moth; see Section 5.3.4) killed by predators can be calculated from: (i) the number of insects entering the pupal stage; (ii) the number of pupae killed by parasitoids, disease, and unknown causes (e.g. weather); and (iii) the number of pupae successfully emerging as adults. In some cases, sampling of the pupal stage may also reveal evidence of feeding by particular types of predators and provide separate estimates of mortality caused by, for example, shrews and ground-dwelling beetles. This simple calculation of predation is possible because the population does not, of course, increase during the pupal stage and because of the non-overlapping (discrete) generations of the insect. Predation is much more difficult to assess in multivoltine insects, such as cereal aphids, with overlapping generations.

In such cases, an alternative to measuring mortality is to calculate a predation "index" of some kind. For example, the research described above on the impact of polyphagous predators on cereal aphids led to more detailed work to identify which of the polyphagous predators found in cereal fields have a significant impact on aphid numbers. This research involved measuring the density of different insect predators during the period when aphid numbers were increasing and dissecting samples of predators to calculate the proportion that had consumed cereal aphids (Sunderland & Vickerman 1980). Different predators were then ranked according to an index calculated by multiplying predator density by the proportion of individuals of the species that had consumed aphids (Table 5.1). This then led to more detailed studies of the beetle species identified as being potentially the most important predators.

Predation rates can also be estimated by direct observations of predation in the field, a technique unsuited to most predatory groups except perhaps spiders (e.g. Sunderland et al. 1986). In addition, serological methods can be used to measure predation. Serological techniques are based on the production of antibodies in rabbits and other mammals against antigens of prey species. These antibodies are then used to detect the presence of particular prey species in the gut contents of predators. One of the most commonly used techniques is ELISA or enzyme-linked immunosorbent assay (Sunderland 1988; Kidd & Jervis 1996; Hagler 2006). For example, Sunderland et al. (1987) developed a predation index for different predator species of cereal aphids based on ELISA. The index is calculated as $P_g d / D_{max}$, where P_g is the percentage of predators testing positive in cereal aphid ELISA tests, D_{max} is the number of days over which cereal aphid antigens are detected in the predator, and d is the density of the predator. Use of this index suggested that spiders were the most important predators of cereal aphids. In a modification of the ELISA system, Hagler and Naranjo (2004) were able to detect the dispersal of commercially bought ladybirds, *Hippodamia convergens* (Coleoptera: Coccinellidae), in cotton fields in Arizona by "tagging" them with rabbit immunoglobulin which could be detected using ELISA.

5.3.3   Density-dependent and density-independent mortality factors

Observations of mortality caused by natural enemies have also been used to identify which, if any, natural enemies have the potential to "regulate" the abundance of particular insect species. As we learned in Chapter 4, density-dependent factors can limit population growth because their proportional impact varies according to population density. Density-dependent factors include predation and competition, and their impact may be manifest by changes in the rates of birth, death, or movement as population density changes (see Figure 4.2). In contrast, the proportional impact of density-independent factors does not change with population density (see Chapters 1 and 2).

The idea that density-dependent factors are needed to regulate animal population abundance (e.g. Nicholson 1933, 1957, 1958) is now widely accepted despite often acrimonious debate among ecologists (Dempster & McLean 1998). However, a few population ecologists, most, notably, working on insects, have argued that density-independent factors are responsible for regulating insect densities (e.g. Andrewartha & Birch 1954; Milne 1957a, 1957b; Den Boer 1991).

Put briefly, the fundamental argument for the importance of density dependence is that populations would increase indefinitely unless their fluctuations were somehow regulated by density-dependent mortality (or natality, emigration, or immigration). The simplest population models demonstrate this (Section 5.4). The counterargument is that in natural (as opposed to mathematical) systems, density-independent factors, particularly weather, are largely responsible for

Table 5.1 Assessment of the importance of different polyphagous species as predators of cereal aphids by the calculation of a predation "index". (From Sunderland & Vickerman 1980.)

	Proportion containing aphid remains at aphid densities of:			Number examined	Mean density of predators (for aphid increase 1–1000/m²)	Predation index*
	1–1000/m² (increase phase)	1000–1/m² (decrease phase)	Limit 1000/m² (increase plus decrease phases)			
Demetrias atricapillus	0.253	0.136	0.230	113	1.23	0.311
Agonum dorsale	0.236	0.336	0.257	653	1.28	0.302
Forficula auricularia	0.278	0.165	0.220	236	0.61	0.170
Tachyporus chrysomelinus	0.051	0.054	0.052	346	2.39	0.122
Tachyporus hypnorum	0.024	0.034	0.260	778	4.50	0.108
Bembidion lampros	0.082	0.167	0.093	989	1.23	0.101
Amara familiaris	0.033	0.019	0.027	255	1.47	0.049
Amara aenea	0.034	(0.00)†	0.034	176	1.42	0.048
Nebria brevicollis	0.086	(0.00)	0.085	531	0.48	0.041
Notiophilus biguttatus	0.040	0.013	0.031	228	0.67	0.027
Asaphidion flavipes	0.048	0.000	0.044	114	0.31	0.015
Amara plebeja	0.016	0.000	0.014	147	0.88	0.014
Harpalus rufipes	0.053	0.056	0.054	147	0.14	0.007
Pterostichus melanarius	0.161	0.073	0.101	346	0.03	0.005
Loricera pilicornis	0.007	0.019	0.011	442	0.27	0.002
Calathus fuscipes	(0.000)	0.098	0.085	47	0.02	0.000

* Values in column 1 multiplied by values in column 5.
† Parentheses denote sample size < 10.

causing, and even limiting, the fluctuations of insect populations. As Andrewartha and Birch (1954) wrote in relation to their classic study of weather on thrips, *Thrips imaginis* (Thysanoptera), on rose: "not only did we fail to find a density-dependent factor but ... there was no room for one". This remark was prompted by the finding (in multiple regression analysis) that weather explained over 80% of the variation in the abundance of thrips (Davidson & Andrewartha 1948a, 1948b). However, as we saw in Section 4.1, even 89% density-independent mortality cannot limit population growth in the absence of some density dependence (see Table 4.2). As well as stimulating the development of mathematical models, the density-dependence debate in the 1950s stimulated the growth of the "life table" approach to the study of insect populations. Most life table work has been carried out in the woods and fields, as described below, though recently very large (more than 100,000 strong) laboratory populations of insects such as mosquitoes have been studied instead (Styer et al. 2007).

5.3.4 Life table studies

Life tables and *k*-factor analysis have been useful tools in the detection of density dependence in insect populations (Varley 1947). One of the best-known examples of the life table approach is the study by Varley and Gradwell (1968) and Varley et al. (1973) of the winter moth on oak trees in Wytham Wood in England (see Chapter 1). Those workers monitored the abundance of adult and larval winter moths for over 15 years and measured the impact of pupal predation, parasitism by *Cyzenis albicans* and other parasitoids, and pathogen (microsporidian) infection. They also quantified the effect of what they called "winter disappearance", mortality between the adult and late larval stage.

The impact of each mortality factor in *k*-factor analysis is expressed as a *k*-value:

$$k_i = \log N_i - \log N_{i-1} \qquad (5.1)$$

where N_i and N_{i-1} are the densities of the population before and after the action of mortality k_i. The *k* value of each factor is calculated (as k_1, k_2, k_3, etc.) and total mortality, *K*, equals k_1, k_2, k_3, etc.

If the values of all *k*-factors are plotted together with total mortality (*K*) for each year of the study,

then the contribution of each *k*-factor can be evaluated. The factor that makes the greatest contribution to total mortality is known as the key factor. This can be assessed visually or evaluated by plotting individual *k*-factors against total mortality, the mortality factor with the greatest slope being the key factor (Podoler & Rogers 1975). For the winter moth, the key factor is winter disappearance (Figure 5.11). Note that the key factor is not necessarily density dependent.

Density dependence is detected by plotting the values of each *k*-factor against population density. A positive slope indicates density dependence, a negative slope indicates inverse density dependence (a decreasing effect on the prey population as density increases) and an absence of a significant relationship indicates density independence. This demonstrated that pupal predation acts as a density-dependent factor on the winter moth (Figure 5.12).

Thus, *k*-factor analysis has shown that winter moth population fluctuations are driven primarily by "winter disappearance" but the population is regulated by pupal predation. This is a critical point – the ecological factor responsible for most of the population change is not necessarily the factor that acts to regulate the population. The largest component of "winter disappearance" – although it includes adult, egg, and larval mortality – is mortality of young winter moth larvae, thought to be mainly determined by the degree of synchrony (or coincidence) between larval emergence and oak bud burst (Hunter 1990, 1992b). It is likely driven primarily by climatic variation, and is density independent.

Detailed instructions for the construction of life tables and *k*-factor analysis have been given by Varley et al. (1973) and Kidd and Jervis (1996). Several reviews have been published summarizing the many insect life table studies that have now been carried out (Dempster 1983; Stiling 1987, 1988; Cornell & Hawkins 1995). Dempster (1983) analyzed 24 datasets and found evidence of density dependence by natural enemies in only three cases. Stiling (1987) examined life table data for 58 species of insect but found evidence of density dependence in only about half of them. In approximately half of the cases where density dependence was detected (27% of all cases) parasitism, predation, or pathogens were responsible. Is density dependence really as rare as these reviews would suggest?

Unfortunately, there are many problems associated with *k*-factor analysis (Dempster 1983; Price 1990).

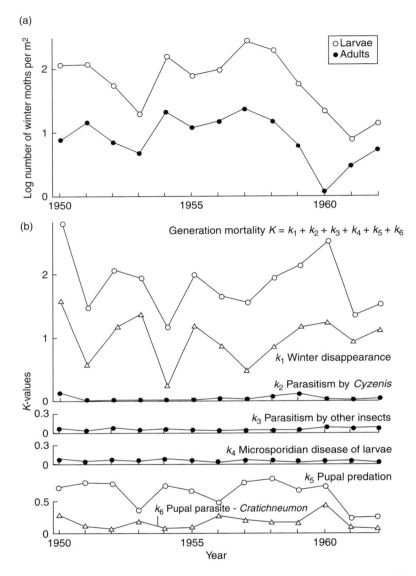

Figure 5.11 Abundance (a) of winter moth larvae and adults in Wytham Wood. England 1950–62 and changes in mortality, and (b) expressed as total generation mortality, K, and k-values for separate sources of mortality. (From Varley et al. 1973.)

First, both the estimation of key factors and the detection of density dependence by regression analysis are statistically flawed. In both cases the variables being analyzed are not independent; for example, k-values are calculated from, and therefore are not independent of, population densities. Alternative tests for the detection of density dependence have been proposed but these may be unnecessarily strict, potentially "missing" density-dependent mortality factors (Hassell et al. 1987). There are several other "technical" problems with k-factor analysis. For example, "traditional" k-factor analysis assumes that each

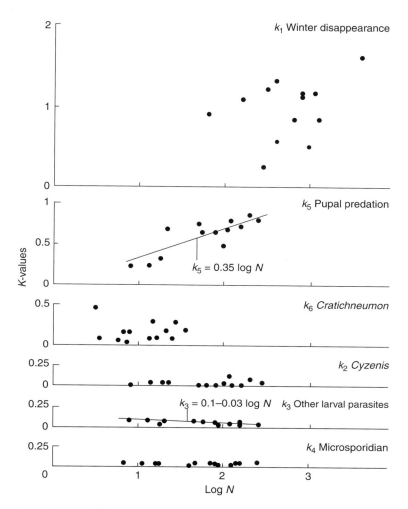

Figure 5.12 Relationships between different winter moth mortalities and population density. (From Varley et al. 1973.)

k-factor acts in sequence, not in parallel. The order in which different mortalities are analyzed may not seem to be important, but Putman and Wratten (1984) found that when they reversed the order of larval starvation and predation in life table analysis of the cinnabar moth, predation, rather than larval starvation, became the apparent key factor.

As well as statistical reasons, there are also ecological reasons underlying the weakness of k-factor analysis. For example, k-factor analysis was not initially designed to detect delayed density dependence (see Chapter 4). We now know that there are

often time lags associated with the operation of the density-dependent processes that regulate populations (Turchin 1990). While time sequence plots generated from k-factor analysis can reveal the action of delayed density-dependent factors through the spiral patterns that they cause (Figure 5.13), there are now better methods for estimating the importance of delayed density dependence. These will be discussed in the section on time-series analysis (Section 5.4.10). Moreover, life tables are really "death tables" and generally categorize sources of mortality only. As such, they may miss density-dependent processes

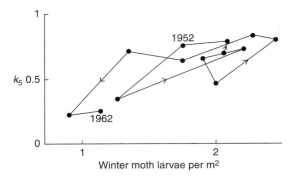

Figure 5.13 Time series plot of pupal predation of the winter moth, the spiral form suggesting a delayed density-dependent component to this mortality. (From Varley et al. 1973.)

operating through birth rates or movement (see Figure 4.2) and may fatally underestimate the roles of resource quality and quantity in population dynamics (Hunter et al. 2000). Methods have been developed by which female choice among plants of different quality can be incorporated into life tables. For example, Preszler and Price (1988) began their life tables of the gall-forming sawfly, *Euura lasiolepis* (Hymenoptera: Tenthredinidae), with eggs in the female ovaries. Females could both fail to initiate galls and, after gall initiation, fail to oviposit within those galls. Consequently, two aspects of female behavior, based upon plant quality, were incorporated into their life tables. Other problems with *k*-factor analysis remain. It does not, on its own, reveal the potentially important influence of spatial density dependence (Hassell 1985a) (Section 5.4.7). Additionally, mortality factors may interact with one another, causing effects on populations that are not additive (Section 5.5).

Some of the problems associated with life tables and *k*-factor analysis are so severe that it has been referred to as a methodological "strait jacket" (Putman & Wratten 1984): *k*-factor analysis clearly should not be used alone to try to understand the population dynamics of insects. However, life tables can be a useful way of comparing, for example, the susceptibility of different types of insect to predation or parasitism. Hawkins et al. (1997) compiled life tables for 78 insect herbivores and found that leaf-miners suffer the greatest levels of parasitism, and gallers, borers, and root-feeders the least. In contrast, exophytic (externally feeding) herbivores experienced the

greatest level of mortality caused by predators and pathogens.

Despite controversy, life table studies and key factor analyses continue unabated. Insect populations studied recently include *Helicoverpa* spp. (Lepidoptera: Noctuidae) in cotton (Grundy et al. 2004), *Nasonovia ribisnigri* (Hemiptera: Aphididae) on lettuce (Diaz & Fereres 2005), and *Bemisia tabaci* (Hemiptera: Aleyrodidae) (Asiimwe et al. 2007) on cassava in Uganda. In this latter example of whitefly populations, the highest rates of mortality across all life stages from egg to adult were from parasitism, then dislodgement, and thirdly predation. While the first and the last factor have at least the potential to be density dependent, dislodgement, where eggs, nymphs, and adults are knocked off the plants by wind and rain, is likely to be density independent. Working with the same insect, *B. tabaci*, but on cotton in Arizona, Naranjo and Ellsworth (2005) discovered that mortality in the fourth nymphal instar was the most important factor accounting for overall generation mortality, attributable mainly to predation by predatory bugs (Hemiptera: Heteroptera) – the key factor.

5.3.5 Experiments on natural enemies and their prey

As an alternative (or complement) to observing natural populations, many ecologists have manipulated experimentally either natural enemy or prey species to quantify the impact of natural enemies on their prey (Hunter 2001a). Experimental manipulation, if carefully done, has the benefit of allowing insect ecologists to infer cause and effect from their results. Of course, experimental approaches have their own set of problems including limitations of timescale (Holt 2000), the turnover of important species within experimental arenas (Leibold et al. 1997), and their general inability to encompass spatial and temporal variation in the strength of ecological factors (Belovsky & Joern 1995; Preszler & Boecklen 1996). These problems are not trivial, yet experimental manipulations provide some of the strongest support for the role that natural enemies can play in the dynamics of prey populations.

The experimental approach takes many forms but most experiments involve the exclusion of natural enemies and a comparison between the abundance of the prey species in exclusion and control "treatments".

Natural enemies may be excluded by cages or barriers (Kidd & Jervis 1996). Cages may be placed over the whole plant, over single branches (sleeve cages) or leaves, or even small parts of a leaf (e.g. "clip-cages"). Barriers to prevent access of natural enemies include greased plastic bands around tree trunks or branches. Exclusion experiments may be used to quantify the impact of the whole natural enemy complex of a single prey species or may focus on the impact of one, or a few, species of natural enemy (Kidd & Jervis 1996). The latter, so-called partial (as opposed to total), exclusion experiments include the use of mesh or gauze cages where the mesh size allows access of some natural enemies but not others (Colfer & Rosenheim 2001). Alternatively, cages or sleeves may be placed over whole plants or branches of trees and bushes to exclude predators such as hoverflies (Diptera: Syrphidae), lacewings (Neuroptera: Hemerobiidae), ladybirds (Coleoptera: Coccinellidae) and parasitoids (Schmidt et al. 2004; Day et al. 2006). The cages or sleeves may be raised slightly above the ground or consist of varying size mesh to selectively to allow predators such as ground beetles (Coleoptera: Carabidae) to move in. Exclusion may also use insecticide treatment to "exclude" unwanted natural enemies from manipulation experiments (Cugala et al. 2006).

For example, Morris (1992) excluded predators from colonies of the aphid *Aphis varians* (Hemiptera/Homoptera: Aphididae) feeding on fireweed (*Epilobium angustifolium*) as part of a complex experiment designed to investigate the effects of predation, interspecific competition (with flea beetles, Coleoptera: Chrysomelidae), and water availability to the host plant. Morris found that coccinellid and syrphid predators had the greatest impact on the abundance of the aphid: in their absence, aphid numbers increased by 10% per day. The other factors had an insignificant impact on aphid abundance.

There are many problems associated with exclusion experiments but the main drawback of these experiments is that the cages may affect the microclimate of the enclosed plant, potentially affecting the performance of the prey species (Kidd & Jervis 1996). There are various possible solutions to this problem, such as "inclusion" cage experiments. In these experiments, known numbers of predators or parasitoids are placed inside the cages (e.g. Dennis & Wratten 1991). Even here, however, the unnatural microclimate may lead to misleading results (Kidd & Jervis 1996).

Moreover, caging experiments often influence the movement behavior of the prey species as well as colonization by predators. Untangling the relative impact of potential cage artifacts on predators and prey can be difficult.

Despite their problems, exclusion experiments can, if carefully done, provide useful information on the impact of natural enemies on prey populations. Other experimental techniques include the use of insecticides to remove natural enemies or even their physical removal (Kidd & Jervis 1996). An alternative experimental approach is to manipulate prey numbers instead of natural enemies. This is particularly useful for measuring the impact of natural enemies in different habitats. For example, Watt (1988, 1990) manipulated the abundance of pine beauty moth, *Panolis flammea* (Lepidoptera: Noctuidae), to measure, amongst other things, the role of natural enemies in different forest habitats. Run in parallel with natural enemy exclusion experiments, this approach identified the role of natural enemies in preventing pine beauty moth outbreaks on Scots pine (Watt 1990) (Figure 5.14). Another approach is to use artificial prey. For example, Speight and Lawton (1976) used laboratory-reared *Drosophila* pupae to assess the effect of weed cover in cereal fields on the foraging of carabid beetles.

Experiments are also carried out to do more than quantify the impact of natural enemies. One interesting example is a "convergence experiment" (Nicholson 1957). In convergence experiments, insect populations are manipulated to artificially high or low densities to determine whether the population returns to an equilibrium density. Although there are many technical problems associated with this type of experiment, it can be useful in detecting density-dependent factors missed in life tables and *k*-factor analyses. For example, Gould et al. (1990) discovered density-dependent mortality in gypsy moth, *Lymantria dispar* (Lepidoptera: Lymantriidae), populations caused by two parasitoids by artificially increasing the numbers of gypsy moth to provide a wide range of densities.

Experiments have also demonstrated how natural enemies can restrict the spread of insect outbreaks. For example, outbreaks of the western tussock moth, *Orgyia vetusta* (Lepidoptera: Lymantriidae), are known to persist for over 10 years, with little tendency to spread into suitable adjacent areas. One possible explanation for this is that natural enemies, more mobile than the moth, which has flightless females and disperses

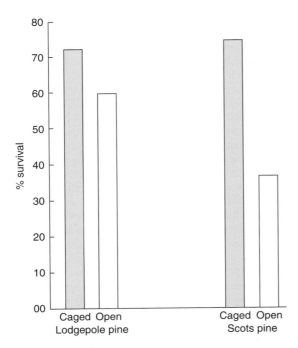

Figure 5.14 The survival of pine beauty moth. *Panolis flammea*, larvae on different host plants in predator exclusion cages and exposed to predation. (From Watt 1989, 1990.)

mainly as passive first instar larvae (Yoo 2006), disperse out from the edges of the outbreak and create a zone around the outbreak area where the ratio of natural enemies to tussock moth is particularly high. This "predator diffusion" hypothesis was tested by establishing experimental populations of tussock moth along a transect from the edge of a tussock moth outbreak into a suitable adjacent habitat where there were few tussock moths (Brodmann et al. 1997; Maron & Harrison 1997). The impact of natural enemies was assessed by measuring the number of eggs and larvae attacked by an egg parasitoid and four species of tachinid larval parasitoids. As predicted, parasitism was greatest in the zone immediately surrounding the outbreak, and it was concluded that the parasitoid did indeed restrict the spread of tussock moth outbreaks. Interestingly, this hypothesis was stimulated by "reaction–diffusion" models, which predict that insects can become patchily distributed in uniform habitats as a result of interactions between mobile natural enemies and their relatively sedentary prey (Maron & Harrison 1997).

This is a particularly good example of the power of experimentation and of how we need models, long-term data, and experimentation to solve problems in insect ecology.

5.4 MODELING PREDATOR–PREY INTERACTIONS

5.4.1 The Lotka–Volterra model

No other aspect of theoretical population ecology has attracted (and is still attracting) more attention than predator–prey modeling (Fan & Li 2007). (In this context we include parasite–host models and host–pathogen models.) We cannot attempt to cover the whole history and complexity of models that have been developed to describe insect populations, and present only some of the main developments in modeling here. We are aware that a few insect ecologists remain skeptical of population models. Nevertheless, models are a useful way of illustrating complex interactions in ecology. They can be used as a formal statement of a hypothesis (perhaps the way we think that a predator interacts with its prey) which can then be tested experimentally. They can be used to simulate particular interactions of interest, such as those between pest insects and their natural enemies on a specific crop plant of economic importance. They can be used to make predictions about population densities at some future date. And, in analytical modeling, they can be used to ask "what if" questions about the basic nature of ecological interactions. Models can be "mechanistic" in which precise mechanisms underlying the interactions (predator foraging behavior, predator satiation) are incorporated explicitly. Models can also be "phenomenological" where the dynamics of the major players are described without reference to the mechanisms underlying those dynamics.

The first population models to describe predator–prey interactions, the Lotka–Volterra equations, were independently produced by Lotka (1925) and Volterra (1931). Such is the lasting impact of these publications that Hutchinson (1978) described Lotka's 1925 book as "one of the foundation stones of contemporary ecology". Hutchinson (1978) and Kingsland (1985) have described the history of early population models and their use, and Gillman and Hails (1997) have provided a thorough introduction to the mathematics of population models.

Lotka and Volterra first chose a model to describe the population dynamics of a prey population growing in the absence of predation. Here, we refer back to the logistic equation that we met in Chapter 4 (equation 4.10). This might be a good time to re-read Sections 4.1 and 4.4.2 to refresh your memory before proceeding.

Recall that the logistic equation takes the form:

$$dN/dt = rN\left[\frac{(K-N)}{K}\right] \qquad (5.1)$$

where N is now the population size of the prey species, r is the per capita exponential growth rate of the prey population, and K is the "carrying capacity" of the environment for prey species N. Lotka and Volterra assumed that, in the presence of a predator, prey population growth rate would be reduced in proportion (b) to the encounters between the predator (P) and prey, such that:

$$dN/dt = rN\left[\frac{(K-N)}{K}\right] - bPN \qquad (5.2)$$

We now need an equation for predator population change. Lotka and Volterra modeled the predator population by assuming that, in the absence of prey, it would decline exponentially at a per capita rate, k:

$$dP/dt = -kP \qquad (5.3)$$

They also assumed that the predator population would increase in proportion, c, to the encounters between predators and prey. This leads to the final equation for the predator population:

$$dP/dt = cPN - kP \qquad (5.4)$$

Taken together to represent the interactions between predator and prey, the Lotka–Volterra equations (equations 5.2 and 5.4) can produce cycles in prey numbers similar to those observed in many insect species (May 1974; Barbour 1985) (e.g. Figures 5.15 and 5.16) and imply that population cycles of prey can be caused by predators. However, realistic cycles are only produced under particular conditions of the Lotka–Volterra model (Gillman & Hails 1997) and the Lotka–Volterra equations have been widely criticized for their lack of realism. Nevertheless, their contribution to theoretical ecology as a stimulus for further research cannot be ignored (May 1974).

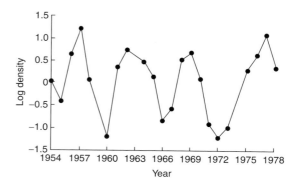

Figure 5.15 Abundance of pine looper moth, *Bupalus piniaria*, pupae in Tentsmuir forest, Scotland 1954–78. (From Barbour 1985.)

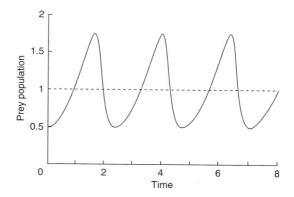

Figure 5.16 Changes in the size of a prey population according to particular conditions of the Lotka–Volterra equation. (From May 1974.)

5.4.2 Discrete population models and density dependence

As we learned in Chapter 4, population models can be constructed either in continuous or discrete time. Continuous time models are suitable for populations with overlapping generations and continuous reproduction, and are best described by differential equations such as those used in the Lotka–Volterra predator–prey equations above. However, discrete time population models are more suitable for the many insects that exhibit synchronous reproduction during discrete breeding seasons, such as univoltine (one generation per year) species. Discrete time models are best described by difference equations that are sometimes easier to construct and understand than models based on differential equations. We met the simplest difference equation in Chapter 4 (see equation 4.2).

Gillman and Hails (1997) described the mathematics of difference equation models. In summary, they usually take the form of an equation describing the abundance of a population (N) at time t, usually written as N_t, as a function of the size of the population in the previous year or generation (N_{t-1}). For example, a population model for an insect with a fecundity of 100 eggs per female (and a sex ratio of 1 : 1) could be described by the equation:

$$N_t = 50N_{t-1} \qquad (5.5)$$

This is analogous to equation 4.2, with $\lambda = 50$. It is called a "first-order" difference equation because abundance at time t is related to abundance one time interval previously ($t - 1$). Second-order difference equations relate abundance to two previous time intervals; for example,

$$N_t = 50N_{t-1} + 50N_{t-2} \qquad (5.6)$$

The logistic population model described in equation 5.1 has several discrete time versions. One of these is:

$$N_t = \lambda N_{t-1}\left(\frac{K - N_{t-1}}{K}\right) \qquad (5.7)$$

We can use equation 5.7 to explore some of the features of difference equations, including the complex dynamics that they are capable of producing. Using the examples in Figure 5.17, where $K = 100$ and 200, increasing the value of λ from 2 to 3.1 results in oscillations between two population densities, or a "two-point limit cycle" (Figure 5.17b), and a further increase in the value of λ to 3.5 results in four-point limit cycles (Figure 5.17c). If λ is increased still further, to 4.0, this results in irregular population fluctuations or "chaotic dynamics" (Figure 5.17d). Chaotic behavior has received much attention in the dynamics of populations and other systems. The main message from models producing chaotic dynamics

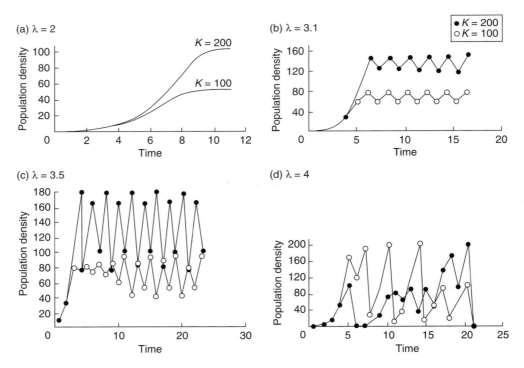

Figure 5.17 The effects of varying λ and K in the discrete logistic equation. (From Gillman & Hails 1997.)

is that it may be wrong to assume that irregular population fluctuations are necessarily caused by irregularly acting factors, such as climatic variability: they may be caused by density-dependent factors. There are no stochastic (irregular) factors operating in equation 5.7 yet, with high values of λ, chaotic dynamics are produced. Why is the discrete time version of the logistic equation capable of producing complex dynamics (Figure 5.17) whereas the continuous time version is not (Figure 4.5)? Discrete time models have a built-in time delay of one generation (between N_t and N_{t-1}) and, as we learned in Section 4.1.1, time delays favor oscillations in population dynamics.

One of the problems with the models described in this section is that, even if they can produce population behavior of the types observed in nature (see Figure 5.15), they are oversimplified and say little about the causes of the population behavior. This problem may be overcome in part by using realistic values for model variables (such as λ), but nevertheless many population ecologists have felt the need to construct more complex models to explore the mechanisms that drive dynamics. The next section considers improvements in modeling predator–prey dynamics that use discrete time models with increasing levels of mechanism. Later sections look at some attempts to model the interactions between insects, their natural enemies, and other factors.

5.4.3 The Nicholson–Bailey model

The first well-known model to describe explicitly the population dynamics of insect prey and their parasitoids was constructed by Nicholson and Bailey (1935). They criticized the predator–prey models of Lotka and Volterra for several reasons including the incorrect assumption that there is an instantaneous response by both predators and prey to encounters between them. Nicholson and Bailey pointed out that the response of natural enemies, particularly parasitoids, to changes in prey density can take some time to occur. For example, effects of prey numbers on the population density of adult parasitoids will likely take one generation to be expressed. Discrete time models are therefore appropriate tools for such modeling.

Nicholson and Bailey attempted to incorporate some degree of ecological realism into their model. None the less, some simplifying assumptions were made, and the most important of these are as follows (Hassell 1981; Kidd & Jervis 1996; Gillman & Hails 1997):

- predators or parasitoids search randomly for their prey;
- predators can consume an unlimited number of prey (once located);
- parasitoids have an unlimited fecundity;
- only one adult parasitoid will emerge from a parasitized host;
- the area, described by Nicholson as the "area of discovery" *a*, searched by a predator or parasitoid is constant (see below); and
- generations of both the natural enemy and its prey are completely discrete and fully synchronized.

It should be noted that the area of discovery referred to above is more or less equivalent to the proportion of the habitat searched by the natural enemy during its lifetime and is used as a measure of the natural enemy's searching efficiency.

Let us start with the equation for the prey. The model begins by assuming exponential growth in the prey population in the absence of any parasitoids:

$$N_{t+1} = \lambda N_t \qquad (5.8)$$

However, when parasitoids are present, not all prey contribute to population growth. Rather, only the proportion surviving the parasitoids contribute. N_t must therefore be multiplied by the proportion surviving, which is assumed to be a negative exponential function of parasitoid (P_t) density:

$$N_{t+1} = \lambda N_t e^{-aP_t} \qquad (5.9)$$

The expression $\exp(-aP_t)$ has host survivorship declining exponentially as the parasitoid population grows. It can also be derived from the Poisson distribution, which is a probability density function that describes random processes such as random encounters between natural enemies and their prey. The reader is referred to Gillman and Hails (1997) for a discussion of probability density functions and their use in models such as the one above.

Now we need an expression for population change in the parasitoid. Here, Nicholson and Bailey assumed that parasitoids converted prey into new parasitoids (parasitoid "birth") with efficiency *c*. Of course, they only produce new parasitoids from hosts that were actually attacked. Given that the proportion of hosts

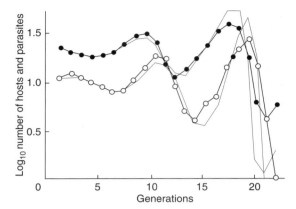

Figure 5.18 Changes in abundance of greenhouse white-fly, *Trialeurodes vaporarium* (●—●), and its parasitoid *Encarsia formosa* (○—○) and as predicted by the Nicholson–Bailey model (solid lines without symbols). (From Burnett 1958; Hassell & Varley 1969.)

surviving is given by $\exp(-aP_t)$, then the proportion attacked must be $1 - \exp(-aP_t)$. The parasitoid equation is therefore:

$$P_{t+1} = cN_t(1 - e^{-aP_t}) \qquad (5.10)$$

The Nicholson–Bailey equations (5.9 and 5.10) predict that there exist equilibrium densities for both the natural enemy and its prey. The equilibrium densities depend upon the particular values of λ and a. However, the equilibria are unstable: even the slightest disturbance leads to cycles of increasing size and the extinction of the parasitoid (see, for example, Hassell 1978) (Figure 5.18). So, as written, the Nicholson–Bailey equations do not do a good job of reflecting reality. Should we just discard them?

5.4.4   Making the Nicholson–Bailey model more realistic

The lack of stability of the Nicholson–Bailey model indicates that it fails to capture some important aspect of predator–prey interactions. Something must be missing from the model which, when added, will improve model stability. Rather than starting again from scratch, we might attempt to modify the Nicholson–Bailey model, based on our understanding of natural systems, to improve its performance. Many

workers have tried to do just that, and some of their findings are reported below.

One ecological factor missing from the Nicholson–Bailey model (which is present in our version of the Lotka–Volterra model in equation 5.2) is density dependence acting on host population growth (Flatt & Scheuring 2004). In other words, there is nothing to limit host growth in the Nicholson–Bailey model except the natural enemy. As we learned in Chapter 4, resource availability places an ultimate limitation on population growth, and can act to regulate populations. The incorporation of resource limitation through density-dependent competition among hosts in the Nicholson–Bailey model does increase the stability of the interaction between predators and their prey (Beddington et al. 1975) (Figure 5.19). This, of course, implies that there is nothing inherently stable in the interactions between insects and their natural enemies – the stability is supplied by intraspecific competition among prey. However, since insects and their natural enemies do persist together for many generations, often at densities below which prey should compete, many ecologists have tried to develop predator–prey models that are inherently stable because of the predator–prey interaction.

5.4.5   Handling time, functional responses, and numerical responses: the effects of prey density

One of the drawbacks of the Nicholson–Bailey model is that it is unrealistic to assume that predators have unlimited appetites and that parasitoids can produce an unlimited number of eggs (Hassell 1981). In addition, it is also unrealistic to assume that the time available to natural enemies to search for prey is not related to prey density (Holling 1959). Holling predicted that as prey numbers increase, and therefore more prey are eaten or parasitized, then the predator or parasitoid spends an increasing amount of time in activities such as capturing, killing, eating, and digesting prey. Holling called this the "handling time" of a natural enemy. The consequence of an increase in handling time as prey density increases is that the number of prey captured per predator does not show a linear (Type I, below) increase as is implicit in the Nicholson–Bailey model.

The relationship between the number of prey killed or parasitized per natural enemy and prey density is known as the functional response, and Holling (1959) described three different types:

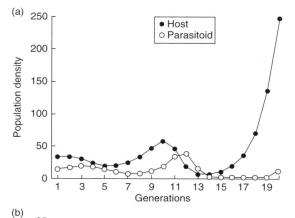

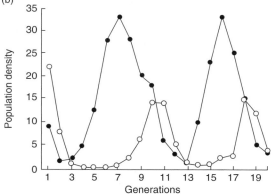

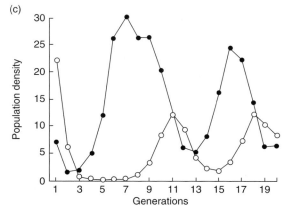

Figure 5.19 The effects of incorporating density-dependence in the Nicholson–Bailey model: (a) without density-dependence; (b) with density-dependence, resulting in cyclical oscillations with upper and lower limits (limit cycles); and (c) with increased density-dependence, resulting in oscillations with decreasing size which approach an equilibrium. (From Kidd & Jervis 1996.)

- Type I: the response is linear up to a plateau (Figure 5.20a);
- Type II: the response rises at a decreasing rate (Figure 5.20b); and
- Type III: the response is sigmoid (Figure 5.20c).

Many predators exhibit a Type II functional response (Schreiber & Vejdani 2006), which arises because, as mentioned above, predators spend an increasing amount of time handling their prey and a decreasing amount of time searching for prey as prey density increases. Holling described the Type II functional response curve with the following equation:

$$N_A = \frac{aTN_t}{1 + aT_H N_t} \qquad (5.11)$$

where N_A is the number of prey attacked per predator, N_t is the density of prey, T is the total time available for searching, T_H is the handling time, and a is the searching efficiency or attack rate of the predator.

Holling developed this equation by carrying out an experiment with a human "predator". This predator was blindfolded and placed in front of a 0.9 m square table on which were pinned a variable number of paper disks 4 cm in diameter. The number of disks found by tapping and removed from the table in a minute was recorded at a range of disk densities. The above equation, derived from this experiment, is therefore known as Holling's disk equation. As the disk equation shows, the Type II functional response depends upon the predator's rate of attack, the time available for the predator to encounter its prey, and handling time. One aspect not explicitly included in the disk equation, but nevertheless important in shaping the functional response, is satiation (Holling 1966). At some point, the number of prey eaten per predator will reach a plateau because the predator is no longer hungry.

A Type III functional response curve can arise in two linked ways, i.e. through learning and through switching. At low densities of a particular prey species, a predator may be considered to be naïve and inefficient at capturing its prey. However, as the density of the particular prey increases and the predator encounters it more frequently, the predator is likely to learn how to find and capture this species of prey more efficiently. Eventually, however, the effects of increasing handling time and satiation with increasing prey density have the same effect as they have in the Type II functional response and produce the sigmoid

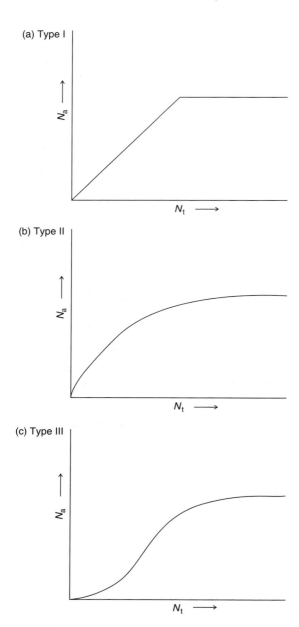

(a) Type I

N_a

N_t

(b) Type II

N_a

N_t

(c) Type III

N_a

N_t

Figure 5.20 Three types of functional response—the relationship between the number of prey killed per predator (or parasitized per parasitoid) (N_a) and prey density: (a) Type I, (b) Type II, and (c) Type III. (After Holling 1959.)

relationship between attack rate and host density. The Type III response can also occur as a result of predator switching, that is, the phenomenon whereby predators swap from one prey item to another as it becomes more profitable to do so. For example, polyphagous predators may switch from prey species A to prey species B as the density of species A decreases and the density of species B increases. In these circumstances a predator might show a Type II response against both prey species together, but a Type III response against them separately.

Laboratory studies have shown that most predators and parasitoids have a Type II functional response (although it should be noted that some experimental procedures may be responsible for producing a Type II response; Van Roermund et al. 1996), a significant number have a Type III response, and the Type I response is very rare (Figure 5.21). Holling considered that the Type II response was most typical of invertebrate predators and the Type III response most typical of vertebrate predators. However, that is clearly not universally true (Figure 5.21).

Exceptions to these categories of functional response curve have been noted, and some researchers have suggested a reclassification or expansion of functional responses (e.g. Putman & Wratten 1984; Schenk et al. 2005). One interesting exception is the response of predators to prey with defensive behavior such as many sawflies. When threatened, sawfly larvae rear backwards, a particularly effective means of defense in these communally feeding species because they all react together. In an experiment on the sawfly *Neodiprion pratti banksianae* (Hymenoptera: Diprionidae) and a pentatomid bug predator, Tostowaryk (1972) found a humped functional response: when the prey reached a certain density, their defensive response to the predator was enough to bring about a decrease in attack rate. Despite this type of exception (which was noted by Holling 1965), Holling's classification has largely stood the test of time and is widely accepted as a useful basis for describing the functional response of most predators and parasitoids.

The impact of the functional response on the interaction between predator and prey depends upon the type of the response (Hassell et al. 1977; Hassell 1978, 1981; Kidd & Jervis 1996). The inclusion of a Type II response in the Nicholson–Bailey model is theoretically destabilizing, but, as handling time is often a small fraction of the total time available (at least for insect predators and prey), the size of the destabilizing effect is likely to be small. A Type III response is theoretically stabilizing because it results in density-dependent predation over lower prey densities.

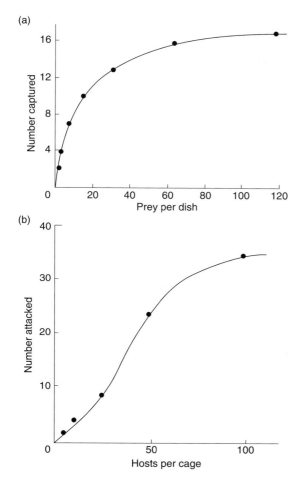

Figure 5.21 Examples of functional responses: (a) concave Type II response for predation by the coccinelid *Harmonia axyridis* on *Aphis craccivora*; and (b) sigmoid Type III response for parasitism of the aphid *Hylopteroides humilis* by the braconid wasp *Aphidius uzbeckistanicus*. (From Mogi 1969; Dransfield 1975; Hassell 1981.)

Natural enemies show, in addition to their functional response, a numerical response. Whereas the functional response is the change in the attack rate of an individual predator or parasite in response to changing prey density, the numerical response is the change in the number of natural enemies in response to changing prey density. Both responses are generally of similar shape and in reality work simultaneously (Omkar & Pervez 2004). For example, *Anagyrus* sp. (Hymenoptera: Encyrtidae) is

a parasitoid of the Madeira mealybug, *Phenacoccus madeirensis* (Hemiptera: Pseudococcidae). In a series of complex laboratory experiments, Chong and Oetting (2006) showed that the largest number of offspring was produced by lone parasitoids (see next section) foraging for hosts in patches containing the lowest parasitoid : host ratio (i.e. higher host densities resulted in higher parasitoid success). As we learned in Chapter 4, densities of insects (including predators) increase as the result of reproduction or immigration. So the numerical response generally measures the effects of prey density on predator recruitment through birth or movement. As with functional responses, we generally recognize Type I, Type II, and Type III numerical responses, based on the shapes of the curves. Notice, however, that the *y*-axis of a numerical response curve is predator density (Figure 5.22c).

The functional and numerical responses of predators to variation in prey density can be combined to calculate an "overall response", measured as the number of prey dying from predation as prey density varies (Kidd & Jervis 1996) (Figure 5.22a). For example, a sigmoid (Type III) overall response can be derived from a Type II functional response and a Type II numerical response (Figure 5.22b).

5.4.6   Mutual interference: the effects of natural enemy densities

Another possible influence on predator–prey dynamics is mutual interference among predators (or parasitoids) (Rogers & Hassell 1974; Hassell 1981; Van Alphen & Jervis 1996). For example, many different types of natural enemy have been observed to disperse following an increase in encounters with other natural enemies (of the same species) or after detecting previous parasitism in hosts. There are also many examples of aggressive and host-marking behavior in female parasitoids. These observations suggest that the searching efficiency of parasitoids and predators declines as the density of parasitoids and predators increases. Laboratory studies have confirmed this (Hassell 1978) (Figure 5.23). In reality, mutual interference is a form of competition among natural enemies. As we might expect, then, mutual interference can enhance the stability of predator–prey interactions in a way that is similar to the stabilizing effects of competition among prey (Section 5.4.4). Mutual interference may be classified as either

(a) Total response

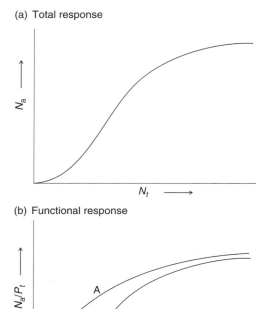

(b) Functional response

(c) Numerical response

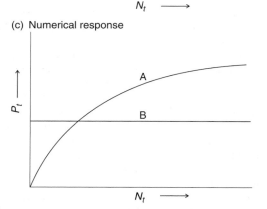

Figure 5.22 Alternative ways of producing a sigmoid Type III total response (a), a combination of the individual functional response (b), and the numerical response (c), where N_a is the number of prey killed by predators. P_t is the density of predators, and N_t is the density of prey. The same total response may be obtained from functional response A or B with numerical response A or B, respectively. (From Hassell 1978.)

direct interference, resulting from behavioral interactions between natural enemies, or indirect interference caused by, for example, superparasitism, decreasing fecundity, or a shift in sex ratio (Visser & Driessen 1991).

Hassell and Varley (1969) produced an empirically derived model for the effect of interference:

$$\log a = \log Q - m(\log P_t) \qquad (5.12)$$

where a is searching efficiency (as discussed above), m is the slope of the line, $\log Q$ is the intercept, and P_t is predator or parasitoid density. Notice that equation 5.12 is the equation of a straight line which, broadly speaking, mimics the responses shown in Figure 5.23. Notice also that, when there is no interference ($m = 0$), $\log a = \log Q$. Q is therefore equivalent to the attack rate in the absence of interference (Van Alphen & Jervis 1996). Equation 5.12 can be incorporated in the Nicholson–Bailey model and shows that the effect of mutual interference on predator–prey dynamics depends on the value of m and the prey rate of increase in the absence of all factors other than predation (λ) (Hassell & May 1973; Hassell 1981). An increase in m leads to an increase in stability, and an increase in λ leads to less stable interactions. The other term in the model above, Q, affects the equilibrium level but not stability.

Hassell et al. (1976) pointed out that it is difficult to predict the impact of interference on particular predator–prey interactions because, although the value of m can be determined from laboratory studies, the value of λ is difficult to acquire. As we pointed out in Chapter 4, λ is not the same as fecundity but includes the effects of birth, movement, and mortality factors other than the interaction being modeled (predation or parasitism). In various predator–prey models, interference seems to enhance stability (Arditi et al. 2004), though it may result in generally low predator densities. However, the application of interference in predator–prey models has been criticized by several researchers (e.g. Free et al. 1977; Hassell 1981). The major criticism is that laboratory studies, conducted in atypically homogeneous environments at atypical natural enemy densities, have exaggerated the value of m and hence the role of mutual interference. Some field studies have also suggested that mutual interference may not be strong (Cronin & Strong 1993).

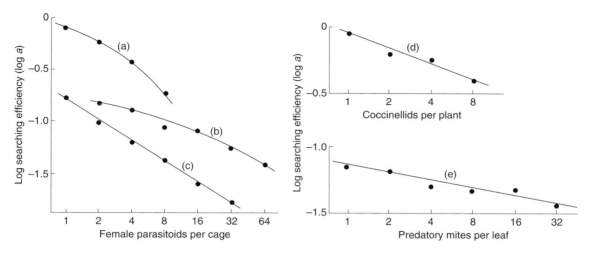

Figure 5.23 Examples of decreasing searching efficiency of predators and parasitoids as their density increases: (a) *Pseudeucoia boche* (Bakker et al. 1967); (b) *Encarsia formosa* (Burnett 1958); (c) *Nemeritis canescens* (Hassell 1971); (d) *Coccinella semtempunctatus* (Michelakis 1973); and (e) *Phytoseiulus persimilis* (Fernando, unpublished in Hassell 1981). (From Hassell 1981.)

5.4.7 Aggregative responses by natural enemies to prey distribution

Most prey populations under natural conditions have a clumped distribution but the Nicholson–Bailey and related models implicitly assume that predators and parasitoids will not respond to this pattern of distribution (Hassell 1981). Instead, these models assume that natural enemies spend the same amount of time in each (equal-sized) patch of habitat irrespective of the density of prey in that patch. Many studies have shown that this is not true and that natural enemies show an aggregative response to the density of their prey, spending a disproportionately large amount of time in patches of high prey density (Figure 5.24).

Spatial variation in host density and the response of natural enemies to it can be modeled by running a modified Nicholson–Bailey model separately for a series of host patches (Hassell 1981). In addition, the distribution of the predators or parasitoids among these patches is modeled using the following equation (Hassell & May 1973):

$$\beta_i = c\alpha_i^{\mu} \qquad (5.13)$$

where β_i is the proportion of the predator population in patch i, c is a constant, α_i is the proportion of the prey population in patch i, and μ is an aggregation

index. The value of μ varies from $\mu = 0$, corresponding to random search, to $\mu = 1$, where natural enemies are distributed in proportion to prey density, to $\mu = \infty$ (infinity), where all predators or parasitoids are concentrated in the highest host density patch and other patches are complete refuges. Values of μ between 1 and ∞ represent increasing levels of predator clumping.

The effect of incorporating the aggregation equation in the Nicholson–Bailey model depends on the value of μ (the natural enemy aggregation index), the prey distribution, and the prey rate of increase as follows (Hassell & May 1973; Hassell 1981):

1 An increase in aggregation by natural enemies in high-density host patches leads to an increase in stability.

2 The more clumped the host distribution, the more likely the interaction is to be stable; but if the host is more or less evenly distributed amongst patches, the interaction will be unstable, irrespective of the degree of aggregation by the natural enemy.

3 An increase in the prey rate of increase tends to decrease stability.

In other words, when natural enemies tend to aggregate in patches where the density of their prey is highest, they will tend to encounter prey at a higher rate than if they were searching randomly. This results in patches with low densities of prey becoming

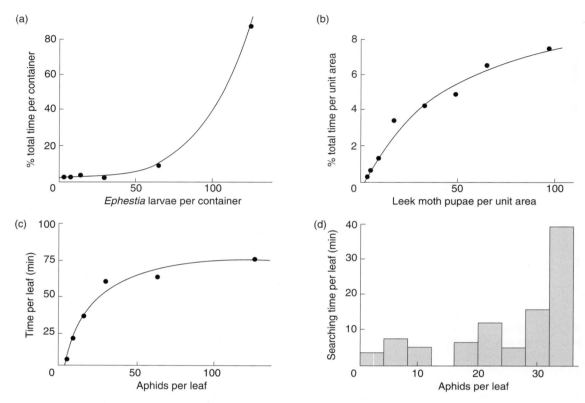

Figure 5.24 Examples of aggregative responses by parasitoids and predators: (a) *Nemeritis canescens*, an ichneumonid parasitoid of flour moth larvae, *Ephestia cautella* (Hassell 1971); (b) *Diadromus pulchellus*, an ichneumonid parasitoid of leek moth pupae, *Acrolepia asseciella* (Noyes 1974); (c) *Diaereliella rapae*, a braconid parasitoid of the aphid *Brevicoryne brassicae* (Akinlosotu 1973); (d) *Coccinella septempunctata*, a predator of *B. brassicae* (M.P. Hassell, unpublished in Hassell & May 1974). (From Hassell & May 1974.)

partial refuges from predation or parasitism, and it is this that contributes towards stability (Hassell 1981). This can be illustrated by comparing the relationship between searching efficiency (a) and predator density (P_t) for natural enemies with (i) a fixed aggregation strategy ($\mu = 1$); (ii) a random search strategy ($\mu = 0$); and (iii) an optimal foraging strategy (Figure 5.25). When predators are following a fixed aggregation strategy, i.e. they are distributed in direct proportion to prey density, their searching efficiency is greater than under a random searching strategy while predator densities are low. However, as predator densities increase, searching efficiency decreases. This effect is similar to the effect of interference on searching efficiency discussed above and has been referred to as "pseudo-interference" (because it may be wrongly attributed to mutual interference; Free et al. 1977). Pseudo-interference is the result of the aggregation of large numbers of natural enemies, not mutual interference, although both are likely to occur in these circumstances. Incidentally, another problem may arise at high enemy (parasitoid) densities: superparasitism. Superparasitism occurs when relatively large numbers of parasitoids are searching for hosts, and lay eggs in hosts already parasitized (Chen et al. 2006).

Real predators and parasitoids are unlikely to remain in high-density host patches when the number of predators or parasitoids reaches a level where their rate of prey capture is poorer than if they were adopting a random searching strategy. Instead, they may adopt an optimal foraging strategy (e.g. Comins & Hassell 1979). Optimal foraging behavior has mostly

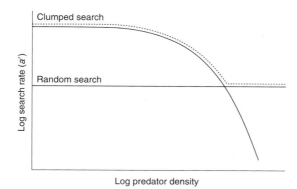

Figure 5.25 The effects of different searching strategies on the relationship between searching efficiency and predator density. The broken line shows the switch between clumped and random search at high densities that an optimally foraging or "prudent" predator would adopt. (From Hassell 1978.)

been studied with vertebrate predators in mind, but it has relevance to invertebrate predators and parasitoids too. Optimal foraging theory predicts that an optimal predator should forage preferentially in patches that are rich in food items and only forage in less profitable patches when the availability of good-quality patches is low (e.g. Royama 1970). In addition, an optimal predator should remain in a patch until its rate of capture falls below the average rate of food capture in the habitat as a whole. Thus in the case illustrated in Figure 5.25, optimally foraging predators may be predicted to change their foraging strategy to minimize the effect of an increase in predator density.

Comins and Hassell (1979) constructed a model for optimally foraging predators and parasitoids. They found that the outcome was qualitatively similar to the model (Hassell & May 1973) developed for natural enemies with fixed aggregation in patches of high host density. The aggregation model appears then to be largely adequate for describing the population dynamics of natural enemies and their prey. The key aspects of that model are that it has patches of prey and predators distributed in space, and incorporates the aggregative behavior of natural enemies. In combination, this leads to spatial density dependence: natural enemies aggregate in areas of high prey density and cause a disproportionate amount of mortality in those patches. This approach to modeling is important not only because it started the development

of other spatially explicit models (see below) but also because it encouraged insect ecologists to include the spatial dimension in considering the temporal fluctuations in the abundance of insects.

5.4.8 The development of spatial models of natural enemy–prey interactions

A further development in modeling predator–prey interactions in space (as well as in time) has been the use of cellular automata models (Comins et al. 1992; Gillman & Hails 1997). In cellular automata models, predator–prey dynamics take place on a grid of cells (e.g. Figure 5.26a) linked by intergenerational dispersal. For example, in the model constructed by Comins et al. (1992), interactions between parasitoids and their prey were modeled in each grid cell using the Nicholson–Bailey model. Dispersal among cells in the square grid ($n \times n$ cells in size) occurred once per generation, and was modeled with the following rules:
1 In a dispersal phase, both a fraction of the adult hosts and a fraction of the adult parasitoids leave the cell in which they emerged and the rest remain and reproduce in it.
2 The dispersing hosts and parasitoids move to the eight cells adjacent to the cell from which they emerged in equal numbers (in most other similar studies, dispersing hosts and parasitoids have been distributed over the whole grid according to a specified rule).
3 Hosts and parasitoids emerging from the cells at the edge of the grid, return to the grid in so-called reflective boundary conditions (for example, individuals that would have moved southeast from the middle of the east side of the grid move one cell south and individuals that would have moved southeast from the southeast corner of the grid stay in that cell).
What were the results of the Comins et al. (1992) model? In small grids (where $n = 10$ cells or less), the parasitoid and its host became extinct within a few hundred generations. However, with larger grid sizes ($n = 15$–30 cells), the model produced three types of pattern that developed in space: "spirals", "spatial chaos", and "crystal lattices" (Figure 5.26). These results are important for two reasons. First, they suggest that predator–prey interactions can generate complex patterns in space even when the environment itself is homogenous. For ecologists, this means that spatial variation in insect densities

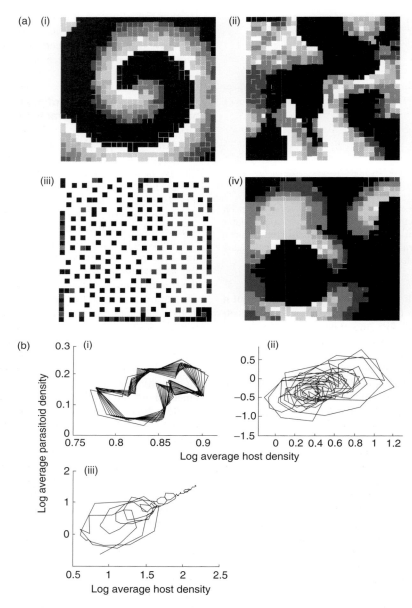

Figure 5.26 Dynamics of an insect host and its parasitoid in a spatial grid of 30 by 30 cells linked by dispersal with Nicholson–Bailey model dynamics ($\lambda = 2$) in each cell (apart from iv—see below). (a) Instantaneous maps of population density with different levels of density of host and parasitoid represented by different levels of shading (black, empty patches; dark shades becoming paler, patches with increasing host densities; light shades; patches with increasing parasitoid densities) resulting in (i) "spirals", (ii) "spatial chaos", (iii) "crystalline structures", and (iv) highly variable spirals from Lotka–Volterra model dynamics. (b) Host–parasitoid density ("phase plane") plots showing dynamics over time with the same parameters as (a). (From Comins et al. 1992; Gilman & Hails 1997.)

does not necessarily result from spatial variation in the environment – predation might be responsible. To date, complex spatial patterns of the types predicted by cellular automata models have been observed in populations of western tussock moth (Maron & Harrison 1997) and larch budmoth (Bjornstad et al. 2002). Second, the Comins et al. (1992) model predicted long-term persistence of the host and parasitoid, despite the unstable nature of the basic Nicholson–Bailey model used in each of the cells. In other words, an interaction that is unstable at a single point in space may actually be stable when dispersal among patches can occur.

5.4.9 Natural enemy–prey models: conclusions

As we pointed out in Section 5.4.1, it is easy to be skeptical of the value of mathematical models in insect ecology. Nevertheless, population models have given us many insights into how insects interact with predators and parasitoids. How else could the relevance of laboratory and field experiments on predators and parasitoids be demonstrated without the use of population models? Lotka and Volterra showed that it was possible to model the interactions between predators and their prey, and since then ecologists such as Nicholson, Holling, Hassell, Comins, and others have worked to inject ecological realism into population models. These ecologists have shown that the long-term persistence of insects and their natural enemies may arise as a result of the behavior of the natural enemies (aggregating in high-density patches of their prey), other density-dependent factors acting on the prey populations, or by the dispersal behavior of natural enemies and their prey (Section 5.4.8). Some of the "analytical" approaches to modeling natural enemy–prey interactions not dealt with here have been discussed by Jervis and Kidd (1996) and Gillman and Hails (1997).

5.4.10 Phenomenological models and time-series analysis

Mechanistic predator–prey models built upon the Nicholson–Bailey skeleton have provided us with an immense amount of information on the potential roles of parasitoid behavior in the dynamics of predator–prey interactions. However, insect ecologists are often limited in the information that they have about a particular species of interest. For example, we may have annual counts of a crop or forest pest, but no information on densities of natural enemies, never mind their functional and numerical responses to variations in pest density. Under these circumstances, it can be difficult or impossible to build accurate mechanistic models with such limited information. Fortunately, there are some statistical techniques that can be used to analyze long-term population counts (or "time series"). Models built from such analyses generally fall under the category of "phenomenological models" because we can study the phenomena without a complete understanding of the potential mechanisms generating the phenomena. The analytical techniques used to build these models fall under the general category of time-series analysis.

Time-series analysis *sensu stricto* (Box & Jenkins 1976) has a long history of use in a variety of disciplines including engineering, social sciences, mathematics, and climatology. We are unable to cover the techniques in detail here, but will focus on its primary use in insect ecology: the detection of long-term patterns in time-series data, particularly population cycles and delayed density dependence (e.g. Berryman 1999). Recall from Section 5.3.4 that ecologists have often been unable to detect density dependence in their analyses of insect life tables, in large part because life tables are not designed to detect delayed density dependence (density dependence operating on a time lag). The application of time-series analysis by Peter Turchin to long-term data of 14 species of forest pest was a key demonstration of the value of the technique. Turchin (1990) showed that eight of the insect species exhibited delayed density dependence, whereas only one of these exhibited direct density dependence using standard techniques. One of his analyses, for the larch budmoth *Zeiraphera diniana* (Lepidoptera: Tortricidae), is shown in Figure 5.27. Notice first that the time series for the larch budmoth appears to show very regular cycles (Figure 5.27a) – we might therefore expect that the population is under the influence of delayed density dependence (see Section 4.1.1). Second, we can look at the "autocorrelation function" (ACF, Figure 5.27b) which describes the statistical relationship (correlation) between population density at time t and densities at various times in the past ($t - 1$, $t - 2$, etc.). Again, we see that the putative population cycle is

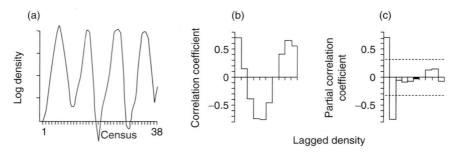

Figure 5.27 Time-series analysis of the larch budmoth, *Zeiraphera diniana*, populations in the Swiss Alps. The time-series (a) and autocorrelation function (b) provide evidence for cyclic dynamics. The partial autocorrelation function (c) suggests that there is density dependence operating with a lag of $t-2$. (From Turchin 1990.)

represented as the rise and fall of the correlation between current density and density in the past. The correlation varies systematically from positive to negative to positive again, reflecting the periodicity of the cycle. Finally, we can study the partial auto-correlation function (PACF, Figure 5.27c) which considers "partial correlations" between current density and density at times in the past. It provides information on the dominant time lag on which density dependence might be operating. More simply, we look for the largest negative value on the PACF, which occurs at time $t - 2$. From this we can infer that delayed density dependence is operating with a time lag of 2 years.

We can go one step further with this analysis. We now know that, if we wish to model the dynamics of the larch budmoth, we need a model that incorporates a delayed density-dependence term. What would that model look like? There are a variety of possible options available to us, but the simplest builds on something we have seen earlier. We saw in Chapter 4 that we could model the dynamics of univoltine insects with discrete time models of the form given in equation 4.2. Recalling that $\lambda = e^r$, we can rewrite equation 4.2 with density dependence:

$$N_t = N_{t-1}e^{r-\alpha N_{t-1}-\beta N_{t-2}} \qquad (5.14)$$

Here, the terms α and β represent the strength of rapid (α) and delayed (β) density dependence operating on the population. Notice that the term β fulfills the requirement of a $t - 2$ lag in density dependence that was detected in the PACF (Figure 5.27c). We can use statistical methods to estimate the values of r, α and β

from the original time-series data, and this is exactly what Turchin (1990) did. His analysis provided the following final model:

$$N_t = N_{t-1}e^{1.2-0.0001N_{t-1}-0.02N_{t-2}} \qquad (5.15)$$

In simple terms, equation 5.15 says that population change of the larch budmoth depends only weakly on density 1 year previously, but strongly on density 2 years previously. A key point here is that it is a phenomenological model. We have no idea what mechanisms underlie the delayed density dependence. It could result from the action of natural enemies (see above) or from induced changes in plant quality (see Chapter 3), and in fact both of these mechanisms have been proposed to explain population cycles in larch budmoth (see below). So the strength of the approach is that it provides a detailed analysis of delayed density dependence and cycles, while the weakness of the approach is that it tells us little about mechanisms. Interestingly, a more recent analysis of the larch budmoth data suggests that both variation in plant quality and parasitism operate on dynamics, but that parasitism is the dominant effect (Turchin et al. 2003). The larch budmoth is used as a case study in Section 5.5 below.

One further advantage of the time-series approach is that it is flexible. More data can be added as they become available, so that numbers of parasitoids or weather variables can be incorporated into the models. In a recent example, the population dynamics of the gall midge, *Taxomyia taxi* (Diptera: Cecidomyiidae), on yew trees was modeled as a function of gall density, parasitoid density, and rainfall using time-series

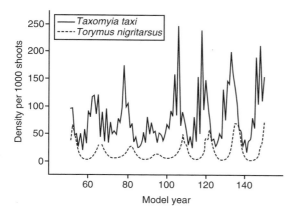

Figure 5.28 A stochastic time-series model of the dynamics of the galling fly, *Taxomyia taxi*, and its parasitoid, *Torymus nigritarsus*, on yew trees in England. The model includes the effects of rainfall and parasitism on galler dynamics, and accurately reflects the 18-year cycles in galler abundance. (From Redfern & Hunter 2005.)

analysis (Redfern & Hunter 2005) (Figure 5.28). However, caution is always necessary with such a non-mechanistic approach. It is based largely on statistical procedures from which we cannot demonstrate cause and effect. In a relatively few cases, spurious detection of delayed density dependence can occur because of regular oscillations in external forces such as weather (Hunter & Price 1998, 2000; Jiang & Shao 2003; Price & Hunter 2005). However, such cases are likely in the minority, and time-series analysis should be seen as a powerful addition to the insect ecologist's tool box, especially when it is used in conjunction with experimental and life table approaches (Hunter 2001a).

5.5 SYNTHESIS: COMBINING THE IMPACTS OF NATURAL ENEMIES AND OTHER FACTORS ON INSECT POPULATION DYNAMICS

5.5.1 The "top-down" versus "bottom-up" debate

Trying to assess the relative importance of regulatory factors from trophic levels above the one under consideration ("top-down") and those from the trophic level

below ("bottom-up") has provided much healthy debate and research in ecology. In a still influential paper written over 45 years ago, Hairston et al. (1960) argued that because the world is covered with a profusion of green plants, herbivores must be having little impact on their host plants, particularly in natural ecosystems. They concluded that herbivores must be kept at insignificant levels by natural enemies. The many instances of successful biological control, observations and experiments on natural enemies, and some of the theoretical models discussed in this chapter have supported this top-down view of population dynamics. However, other ecologists have also argued that plants play the dominant role in determining herbivore numbers because they vary so much in quality (see Chapter 3) and set a resource limit, or ceiling, to population growth (see Chapter 4) (Dempster & Pollard 1981). Insect numbers cannot increase past the number theoretically sustained by this ceiling and will be increasingly affected by intraspecific competition as they approach the ceiling.

Dempster and Pollard (1981) argued that there was little evidence to support the models that predict regulation around an equilibrium density (see examples above). One of the examples they gave was the cinnabar moth, *Tyria jacobaeae* (Lepidoptera: Arctiidae), whose abundance generally parallels the biomass of its host plant, ragwort (Figure 5.29a). The population dynamics of many species appear to fit the "ceiling" model, sometimes in subtle ways. For example, Auerbach (1991) found that natural enemies played an insignificant role in the population dynamics of the aspen blotch miner, *Phyllonorycter salicifoliella* (Lepidoptera: Gracillariidae) and concluded that the abundance of this insect each year approached a ceiling set by the numbers of young leaves available to the adult stage during oviposition. In this case, therefore, the ceiling was not set by the total amount of foliage that might be wrongly assumed to be available to this insect, but only the number of nutritious, young leaves. Dempster and Pollard (1981) extended their argument to include the abundance of natural enemies, suggesting that their numbers too are determined by a ceiling set by the abundance of their insect hosts (Figure 5.29b). There are, however, many examples showing that abundance is not always determined by resource limitation.

Another way that the interaction between herbivores and their host plants may regulate insect

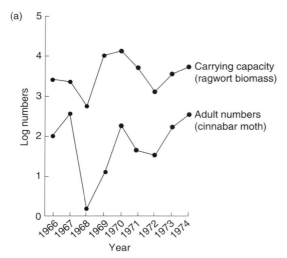

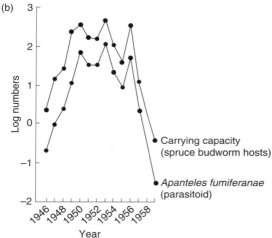

Figure 5.29 The influence of resource limitation on herbivores and natural enemies: (a) abundance of cinnabar moth *Tyria jacobaeae* adults and biomass of its host plant, ragwort, the determinant of the carrying capacity of its habitat; and (b) abundance of the parasitic wasp *Apanteles fumiferanae* and its host, spruce budworm (*Choristoneura fumiferana*). (From Dempster & Pollard 1981.)

numbers is through density-dependent changes in birth and mortality rates brought about by herbivore-induced changes in plant quality (see Chapter 3 and Section 4.5.2). This possibility was first suggested by Haukioja (1980), who argued that the population cycles of insects such as the autumnal moth, *Epirrita*

autumnata (Lepidoptera: Geometridae), in Scandinavia are the result of an insect-induced deterioration in the quality of birch as the abundance of the autumnal moth increases. Host plants may also play a significant role in determining insect numbers through, for example, phenological asynchrony and stress-mediated changes in plant quality (see Chapter 3). The value of insect-resistant crop varieties (see Chapter 12) also shows that bottom-up factors can determine insect numbers. Life tables (Section 5.3.4) may be used to assess the relative importance of natural enemies and plant factors as causes of insect mortality. Cornell and Hawkins (1995) examined 530 life tables for 124 insect herbivores and found that natural enemies were a more frequent cause of mortality for immature herbivores (48% of cases) than plant factors (9%). However, they suggested that life table studies have probably underestimated the importance of plant factors.

5.5.2   Combined effects and multiple equilibria

The top-down versus bottom-up debate as presented above was a dichotomy between regulation of insect numbers by natural enemies or by the host plant, respectively. Clearly, neither model applies in all cases. More importantly, we now recognize subtle ways in which the effects of plants and enemies can combine to determine the abundance of insect herbivores, and in particular, over varying spatial scales (Gripenberg & Roslin 2007). In the majority of cases, there seems to be a role for both top-down and bottom-up effects (Hunter & Price 1992; Bonsall et al. 1998; Raworth & Schade 2006). We can combine the effects of variable plant quality and natural enemies into graphic (and mathematical) models that explore population regulation and multiple equilibria (Hunter 2001a). An example of such a graphic model is shown in Figure 5.30. In this simple case, we have ignored immigration and emigration, and focused only on rates of mortality and natality (birth). Notice first, in Figure 5.30a, that mortality acting upon the herbivore can result from both natural enemies and competition for resources. Natural enemies are assumed to act in a density-dependent fashion over lower ranges of herbivore density (increasing portion of the predation curve) and in an inverse density-dependent fashion over higher ranges of herbivore density (decreasing portion

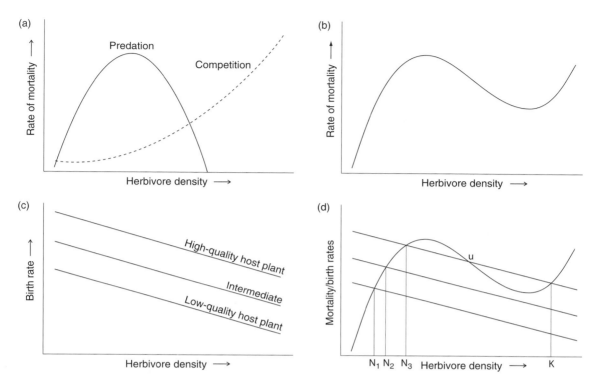

Figure 5.30 Model of the combined role of different factors to determine equilibrium densities in insect herbivores: (a) separate relationships between the mortality caused by predation and competition and herbivore density; (b) combined mortality due to predation and competition; (c) the effects of host-plant quality and herbivore density on herbivore birth rate; and (d) the combined role of mortality and birth rate. Note that, even when predation maintains the herbivore population below its carrying capacity, variation in food quality still causes variation in equilibrium density (N_1, N_2, and N_3). N_1, N_2, N_3, and K are theoretically stable equilibria and u is an unstable equilibrium. For further details, see text. (From Hunter 1997.)

of the predation curve). This switch from density dependence to inverse density dependence is predicted to occur as a result of satiation, handling time, or reproductive limitation (Section 5.4.5). In contrast, the competition curve shows increasing strength at higher herbivore densities, as we might expect for resource limitation when herbivore numbers get large. The combined effects of predation and competition on herbivore mortality are shown in Figure 5.30b, which is simply the sum of the two previous curves. Now we can deal with birth rate (Figure 5.30c). We assume that birth rate is density dependent, so that it declines as herbivore densities increase. Notice also that there are plants of three different qualities, with higher birth rates on the high-quality plants. In Figure 5.30d, we can combine birth rates

and death rates together to explore population equilibria for the herbivore.

The key to understanding Figure 5.30d is that, whenever the death rate line intersects the birth rate lines, we have an equilibrium. This occurs simply because, if death rates equal birth rates, then the population does not change. First consider the birth rate lines for plants of low and intermediate quality. These lines only cross the death rate curve at one point each (N_1 and N_2) and both are to the left of the top of the natural enemy curve (in the density-dependent portion). This means that, on plants of low or intermediate quality, there is only a single equilibrium each, where the populations are regulated by the natural enemy. There is an important point here – even though the herbivore population is regulated by predation, a change in plant quality still causes a change in the size of the equilibrium population. Note that the equilibrium for the intermediate quality plant (N_2) is higher than that of the low-quality plant (N_1). The general message from this comparison is that the equilibrium population size is set by both top-down (predation) and bottom-up (plant quality) forces, even when the population is regulated by the natural enemy.

Now consider the birth rate line for the high-quality host plant on Figure 5.30d. It intersects the death rate curve at three points, generating three equilibria. There is a low-density equilibrium (N_3) where the herbivore population is regulated by the natural enemy. There is a high-density equilibrium (K) where the population is limited by competition for resources – it is in the inverse density-dependent portion of the predation curve, where regulation by natural enemies is no longer possible. Finally, there is an unstable equilibrium (u) which lies between the two stable equilibria. Why is the middle equilibrium unstable? Moving slightly to the right of u, notice that the birth rate exceeds the death rate. When there are more births than deaths, the herbivore population will increase until it reaches the new equilibrium, K. Likewise, moving slightly to the left of u, the death rate exceeds the birth rate, and so the population will decline until it reaches the lower equilibrium, N_3. The upper and lower equilibria (K and N_3) are both locally stable because they act as attractors. Slight deviations to the right or left of K or N_3 will cause the populations to return back to those equilibria.

The existence of multiple equilibria provides the potential for populations to switch between predator regulation and regulation by competition for resources. For example, a herbivore population on high-quality plants might be kept at low density (N_3) by natural enemies for some extended period. However, if immigration of herbivores increased their density above the unstable equilibrium u, the population would then increase to the upper equilibrium, K. Changes in weather conditions, changes in plant quality, or other factors that influence the densities of herbivores and their enemies might cause the herbivore population to "switch" from one equilibrium density to the other. Insect ecologists now believe that some herbivore outbreaks occur as a result of switching between upper and lower equilibrium densities. For example, grasshopper populations in the grasslands of the United States are thought to vary between predator

regulation and competition for resources based on the amount of rainfall (Belovsky & Joern 1995). As a final "thought experiment", imagine that weather conditions caused the high-quality plants in Figure 5.30d to become of extremely high quality. The birth rate line might rise vertically on the graph such that there was a single equilibrium to the far right (at high density) where only regulation by resource limitation was possible. This seems to be the case in some years for aphids living on milkweed plants in the southeastern United States (Helms et al. 2004).

So what are the take-home messages from such analyses? First, equilibrium densities of insect populations are set by both top-down and bottom-up forces, even when one or the other is the dominant force regulating the population. Second, the existence of multiple equilibria suggests that populations might vary from predator regulation to resource limitation depending upon spatial and temporal variation in environmental conditions. The top-down versus bottom-up view of herbivore population dynamics is a false dichotomy. It is much more interesting and informative to explore the interactions between predation pressure and resource limitation as they act on insect population dynamics. The following sections describe a few examples where insect ecologists have tried to tease apart the dynamics of insect populations, using a range of approaches, to identify how different ecological factors interact to determine insect population change.

5.5.3   Spruce budworm

In the previous section, we suggested that changes in environmental conditions, particularly plant quality, could act to shift insect populations from predator limitation (top-down) to resource limitation (bottom-up) (see Figure 5.30d). Outbreaks of the spruce budworm, *Choristoneura fumiferana* (Lepidoptera: Tortricidae), appear to provide an example of exactly that (see Plate 5.2, between pp. 310 and 311). Outbreaks become increasingly likely as forests age (Peterman et al. 1979), in large part because of increases in spruce budworm recruitment (realized birth rates). In young, or "immature", forests the recruitment rate is low because forest conditions are relatively unfavorable for the spruce budworm. As the forest ages, it becomes more suitable for the spruce budworm but outbreaks are prevented by predation unless large numbers of moths immigrate into the forest ("intermediate" forest). Eventually, however, the ability of predators to prevent an outbreak disappears ("mature" forest) because recruitment rates are simply too high for regulation by predators. In general, when an outbreak does occur in a mature forest, its impact is strongly constrained by forest characteristics (stand composition and structure) (Bouchard et al. 2006). Thus, over decadal timespans, we see a change from top-down regulation to bottom-up regulation, contingent upon the age and composition of the forest. Spruce budworm outbreaks can be sufficiently severe to significantly influence forest succession (Sturtevant et al. 2004), and in fact to reset the successional stage of the forest to one that again favors regulation by natural enemies. Long-term cyclic dynamics in the system are ultimately driven by the interactions among forest age, insect herbivore recruitment rates, and natural enemies.

5.5.4 Larch budmoth

The larch budmoth, *Zeiraphera diniana* (Lepidoptera: Tortricidae), shows 8–10-year population cycles in the Swiss Alps (Turchin et al. 2003; see also Chapter 1) (Figure 5.31). It has been suggested that regional outbreaks begin in local, specific, foci from which they spread into adjoining areas, the so-called "epicenter hypothesis" (Johnson et al. 2004). Several factors have been suggested to cause population cycles, including delayed density-dependent parasitism and induced plant resistance (Benz 1974; Baltensweiler & Fischlin 1988). Several parasitoids attack the larch budmoth; different species appear to attack the budmoth at different stages of the population cycle (Delucchi 1982). Feeding damage by the larch budmoth results in a decline in larch needle quality and this is thought to cause an increase in larval mortality. Larch budmoth damage also results in a decline in the quantity of available foliage, and indeed the overall growth patterns of larch (Nola et al. 2006).

As we learned in Section 5.4.10, phenomenological models of larch budmoth dynamics are useful for describing the dynamics of the system, but do not provide information on the factors and mechanisms driving the population cycles. Van den Bos and Rabbinge (1976) constructed a simulation model to try to explain the cycles of the larch budmoth. They incorporated the three factors mentioned above in

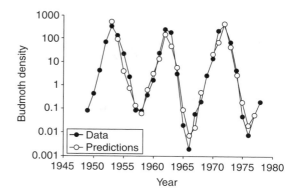

Figure 5.31 Observations and model predictions of population changes in the larch budmoth, *Zeiraphera diniana*, in the Swiss Alps. The model includes only the interaction between the budmoth and its parasitoids (see Table 5.2 for model choices). (From Turchin et al. 2003.)

their model: parasitism, changes in host-plant quantity, and induced changes in host-plant quality. They concluded that all factors were needed to explain the population cycles. More recently, Turchin et al. (2003) explored the mechanisms driving larch budmoth dynamics in some detail. They began by developing five different mechanistic models (Table 5.2) which varied from effects of plant quality alone, to effects of parasitoids alone, to combinations of plant quality and parasitism. Although the models in Table 5.2 may look complicated, do not be discouraged. They are all modifications of models that you have seen in preceding sections, and can be readily understood with reference to a detailed text on population modeling (e.g. Gillman & Hails 1997). The key point is that the models were built before the data were analyzed so that the investigators could be sure of the mechanisms inherent in each model. The data were then fit to each model in turn using non-linear time-series analysis, and the fits of the models compared to the real data. The model containing only the parasitoid–host interaction provided an excellent fit to the data, explaining about 90% of the variation in larch budmoth densities over time. The full tri-trophic model, including both parasitoids and plant quality, provided only a marginal improvement. These analyses suggest that, while variation in plant quality and parasitism both contribute to larch budmoth population dynamics, the effect of the parasitoids is much stronger than that of plant quality.

Table 5.2 Five alternative models used to analyze long-term data of larch budmoth *Zeiraphera diniana* populations in the Swiss Alps. While the full tri-trophic model was a significantly better fit to the data, the relatively simple parasitoid–host model explained about 90% of the variance in budmoth densities over time. (From Turchin et al. 2003.)

Model and parameters	Equation
Plant quality I (the "non-linear" version) N_t: budmoth density	$N_{t+1} = \lambda_0 N_t \frac{Q_t}{\delta + Q_t} \exp[-gN_t]$
Q_t: plant quality index	$Q_{t+1} = (1-\alpha)\left(1 - \frac{N_t}{\gamma + N_t}\right) + \alpha Q_t$
Plant quality II (alternative N_t equation, the "linear" version) N_t: budmoth density L_t: needle length	$N_{t+1} = N_t \exp[u + vL_t - gN_t]$
Parasitoid–host N_t: budmoth density	$N_{t+1} = \lambda_0 N_t \exp\left[-gN_t - \frac{aP_t}{1 + ahN_t + awP_t}\right]$
P_t: parasitoid density	$P_{t+1} = \phi N_t \left\{1 - \exp\left[-\frac{aP_t}{1 + ahN_t + awP_t}\right]\right\}$
Tri-trophic I (full model) N_t: budmoth density	$N_{t+1} = \lambda_0 N_t \frac{Q_t}{\delta + Q_t} \exp\left[-gN_t - \frac{aP_t}{1 + ahN_t + awP_t}\right]$
P_t: parasitoid density	$P_{t+1} = \phi N_t \left\{1 - \exp\left[-\frac{aP_t}{1 + ahN_t + awP_t}\right]\right\}$
Q_t: plant quality index	$Q_{t+1} = (1-\alpha)\left(1 - \frac{N_t}{\gamma + N_t}\right) + \alpha Q_t$
Tri-trophic II (simplified "linear" version) N_t: budmoth density P_t: parasitoid density L_t: needle length	$N_{t+1} = N_t \exp\left[u + vL_t - \frac{aP_t}{1 + awP_t}\right]$

5.5.5 Lime aphid

Although building detailed models of the population dynamics of univoltine insects such as the spruce budworm and the larch budmoth is not easy, it is often harder to analyze the even more complex dynamics of multivoltine insects such as aphids. One example is provided by the work of Barlow and Dixon (1980) on the lime aphid, *Eucallipterus tiliae* (Hemiptera: Aphidae) (see Plate 5.3, between pp. 310 and 311). The fluctuations of this aphid were monitored for 8 years on lime (*Tiliae* × *europaea*) trees in Glasgow, Scotland. The lime aphid has four or five generations per year from fundatrices, which emerge from eggs in April–May, through to oviparae, which lay the next cohort of eggs in September–October. The

generations of the lime aphid rapidly overlap and the population assessments carried out by Barlow, Dixon, and coworkers (Figure 5.32a) suggest very complicated population dynamics with no consistent pattern of within-year fluctuations in abundance. However, patterns in the dynamics of the lime aphid emerge when the trends in the abundance of fundatrices and oviparae are examined. There is a clear positive relationship between the number of oviparae in one year and the number of fundatrices the following year. In contrast, there is a negative relationship between the number of fundatrices in one year and both the number of oviparae later that year and the number of fundatrices the following year (Figure 5.32b).

Barlow and Dixon constructed a simulation model to try to explain the within-year and between-year

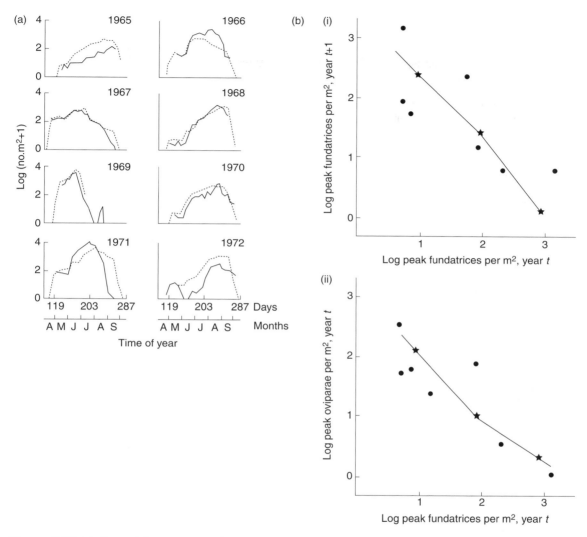

Figure 5.32 (a) Observed fluctuations in abundance of the lime aphid from 1965 to 1972 (solid lines) and simulated population density (broken lines). (b) Relationships between: (i) lime aphid densities in successive years (numbers of fundatrices, aphids which emerge from eggs in the spring); and (ii) different aphid generations within the same year (fundatrices and oviparae, aphids which lay eggs in the fall). Observed data are shown as open circles, and simulation model results are shown as solid lines with stars. (From Barlow & Dixon 1980.)

fluctuations in lime aphid numbers. They considered various factors including:

- predation by the two-spot ladybird, *Adalia bipunctata* (Coleoptera: Coccinellidae);
- predation by the black-kneed capsid, *Blepharidopterus angulatus* (Hemiptera: Miridae);

- the effects of plant quality (specifically, seasonal variation in plant nitrogen concentration) on growth, development, and reproduction;
- the effects of temperature on the emergence of fundatrices in the spring and the development rate of aphid nymphs; and

• the effects of aphid density (measured in different ways, see below) on reproduction and emigration.

The resulting full-factor model was very successful in simulating the within-year population fluctuations of the lime aphid (Figure 5.32a). Barlow and Dixon then examined the role of each of the major factors affecting the lime aphid by removing each factor in turn from their model. They also examined the role of each factor by running the model with each of these main factors alone. They assessed the impact of the removal or inclusion of each factor by looking at the effect it had on the relationship between the year-to-year changes in the numbers of fundatrices. The factors investigated in this way included the effect of adult density on flight (and hence emigration), the effect of nymphal density on flight, and predation.

When each factor was removed in turn, some factors, such as the effect of current adult density on emigration, were found to have little impact on the year-to-year dynamics of the lime aphid (Figure 5.33). Two factors did have an impact: the effect of nymphal density on emigration and the effect of predation. However, the former appears to affect the lime aphid only at high densities (Figure 5.33b), whereas the latter (predation) appears to have an effect at all densities (Figure 5.33e). When each factor was included alone in turn (Figure 5.34), predation was found to be the only factor that reasonably predicted the year-to-year dynamics of the lime aphid. No factor other than predation was capable of regulating aphid densities alone. However, predation had this regulatory effect only at low and medium densities (Figure 5.34e): at high densities, predation alone was incapable of regulating lime aphid numbers.

From their simulation studies, Barlow and Dixon concluded that different factors interact to regulate lime aphid numbers, and that the relative importance of these factors varies according to the abundance of the lime aphid. This conclusion is very similar to that described in Section 5.5.2.

5.5.6 Insect–plant–natural enemy interactions

The above examples show how different factors (predation, parasitism, competition, variation in

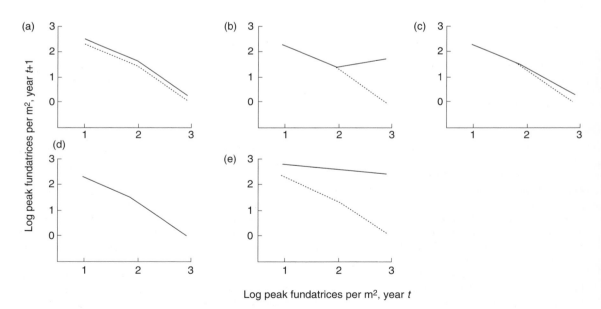

Figure 5.33 The effect of removing different factors from the lime aphid model on the relationship between lime aphid (fundatrix) abundance in successive years: (a) impact of adult density on flight; (b) impact of nymphal density on flight; (c) changes in adult weight; (d) Impact of cumulative aphid density on flight; and (e) predation. In each case the new relationship is shown as a solid line and the original relationship as a broken line. (From Barlow & Dixon 1980.)

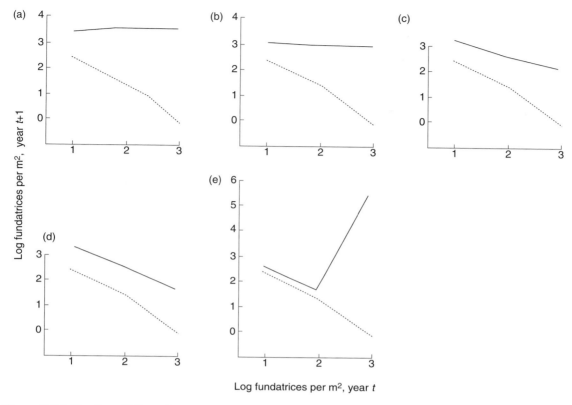

Figure 5.34 The effect of different factors from the lime aphid model acting in isolation on the relationship between lime aphid (fundatrix) abundance in successive years: (a) impact of adult density on flight; (b) impact of nymphal density on flight; (c) changes in adult weight; (d) impact of cumulative aphid density on flight; and (e) predation. In each case the new relationship is shown as a solid line and the original relationship as a broken line. (From Barlow & Dixon 1980.)

host-plant quality) together determine insect population dynamics. One idea that has emerged from studies of plant–herbivore–enemy interactions is the concept of contingency. The idea is a simple one – rather than the effects of one factor (say, natural enemies) on an insect herbivore being consistent in time and space, the effect is contingent upon variation in another ecological factor. For example, the efficacy of a natural enemy may vary depending upon traits expressed by the host plant on which the herbivore is feeding (Price et al. 1980). Such contingent interactions among three trophic levels have been termed "tri-trophic interactions". We learned in Section 3.3.7 that natural enemies can respond to plant traits, including chemical cues, in order to locate their insect herbivore prey. We should therefore not be surprised

that variation in plant traits can influence the success of enemy foraging.

Such is the dependence of parasitoids on plant cues that even a slight genetic change in the plant may result in a decline in parasitism. For example, Brown et al. (1995) demonstrated that very slight genetic changes in host plants can affect the impact of several natural enemies. They measured the effects of two different parasitoids and a mordellid beetle predator on the goldenrod ball gallmaker, *Eurosta solidaginis* (Diptera: Tephritidae), living on *Solidago altissima*, *S. gigantean*, and host races derived from the two ancestral plant species. Observations on the behavior of one species of wasp, *Eurytoma obtusiventris* (Hymenoptera: Eurytomidae), showed that it preferred to search on the ancestral plants rather than on the derived plant races and that parasitism was much higher on the former (30.5%) than on the latter (0.4%). In this case, rates of parasitism are contingent upon genetic variation among plants, and it provides an excellent example of a tri-trophic interaction.

When insect herbivores colonize plants on which enemies do not forage (such as *Eurosta solidaginis* above), they are considered to occupy "enemy-free space" (Jeffries & Lawton 1984; Heard et al. 2006). The colonization of enemy-free space can have a dramatic impact on insect abundance. For example, the pine beauty moth, *Panolis flammea* (Lepidoptera: Noctuidae), feeds on the native Scots pine (*Pinus sylvestris*) in Scotland but colonized lodgepole pine (*Pinus contorta*) after it was introduced as a plantation tree from Canada and the USA. Pine beauty moth is found in low numbers on Scots pine but regularly reaches "outbreak" densities on lodgepole pine. This pattern occurs probably because there are fewer natural enemies in lodgepole pine plantations than in Scots pine forests and parasitic wasps, at least, are poorer at locating their prey on lodgepole pine (Watt et al. 1991).

The physical structure of plants may also affect prey location or prey capture by natural enemies. For example, the parasitoid, *Encarsia formosa* (Hymenoptera: Aphelinidae), is poorer at finding and killing greenhouse whitefly, *Trialeurodes vaporariorum* (Homoptera: Aleurodidae), on cucumber varieties with hairy leaves (Van Lenteren et al. 1995). This is one example of a potential pitfall of selecting plant varieties for resistance to insect pests. Another example is the reduction in parasitism and predation on the wild tomato variety PL 134417, which has glandular trichomes containing the methyl ketones 2-tridecanone and 2-undecanone (Kennedy et al. 1991; Farrar et al. 1992). This variety is resistant to pests such as the tobacco hornworm, *Manduca sexta* (Lepidoptera: Sphingidae), *Helicoperva* (*Heliothis*) *zea* (Lepidoptera: Noctuidae), and the Colorado beetle, *Leptinotarsa decemlineata* (Coleoptera: Chrysomelidae). However, predators and parasitoids such as *Archytas marmoratus* and *Eucelatoria bryani* (Diptera: Tachinidae) are less effective on PL 134417 than on tomato varieties that are susceptible to pest insects. These tachinid parasitoids both attack *H. zea* but have different life history characteristics. *E. bryani* lays its eggs directly into the host, but *A. marmoratus* gives birth to special larvae, called planidia, which attach themselves to passing *H. zea* larvae, penetrate the host

cuticle, and develop in the host pupae. The glandular trichomes of the wild tomato kill *A. marmoratus* planidia and the larval development of both species of parasitoid is deleteriously affected by the methyl ketones consumed by their prey. Experimental removal of the trichomes eliminates the negative effects on these and other natural enemies. Predation and parasitism on hybrid plants with trichomes which do not contain the methyl ketones is intermediate between the susceptible and PL 134417 varieties, indicating that the negative effect of the glandular trichomes (in PL 134417) is partly physical and partly chemical (see also Chapter 12).

Other aspects of the physical structure of host plants that may affect insect natural enemies include bark texture and bark hardness, both of which influence the density of parasitic wasps that attack the bark beetle, *Ips typographus* (Coleoptera: Scolytidae), on spruce (Lawson et al. 1996). Stem toughness may have a general role to play in variation in parasitism rates among host plants. For example, when *Borrichia frutescens* plants are fertilized or shaded, rates of egg parasitism of the plant-hopper *Pissonotus quadripustulatus* increase because plant stems are softer for parasitoids to penetrate in search of eggs (Moon et al. 2000). In contrast, high salinity reduces parasitism because stems become tougher (Moon & Stiling 2000). These two examples from *Borrichia* plants provide evidence that variation in the abiotic environment (shade, nutrients, salinity) can drive variation in plant quality and the form of tri-trophic interactions.

Another general effect of plant structure on natural enemies is the effect of gall size on parasitism of galling insects. Stiling and Rossi (1996) found that parasitism of the gall midge *Asphondylia borrichiae* (Diptera: Cecidomyiidae) on *Borrichia* decreased as gall size increased, largely because the most abundant parasitoid was unable to parasitize insects within large galls. As gall size was determined by plant clone and site conditions, these factors also indirectly determined parasitism. Variation in gall size also influences parasitism of *Eurosta solidaginis* (Diptera: Tephritidae) on goldenrod (Abrahamson & Weis 1997) and parasitism of *Euura lasiolepis* (Hymenoptera: Tenthredinidae) on the Arroyo willow (Price & Hunter 2005).

Perhaps the simplest plant–insect–natural enemy interaction is the effect of plant quality on herbivore quality, and subsequent effects on enemy performance (see Hunter 2003a for a recent review). For example, several entomologists have suggested that insects on relatively poor-quality host plants tend to suffer higher rates of predation than insects on better quality host plants, because of their slower development rates (Price et al. 1980). This has been termed the "slow growth, high mortality hypothesis" (Clancy & Price 1987; Cornelissen & Stiling 2006). Haggstrom and Larsson (1995) tested this hypothesis by comparing the mortality caused by predators of the willow-feeding leaf beetle *Galerucella lineola* (Coleoptera: Chrysomelidae) on *Salix viminalis* and *S. dasyclados*. *G. lineola* develops faster on *S. viminalis* than on *S. dasyclados*, and, in support of the slow growth, high mortality hypothesis, total predation during the larval period and the daily larval predation rate was found to be greater on *S. dasyclados* than on *S. viminalis*. Loader and Damman (1991) obtained similar results in a study of the cabbage white butter-fly *Pieris rapae* (Lepidoptera: Pieridae) on collard plants experimentally treated to be either rich or poor in leaf nitrogen concentration. Not all studies have given support to the slow growth, high mortality hypothesis. Leather and Walsh (1993) found that predation of pine beauty moth larvae was greater on faster growing larvae on better quality host plants. Leather and Walsh compared predation on different provenances of lodgepole pine and, as different provenances are known to emit different mixtures of monoterpenes,

perhaps predators were responding to the chemical composition of the host plants. In reality, there are as many examples contradicting the slow growth, high mortality hypothesis as there are in support of it (Hunter 2003a) but the key point remains – there are often dramatic differences in the mortality imposed by natural enemies on insect herbivores on different host plants. It is likely that physical, chemical, and nutritional factors all contribute to this pervasive pattern.

It is not only insect natural enemies that are prone to feel the effects of tri-trophic interactions. The efficacy of pathogens that attack insects can also vary with host-plant quality. For example, the gypsy moth nuclear polyhedrosis virus (GMNPV) is a major natural enemy of gypsy moth, *Lymantria dispar* (Lepidoptera: Lymantriidae). Defoliation caused by gypsy moth larvae is known to induce chemical changes in the foliage of its host plants that, in turn, reduce gypsy moth performance (Rossiter et al. 1988; see also Chapter 12). However, the chemical composition of oak foliage also affects GMNPV (Keating et al. 1990; Schultz et al. 1990). High foliar tannin concentrations inhibit GMNPV, increasing gypsy moth resistance to the pathogen (Figure 5.35). Thus what may appear to be separate interactions between the gypsy moth and its host plants, and between the gypsy moth

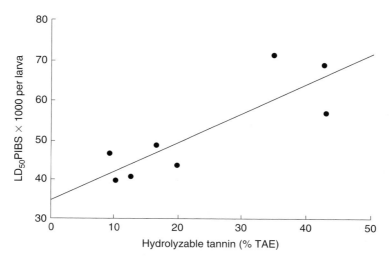

Figure 5.35 Relationship between the concentration of hydrolyzable tannins in oak leaves from eight different locations and the susceptibility of gypsy moth larvae to GMNPV on leaves from the same locations (as measured by LD_{50} tests). TAE, tannic acid equivalents. (From Schultz et al. 1990.)

and one of its natural enemies, is a more complex interaction involving all three. This interaction has been explored in a model by Foster et al. (1992).

5.5.7 Interactions among natural enemies and intraguild predation

In Section 4.6, we described competitive interactions among insect predators. In recent years, a more complex form of interaction among predators, intraguild predation (Polis et al. 1989), has received increasing attention. A simple form of intraguild predation is shown in Figure 5.36. Predator 1 and

predator 2 both consume the same kind of prey, and might be expected to compete for that prey resource. However, predator 2 (the intraguild predator) also consumes individuals of predator 1 (the intraguild prey). Both are nominally in the predator guild, hence the term intraguild predation. In reality, predator 2 is an omnivore, feeding at more than one trophic level. Undoubtedly, it makes sense to feed in this way; supplementing their diet with easily caught and consumed prey in the same guild increases nitrogen uptake and its consequent advantages (Matsumura et al. 2004). Interactions of this type are much more common than was once thought (Denno & Fagan 2003) (Figure 5.37), but their ecological consequences are not yet fully clear (Rosenheim 1998). Will the intraguild predation reduce competition for the common prey? What will be the effects on the dynamics of the prey population? These are important questions both for ecological theory and for applied entomology. For example, if the prey in question is an agricultural pest, and the two predators are potential biological control agents, will the suppression of the prey population be diminished or enhanced by the intraguild predation?

There are at least six potential outcomes of intraguild predation. The first is simply that the three species cannot coexist at a stable equilibrium. This has been explored using analytical models (Holt & Polis 1997) which suggest that coexistence is only possible when the intraguild prey (predator 1 in Figure 5.36) is a superior competitor for the basal prey resource. The other five potential outcomes for the prey population are shown in Figure 5.38, from work by Ferguson and

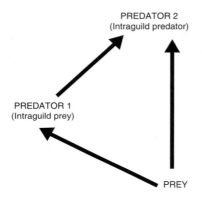

Figure 5.36 A simple illustration of intraguild predation. Predator 1 and predator 2 compete for the same prey item. However, predator 2 also feeds upon predator 1.

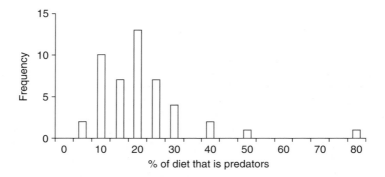

Figure 5.37 The frequency of intraguild predation (percentage of diet that is other predators) of spiders. Almost all spiders incorporate other predators as well as herbivores in their diets. (From Denno & Fagan 2003.)

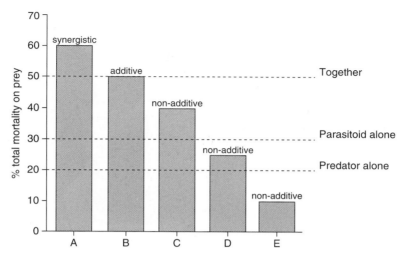

Figure 5.38 Potential outcomes of intraguild predation for the mortality of a shared prey species. Effects can be synergistic (A) such that combined mortality is greater than the sum of the enemies acting alone. Effects can be additive (B), so that combined mortality is simply the sum of the enemies acting alone. Effects can also be non-additive (C–E) such that combined mortality is lower than the sum of the enemies acting alone. (From Ferguson & Stiling 1996.)

Stiling on intraguild predation between a predator and a parasitoid. Possible outcomes range from a higher than expected mortality on the prey population based on the individual effects of predators, to a much lower than expected mortality. Synergistic effects between the enemies result in higher than expected mortality. Additive effects result when total mortality is simply the sum of mortalities imposed by the enemies acting alone. Non-additive effects result when total mortality is less than the sum of mortalities imposed by the enemies acting alone. In the study by Ferguson and Stiling (1996), they found that the combined mortality of a coccinellid predator and a parasitoid on the aphid *Dactynotus* (Hemiptera/Homoptera: Aphididae) was less than that imposed by the parasitoid alone. In other words, suppression of the herbivore population was compromised by the presence of the intraguild predator. In an extreme but surprisingly common case, predators such as ladybirds are actually cannibalistic (Dixon 2007). This in fact reduces the population density of a particular predator, enabling more than one species to coexist.

It is now clear that intraguild predation can have varying effects upon herbivore population size (Rosenheim 1998; Eubanks & Denno 2000). Most likely, effects seen at the herbivore level will depend upon the strength of intraguild predation (Moran et al. 1996). If the intraguild predator feeds preferentially upon the herbivore, then suppression of the herbivore population will likely be sustained. In contrast, if the intraguild predator feeds preferentially on the intraguild prey, herbivore suppression should be diminished. For example, Rosenheim et al. (1993) studied the predatory community associated with the aphid *Aphis gossypii* (Hemiptera/Homoptera: Aphididae) on cotton. They found that generalist hemipteran predators (*Zelus* and *Nabis* species) consumed larvae of the green lacewing, *Chrysoperla carnea* (Neuroptera: Chrysopidae), also a major predator of *A. gossypii*. Experiments showed that predation of lacewing (intraguild prey) by either hemipteran species (intraguild predators) could result in cotton aphid outbreaks. However, in the same cotton system, Colfer and Rosenheim (2001) demonstrated that interactions between the coccinellid *Hippodamia convergens* (Cdeoptera: Coccinellidae) and an aphid parasitoid actually improved suppression of the cotton aphid populations. When several predator species attempt to exploit a single prey species, intraguild predation, and its combined effect on the prey population, can become complex, as shown by the example in Figure 5.39 (Rosenheim 2005). Here, predatory interactions

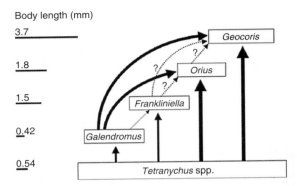

Body length (mm)

3.7

1.8

1.5

0.42

0.54

Figure 5.39 A size-based ladder of intraguild predation among arthropods that prey on spider mites, *Tetranychus* spp., in California cotton. Arrows point from prey to predator, and the width of the arrow is approximately scaled to the strength of the effect of the predator on the prey population. Dashed arrows with question marks indicate interactions that have not been studied. Also shown are the body lengths of the adult female stages of each species. (From Rosenheim 2005.)

Figure 5.40 Intraguild predation in action. The wolf spider, *Pardosa littoralis*, is consuming a mirid egg predator, *Tytthus vagus*. Both species are predators on the plant-hopper, *Prokelisia dolus*. (Courtesy Claudio Gratton, from Finke & Denno 2002.)

among species are a function of body size, and the overall message is that potential suppression of the spider mite pest, *Tetranchychus urticae*, is difficult to predict.

To add to the complexity, recent studies suggest that the strength of intraguild predation varies with plant quality and vegetation structure. For example, both mirid egg predators, *Tytthus vagus* (Hemiptera/ Heteroptera: Miridae), and wolf spiders, *Pardosa littoralis*, feed upon the plant-hopper, *Prokelisia dolus* (Hemiptera/Homoptera: Delphacidae), in coastal salt marsh communities in the eastern United States. The wolf spider also feeds on the egg predator (Figure 5.40), an example of intraguild predation. Work by Finke and Denno (2002) has demonstrated that the effects of predation on the herbivorous plant-hopper depend upon vegetation structure. In structurally simple habitats, without an accumulation of cordgrass thatch, the predators interact antagonistically, with spiders consuming mirids. As a result, predation pressure on the plant-hopper is reduced and their populations grow. In contrast, structurally complex habitats with an accumulation of thatch provide spatial refuges for the mirid egg predator. The intensity of intraguild predation is reduced, thereby increasing the combined effectiveness of the predators in suppressing plant-hopper populations. In Section 5.5.6 we described various ways in which plant

quality can influence the efficacy of natural enemies interacting with their herbivorous insect prey. It now seems likely that the intensity of predator–predator interactions, and intraguild predation, will vary based on the quality and architecture of plants.

The dichotomy between plant-based and enemy-based effects on herbivore populations is no longer workable in an age where we recognize the importance of complex interactions within and among trophic levels, the impacts of climatic variation on species interactions, and the preponderance of multi-trophic interactions in natural and managed systems. The examples that follow are designed to provide readers with a necessarily brief taste of the ecological complexity that can influence plant–insect herbivore–natural enemy interactions.

First, other members of ecological communities can have significant effects upon insect predator–prey relationships. In Section 3.3.4, we described how endophytic fungi can produce toxins that confer resistance to certain species of plants. However, endophytes may also have detrimental effects upon the natural enemies of these herbivores. Bultman et al. (1997) found that parasitoids reared on fall army-worm, *Spodoptera frugiperda* (Lepidoptera: Noctuidae), that had fed on tall fescue infected by the endophytic fungus *Acremonium coenophialum* had a lower survival rate compared with those reared on fall armyworm that had fed on uninfected grass. However, as the endophytic fungus also has a direct negative effect on the herbivore (through the alkaloids produced symbiotically by the grass and the fungus), it is unclear what the net effect of the fungal infection on the herbivore might be.

Likewise, the presence or absence of one species of herbivore may influence rates of predation or parasitism upon a second species of herbivore. Consider Figure 5.41. An increase in the density of herbivore 1 might cause an increase in the density of the predator which then causes a decline in the density of herbivore 2. Such effects have been termed "apparent competition" (Holt 1977) because herbivore populations appear to be negatively affected by one another. However, the mechanism operates through a shared natural enemy rather than a limitation of resources for the herbivores. As an example, Evans and England (1996) showed that the presence of pea aphids can lead to an increase in parasitism rates on alfalfa weevils, probably because parasitic wasps benefit from the honeydew that the aphids provide. Evans and England also showed that the addition of a predator (the seven-spot ladybird, *Coccinella septempunctata*; Cdeoptera: Coccinellidae) led to increased predation of alfalfa weevils but also led to a decrease in weevil parasitism, presumably because the predator also reduced the number of aphids and hence the honeydew supply for parasitoids.

Many other external factors can affect the relationship between natural enemies and their prey. Drought stress is often considered to have an indirect effect on insect herbivores (see Chapters 3 and 12). Drought stress may also have an effect on natural enemies. Godfrey et al. (1991) found that natural enemies were more abundant in drought-stressed than in irrigated

corn (*Zea mays*). What caused this is unknown but it may be that drought-stressed plants emit volatiles similar to those they emit after attack by insect herbivores and which are known to attract natural enemies (see Section 3.3.7).

Forest fragmentation may lead to an increase in the severity of pest outbreaks because of a decrease in the impact of natural enemies (Schmidt & Roland 2006). Roland et al. (1997) studied the effect of parasitism on the forest tent caterpillar, *Malacosoma disstria* (Lepidoptera: Lasiocampidae), a major defoliator of North American boreal forest dominated by trembling aspen, sugar maple, and balsam poplar (Cooke & Lorenzetti 2006). The major parasitoids of *M. disstria* are the flies *Patelloa pachypyga* (Diptera: Tachinidae) and *Arachnidomyia aldrichi* (Diptera: Sarcophagidae). *P. pachypyga* lays its eggs on aspen foliage and *A. aldrichi* lays its eggs directly on the cocoon of *M. disstria*. Parasitism by *P. pachypyga* (but not by *A. aldrichi*) is lower in fragmented than in continuous forest (Roland & Taylor 1997; Roland et al. 1997). The lower abundance or efficiency of the tachinid parasitoid in forest fragments may be due to a preference for the relatively cool, humid conditions of continuous forest stands. Some parasitoids are known to prefer such conditions; others prefer warm, dry microclimates (Weseloh 1976). Other factors may be at least partly responsible for the prolonged outbreaks of *M. disstria* in fragmented forest. For example, virus transmission may be lower in forest patches than in continuous forest (Roland & Kaupp 1995) and the fecundity of *M. disstria* may be greater in fragmented than in continuous forest (Roland et al. 1997).

Some trees experience more damage by pests in urban habitats, such as roadsides, than in forested habitats (Speight et al. 1998). This may be due to a range of factors, including differences in the abundance of natural enemies. Hanks and Denno (1993a), for example, found that generalist predators, such as phalangids, earwigs, and tree crickets, were less abundant on trees along roadsides and in parking lots than in forested areas. Studies on experimental cohorts of the armored scale insect, *Pseudaulacaspis pentagona* (Hemiptera: Diaspididae), showed that predator-induced mortality was greater on forest trees, and Hanks and Denno concluded that natural enemies and plant–water relations (indirectly affecting scale insect performance through host-plant quality; see Chapter 3) jointly explained the greater abundance of scale insects in disturbed habitats.

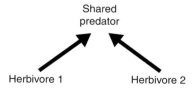

Shared
predator

Herbivore 1   Herbivore 2

Figure 5.41 Diagram illustrating apparent competition. Negative associations between populations of two herbivores can result from sharing a natural enemy rather than from competition for limited resources.

The above examples are given to illustrate the effect of habitat characteristics on the impact of natural enemies. It should not be assumed from these examples that all forest pests are similar to *M. disstria*. The outbreaks of other major forest pests, such as the spruce budworms *Choristoneura fumiferana* and *C. occidentalis* (Lepidoptera: Tortricidae) are most severe in continuous forests (e.g. Swetnam & Lynch 1993). In an additional example, Wesolowski and Rowinski (2006) have suggested that densities of caterpillars of winter moth, *Operophtera brumata* (Lepidoptera: Geometridae), are much reduced in fragmented forests, especially those surrounded by conifer woodland, compared with continuous oak woodland. As we have seen elsewhere, early-stage winter moth larvae disperse on silken threads from high-density populations. In forest fragments, dispersal leads to high mortality because the larvae are unable to locate new, suitable host tree species. It should also be pointed out that not all studies have found that insects are more abundant on trees in urban than in natural (or seminatural) habitats (e.g. Nuckols & Connor 1995).

Perhaps the most important habitat feature from the point of view of the impact of natural enemies on insect herbivores is plant diversity. There are several reasons why monocultures are more susceptible to pest outbreaks than are mixed-species agricultural or forest crop stands, but one main reason appears to be that an increase in plant diversity leads to an increase in the diversity and abundance of natural enemies. Such is the importance of plant diversity for promoting the action of natural enemies that it forms an important basis for many pest management strategies such as intercropping (see Chapter 12). For example, Schellhorn and Sork (1997) found that the abundance of natural enemies (coccinellids, carabids, staphylinids) of collard-feeding pests was greater in polycultures (collards plus weeds) than in monocultures (collards alone). Similarly, Hooks et al. (2006) found that biological control agents such as spiders and parasitoid wasps of the cabbageworm, *Artogeia rapae* (Lepidoptera: Pieridae) in Hawaii caused greater mortality in mixed cropping systems than in monoculture. However, things may not be as simple as these two examples suggest. Wilby et al. (2006) reported complex and contradictory results from studies on the relationships between arthropod diversity and rice pest communities in Vietnam.

5.6 CONCLUSION

The above examples show how different factors – predation, parasitism, the amount and quality of the food resource, and the abiotic environment – interact to determine insect population dynamics. These examples provide ample evidence that, to understand the fluctuations of insect populations, we need to use all of the tools available to the ecological entomologist. Population monitoring and life table analysis, experimental techniques, and population modeling can all contribute in important and synergistic ways to our understanding of insect population change. Ecologists have been trying to explain the dynamics of insects and their natural enemies using models for over 70 years. We hope that we have demonstrated that even some simple models have provided insight into natural enemy–prey interactions. In recent years, insect ecologists have uncovered more and more of the complex interactions involving insects, their host plants, their natural enemies, and other factors. Surprisingly few models have been constructed to explore these interactions. Those that have been built (Section 5.4) have surely demonstrated that we should not ignore the complexity of insect population dynamics and that we can effectively use models to explore these dynamics.

However, it may be said that even after detailed research, we do not always arrive at a complete understanding of the population dynamics of the species studied across its full geographic range. A good example of how some species, at least, have different population dynamics in different parts of their geographic range is the winter moth. As discussed above, work on the winter moth, *Operophtera brumata* (Lepidoptera: Geometridae), in Wytham Wood in England showed that it was regulated by pupal predation and that one of its parasitoids, *Cyzenis albicans* (Diptera: Tachinidae), had no significant role in its population dynamics (Section 5.3.4). In Canada, however, it was the introduction of this parasitoid that brought about the dramatic reduction of the winter moth *O. brumata*. Subsequent studies on its natural enemies suggested that parasitism and predation acted together to cause this reduction in winter moth densities, but the fact remains that the dynamics of this insect vary across its geographic range. Its dynamics also appear to vary from host to host. In the UK, the winter moth has recently emerged as a pest of Sitka spruce and heather (see Plate 5.4, between pp. 310 and 311). Hunter et al. (1991) failed to find that any of the factors that significantly affected the winter moth on oak affected it on Sitka spruce. Its dynamics on heather also appear to be different (Kerslake et al. 1996).

The studies discussed above show that insect populations tend to vary as the result of interactions among a multiplicity of factors. Are there therefore a million types of population dynamics? The answer is probably no (Lawton 1992). The types of population dynamics generally fall into a few major categories, even if more than one type of dynamic behavior can be found in a single species. The reason for this conclusion is that underlying the complexity of population dynamics is a relatively simple set of rules or principles (Berryman 1999) from which a diversity of dynamics can emerge. We have described some of these principles in Chapters 4 and 5 (e.g. exponential growth, negative feedback, and density dependence) and encourage readers to explore the rest; it will be worth the effort.

Chapter 6

EVOLUTIONARY ECOLOGY

6.1 INTRODUCTION

There are many ecological processes that are impossible to understand without considering evolution. Ecologists are rather prone to examining species of animals and plants at one point in time, and often also at one point in space. In fact, these species consist of a series of populations that may differ genetically and hence possess very different ecological properties (Bradshaw 1983). Populations of insects, as with all other taxa, are subject to natural selection, and all the topics that we discuss in the book, whether they be insects and climate, insects and plants, or insects and other animals or pathogens, are concerned with the results of evolution. Evolution is essentially about adaptation to environment, or fitness (Bulmer 1994), and the ways in which organisms develop, reproduce, and die have been shaped by natural selection (Nylin & Gotthard 1998; Southwood 2003). Fitness considers the relative ability of individuals or their progeny to survive under various constraints imposed by the abiotic and biotic environment in which they find themselves. In this chapter we discuss some of the mechanisms of evolutionary ecology pertaining to insects, emphasizing particular examples to illustrate the complexities achieved as solutions to the problem of maximizing fitness.

6.2 LIFE HISTORY STRATEGIES

Because insects comprise such a vast number of species, it is not surprising that we can discover an enormous range of life history types. Even within a species, different individuals in different places or times may exhibit distinct life histories. Larvae of butterflies, for example, are adapted for feeding and growth, whereas their adults are specialized for dispersal and reproduction. The former activity may take some time, depending on food supply and climate, whereas the latter stage may be very much shorter. Adult aphids of a single species may vary in the incidence of winged forms (alatae), again depending on food supply and time of year, whereas other aphids alter the incidence of males and hence sexual reproduction in their life histories. Some adult insects never reproduce at all, as is the case with worker bees. Many species may produce enormous numbers of offspring, but others only a very few. Whatever the life history strategies evolved by insects, these may also vary within a species in different localities.

The life history strategy selected by evolution for a particular set of ecological circumstances thus varies tremendously. Some insects have evolved extremely rigid and non-adaptable strategies, as in the case of stick and leaf insects (Phasmida), where many species are host-plant specialists, wingless as adult females, and reproduce only asexually. Others, however, are much more flexible in their life history patterns within one locality. The pea aphid, *Acyrthosiphon pisum* (Hemiptera: Aphididae), and its wasp parasitoid, *Ephedrus californicus* (Hymenoptera: Aphidiidae), both vary their life history strategies (in terms of variable numbers of offspring, lifespan, and development time) in response to variations in the quality of the aphids' host plant (Stadler & Mackauer 1996).

6.2.1 *r*- and *K*-selection

All organisms have to make an evolutionary "choice" when it comes to determining whether to produce

Table 6.1 Some characteristics of *r*- and *K*-selected insects. (From MacArthur & Wilson 1967; Southwood 1977; McLain 1991; Maeta et al. 1992.)

Characteristic	*r*-selected	*K*-selected
Body size in general	Small	Large
Body size male vs female	Larger females	Smaller females
Colonization ability	Opportunistic	Non-opportunistic
Development rate	High	Low
Dispersal ability	High	Low
Egg size	Small	Large
Fecundity	High	Low
Investment in each offspring	Low	High
Longevity	Short	Long
Occurrence in succession	Early	Late
Population density	Fluctuating, sometimes widely	Stable
Intraspecific competition	Often "scramble"	Often "contest"
Role of density dependence	Unimportant	Important
Population "overshoot"	Frequent	Rare
Sex ratio	Female biased	Normal

very large numbers of offspring in whom very little investment is made per individual, or alternatively to have rather few offspring, but to ensure that each one has the best possible chance of surviving to adulthood. Ecologists have pigeonholed each strategy into *r*- or *K*-selection, and though these concepts have been criticized in recent years (Reznick et al. 2002), they contain such important themes of density-dependent regulation, resource availability, and so on, that we consider them extremely useful in describing how a particular population or species of insect has evolved and adapted to particular and local conditions. Table 6.1 provides some of the characteristics of both types, though it must remembered that they are merely two opposing ends of a theoretical spectrum of reproductive strategy, with many intermediates being found in the real world. From the table, it is clear that *r*-selected insects are likely to be unpredictable in their ecologies, to undergo "boom and bust" population cycles, to produce large numbers of offspring, and essentially to overexploit their habitats and leave them in ruins. Individuals of *r*-selected species run the high risk of an early death (Davis 2005), and many species are small or even tiny (Thorne et al. 2006). Classic examples include locusts and aphids (see Chapter 1). *K*-selected insects, on the other hand, are more often regulated by density-dependent factors such as intraspecific

competition and natural enemies, and they have relatively few offspring and avoid overexploiting their habitats. Some species are characterized by having giant eggs, low fecundities, and prolonged lifespans (Maeta et al. 1992). Examples include tsetse flies, carpenter bees, and many (though by no means all) forest Lepidoptera.

r-selected species often dominate in habitats that are relatively hard to find in time and/or space, or exist for only short periods of time. Thus, for example, work in the Apennine region of Italy found that 70% of dung beetle (Coleoptera: Scarabaeoidea) species living in cattle dung exhibited *r*-selected characteristics (Carpaneto et al. 1995. Dung pats are scattered fairly randomly around grassland, and remain fresh and hence suitable for colonization by beetles for only a short time, so that insects wishing to exploit them have to have evolved efficient and rapid searching mechanisms. *r*-selected insects also tend to be early colonizers of new or disturbed habitats, by virtue of their essentially opportunistic and pioneer characteristics. As mentioned above, aphids (Hemiptera: Aphididae) and their relatives, leafhoppers (Hemiptera: Cicadellidae), for example, are highly mobile and have high reproductive potentials; they are typically *r*-selected, and in one set of experimental manipulations they appeared early in the recolonization of apple previously emptied of their insect fauna by insecticide spraying (Brown 1993). It has also been suggested that *r*-selection is an ancestral trait in some insects, and *K*-selection becomes more widespread as evolution progresses. Kranz et al. (2002) presented the example of the evolution of eusociality in gall-forming thrips (Thysanoptera) species. Figure 6.1 shows that solitary species, considered to represent part of an ancestral lineage to social species, exhibit a much higher brood size than their eusocial relatives.

As with most things ecological, life is never simple, and some species may show traits of both *r*- and *K*-selection at different times or in different locations. A single species may exhibit *r*- or *K*-selected traits depending on its population density. Low population densities of the bean weevil, *Acanthoscelides obtectus* (Coleoptera: Curculionidae), showed *r*-selected characteristics of high fecundity, earlier age at first reproduction, and higher growth rates when compared with high-density populations, which appeared to be more *K*-selected (Aleksic et al. 1993). In the case of the mosquito *Anopheles messeae* (Diptera: Culicidae),

populations showed a seasonal conversion from *r*-selected traits to *K*-selected ones in periods of so-called population "prosperity" (Gordeev & Stegniy 1987), when resources were plentiful. Some of the characteristics mentioned in Table 6.1 will now be considered in greater detail.

6.2.2 Survivorship

Survivorship (or perhaps more accurately fatalityship) describes the rates at which organisms die as they progress through a generation. In theory, animals including insects exhibit one of three types of survivorship curve, as shown in Figure 6.2. These distributions may have limited real-world use, as most insects do not really conform to theory. Nevertheless, there is merit in considering a basic difference between a species or population wherein most of the individuals die at a very young age, compared with those where most individuals live to be old (Type III vs. Type I). The Type II survivorship curve describes a situation where no particular life stage shows significantly higher mortalities than any other. This pattern, where constant mortality occurs across all ages, is more characteristic of birds and other vertebrates than insects (Strassman et al. 1997). However, it should be noticed that the *y*-axis is normally depicted on a logarithmic scale; back-transformation reveals that population declines illustrated by survivorship curves are largest in early stages even when the

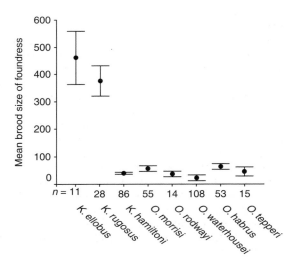

Figure 6.1 The mean brood size and 95% CI bars of the foundress for *Kladothrips ellobus, K. rugosus* (solitary species in the ancestral lineage to the eusocial species), *K. hamiltoni, Oncothrips waterhousei, O habrus,* and *O. tepperi* (eusocial species). *O. morrisi* (which has a non-dispersing, fully reproductive fighting morph), and *O. rodwayi* (secondarily solitary). *n*, number of galls for each species in analyses. (From Kranz et al. 2002.)

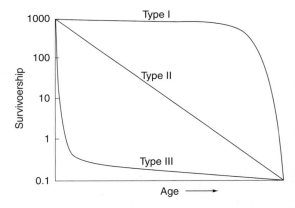

Figure 6.2 Three general types of survivorship curve. (From Begon et al. 1990.)

curve is a straight line (Type II). What really counts is the steepness of the slopes over a particular life stage, though it must be borne in mind that steepness may be at least partly related to environmental conditions such as temperature (Legaspi & Legaspi 2005) rather than an innate result of insect evolution. None the less, three examples are given, one for each type of survivorship curve.

Figure 6.3 shows the mean percentage survival over three generations of several life stages of the lesser cornstalk borer (LCB), *Elasmopalpus lignosellus* (Lepidoptera: Pyralidae) (Smith & Johnson 1989). The larvae of this species can cause serious economic damage when they tunnel in the leading shoots of maize and other crops, causing dead-heart symptoms. Like many of its family, LCB is not particularly host plant-specific, and in this particular example, the so-called cornstalk borer is in fact acting as a pest of peanuts in Texas. From the graph, it is clear that no life stage through a generation exhibits particularly greater mortality than any other. There is just a suggestion that eggs and pupae suffer least, having the shallowest gradients. This is often the case, where quiescent, immobile, and relatively defenceless forms such as these are placed by the previous life stage (ovipositing females or late-stage larvae) in safe and protected habitats. In fact, for borers such as LCB, the larvae are also fairly well protected from external influences including high levels of parasitism and disease, reflected in the roughly Type II curve.

Aphids more often than not show complex and specialist life history strategies, optimized for reproductive and host exploitation (Section 6.3). *Sitobion avenae* (Hemiptera: Aphididae), the grain aphid (see Plate 6.1, between pp. 310 and 311), is a notorious pest of winter and spring cereals in northwestern Europe, which, given the right weather conditions (dry and warm) when the crop enters the milk-ripe stage after flowering, can cause losses to wheat extending to several tonnes per hectare. All aphids at this time of the year are parthenogenetically reproducing females, each of whom has a maximum lifespan in the laboratory of around 10–12 weeks (Thirakhupt & Araya 1992). Under these ideal conditions, most females survive the nymphal stage and enter adulthood with little significant mortality (Figure 6.4). Heavy losses occur only when the post-reproductive stage is reached after a couple of months. It is important to realize that this type of survivorship is unlikely to be encountered in the field, because natural enemies are likely to exact a toll. However, the importance of parasitoids, predators, and pathogens in the life of aphids and their relatives is often limited (see Chapter 12), so that Type I survivorship curves may still operate. Even parasitoids themselves follow this type of strategy on occasion. Adults of *Cotesia flavipes* (Hymenoptera: Braconidae), a parasitoid of another pyralid moth related to the cornstalk borer discussed above, also show a Type I survivorship curve, where most individual females live until the end of their normal lifespan, in this case a mere 24 h (Wiedenmann et al. 1992).

Finally, some insects produce enormous numbers of offspring, a high proportion of which are lost early in life. Scale insects (Hemiptera: Coccidae) provide some good examples. The horse-chestnut scale, *Pulvinaria regalis*, may lay several thousand eggs, according to the size of the female. As Figure 6.5 shows, mortality during the egg stage is fairly minor (M.R. Speight, unpublished data), but when the eggs hatch to produce first instar nymphs, called crawlers, very large losses may occur. The mobile crawlers disperse away from the egg masses from which they originated on the main trunks of both host and non-host trees, and it is their job to locate leaves of limes, horse-chestnuts, or sycamores on which older nymphs will feed during the summer months. Many crawlers merely move to the foliage of the trees on which they hatched, but the majority are blown off their hosts and lost on the wind, very few indeed ending up by chance on a

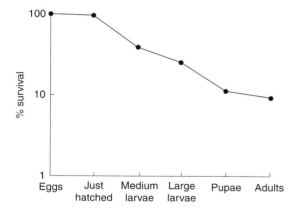

Figure 6.3 Survival of cornstalk borer *Elasmopalpus lignosellus* (mean of three generations). (From Smith & Johnson 1989.)

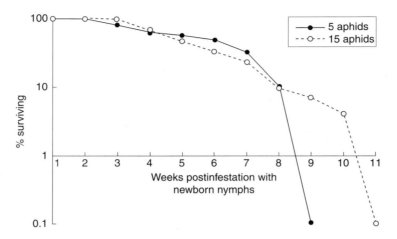

Figure 6.4 Survivorship curves of two colonies of the cereal aphid, *Sitobion avenae*. (From Thirakhupt & Araya 1992.)

new suitable host tree (Speight et al. 1998). These huge dispersal losses are analogous to those seen in the case of sessile marine invertebrates that consign their young larval stages to the vagaries of ocean currents, in the hope that at least a few will colonize new and distant substrates. Insects with heavy mortalities early in a generation such as these, though appearing in disturbingly high numbers in the egg and early larval instar stage, may soon be reduced to relatively low population densities. In Texas for

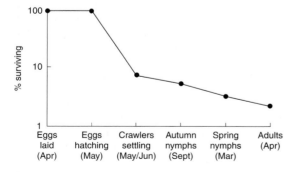

Figure 6.5 Survival of the horse-chestnut scale, *Pulvinaria regalis*, through an annual generation. (M.R. Speight, unpublished observations.)

example, the notorious cotton budworm, *Helicoverpa* (*Heliothis*) *zea* (Lepidoptera: Noctuidae), suffers its highest generation mortality in the egg stage, with more than 97% of all the budworm dead by the fifth instar (Pustejovsky & Smith 2006) with no intervention from the pest manager.

So, evolution has produced several types of survivorship strategy wherein the rates of mortality vary according to the life stage. Larvae or nymphs tend, of course, to be the longest lived life stage. Cossid moth larvae (see Plate 6.2, between pp. 310 and 311), the famous witchetty grubs of the Australian outback, may live for many years, slowly growing inside trees. An added complication, however, is that different life stages of a species may show tendencies towards various shapes of survivorship curve at certain times of year, during different seasons (Maharaj 2003), or within one life stage. It is therefore important to treat eggs, larvae, and adults, for example, as separate stages wherein survival may show different patterns. Larvae of the caddisfly, *Rhyacophila vao* (Trichoptera: Rhyacophilidae), predate other invertebrates in small streams in Canada (Jamieson-Dixon & Wrona 1992). Their first winter is spent as the first three instars, developing to the fourth instar by summer and the fifth by fall. The second winter is spent as fifth instars, with new adults appearing in the summer of the second year. Figure 6.6 illustrates survivorship of a cohort of larvae in the third instar in late spring,

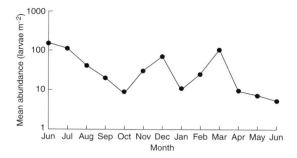

Figure 6.6 Survival of a cohort of larvae of the caddisfly, *Rhyacophila vao*. (From Jamieson-Dixon & Wrona 1992.)

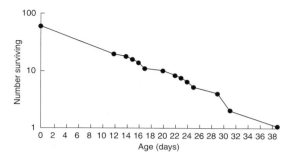

Figure 6.7 Survival of adult female damselfly, *Pyrrhosoma nymphula*. (From Bennett & Mill 1995.)

and shows an exponential decline in numbers until winter. Higher mortalities in the active months are not surprising; losses through predation, competition, and possibly starvation are likely to operate when the stream temperatures are relatively high, and a species such as this with a long larval stage through seasonally variable climatic conditions would be expected to show a Type III survivorship curve.

Insect species with a much shorter time spent in a particular life stage are less likely to exhibit this type of survivorship. Remaining with freshwater insects, though this time terrestrial adults, the damselfly *Pyrrhosoma nymphula* (Odonata: Zygoptera) has a total adult lifespan of around 20 days in the UK (Bennett & Mill 1995) (Figure 6.7). Within this timeframe, the insects are reproductively active for a mean of 6.7 days only, and before this, young adults of both sexes remain away from water until sexually mature. It is thought that this tactic has evolved to protect them from predation before they have had a chance to reproduce, and the survivorship curve reflects the fact that prereproductives survive as well as, if not better than, mature adults.

Finally, differing life history strategies may be used by closely relayed species to avoid interspecific competition. In Hawaii for example, melon flies and fruit flies that feed on essentially the same substrates exhibit very different life history strategies (Vargas et al. 2000). *Bactrocera cucurbitae* shows late onset of reproduction, a longer lifespan, and a lower rate of increase, whereas *Ceratitis capitata* (Diptera: Tephritidae) shows early reproduction, a short lifespan, and a high intrinsic rate of increase.

6.3 SEXUAL STRATEGIES: OPTIMIZING REPRODUCTIVE POTENTIAL

As discussed above, many insects appear to maximize their reproductive potentials, and this is manifested by the evolution of mechanisms that increase their fecundities. Two of the most important functions of adult insects are to disperse and reproduce, though these two can be mutually incompatible such that there may be a tradeoff between dispersal and reproduction (Lock et al. 2006). Dispersal can be especially important for insects whose normal habitats are ephemeral or extremely patchy. Freshwater insects are again a good example of this. It is crucial, for example, for pond skaters or striders (Hemiptera/Heteroptera: Gerridae) (see Plate 6.3, between pp. 310 and 311) to move from pond to pond. But there are costs involved. Simply put, the further an individual disperses, the lower its fecundity is likely to be (Hanski et al. 2006), since energy for one activity uses up energy which could be used for the other. Gu et al. (2006) studied the tradeoff between mobility and reproductive fitness in the codling moth, *Cydia pomonella* (Lepidoptera: Tortricidae). This insect is one of the most serious pests of orchards in the world (see Chapter 12), and it is known to have two distinct strains (genotypes) typified by different combinations of physiological, morphological, and behavioral traits in terms of dispersal capacity. The mobile strain shows a reduced survival with age, and also a reduced maximum fecundity when compared with the sedentary strain (Figure 6.8a), indicating the "cost" of being

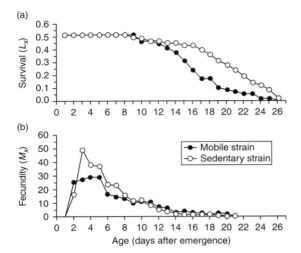

Figure 6.8 The age-specific survivorship (L_x) and fecundity (M_x) of female adults in the mobile and sedentary strains of *Cydia pomonella*. (From Gu et al. 2006.)

mobile in order to move to pastures (or orchards) new. Some of the best examples of insects that have evolved specialized morphs for dispersal versus reproduction are aphids (Braendle et al. 2006).

Two extreme examples of evolutionary tradeoffs are the development of flightlessness and of asexuality. Both mechanisms fulfill other purposes too, which will also be discussed here for convenience.

6.3.1 Flightlessness

Loss of the ability to fly has occurred in nearly all orders of winged insects, many times within most of these, and perhaps hundreds of times in the Hemiptera and Coleoptera (Andersen 1997). Adult insects do not always have wings (see Plates 6.4–6.6, between pp. 310 and 311). Figure 6.9 shows the major orders of insects and the percentages of each that exhibit flightless adults (Wagner & Liebherr 1992). In most cases, the appearance of flightlessness must be a secondary event, since the ancestral state of the Pterygota is to have wings as an adult, and around 5% of so-far described species are now wingless (Whiting et al. 2003). In some cases, whole species may be flightless, alternatively adults may be polymorphic for flight ability or winglessness (Carroll et al. 2003),

or one sex (usually the male) flies whilst the other cannot (Gade 2002). As Wagner and Liebherr (1992) pointed out, the evolution of wings is heralded as the most important event in the diversification of insects, yet as the figure shows, the loss of wings and flight has occurred in almost all the pterygote orders, albeit rather late in most cases. Figure 6.10 shows a phylogeny of ball-rolling dung beetles in the tribe Scarabaeini (Coleoptera: Scarabaeidae) (Forgie et al. 2005). It seems that in this case, flightless species (depicted by dashed lines at the bottom of the figure) are monophyletic, highly derived, and originating very late in the lineage.

Loss of the ability to fly seems in part to be related to season, such as short-winged morphs of the mole cricket *Gryllotalpa orientalis* (Othoptera: Gryllotalipdae) that occur during winter, whilst long-winged morphs appear in spring and summer to disperse (Endo 2006). Alternatively, flight ability may be related to the stability of the habitat in which insects live (Hunter 1995). It is argued that if dispersal from one habitat to another is not required for the long-term survival of populations, because a single habitat has predictable and long-lasting abiotic and biotic factors to which the species is well adapted, then there is no need to disperse (Wagner & Liebherr 1992; Carroll et al. 2003). Roff (1990) has reviewed reports in the literature of the occurrence of flightlessness in insect groups from various habitats. These include woodland, desert, surfaces of sea and snow, caves, bird and mammal body surfaces, and hymenopteran or termite nests. Woods, deserts, and caves, for example, are clearly permanent ecological fixtures, at least until the fragmentation of forests, for example, because of human activities, occurs on a large scale, whereupon it might be expected that flightless species will be the first to be endangered. Habitats with less than average incidence of flightless species are predictably changeable or ephemeral ones, such as the seashore and fresh water (both its margins and the water body itself) (Roff 1990). So, why should insects lose their supposedly most important trait as soon as it is no longer required? Clearly, there must be a cost involved in having full wings and using them to fly; this cost can be related to reproductive potential.

So, here is undoubtedly a tradeoff between flight capability and reproductive success. As mentioned above, where adults are flightless, it tends to be the females that exhibit this trait rather than the males (Wagner & Liebherr 1992). One order that contains

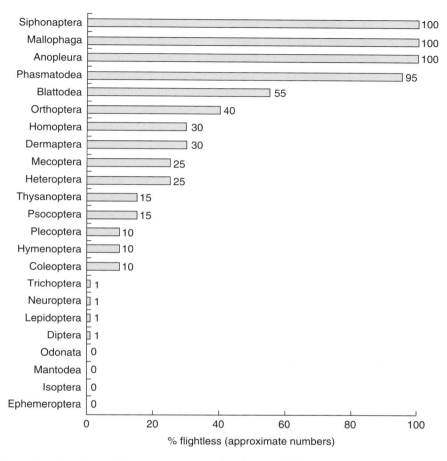

Figure 6.9 Percentage of species within temperate insect orders that are flightless. (From Wagner & Liebherr 1992.)

many flightless species is the Phasmida (stick and leaf insects). Many female adults have wings that are very much reduced or, in some cases, completely absent, yet many also have males with full wings, who fly readily. The Lepidoptera are not renowned for a high incidence of flightlessness, but in woodland habitats several families, such as the Lymantriidae and Geometridae, do have flightless species. In the case of the latter family at least, it has been shown that those species with reduced wings show a significantly higher maximum fecundity than fully winged species (Hunter 1995) (Figure 6.11).

Females, of course, have to invest much greater resources into reproduction than do males, which frequently means that they are much bigger or heavier than their males. Figure 6.12 shows this in the case of the grasshopper, *Phymateus morbillosus* (Orthoptera: Acrididae) (Gade 2002). Adult males and females of this species look very similar at first glance, but as the table shows, males are in fact significantly smaller and lighter, and their wing-loading is therefore much lower. Thus, males are able to fly for up to a minute at a time, and cover many meters. Females on the other hand are unable to fly, probably because they cannot produce enough lift.

Even male insects, however, exhibit an evolutionary "choice" to be winged (macropterous) or non-winged and hence flightless (brachypterus) on

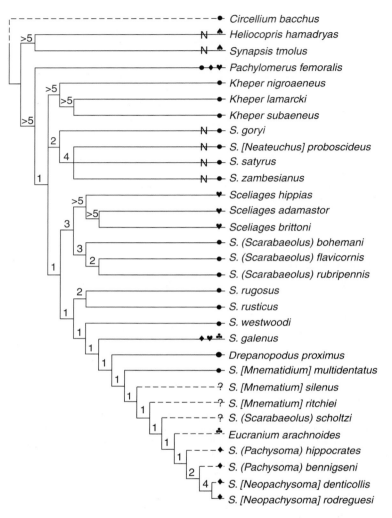

Figure 6.10 Phylogeny of the Scarabaeini (Coleoptera: Scarabaeidae). Total characters data set. NONA, unweighted, strict consensus of three trees ($CI_1 = 0.25$, $RI_1 = 0.50$, length = 1835). Branch supports (Bremer decay indices) provided above nodes. Flightless taxa (dashed branch). Nocturnal taxa (N), S., *Scarabaeus*. Rolling ●; tunnelling ♠; pushing ♥; dragging ♦; carrying ♣; unknown ? (From Forgie *et al.* 2005.)

occasion. Socha (2006) studied the mating success of adult male firebugs, *Pyrrhocoris apterus* (Hemiptera: Pyrrhocoridae). When exposed to intraspecific competition, 5–7-day-old flightless males were three to five times more successful at finding a mate and copulating with her than winged ones (Figure 6.13). Possessing wings can be a definite disadvantage at times.

Many species of Hemiptera exhibit both macropterous and flightless brachypterous morphs within one

species. Some individuals within a species may even keep their wings, but lose their flight muscles, as with the soapberry bug, *Jadera haematoloma* (Hemiptera: Rhopalidae) (Carroll et al. 2003). Another example of this type exists in northern European populations of the water strider, *Gerris thoracicus* (Hemiptera: Gerridae). Both morphs start off as long-winged forms that are able to fly before the reproductive period, but one form histolyse (breakdown) their flight

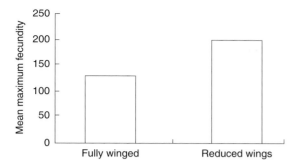

Figure 6.11 Simple comparison of mean maximum reported fecundity of woodland geometrid moths according to their winged status. The means are significantly different. (From Hunter 1995.)

muscles at the start of reproduction, and hence lose the power of flight (non-flyers) (Kaitala 1991). The other form maintains its ability to fly throughout adult life (flyers). In experiments, both were maintained in one of two regimes, one where food (insect prey) was abundant, and the other where it was limited to three *Drosophila* per adult per day. Figure 6.14 shows that no matter what the food regime, the non-flyer morph had a higher fecundity and reproductive rate (measured as number of eggs laid per day) compared with the flyers, though both variables were also significantly affected by the quantity of food available to female insects (Kaitala 1991). The ability

to change reproductive strategy via alterations in ecology or morphology (known as phenotypic plasticity) according to environmental conditions enables an animal to perform "tradeoffs" between such things as longevity, fecundity, and reproductive rate. In the example of water striders, flightless females are able to be more reproductive at low food levels, but clearly are unable to move between distant and unpredictable habitats. Interestingly, even if flight has been abandoned, dispersal is still a crucial part in the lives of many insects, and in one example it seems that insects with wings are more efficient, even "determined", walkers than wingless morphs in the same species (Socha & Zemek 2003) (Figure 6.15).

Other Hemiptera show more complex phenotypic plasticities, via changes in the occurrence of winged (alate) and wingless (apterous) morphs in response to variations in food supply and/or seasons. Aphids (Hemiptera: Aphididae) such as the grain aphid, *Sitobion avenae*, produce apterous offspring when food is both abundant and high in organic nitrogen, as occurs during the milk-ripe stage of cereal growth, when the grain is forming in early summer. Embryos can be found developing in the bodies of relatively young female nymphs, and their volume increases exponentially as their mother matures. Third instar nymphs are to be found around the fourth day of life, and they become adult on the seventh day in apterous morphs, whereas alatae do not become adult until the eighth day (Newton & Dixon 1990). Not only is there a tradeoff between being able to fly and development time, but, as can be seen from Figure 6.16, the

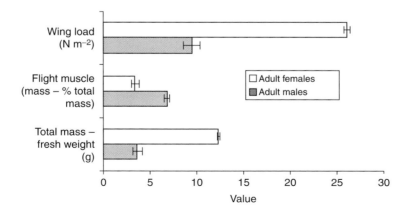

Figure 6.12 Mass and morphometrics for male and female *Phymateus morbillosus*. (From Gade 2002.)

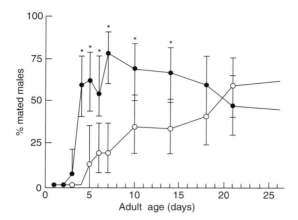

Figure 6.13 Changes in the percentage of mated males of two-wing morphs during the first 28 days of adult life. ●, brachypterous morph; ○, macropterous morph. Means ± 95% confidence intervals are shown. (From Socha 2006.)

embryos of apterous females grow faster. Also, the total fecundity of apterae is greater than that of alatae. Hence, winged morphs initially produce smaller and fewer offspring than non-winged ones, though alate females are able to catch up with their apterous sisters after 4 days of reproduction (Newton & Dixon 1990).

It is clear that if the habitat is able to provide sufficient quantities of high-quality food, then a better strategic "bet" for aphids is to remain wingless. Only when this food declines as the season progresses, or intraspecific competition intensifies, is there a need to move, and hence having wings becomes an advantage (Leather 1989). One final problem for apterous aphids involves their reduced capacity to avoid being eaten by predators or attacked by parasitoids. Certainly, aphids such as the pea aphid, *Acyrthosiphon pisum* (Hemiptera: Aphididae), produce an increasing proportion of winged morphs when exposed to natural enemies such as hoverflies, lacewings, and parasitic wasps (Kunert & Weisser 2005). Failing this, apterous individuals quite sensible drop from the host plant to the ground as enemies approach (Gish & Inbar 2006), whereupon they then have the challenge of finding their way back to a suitable plant. It seems that mature aphids are particularly good at this, being able to find their way back to a plant from 12 or 13 cm away in around 40 seconds.

Finally, we have assumed that all insects in the Pterygota have become wingless as a secondary adaptation (hence the name). Only one instance, that of some stick and leaf insects, has so far been reported where the ancestral state is thought to be flightlessness, with wings returning later in evolution (Klug &

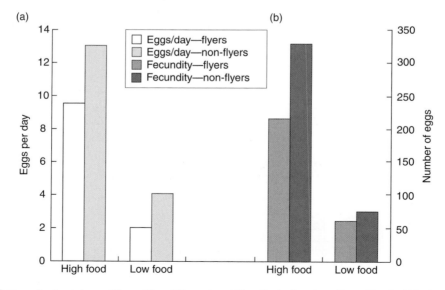

Figure 6.14 Reproductive rates and fecundities of the water strider, *Gerris thoracicus*. (From Kaitala 1991.)

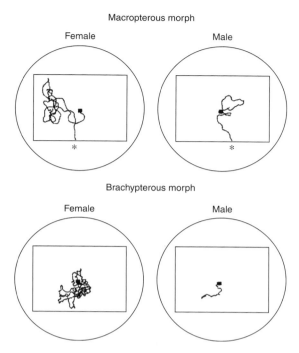

Macropterous morph

Female   Male

*

Brachypterous morph

Female   Male

Figure 6.15 Examples of complete locomotory tracks of macropterous and brachypterous females and males of *Pyrrhocoris apterus*. Circles represent arena walls, rectangles indicate observation zones, and small solid squares mark the start of the tracks. ∗ Denotes that macropters abandoned observation zone before time limit of 1 h. (From Socha & Zemek 2003.)

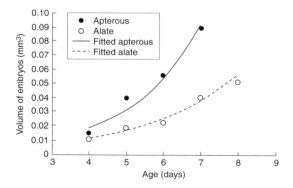

Figure 6.16 Volume of largest embryos in relation to age of alate and apterous mothers of *Sitobion avenae*. (From Newton & Dixon 1990.)

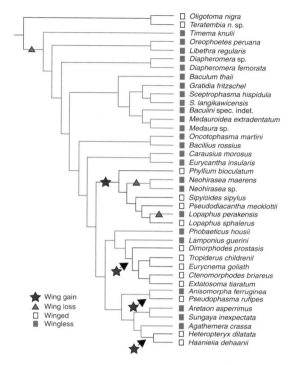

☐ *Oligotoma nigra*
☐ *Teratembia* n. sp.
■ *Timema knulii*
■ *Oreophoetes peruana*
■ *Libethra regularis*
■ *Diapheromera* sp.
■ *Diapheromera femorata*
■ *Baculum thaii*
■ *Gratidia fritzschel*
■ *Sceptrophasma hispidula*
■ *S. langikawicensis*
■ *Baculini* spec. indet.
■ *Medauroidea extradentatum*
■ *Medaura* sp.
■ *Oncotophasma martini*
■ *Bacillius rossius*
■ *Carausius morosus*
■ *Eurycantha insularis*
☐ *Phyllium bioculatum*
■ *Neohirasea maerens*
■ *Neohirasea* sp.
☐ *Sipyioides sipylus*
☐ *Pseudodiacantha mecklottii*
■ *Lopaphus perakensis*
☐ *Lopaphus sphalerus*
■ *Phobaeticus housii*
■ *Lamponius guerini*
☐ *Dimorphodes prostasis*
☐ *Tropiderus childrenil*
☐ *Eurycnema goliath*
☐ *Ctenomorphodes briareus*
☐ *Extatosoma tiaratum*
■ *Anisomorpha ferruginea*
☐ *Pseudophasma rufipes*
◩ *Aretaon asperrimus*
■ *Sungaya inexpectata*
■ *Agathemera crassa*
☐ *Heteropteryx dilatata*
☐ *Haanieiia dehaanii*

★ Wing gain
▲ Wing loss
☐ Winged
■ Wingless

Figure 6.17 Character mapping of wing types on phasmid phylogeny. Parsimony optimization of winged (blue) and wingless (red) states for male phasmids on the optimization alignment topology. This reconstruction requires seven steps with four wing gains and three losses; DELTRAN optimization requires five wing gains and two losses. Maximum likelihood reconstruction produces similar results. (From Whiting et al. 2003.)

Bradler 2006). Some of the Phasmatodea appear to have arisen as wingless forms, and those species that now have wings (no more than 40% of total) have evolved them secondarily, as illustrated in Figure 6.17 (Whiting et al. 2003). Furthermore, as the diagram shows, wings actually evolved separately and independently in various species of phasmid.

6.3.2 Asexuality

Sexual reproduction may be rare or non-existent in some organisms (Halkett et al. 2005). However, others are able to maintain both sexual and asexual forms side by side as it were, with gene flow taking place between the two populations. Each form appears at different times of the year, for example with over-wintering forms such as eggs produced sexually; whilst rapidly reproducing forms in spring and early summer, when plants are most nutritious, are produced by asexual means (McCornack et al. 2005).

In many insect species, males are unknown or non-functional, and offspring are produced parthenogenetically without fertilization. Parasitic wasps, aphids, stick insects, and sawflies, for example, all have asexual species, or morphs of species. The evolution of sex in any organism is of course a process by which genetic diversity, hybrid vigor, and outbreeding are maintained or promoted. Sexual reproduction is said to enable organisms to adapt to changing environments, to move into new niches, to avoid parasites, and to act as a "sieve" to remove deleterious genes and mutants (Fain 1995). However, it may be that these normally admirable traits are, on occasion, not desirable or adaptively significant, that is, in some situations, males may be counterproductive or simply a waste of resources. Let us take a theoretical example where normally a population consists of roughly equal numbers of males and females. Fifty percent of this population is necessarily non-productive, in that the males consume resources, or merely take up space, without producing offspring. Reducing the number of males, or eradicating them altogether, should in theory dramatically increase the reproductive potential of the population; in fact, double it (Keeling & Rand 1995). All that would be lost would be genetic diversity, itself not always a desirable trait, especially if hard-earned (in evolutionary terms) specialization such as host-specific parasitism or herbivory are to be maintained unchanged in offspring. Hamilton et al. (1990) suggested that the main reason for the evolution of sexual reproduction is linked to resistance to parasites and pathogens, via a continuing adaptability in the host, but this supposed advantage cannot operate in all cases, otherwise asexuality in higher animals, including insects, would not be so widespread.

It certainly seems that there is a survival "cost" to mating. Figure 6.18 shows that females of the butterfly *Colias eurytheme* (Lepidoptera: Pieridae) that have mated have substantially shorter lifespans than virgins (Kemp & Rutowski 2004). It has been suggested that male ejaculates may be toxic to females, but that the tradeoffs of sexual reproduction for the butterflies, apart from the factors mentioned above,

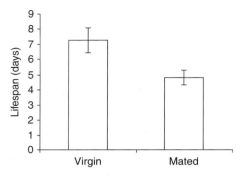

Figure 6.18 Lifespan of virgin versus mated *Colias eurytheme*. (From Kemp & Rutowski 2004.)

come from nuptial "gifts" in the form of nutritious spermatophores that males give to females, thus increasing their reproductive success. Nevertheless, there is no doubt that remaining unmated can make you live longer.

A large number of species, again in the Hemiptera, reproduce predominantly, if not exclusively, by asexual means, and Table 6.2 summarizes all such possibilities (Wilson et al. 2003). Aphids have a widespread tendency to lose the sexual phase of reproduction, and they frequently exhibit cyclical parthenogenesis where their life history strategy has evolved to separate sexual reproduction from the maximization of biomass and offspring production (Wilson & Sunnucks 2006). Both activities involve specializations, ecologically and morphologically, to gain efficiency, and these strategies may in fact evolve more than once within one species, as shown for the bird cherry oat aphid, *Rhopalosphium padi* (Hemiptera: Aphididae), in Figure 6.19 (Delmotte et al. 2001). Assuming that completely asexual lineages cannot revert to sexuality, having lost the mechanisms for it, then the cladogram shows that entire asexuality has appeared three times. These mechanisms also allow host-plant alternation, a phenomenon known as heteroecy. An excellent example of an aphid that undergoes both cyclical parthenogenesis and heteroecy is the black bean aphid, *Aphis fabae* (Hemiptera: Aphididae) (see Plate 6.7, between pp. 310 and 311), whose life cycle is illustrated in Figure 6.16 (Blackman & Eastop 1984). Some of the terms relating to different aphid morphs need explanation: "fundatrix" is the first parthenogenetic generation; "virginoparae" are alate or apterous females who produce live female offspring

Table 6.2 Aphid life cycles: all four life cycles can be found in a single aphid species. (From Wilson et al. 2003.)

	Sexual			Asexual	
	Cyclical-parthenogenesis	Intermediate strategies		Male-producing obligate parthenogenesis	Obligate parthenogenesis
Spring	Foundresses hatch from sexual overwintering eggs Numerous parthenogenetic generations	Foundresses hatch from sexual overwintering eggs Numerous parthenogenetic generations	Continuous parthenogenetic reproduction	Continuous parthenogenetic reproduction	Continuous parthenogenetic reproduction
Summer					
Fall (shorter days and cooler temperatures)	Production of males and sexual females Males and sexual females mate and lay overwintering eggs	Production of males and sexual females Males and sexual females mate and lay overwintering eggs		Production of some males that can mate with sexual females produced by CP and I clones*	
Winter	Eggs diapause overwinter and hatch in spring	Eggs diapause overwinter and hatch in spring	Continuous parthenogenetic reproduction through winter	Continuous parthenogenetic reproduction through winter	Continuous parthenogenetic reproduction through winter

*CP, cyclically parthenogenetic; I, intermediate strategies

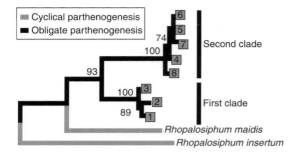

Figure 6.19 Rooted neighbor joining tree based on Kimura two-parameter distances of 10 mitochondrial haplotypes (1199 bp comprising cytochrome *b*). The numbers at the nodes are bootstrap percentages (1000 replicates). Changes in reproductive mode are mapped along the neighbor joining tree, assuming that sexual reproduction is the ancestral reproductive mode and that reversion to sexuality is impossible. (From Delmotte et al. 2001.)

parthenogenetically; "gynoparae" are alate parthenogenetic females who migrate to the primary host and produce egg-laying, sexual, female offspring. It should be noted that all offspring are produced viviparously, allowing the shortest possible generation times.

Thus it can be seen that aphids such as this are able to maximize their exploitation of the host plant at the best times of year, when the food quantity and quality is at its height, and weather conditions allow for rapid development rates. No wastage of resources by non-productive males is allowed. Sexual reproduction prepares the population for winter, providing the scope for genetic mixing so that next year's generations will, if required, be able to cope with environmental changes. On another point, eggs tend to be most resistant to winter cold, and are thus again an insurance policy for the aphids. Some authors such as Delmotte et al. (2001) have called this "bet-hedging", where facultative asexual forms can revert to sexuality when the varying demands of dispersal and overwintering dominate (Halkett et al. 2006) (Figure 6.21).

Some aphids take this to even further extremes; for example, the green spruce aphid, *Elatobium abietinum* (Hemiptera: Aphididae). The British population never has males (anholocyclic life cycle), whereas the population in continental Europe uses sexual reproduction in the fall (holocyclic life cycle) to produce the overwintering adult stage. The onset of fall and then winter in the UK does not appear to be a sufficiently strong stimulus to trigger the production of males.

There is also an advantage in overwintering in the more cold-susceptible adult stage (see Chapter 2), in that reproduction can ensue at the earliest possible opportunity in the following spring. Judging by the seriousness of *Elatobium* as a pest of spruce in the UK, the loss of sexual reproduction has not hampered their ability to fully exploit their host.

6.4 LIFE HISTORY VARIATIONS WITH REGION

As illustrated with *Elatobium* above, within a species' range, the dominant life history strategy may also vary according to the geographic location of a particular population, based predominantly on the local environmental conditions. For example, worker castes in the ant *Trachymyrmex septentrionalis* (Hymenoptera: Formicidae) differ in their mean sizes (Beshers & Traniello 1994). Colonies in Long Island, New York State, have larger individuals on average than those in Florida. It is suggested that the northerly colonies have evolved a life history strategy adapted to survive temperate winters, whereas those in the subtropical south have experienced different selection pressures and have adapted for rapid colony growth in the absence of climatic constraints. In Western Europe, the grayling butterfly, *Hipparchia semele* (Lepidoptera: Satyridae), shows significant regional differences in several life history features, including longevity, fecundity, egg-laying rates, and egg size (Garcia-Barros 1992). This species feeds on grasses as a larva, and is commonly found in heathland, sand dunes, and along the edges of cliffs. In the north of the region, there is a higher concentration of egg production early in the life of the adult female, and smaller eggs are produced, relative to butterfly populations in Mediterranean areas. Adult females also live longer in the latter region. Adverse climatic conditions expected in more northerly climes mean that adult females must reproduce faster and earlier before the shorter warm season is over. Of course, these examples describe extremes of what is likely to be a continuum of strategies where a species range is continuous, or clinal. Where a single species is necessarily split into more isolated populations by natural barriers or manmade ones, demes may be generated where population characteristics may be discontinuous (Section 6.6).

Back with aphids, we have seen how populations of the spruce aphid *Elatobium abietinum* seem to

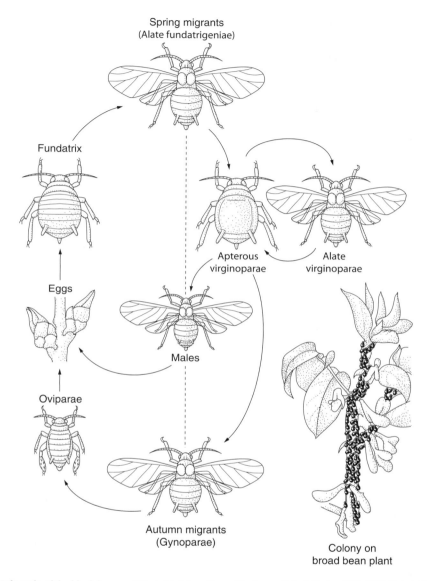

Figure 6.20 Life cycle of the black bean aphid, *Aphis fabae*. (From Blackman R.L. & Eastop V.F. (1984) Aphids on the World's Crops, An Identification Guide. John Wiley & Sons, New York.)

lose their holocyclic lifestyle as they move into the more northerly climes of the UK. However, most parthenogenetic individuals seem unable to survive temperatures below −5 or −10°C (Dedryver et al. 2001), so it may be that continental winter temperatures in Europe may require the use of eggs produced by sexual reproduction as the only overwintering life stage to be able to survive the cold (see also Chapter 2). Since only sexual females can lay cold-resistant eggs (Vorburger et al. 2003), asexuality would not survive in such conditions (Simon et al. 2002).

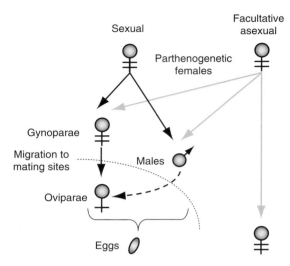

Figure 6.21 Schematic representation of the fall switch to sexual reproductive mode for the two main reproductive strategies in *Rhopalosphium padi*. (From Halkett et al. 2005.)

6.5 COEVOLUTION

Darwin did not use the word coevolution in *On the Origin of Species* (1859), but he did use the word coadaptation to describe reciprocal evolutionary change between interacting organisms (Thompson 1989). For example, in discussing the evolution of interactions between flowers and pollinators, Darwin wrote, "Thus I can understand how a flower and a bee might slowly become, either simultaneously or one after the other, modified and adapted in the most perfect manner to each other, by continual preservation of individuals presenting mutual and slightly favorable deviations of structure" (Darwin 1859, p. 95). As we have come to expect, Darwin provided us with a fundamental description of, and a mechanism for, interactions between organisms that result in reciprocal changes in traits (morphology, behavior, physiology) over evolutionary time, and his observations have generated the field of study that we now call coevolution.

It is important that we have a clear understanding of the concept of coevolution before we apply it to the ecology of insects. As Janzen (1980) has pointed out, misuse of the concept of coevolution is common,

and he offers this definition: "an evolutionary change in a trait of the individuals in one population in response to a trait of the individuals of a second population, followed by an evolutionary response by the second population to the change in the first." It is this reciprocal change (A affects B, which affects A again) that distinguishes coevolution from simple adaptation by organisms to the abiotic and biotic environment. For example, an insect herbivore that has the ability to detoxify certain secondary metabolites in the tissues of its host plant is not necessarily "coevolved" with that plant. The secondary metabolites might be present for a variety of reasons (see Chapter 3), or the herbivore may have had its detoxification mechanisms in place before encountering the host plant in question. Janzen (1980) also offered a definition for "diffuse coevolution" in which the populations of A or B or both are actually represented by an array of populations that generate a selective pressure as a group. Potential examples include some mimicry complexes among multiple species of butterfly or some insect pollinator interactions with flowering plants: the floral adaptations of most flowering plants probably represent the result of selection from a number of different pollinator species. In both kinds of coevolution (diffuse and pairwise), the key requirement is repeated bouts of reciprocal genetic change specifically because of the interaction. As reciprocal bouts of genetic change are not easily observed in natural populations, the study of coevolution requires careful detective work, combining the disciplines of ecology, systematics, and genetics (e.g. Hougen-Eitzman & Rausher 1994; Iwao & Rausher 1997). More complex subdivisions of the coevolutionary process, particularly subdivisions of diffuse coevolution, are possible (Thompson 1989, 1997) but are beyond the scope of this text. Interested readers are strongly encouraged to read Thompson's (1994) excellent text on the subject.

Reciprocal evolutionary change can occur between populations of organisms that interact with one another in a variety of ways (see, for example, Figure 4.3). Competitive, parasitic, and mutualistic interactions can all provide grist for the coevolutionary mill. In the following sections, we describe potential examples of antagonistic coevolution (insect herbivores on their host plants) and mutualistic coevolution (flowering plants and their insect pollinators, and ant–plant mutualisms).

6.5.1 Antagonistic coevolution

In Chapter 3, we described in some detail the myriad of physical and chemical traits exhibited by plants that can influence the preference (host choice) or performance (growth, survival, reproduction) of insect herbivores. We were careful to point out that simply demonstrating that a plant trait exhibits antiherbivore activity is not sufficient evidence to conclude that the trait is an evolved defense. Many physical and chemical features of plants appear to serve a variety of functions, and tight evolutionary associations between specific plant traits and specific insect herbivores are difficult to demonstrate unequivocally. However, there is an almost devout belief among many insect ecologists that the diversity of chemical structures found in the tissues of plants is in large part the result of selection by insect herbivores. If this is true, we might expect to find that a subset of these selective events has been reciprocal and, hence, coevolutionary, so that speciation in particular insect groups is closely coupled with speciation in their host plants (Brandle et al. 2005). This is the basis of what has been described as the "evolutionary arms race" between insect herbivores and their host plants. In its simplest form, the arms race can be characterized as follows: one or more individuals within a plant population develop a new, genetically based defensive trait by random mutation or recombination. Individuals with this trait suffer lower levels of herbivory than their conspecifics. Low levels of herbivory are associated with higher rates of survival or fecundity, and the proportion of individuals carrying the novel defense increases over time by the process of natural selection. Subsequently, within the herbivore population, one or more individuals develop a genetically based ability to breach the novel plant defense. These insects are at an advantage over their conspecifics, and so the ability to breach the plant defense spreads through the insect herbivore population. At some future point, yet another novel plant defense emerges, and the process of "escalation" begins again.

An important parallel to this coevolutionary process was suggested by Ehrlich and Raven (1964), who argued that coevolution between plants and insects could lead to speciation events and adaptive radiation of taxa. This has subsequently come to be known as "escape-and-radiation coevolution" (Thompson 1989). Ehrlich and Raven envisaged five steps to this type of coevolution, as follows:

1 Plants produce novel secondary compounds through mutation and recombination.

2 The novel chemical compounds reduce the palatability of these plants to insects, and are therefore favored by natural selection.

3 Plants with these new compounds undergo evolutionary radiation into a new "adaptive zone" in which they are free of their former herbivores. Speciation results in a new taxon of plants that share chemical similarity.

4 A novel mutant or recombinant appears in an insect population that permits individuals to overcome the new plant defenses.

5 These insects then enter their own new "adaptive zone" and radiate in numbers of species onto the previously radiated plant species containing the novel compounds, thereby forming a new taxon of herbivores.

Ehrlich and Raven's view of coevolution, then, results in speciation of both plants and insects rather than a tight coevolved interaction between a specific species of plant and a specific insect herbivore.

Ehrlich and Raven (1964) amassed an extraordinary amount of information to support their view of escape-and-radiation coevolution. Using the associations between major taxa of butterflies and their host plants, they argued convincingly that most butterfly taxa are each restricted to using a few related families of plants. Examples of major butterfly–host associations include Papilionidae on Aristolochiaceae, Pierinae on Capparidaceae and Cruciferae, Ithomiinae on Solanaceae, and Danainae on Apocynaceae and Asclepiadaceae (Table 6.3). In addition, when taxonomic associations between butterflies and their hosts are "broken", it may be because the plants are phytochemically similar. For example, the plant families Rutaceae and Umbelliferae share some attractant essential oils (Dethier 1941), which may be responsible for an apparent shift of some Rutaceae-feeding groups of Papilio to Umbelliferae. Ehrlich and Raven's paper has had a profound impact on the fields of evolutionary and population ecology, and is a "must read" for any serious student of insect ecology.

What evidence is there that coevolution in general is a powerful evolutionary force, and that Ehrlich and Raven's hypotheses in particular are correct? Surprisingly, few studies have emerged in full support of escape-and-radiation coevolution (Smiley 1985; Thompson 1989), perhaps because detailed and

Table 6.3 Some relationships between butterfly taxa and plant taxa discussed by Ehrlich and Raven (1964). (From Price 1996.)

Butterfly family	Subfamily	Tribe or lower taxon	Location	Plant family or higher taxon
Pieridae	Dismorphiinae			Leguminosae (exclusively)
	Coliadinae			Leguminosae (mostly)
	Pierinae	Pierini	Temperate	Cruciferae
			Tropical	Capparidaceae
		Euchloini	Temperate	Cruciferae
			Tropical	Capparidaceae
Nymphalidae	Ithomiinae			Solanaceae
	Danainae			Apocynaceae
				Asclepiadaceae
	Morphinae	11 Genera		Monocotyledons
		Morpho		Dicotyledons
	Satyrinae			Graminae, Cyperaceae
	Charaxinae			Ranales and others

unequivocal phylogenies of both phytophagous insects and their host plants have emerged only recently. One possible example in support of Ehrlich and Raven was presented by Berenbaum (1983), who described the diversification of coumarin-containing plants that followed the increasing modification of the coumarin molecule. Simple coumarins are small molecules (Figure 6.22a) found in many families throughout the plant kingdom. A modification of this molecule, hydroxycoumarin, occurs in about 30 plant families. Linear furanocoumarins (Figure 6.22b) are found in eight plant families, but most genera are within the Rutaceae or Umbelliferae. Finally, the angular furanocoumarins (Figure 6.22c) occur in only 13 plant genera (including two in the Leguminaceae and 11 in the Umbelliferae). One striking fact is that the diversity of plant species containing coumarin-based molecules increases as the complexity of the molecule increases (Figure 6.23). Genera containing angular furanocoumarins are the most speciose, followed by genera containing linear furanocoumarins, and so on. These data suggest that the evolution of phytochemical complexity may indeed shift plants into a new "adaptive zone" where radiation can occur. Even more striking, the number of species of insect herbivore associated with each chemical type also increases with chemical complexity (Figure 6.24). The diversity of *Papilio* butterfly species (see Plate 6.8, between pp. 310 and 311) and moths in the tribe Depressarini increases with the complexity of the coumarin structure. This would suggest that insects that have overcome each of the new coumarin forms have entered their own adaptive zone.

Not everyone agrees that the evolution of furanocoumarin-containing plants, and their associated insect herbivores, provides support for Ehrlich and Raven's hypotheses. Thompson (1986) has argued that differences in the way that species are lumped and split in different phylogenies can change the conclusions of the analyses. Moreover, "parallel cladogenesis" (in this case, the congruent development of taxa of insect herbivores with taxa of their host plants) is not sufficient evidence to demonstrate that escape-and-radiation coevolution has operated. Secondary plant metabolites can serve a large variety of functions (see Chapter 3), and the elaboration of chemical structures, and the radiation that may follow, need not be insect generated. If insects were not the major force leading to plant diversification, then adaptation and radiation by insect herbivores onto plants with novel compounds would be simple evolution, not coevolution. Although recent advances in molecular techniques have provided us with much more accurate insect and plant phylogenies, the precise mechanisms leading to parallel cladogenesis may always be open to question. Below, we provide some examples where parallel cladogenesis has been explored using molecular phylogenies and we consider the degree to which they support the theory of escape-and-radiate coevolution. We add the caveat

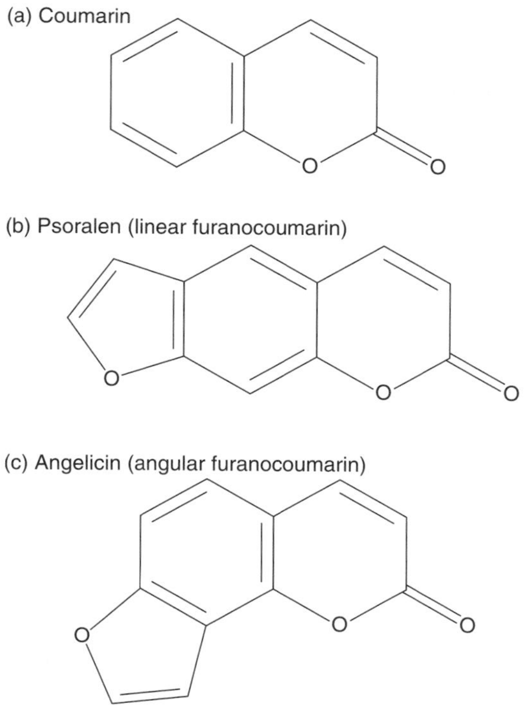

(a) Coumarin

(b) Psoralen (linear furanocoumarin)

(c) Angelicin (angular furanocoumarin)

Figure 6.22 Increasing complexity of (a) the basic coumarin structure found in many plant families, through (b) linear furanocoumarins found in eight plant families, and (c) angular furanocoumarins found in two plant families.

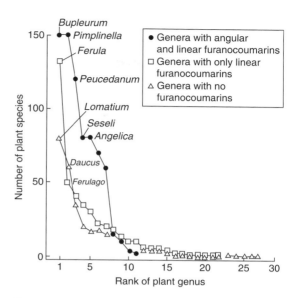

Figure 6.23 Number of the plant species per genus in the genera of Umbelliferae with different chemistries. There are generally more species per genus in those genera whose species contain angular and linear furanocoumarins. (After Berenbaum 1983; from Strong et al. 1984.)

that parallel cladogenesis is necessary, but not sufficient, to infer that coevolution has taken place.

It is first important to note that comparative molecular phylogenies of insects and their host plants do not always support the existence of parallel cladogenesis. For example, phylogenetic trees of European flies in the genus *Urophora* (Diptera: Tephritidae) are not congruent with the phylogenies of their thistle host plants (Brandle et al. 2005) (Figure 6.25). In other words, there exists no tight linkage between speciation events in the thistles and speciation events in the flies. Similarly, a molecular phylogeny of 77 leaf-mining *Phyllonorycter* (Lepidoptera: Gracillariidae) moths reveals that levels of cospeciation between leaf-miners and their host plants are not greater than those expected by chance, despite the physical intimacy of the relationship (Lopez-Vaamonde et al. 2003). Counter examples to parallel cladogenesis and

coevolution are important because they allow us to gauge more accurately the relative importance of coevolution and "standard" evolution in the development of insect–plant relationships. In fact, an increasing number of studies using phylogenetic analysis suggest that cospeciation events are rare in plant–insect systems (Percy et al. 2004). Rather, the partial congruence between plant and insect phylogenies suggests that there is a high frequency of host switching (to related plants) and tracking of plant traits by insects rather than tight cospeciation and parallel cladogenesis. For example, some phylogenetic analyses have suggested high levels of parallel cladogenesis between psyllids (Hemiptera) and legumes on the Atlantic Macaronesian Islands. However, the use of molecular clocks reveals that the main legume radiation occurred some 8 million years ago whereas the insect radiation occurred about 3 million years ago (Percy et al. 2004). Clearly, speciation events in the legumes and psyllids were not reciprocal, as required by escape-and-radiate coevolution.

The studies by Judith Becerra and colleagues have provided one of the most complete investigations

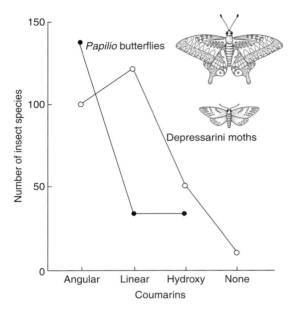

Figure 6.24 Number of species in two lepidopteran groups associated with Umbelliferae differing in their coumarin-based defences. (After Berenbaum 1983; from Strong et al. 1984.)

of the factors underlying the evolution of host use by phytophagous insects. Becerra's work distinguishes among three possible routes to host-plant use by insect herbivores (Becerra & Venable 1999). First, the evolution of host use by insects may be driven primarily by geographic overlap. Simply put, insects may be more likely to adapt to plants that are in the same geographic range as their current host plants. Second, coevolution and parallel cladogenesis may drive patterns of host use by insect herbivores. Finally, chemical similarity among potential host plants, irrespective of the relatedness of those plants, may determine host use by insects over evolutionary time. Distinguishing between the second and third hypotheses is important. While related plants are likely to be chemically similar to one another, chemical similarity need not follow exactly the same pattern as that of genetic relatedness. In order to distinguish among the three hypotheses, accurate molecular phylogenies are required for both insects and plants, in addition to information on the historical distribution of organisms.

Becerra and Venable (1999) accomplished these comparisons using data for the chrysomelid beetle, *Blepharida*, on its host plants in the genus *Bursera*. This ancient insect–plant interaction is more than 100 million years old, probably dating back to before the separation of Africa from South America (Becerra 2004). *Bursera* is defended by terpene-containing resins, and many species of *Blepharida* are monophagous. While there exists some degree of congruence between the molecular phylogenies of the beetles and their hosts, there is also some divergence (Figure 6.26). Moreover, geographic overlap between plants and herbivores does not adequately explain host use by beetles. The one exception to this is the beetle species, *Blepharida alternata* (Coleoptera: Chrysomelidae), which is unusually polyphagous. For this species, geographic overlap of hosts appears to have been important. In general, however, chemical similarity among host plants provides the best explanation of the evolution of host use by *Blepharida* (Table 6.4). Related insects appear to use hosts that are chemically similar rather than those plants that are most closely related phylogenetically. That being said, coevolution may still be operating at some level between *Blepharida* and *Bursera*. Molecular clocks have been used to date the defensive adaptations of *Bursera* and the counteradaptations of *Blepharida* (Becerra 2003). Results show that plant defenses, and insect counter-adaptations, evolved roughly in synchrony. For example, some species of *Bursera* hold their resins under pressure so that they "squirt" the herbivores when the resin canals are broken (see Chapter 3). The evolutionary age of this adaptation is approximately the same as the ability of some beetle species to disarm the defense. Overall, there appears to be good evidence for some degree of coevolution between *Bursera* and *Blepharida*, but it has not resulted in the tight parallel cladogenesis that we might have expected. Rather, chemical similarity trumps phylogenetic relatedness in driving the evolution of host use by *Blepharida*.

Evidence of tight coevolution at the species level between insect herbivores and their host plants is also not as common as we might once have thought. Spectacular examples do exist, such as the evolution of egg-mimic structures on *Passiflora* vines that are thought to deter *Heliconius* butterflies (see Plate 6.9, between pp. 310 and 311) from laying eggs (Gilbert & Raven 1975) (Figure 6.27). There seems little doubt that the egg mimics are a defense directed at the butterflies and that the energy-efficient flight and

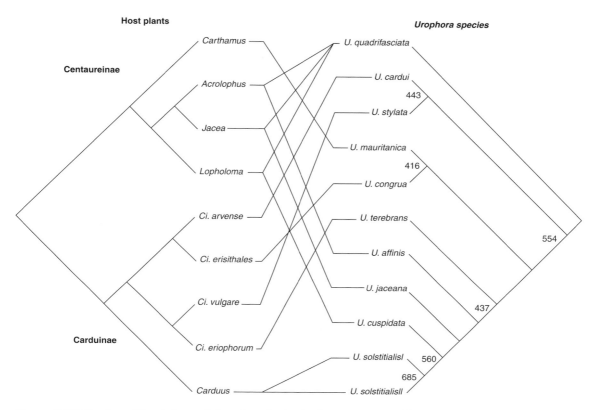

Figure 6.25 Topology of the phylogenetic hypothesis for 11 *Urophora* taxa generated by maximum likelihood from allozyme frequency data (consensus tree from 1000 bootstrap samples). The tree was rooted with *Myopites blotii*. Bootstrap support is shown for all nodes (frequency > 40%). The host-plant tree was derived from allozymes as well as cladistic analyses of morphological data. *Lopholoma, Jacea,* and *Acrolophus* are subgenera of the genus *Centaurea*. Hosts and associated *Urophora* taxa are connected by lines. (From Brandle et al. 2005.)

inspection responses of the butterflies have resulted from selective pressures imposed by the vine (Gilbert 1975). With such enthralling natural history stories, it is not surprising that many insect ecologists embraced coevolution as a dominant force in insect–plant interactions, and were less careful in applying the concept to their own data than they should have been (Janzen 1980). In reality, we might expect tight coevolution between insect herbivores and their host plants to be relatively rare for several reasons.

Perhaps most importantly, many different kinds of organisms can attack plants, including bacteria, fungi, viruses, nematodes, molluscs, mammals, birds, and reptiles. We should not expect perfect adaptation or counter-adaptation between insects and plants

when such a diverse group of consumers have the potential to injure many plant species (Poitrineau et al. 2003). Second, different insect species on the same host plant can exert selection pressures in different directions. For example, tannins appear to deter leaf-chewing insect herbivores on red oak (*Quercus rubra*) in Pennsylvania while, at the same time, attracting gall-forming and leaf-mining insects (Hunter 1997). Clearly, the advantages of producing tannins will vary depending on the relative densities of leaf-chewing and endophagous herbivores. Third, it is not clear that most insects exert much of a selective force on most plants most of the time (Strong et al. 1984). Rather, strong selection imposed on plants by insects may be relatively rare and episodic in natural

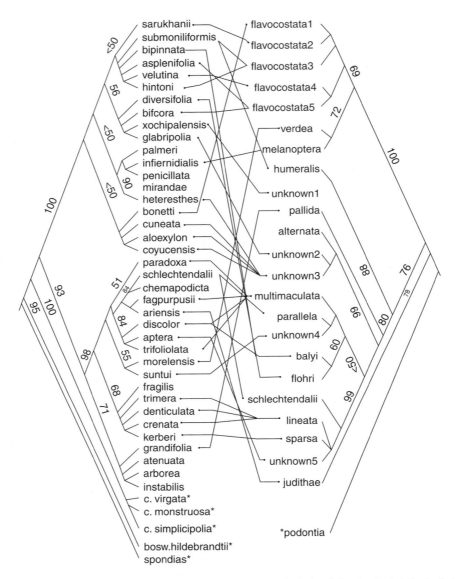

Figure 6.26 Feeding associations between *Bursera* host plants (left) and *Blepharida* beetles (right). Both phylogenies were reconstructed by using the nucleotide sites of the nuclear ribosomal DNA internal transcribed spacer sequences. Asterisks indicate outgroups, and the numbers above branches are bootstrap percentages. (From Becerra & Venable 1999.)

systems, and the links between episodic selection and plant diversification are not well established. Fourth, insects respond to traits other than defensive chemistry when selecting their host plants. For example, diversification and evolutionary patterns of

dietary breadth of seed beetles in the genus *Stator* (Coleoptera: Chrysomelidae, Bruchinae) appear more closely linked to oviposition substrate than to defensive chemistry (Morse & Farrell 2005). Finally, related to the second and third points just mentioned, the

Table 6.4 Correlation of *Blepharida's* phylogeny (topology B) with host biogeography, host phylogeny, and host chemistry (topology A). Analyses are reported with and without including the generalist, *Blepharida alternata*. (From Becerra & Venable 1999.)

Topology A	Topology B	DC		PDC				
				Discounting correlation with plant chemistry		Discounting correlation with plant phylogeny		Discounting correlation with plant biogeography
Excluding *Blepharida alternata*								
Plant biogeogram	Insect phylogeny	0.89	$p = .055$	0.91	$p < .1$	0.90	$p < .1$	–
Plant phylogeny	Insect phylogeny	0.89	$p = .013$	0.91	$p < .1$	–	0.88	$p < .05$
Plant chemogram	Insect phylogeny	0.73	$p < .001$	–	0.76	$p < .05$	0.77	$p < .05$
Including *Blepharida alternata*								
Plant biogeogram	Insect phylogeny	0.92	$p = .038$	0.92	$p < .05$	0.92	$p < .05$	–
Plant phyiogeny	Insect phylogeny	0.90	$p = .005$	0.94	$p < .1$	–	0.92	$p < .05$
Plant chemogram	Insect phylogeny	0.83	$p < .001$	–	0.84	$p < .05$	0.87	$p < .05$

DC, distortion coefficient; PDC, partial distortion coefficient.

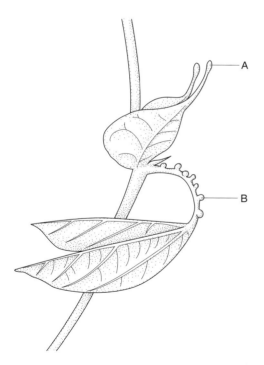

Figure 6.27 Mimetic *Heliconius* eggs on *Passiflora* (A) and extrafloral nectaries (B). (From a photograph in Gilbert & Raven 1975, reprinted from Strong et al. 1984.)

"insect–plant interface" is highly variable in space and time. Different plant (and insect) populations separated in space are under different selective pressures and unlikely to be on exactly the same "coevolutionary trajectory" at the same time (Thompson 1997). Similarly, a single population of insects and plants may interact to varying degrees at different times depending upon variation in the biotic and abiotic environment. The idea that the strength of coevolutionary interactions can vary in space and time has come to be known as the geographic mosaic theory of coevolution (Thompson 1997). Under this theory, interaction traits expressed by plants and herbivores match at coevolutionary hotspots, where selection is reciprocal, and mismatch at coldspots, where reciprocity is not a factor.

There are examples to support the geographic mosaic theory of coevolution from both antagonistic and mutualistic (below) interactions between insects and plants. For example, the effects of seed predation on cone and seed evolution in lodgepole pine, *Pinus contorta*, in the United States forms a geographic mosaic (Siepielski & Benkman 2004). Natural selection by red crossbills results in the evolution of larger lodgepole pine cones with thicker distal scales and, in the absence of squirrels, more seeds with a greater ratio of seed mass to cone mass. However, in the Little

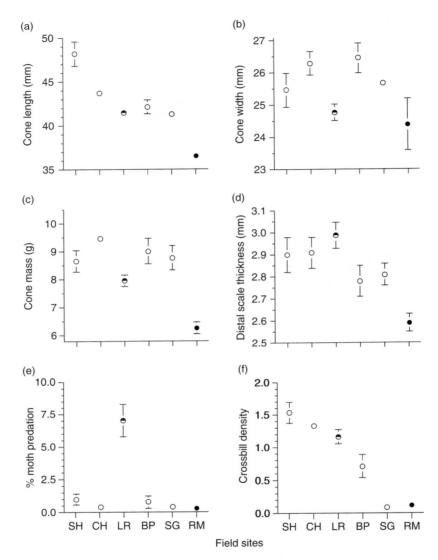

Figure 6.28 Four cone traits under selection by moths (a–c) or crossbills (d) in the Little Rocky Mountains, percent damage by moths (e), and the density of crossbills (f). The acronyms along the x-axis represent field sites. Only RM supports squirrels. Notice that site LR suffers high moth damage, associated with low cone width (b) and mass (c). (From Siepielski & Benkman 2004.)

Rocky Mountains, seed predation by cone borer moths, *Eucosma recissoriana* (Lepidoptera: Pyralidae), results in selection for smaller cones with fewer seeds (Figure 6.28). In other words, there is no single coevolutionary trajectory between lodgepole pine and its seed predators. The coevolutionary process varies in space depending upon the identity of the species within the local community. A similar conclusion was reached by Zangerl and Berenbaum (2003) in their study of coevolution between wild parsnip plants, *Pastinaca sativa*, and the parsnip webworm, *Depressaria pastinacella* (Lepidoptera: Oecophoridae).

Wild parsnips produce chemical defenses called furanocoumarins, which can be detoxified by the webworm's cytochrome P450 monooxygenases. Zangerl and Berenbaum examined 20 populations of wild parsnip and webworms in Illinois and Wisconsin for phenotype matching between plant defenses and insect detoxification systems. Twelve of the populations displayed phenotype matching, consistent with hotspots of coevolutionary interaction. In the other eight populations, the presence of an alternative host plant, cow parsnip, appears to have disrupted the tight interaction between wild parsnip and parsnip webworms (Zangerl & Berenbaum 2003).

What is clear is that attitudes have changed concerning the "evolutionary arms race" between insect herbivores and their hosts. In 1964, Ehrlich and Raven suggested: "The plant–herbivore interface may be the major zone of interaction responsible for generating terrestrial organic diversity." In 1984, 20 years later, Strong et al. concluded: "Coevolution most certainly does not provide a general mechanism to explain the contemporary structure of phytophagous insect communities." Today, we recognize that hotspots (and coldspots) of antagonistic coevolution exist, and that variability in the biotic and abiotic environment determine where and when coevolution will be strong. In the final analysis, insect and plant coevolution has been proceeding apace for a very long time. To quote Schoonhoven (2005), "clearly insects and plants have, over the millennia, formed a biological partnership that has flourished to mutual benefit". It is this concept of *mutual* benefit that we shall consider next.

6.5.2 Mutualistic coevolution

There is little doubt that mutualistic associations are common in insects. Table 6.5 provides an estimate of the number of species of insects in the British Isles that are involved with mutualistic interactions. Overall, about 36% of the British insect fauna have at least one mutualist association, and many have more than one (Price 1996). Although any estimation technique is open to criticism, it seems clear that mutualism is a pervasive feature of the ecology and evolution of insects. The success of some insect groups is almost certainly dependent upon their associations with other species. For example, flagellate protozoan or bacterial symbionts in the guts of termites (Isoptera)

Table 6.5 An estimate of the numbers of insects involved with mutualistic microorganisms in the British Isles. (Based on data from Buchner 1965; Price 1996.)

Order	Species	No. with mutualists
Orthoptera	39	8
Phthiraptera	308	308
Thysanoptera	183	183
Heteroptera	411	288
Homoptera	976	891
Lepidoptera	2233	1116
Coleoptera	2844	709
Hymenoptera	6224	2874
Diptera	3190	811
Siphonaptera	47	47

have permitted this order of insects to exploit a high-cellulose diet that is unsuitable for most other insect species (Bignell et al. 1997). The association is sufficiently well developed that many termites have modified guts with anaerobic microsites to enhance cellulase activity. Presumably, this association has facilitated the radiation of termites into the wide variety of environments in which they are found today (see Section 6.5.5 for more details). The role of termites in terrestrial ecosystems is considered in detail in Chapter 8.

The "intraorganismal" mutualisms of termites and other insects, where the associated species actually live within the body of the insect, almost inevitably require adaptive adjustments by both species in the association, and are therefore amongst the most reliable examples of coevolution (Takiya et al. 2006; see also Section 6.5.5). Many groups of Hemiptera (cicadas, plant-hoppers, aphids, etc.) have specialized organs called mycetocytes in which mutualistic microorganisms live. The presence of these organs in even very primitive Hemiptera has led some researchers to suggest that the whole taxon was able to radiate onto plants and suck plant juices because their symbionts provided essential nutrients absent from the plants (Buchner 1965; Price 1996). If true, this would be one of the most vivid examples of adaptive radiation driven by coevolution. Aphid–symbiont interactions are considered in more detail later in this chapter. Nutrition mutualists are also found in phytophagous Lepidoptera and parasitic fleas (Siphonaptera) and lice (Phthiraptera).

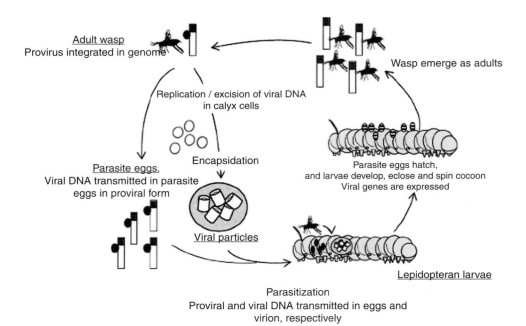

Adult wasp
Provirus integrated in genome

Wasp emerge as adults

Replication / excision of viral DNA
in calyx cells

Encapsidation

Parasite eggs.
Viral DNA transmitted in parasite
eggs in proviral form

Parasite eggs hatch,
and larvae develop, eclose and spin cocoon
Viral genes are expressed

Viral particles

Lepidopteran larvae

Parasitization
Proviral and viral DNA transmitted in eggs and
virion, respectively

Figure 6.29 Polydnavirus and endoparasitoid wasp cycles. Viral DNA is integrated in the genome of the wasp. The viral DNA is chromosomally transmitted to the next generation as a provirus. In ovarian cells circular DNA molecules produced from the provirus are packaged to form viral particles. During oviposition the wasp injects viral particles to the lepidopleran host. Virus particles enter host cells and viral genes are expressed but the virus genome does not replicate in the host. Virus gene products allow successful development of the parasitoid larvae that emerge out from the parasitized insect, spin their cocoons, and emerge as adults to mate and search for new insect hosts. (From Dupuy *et al.* 2006.)

One fascinating "endosymbiotic" relationship that has gained increasing attention in recent years is the association between some parasitic wasps (Hymenoptera: Ichneumonidae and Braconidae) and polydnaviruses (PDVs). Ichneumonids and braconids are (mainly) endoparasites of other insects (see Chapter 5) and, as such, must deal with the immune systems of their hosts. These parasitoids have adopted viral mutualists that suppress the cellular immune response of their host insect. Figure 6.29 depicts a general life cycle (Dupuy et al. 2006). The PDVs (or virus-like particles, VPs) are held within the lumen of the oviducts of female wasps. PDVs enter the host insect, often a lepidopteran larva, when the female wasp oviposits within the body of the host. The PDVs have been detected in host hemocyctes (Han et al. 1996), where they apparently attack the immune system. Because (i) successful parasitism appears much reduced in the absence of PDVs, and (ii) the PDVs replicate primarily in the wasps, the association

appears truly mutualistic. To what degree the association has led to coevolution and parallel cladogenesis is as yet unclear, but the association is undoubtedly ancient, with its origins proposed between 75 and 140 million years ago (Webb et al. 2006).

Mutualisms between insects and other taxa that are not intraorganismal (i.e. one mutualist does not live permanently within the other) may be less tightly coevolved. In fact, while searching for examples of tight coevolution that included insects for this text, we were struck by the repetition of four or five well-known examples (e.g. yucca moths, *Heliconius* butterflies, fig wasps, etc.) from one review to the next. Although we commit the same sin below, we suggest that the lack of diversity of examples of tight coevolution among free-living mutualists historically chosen by researchers reflects a paucity of such relationships in natural systems. Most free-living insects, or their free-living mutualists, are probably not in tight coevolutionary relationships with one another. There are, of course, some exceptions described below. But we believe that the best examples of the coevolution of insects with mutualists are intraorganismal (see above).

"Looser" mutualisms between individuals or populations of insects and their symbionts, resulting from diffuse coevolution, are perhaps more common than tight species-to-species coevolution. The pollination of most insect-associated angiosperms (flowering plants) probably falls into this category, having evolved via plants "co-opting" non-specialized insect visitors into flower feeding in order to pick up pollen (Bronstein et al. 2006). This in turn exerts strong evolutionary selection pressure on both plants and insects to develop highly complex morphological, phenological, behavioral, and chemical traits. Many flowering plants use insects such as bees as "flying genitalia" for reproduction, providing nectar rewards in return for the transfer of gametes. However, the majority of flowering plants are visited by more than one species of insect and most insect pollinators visit more than one species of plant. Diffuse coevolution may be the best way to view the adaptive changes in flowers and insects that benefit both the pollinator and pollinated species. Roubik (1992) has suggested that there is very little tight coupling between bees and flowering trees in the New World tropics, and has found over 200 pollen species within bee colonies at a single site. As Feinsinger (1983) remarked, "most plants and pollinators move independently over the landscape, not matched in pairs". Indeed, many pollinating insects appear adapted specifically to account for spatial and temporal variation in flowering plants. Inoue and Kato (1992) provided compelling evidence that variation in the body size of bumblebees, both within a single colony at one point in time and from the start to the end of the pollinating season, allows the bees to visit a wide array of plant species. Their studies of five species of *Bombus* (Hymenoptera: Apidae) in Japan have demonstrated remarkable plasticity in proboscis length and head capsule width within a species. Although many pollination biologists have reported the averages of morphological features in insect pollinators that match the morphology of flowers (e.g. Figure 6.30), it is the variation around these averages that provides the best evidence for diffuse coevolution between flowers and pollinators (Figures 6.31 and 6.32). That being said, not all plants are simply "open for business" for any passing pollinator. Even when a plant species as a whole receives visits from many insect species, there can be variation among plant populations, or among individual plants within populations, in the pollinators that visit. In studies of the insect-pollinated shrub, *Lavandula latifolia*, some plant populations, and some individual shrubs, receive a relatively small subset of all potential pollinators (Herrera 2005). In other words, at the local scale, some insect-pollinator systems may not be quite as "diffuse" as some others.

Of course, there are a few examples of insect pollinator–plant mutualisms that are highly specialized

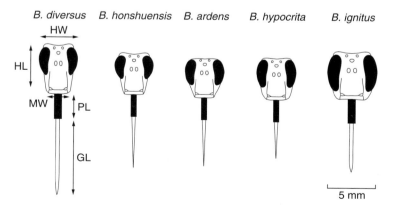

Figure 6.30 Average sizes of the heads of workers of five bumblebee species in Japan. These averages hide the considerable variation in morphology that occurs within species. (From Inoue & Kato 1992.)

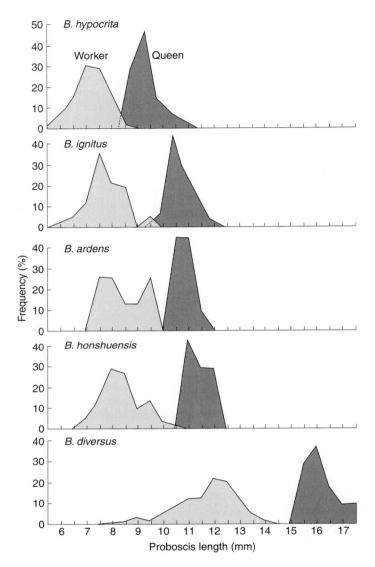

Figure 6.31 Frequency distributions of proboscis length of five bumblebee species. (From Inoue & Kato 1992.)

and, presumably, tightly coevolved. Gilbert (1991) has suggested that the pollination of *Anguria* and *Gurania* vines by *Heliconius* butterflies represents the result of reciprocal evolutionary change. Male vines flower throughout the year, even when females are not in flower, providing a constant source of nectar (and pollen) for the long-lived butterflies. The combination of longevity and the ability of *Heliconius* to learn specific routes through the forest provides an evolutionary incentive for the vines to offer a consistent reward to visitors. In fact, a single male inflorescence of *Anguria* may produce about 100 flowers consecutively. Even though each flower lasts only a day or so, the inflorescence will attract *Heliconius* for several months. Butterflies increase the probability of seed set and are predictable visitors over extended periods

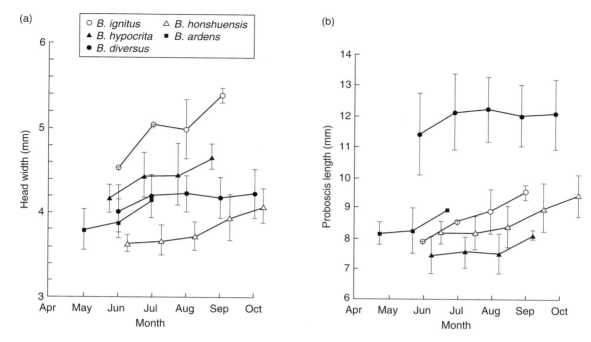

Figure 6.32 Seasonal changes in head width (a) and proboscis length (b) of workers of five bumblebee species in Japan. (From Inoue & Kato 1992.)

while the plants provide nutritious nectar and pollen that presumably improve both the fecundity and longevity of *Heliconius*. Various species of *Heliconius* differ only slightly in wing coloration and pattern, and closely related species seem to readily hybridize in nature, and hybrid traits thus produced are able to generate reproductive isolation between species (Mavarez et al. 2006). Note also that most *Heliconius* host plants are toxic, containing cyanogenic glycosides. The butterfly larvae sequester these toxins from their plant food, and pass them on to their adults to use them as defenses against their own predators. So important is this chemical defense in the adult *Heliconius* (both of course for the insect itself but also for the plant that originally produced the toxin − without the butterfly, there would be no pollination) that males actually transfer plant-derived cyanogens to the female in the "nuptual gift" (see Chapter 7) at mating (Cardosa & Gilbert 2007).

Arguably the most remarkable example of plant–pollinator coevolution is that of the fig wasps (Hymenoptera: Agaonidae) and fig trees *Ficus* (Galil &

Eisikowitch 1968). With few exceptions, each species of fig in the genus is pollinated by a different species of agaonid wasp, and the near-perfect match between fig species and wasp species is an astonishing example of parallel radiation. Figs produce false fruits (Figure 6.33) within which male and female flowers are produced. Initially, an opening at the top of the inflorescence, called the ostiole, opens in synchrony with female flowering. Female wasps enter through the ostiole and pollinate some female flowers while ovipositing in, and pollinating, others. Flowers that receive pollen only (usually long-styled flowers) develop to produce seeds, whereas flowers that receive both wasp eggs and pollen (usually short-styled flowers) become galls with developing wasps inside. Male wasps, wingless with modified, elongated abdomens, emerge from the galls first and mate with females while they are still within the galls. The males then bore exit holes in the side of the inflorescence. When the mated females finally emerge from their galls, it is in synchrony with male flowering. Female wasps collect pollen from the male flowers

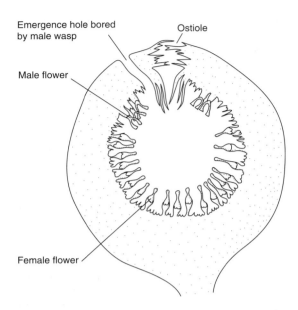

Emergence hole bored
by male wasp

Ostiole

Male flower

Female flower

Figure 6.33 Cross-section of a fig inflorescence. (After Galil & Eisikowitch 1968; Price 1996.)

Figure 6.34 Female yucca moths (*Tegeticula*) pollinating their *Yucca* host. (Courtesy Olle Pellmyr.)

and store it in specialized pouches on the coxae of their front legs. They then leave the inflorescence through the hole bored by the males, and fly in search of another fig to start the process over again. Almost every aspect of inflorescence morphology and phenology (timing of development) appears adapted to the morphology and phenology of the fig wasps (and vice versa). The intricacy of the coevolved relationship may arise, in part, because the fig wasps are indeed "endosymbionts" for at least a portion of their life cycle, living as they do within the false fruits of the fig trees. As we suggested earlier, intraorganismal mutualisms are likely to be more tightly coevolved than those of free-living organisms.

As in all mutualisms, there are costs and benefits to the organisms involved in the mutualistic interaction. For example, when fig wasps form galls in the false fruits of figs it presumably imposes some cost to fig fitness. The balance between costs and benefits is important because, if costs begin to outweigh benefits for one of the partners, the interaction can change from mutualism to parasitism. Given that the costs and benefits of mutualism are likely to vary with environmental conditions, we might expect

to see spatial and temporal variation in the outcome of mutualistic interactions. This phenomenon is understood nowhere better than in the interaction between *Yucca* plants and yucca moths (Figure 6.34). *Yucca* and true yucca moths (primarily the genera *Tegeticula* and *Parategeticula*) participate in an obligate mutualism in which the *Yucca* are pollinated by the moths, while moth larvae consume a portion of the seeds produced by the *Yucca* plants. In a typical *Tegeticula* life cycle, female moths eclose and mate with a male in the *Yucca* flowers. Females then collect pollen with specialized tentacles and oviposit into a *Yucca* ovary. After oviposition, females actively apply pollen to the stigmatic cavity, ensuring that there will be developing seeds for the larvae to feed on. The larvae feed until the fifth instar and then drop into the soil to diapause and pupate. The yucca moths therefore act as both plant pollinators and floral parasites of their *Yucca* hosts. Not all moth species that pollinate *Yucca* plants are active pollinators. For example, there are two species of *Tegeticula* that lack the specialized tentacles of the other yucca moths, and perform no active pollination services. These species are considered to be cheaters in the mutualism. Even those species of yucca moths that actively pollinate their hosts may cheat by laying a large number of eggs within *Yucca* flowers, imposing greater costs on *Yucca* fitness than benefits through pollination. Some *Yucca* appear able to punish such cheaters by abscising flowers that contain too many moth larvae.

A related group of moths, in the genus *Greya*, provide a fascinating example of the dynamic balance between mutualism and parasitism. We learned earlier that coevolutionary processes occur in geographic mosaics, and the same is true for the interaction between the moth *Greya politella* (Lepidoptera: Incurvariidae) and its host plant, *Lithophragma parviflorum*. *Greya* does not actively pollinate *Lithophragma*: pollination is passive, and occurs during oviposition for seed parasitism. Results from studies by Thompson and Cunningham (2002) demonstrate that the interaction can be strongly mutualistic in some habitats, but commensal or antagonistic in neighboring habitats (Table 6.6). The relationship becomes antagonistic when the fitness gained by *Lithophragma* plants through pollination services is less than the fitness loss incurred through seed predation. This depends largely on the availability of other pollinator species in the habitat. When other pollinators are abundant, the fitness losses to seed parasitism are not balanced by fitness gains because *Lithophragma* does not need *Greya* for pollination. In contrast, when other pollinator species are rare, *Lithophragma* relies upon *Greya* for pollination, and the costs of seed loss are more than balanced by the benefits of pollination.

Pollination mutualisms that also include floral parasitism occur in a number of other systems (Hartmann et al. 2002; Kawakita et al. 2004) and, whenever pollination is accompanied by seed consumption, there is likely to be evolutionary tension between mutualism and parasitism. There is some evidence to suggest that pollination mutualisms of the type represented by *Yucca* and yucca moths may evolve from parasitic relationships (Westerbergh & Westerbergh 2001). For example, the geometrid moth, *Perizoma affinitatum* (Lepidoptera: Geometridae), is a seed predator of the plant *Silene dioica*. Although the moth does not actively pollinate its host in the same way that most *Tegeticula* species do, the transfer of pollen among flowers during oviposition is quite substantial (Westerbergh 2004). As a result, the net effect of the seed predator on *S. dioica* depends upon the availability of other pollinators in the habitat, in a manner similar to that of the *Greya–Lithophragma* interaction. If fewer than 60% of *Silene* flowers are pollinated by other pollinator species, the net effect of *Perizoma* is actually positive, despite the seed predation. Whether this system provides a model for the evolution of mutualism from parasitism has yet to be determined. It is clear, however, that the fine line that exists between mutualism and parasitism results in fascinating and complex evolutionary interactions across geographically heterogeneous landscapes.

As we have seen, there is no advantage to a plant to attract insects to its flowers if pollen is not transferred, but there are many examples of "cheaters" on pollination mutualisms (Bronstein et al. 2006). In these cases, insects from a variety of groups steal the nectar from flowers but fail to contact pollen in the process. These nectar robbers may chew holes in the flowers, bypassing the anthers and stigmas. Ironically, however, even this type of cheating may have a happy ending; the flowers of *Linaria vulgaris*, a perennial herb in Colorado, are attacked by the bee *Bombus occidentalis* (Hymenptera: Apidae) which steals nectar by biting holes in the base of flowers.

Table 6.6 Effects of *Greya politella* oviposition on the probability of floral development in *Lithophragma parviflorum*. Numbers are the ratio of the percentage of developed capsules containing eggs to the percentage of aborted capsules containing eggs. Significant effects (in bold) are greater than 1 when pollination benefits outweigh seed consumption costs, and less than 1 when costs exceed benefits. (Data Thompson & Cunningham 2002.)

| Population | Effect of *Greya* oviposition on floral development | | |
	1997	1998	1999
Mutualistic effect of *Greya*			
Berg	1.1	**3.0**	**1.8**
Meadow	–	**2.4**	1.6
Saddle	0.8	**20.5**	**9.2**
Turnbull	1.0	1.0	2.4
Antagonistic effect of *Greya*			
Rapid	**0.5**	0.0	1.3
Salmon	–	**0.4**	–
South Fork	–	**0.6**	0.4
Wenaha	2.1	**0.4**	1.6
No effect of *Greya*			
Granite	0.5	1.1	1.0
Keahing	1.1	1.1	1.2
Selway	–	0.9	–
Wind	–	0.7	1.0

This in effect creates extrafloral nectaries (see below) which attract ants that in turn reduce the numbers of flower- and seed-eating beetles (Newman & Thomson 2005).

6.5.3   Ant–plant associations I: myrmecophytism

In Chapter 3, we described a special kind of "plant defense" in which certain plants provide incentives (food and shelter) to ants. The ants, in turn, are thought to remove insect herbivores from the plants and so reduce tissue damage. Ant–plant associations have arisen independently in many plant taxa, and are often thought to represent good examples of coevolution. For example, Janzen (1966, 1967) described the close association between species of *Acacia* in Central America and ants in the genus *Pseudomyrmex*. So-called Beltian bodies on the leaf tips of *Acacia* provide protein rewards to foraging ants. In addition, hollow swollen thorns provide shelter (domatia), and extrafloral nectaries provide sources of sugar (see Figure 3.21). Plants that provide such rewards for ants are termed "myrmecophytic", or "ant-loving". Myrmecophytism occurs in over 100 genera of plants, though it is almost entirely restricted to tropical regions of the world (Gaume et al. 2005).

Plants in the genus *Cecropia* in Central America also provide domatia and food rewards for ants. In this case, the extrafloral nectaries produce glycogen, a highly branched polysaccharide that is normally found in animal tissues such as our own liver and muscle. Glycogen can be broken down rapidly by animals to release glucose. The production of an animal sugar by a plant is a remarkable adaptation that is hard to explain in any context except the benefits that must be received by attracting foraging ants. It seems obvious that ants benefit from the relationship, but what about the plants? In one study, Vasconcelos and Casimiro (1997) have shown that *Azteca alfari* ants influence the foraging of leaf-cutting ants, *Atta laevigata* (Hymenoptera: Formicidae; note that all ant species are in the family Formicidae (order Hymenoptera)), among *Cecropia* species. Leaf-cutting ants cut foliage from plants and use the leaf fragments to grow fungus gardens within their nests. Although the leaf-cutters are actually fungivorous, they can be viewed as herbivores from the plant's perspective. In the presence of the predaceous *A. alfari*,

the leaf-cutting ants are forced to utilize less-preferred species of *Cecropia*. Similarly, Fonseca (1994) demonstrated a clear benefit to the plants as a result of hosting ants. Experimental removal of *Pseudomyrmex concolor* from the myrmecophytic plant species *Tachigali myrmecophila* resulted in a 4.3-fold increase in herbivore densities, and a 10-fold increase in levels of leaf damage. Leaf longevity was two times higher on plants occupied by ants than on unoccupied plants, and apical growth was 1.6-fold higher on occupied plants. Ant–plant associations such as these can vary from very loose to very tight and complex (Dutra et al. 2006). The efficiency of protection by ants varies with plant and insect species (Dejean et al. 2006). As an aside, ant-plants may also reduce their investment in "traditional" antiherbivore traits such as chemical defenses (Janzen 1966; Nomura et al. 2000; Eck et al. 2001).

However, not all studies have documented clear benefits to plants of hosting ants. For two Amazonian species of *Cecropia*, herbivory levels are sometimes low, and do not differ when ants are excluded from the plants (Faveri & Vasconcelos 2004). In this case, plants are paying the costs of the mutualism (providing food and shelter) without receiving obvious benefit. It may be that attack by herbivores is unpredictable, and it pays to host ants for those times when herbivory rates increase. We might also expect natural selection to favor plants that only pay the costs of mutualism when there are clear benefits to be gained. Under this scenario, we might expect plants to increase the reward that they offer to ants once herbivores begin to attack. Is there any evidence that ant recruitment is induced by herbivore damage? In Section 3.3.7 we described the volatile signals that some plants use to attract the natural enemies of herbivores. A number of studies (Bruna et al. 2004; Christianini & Machado 2004; Romero & Izzo 2004) have shown recently that ants can also respond to signals produced from damaged leaf tissue. For example, damage to young leaves of the ant-plant, *Hirtella myrmecophila*, in the central Amazon, results in the recruitment of *Allomerus octoarticulatus* (Hymenoptera: Formicidae) ants to the site of damage (Romero & Izzo 2004) (Figure 6.35). Here, the reward for the ants is presumably increased predictability of herbivores as food.

There is also evidence to suggest that some extrafloral nectar production is induced by herbivore activity. The North American tree, *Catalpa bignonioides*,

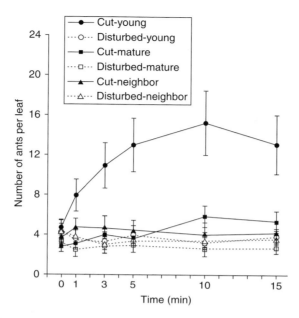

Figure 6.35 The effects of damage to the ant-plant *Hirtella myrmecophila* on ant recruitment to sites of damage. (From Romero & Izzo (2004).)

uses extrafloral nectar to attract the ant, *Forelius pruinosus*, in response to damage by the sphingid moth, *Ceratomia catalpae* (Lepidoptera: Sphigidae). When caterpillars attack *Catalpa* leaves, sugar production from the extrafloral nectaries on those leaves increases by two- to three-fold within 36 h (Ness 2003) (Figure 6.36a). As a result, ant attendance on damaged leaves increases dramatically (Figure 6.37b). In this case, the presumed costs of higher nectar production are only paid when herbivores damage the plants and there are benefits to be gained from herbivore removal. However, there is some evidence that the ants themselves, rather than herbivores, can induce increased rewards from ant-plants. Ants appear able to induce the formation of twig-like domatia on the Amazonian tree, *Vochysia visimaefolia* (Bluthgen & Wesenberg 2001). If ants induce greater rewards in the absence of increased herbivory, it may be a form of plant exploitation by ants. Note, however, that the use of domatia can depend on the age of the trees in question. In the case of *Cordia alliodora*, a South and Central American tree species, Trager and Bruna (2006) found that only older (5 or more years of age) trees had ants in domatia that defended against defoliation by lepidopteran larvae.

Myrmecophytic plants may gain additional advantages from the ants that they attract beyond the removal of insect herbivores. In several cases, ants have been shown to prune back plants adjacent to the one on which they live. Pruning of neighbors may reduce competition between their host plant and surrounding plants for light and nutrients. However, ants may also benefit by pruning. Studies of *Triplaris americana* plants in Peru and their associated *Pseudomyrmex dendroicus* ants have shown that pruning activity reduces invasions by *Crematogaster* ants by physically separating adjacent plants. *Crematogaster* invasions inhibit *Pseudomyrmex* feeding activity, and the invaders can steal *Pseudomyrmex* broods and usurp their nests (Davidson et al. 1988). This example illustrates the general point that we have to be careful when we make assumptions about the benefits of apparent mutualisms. In this case, pruning of neighbors may actually be more important for the ants than for their plant hosts.

Indeed, the fact that *Crematogaster* ants will readily utilize the nests of *Pseudomyrmex* on *T. americana* suggests that the mutualism between the plant and *Crematogaster* may not be especially tight. We do not really know whether the plant benefits more

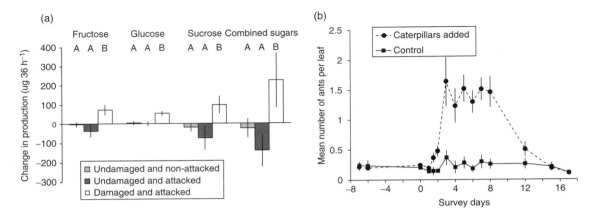

Figure 6.36 (a) The effects of foliar damage to *Catalpa bignonioides* on sugar production by extrafloral nectaries. (b) The effects of caterpillar feeding on *Catalpa bignonioides* on ant recruitment to attacked leaves. Caterpillars were added to the treatment plants after day 0, and removed after survey day 4. Control plants lacked caterpillars. (From Ness (2003).)

from the presence of one ant species than the other. Previous examples described in this chapter have taught us to be suspicious of inferring coevolution by simply observing current ecological relationships, and ant–plant associations should be no exception. Are ant–plant mutualisms really tightly coevolved? Certainly, there is reason to doubt that ant–plant associations have arisen by escape-and-radiate coevolution (above). For example, the *Leonardoxa africana* (Leguminosae: Caesalpinioideae) complex in Africa is composed of four closely related plant species. DNA sequence data have shown that the interactions of two of the four plant species with the ants *Aphomomyrmex afer* and *Petalomyrmex phylax* appear to have arisen independently and are not the result of cospeciation (Chenuil & McKey 1996). They further suggest that specific ant–plant associations originate by "ecological fitting of preadapted partners", not escape-and-radiate coevolution. Likewise, there is no strong evidence of cospeciation to be found in the analysis of molecular phylogenies of *Crematogaster* ants and their *Macaranga* hosts in Southeast Asia (Quek et al. 2004). Rather, myrmecophytism appears to have evolved independently multiple times within the genus *Macaranga* (Blattner et al. 2001; Davies et al. 2001). Indeed, some plants can host a variety of different ant species in space or time. For example, the tree *Barteria nigritana*, from Lower Guinea, hosts

a series of ant species in predictable order as plants age (Djieto-Lordon et al. 2004). There is no tight relationship with a single ant species. Similarly, the two neotropical myrmecophytes, *Cordia nodosa* and *Duroia hirsute*, can associate with ants in the genera *Azteca*, *Allomerus*, or *Myrmelachista*. In this system, *Azteca* and *Allomerus* provide the greatest protection from insect herbivores while *Myrmelachista* provides the greatest protection from encroaching vegetation (Frederickson 2005). The relative benefits to the plants of associating with the different ant partners will therefore likely depend on the relative costs of herbivory and plant competition to plant fitness.

As with pollination mutualisms there can be a fine line between beneficial and deleterious effects of ants on plants. In some cases, foraging ants may not only discourage insect herbivores, but insect pollinators as well. For example, the semimyrmecophyte, *Humboldtia brunonis*, in southern India produces extrafloral nectaries that are close to the real flowers. Foraging ants discourage visitation by pollinators, and actually reduce fruit production by the plant (Gaume et al. 2005). One ant species, *Crematogaster dohrni*, visits the real flowers as well as the extrafloral nectaries, and can damage the flowers before pollination. The damage can be sufficiently severe that the ant castrates the plant, a clear switch from mutualism to parasitism. Can plants "punish" ants that cheat on

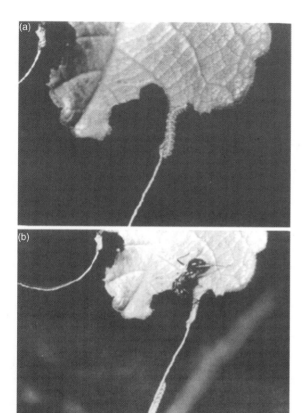

Figure 6.37 Larvae of *Eunica bechina* (a) build frass chains as they feed, and (b) retreat down these chains when harrassed by ants. (From Freitas & Oliveira 1996.)

the mutualism, in the same way that *Yucca* plants can punish yucca moths by floral abscission (see above)? The ant-plant *Hirtella myrmecophila* is also susceptible to castration by its ant partner, *Allomerus octoarticulatus*. In this system, the plant produces leaf pouches as domatia for ants. *Hirtella* drops the domatia from older leaves, which are less susceptible to herbivores than are younger leaves, and flower castration is negligible on older branches when compared to younger branches. As a result, fruit set is primarily restricted to older branches, minimizing the effects of cheating by ants (Izzo & Vasconcelos 2002).

Even if ant–plant associations are not all tightly coevolved, they provide dramatic examples of mutualism and adaptation, and are remarkably common. In the Pasoh Forest Reserve of peninsular Malaysia, for example, 91 of 741 woody plants examined had extrafloral nectaries (12.3% of species and 19.3% of vegetation cover). The nectaries occurred in 47 genera and 16 plant families. The interactions between the plants with extrafloral nectaries and ants appeared to be rather facultative and non-specific (Fiala & Linsenmair 1995). A continuum of myrmecophytism, from weakly facultative to obligate, is shown by the paleotropical tree genus *Macaranga* (Euphorbiaceae) (Fiala & Maschwitz 1991). All *Macaranga* species provide food for ants in various forms (extrafloral nectaries or fat bodies), but the obligate myrmecophytes start producing food rewards at a younger age and may produce more. In addition, only the obligate myrmecophytic *Macaranga* offer nesting spaces (or domatia) inside internodes in the stem of the plant, which become hollow because of the degeneration of the pith. Non-myrmecophytic trees retain solid stems with a compact and wet pith. The stem interior of some "transitional" species remains solid, but the soft pith can be excavated: provision of nesting sites may be the most important step toward obligate myrmecophytism in this genus of trees.

As might be expected, some insect herbivores have developed a variety of adaptations to counter the effects of ants on plants. For example, although larval densities of *Polyhymno* sp. (Lepidoptera: Gelechiidae) are lower on ant-defended individuals of *Acacia cornigera* in Mexico than on ant-free individuals, larvae gain some protection from the ant *Pseudomyrmex ferruginea* by constructing sealed shelters made from the pinna or pinnules of *Acacia* leaves. As a result of this protection, the larvae of *Polyhymno* can sometimes reach densities that can defoliate and kill their host plants (Eubanks et al. 1997). Likewise, the Brazilian savanna shrub *Caryocar brasiliense* has nectaries that attract ants, and oviposition by the butterfly *Eunica bechina* (Lepidoptera: Nymphalidae) is deterred by the presence of real and experimental (rubber) ants. Although larval mortality is strongly affected by ant visitation rates to plants, the butterfly larvae have adopted a remarkable defense to reduce predation pressure. They build stick-like structures from their own frass (fecal pellets) and retire to the end of these frass chains when harassed by the ants (Freitas & Oliveira 1996) (Figure 6.37).

6.5.4 Ant–plant associations II: myrmecochory

A second form of ant–plant association that can be described as a mutualism is myrmecochory – the dispersal of seeds by foraging ants. In the same way that plants can exploit pollinators as flying genitalia, so they can also exploit ants as mobile gardeners. By providing rewards for ants on the surface of seeds, some plants can benefit from the movement of their propagules across distances greater than would otherwise be possible. In such cases, the phenology of seed production appears to coincide with periods of greatest ant activity (Oberrath & Bohning-Gaese 2002).

Nakanishi (1994) distinguished between three kinds of myrmecochory. First, there is myrmecochory with autochory. In this case, the seed pods of plants release the seeds explosively so that seeds are already deposited some distance from the parent plant, and are then dispersed over greater distances by foraging ants. Second, there is myrmecochory with vegetative reproduction: plants reproduce both vegetatively and sexually, with ants transporting the sexual propagules (seeds). Finally, there is "pure" myrmecochory, where the only dispersal of seeds away from the parent plant relies on ants. The last of these depends most heavily on ants to disperse seeds away from the parent, and we might expect plants that rely on pure myrmecochory to provide the greatest reward to their ant mutualists.

How do plants reward ants for seed dispersal? In some cases, ants are rewarded simply by the fleshy fruit around some small seeds (Passos & Oliveira 2003). In such cases, ants are not the only potential dispersal agents of fleshy seeds, and the interaction is facultative. In contrast, plants that have evolved specifically for seed dispersal by ants generally have detachable protrusions on their seed surfaces called elaiosomes (Gammans et al. 2005) (Figure 6.38). Elaiosomes are high in lipids and fatty acids, and can also contain proteins. Foraging ants return seeds to their nest, remove the reward, and the seeds are left to germinate, often underground in soil that has been aerated by ant activity. It has been suggested that the chemistry of elaiosomes influences their attractiveness to ants and subsequent ant behavior.

Figure 6.38 (a) Elaiosome on a seed of *Croton priscus*. (b) Ant worker in the genus *Pheidole* removing an elaiosome from *Croton priscus*. (Photographs courtesy I.Sazima, property of the Association of Tropical Biology, Kansas, USA.)

For example, a comparison of the chemistry of the elaiosomes from three species of *Trillium* found significant variation in their protein, neutral lipid, and fatty acid contents (Lanza et al. 1992). Results suggested that oleic acid stimulated ants to pick up the seeds and that linoleic acid stimulated ants to carry the seeds to the nest. Variation in these compounds, and in lipid to protein ratios, appeared to explain the relative dispersal success of the three *Trillium* species. Elaiosomes can be extremely important for ant nutrition. In a labeling experiment with

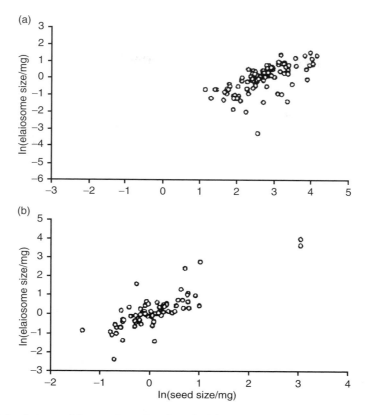

Figure 6.39 Covariation between elaiosome size and seed size for (a) 109 *Acacia* species, and (b) 98 other species of plant in Australia. (From Edwards et al. 2006.)

stable isotopes, Fischer et al. (2005) demonstrated that elaiosomes can provide about 87% of daily nitrogen and 79% of daily carbon to ant larvae.

Plants have to reward their seed-dispersing ants appropriately, and, as might be expected, larger seeds have larger elaiosomes (Edwards et al. 2006) (Figure 6.39). In this Australian example, heavier seeds had heavy elaioasomes attached, across broad plant taxonomic groupings. Slopes in these relationships greater than one, by the way, indicated that ants demand proportionately larger rewards to remove larger seeds. As an aside, it has been suggested that the capitula (small protrusion) on the eggs of some stick insects mimics the elaiosomes of seeds, and stimulates ants to collect them. Eggs that were buried in ants' nests were found to suffer lower rates of parasitism than unburied eggs (Hughes & Westoby 1992).

Do plants really benefit from elaiosome production and ant dispersal? In the fynbos shrub lands of South Africa, removal of the elaiosomes from seeds of *Leucospermum truncatulum* by ants reduces the rate of seed predation by rodents, but has little effect on the germination success of the surviving seeds (Christian & Stanton 2004). In contrast, Horvitz and Schemske (1994) found that seeds with rewards had 1.6-fold higher emergence and were dispersed on average three-fold farther than seeds without rewards. However, other biotic and abiotic factors also influenced seed dispersal and germination so that the value to the plant of elaiosome production is not likely to be equivalent in every environment. The tropical pioneer tree *Croton priscus* (Euphorbiaceae) has explosive seed capsules that are dispersed some distance (about 3 m) ballistically. However, they still produce elaiosome-bearing seeds (myrmecochory with autochory; Figure 6.38a). Research by Passos and Ferreira (1996) found that ants remove such seeds at a rate of about 88% per day and move them 1–2.5 m farther than for autochory alone. Seedlings of *C. priscus* are often found on the refuse piles left by foraging ants, suggesting that ant dispersal does result in germination (Passos & Ferreira 1996). In a comparison of 10 different ant-dispersed plant species, Peters et al. (2003) reported that seeds with larger elaiosomes had significantly higher removal rates than seeds with smaller elaiosomes. Assuming a positive relationship between seed removal and subsequent germination success, investment in large elaiosomes appears to pay off. In an experiment in mountain grasslands, Dostal (2005) reported that 63.8% of seeds with elaiosomes were removed by ants within a 39 h period, in comparison to only 10.9% of seeds without elaiosomes. However, seed removal in this system did not contribute significantly to the build up of the seed bank. Indeed, the dispersal of seeds by ants may not always be entirely beneficial. Bond and Stock (1989) showed that *Leucospermum conocarpodendron* seeds dispersed by ants after fire in the Cape fynbos ecosystem of South Africa were generally moved to nutrient-poor sites in comparison with seeds that were passively dispersed from the plant.

If elaiosome production and the dispersal of seeds by ants is the result of coevolution, it is probably diffuse rather than tight coevolution. There do not appear to be many examples of obligate relationships between a single species of plant and a single ant disperser. Rather, ants will disperse seeds from many plant species, including novel species in the environment. For example, Pemberton (1988) described myrmecochory between plants introduced to North America (e.g. leafy spurge, *Euphorbia esula*) and ants native to North America (e.g. *Formica obscuripes*). As there cannot have been a long-term evolutionary relationship between the introduced plant and native ant, this demonstrates the facultative nature of the mutualistic interaction. That being said, not all ant species are equally effective at dispersing the seeds of myrmecochorous plants. This has become of increasing concern as native ant communities become disturbed or displaced by exotic ant invasions (Zettler et al. 2001). For example, the Argentine ant, *Linepithema humile*, is an exotic invader on several continents of the world. In California, Argentine ants

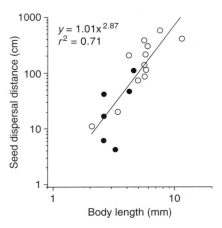

Figure 6.40 The relationship between ant body length and the distance that seeds are dispersed by ants. The closed circles represent sites invaded by exotic ants while open circles represent uninvaded sites. Notice that invaded sites have smaller ants that carry seeds shorter distances. (From Ness et al. (2004).)

displace the common harvester ant, *Pogonomyrmex subnitidus*, a species important for the dispersal of the tree poppy, *Dendromecon rigida*. The Argentine ant is too small to carry the relatively large seeds of the tree poppy, and does not act as an effective dispersal agent (Carney et al. 2003). The Argentine ant has also invaded the fynbos of South Africa, and appears to reduce the recruitment of large-seeded species into the plant community (Christian 2001; see also Section 8.6 and Figure 8.23). Again, the effect appears to result from the small size of Argentine ants relative to the native ant community, and the inability of Argentine ants to transport large seeds. This appears to be a common phenomenon: invasive ant species are often small, yet displace larger native ants, and reduce the dispersal of large-seeded plants (Ness & Bronstein 2004; Ness et al. 2004) (Figure 6.40). These results suggest that even relatively loose mutualisms, such as those between ants and myrmecochorous plants, are subject to disruption, with significant consequences for community dynamics.

6.5.5 Insect–microbe interactions

Associations between organisms can take many forms as discussed in various other parts of this book. Two of these are parasitism and mutualism, in fact only parts of a rather continuous spectrum. Insects like all organisms have many parasites, some of which form the basis of biological control, as discussed in detail in Chapter 12, but here we present the curious tale of the parasitic bacterium, *Wolbachia*.

Various unrelated bacteria appear to have deleterious effects on male animals (Zchori-Fein & Perlman 2004). Members of the genus *Wolbachia* belong to the Alpha-Proteobacteria (Mitsuhashi et al. 2004) and form intracellular associations with insects and many other arthropods. These associations cause a variety of reproductive abnormalities, which include cytoplasmic incompatibility, feminization of males, induction of parthenogenesis, and, most severely, the killing of males. In essence, all of these factors serve to enhance the success of transmission of the bacterium to new host generations via eggs (Moreau & Rigaud 2003) by increasing the number of females in a host population at the expense of males (who after all just waste resources). Host sex ratios are particularly skewed to favor females (Jiggins et al. 2002) because male gametes tend to be too small to accommodate the parasitic bacteria within them, and hence being associated with males can be considered to be an evolutionary dead end (Koivisto & Braig 2003). On the other hand, a large number of insects from many orders have evolved mutually beneficial, as opposed to deleterious, associations with microorganisms including fungi, bacteria, and protozoans. This coevolution has arisen many times, and is predominantly an adaptation to promote the utilization of difficult or nutrient-poor food resources, such as plant sap, wood, and leaf material. The symbiotic microorganisms may exist outside the insect's body (as with fungus-gardening ants mentioned above), inside the gut of the insect (as with certain termites), or inside the cells of the insect (as in the case of aphids). The mechanisms of this coevolution and its results will be discussed in detail as they occur in various insect groups.

Coleoptera

Various families of beetle have evolved complex symbioses with microorganisms. Some, such as the chafers and rhinoceros beetles (Coleoptera: Scarabaeidae), possess colonies of methane-producing bacteria in their hindguts, which break down plant material consumed by the host (Hackstein & Stumm 1994).

Weevils (Coleoptera: Circulionidae) have developed specialized organ-like structures called mycetomes, which contain enterobacteria (Campbell et al. 1992). In the case of the grain weevil, *Sitophylus oryzae*, a widespread pest of stored cereals, the symbionts occur in the ovarioles of the female (Grenier & Nardon 1994), from where they may be transferred to the offspring via the eggs. Woodworm, museum beetles, and so-called cigarette beetles (Coleoptera: Anobiidae) are all able to feed on dead organic material, from wood to skins and even wool. They also possess mycetomes that open into the alimentary tract between the fore- and midgut, which in the case of wood-boring species contain yeast-like symbionts including *Simbiotaphrina buchneri* and *S. kochii* (Noda & Kodama 1996). These yeasts are thought to aid in larval nutrition (Suomi & Akre 1993) and also to detoxify plant-derived toxins such as tannins (Dowd & Shen 1990). Museum beetles, *Anthrenus* spp. (Coleoptera: Dermestidae), are notorious destroyers of preserved animal material from skins, hides, mummified bodies, and insect collections (Figure 6.41). Trivedi et al. (1991) examined the ability of *Anthrenus* larvae to digest wool fiber, a particularly difficult food source. The larvae (so-called "woolly bears") bite off wool fragments in 20–100 mm lengths. Digestion of the outer scales of the wool occurs in their sack-like midgut. In the ileum, large numbers of bacterial cells help to digest the keratin fibrils, resulting in complete structural disintegration of the wool fiber. Actively feeding amoeba-like protozoans gradually replace the bacterial flora in the posterior ileum and rectum, where digested wool components and microbial biomass become compacted because of the absorption of nutrient fluid from the hindgut. The final compact fecal pellets appear as dust – anyone who has had an infestation of museum beetle in their prized collection of insects or shrunken heads will vouch for the efficiency of this symbiotic interaction.

One of the most widespread beetle–microbe symbioses involves two families of wood- and bark-borers, the bark beetles (Coleoptera: Scolytidae) and the ambrosia beetles (Coleoptera: Platypodidae). Both families contain large numbers of both tropical and temperate species that are frequently serious pests of trees, both standing and fallen or felled. The beetles themselves can kill trees, especially those that are already in some way debilitated by fire, windblow, drought, or defoliation, by extensive larval boring between the bark and sapwood, which effectively

Figure 6.41 Insect specimen destroyed by *Anthrenus*. Larvae and adults can be seen. (Courtesy P. Embden.)

their own nutrition, but they rely on the beetles to carry them from host to host, so much so that some fungal species are unable to survive on their own. This feeding habit in platypodids is known as xylomycetophagy, where the ectosymbiotic "ambrosia" fungi form the major part of the food of both larval and adult beetles (Beaver 1989). As mentioned above, the beetles themselves are frequently important forest pests, but the fungi can also cause serious wood degradation. Blue-stain ascomycete fungi in the genera *Ceratocystis* and *Ophiostoma*, for example, can render timber unmarketable (Jankowiak 2005) (see Plate 6.10, between pp. 310 and 311).

To vector their fungal symbionts efficiently, many species of beetle have evolved specialized structures called mycangia in which the fungi proliferate (Cassier et al. 1996). Mycangia have openings to the outside of the prothorax of adult beetles, ready to inoculate new host trees when the beetles tunnel into their bark and wood to deposit their eggs. Fungal taxonomy is, as always, complex and seemingly incomplete, at least to an entomologist, but various genera are commonly found in association with these mycangia, including *Graphium* spp. (Cassier et al. 1996), *Ceratocystiopsis* spp. (Coppedge et al. 1995), *Monacrosporiurn* spp. (Kumar et al. 1995), and *Ophiostoma* spp. (Fox et al. 1992). The coevolution of these highly successful (and frequently obligate) beetle–fungus associations is clearly ancient; amber from the Dominican Republic, dated at 25–30 million years old (see also Chapter 1), contains well-preserved adult platypodids wherein the mycangia are readily identifiable, still containing spores of symbiotic fungi (Grimaldi et al. 1994). The fungi themselves are highly adapted to being disseminated by beetles. They produce sticky spores at the tips of elongate fruiting bodies which erupt from the sapwood into the larval and pupal galleries of the beetles, enabling the insects to acquire the spores either on their bodies or in their mycangia before emerging as young adults to fly to new host trees (Zhou et al. 2007).

The role of these fungal symbionts in the life of the beetles has been investigated in detail. In Figure 6.42a, it can be seen that bark beetles in the genus *Ips*, globally widespread pests of pine and spruce especially, have significantly shorter development times when their fungi (*Ophiostoma* spp.) are present (Fox et al. 1992). Not only that, but as Figure 6.42b shows, the length of the tunnels excavated by bark beetle larvae are significantly shorter when fungi

ringbarks or girdles the host. Some species cause extra problems by way of the maturation feeding activities of young adults before mating; Dutch elm disease is a classic example of this, which we discuss fully in Chapter 11. Almost all scolytids and platypodids have evolved intimate symbiotic associations with various types of fungi, which assist the beetles by breaking down wood material, aiding in the rotting process by rendering the host trees susceptible to further beetle attack, or, as in the case in some of the ambrosia beetles, providing a direct source of food for the growing larvae. These ambrosia symbioses have evolved seven times, according to Mueller et al. (2005), and in the main the fungal symbionts grow externally to the beetles, using wood substrates for

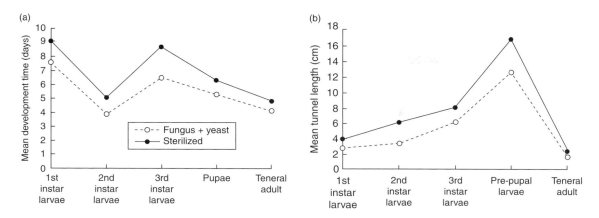

Figure 6.42 Mean development time (a) and tunnel length (b) of various life stages of the bark beetle *Ips paraconfusus* with and without symbionts (all comparisons within a life stage significantly different at $P = 0.05$ level except adult). (From Fox et al. 1992.)

are available, indicating that these symbionts reduce the need for larvae to feed extensively on what can only be described as a suboptimal diet. The number of beetles in a population that carry fungal symbionts, and the concentration of the microbe in each insect, can be variable. It was said that in the case of *Scolytus* spp. (Coleoptera: Scolytidae), the beetle vectors of Dutch elm disease, one in two newly adult beetles visiting a new and healthy tree to maturation feed would carry the fungus *Ceratocystis* (C.J. King, personal communication). In another world-renowned genus of bark beetle, *Dendroctonus*, there is a highly significant relationship between the number of female beetles carrying the symbiotic fungus and both the weight of the beetles and their fat content (Coppedge et al. 1995) (Figure 6.43). Both these factors are directly correlated with the survival of the beetles and their reproductive fitness.

Unlike bark beetles, where the larvae may not feed directly on the fungus, the development of a brood of ambrosia beetle larvae is dependent on the growth of their symbiotic fungi. In some species, such as *Xyleborus mutilatus* (Coleoptera: Scolytidae), adult females utilize the whole gallery system as space available for culturing the fungus which their larvae then eat (Mizuno & Kajimura 2002). It is not surprising, therefore, to find that the total length of gallery systems is directly and positively related to the number of offspring – more space, more food. A mother beetle can determine the number of eggs she produces in response to the amount of ambrosia fungus, and

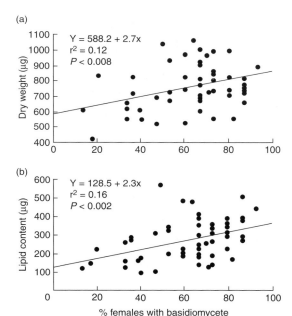

Figure 6.43 Regressions of adult dry weight (a) and lipid content (b) of the bark beetle *Dendroctonus frontalis* against percentage of females carrying a basidiomycete fungal symbiont. (From Coppedge et al. 1995.)

expand her tunnels appropriately. In Figure 6.44 it can be seen that adult females are even able to produce at least one male egg per tunnel branch after

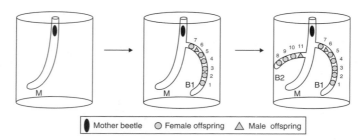

Figure 6.44 A schematic model of reproduction of *Xyleborus pfeili* in relation to gallery construction. The numbers mean the order of oviposition (emergence) in offspring. M, main gallery, B1–2: branch tunnels. (From Mizuno & Kajimura 2002.)

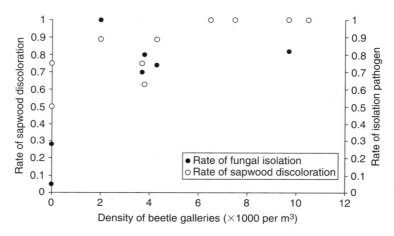

Figure 6.45 Relationships between the density of ambrosia beetle galleries and the rates of isolation of the fungus *Raffaelea quercivora* plus the sapwood discoloration it causes. (From Kinuura & Kobayashi 2006.)

laying several females, so that insemination, albeit by a sibling, is ensured. There is no doubt that the density of ambrosia beetles and/or the intensity of their attack on trees are closely related to fungal infection of the hosts. Kinuura and Kobayashi (2006) studied the attacks on Japanese oak trees (*Quercus crispula*) by the ambrosia beetle *Platypus quercivorus* (Coleoptera: Platypodidae). These beetles excavate tiny tunnels or galleries known as pinholes in the sapwood of host trees, inoculating the trees with the blue-stain fungus, *Raffaelea quercivora*. In Figure 6.45 it can be seen that the number of beetle galleries (pinhole tunnels) is closely related to fungal attacks and the resulting damage.

Hemiptera

Symbiotic relationships between members of the order Hemiptera and microorganisms have been studied most intensively in the aphids (Hemiptera: Aphididae). These insects possess a wide variety of genetically distinct microbial symbionts. In some species of aphid, these associations appear to be obligate, whilst in others they seem to be facultative (Simon et al. 2003). Most aphids possess obligate symbiotic bacteria in the genus *Buchnera* (Wilkinson & Douglas 1995), though there are a variety of secondary or accessory bacterial types that also occur (Haynes et al. 2003), including Gamma-Proteobacteria such as *Serratia symbiotica* and *Hamiltonella defensa* (Douglas et al. 2006). These intracellular symbionts are located in the cytoplasm of structures known as mycetocytes (or bacteriocytes), large cells in the aphids' abdomen (Fukatsu 1994; Braendle et al. 2003). The association is thus known as "mycetocyte symbiosis" (Douglas 1998). As well as occurring in the mycetocytes, for the definition to be upheld, the microorganisms have to be maternally inherited (vertically transmitted), and both the insect and the microorganisms have to require the association. In fact, many groups of insects exhibit mycetocyte symbioses; most of these insects live on nutritionally poor or unbalanced diets such as vertebrate blood, wood, or phloem sap (Douglas 1998). For aphids, this last diet is normally deficient in essential amino acids, which the bacteria are able to synthesize (Liadouze et al. 1995; Birkle et al. 2002), and studies of aphid fossils using rRNA molecular clocks suggest that this association may have been established as much as 160–280 million years ago. *Buchnera*, which may represent as much as 2–5% of the total aphid biomass (Whitehead & Douglas 1993), is thought to be related to well-known enterobacteria (van Ham et al. 2003) such as *Escherichia coli*, indicating a free-living and monophyletic origin of the symbionts (Harada & Ishikawa 1993). *Buchnera* is obligatedly transmitted from adult female aphids to their offspring via the maternal ovaries into the new embryos (Cloutier & Douglas 2003). The problem with this system is that transfer of symbionts from mother to offspring involves a "bottleneck", as illustrated by Figure 6.46 (Wilkinson et al. 2003), so that only a small amount of genetic material from the maternal bacterial population is transferred to the offspring.

Much research has been carried out to elucidate the advantages conferred on aphids by their bacterial symbionts. The roles of the various secondary bacterial symbionts remain unclear, and it appears that their associations with aphids are indeed polyphyletic (Tsuchida et al. 2006). However, *Buchnera* is undoubtedly an essential component of an aphid's ability to feed on phloem sap. Phloem sap, as we mentioned in Chapter 3, is a far from ideal food source for animals. It is rich in sugars, of course, but carries an inadequate supply of essential amino acids, as Figure 6.47a shows (Douglas 2006). Many amino acids essential to aphids such as leucine, lysine, and methionine are in very low concentrations in sap, but *Buchnera* is able to supply these and other amino acids in relatively large quantities (Figure 6.47b). Using the relatively

Figure 6.46 Transfer of *Buchnera* (dark granules) from adult aphid tissues (on right) to embryo (on left). (From Wilkinson et al. 2003.)

simple system of treating the insects with antibiotics that kill the bacteria, aposymbiotic aphids can be created, and Figure 6.48 shows that pea aphids, *Acyrthosiphon pisum* (Hemiptera/Homoptera: Aphididae), deprived of *Buchnera* grew very slowly relative to "normal" ones and that both the number of embryos and the size of each in developing females are much reduced (Douglas 1996). In fact, it seems that embryos are more dependent on symbiotic bacteria than are maternal tissues, probably because of limitations caused by the absence of certain amino acids such as tryptophan and phenylalanine. There does in fact appear to be a lot of variation in the ability of *Buchnera* strains to synthesize amino acids such as tryptophan in different aphid clones (Birkle et al. 2002).

Isoptera

The ability to digest cellulose is rare in most higher animals, including insects. This cellulytic capacity may be uncommon simply because it is rarely an advantage (Martin 1991) in that cellulose digestion is usually mediated by microorganisms. Even termites (Isoptera), wherein some independent cellulose digestion may occur, rely in the main on a battery of microbial symbionts to break down the celluloses and hemicelluloses in their food – mainly dead wood and other plant tissues (Figure 6.49). Some termites have a complex and varied gut microflora, consisting of various bacteria and yeasts including *Bacillus*, *Streptomyces*, *Pseudomonas*, and *Acinetobacter* (Schaefer et al. 1996). Indeed, in others, such as *Coptotermes*

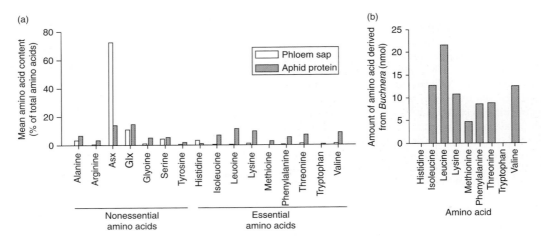

Figure 6.47 The nitrogen barrier to phloem sap utilization: amino acid relations of the pea aphid *Acyrthosiphon pisum* line LL01 feeding on *Vicia faba*. (a) Amino acid content of *V. faba* phloem sap and aphid protein (excluding the non-essential amino acids cysteine and proline, which cannot be quantified by the method adopted). Asx, aspartic acid and asparagine; glx, glutamic acid and glutamine. (b) Amino acids derived from *Buchnera* symbionts during the 2 day of the final larval stadium, as calculated from the difference between amino acids required for protein growth and acquired by feeding on plant phloem sap. (Unpublished data of L.B. Minto, E. Jones, & A.E. Douglas; and data from Wilkinson & Douglas 2003.) (From Douglas 2006.)

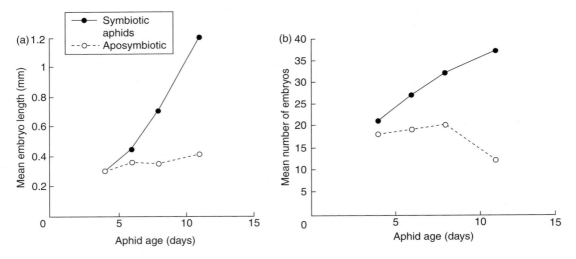

Figure 6.48 Size (a) and number (b) of embryos in nymphs of the pea aphid, *Acyrthosiphon pisum*, with and without their endosymbiotic bacteria. (From Douglas 1996.)

spp., protozoans including *Pseudotrichonympha* are essential for wood degradation in the gut, and though there is still some controversy, it is thought that the hindgut protozoan fauna of termites such as

Mastotermes actively produce cellulases (Watanabe et al. 2006). Certainly, termites which had these protozoans eliminated showed a 30% reduction in their wood-attacking activities, though they were

Figure 6.49 Worker termits consuming dead leaves.

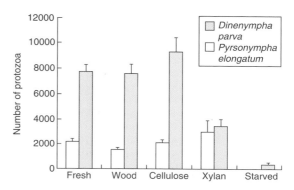

Figure 6.50 Numbers of protozoans of two species in the hindgut of the termite *Reticulitermes speratus* maintained on different diets for 20 days. (From Inoue et al. 1997.)

readily able to pick up new symbionts from fresh worker termites in the neighborhood (Yoshimura et al. 1993). As all endosymbiotic microbes in termites inhabit the gut of their hosts, the insects usually lose their symbionts when they empty their guts before a molt (ecdysis) (Lelis 1992), and hence have to pick them up again in the next instar from surrounding termites. This is thought to be one of the reasons for the evolution of sociality in termites, which ensures that there is always someone close by from whom to reacquire the essential symbionts. It may thus be possible to trace the ancestry of termites back to primitive woodroaches (Dictyoptera: Cryptocercidae), which also have cellulytic protozoans in their hindguts (Thorne 1991).

Termites in the family Rhinotermitidae, such as *Reticulitermes* spp., have been found to support at least 11 species of hypermastigote flagellate protozoans in their hindguts, many of which seem to perform different functions under the general heading of wood degradation. In experiments in Japan, Inoue et al. (1997) fed termites with different diets and assessed the population densities of various protozoan species in their guts after 20 days. Figure 6.50 illustrates the effect on just two protozoan species, comparing different diets with the natural (freshly caught) situation. Wood and pure cellulose seem able to maintain high densities of protozoans, but xylan, a secondary compound found in wood material alone reduces the numbers of one species significantly, and starvation virtually or completely wipes out the symbionts. Both termite and symbiont clearly require adequate mixtures of food derived from natural

wood material, and these obligate associations appear to be the result of close coevolution (Hongoh et al. 2005). Some Rhinotermitidae even have the facility to fix atmospheric nitrogen using bacteria in their hindguts (Curtis & Wailer 1995), but the most complex insect–microbe relationships in the Isoptera are found in the so-called higher termites, such as *Macrotermes* spp., which culture symbiotic "gardens" of the fungus *Termitomyces* (Bignell et al. 1994). The oldest fossil fungus garden produced by termites so far found was discovered in 7-million-year-old sandstones from the Chad Basin in Africa (Duringer et al. 2006). Young worker termites eat dead plant material encountered during their foraging activities away from the colony, but also consume the conidia of this fungus when back in the nest. The conidia contain cellulase enzymes, which may help to break down plant tissues in the termite gut. Fungus gardens also break down composted plant debris directly, which older workers feed to the juveniles, the latter receiving food in which cellulose, hemicellulose, pectin, and lignin are all already significantly degraded (Bignell et al. 1994). Nests of species such as *Odontotermes* spp. and *Macrotermes* spp. consist of underground fungus-comb chambers, the fungus gardens, built within the soil profile and connected by galleries. Put simply, the fungus garden is a type of digestive system exterior to the insect's body, which break down cellulose. This process can result in considerable metabolism and hence release of gases such as carbon dioxide. As Figure 6.51 shows, termite mounds (see Plate 6.11, between pp. 310 and 311) with active

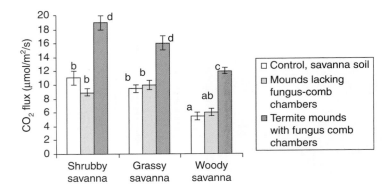

Figure 6.51 In situ soil respiration rates in three substrates according to the presence of termite fungus comb. (From Konate et al. 2003.)

fungus gardens produce significantly more carbon dioxide than those without (or indeed, bare soil) (Konate et al. 2003). The authors suggest that this symbiosis between termites and fungi can represent nearly 5% of the total above-ground net primary production in a West African tropical savanna ecosystem.

One of the most complex problems in the evolutionary ecology of termites and their various symbionts involves the carbon to nitrogen balance of their food relative to their own tissues (Higashi et al. 1992). Dead wood contains less than 0.5% organic nitrogen, whereas termite tissues range from 8% to 13% (fresh weight). Wood is also very much richer in carbon-based celluloses and hemicelluloses so that the C : N ratio can range from 350 : 1 to 1000 : 1; very much higher than that required by the insects. Termites have evolved two solutions to this problem. One involves adding nitrogen, the other, deleting carbon (Higashi et al. 1992). Figure 6.52 shows pathways that may solve these problems, and illustrates the roles of microbial symbionts. Those species that add nitrogen to their food may do so either by using nitrogen-fixing bacteria to fix atmospheric nitrogen (route N-1 in Figure 6.52), or by relying on other bacteria that can synthesize amino acids from inorganic nitrogen excreted as urea by the insects themselves (route N-2a). Excreted nitrogen may also be used as manure for termite fungal gardens, with the termites eating the fungi (route N-2b). Other termites get rid of carbon instead. Symbiotic methanogenic bacteria release excess carbon from wood material in the form of methane (route C-1), or once again, the fungi in the

fungus gardens can export carbon to the atmosphere in the form of carbon dioxide during their respiration (route C-2) (Higashi et al. 1992). From these observations, it is hypothesized that two evolutionary strategies have been followed by termites. Some possess only carbon-eliminating symbionts, and live entirely inside wood, whereas the second group possess both types of symbionts and can deal with carbon- and nitrogen-balancing routes. These species are able to forage outside their nests, and have greater productivity and larger colonies. The fact that colonies can become larger when both carbon and nitrogen routes are available also means that this group of termites can possess truly sterile worker castes, wherein the likelihood of having to take over reproduction is very low because of the large number of individuals in the colony.

Hymenoptera

The ant family Formicidae contains a tribe called the Attini, which are unique among the ants in that they depend on externally cultured symbiotic fungi for food for their larvae. In fact fungi are but one of a whole array of microbes associated with these ants. Van Borm et al. (2002) found everything from flavobacteria and proteobacteria to Gram-positive bacteria in one ant species. Ant "fungiculture" arose 50–60 million years ago (Poulsen & Boomsma 2005). The fungi, *Leucoagaricus gongylophorus*, hydrolyze plant polysaccharides which form a substantial part of the ants' diet (Silva et al. 2003), though the fungus

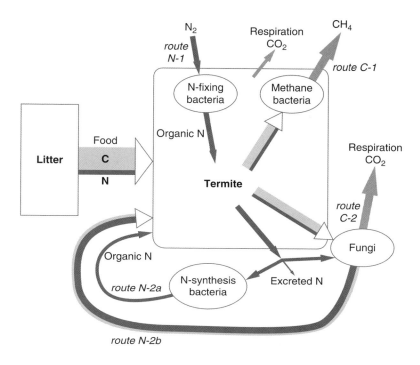

Figure 6.52 Summary of possible carbon–nitrogen balance mechanisms used by various species of termites (see text). (From Higashi et al. 1992.)

may not itself be capable of metabolizing cellulose (Abril & Bucher 2002). In return, the fungi derive dispersal, nourishment, and, as we will see later, freedom from their own microbial parasites. The most advanced group of these insects are known as the leaf-cutter ants from genera including *Acromyrmex* and *Atta*, which together constitute 39 New World species (Cherrett et al. 1989) (see Plate 6.12, between pp. 310 and 311). Leaf-cutters are able to utilize enormous quantities of plant material – not just leaves, but also stems, fruit, and flowers cut from living plants – as substrates on which their fungal symbionts, homobasidiomycete fungi in the order Agricoles (Hinkle et al. 1994), grow inside the ant colony producing a "fungal garden" (see Plate 6.13, between pp. 310 and 311). Because of these activities, leaf-cutter ants are considered to be extremely serious forest and farm pests in Central and South America. Indeed, Cherrett et al. (1989) suggested that they consume more grass than cattle, and Wint

(1983) reported that up to 80% of defoliation seen in rainforests in Panama was caused by leaf-cutters.

Clearly, this obligate symbiosis works very well. The basic biology is simple enough. Ants themselves specialize in the degradation of low molecular weight substrates such as oligosaccharides, whereas the fungi deal with higher molecular weight polysaccharides (Richard et al. 2005). The fungi produce hyphae that terminate in apical swellings called gongylidia (Maurer et al. 1992) that are eaten by adult ants and fed by them to their larvae. These nutrient-rich structures are "designed" for easy harvesting by the farmer ants (Mueller et al. 2005), and these authors are convinced that this association is an excellent example of "agriculture in insects". The gongylidia appear to have no function except to provide ant "rewards" (Bass & Cherrett 1995), and occur grouped together in bite-sized chunks called staphylae (Cherrett et al. 1989). The greater the number of gongylidia in an ant colony, the more workers

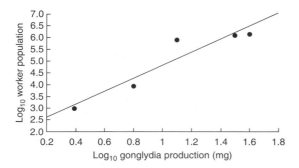

Figure 6.53 Relationship between the productivity of symbiotic fungi (measured by gongylidia production) and the maximum number of workers in leaf-cutter ant colonies. (From Cherrett et al. 1989.)

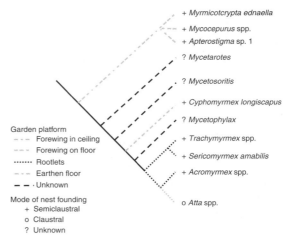

Figure 6.54 The use of fungal garden platforms and nest-founding modes plotted on a phylogenetic hypothesis for Attini. (From Fernandez-Marin et al. 2004.)

can be supported (Figure 6.53). This relationship predominantly works via the larvae, which use the staphylae as a protein-rich source of plant-derived nutrients. The insects use the fungus to defeat physical and chemical plant defenses, and the fungus gains because the chewing activities of the ants allow easy penetration of plant material by their hyphae (Cherrett et al. 1989). Despite being rather widespread in terms of species, ant–fungus associations are very specific, and experiments have shown that cross-inoculation of one ant species with the fungus found in another, albeit closely related, species, results in declines in numbers of worker and reproductive ants produced by the recipient colony, as well as a general reduction in the biomass of the fungus garden itself (Mehdiabadi et al. 2006).

The establishment of a new ant colony complete with fungal symbionts is tricky. The single queen (so-called foundress) needs to transport, nourish, and cultivate the fungus on her own, and the methods used vary between species (Fernandez-Marin et al. 2004). Figure 6.54 shows the relationships between different nest-founding modes within the leaf-cutter ant tribe (Attini), and needs some explanation. Claustral foundation in the genus *Atta* is where the newly mated queen establishes a colony without foraging – she places the new fungus garden directly on the chamber floor and nourishes it and her brood from her own body reserves. Semiclaustral foundation, on the other hand, as in the genus *Acromyrmex*, is where the queen forages for substrates on which to grow the fungus and nourishes her broods directly

from the growing fungus. The figure also shows that some leaf-cutter species such as *Mycocepurus* suspend their shed forewings from the chamber ceiling on which to establish the fungus garden, thus keeping it well clear of the chamber floor and hence isolating it from soil microorganisms that might infected and contaminate the culture.

All is not entirely well in the ants' garden – there is also a villain of the piece in the form of a genus of specialized, highly pathogenic microfungi that attack the ants' fungal cultures (Gerardo et al. 2006). This pathogen, *Escovopsis*, has only ever been found in nests of attine ants, and once established in an ant colony, it can cause immense devastation. Such pathogen infections of the fungus garden has led to the development of grooming or "weeding" behaviors in the ants, who actively remove parasite-infected parts of the fungus garden in order to keep the majority clear of infection (Currie & Stuart 2001). In Figure 6.55 it can be seen that strenuous activity on the part of *Atta* workers ensues when the potentially devastating fungal parasite *Escovopsis* is added to fungus colonies. The ant "rapid response team" removes infected fungus garden and dumps it outside the nest. The pinnacle of fungus garden protection is the coevolution of a highly complex mutualism within leaf-cutter ant nests involving a filamentous bacterium

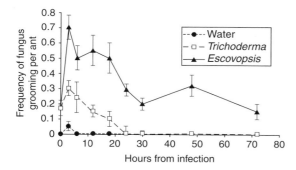

Figure 6.55 Behavior of worker leaf-cutter ants (fungus grooming) in prescence of fungus-garden pathogens. (Currie & Stuart 2001.)

(actinomycetes) that the ants use to combat *Escovopsis* (Currie et al. 2003). This third mutualist grows on the ants' cuticles, and produces an antibiotic that destroys *Escovopsis* (Poulsen et al. 2002). A worker ant picks up *Pseudonocardia* (Poulsen et al. 2002) within a few days of emerging from the pupa, and the bacterial cover on its body exponentiates so that after 10–15 days, its entire cuticle is covered. There is clearly a metabolic cost to this association, in that ants with *Pseudonocardia* on their bodies eat more fungus, and show significantly increased respiration rates, than those with no bacteria on their bodies, but the obvious tradeoff is that the all-important fungus garden is protected for the good of the whole ant colony.

6.6 SEQUESTRATION OF PLANT SECONDARY METABOLITES

In Chapter 3, we described some of the secondary metabolites in plant tissue that can influence the foraging behavior and feeding success of insect herbivores. At least some of these secondary metabolites are likely to serve a defensive function in plant tissue and, earlier in this chapter, we described potential evolutionary and coevolutionary associations between insects and plants that are chemically based. Here, we explore one further evolutionary response of insects to plant secondary metabolites – the ability of some herbivores to retain and utilize plant chemicals for their own defensive purposes (Rowell-Rahier & Pasteels 1992). This process is known as sequestration. Sequestered plant defenses can make insect

herbivores less attractive, palatable, or nutritionally valuable to parasitoids (Campbell & Duffy 1979; Greenblatt & Barbosa 1981), invertebrate predators (Dyer & Bowers 1996; Narberhaus et al. 2005), and vertebrate predators (Brower & Brower 1964; Rothschild 1973). The number of plant toxins sequestered by individual insect herbivores can be dazzling. For example, Wink and Witte (1991) have identified 31 different quinolizidine alkaloids in the aphid *Macrosiphum albifrons* and 21 in *Aphis genistae* (Hemiptera: Aphididae). The concentrations of alkaloids in the tissues of these aphids can exceed 4 mg/g fresh body weight.

Of course, not all chemical defenses found on or in the bodies of insect herbivores are sequestered from plants (Soe et al. 2004). For example, phytophagous leaf beetles in the genus *Oreina* (Coleoptera: Chrysomelidae) have a variety of defensive strategies (Pasteels et al. 1995) (Table 6.7). All beetles within the genus produce pronotal and elytral secretions that appear defensive in function. Most of the species secrete cardenolides, but these are synthesized *de novo* from cholesterol, and are not derived from the plants on which *Oreina* beetles feed. In contrast, *Oreina cacaliae* secretes nitrogen oxides of pyrrolizidine alkaloids (PAs) and no cardenolides. The PAs are derived from their host plants. Between these two extremes are other species of *Oreina* that can both synthesize and sequester their defensive secretions. These intermediate species can exhibit remarkable variation in their defenses both within and among populations. This variation probably arises from differences in the local availability and toxicity of their host plants, but allows for considerable flexibility in defense.

We should stress that sequestration is an active process. Not all plant secondary metabolites are retained, and those that are sequestered can be modified to varying degrees. For example, the polyphagous grasshopper *Zonocerus variegates* (Orthoptera: Acrididae; see Plate 6.14, between pp. 310 and 311) sequesters only two (intermedine and rinderine) of the five PAs from the flowers of the tropical weed, *Chromolaena odorata*. About 20% of the PAs that are sequestered are converted to lycopsamine and echinatine by chemical inversion at one of the carbon bonds of the compounds (Biller et al. 1994). In other words, insects that make use of plant toxins for defense have the capacity to sequester certain compounds, excrete others, and sometimes modify those that they keep (Hartmann et al. 2003).

Table 6.7 Sequestered and autogenous defensive secretions in chrysomelid beetle genus *Oreina*. (Data from Rowell-Rahier Pasteels 1992; Pasteels et al. 1995.)

Species	Sequestered secretions	Autogenous secretions
O. bifrons	—	Cardenolides
O. gloriosa	—	Cardenolides
O. speciosa	—	Cardenolides
O. variabilis	—	Cardenolides
O. speciosissima	Pyrrolizidine N-oxides	Cardenolides
O. elongata	Pyrrolizidine N-oxides	Cardenolides
O. intricata	Pyrrolizidine N-oxides	Cardenolides
O. cacaliae	Pyrrolizidine N-oxides	—

Moreover, not all compounds sequestered by insects from plant tissue are necessarily toxic or defensive. For example, poplar hawkmoth and eyed hawkmoth larvae (*Laothoe populi* and *Smerinthus ocellata*, respectively (Lepidoptera: Sphingidae)) sequester chlorophylls and carotenoids (pigments) from their hosts. The pigments are translocated to the integument, and provide for accurate color matches between the insect larvae and the plants on which they are feeding (Grayson et al. 1991). Although this sequestration is defensive, presumably helping to protect larvae from visually hunting predators, it does not rely on plant toxins. Similarly, the Lycaenidae, the second largest family of butterflies, appear to use excretion rather than sequestration as the dominant mode of coping with toxic host plants. However, many larvae sequester flavonoids from their hosts, which are later concentrated as pigments in the adults' wings, and are thought to play a role in visual communication (Fiedler 1996; Burghardt et al. 2001).

None the less, there is clear evidence that toxins sequestered from plants can act as powerful deterrents to their natural enemies. For example, the leaf beetle *Chrysomela confluens* (Coleoptera: Chrysomelidae) produces a salicylaldehyde-based defensive secretion that is effective against generalist predators. In an experiment by Kearsley and Whitham (1992), some larvae were "milked" of their defensive secretions, whereas other larvae were left with their secretions intact. Only 7% of the unmanipulated larvae were attacked by ants, and none suffered serious injury. In contrast, 48% of the milked larvae were attacked

by ants, and two-thirds of these were killed. The salicylaldehyde secretion is produced from salicin, a precursor present in the leaves of host plants consumed by *C. confluens*. It is interesting to note that the conversion of salicin to salicylaldehyde liberates glucose, which may act as a source of energy for the larvae. If there are any costs associated with sequestration and secretion, they may be offset by the energy gained from the liberation of glucose. In this case, at least, larvae appear to obtain their defense "for free". In a similar study, Dyer and Bowers (1996) demonstrated that *Junonia coenia* (Lepidoptera: Nymphalidae) larvae sequester iridoid glycosides from their host plants. In their experiments, larvae fed on diets with high concentrations of iridoid glycosides, and individual larvae with high rates of sequestration, were more likely to be rejected by (and survive attack by) ant predators. Interestingly, the diets that provide insect herbivores with chemical protection against natural enemies are not always the diets that are most favorable nutritionally. This can lead to diet mixing, whereby insect herbivores consume both palatable plants (for nutritional gain) and chemically defended plants (for defensive sequestration). For example, the generalist caterpillar, *Estigmene acrea* (Lepidoptera: Arctiidae) grows larger on *Viguiera dentata* (Asteraceae) than on *Senecio longilobus* (Asteraceae) (Singer et al. 2004a) (Figure 6.56a). However, late instar *E. acrea* larvae prefer *S. longilobus* in choice tests. *S. longilobus* produces PAs that provide larvae with some protection from their parasitoids (Figure 6.56b). The balance between benefits of growth (food quality) and defense (sequestration) appears to maintain the

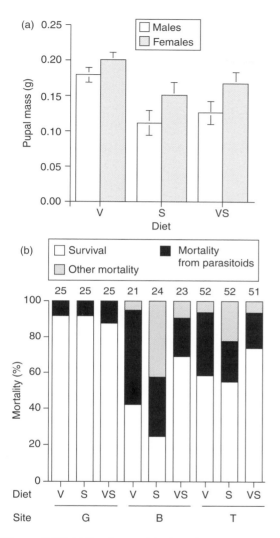

Figure 6.56 (a) Pupal mass of the Arctiid, *Estigmene acrea*, fed diets of *Viguiera dentata* (V), *Senecio longilobus* (S), and a mixture of the two plants (VS). (b) Mortality of the Arctiid, *Estigmene acrea*, feeding on diets of *Viguiera dentata* (V), *Senecio longilobus* (S), and a mixture of the two plants (VS) at three field sites (G, B, T). At sites B and T, mixed diets result in lower overall mortality. Mixed diets include highly nutritious food (V) and protection from parasitoids (S). (From Singer et al. 2004a.)

use of diet mixing by these caterpillars. The same tradeoff between food quality and defense also favors diet mixing in other species of Arctiid moths (Singer et al. 2004b).

One of the most extensively studied cases of sequestration by an insect herbivore is that of the monarch butterfly, *Danaus plexippus* (Lepidoptera: Danaidae). Monarchs sequester cardenolides from their milkweed (*Asclepias*) host plants. Cardenolides are toxic, bitter-tasting steroids that attack the sodium–potassium–adenosine triphosphatase (ATPase) enzyme system of most animals, and they have been shown to protect adult (and perhaps larval) monarchs from their natural enemies (Brower et al. 1988). Cardenolides are sequestered by monarch larvae (see Plate 6.15, between pp. 310 and 311) that feed on *Asclepias*, and are then concentrated into the exoskeletal tissues of adults. The insensitivity of monarchs to cardenolides is thought to result from a single amino acid replacement (asparagine replaced with histidine) at position 122 in the α-subunit of the Na^+–K^+–ATPase system (Holzinger & Wink 1996). To understand the cardenolide–monarch–predator interaction, we must appreciate that (i) *Asclepias* varies in the concentration of cardenolides in its tissues, and (ii) the movement of monarchs in space and time dictates the type of *Asclepias* to which they are exposed and their subsequent level of protection from predators.

Asclepias species vary in toxicity at a variety of spatial scales. For example, *Asclepias syriaca*, the species on which about 92% of larvae feed in the northern USA, exhibits large-scale variation in cardenolide concentration across its range from North Dakota east to Vermont, and south to Virginia (Malcolm et al. 1989) (Figure 6.57). At very fine spatial scales, cardenolides within individual plants are inducible (see Chapter 3), increasing in concentration following defoliation by monarchs (Malcolm & Zalucki 1996) and feeding by aphids (Martel & Malcolm 2004) (see Plate 6.16, between pp. 310 and 311). Induction therefore will generate small-scale variation among individuals within populations. In addition, small- and large-scale variations in cardenolide concentration are superimposed upon seasonal variation, and all levels of variability are likely to influence the monarch–enemy interactions because of diversity in the power of sequestered defenses.

Perhaps the most striking example of variation in the cardenolide–monarch–milkweed interaction comes from studies of variation among milkweed species. Milkweeds used by monarchs in the southern USA have much higher concentrations of cardenolide than the species *A. syriaca*, used by most larvae in the

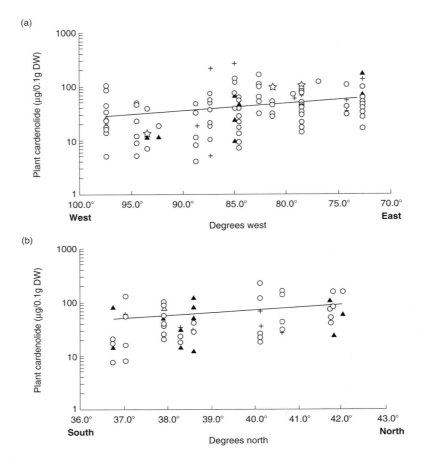

Figure 6.57 (a) Longitude and (b) latitudinal variation in the cardenolide concentrations of *Asclepias syriaca*. DW, dry weight. (From Hunter et al. 1996.)

north (Hunter et al. 1996) (Table 6.8). The migration of monarchs among species of *Asclepias* that vary in toxicity brings them in and out of "zones of susceptibility" to vertebrate natural enemies. Monarchs become susceptible to vertebrate predators (e.g. mice, black-backed grosbeaks, black-headed orioles) if their total cardenolide concentration drops below 121 g per butterfly (or 57 g/0.1 g dry weight of butterfly; Fink & Brower 1981). If we start a typical year in January, we can follow the movement patterns of monarchs, the chemistry of the hosts on which they feed, and changes in monarch cardenolide defenses over time (Figure 6.58). In January, almost the entire monarch fauna east of the Rocky Mountains overwinters in

Table 6.8 Cardenolide contents of the commonest *Asclepias* species used by monarch butterfly larvae in southern and northern USA. (Data from Hunter et al. 1996 and sources therein.)

Asclepias spp.	Mean cardenolide content (μg/g leaf)
A. syriaca	50
A. humistrata	417
A. viridis	376
A. asperula	886

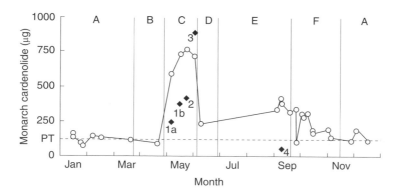

Figure 6.58 Annual cycle of mean cardenolide concentrations (μg per butterfly) in populations of *Danaus plexippus*. Open circles are cardenolides in butterflies; solid diamonds are cardenolides in *Asclepias* hosts. 1a, *A. viridis* in Texas and Louisiana; 1b, *A. viridis* in Florida; 2, *A. humistrata* in Florida; 3, *A. asperula* in Texas; (4) *A. syriaca* in northern USA. Temporal zones are: A, overwintering in Mexico; B, northward spring migration to southern USA; C, spring breeding in southern USA; D, spring generation arriving in northern USA; E, summer breeding in northern USA; F, fall migration from northern USA to Mexico. For further details, see text. (From Hunter et al. 1996.)

high-altitude fir forests in central Mexico. These adults developed from larvae feeding in the northern USA on *A. syriaca*. Because *A. syriaca* has relatively low levels of cardenolides, and because adult butterflies apparently lose cardenolides as they travel south, the adults overwintering in Mexico in January are not well protected from predators, and suffer high rates of vertebrate predation (Malcolm & Zalucki 1993). This is seen in Figure 6.58A where cardenolide concentrations in the butterflies are just at the threshold of susceptibility to vertebrate predators.

During March and April (Figure 6.58B), the surviving adult monarchs migrate northwards, into southern USA (e.g. Florida, Louisiana, Texas) and produce the first larvae of the year. These larvae feed on southern species of *Asclepias* (e.g. *A. viridis*, *A. asperula*, *A. humistrata*), which have much higher cardenolide concentrations (Table 6.8), and the adult population arising from these larvae has high levels of sequestered cardenolides (Figure 6.58C). These new spring adults, well protected against their vertebrate predators, then migrate to the northern USA (Figure 6.58D). They begin breeding and ovipositing on *A. syriaca*, and produce another three generations. These northern generations (Figure 6.58E) are less well protected from predation, although sequestration can concentrate cardenolides to levels higher than those in the host plant. However, as the adults produced in the north migrate south to Mexico for

the winter (Figure 6.58F), they lose cardenolides, becoming increasingly susceptible to predators (Figure 6.58A, right-hand side).

The migration of monarchs, then, exposes them to *Asclepias* species that vary in cardenolide content, and there is a dynamic interaction between monarchs, cardenolides, and vertebrate predators over the course of a single year.

Although most studies of the natural enemies of monarchs have focused on vertebrate predators, one study has shown that insect parasitoids are also influenced by the cardenolides sequestered by monarchs. The number of adult tachinid parasitoids (Diptera: Tachinidae) emerging from parasitized monarchs declined as the cardenolide concentration of the *A. syriaca* on which the larvae were feeding increased (Hunter et al. 1996). Because tachinid larvae hatch outside their hosts and then penetrate the cuticle, the reduced number of emerging parasitoids could have resulted from either reduced rates of penetration or lower levels of survivorship of parasitoids within host larvae. Interestingly, the probability of parasitism was unrelated to cardenolide concentration, so the cardenolides did not protect larvae from tachinids directly. Rather, the sequestered compounds simply reduced the number of parasitoids emerging. This suggests that the cardenolide–monarch–parasitoid interaction has a greater effect on the third trophic level (tachinids) than the second (monarchs).

We pointed out above that the cardenolides sequestered by monarchs were concentrated into exoskeletal tissues. It makes intuitive sense that insects are more likely to survive attack by enemies if they concentrate their toxins in the tissues first encountered by those enemies: bitter tasting or noxious compounds can produce a rapid rejection response by both vertebrate and invertebrate predators. Of course, over evolutionary time, we might expect the concentration of toxins to tissues that maximize the inclusive fitness of the sequestering insect. For example, the painted grasshopper *Poecilocerus pictus* (Orthoptera: Pyrgomorphidae) sequesters cardenolides from the milkweed *Calotropis gigantea* (Pugalenthi & Livingstone 1995). Concentrations are particularly high in the metathoracic scent gland, which probably serves to warn predators even before contact is made. However, concentrations are also high in the ovaries and eggs. The painted grasshopper may be providing protection to its unborn offspring (and neonate nymphs) from predators and parasites by including cardenolides with other egg provisions. This is an excellent example of a maternal effect (Rossiter 1994, 1996) in which the environmental experience of the parent (in this case, cardenolides consumed by the mother) is passed on to the offspring and influences their performance. Because the success of offspring is a primary component of fitness, it should be no surprise to see sequestered compounds concentrated in tissues that protect reproductive investment. A further example is provided by the generalist moth, *Estigmene acrea* (Lepidoptera: Arctiidae), which sequesters PAs from its host plants (see above). Some males will transfer most of their sequestered PAs to females during copulation. The females then transfer the PAs to their eggs (Hartmann et al. 2004). PAs in arctiid eggs have been shown to protect the eggs from predation by lacewing larvae (Eisner et al. 2000) (Figure 6.59).

One common feature of insects that are protected by toxins (or venoms, such as bees and wasps) is bright "aposematic" coloration. Aposematism is the combination of warning coloration and toxicity to predators, and has been studied for years by entomologists and evolutionary biologists. The hypothesis, at its most basic, states that aposematic insects gain some protection from visually hunting predators that can learn to ignore distasteful prey. The subject is too broad to cover in detail here, but Owen's (1980) text provides a useful introduction to the topic. We restrict our comments to a couple of general points.

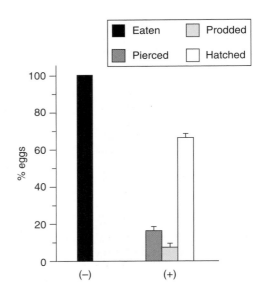

Figure 6.59 The fate of arctiid moth eggs without (−) and with (+) pyrrolizidine alkaloids under predation pressure from lacewing larvae. All eggs without chemical protection are consumed by lacewings. (From Eisner et al. 2000.)

First, there is no direct link between aposematism and sequestration (Pasteels & Rowell-Rahier 1991). It should be recalled that chemical defenses in insects can be autogenous (self-generated) and so not all insects that display warning colors have sequestered compounds from plants. For example, although aposematism and chemical defense are both common in the Chrysomelinae (Coleoptera: Chrysomelidae), sequestration is the exception, not the rule, in this group of insects. Sequestration appears to be a secondary trait derived from aposematic, autogenously defended, ancestors.

Second, aposematic coloration has traditionally been considered to lead to two kinds of mimicry by other species of insect. Batesian mimicry is when a palatable mimic comes to resemble an unpalatable, aposematic model, and so gains protection from predators that have learnt to avoid the model insect. Mullerian mimicry, on the other hand, is when a series of unpalatable aposematic insects come to resemble one another, so reinforcing the message of unpalatability (or danger) to potential predators (see Plates 6.17 and 6.18, between pp. 310 and 311).

However, mimetic relationships do not always fall neatly into these categories, and can be difficult to untangle (Ritland & Brower 1993). For example, the viceroy butterfly, *Limenitis archippus* (Lepidoptera: Nymphalidae), has been considered a classic example of a Batesian (palatable) mimic of two aposematic danaid models (the monarch, *Danaus plexippus*, and the queen, *D. gilippus berenice*). However, experiments with captive red-winged blackbirds indicate that *L. archippus* may be as unpalatable as *D. gilippus*, and that the mimetic complex in Florida (USA) may be dynamic, shifting along a continuum from Batesian to Mullerian mimicry. Queens and monarchs feed on a variety of hosts, and therefore vary in palatability in space and time. The relative abundance of the three species varies in space and time also, and viceroy coloration varies from monarch-like to queen-like, with intermediate forms (Ritland & Brower 1993). The precise division of species among Batesian and Mullerian mimicry may be, in reality, more complicated than it first appears. There are many other kinds of mimicry in the insect world that we do not have space to consider here. Again, we direct interested readers to an introductory text on the subject (Owen 1980).

6.7 DEME FORMATION AND ADAPTIVE GENETIC STRUCTURE

In previous sections of this chapter, we have described the evolutionary responses of insects to their biotic and abiotic environment. It is important to recognize, however, that environmental variation generates a series of habitat mosaics, and that the evolutionary responses of insects occur in a patchy environment. Because the environment is patchy, insect populations can frequently become structured into discrete genetic groups or "demes". Demes can be thought of as groups of individuals that show marked genetic similarity because gene flow within the group exceeds gene flow among groups. Of course, restricted gene flow and deme formation can occur for many reasons, including stochastic (chance) events that isolate individuals from others in the population, or environmental factors that isolate populations from one another (Cognato et al. 2003). However, deme formation can also result from adaptation to local environmental conditions, and this is called adaptive genetic structure (Mopper 1996).

We have seen in this and earlier chapters that host-plant quality can vary markedly in space. Quality can vary among populations of plants, among individuals within a population, and even among different parts (e.g. branches) of the same plant (Denno & McClure 1983). If insect herbivores become locally adapted to such variation in their host plants, then adaptive genetic structures can develop. Measuring such an adaptive genetic structure is important, because it helps us to understand interactions between insects and plants in ecological and evolutionary time, and may provide information on the routes of speciation.

The study of deme formation within populations of insect herbivores came to prominence in the early 1970s, when Edmunds was studying populations of black pineleaf scale, *Nuculaspis californica* (Hemiptera: Diaspididae) on its host *Pinus ponderosae*. In a paper describing the ecology of this insect–plant interaction, Edmunds (1973) noted that "scale populations apparently become adapted to specific host individuals", and his observations paved the way for a classic study of deme formation in this system that was later published in *Science* (Edmunds & Alstad 1978). In this study, scale insects from 10 infested trees were transferred onto 10 uninfested trees, and the performance of the insects compared between natal (original) and novel hosts. The survival of scale insects was higher on the original host tree than on the trees to which they were transferred, and Edmunds and Alstad concluded that scales had become locally adapted to the individual tree on which they lived. Although there was some criticism of the methods and conclusions in this paper (e.g. Unruh & Luck 1987), subsequent studies comparing variation in enzymes produced by scales ("allozyme markers") have confirmed that there is genetic structure in *N. californica* populations among trees, and among individual branches within trees (Alstad & Corbin 1990).

Genetic structure within populations of phytophagous insects now appears to be widespread (Mopper & Strauss 1998). Deme formation has been documented in such diverse taxa as Coleoptera, Hemiptera, Thysanoptera, Lepidoptera, and Diptera. Genetic differentiation can even occur between pest insects in agricultural fields and their conspecifics on native plants in field margins (Vialatte et al. 2005). What factors could cause insect herbivores to exhibit genetic differentiation among individual plants, or even different parts of the same individual? Alstad (1998) described three possible sources of genetic structure

> **Table 6.9** Three hypotheses on the origin of genetic structure within insect populations.
>
> 1 *Intrinsic local adaptation hypothesis*
> Genotypic variation among individual plants influences their quality as hosts for insects
> Genotypic variation among individual insects results in some individuals outperforming others on particular host
>   genotypes
> Natural selection favors the insect genotype(s) most suitable for the genotype of each individual plant
> Reduced gene flow or strong selection maintains differences in the genotypes of insects among individual plants
>
> 2 *Extrinsic local adaptation hypothesis*
> Local variation in the environment (light, temperature, nutrients, water, etc.) generates variation in quality both within
>   and among plants
> Genotypic variation among individual insects results in some individuals outperforming others on particular host
>   phenotypes or on parts of individual plants that vary in quality
> Natural selection favors the insect genotype(s) most suitable for the phenotype of each plant or plant part
> Reduced gene flow or strong selection maintains differences in the genotypes of insects among plants or plant parts
>
> 3 *Drift hypothesis*
> The insect population is characterized by low levels of gene flow among plants or plant parts
> Isolated groups of insects lose certain alleles whereas others become fixed because of random drift
> The population as a whole develops genetic structure by non-adaptive processes

(Table 6.9). The intrinsic local adaptation hypothesis suggests that, over evolutionary time, insects become adapted to the genotype of the particular plant on which they live, because plant genotype is the principal factor influencing plant quality (defenses, nutritional quality) for insects. In contrast, the extrinsic local adaptation hypothesis suggests that environmental effects on plant quality (e.g. nutrient availability, sun vs. shade, etc.) are dominant in determining the phenotypic traits of plants that are important to herbivores. These two hypotheses may appear similar because they both operate through phenotypic traits of the plant. However, the extrinsic local adaptation hypothesis allows for differences in quality within plants that are not genetically based. For example, if different branches of the same tree are growing in different environments (e.g. sun vs. shade) their qualitative differences can contribute to genetic structure in the insect population. Under both hypotheses, however, genetic structure is considered to arise from adaptive genetic change in the insect population. Finally, the drift hypothesis suggests that limited gene flow can result in genetic structure by chance alone. In other words, isolated groups of insects become genetically distinct simply by the random loss of alleles that are shared by the global population. In this last case, demes form not by adaptation, but by "neutral evolution".

All three hypotheses leading to genetic structure probably require a couple of conditions in common. First, the plant must be long-lived relative to the insect herbivore. The development of genetic structure by insects on annual plants seems unlikely because their hosts live for only 1 year. Few, if any, insect species are likely to have a sufficient number of generations within that year for a distinct genetic structure to develop by either adaptation or drift: they will broadly resemble a cross-section of the entire population. In contrast, on long-lived tree species, phytophagous insects can produce literally hundreds of generations on the same individual plant, providing the potential for adaptation by the herbivore. Second, deme formation seems more likely when gene flow is low among the fragmented groups of individuals. If sexual reproduction is common among individuals from different groups, then they are unlikely to develop genetic dissimilarity. Restricted gene flow can occur for a number of reasons, including low rates of insect dispersal within and among plants, or phenological asynchrony (differences in developmental timing) among groups living on different hosts.

What evidence is there in support of adaptive versus neutral theories of deme formation? Virtually no studies have compared the extrinsic versus intrinsic local adaptation hypotheses. One reason for this is the very real difficulty in estimating the relative contributions

Table 6.10 Experimental tests of the adaptive deme formation hypothesis. (After Mopper 1996. Data from: 1. Edmunds and Alstad 1978; 2. Rice 1983; 3. Wainhouse and Howell 1983; 4. Karban 1989; 5. Hanks and Denno 1993*b*; 6. Unruh and Luck 1987; 7. Cobb and Whitham 1993; 8. Mopper et al. 1995; 9. Stiling and Rossi 1998; 10. Ayres et al. 1987.)

Insect	Host plant	Mobility	Deme formation
Black pineleaf scale[1] *Nuculaspis californica*	Ponderosa pine *Pinus ponderosae*	Sessile	Yes
Black pineleaf scale[2] *Nuculaspis californica*	Sugar pine *Pinus lambertiana*	Sessile	No
Beech scale[3] *Cryptococcus fagisuga*	Beech *Fagus sylvatica*	Sessile	Yes
Thrips[4] *Apterothrips secticornis*	Seaside daisy *Erigeron glaucus*	Sessile	Yes
Armored scale[5] *Pseudaulacaspis pentagona*	Mulberry *Morus alba*	Sessile	Yes
Needle scale[6] *Matsuccoccus acalyptus*	Pinyon pine *Pinus edulis*	Sessile	No
Needle scale[7] *Matsuccoccus acalyptus*	Pinyon pine *Pinus edulis*	Sessile	No
Leaf-miner[8] *Stilbosis quadricustatella*	Sand-live oak *Quercus geminata*	Dispersive	Yes
Gall midge[9] *Asphondilia borrichia*	Sea oxeye daisy *Borrichia frutescens*	Dispersive	Yes
Geometrid moth[10] *Epirrita autumnata*	Mountain birch *Betula pubescens*	Dispersive	No

of genotype and environment to variation in qualitative traits (e.g. nutrition, defense) of long-lived plants in natural populations (Klaper & Hunter 1998). Although new statistical techniques combined with molecular studies (e.g. microsatellite DNA markers) are making this feasible, it has been extremely difficult to say with any certainty whether differences among trees in natural populations are genetic or environmental. However, if we combine the intrinsic and extrinsic hypotheses together into an "adaptive" model, we can compare it with the neutral (drift) model of deme formation. The adaptive model argues that, as insects repeatedly colonize a natal host, they become more successful at exploiting it and less successful at exploiting novel conspecific hosts. After several generations, natural selection produces fine-scale demic structure in the insect population. If this is true, transplanted insects will fare worse on novel hosts than on natal hosts. In contrast, the neutral

model predicts that genetic structure arises from drift, and is not adaptive. If true, the performance of transplanted insects will not differ between novel and natal hosts.

Mopper (1996) and Mopper and Strauss (1998) have collected together many of the experimental tests of deme formation in insect herbivore populations. Of 10 field experiments reported by Mopper (1996), six support the adaptive model of deme formation and four do not (Table 6.10). Like many of the other topics covered in this book, the jury is still out on which forces, selective or stochastic, generate genetic structure in insect herbivore populations. However, compelling examples of adaptive deme formation do exist. Karban (1989) found that thrips, *Apterothrips secticornis* (Thysanoptera), displayed higher growth rates on natal plants (*Erigeron glaucus*) than on novel plants of the same species. His experiment was particularly well designed because all the thrips were kept in a common environment on novel hosts for several generations beforehand. By doing this, Karban made sure that differences in performance among hosts were due to the genotype of the insects and not to their immediate environmental history.

One interesting fact to emerge from Mopper's (1996) review is that deme formation may not be linked to the potential of insects to disperse. Three of the 10 insect species examined were dispersive, and two of these exhibited deme formation (Table 6.10). This is surprising because dispersal ability is likely to be related to gene flow, and high levels of gene flow should eliminate genetic structure. One possible explanation is that strong selection pressure maintains genetic structure, even in the presence of significant gene flow. Introduced, "novel" genotypes might be removed rapidly if host quality exerts considerable selection on insect genotype. Of course, to demonstrate that selection is mediated by host quality, it is necessary to show that survival or fecundity of "novel" insects is lower than that of "natal" insects because of plant attributes. Remarkably few studies of deme formation in insects have attempted to show that mortality or fecundity is directly linked to plant traits. The importance of this is aptly demonstrated by an experimental study of deme formation in the leaf-mining moth, *Stilbosis quadricustatella* (Lepidoptera: Cosmopterigidae), on oak (Mopper et al. 1995) (Figure 6.60). Leaf-miners are excellent insects for such studies because they leave evidence of the causes of mortality behind in the mine. Plant-related mortality

can be separated from predation or parasitism by examining damage (or lack thereof) to the mine.

Mopper and her colleagues transferred *S. quadricustatella* among: (i) natal and novel sites; (ii) natal and novel oak species (*Quercus geminata* and *Q. myrtifolia*); and (iii) natal and novel *Q. geminata* individuals within a site (Figure 6.61). They found that, as predicted by the adaptive deme formation hypothesis, there was significantly lower plant-mediated mortality in the natal treatment groups at all three spatial scales (site, species, individual). Apparently, there is genetic structure in the leaf-miner population at a variety of scales, and host quality is directly responsible for some larval mortality. However, despite reduced plant-mediated mortality on natal trees, rates of overall survival were not different. Mortality from natural enemies was actually lower on novel trees than on natal trees, and it compensated for the effects of tree quality on survival (Figure 6.61b). If the investigators had not divided mortality among host and enemy effects, overall survival would have been identical, and the hypothesis of adaptive deme formation rejected. In reality, leaf-miners exhibited adaptive deme formation on oaks by their 10th generation on an individual tree and, by 40 generations adaptive genetic structure was strong (Mopper et al. 2000).

In several of the studies in which adaptive deme formation has been demonstrated, endophagous (internally feeding) insects have been used in the experiments. Galling and mining insects (see Chapter 3) are intimately associated with their host plants, and it has been suggested that they are more likely to adapt to the quality of individual hosts over time than are insects on the plant surface (Yukawa 2000). External feeders may be more prone to unpredictable environmental changes that limit the insects' responses to the "constant" nature of their host plant. Endophagous insects may be somewhat "buffered" from the environment. For example, adaptive deme formation has been demonstrated for the gall-forming midge, *Asphondylia borrichia* (Diptera: Cecidomyiidae), on sea oxeye daisy, *Borrichia frutescens* (Stiling & Rossi 1996, 1998). In this case, gall size is significantly larger on natal than on novel plants, and adult flies prefer natal to novel plants for oviposition. However, it is clear that we need many more studies before we can generalize about links between feeding mode and deme formation. Indeed, the study of genetic architecture in insect herbivore populations still suffers from a lack of experimental studies, even on

Figure 6.60 Two lepidopteran leaf mines (*Stilbosis quadricustatella*) on the sand-live oak, *Quercus geminata*. (Courtesy of S. Mopper.)

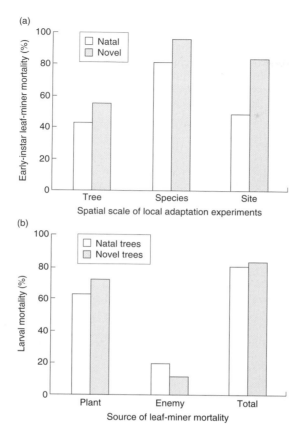

Figure 6.61 (a) Experimental transfer of *Stilbosis quadricustatella* between natal and novel environments demonstrates genetic structure at three spatial scales (tree, species, and site). (b) Larval mortality of *S. quadricustatella* transferred between natal and novel host plants. Although host-mediated mortality is lower on natal hosts, predation is higher. Overall, levels of mortality do not differ between natal and novel hosts. (From Mopper et al. 1995.)

those taxa such as scale insects from which the original hypotheses of deme formation were developed. Further work on scale insects of red (*Pinus resinosa*) and Scots (*P. sylvestris*) pine suggest that there is no local adaptation to individual trees within species, but that there is some local adaptation among pine species (Glynn & Herms 2004). Turning full circle, Alstad now appears to favor a neutral drift model, and not an adaptive model, for deme formation in the black pineleaf scale insect. He has concluded (Alstad 1998), "after 20 years field research with black pineleaf scale and ponderosa pine, the neutral drift and the extrinsic local adaptation hypothesis

remain viable. The intrinsic local adaptation hypothesis is dead." We need much more research to determine how often, and by what mechanisms, insect herbivore populations develop genetic structure.

6.8 EXTREME WAYS OF LIFE

It should be clear by this stage in the chapter that evolution, through the process of natural selection, can result in adaptation by insects to a wide variety of environments. In this final section, we discuss some examples where insects have adapted to habitats that may be considered to be extreme, in that they present particularly difficult conditions for insects and hence demand very special modifications to ecology, physiology, and behavior. These extremes will be considered under the headings of salinity, temperature, toxic environments, and caves.

6.8.1 Salinity

Living in fresh water for at least part of the life cycle is a common feature for some insect orders, but it is perhaps initially surprising to find that water containing appreciable quantities of salt is almost entirely unexploited by insects. Insects are certainly found living in close association with the sea. For example, shore flies (Diptera: Ephydridae) are common in salt marsh habitats, exhibiting high abundance and species richness (Kubatova-Hirsova 2005) (Figure 6.62). True sea water is defined as containing in the region of 35 ppt (parts per thousand) dissolved sodium chloride. Water with 8–35 ppt is known as brackish, and with anything below 8 ppt is thought of as fresh. Almost all aquatic insects are to be found only in fresh water, but a few species such as brine flies have evolved the ability to withstand relatively high levels of salinity. *Ephydra hians* (Diptera: Ephydridae) is a species commonly found in warm, saline, alkaline lakes in Canada, where its larvae graze on microbial mats consisting mainly of photosynthetic filamentous cyanobacteria (Schultze-Lam et al. 1996). Other species of *Ephydra* are thought to be fairly tolerant of salinity levels, but some brine flies may be so well adapted to these conditions that they are unable to tolerate lower salinities. *Paracoenia calida* (Diptera: Ephydridae) is endemic to saline springs in northern California, with salinity levels of 22 ppt. It is suggested that its physiology has become highly adapted to these waters, and specializations for life under such severe and rare conditions may account for the endemism (Barnby & Resh 1988). Some beetles have managed to partition the resources according to salinity and actual divide them up between related species. In Figure 6.63 two species of tiger beetle (Coleoptera: Cicindellidae) found in inland salt marshes in Nebraska, USA, reduce their niche overlap by separating oviposition sites in different salinity levels. Neither species will lay eggs in zero salinity, a degree of specialization to the saline environment, whilst neither can cope with very salty conditions.

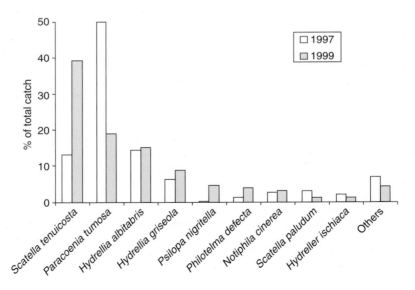

Figure 6.62 Percentage of each species representing more than 2% of shore flies collected in each study year. (From Kubatova-Hirsova 2005.)

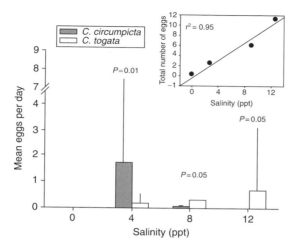

Figure 6.63 Oviposition by adult *Cicindela circumpicta* and *C. togata* as a function of experimental salt concentrations. Bars represent the mean number of eggs $\pm$ SE per sampling period ($n = 8$). For each salinity, a Student's *t*-test was performed (*P*-values are shown above the bars). The inset shows the linear regression of the total number of *C. togata* eggs per condition versus salt concentration. Neither species laid any eggs in the zero salt treatment. (From Hoback et al. 2000.)

In the middle, however, salinity is used as an ecological partitioning mechanism, with one species, *Cicindela circumpicta*, clearly preferring egg-laying sites in less saline conditions to *C. togata* (Hoback et al. 2000).

The sea covers somewhere in the region of 70% of the Earth's surface, but there are almost no insect species at all that could be considered to be marine, in the strict sense of the word. Certainly, many species can be found in the intertidal zone, in the strand line, and in salt marshes, but in almost all of these cases, the insects that live there are able to do so by avoiding much contact with sea water. Along the North Sea coast of Germany, for example, 310 species of Lepidoptera from 29 families were found inhabiting salt marsh ecosystems (Stuening 1988), but most are found in salt-tolerant but terrestrial plant communities dominated by grasses in the upper reaches where there is little tidal influence. Even then, some insects do expose themselves to extreme salinities. Tropical mangrove mud, for example, can become hypersaline when a long way from the refreshing tides and subject to water evaporation, and salinities as high as 89–160 ppt occur regularly (Olaffson et al. 2000). Amongst the more "normal" marine invertebrates such as copopod crustacea found in these

habitats in Zanzibar, chironomid midge larvae (Diptera: Chironomidae) also occurred in some numbers. Not surprisingly, their abundance was negatively correlated with salinity, but, nevertheless, living for long periods in salinities over 100 ppt must be some sort of record for an insect.

The intertidal (littoral) region is, of course, where the land meets the sea, and habitats within the region, along with the organisms that live there, are covered by sea water for varying amounts of time, depending on where they live in relation to the low tide mark. Inventories of insects found in the intertidal zone can include various major orders, with beetles (Coleoptera) and flies (Diptera) dominating (Camus & Barahona 2002). Highest up the intertidal zone is the strandline, or seaweed zone, where insects proliferate. Certain rove beetle (Coleoptera: Staphylinidae) are endemic to these areas (Ahn et al. 2000), and seaweed flies such as *Coelopoda frigida* (Diptera: Coelopidae) lay their eggs amongst washed-up seaweeds such as *Fucus* spp., and they are now so well adapted to this larval food supply that the mere presence of the alga induces egg laying (Dunn et al. 2002). Further down the intertidal region, life becomes a tradeoff between new and relatively empty (at least of other insects) habitats, and increasingly prolonged exposure to sea water. Two basic strategies have evolved, one to try to avoid too much contact with sea water, and the other to tolerate it, indeed on occasion to embrace it.

Avoidance involves mechanisms to hide from the sea when the tide is in, and to only become active when the tide goes out. So, for example, when the sea covers salt marshes, insects such as the beetle *Bledius spectabilis* (Coleoptera: Staphylinidae) may be able to avoid coming into contact with water by constructing burrows into which they retreat. When the tide comes in, surface tension blocks the narrow neck of the burrow, preventing flooding (Wyatt 1986). Similarly, beetles such as *Bryothinusa* and *Neochthebius* (Coleoptera: Staphylinidae and Hydraenidae) live amongst small pebbles which are covered at high tide (Ohba 2003). Their bodies are densely covered with both long and short hairs, as well as with water-repellent oils. Once submerged, these insects are extremely waterproof, and as long as they can survive low oxygen conditions until the tide recedes again, they never actually come into direct contact with sea water.

The other tactic for living between the tides is to "set up home" there, which is what various species of Diptera have accomplished most successfully. Figure 6.64 shows a principal components analysis (PCA) biplot of several invertebrate communities that established themselves in small artificial "ponds" set up in American salt marshes (Stocks & Grassle 2001). More typical marine animals such as polychaetes, sea anemones, and gastropod molluscs soon established themselves in these "mesocosms", and chironomid midge larvae exhibited communities all of their own in these new habitats.

One of the most amazing associations between insects and other intertidal, properly marine, organisms, has been studied by Harley and Lopez (2003). Adult flies in the genus *Oedoparena* (Diptera: Dryomizidae) lay their eggs on the shells of barnacles such as *Balanus* spp. in the intertidal zone of Washington, USA (see Plate 6.19, between pp. 310 and 311). Hatching larvae immediately enter the hosts' mantle cavity and consume the barnacle inside, whereupon they have to migrate (when the tide is out to avoid being washed away), to at least two more prey before pupating finally inside the shell of their last meal. It can be seen from Figure 6.65 that attacks of the predator are seasonal, with peak abundances in late

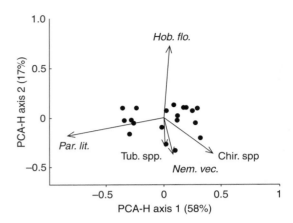

Figure 6.64 Principal components analysis (PCA) biplot of several invertebrate communities in small "ponds" in salt marshes. *Hob. flo., Hobsonia florida* (polychaete); *Par. lit., Paranalis litoralis* (oligochaete); Tub. spp, Tubificidae spp. (oligochaete); *Nem. vec., Nematosteva vecrensis* (anemone); Chir. spp, Chironomus spp. (midge larvae). (From Stocks & Grassle 2001.)

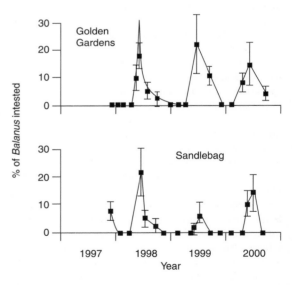

Figure 6.65 Frequency of infestation by larvae of *Oedoparena* spp. of adults of *Balanus glandula* at two sites in Washington. Means ± SE are shown. (From Harley & Lopez 2003.)

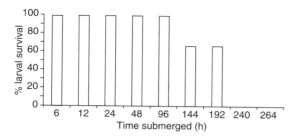

Figure 6.66 Survival of *Oedoparena* larvae following submergence in seawater for varying periods of time. (From Harley & Lopez 2003.)

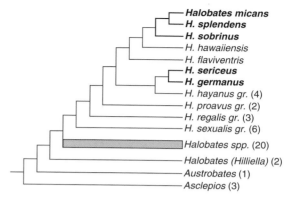

Figure 6.67 Phylogeny of *Halobates* spp. (simplified after Damgaard et al. 2000). Branches leading to oceanic species are highlighted and species names are shown in boldface. Number of species are indicated for species groups. Shaded branch denotes the clade composed of 20 coastal species. (From Andersen et al. 2000.)

spring and early summer, and that barnacle mortality can reach over 20%. Clearly, *Oedoparena* larvae must be able to survive emergence in sea water for the part of the tidal cycle that is underwater, and also withstand high temperatures inside the shells of dead barnacles on hot summer days when the tide is out. Their staggering ability to survive submerged in full sea water is illustrated in Figure 6.66. Never in the wild would they be expected to have to live for up to 100 h constantly in this condition, but clearly they can. Their entire larval life is spent below the high tide mark, as indeed is the pupal stage too, since the final instar larvae spend the winter in empty barnacle shells. This must rank as one of the most extreme adaptations of insects to living in the sea, without becoming truly marine.

There are very few truly marine insects. Indeed, if "marine" implies that the species lives in the sea (i.e. below the surface), then there are just about none, though just a handful occupy the sea–air interface by exploiting the surface of the sea. These are the sea-skaters or ocean-striders, belonging to the genus *Halobates* (Hemiptera: Gerridae), of which only five species have colonized the open ocean (Andersen 1991), all of them in the tropics. The phylogeny of species within the genus *Halobates* is shown in Figure 6.67 (Andersen et al. 2000). It seems that oceanic life has evolved more than once, as indicated by the bold lines in the diagram (Damgaard et al. 2000). All these species would appear to have evolved from freshwater gerrids, of which there are plenty of species, via a coastal, intertidal, or estuarine habitat. The earliest fossils of sea-skaters are about 45 million years old (Andersen et al. 1994), and show very

similar characteristics to modern species. The adults are always wingless, presumably reflecting the stable nature of marine habitats (Andersen 1999), and they feed on dead bodies of arthropods floating on the surface of the sea. Eggs are laid on whatever hard substrates are available; floating seaweed and albatross tails have even been suggested. Because these oviposition sites must be highly ephemeral, sea-skaters are able to make the best of what they find in the open ocean. Thus Cheng and Pitman (2002) were able to find over 800 *Halobates sobrinus* surrounding a single egg mass containing an estimated 70,000 eggs on a plastic 4 L milk jug, floating in the eastern Pacific. Some species have been found more than 150 km from shore (Cheng et al. 1990), but others are more coastal, especially in the nymphal stages. Figure 6.68 represents the distributions of nymphs and adults of *H. fijiensis* in relation to the coastline in Fiji, and it can be seen that most juveniles are found close to the seaward edge of mangroves, and generally confined to pools high on the shore when the tide recedes. Adults, on the other hand, can be found some distance out to sea in sheltered bays, though sparsely distributed (Foster & Treherne 1986). Thus potential competition for food and space between adults and offspring is likely to be reduced.

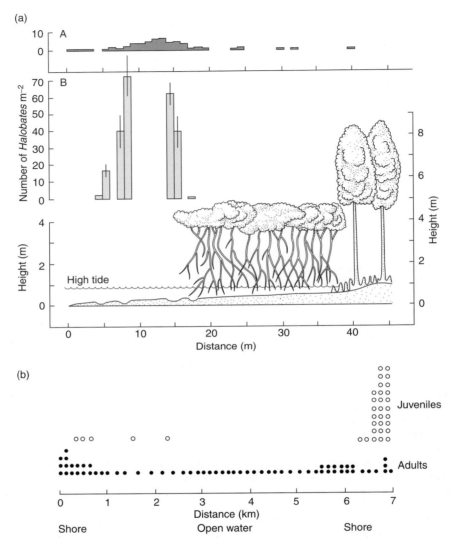

Figure 6.68 (a) Distribution of juvenile *Halobates fijiensis* along upper shore and mangrove transects at (A) high tide and (B) low tide ($\pm$ SE). Numbers were counted in five 1 m squares every 1 m. (b) Distribution of adult and juvenile *H. fijiensis* along an open-water transect across Lucacala Bay, Fiji. (From Foster & Treherne 1986.)

6.8.2 Temperature

We discussed the multivarious effects of temperature on insects in Chapter 2. Exceedingly high temperatures that approach the point of protein denaturation must represent an upper limit to animal distribution, but, nevertheless, some insects live in very hot water. The brine fly *Paracoenia calida* mentioned above occurs in high densities in water that is not only very salty, but where the water at its spring source reaches up to 54°C (Barnby & Resh 1988). Eggs are laid near the source to remain warm, and the larvae, which

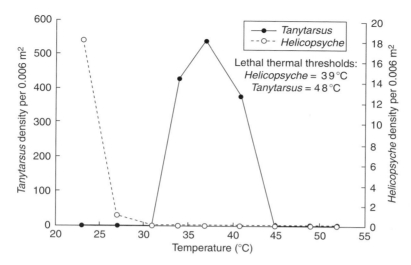

Figure 6.69 Distribution of larvae of a chironomid midge, *Tanytarsus* sp., and a caddisfly, *Helicopsyche borealis*, according to the thermal gradient in a stream emanating from a hot spring. (From Lamberti & Resh 1985.)

are to be found in all but the hottest water, avoid the problem of low oxygen tensions in warm water by using atmospheric oxygen for respiration. Water temperatures can be responsible for the distribution of insects according to thermal gradients in hot springs (Figure 6.69). In a study on streams supplied by hot springs in California (Lamberti & Resh 1985), no macroinvertebrates were to be found in water with temperatures equal to or above 45°C, but the highest density (mainly midges (Diptera: Chironomidae)) occurred at 34°C, with maximum species richness at a much more tolerable 27°C. Both midges and caddisflies (Trichoptera) were commonest at sites that were several degrees below their lethal thermal thresholds, which suggests some leeway in the system should temperatures suddenly rise.

Temperature extremes are more common on land, where the buffering effects of water are absent. The tropics are of course dominated by high temperatures, so that tropical insects such as termites exhibit very high critical thermal maximum temperatures (Ct_{max}). Various genera such as *Coptotermes* and *Reticulitermes* have Ct_{max} values between 49 and 50°C (Hu & Appel 2004). Deserts can range from scorchingly hot to well below freezing within a single 24 h cycle, and insects have adaptations to tolerate both extremes. According to Gehring and Wehner (1995), ants in the genus *Cataglyphis* (Hymenoptera: Formicidae) are amongst

the most thermotolerant land animals known. They forage in the desert at temperatures above 50°C, when their thermal lethal temperatures are a mere 53–56°C, depending on species. It seems that the ants are able to perform this feat by accumulating special heat shock proteins in their tissues before venturing out.

The ability of many insects to survive periods of freezing by supercooling is described in Chapter 2, and it can be seen that temperatures several degrees below zero can often be survived by changes in cell chemistry induced by exposure to low but not lethal temperatures. One example of extreme cold survival comes from *Onychiurus*, a genus of springtail (Collembola) living in Spitsbergen. Monthly mean air temperatures in this region exceed 0°C for only 4 months of the year (June to September; Strathdee & Bale 1998), and temperatures in the soil can be expected to be appreciably lower most of the time. Strictly speaking, of course, springtails are no longer considered to belong within Insecta, but as hexapods their evolutionary ecology might be expected to parallel that of true insects. These animals are exposed to soil temperatures as low as −29.6°C, and temperatures that remain below zero for up to 289 days. In fact, the animals may be encased in ice for an astonishing 75% of the year (Coulson et al. 1995).

Finally, there appears to be a perhaps unsurprising latitudinal influence on the ability of insects to survive

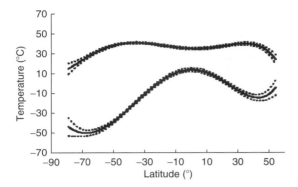

Figure 6.70 Best-fit polynomial regression lines (± 95%) showing the relationship between latitude and the absolute maximum (top line) and absolute minimum (bottom line) temperatures across the New World (negative latitudes are north of the equator). (From Addo-Bediako et al. 2000.)

extreme thermal limits (Addo-Bediako et al. 2000), due to greater variations in climatic conditions in these regions. From Figure 6.70 it is clear that minimum temperatures in the southern and especially northern hemispheres reach very low temperatures, and indeed, the lower bounds of supercooling points and lower lethal temperatures of insects do decline with latitude.

6.8.3 Toxic environments

A large number of chemicals derived from human activities of one sort or another build up in the tissues of insects, which are exposed to the toxins (Nummelin et al. 2007). Many types of chemical are produced as byproducts of manufacturing, agrochemical, and power generation industries. These unwanted by-products include atmospheric pollutant gases and petroleum wastes in water and soil. Other "toxic chemicals" are actually synthesized on purpose, such as fuels including petrol and jet or even rocket fuel. Some of these compounds end up in the environment, where they can have dire consequences for insects and other animals. Not surprisingly, residues of pollutants including petroleum hydrocarbons, ammonia, and metals in river sediments in Indiana, USA, were all toxic to *Chironomus* midge (Diptera: Chironomidae) larvae (Hoke et al. 1993). Even more

understandable is the lethal effect of aircraft fuels. Chironomid adults of various species have been seen to be overcome by fumes during the refueling of helicopters, and dead adult midges have actually been found clogging up the filters in the fuel lines of such aircraft belonging to the Irish Air Corps (O'Connor 1982). The evolution of fuel tolerance in insects is somewhat hard to imagine, though there is, in fact, some evidence suggesting that not all insects are equally affected. In the case of standard, petroleum-derived jet fuel as used by the US Air Force, the order of decreasing susceptibility was shown by earwigs, rice weevils, flour beetles, ladybirds, and finally cockroaches (Bombick et al. 1987) – cockroaches have apparently adapted to survive almost anywhere. Fuel-derived toxins may also have sublethal effects on insects, which, though the exposure may be rather short-lived, can have serious effects on population survival in the longer term.

Undoubtedly, insect ecologies and physiologies have had little time to adapt, even if it were possible, to the presence of solid rocket fuel components in their environment. None the less, honey bees, *Apis mellifera* (Hymenoptera: Apidae), in the vicinity of the Kennedy Space Center, Cape Canaveral, Florida, are exposed to hydrogen chloride (HCl) gas, a hazardous exhaust component of space shuttle launches. In experiments, Romanow and Ambrose (1981) found that HCl caused depressions in brood and honey production, elevations in aggressive behavior by the bees, and increases in the incidence of both viral and fungal stress-related diseases. Quite how the evolutionary ecology of bees would handle this problem given time and sufficient selection pressure is uncertain.

As suggested above, there is no doubt that certain groups or species of insects are better able to tolerate toxins such as petrochemicals than others. One example concerns a train derailment in New York in 1997 where around 26,500 L of diesel fuel were spilt into a trout fishing stream (Lytle & Peckarsky 2001). Slicks of diesel were seen floating 16 km downstream within 24 h of the accident, and 92% of all fish in the river system died. Figure 6.71 shows what happened to the dominant invertebrate populations. The density of all invertebrates was clearly severely reduced immediately after the spill, and the effect was still in evidence in the following spring (Figure 6.71a). However, those insects that did survive were dominated by the so-called riffle beetle,

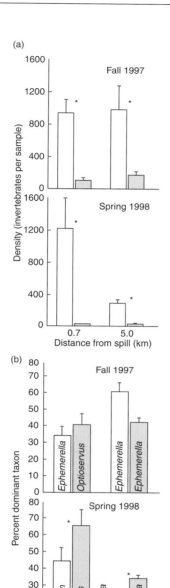

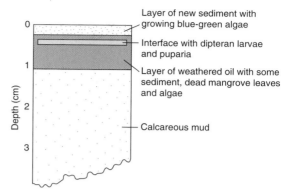

Figure 6.72 Diagramatic cross-section of the soil profile from a mangrove oil spill in the United Arab Emirates, showing the location of oil tolerant fly larvae and pupae. (From Erzinclioglu et al. 1990.)

Figure 6.71 (a) Density of invertebrates, and (b) percentage of dominant taxa (mean ± SE) from 1-min kick samples at reference (clear; white bars) and impact (polluted; grey bars) sites, 2 weeks after the spill (Fall 1997) and 3–4 months after the spill. (From Lytle & Peckarsky 2001.)

Optioservus (Coleoptera: Elmidae) (Figure 6.71b). This species is known to be tolerant of petrochemicals, and as the figure shows, this species comprised more than 40% of all the surviving invertebrates close to the spill. Further away, and as time after the spill lengthens, the more naturally occurring mayfly nymph *Ephemerella* (Ephemeroptera: Ephemereillidae) prevails. Quite why or how the beetle has become tolerant of these types of toxins is less easy to explain.

A very few insects have actually adapted to crude oil as a habitat. Maggots of the fly *Psilopa petrolii* (Diptera: Ephydridae) were found before World War II living in crude oil spillages in California, where they fed on other insects that fell into the oil (Erzinclioglu et al. 1990). It is not known how this insect avoids or tolerates the toxic effects of the oil, but here is clearly an unoccupied niche ripe for colonization. Much more recently, similar dipteran larvae have been discovered in intertidal sediments in the Arabian Gulf, living in the interface between a layer of oil derived from fuel oil spillages and a new thin overlying layer of blue-green algae (Figure 6.72) (Erzinclioglu et al. 1990). To make matters worse, these oily sites also experienced periodic inundations of highly saline sea water.

Other anthropogenic toxins may have less direct effects. A wealth of literature describes the many effects of acid rain on animal and plant communities, either by affecting the insects directly or by changing the characteristics of their host plant. Many examples

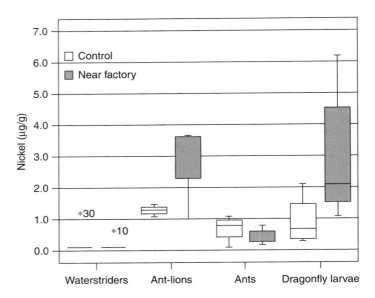

Figure 6.73 Box plot of nickel concentration of four insect groups near a steel factory compared with a control. (From Nummelin et al. 2007.)

are available of both positive and negative influences (see Chapters 3 and 8), and only one is presented here. In northeast France, many mountain streams have become acidified because of precipitation in their catchment areas depositing atmospheric pollutants such as sulphur oxide and nitrogen oxide gases. pHs as low as 4.9 have been recorded, with associated reductions in bicarbonate ions and elevated aluminum concentrations (Guerold et al. 1995). The populations of certain aquatic insects were seriously depleted by these conditions. Mayflies (Ephemeroptera) were least abundant in the acidic waters, while stoneflies (Plecoptera) were the dominant insect group. The latter have adapted to more acidic conditions, and as environmental change progresses it will be these types of insects that will survive more readily (see also Chapter 8).

Animals may themselves not be affected particularly adversely by environmental pollutants such as heavy metals, but their bodies may exhibit relatively high levels of such toxins. In Figure 6.73, Nummelin et al. (2007) show how certain types of insects, specifically dragonfly nymphs (Odonata/Anisoptera) and ant-lions (Neuroptera: Myrmeleontidae), collected in the vicinity of a steelworks have significantly higher

concentrations of nickel in their tissues compared with controls. Other species in this study such as water striders (Hemiptera/Heteroptera: Gerridae) did not show high levels of nickel, but levels in their bodies were found to be good indicators of iron and manganese levels in the environment.

6.8.4 Caves

Caves in general comprise a variety of habitats, depending on how much of their lives animals spend underground, and how specialized they have become. Troglobytes are the most extremely adapted, and, within the insects, consist of a wide variety of orders and families, including ants, flies, cockroaches, and crickets (Reeves & McCreadie 2001; Roncin & Deharveng 2003), and in particular, beetles (Deuve 2002; Trezzi 2003; Peck & Thayer 2003). Some insects only enter caves in search of food supplies provided by bat guano or carrion. Others, however, spend their entire lives as parts of the cave ecosystem where the environments are typified by scarce but fairly constant resources and conditions. Beetles in the families Paussidae, Carabidae, and Staphylinidae are common, the two latter groups being dominated by carnivores and predators. Cholevids are especially diverse in caves, where they mainly survive as saprophages and detritivores (Moldovan et al. 2004). Some species show modifications to mouthparts when compared with relatives that live above ground, such as mandibles and associated hairs and other appendages clearly adapted for a new semiaquatic way of life, only really suited for permanent cave dwelling. An extreme habitat within these already extreme habitats is known as "cave hygropetric" (Sket 2004). Most cave walls are permanently moist or wet, if not running in water, and more specialist cholevid beetle species can be found living amongst other fairly generalist predators and scavengers from millipedes to isopods and even leeches.

Chapter 7

PHYSIOLOGICAL ECOLOGY

7.1 INTRODUCTION

Physiology can be thought of as the internal processes of life – we know them as processes such as digestion, excretion, water balance, and so on. Physiological ecology, then, describes those physiological mechanisms that are fundamental to the ways in which organisms interact with their environment, including other organisms. In this chapter, we consider some of the important physiological processes that make insects so successful in so many different environments. The topics that we cover here are far from an exhaustive list – there are whole books on the subject. Moreover, while we present the material in discrete categories, there is considerable overlap among the mechanisms and processes that we describe. Our purpose here is to introduce readers to the subject of physiological ecology in the hope that they may follow up specific interests by consulting more detailed texts. Our approach is this chapter is to open with feeding by juveniles and to close with reproduction by adults, targeting a few of the many physiological processes that occur along the way during insect development.

7.2 FOOD AND FEEDING IN JUVENILES

The major chemical constituents of the insect body were introduced in Chapter 3. As in all animals, nitrogen-rich proteins are the basic building blocks of tissue and muscle, carbohydrates provide the energy (fuel), while fats tend to act as fuel storage. Because proteins are the major building blocks of tissue, juvenile insects require relatively large amounts in their diets, and proteins may often represent a limiting resource. In Chapter 3, we described how plants have relatively low levels of organic nitrogen in their tissues so that herbivorous insects are particularly constrained by levels of plant nitrogen. In fact, insects employ many methods to increase the levels of protein derived from their diet, and to avoid waste where possible. It is therefore not surprising that many juvenile insects, especially those that have to exist on low-protein diets, regularly eat their own exuviae (shed cuticle) (Mira 2000). Except for the sclerotin component (see Chapter 1) of insect cuticle, a large amount is digestible, freeing up the constituent chemicals for future use. In this "waste not, want not" tactic, cockroaches, for example, have been shown to resorb up to 58% of the nitrogen present in the exuviae.

In general terms there is nothing particularly unusual about the protein, carbohydrate, and fat (lipid) content of the insect body, nor its required diet content, except when ecologies have become very specialized. For example, some blood-feeding insects such as tstetse flies (Diptera: Glossinidae) possess abundant proteases (enzymes which digest proteins), but carbohydrases (for digesting carbohydrates) are virtually absent. Adult butterflies, on the other hand, which feed mainly on sugary nectar, only possess invertases in any quantity, which are used to digest particular kinds of sugar.

Most insects tend to be composed predominantly of proteins, with less carbohydrates and lipids per unit body mass. As Figure 7.1 shows for adult *Drosophila*, absolute concentrations of the three chemical groups increase linearly with body size, and protein levels are highest across all ranges of body mass (Marron et al. 2003). Notice that females are, on the whole, bigger

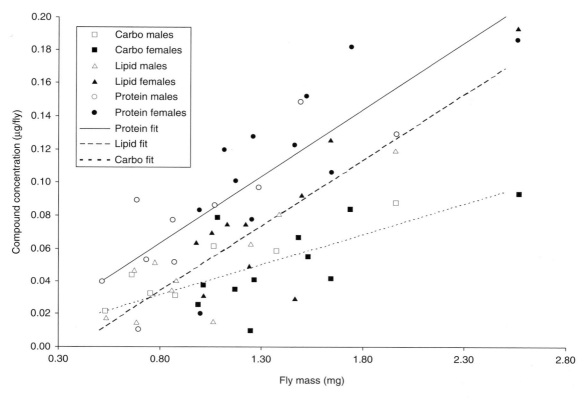

Figure 7.1 Carbohydrate, lipid, and protein levels in adults of 16 species of mesic *Drosophila*. (From Marron et al. 2003.)

and heavier than males, a feature common within the Insecta and elsewhere in the animal kingdom, depending on latitude (Blankenhorn 2000). It is also crucial that a large proportion of the food eaten by an insect nymph or larva is converted into growth – a low level would be wasteful and inefficient. In experiments with the caterpillars of various species of *Heliothis* (Lepidoptera: Noctuidae), Lee et al. (2006) found that the amount of organic nitrogen eaten correlated with nitrogen growth up to an asymptote (Figure 7.2), with somewhere in the range of 20% to 40% efficiency. Carbohydrate assimilation efficiency ranged from 10% to 20%.

The majority of insects do most if not all of their growth as juveniles (nymphs or larvae). Consequently, the quantity and quality of food that a juvenile insect is able to acquire have crucial implications for its adult life, in particular on fecundity and reproduction. The length of the juvenile stage, the number of instars,

and the nutritional value of the food help to determine the size (weight) of the pupa, and hence the emerging adult. However, the length of time spent as a juvenile represents a tradeoff between benefits (longer to eat and grow) and costs (risk of mortality, declining environmental conditions, etc.).

We might expect insects to exhibit physiological mechanisms that maximize their growth rate per unit time, particularly on poor-quality diets. For example, wood is very low in nutrients, especially the all-important organic nitrogen. Larval longhorn beetles are all wood- or bark-borers, and can be extremely serious secondary pests of stressed or otherwise debilitated trees (see Chapter 12). Some longhorns appear to possess the rare (in eukaryotes) enzymes required to break down cellulose. Moreover, they are very sensitive to variation in the protein content of their diet. For example, larvae of *Morimus funereus* (Coleoptera: Cerambycidae), an oak-feeding

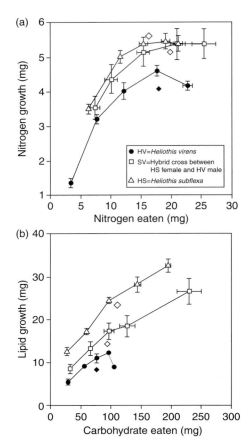

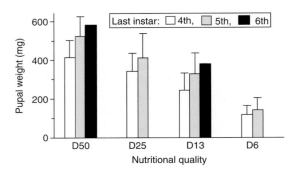

Figure 7.3 Body weight of *Psacothea hilaris* pupae reared on diets of sequentially lowered nutritional quality (D50, D25, D13, D6). Pupae were classified based on the larval instars they completed. (From Shintani et al. 2003.)

Figure 7.2 Utilization plots describing the bivariate means (±1 SE) of (a) nitrogen growth versus total nitrogen consumption and of (b) lipid growth versus total carbohydrate consumption in the no-choice test. Growth data presented for SV and HS caterpillars are corrected by multiplying the factor of 0.980 and 1.579 to each. This is done to account for their difference in initial fresh mass relative to HV caterpillars. In each genotype, the points from left to right are 7:35, 14:28, 21:21, 28:14, and 35:7 protein : carbohydrate (P:C) diets for the nitrogen utilization plot whereas such order is reversed for the lipid plot. Diamond symbols indicate the mean reference values for animals that are free to choose between 35:7 and 14:28 P:C diets. (From Lee et al. 2006.)

longhorn beetle from Eastern Europe, can change their proteolytic activity depending upon the nitrogen content of their diet (Ivanovic et al. 2002). Insect species with a narrow range of food types or host species (specialists) tend to self-select lower protein diets than closely related species with a broader range of diets (non-specialists) (Lee et al. 2006). In addition, it is predicted that nutritional regulation abilities are less important in animals such as insects with narrower diet breadths (Despland & Noseworthy 2006).

The growth of insect larvae and the length of the larval stage are critically dependent on food quality. In the study illustrated in Figure 7.3, Shintani et al. (2003) reared another wood-boring beetle, the yellow-spotted longhorn beetle, *Psacothea hilaris* (Coleoptera: Cerambycidae), on diets of different quality. Artificial diets were comprised of commercial insect diet (originally designed for silkworms) mixed with varying amounts of indigestible cellulose powder. Increasing the ratio of cellulose in the diet (from left to right) caused a clear reduction in the mass of the pupae. Furthermore, the number of larval instars completed before pupation also declined. In the case of the poorest diet (D6:94), only the fourth and fifth instars were reached, with pupal mass averaging a mere 25% of that attained on a better diet (D50:50). The implications (generally lower fitness) of reduced pupal, and hence adult, weight are considered later in this chapter. In experiments with lepidopteran larvae, in this case those of the tobacco hornworm, *Manduca sexta* (Hemiptera: Sphingidae), Ojeda-Avila et al. (2003) discovered that larvae on low-sucrose, high-protein diets grew most rapidly, and produced the largest pupae and adults (Figure 7.4). The amount of food consumed (dry weight) was positively and linearly related to the growth of larvae (fresh weight) and, of course, older larvae (fourth instar) ate more

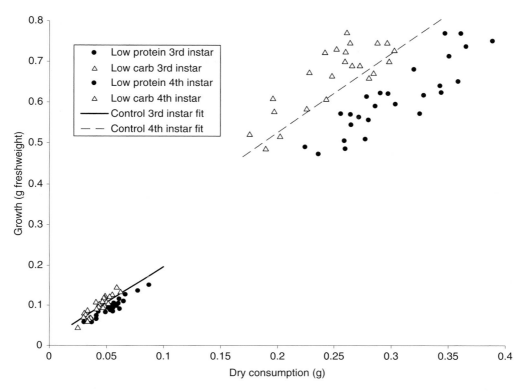

Figure 7.4 Fresh weight growth of two larval instars of tobacco hornworm, *Manduca sexta*, feeding on three types of diet. (From Ojeda-Avila et al. 2003.)

and grew more than younger ones (third instar). Most significantly, insects on low-protein diets grew substantially more slowly for any given consumption rate, another illustration of the vital importance of protein in the diets of insect larvae.

Diet quantity and quality also influence the growth and fitness of carnivorous insects. In fact, one of the most efficient types of feeding in this context is cannibalism (Mayntz & Toft 2006), since the prey can be considered to be ideally matched to the predator's nutrition requirements. Ladybirds (Coleoptera: Coccinellidae) despite their (often undeserved) fame as biological control agents of aphid pests (see Chapter 12), are fairly polyphagous, and will eat a variety of animal material as long as it is of the right size. In fact, some ladybirds are rather accomplished cannibals; hatching larvae will eat the eggs of their siblings all around them rather than expend risk and effort to find aphids for themselves. As with plant

material, the nutrient content of live prey varies considerably, with a resultant influence on the predators. For example, Specty et al. (2003) studied the effects of diet quality on the growth of the ladybird *Harmonia axyridis*. They compared the final mass of various ladybeetle life stages fed on "normal" food, i.e. the aphid *Acyrthosiphum pisum* (Hemiptera/Homoptera: Aphididae), with those fed eggs of the flour moth *Ephestia kuehniella* (Lepidoptera: Pyralidae). Biochemical analyses showed that *E. kuehniella* eggs were twice as rich in amino acids than the bodies of *A. pisum* (12% fresh weight as oppose to 6%), and three times richer in lipids. Conversely, the aphids were one and a half times richer in glycogen than moth eggs. When fed on moth eggs, ladybirds ended up significantly heavier as both male and female adults than when fed on aphids (Figure 7.5); these differences mainly appeared during the pupal and adult stages. Development time to adulthood was the

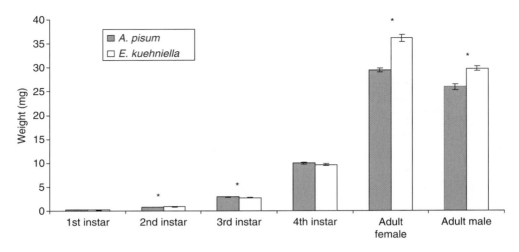

Figure 7.5 Weight at molting of various life stages of the ladybird *Harmonia axyridis* fed on a natural diet (*Acyrthosiphum pisum*) and substitute food (*Ephestia kuehniella* eggs). Asterisks signify a significant difference at $P < 0.05$. (From Specty et al. 2003.)

same for both diets, but total mortality over the generation dropped from around 24% on aphids to a mere 3.3% when fed on moth eggs. Clearly, ladybeetles benefited from feeding on eggs rather than aphids, and we might wonder why ladybirds persist in eating aphids when moth eggs would appear to be better. Of course, in the field, aphids are much more readily available, and are highly abundant and self-replicating, when compared with the occasional clutch of lepidopteran eggs. This illustrates a tradeoff between the best nutrients and the highest ecological availability of nutrient "packets".

Food supply for juvenile insects is especially problematic if they are totally dependent on adults to feed them. Honeybees, *Apis mellifera* (Hymenoptera: Apidae), are a case in point, where the larvae are actually imprisoned within cells in the hive and fed by the workers. Pollen is the main nitrogen source for growing bee larvae, consisting of between 10% and 25% protein (Keller et al. 2005). Once collected by the foraging workers, pollen is stored in special cells within the colony from where it can be fed to the juveniles. Schmickl and Crailsham (2001) manipulated the pollen stores within a honeybee colony using pollen traps, which removed pollen from returning worker bees before they could deposit their cargo, and/or by using water sprays to simulate heavy

rain which deterred the workers from foraging at all. There was a steady decrease of pollen stores within hives during non-foraging periods, and a steady increase during foraging periods (Figure 7.6a). Pollen supply during non-foraging periods got critically low as the larvae aged and grew, and their pollen demand increased. One result of this is shown in Figure 7.6b. During non-foraging periods, 3-day-old larvae were the most likely to die, mainly due to cannibalism by the workers who then fed them to older larvae until capping (when the larvae are sealed off in their cells to pupate). Capping occurs significantly earlier in non-foraging periods, so that workers can still be produced even during low food periods. Since less investment has been made by the colony in younger larvae, they can be sacrificed for the good of the colony, and replaced relatively easily when foraging for pollen and nectar can be reinstated.

Given that protein availability appears to be limiting for juvenile insects in general, we might predict that a sufficient supply of organic nitrogen is particularly important for female insects. The protein content of insect eggs is quite high, and egg lipoproteins provide most of the resources for embryonic and early larval development. We have already pointed out that female insects are generally larger and heavier than males, presumably as an adaptation

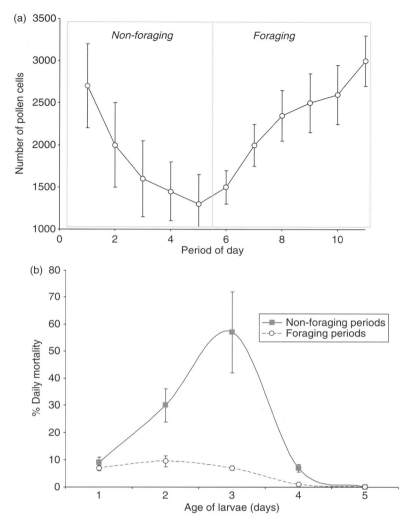

Figure 7.6 (a) Number of pollen cells in a colony of honeybees, *Apis mellifera*, comparing non-foraging with foraging periods, (b) Daily mortality of bee larvae when deprived of pollen (non-foraging periods). (From Schmickl & Crailsham 2001.)

to increase reproductive output. We might therefore expect differences in the nutritional physiology of male and female insects. Telang et al. (2001) studied caterpillar foraging by the tobacco budworm, *Heliothis virescens* (Lepidoptera: Noctuidae), wherein the insects were provided with a choice of diets containing varying percentages of protein versus carbohydrate and their consumption measured. As the ratio of protein to carbohydrate increased (i.e. more protein, less

carbohydrate), the amount of dry food eaten decreased roughly linearly (Figure 7.7), so that much less food was consumed on high-quality diets than on low-quality diets. The figure also shows that females consumed marginally more food than did males. Further work demonstrated that females accumulate more protein and carbohydrate over the whole larval instar than do males, and that they are able to regulate food intake according to protein content.

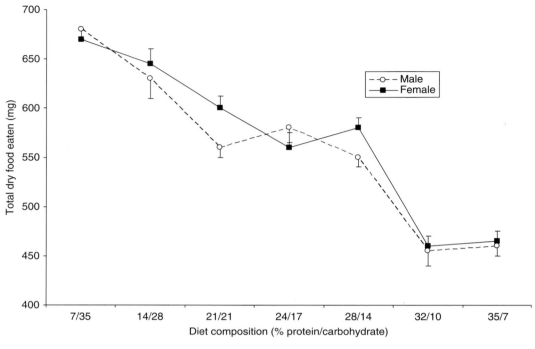

Figure 7.7 Dry weight of food eaten by larvae of the tobacco budworm, *Heliothis virescens*, according to the protein/carbohydrate percent dry weight in the diet. (From Telang et al. 2001.)

In other words, female caterpillars are substantially more likely to seek out and utilize high-protein diets than are males. In the field, late instar *H. virescens* larvae can be found feeding on cotton flowers and anthers, rather than leaves, in order to maximize their protein intake.

7.3 FOOD AND FEEDING IN ADULTS

We have seen that the quantity and quality of food available to insect nymphs and larvae have important consequences for growth and final size of pupae and adults. In turn, adult size is often a good predictor of reproductive success (see later). Sometimes, however, the relationship between pupal/adult size and subsequent fecundity is modified by the resources available to young adults prior to, or during, reproduction. Put simply, the quantity and quality of adult food can influence significantly at least some aspects

of reproductive performance. Two very different examples will illustrate this concept.

First, many young adult insects from butterflies to hover flies, from parasitic wasps to wood-boring beetles, feed on flowers. This maybe to enhance water uptake or to acquire essential food supplies such as sugars for fuel and proteins for egg production. The larvae of longhorn beetles (Coleoptera: Cerambycidae) are important wood-borers (Speight & Wylie 2001), and the genus *Phoracantha* contains some of the most notorious pest species of all. *Phoracantha recurva* and *P. semipunctata* are extremely serious pests of eucalyptus trees all over the tropics and subtropics (see Chapter 12) whose larvae tunnel under the bark, and later into the timber of stressed host trees, causing much death and dieback. Female fecundity in both species is clearly related to the source of pollen in the young adults' diet (Figure 7.8). Bee pollen from the deserts in the southwestern USA, or from British Columbia in Canada, is a poor food

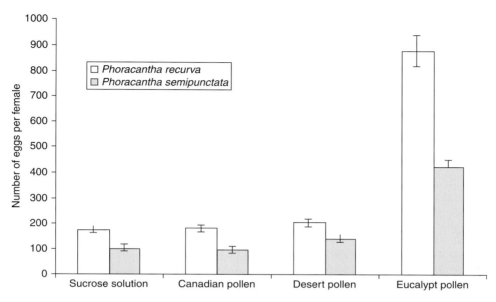

Figure 7.8 Mean lifetime fecundities for two species of eucalyptus longhorn beetle, *Phoracantha* spp., fed on different diets. (From Millar et al. 2003.)

source when compared with pollen from eucalyptus trees (Millar et al. 2003). For example, *P. recurva* females that fed on eucalyptus pollen laid between four and eight times more eggs than those on other diets. In addition, feeding on eucalyptus pollen increased the longevity of *P. semipunctata* females by as much as 70% over the other diets. It seems likely that the effects of diet source on *Phoracantha* beetles are a result of feeding stimulation (hence the response to the specific host-plant pollen), and/or differences in nutritional quality of the various pollens. It is certainly the case that pollen feeding is very important in insect ecology, providing insects with sources of amino acids, proteins, carbohydrates, lipids, sterols, and vitamins.

A completely different ecology which also demonstrates the effects of food quality on the reproductive capacity of adult insects is exemplified by the burying beetles. Adult beetles such as *Nicrophorus orbicollis* (Coleoptera: Silphidae) feed on carrion, and the larvae of other carrion insects, prior to laying eggs. During this time, body weight increases, as does the concentration of juvenile hormone (see below) and ovary mass. If adult females are starved for any reason (such as the lack of suitable dead bodies),

their juvenile hormone titer stays low and their ovaries remain small (Trumbo & Robinson 2004). In order to investigate the effect of food supply on reproductive success, postemergent female beetles were caged with either mealworm (Coleoptera: Tenebrionidae) larvae (a food supply that they would rarely if ever utilize in the wild) or blowfly larvae (Diptera: Calliphoridae) (a usual and preferred food supply, occurring as they do on carrion). Beetles on the less-preferred mealworm diet laid around 33% fewer eggs, which were nearly 20% lighter, than did beetles feeding on the preferred blowfly larvae (Trumbo & Robinson 2004) (Figure 7.9). Females feeding on mealworms also produced offspring that exhibited delayed emergence from the pupal stage. Delayed emergence can impose a significant cost on a rare and ephemeral food source such as an animal corpse. Overall, then, undernourished females produced fewer eggs, smaller eggs, and offspring with delayed development. The lesson is that food quality can be important to the success of adult insects as well as to that of larval and nymphal stages.

One final way to enhance the food intake of young adult females is to utilize the services of male partners. This can take the form of nuptial gifts, where the male

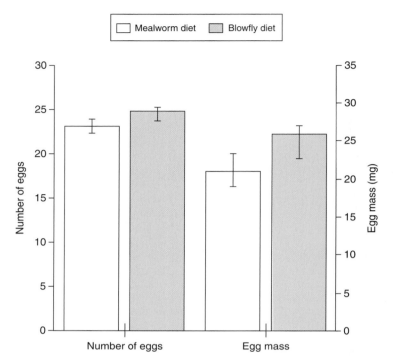

Figure 7.9 Mean number of viable eggs, and egg weight, for the burying beetle, *Nicrophorus orbicollis*. showing that blow fly larvae were preferred over mealworms (both sets of means were significantly different at $P < 0.05$). (From Trumbo & Robinson 2004.)

donates protein-rich material to his mate in order to enhance their offspring's chances of survival. One example concerns the katydid, *Isophya kraussii* (Orthoptera: Tettigoniidae). Most katydid species are herbivorous, and hence mechanisms to increase protein intake are especially important since their diet is particularly deficient in this type of nutrient (see Chapter 1). During copulation, the male transfers a spermatophore (the nuptial gift) to the female, which comprises the ampullae that contain the sperm and the so-called spermatophylax. The latter is a protein-rich substance produced by glands in the male reproductive tract, which recipient female katydids assimilate to enhance egg production (Voigt et al. 2006). Note that the female habit of eating the male during copulation, as in some praying mantids, does not seem to enhance protein intake significantly and is therefore of little use for producing more offspring (Maxwell 2000).

7.4 METABOLISM

Once food of whatever quantity and quality has been ingested and digested, components of the digested food are utilized to produce energy via respiration, a process requiring oxygen and producing carbon dioxide (CO_2). As a consequence, the metabolic rates of organisms can be estimated either by oxygen consumption or by CO_2 evolution. Figure 7.10 illustrates a general relationship between body size and standard (or resting) metabolic rate for individuals from five groups of arthropods (ants, beetles, mites, spiders, sun spiders; Lighton et al. 2001). Metabolic rate increases as body size gets bigger. Indeed, body size is a better predictor of metabolic rate than is taxonomic position within the arthropods. However, within insect lineages, the same pattern does not always hold, with larger insects exhibiting lower resting metabolic rates than smaller ones (Table 7.1).

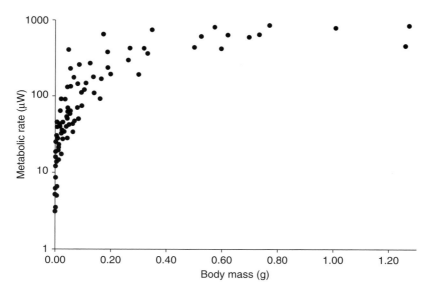

Figure 7.10 Metabolic rates of individuals within five groups of terrestrial arthropods in relation to body size. (From Lighton et al. 2001.)

Here, metabolic rate has been measured in five species of dung beetle from southern Africa (Duncan & Byrne 2000). At a constant temperature, heavier beetles have lower metabolic rates than lighter beetles. Notice

Table 7.1 Standard metabolic rate (mean ± SD) of telecoprid beetles at 20°C.

Species	Mass (g)	Metabolic rate (W kg⁻¹)
Sisyphus fasiculatus	0.136 ± 0.012	1.789 ± 0.38[a]
Scarabaeus rusticus	0.564 ± 0.131	0.987 ± 0.60[b]
Anachalcos convexus	1.421 ± 0.35	0.719 ± 0.21[b]
Scarabaeus flavicornis	0.322 ± 0.053	1.019 ± 0.27[b]
Circellium bacchus	7.285 ± 2.93	0.331 ± 0.1[c]

Means with the same letter are not significantly different ($P < 0.05$: ANOVA with Tukey's multiple range test).

particularly the smallest beetle species with exponentially higher metabolic rates than the slightly larger ones.

However, metabolic rate is influenced by more than simply body size (Nespolo et al. 2005). Insects, of course, like almost all animals on the planet, are poikilothermic and ectothermic, and hence largely unable to regulate their internal temperatures independently of the outside environment. As a result, insects that are exposed to variation in environmental temperatures undergo substantial changes in metabolic rate, even while inactive. Figure 7.11 illustrates how the resting metabolic rate (RMR) of *Rhyzopertha dominica* (Coleoptera: Bostrychidae), the lesser grain-borer, varies with temperature (Emecki et al. 2004). This insect is a serious pest of stored grains, especially when temperatures in storage silos are raised. CO_2 production (as a measure of RMR) increases fairly linearly up to 30°C, but there is some suggestion of an increased CO_2 production at higher temperatures. High metabolism, of course, facilitates rapid growth rates and low generation times as long as copious amounts of food are available, never a problem in a grain store. In other words, high metabolic rates at high temperatures exacerbate the pest status of the lesser grain-borer.

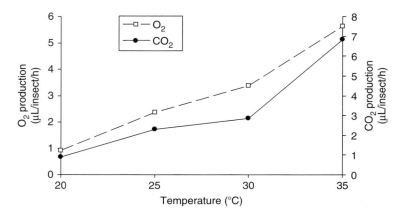

Figure 7.11 Respiration rates of adult *Rhyzopertha dominica* in normal air at different temperatures. (From Emecki et al. 2004.)

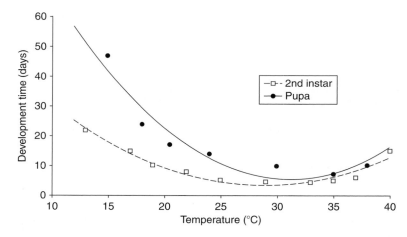

Figure 7.12 Development time in days of second instar and pupae of pea weevil *Bruchus pisorum* according to differing conditions of constant temperature. (From Smith & Ward 1995.)

Many insects actually show non-linear relationships among temperature, metabolic rate, and development time. As a result, there is usually an optimal temperature for development, above and below which development times increase. An example of this is shown in Figure 7.12, which illustrates the development times of two stages in the life cycle of the pea weevil, *Bruchus pisorum* (Coleoptera: Bruchidae). In this case, development time reaches a minimum at about 30°C, and actually rises again as conditions become warmer still (Smith & Ward 1995).

We might expect that insects living in difficult or extreme habitats (see Chapter 6) might have modified metabolic rates that minimize the physiological risks of their habitats. So, insects living in deserts, where food and especially water supply are very limited, might be expected to exhibit reduced metabolic rates for a given external temperature. Duncan et al. (2002) studied the activity and metabolic rates of nine species of tenebrionid beetle from the Negev Desert in Israel. There were no differences in standard metabolic rate among the nine species, though the nocturnal ones showed a form of continuous respiration compared with the diurnally active ones, who exhibited discontinuous gas exchange cycles (see below), presumably to conserve water. More importantly, there were no real differences between the metabolic rates of desert dwellers and their more temperate relatives. The Negev beetles apparently use ecology and behavior to manage their water and food problems rather than metabolic adaptation. However, the ability to change metabolic rate under difficult environmental conditions has been reported in other insects. For example, the alder leaf beetle, *Agelastica alni* (Coleoptera: Chrysomelidae), reduces its metabolic rate under oxygen deprivation (Kolsch et al. 2002) (Figure 7.13). In this study, heat flow rate (another measure of metabolic rate) declined when normal air was replaced with nitrogen. Flexibility in metabolism is likely to afford insects some physiological buffering from environmental variation.

Finally we must also think about what happens to metabolic rate as insects perform different activities. Resting or standard metabolic rate is the usual estimate, because it is the easiest to measure in experiments. When it is possible to measure metabolic rate in mobile insects, it is clear that things speed up considerably. According to Vogt et al. (2000), the fire-ant *Solenopsis invicta* (Hymenoptera: Myrmicidae) increases its metabolic rate between 38- and 50-fold when flying as compared with resting. Table 7.2

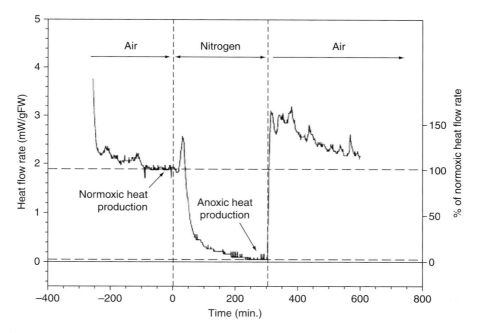

Figure 7.13 Effect of hypoxia/anoxia on the heat flow rate of *Agelastica* beetles at 21.7°C. (From Kolsch et al. 2002.)

Table 7.2 Mean metabolic rates (ml CO_2/g/min) for adult *Phoracantha semipunctata* whilst at rest, running, or flying. All measurements carried out at 20°C. (From Rogowitz & Chappell 2000.)

Metabolic rate at rest		Metabolic rate running		Metabolic rate flying	
Male	0.010 ± 0.004	Male	0.069 ± 0.004	Sex not	
Female	0.009 ± 0.001	Female	0.054 ± 0.003	specified	0.300 ± 0.02

shows the results of activity experiments on the long-horn beetle, *Phoracantha semipunctata*. Apart from the slight differences between sexes, the major finding is that the metabolic rate increases five times or so from rest to running, and six times or so from running to flying (Rogowitz & Chappell 2000). Even then, these authors suggest that the metabolic rate during flight for their beetle is substantially less than for equivalently sized Lepidoptera, perhaps as a consequence of wing mechanics. In general, Lehmann and Heymann (2006) suggest that compared to other types of locomotion, the high power demands for flight requires that metabolic activity increases up to 10–15 times over resting metabolism.

7.5 RESPIRATION

In addition to chemical fuels, all insects need to exchange gases to power metabolism. Insects take in oxygen in order to burn chemical fuel, and release CO_2 as a product of metabolism. This process of gas exchange is called respiration. As described in Chapter 1, most insects respire through a system of tubes called tracheae. The tracheae connect to the outside world via the spiracles, special valves under muscular control that serve to isolate the internal tissues from the environment. The tracheae form a hugely complex network of largely hemolymph-filled tubes, which terminate in the tracheoles, tiny tubes that reach less than 1 μm in diameter. As the photograph in Figure 7.14 illustrate, tracheae and tracheoles permeate all parts of the insect body, including the legs and wings. They are surrounded by rings of chitin that support the tubes and help to prevent them from collapsing when the non-compressible hemolymph is removed

during periods of high metabolic activity. Tracheae form part of the complex insect circulation system, illustrated in Figure 7.15. The direction of hemolymph flow through the circulatory system can actually alternate through the process of heartbeat reversal (Hertel & Pass 2002). Although insects have a main "heart", there is often a well-developed system of auxiliary pumping organs, especially in the extremities, that can be under autonomous nervous control.

The process of respiration takes several forms (Westneat et al. 2003). First, passive gas diffusion occurs both ways, into and out of the tissues surrounding the tracheae, which contain hemolymph at rest. Passive diffusion is a relatively slow process, especially in large insects (Harrison et al. 2005). However, bigger insects do seem to possess higher capacities for respiratory gas diffusion, as shown by Figure 7.16. Here, Harrison et al. show that the diffusion capacity of tracheae increases with body size

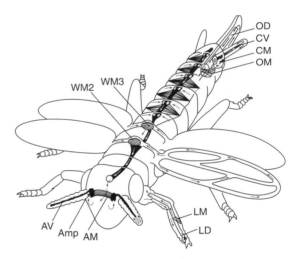

Figure 7.15 Diagram of an idealized insect with the maximum possible set of circulatory organs. Central body cavity: dorsal vessel composed of anterior aorta and posterior heart region. Antennae: pulsatile ampullae (Amp) with ostia connected to antennal vessels (AV), pumping muscle associated with ampullae (AM). Legs: pulsatile diaphragm (LD) and associated pumping muscle (LM). Wings: dorsal vessel modification with ostia in mesothorax (WM2) and, separate pumping muscle in metathorax (WM3). Cerci: cercal vessels (CV) with basal suction pump (CM). Ovipositor: each valvula with non-pulsatile diaphragm (OD) and basal forcing pump (OM). Arrows indicate direction of hemolymph flow. (From Hertel & Pass 2002.)

Figure 7.14 Insect tracheae showing chitinous rings. (Courtesy John Kimball.)

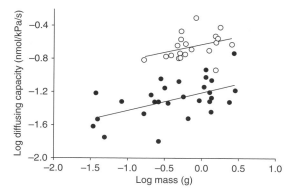

Figure 7.16 Log diffusing capacity versus log mass for the sixth (●) and 10th (○) dorsal transverse tracheae. (From Harrison et al. 2005.)

(mass). Both the dorsal transverse tracheae (DTT) that transport gases from the sixth spiracle to the midgut (sixth DTT) and the DTT that transport gases from the 10th spiracle to the hindgut (10th DTT) show higher capacities in bigger insects. This system does have limitations, however. During activity, osmotic pressures withdraw the fluid from the tracheae leaving air spaces instead where gaseous diffusion can occur more swiftly. Gas movement in the tracheae can be assisted by pumping of the dorsal heart and by muscular contractions in the abdomen. In addition, special air sacs, and accessory or auxiliary hearts in the appendages, assist the pumping process. Routine movements of locomotion also pump gases round the system via these air sacs. This last process is rather elegant, whereby active movements such as flight, requiring the highest rates of gas exchange, automatically generate bellows actions to achieve maximal gas exchange. Finally, there appears to be a system of active tracheal compression, especially in the head and thorax of certain insects, which undergoes a rapid cycle completed in less than 1 s.

Figure 7.17 illustrates a sequence of four dorsoventral images taken of the head and thorax region of the beetle *Platynus decentis* (Coleoptera: Carabidae) using X-ray imaging (Westneat et al. 2003). The white arrowhead in image A shows the resting state, where the tracheal tubes are expanded, while the same marker in C shows maximal compression. This type of breathing may have played an important role in the evolution of running and flying by early

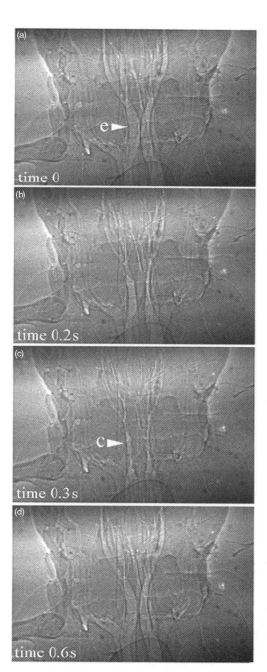

Figure 7.17 Respiration by tracheal compression in the head and thorax of the beetle *Platynus decentis* (dorsoventral views (head up)). (From Westneat et al. 2003.)

insects in terrestrial ecosystems. Moreover, a well-developed system for circulation and respiration may have paved the way for the provision of the high levels of oxygen that are required by the complex nervous, sensory, and feeding systems ubiquitous in modern insects.

In insects, gas exchange generally takes place in a series of discrete phases. These so-called discontinuous gas exchange cycles (DGCs) are widespread in insects, as well as in other arthropods from ticks to harvestmen (Chappell & Rogowitz 2000). DGCs are also described as constriction–flutter–open (CFO) cycles (Vanatoa et al. 2006), wherein a sequence of three distinct phases of spiracle function occur during a DGC. The closed phase (C) is when the spiracles are closed and gas exchange is negligible, and oxygen is taken out of the tracheae into the tissues for respiration. At this stage, however, CO_2 remains dissolved in the tissues, so that the net effect is that intratracheal pressures drop. Next comes the flutter phase (F) when the spiracles open minimally and

fleetingly, and most gas flow is into the low-pressure tracheae from the outside. Finally comes the open phase (O) when the spiracles open fully and gas exchange (oxygen in and CO_2 out) occurs, sometimes assisted by pumping actions of body musculature. The open phase may be rather infrequent, with many hours separating each one, depending on habits and habitat, and, of course, the ambient temperature.

Figure 7.18 illustrates DGCs from the eucalyptus longhorn beetle, *Phoracantha semipunctata*, under different temperatures (Chappell & Rogowitz 2000). The RMR is expressed as the mean VCO_2, a measure of CO_2 production per hour. The large irregular peaks show the O phase, which in the case of *Phoracantha* take place roughly every 30 min at 10°C, but increase rapidly in frequency at higher temperatures. The much smaller but very frequent oscillations observed between the large peaks represent a combination of closed and flutter (C + F) phases. It is important to

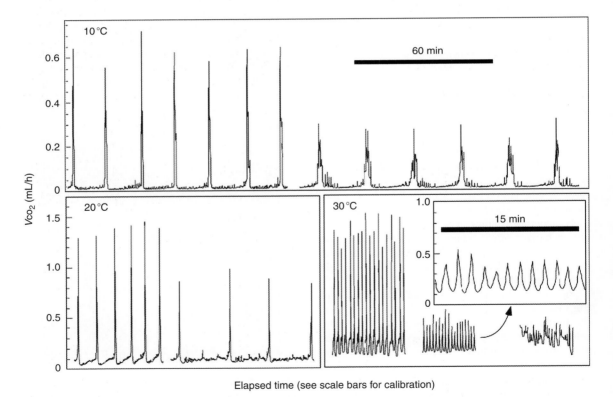

Elapsed time (see scale bars for calibration)

Figure 7.18 Examples of discontinuous gas exchange cycles from *Phoracantha semipunctata* at three ambient temperatures. (From Chappell & Rogowitz 2000.)

notice the decrease in duration of the C + F phases in relation to the O phase at high temperatures. Temperature-regulated metabolic rate requires much more frequent and relatively massive gas exchange at higher ambient temperatures. What are the ecological pressures that have led to the evolution of DGCs in insects? There are several alternative explanations. For example, it has been proposed that DGCs have evolved as a water conservation mechanism because most water loss from insects takes place when the spiracles are open (see below). A short and infrequent O phase helps to prevent these losses, since in the C phase there is little or no scope for water egress, and in the F phase the net movement of gases is inwards. A second hypothesis suggests that DGCs help insects to live in hypoxic (low oxygen) habitats. For example, while Chappell and Rogowitz carried out these experiments on adult longhorn beetles that live externally, their larvae and pupae spend all their lives (save the very first larval instar) under bark or wood where oxygen levels would likely be low.

Recall from Chapters 1 and 2 that some life stages of insects are much more quiescent than is the adult stage, and that diapause is a common life history trait expressed by insects. However, even insects in diapause need to respire, and here again, respiration patterns show bursts and flutters of CO_2 emission (Tartes et al.

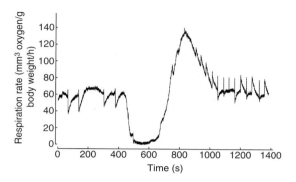

Figure 7.19 Respiration pattern with one large CO_2 burst and microbursts (sharp spikes) during flutter in diapausing *Pieris brassicae*. (From Tartes et al. 2002.)

2002) (Figure 7.19). Butterfly pupae might be expected to have very low metabolic rates (though they likely increase with outside temperature, especially towards spring), and *Pieris brassicae* (Lepidoptera: Pieridae) exhibits a pattern of CO_2 output with a single large burst occurring once every 10–16 h. Microbursts of air intake and CO_2 output occur irregularly but more frequently. Beetles illustrate similar behaviors during low-temperature diapauses. As Figure 7.20 depicts, adult Colorado beetles, *Leptinotarsa decemlineata*

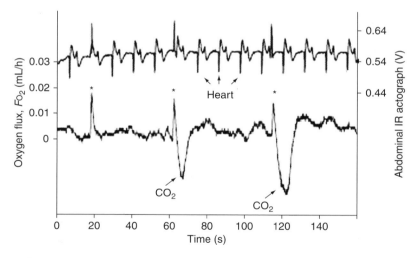

Figure 7.20 Gas exchange microcycles (flutter) in a diapausing Colorado potato beetle at 0°C. The upper trace is an infrared (IR) actographic recording, the lower trace is a simultaneous recording by the respirometer-actograph. Abrupt air intakes into the tracheae, marked with asterisks, are visible on both traces; CO_2 microbursts are visible only on the respirometric recording. Note that the heart is beating. (From Vanatoa et al. 2006.)

(Colopeoptera: Chrysomelidae), show a rhythmic heart beat at 0°C, but experience abrupt air intakes every 60 s or so (Vanatoa et al. 2006). The ecological significance of discontinuous patterns of gas exchange in diapausing insects probably relates as before to a combination of water loss avoidance and the potential for anoxic conditions.

Whatever the specific type of respiration, continuous or otherwise, the rate at which oxygen is consumed (and hence metabolism proceeds) is fundamentally linked to activity and temperature. However, the relationships among ambient temperature, metabolic rate, and respiration are not always straightforward in insects. Despite the textbook definitions, many insects are in fact able to show some degree of endothermy

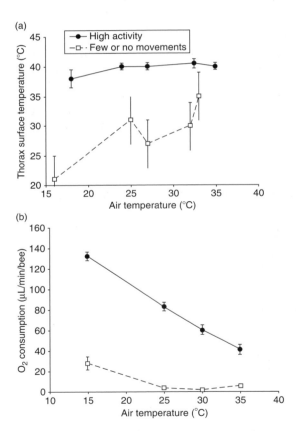

Figure 7.21 (a) Temperature on the surface of the thorax of foraging-age bees in relation to activity level. (b) Mean oxygen consumption of foraging-age bees in relation to activity level. (From Stabentheiner et al. 2003.)

(or at least, heterothermy), in that they can adjust the temperature of their tissues in order to keep their muscles warmer than the outside temperatures before and during flight. Bees, for example, have to generate metabolic heat to achieve a minimum temperature to activate their flight muscles so that they are able to forage and pollinate plants (Nieh et al. 2006). Respiration rate, measured by oxygen consumption, was studied in honeybees, *Apis mellifera* (Hymenoptera: Apidae), by Stabentheiner et al. (2003). Bees are known to be ectothermic at rest in the hive. However, they are able to switch to endothermy to regulate nest temperatures and to prevent chill or even freezing in swarms. They may also exhibit endothermy during external foraging trips, so that they can collect essential pollen and nectar on days when it would be simply too cold to fly if ectothermic. These plastic responses to temperature variation and the ability to "turn on" endothermy must have substantial implications for metabolic rate and respiration when bees need to be active in relatively cold conditions. During high levels of activity, when bees are performing endothermically, the surface of their thorax can be an incredible 20°C warmer than the air around it (Figure 7.21a). This elevated body temperature has huge ecological advantages, but it comes at a considerable cost in terms of oxygen consumption (Figure 7.21b), metabolic rate, and fuel use. In fact, endothermic bees actually show a significant decrease in oxygen consumption as temperatures rise because they do not have to work so hard to maintain body temperatures in excess of ambient.

7.6 EXCRETION

So far in this chapter, insects have consumed food, taken in oxygen to burn the fuel that drives metabolism, and released carbon dioxide. Now, the waste material generated by the combination of these three processes has to be disposed of. Given that nitrogen forms such an important part of insect body tissues, it should be no surprise that nitrogen is turned over, and excreted, during growth and maintenance. There three main types of nitrogen-based excretory products – ammonia, urea, and uric acid – in combination with more minor products such as allantoin, creatine, and amino acids. Whether an animal excretes predominantly ammonia, urea, or uric acid depends upon a number of factors, predominantly environmental, and centers on the relative toxicity of

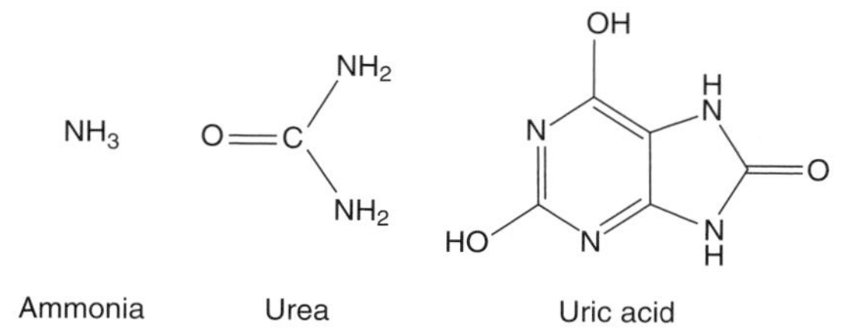

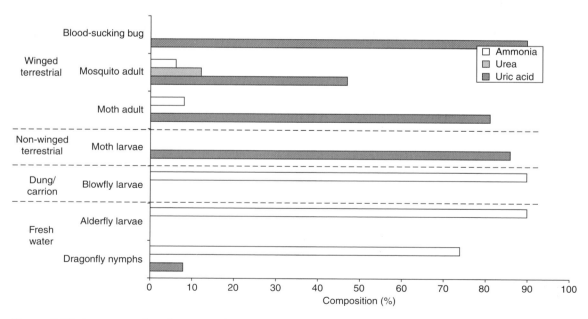

Figure 7.22 Chemical structures of ammonia, urea, and uric acid.

ammonia when it is concentrated in body tissues (Wright 1995). If we look at the chemical structures of ammonia, urea, and uric acid (Figure 7.22) we can see that the nitrogen : hydrogen ratio decreases markedly from uric acid, through urea, to ammonia. At the same time, molecular complexity increases from ammonia to uric acid. In combination, this means that ammonia is the most toxic but the least energy-expensive to make, while uric acid is the least toxic but the most energy-expensive to make. Should insects opt for cheap, toxic waste products or expensive, non-toxic ones? Certainly, some insects produce copious quantities of ammonia. Soil-feeding

termites such as the genus *Cubitermes* (Isoptera) for example have been shown to accumulate huge quantities of the chemical in their mounds, derived from high concentrations in the hindguts of individual insects (Ji & Brune 2006); but more generally, the type of excretory compounds produced by insects appears to depend largely upon the availability of water in their particular environment.

The process of converting ammonia to urea is called ureotelism and the final step, from urea to uric acid, is uricotelism. At each stage in the process, less water is needed for chemical synthesis and, most importantly, the waste product needs less water for its disposal. Toxic ammonia must be flushed from the body with considerable amounts of water, whereas non-toxic uric acid can be disposed of in almost solid form. In simple terms, it takes roughly 0.5 L of water to eliminate 1 g of nitrogen as ammonia, 0.05 L for urea, and a mere 0.001 L for uric acid. Unsurprisingly, then, only those insects that live in fresh water, or a similar highly aqueous environment such as dung or carrion, can excrete ammonia (Figure 7.23). Those species living away from copious supplies of water have to pay the extra energetic burden and produce much less soluble compounds, usually uric acid. A glance at Figure 7.23

Figure 7.23 Excretory products from insects living in different habitats.

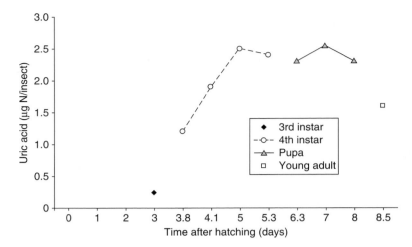

Figure 7.24 Uric acid accumulation during the development of *Aedes aegyptii*. (From von Dungern & Briegel 2001.)

confirms that aquatic insects excrete predominantly ammonia whereas terrestrial insects excrete primarily uric acid. Mosquitoes such as *Aedes aegyptii* (Diptera: Culicidae), with freshwater larval stages and terrestrial adult stages, provide a compelling example. The amount of uric acid produced is minimal before the third larval instar, but rises to a plateau during the fourth instar (von Dungern & Briegel 2001) (Figure 7.24). Uric acid concentrations then stay fairly constant during the pupal stage, and remain relatively high into adulthood. The drop in uric acid concentrations in young adult mosquitoes is related to the disposal of waste accumulated during the pupal stage in the meconium as the adult emerges. In fact, though not shown in the figure, uric acid levels stay fairly high in the adult as it feeds, but do vary according to the level of blood meals taken.

By now it should be clear that excretion by animals, including insects, is intimately associated with their water relationships. We consider these next.

7.7 WATER RELATIONS

Water availability is thought to be one of the most important abiotic variables influencing the distribution of animal species, especially small ones like insects (Addo-Bediako et al. 2001). As suggested in Chapters 1 and 2 of this book, one of the fundamental keys to the inordinate success of the insects is the

waterproofing that enables them to spend prolonged periods of time in highly desiccating environments. Just to take an extreme example, let us imagine a small insect, with a high surface area to volume ratio, flying through the air in the heat of the tropical day. First, the air temperature is high resulting in a high metabolic rate (even if the insect is at rest) and a very drying atmosphere. Air is flowing over the insect's body as it flies, promoting evaporation from the body surface. The high metabolic and respiratory demands of flight require copious supplies of oxygen to be taken in through the spiracles, necessitating frequent open phases during which water can be lost. With all of these potential routes of water loss, it seems highly unlikely that insect flight could have evolved without extremely efficient water management. We have met some of the mechanisms of water retention already; they include waterproof cuticles, discontinuous gas exchange, and tolerance to desiccation.

For the majority of insects, respiratory water loss seems to be less important than cuticular water loss (Chown 2002) (Table 7.3). We can therefore assume that, in the majority of cases, respiratory adaptations such as discontinuous gas exchange are sufficient to minimize respiratory losses. What about cuticular losses of water? The outermost layer, the epicuticle, is a highly complex structure at the micro (or even nano) level. It is covered with a lipid or wax layer, which is in turn overlaid by a cement layer. These layers can be incredibly thin – the electron

Table 7.3 Proportional contributions (%) of respiratory and cuticular water loss to total water loss in various insect species. All measurements taken at 24 or 25°C. (From various sources in Chown 2002.)

Insect species	Group or common name	% Respiratory loss	% Cuticular loss
Periplaneta americana	Cockroach	13	87
Melanoplus sanguinipes	Grasshopper	15	85
Romalea guttata	Grasshopper	3	97
Aphodius fossor	Dung beetle	5	95
Scarabaeus galenus	Scarab beetle	7	93
Scarabaeus gariepinus	Scarab beetle	23	77
Camponotus vicinus	Carpenter ant	2	98
Pogonmyrmex rugosus	Harvester ant	2.4	97.6
Average		8.8	91.2

micrographs shown in Figure 7.25 of coleopteran cuticle (Hariyama et al. 2002) illustrate that the entire epicuticle is much less than 1 μm thick. The uppermost dark layer in Figure 7.25b indicates the relative position and size of the wax and cement layers. The wax layer is fragile and its main function is water retention. It consists of bipolar molecules with hydrophilic and hydrophobic ends. The latter face outward and prevent water leaving the insect body. The cement layer protects the wax layer, the molecules of which can be disoriented by heat or abrasion. For example, water loss rates in various castes of desert ants were increased two or even three-fold in individual workers exposed to soil during nest excavation work (Johnson 2000). Abrasion of the cuticle by soil and sand grains damages the cement and underlying wax layers, and decreases the density of hydrocarbons. Indeed, apocryphal stories suggest that throwing handfuls of road dust into flour reduces the incidence of insect grubs, a research project waiting to be carried out by an enthusiastic reader.

The detailed biochemistry of cuticular waterproofing is beyond the scope of this book, but it is indeed complex. Major types and relative quantities of hydrocarbons not only vary among insect species, but also within a single life cycle, so that eggs, larvae, ands adults can exhibit different biochemical "signatures" from their waterproofing structures (Nelson & Charlet 2003). More important for the ecology of insects is the efficiency of cuticular waterproofing, which is largely temperature dependent (Figure 7.26). In this example, there is a clear transition temperature (or critical temperature; Gefen & Ar 2006) at which the wax

layer breaks down almost completely, in this case at about 40°C (Ramlov & Lee 2001). Until this temperature is reached, the insect shows a remarkable ability to maintain low levels of water loss through its cuticle.

As ecologists, it should be no surprise to us that the degree of cuticular waterproofing is linked to habitat. Insects that live in moist environments tend not to show such extreme waterproofing as those that live in more arid places (Figure 7.27). Ramlov and Lee (2001) illustrate that cuticular permeability can be up to 10 times greater for insects that live in dry, desiccating conditions when compared with those in moist habitats. Notice also that dead insects show much the same levels of permeability as live ones – waterproofing is a simple chemical process which once established does not require a living animal to maintain it (until it needs repairing).

Of course, if and when water loss does occur, insects have evolved various systems for mitigating the problems that result. The solutions can be quite simple, such as drinking when water is available, coupled with tolerating desiccation when it is not. Desert beetles represent an excellent system for studying the combination of these tactics. Adaptations to extreme conditions such as deserts are discussed further in Chapter 6, but for now let us take the example of *Stenocara gracilipes* (Coleoptera: Tenebrionidae) from the Namib Desert in southwestern Africa. Desert tenbrionids form a major component of an ecological guild known as macrodetritivores (Ayal 2007), who are responsible for a great deal of plant decomposition via microbial digestion in their guts. To perform this feat, these beetles have to actively forage for sparse plants in

(a)

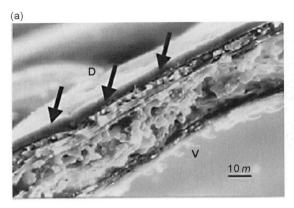

(b)

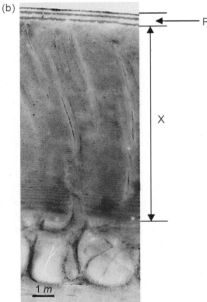

Figure 7.25 Scanning (a) and transmission (b) electron micrographs sectioned with the sagittal plane. The epicuticle (P) just below the dorsal surface was distinguished from the exocuticle (x) by the difference of thickness of each lamination. Five layers were observed in the epicuticle, and a multi-layer in the exocuticle. Arrows indicate the cuticle which includes the epi- and exocuticle. D, dorsal side. V, ventral side. (From Hariyama et al. 2002.)

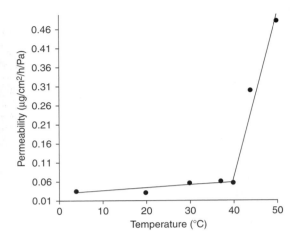

Figure 7.26 Cuticular permeability of third instar larvae of the gallfly *Eurosta solidaginis* according to temperature. (From Ramlov & Lee 2001.)

the desert sands and in doing so, run the risk of severe water shortages. Various mechanisms have been evolved to minimize such risks.

First of all, *S. gracilipes* produces copious quantities of wax from cuticular pores to enhance its waterproofing – hence its common name of wax-blooming beetle. Metabolic water production (water derived from food metabolism) can only provide a certain amount of the insect's internal needs. Without rain to provide drinking water, problems of osmoregulation can ensue, including serious shifts in hemolymph osmolality and volume, allied to changes in sodium, potassium, and chloride concentrations in the body fluids. As the number of days without water increases, and desiccation progresses, the beetles experience a significant reduction in water content (Figure 7.28a). Body water is almost instantly replenished when they are able to drink. However, the changes in hemolymph osmolality and potassium concentration do nothing so dramatic – there is some fluctuation but, during dehydration and subsequent rehydration, their levels stay much more constant (Naidu 2001) (Figure 7.28b). As Naidu states, *S. gracilipes* "demonstrates an exquisite capacity" to osmoregulate under conditions of acute water shortage and abundance. The only remaining question is where does the drinking water come from?

The Namib Desert is characterized by high winds, extreme temperatures, and virtually no rainfall whatsoever. However, dense early morning fogs frequently roll in from the adjacent Atlantic Ocean, and *Stenocara* adults stand with their bodies tilted forward into the wind seemingly collecting droplets of water on their bodies to drink (Parker & Lawrence 2001). Ironically,

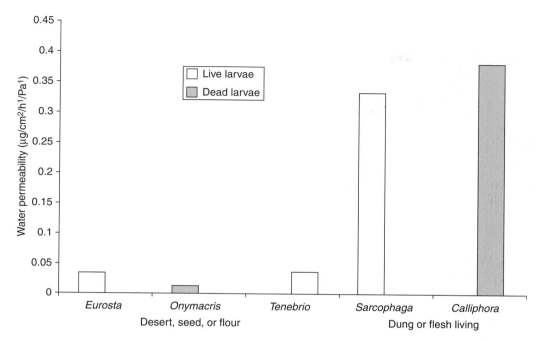

Figure 7.27 Water permeability of the cuticle of various species of larvae from different habitats. (From Ramløv & Lee 2001.)

it seems that the secret of droplet formation lies with the absence of wax on small bumps or points on the elytra that become hydrophilic. These bumps, 10 μm in diameter, are surrounded by very waxy troughs (Parker & Lawrence 2001) (Figure 7.29). Minute droplets of water from the fog grow on the hydrophilic seeding bumps or points. Eventually, the droplets coalesce into a steady stream of water on the elytra, which runs down to the mouth, providing sufficient water to sustain the beetles in the complete absence of rainfall. These mechanisms are so sophisticated that technologists and engineers have been trying to produce artificial versions of them (Zhai et al. 2006). According to these authors, potential applications of such surfaces include water harvesting systems, controlled drug-release coatings, open-air microchannel devices, and lab-on-chip devices.

7.8 DEVELOPMENT: ECDYSIS AND PUPATION

There are few aspects of insect physiology that are more fundamental to the ecology of insects than the growth and molting of juvenile stages, and their subsequent maturation into adulthood. Separating the ecological roles of juveniles (growth) from those of adults (sex, dispersal) is one of the great successes of insects. Exactly how does it happen? At its most simple, juvenile insects grow through nymphal or larval instars, shedding their skin (ecdysing) as they go, eventually turning directly into an adult in the exopterygotes (hemimetabolous insects), or entering a pupal stage that metamorphoses into the adult in the endopterygotes (holometabolous insects). The enormous advantages given to larvae in terms of specialization, lack of competition with adults and so on, cannot be overstated. However, there are at least two basic physiological problems with this otherwise wonderful system. First, insects need to shed their skin, or molt, at regular intervals in order for the juvenile to grow. Second, there are profound and complex morphological and physiological changes that must take place between the juvenile and adult stages.

Both molting and maturation are under the control of a suite of hormones, two of the most important being juvenile hormone (JH) and ecdysone (ecdysteroids). Barnes et al. (2001) label JH the "status quo

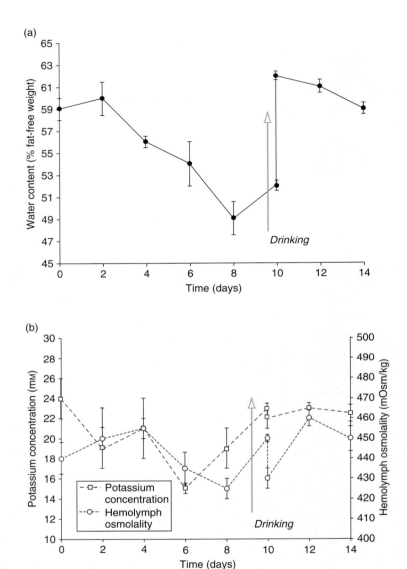

Figure 7.28 (a) Effects of dehydration and rehydration on (a) water content by weight in the desert beetle *Stenocara gracilipes*, and (b) on hemolymph osmolality and potassium concentration in *S. gracilipes*. (From Naidu 2001.)

hormone'' because, when it is present in the insect, each molt results in another nymph or larva (the next instar). When JH levels are reduced, molting results in the appearance of the adult directly in exopterygotes, or via pupation in endopterygotes. Ecdysone, on the other hand, is itself under the control of another

hormone, prothoracicotrophic hormone (PTTH). When in relatively high concentrations, ecdysone stimulates the juvenile insect to molt. Figure 7.30 illustrates the levels of JH and ecdysone during the life cycle of a typical endopterygote, and show how the two hormones interact in the growth and maturation of insects. During the

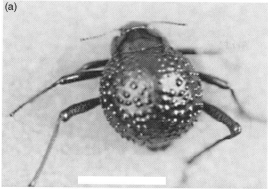

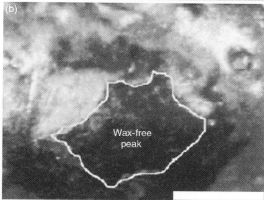

Wax-free
peak

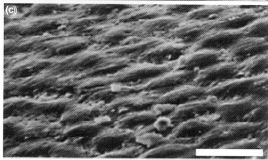

Figure 7.29 The water-capturing surface of the fused overwings (elytra) of the desert beetle *Stenocara* sp. (a) Adult female, dorsaf view; peaks and troughs are evident on the surface of the elytra. (b) A "bump" on the elytra, stained with Red O for 15 min and then with 60% Isopropanol for 10 min. a procedure that tests for waxes. Depressed areas of the otherwise black elytra are stained positively (waxy, light), whereas the peaks of the bumps remain unstained (wax-free, black). (c) Scanning electron micrograph of the textured surface of the depressed areas. Scale bars: a, 10 mm; b, 0.2 mm c, 10 μm. (From Parker & Lawrence 2001.)

pupal stage, high levels of ecdysone and very low concentrations of JH combine to stimulate the pupa to perform its final molt into the adult.

7.8.1 Ecdysis

Each nymphal or larval instar culminates in one or more "pulses" of ecdysone (in fact these are pulses of its active derivative, 20-hydroxyecdysone; Gilbert 2004). These pulses cause changes in the "commitment" of epidermal cells, and stimulate them to produce enzymes that digest and recycle parts of the cuticle. The sequence of diagrams in Figure 7.31 illustrates the molting process. The major component of the aptly named molting fluid is the enzyme chitinase, which causes the degradation of chitin in the old cuticle, separating it from the epidermal cells. The old cuticle is then cast off within 24 h or so (Zheng et al. 2003). It is important to note that the linings of the tracheae, foregut, and midgut are discarded and replaced at each molt along with the outer exoskeleton. The insect therefore loses anything residing in these parts of the gut at every molt. Gut symbionts for example (see Chapter 6) need to be replaced by termites and cockroaches after each ecdysis.

7.8.2 Pupation and wing development

JH plays additional, complex roles during insect development, such as influencing the growth of wings and other appendages (mouthparts, legs, genitalia, etc). Though unseen on the outside of the juvenile animal, insect larvae begin to develop wings in later instars from structures known as imaginal disks. These imaginal disks grow constantly and exponentially during late larval stages and into the pupal stage. Finally, they cease to grow, and subsequently differentiate into adult appendages such as wings (Miner et al. 2000). Much of the growth of imaginal disks seems to be related to the general growth of the larval body and, in starved insects, disk growth ceases fairly rapidly. Differentiation of disks into complex organs such as wings is influenced by JH (Miner et al. 2000) (Figure 7.32). The continued presence of JH in late-stage larvae or pupae inhibits further differentiation of the imaginal disks. The combination of disk growth rate related to body size and food supply, and the regulation of imaginal disk size by JH, ensures that

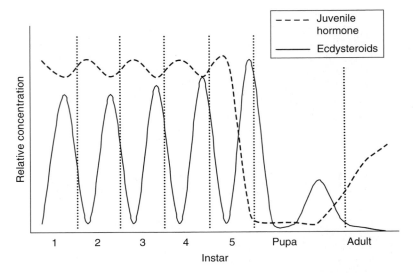

Figure 7.30 Variations in juvenile hormone and ecdysone during life cycle of a typical holometabolous insect. (From Barnes 2001.)

insects develop to adults with wings perfectly proportioned to the size of the rest of the body. Relatively high concentrations of JH repress the development of other adult characteristics, such as eyes (Allee et al. 2006).

Though by its very name, JH is usually associated with juvenile development in insects, it can also play a role in the adult lives of insects. It helps to regulate oocyte and egg development, and has also been shown to influence various types of behavior in social insects (Scott 2006). Aggression in honeybees, for example, is associated with high JH concentrations; guard bees in colonies are highly aggressive in order to defend the colony against intruders and predators, and they show significantly higher levels of JH compared with equivalent worker bees.

7.9   REPRODUCTION

Once an insect reaches adulthood, its final set of missions is to disperse, find a mate (or mates), and seek a suitable habitat in which to place its offspring. Reproductive ecology, behavior, and physiology of insects encompass enormous areas of research, and include some of the most complex but elegant systems on Earth. Elsewhere in this book, we discuss various aspects of insect reproduction, including mate selection, sperm competition, oviposition, asexuality, pheromones,

and so on. Here, we conclude our introduction to physiological ecology with the final moments of the insect life cycle: fecundity and the production of eggs.

7.9.1   Eggs and fecundity

Lifetime fecundity is one fundamental indicator of an animal's reproductive success. Put simply, the more eggs that an adult female produces during her lifetime, the higher the probability should be that at least some of her offspring will live to reproduce. In reality, the quality of eggs as well as the quantity of eggs will influence offspring survival (Rossiter 1994). As such, insects of a given size can generally either produce many small eggs with few provisions for their offspring, or fewer large eggs that contain more resources.

Tradeoffs between egg number and egg size emerge primarily because fecundity is a hugely expensive operation in terms of time, energy, and resources. For most insects, it is a once in a lifetime event after which the adult will die. The energy and resources available for egg production are largely a function of the size of the adult laying the eggs (Riddick 2006), which in turn is related to pupal size (Bauerfeind & Fischer 2005), itself a function of food collected and assimilated by the juvenile (nymphal or larval) stage. In some cases, such relationships can be seen

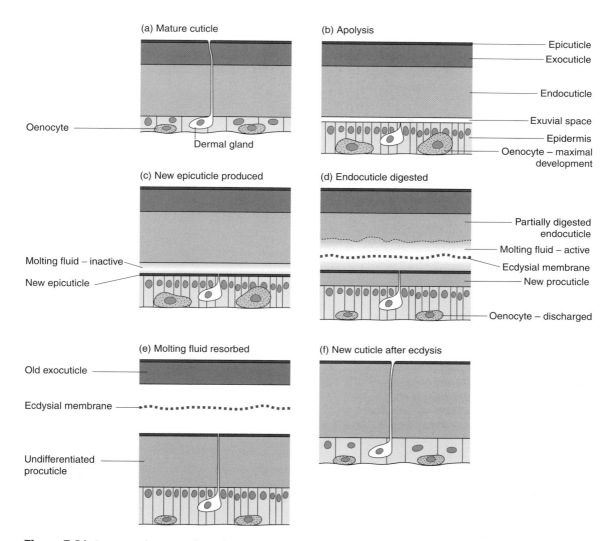

Figure 7.31 Sequence of stages in the molting of an insect cuticle. (From Zheng et al. 2003.)

to be inversely density dependent, so that individual fecundity decreases as population density increases (Gur'yanova 2006). Insect fecundity is also frequently correlated with body size, illustrated by examples from two distinctly different insects (Figures 7.33 and 7.34). *Anagyrus kamali* (Hymenoptera: Encyrtidae) is a parasitic wasp whose preferred host is the hibiscus mealybug, *Maconellicoccus hirsutus*. *Pulvinaria regalis* (Hemiptera: Cocccidae), in contrast, is a sap-feeding scale insect, related rather closely to mealybugs.

Notice that neither of the relationships between body size and fecundity are linear. Instead, both are sigmoidal, indicating that there is a maximum potential fecundity for both insects. No matter how large a female becomes, her egg load is finite. Also note the huge difference in the maximum fecundity of the two species of insect. The parasitoid has an average maximum fecundity of around 150 eggs per female (Sagarra et al. 2001), while that of the scale insect exceeds 2500 (Speight 1994).

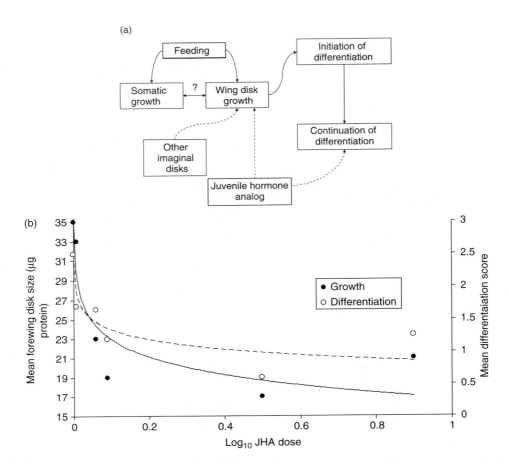

Figure 7.32 (a) Summary diagram of the control of growth and differentiation of the wing disks of *Precis coenia*. (Dashed lines, negative effect). (b) Effect of the juvenile hormone analog (JHA) methoprene on growth and differentiation of the forewing disk in final stage larvae of the butterfly *Precis coenia*. (From Miner et al. 2000.)

Some species of insect can modify batch or clutch size depending upon the quality of the environment in which they are laying eggs. Fewer but larger clutches may be preferred over more frequent but smaller ones, according to the likely benefits offered by a particular oviposition site. Such is the case of the leaf-mining moth, *Paraleucoptera sinuella* (Lepidoptera: Lyonetiidae). This tiny moth feeds as a larva on *Populus* species (poplars and aspen) as well as on *Salix* (willow) from Western Europe to Japan. First, the total number of eggs laid per female varies considerably with host-plant species (Kagata & Ohgushi 2001). In addition, the number of eggs laid per clutch also varies among host plants. Most clutches on *Populus* contain five or six eggs,

whereas on *Salix* clutches are generally half that size (Figure 7.35). It seems that ovipositing adults can choose the numbers of eggs to lay in one batch according to perceived food quality and quantity. Scattering eggs more widely and less densely would appear to be a better bet on less than ideal host-plant species (Kagata & Ohgushi 2002).

Host plants can therefore influence insect reproduction and this seems particularly important for specialists such as the beetle *Leptinotarsa undecimlineata* (Coleoptera: Chrysomelidae), related to the notorious Colorado beetle, *L. decemlineata*. Unlike its more familiar cousin, which eats a variety of plant species, *L. undecimlineata* is a monophagous species

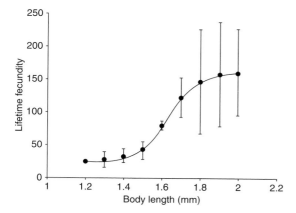

Figure 7.33 Lifetime fecundity in the parasitoid wasp *Anagyrus kamali* related to body size. (From Sagarra et al. 2001.)

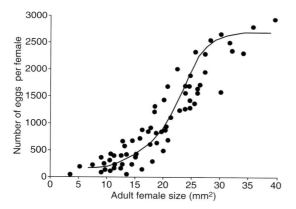

Figure 7.34 Adult female size (surface area) in relation to fecundity of horse chestnut scale, *Pulvinaria regalis*. (From Speight 1994.)

eating only a limited number of species of *Solanum* in the Neotropics, in particular, *Solanum lanceolatum* (Lopez-Carretero et al. 2005). Oocytes fully develop in their precursor germaria when the beetle is fed on its normal host plant, but when the non-host *S. myriacanthum* is provided as food, oocyctes do not develop until the host plant is switched back to *S. lanceolatum*. In this way, host specialization is maintained, since eggs are unlikely to be laid on plants which are the "wrong" species for the specialist insect.

At the physiological level, what determines the number and size of eggs that female insects produce?

Oocytes are essentially the precursors of eggs – ova are produced when oocytes undergo cell divisions and combine with yolk material. Hence the number of oocytes in a female insect, and their size, are good indicators of fecundity, potential offspring survival, and general reproductive success to come. In turn, oocytes and the eggs produced from them are influenced by energy consumption, food supply, and, in particular, the size of the abdominal fat body. Indeed, food quality and quantity can have a significant effect on insect fecundity, so that adult mirid bugs, *Macrolophus caliginosus* (Heteroptera: Miridae), produced many more eggs when both the nymphs and adults were provided with live, natural food (the eggs of the flour moth, *Ephestia kuehniella* (Lepidoptera: Pyralidae)) (Vanderkerkhove et al. 2006) (Figure 7.36). The fat body is used in insects for lipid and protein storage, and for the production of egg yolk proteins. As such, it is a useful index of insect nutritional and reproductive status. In female paper wasps, *Mischocyttarus mastigophorus* (Hymenoptera: Vesidae), those with large fat bodies produce significantly larger oocytes than did wasps with small fat bodies (Markiewicz & O'Donnell 2001) (Figure 7.37). Moreover, the condition of the fat body is itself correlated negatively with foraging rate, and positively with the amount of time that wasps spend in the nest. In other words, female wasps, particularly queens, are able to increase their reproductive capacity by reducing energy-consuming (and worker-like) tasks such as foraging. If energy is not wasted in flight, greater fat stores can be accumulated, increasing an individual's potential for reproduction. Again, there exits an evident tradeoff between reproductive capacity and other essential adult activities.

Given that tradeoffs between competing adult functions are common, it is interesting to consider what fraction of the resources, accumulated during the larval stage, is allocated to egg production. Bagworms (Lepidoptera: Psychidae) provide a useful model insect with which to answer this question. Bag-worm larvae construct cases (bags) from leaf fragments which they line with silk. They then wander over their host plant, carrying their case with them, rather in the manner of hermit crabs or caddisfly larvae. Pupation takes place in the bag and, in species such as the oil palm bag-worm, *Metisa plana*, wingless (apterous) females also complete all of their reproductive activities within the bag. In this Malaysian bagworm species, the mean weight of female pupae is

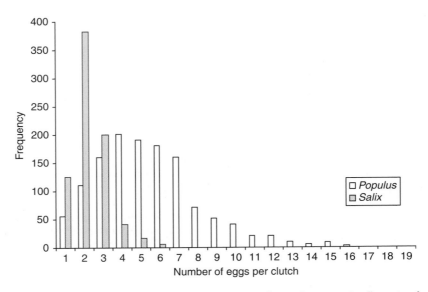

Figure 7.35 Frequency distribution of clutch size in the leaf-mining moth, *Paraleucoptera sinuella*, on two host-plant species. (From Kagata & Ohgushi 2002.)

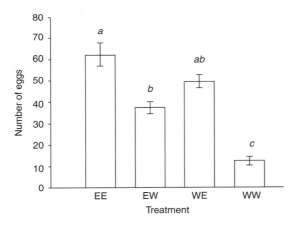

Figure 7.36 Mean number ± SE of eggs oviposited during an 11-day period for female mirid bugs fed *Ephestia kuehniella* eggs as nymphs and adults (EE), *E. kuehniella* eggs as nymphs and artificial diet 2 as adults (EW), artificial diet 2 as nymphs and *E. kuehniella* eggs as adults (WE) and artificial diet 2 as nymphs and as adults (WW). Bars with the same letter are not significantly different (P>0.05; Mann–Whitney U-test). (From Vanderkerkhove et al. 2006.)

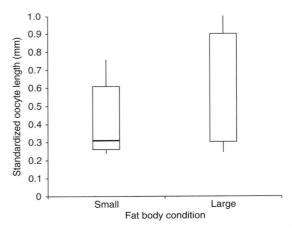

Figure 7.37 Boxplots of mean oocyte length in the paper wasp *Mischocyttarus mastigophorus* according to fat body size. The means are significantly different at $P < 0.05$. (From Markiewicz & O'Donnell 2001.)

almost identical to that of the resulting unmated female adults, showing no appreciable loss of weight or resources (Rhainds & Ho 2002) (Figure 7.38). Incredibly, the weight of the egg mass laid by the same females is about two-thirds of the virgin adult weight. In other words, female bag-worms allocate about two-thirds of the resources which they accumulate as larvae (and which produces pupae of a certain size) into egg production. This is an impressive figure, consistent with the prediction that natural selection

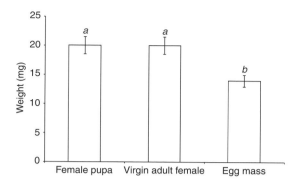

Figure 7.38 Weights of three successive life stages of female bag-worm, *Metisa plana*, from oil palm in Malaysia. Means with the same letter are not significantly different at $P < 0.05$. (From Rhainds & Ho 2002.)

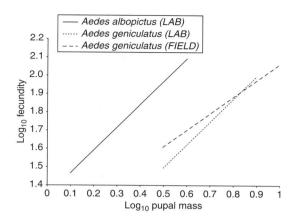

Figure 7.39 Relationships between size of pupae and resulting adult female fecundity in two species of mosquito (fitted regression lines only.) (From Armbruster & Hutchinson 2002.)

will favor traits of those individuals who make the greatest relative contribution to the next generation. It is, perhaps, therefore a shame that in this species of moth, huge mortalities occur at the very early larval stage due to aerial dispersal of very young larvae away from the maternal tree.

Given that adult insect size is usually related to female fecundity, and that adult holometabolous insects emerge from pupae, we might reasonably expect that pupal mass will correlate positively with adult fecundity. So it is in the case of mosquitoes in the genus *Aedes*, well-known vectors of many arboviruses (see Chapter 11). Armbruster and Hutchinson (2002) found that the weight of mosquito pupae correlated well with the fecundity of adult female flies emerging from the pupae (Figure 7.39). In addition, there were some interesting differences between the species of *Aedes* in their allocation to egg number. For example, *Aedes albopictus* (Diptera: Culicidae) is a smaller mosquito than *A. geniculatus* but the fecundity range for the two species is very similar. In other words,

on a per unit body mass basis, *A. albopictus* produces significantly more eggs than does *A. geniculatus*.

7.10 CONCLUSION

We hope that we have given you a small taste of the many fascinating mechanisms by which the physiology of insects contributes to their success in variable environments. We encourage you to explore in more depth the topics that we have introduced here, and the many others that we lacked space to include. In many ways, physiological ecology provides a natural interface between the molecular processes that define life, and the ecology and behavior of insects in ecosystems. At this point in the book, we proceed to a very different scale of organization, and consider how insects participate in the ecosystem processes that occur on Earth.

Chapter 8

INSECTS IN ECOSYSTEMS

8.1 WHAT IS ECOSYSTEM ECOLOGY?

So far in this book, we have concentrated on inter-actions between one or a few species of insects and their abiotic and biotic environments; our focus has been at the population and community level. In this chapter, we look at the importance of insects at the broader ecological scale of entire ecosystems.

We can define an ecosystem as all the interactions among organisms living together in a particular area, and between those organisms and their physical en-vironment (Tansley 1935). The foundation of ecosys-tem ecology is that ecosystems can be represented by the flow of energy and matter from one subsystem to the next. Figure 8.1 shows how energy from the sun is trapped by plants, or "primary producers", in both terrestrial and aquatic systems, and is passed on to primary consumers (herbivores), and then on to sec-ondary and tertiary consumers. Subsequently, the energy reaches the decomposers and, finally, is either "trapped" (at least temporarily) in matter that is diffi-cult to decompose or dissipated as heat. In fact, energy is lost to heat during each exchange between trophic levels (see below). Ecosystem ecologists study the movement of energy and matter among components in this way. Although insects are never primary pro-ducers, they are important primary, secondary, and even tertiary consumers in some ecosystems. Moreover, they are fundamental participants in the decomposition process in terrestrial and aquatic eco-systems. Historically, ecosystem ecologists have paid less attention to the identity of a particular individual or species than its "functional role" in the environ-ment (decomposer, producer, etc.). That approach

has changed somewhat in recent years with the rec-ognition that some species, so-called ecosystem engin-eers (Jones et al. 1994; Hastings et al. 2007), may play a more important role in ecosystems than do most other species. There is also some evidence to suggest that the diversity of organisms in ecosystems can influence energy flows and the cycling of matter (Tilman et al. 1994; Loreau et al. 2001). As a conse-quence of these developments, we see far greater integration among the disciplines of population, com-munity, and ecosystem ecology than has existed previously (Jones & Lawton 1995; Hunter 2001d; Madritch & Hunter 2002); it is an exciting time to be an ecosystem ecologist who is interested in insects.

When we look at a complex ecosystem such as a tropical forest or temperate lake, we know that all of the species in it are dependent on the availability of energy and matter (e.g. nutrients) flowing into the system. We ought to be able to chart the input and output of energy and materials, and see how they are divided up to support the diversity of organisms in the system. Specifically, we ought to be able to assess the importance of insects in the transformation and flow of energy and matter.

8.2 A FEW FUNDAMENTALS OF ECOSYSTEM ECOLOGY

Before exploring the role of insects in ecosystems, we should describe a few important rules or laws that govern the movement of matter and energy through systems. These rules set fundamental limitations on the structure (biological entities) and function (energy

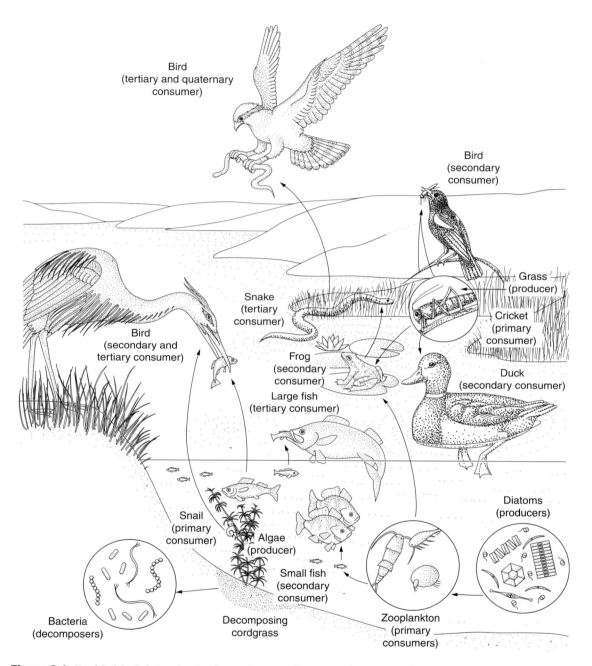

Figure 8.1 Simplified food chains showing how primary producers in both aquatic and terrestrial systems trap energy from the sun. The energy is then passed through subsequent trophic levels. (From Raven et al. 1993.)

and matter flows) of ecosystems and, by extension, the potential effects of insects on ecosystem processes.

8.2.1   Energy and ecosystems

Where does the energy come from to drive ecosystems? What is the source of all the energy required for organisms, including insects, to grow, move, reproduce, and repair damaged tissues? Energy exists in several forms: heat; radiant energy (electromagnetic radiation from the sun); chemical energy (stored in the chemical bonds of molecules); mechanical energy; and electrical energy. One fundamental rule governing energy, called the first law of thermodynamics, says that energy cannot be created or destroyed. It can be converted from one form to another, or moved from one system to another, but the total amount never increases or decreases. That means that when the universe was created about 15 billion years ago, the amount of energy existing then equals the amount of energy existing now. The energy has changed form over and over again in that time, but nothing has been gained or lost.

So, if organisms cannot create the energy that they require to live, they have to transform it from some existing source. Plants (or, strictly speaking, "autotrophs" including some bacteria and Protista) are critical to this process in almost all ecosystems. They trap energy from the sun, converting electromagnetic energy into chemical energy by photosynthesis. The chemical energy is stored in the molecules that make up plant tissue. Herbivores, including many of the phytophagous insects discussed in Chapter 3, can convert this chemical energy into the mechanical energy of flight. Secondary and tertiary consumers such as insect predators and parasitoids (see Chapter 5) then utilize or convert the chemical energy stored in herbivores for their own growth, movement, and reproduction.

However, the transformation of energy from one form to another always involves some loss of "useful" energy. For example, when a grasshopper transforms the chemical energy of plant tissue into mechanical energy to leap away from a mantid predator, some of the chemical and mechanical energy is changed into heat energy, and "lost" to the plant–herbivore–predator system. That does not mean it disappears (remember the first law of thermodynamics); it is still in the environment, but in a form that biological life cannot

readily reuse. This is the second law of thermodynamics: during the transformation of energy from one form to another, there is a continual loss of energy from "useful" or reusable forms, to "less useful" or non-reusable forms (heat). That means that the amount of useful energy available to do "work" in the universe decreases over time, even if the total amount of energy stays the same.

The first and second laws of thermodynamics are therefore critical to the way that ecosystems are put together. First, the amount of energy entering the ecosystem sets a limit on the total productivity of the system, because the system cannot generate energy of its own. Second, the loss of useful energy to heat sets a limit on how many transformations, or retransformations, can take place. The losses to heat at each stage mean that, eventually, all the useful energy will have been used up. This will set a limit to the number of different transformations or "trophic levels" (from plants to herbivores to carnivores to decomposers) that can take place. In fact, the transformation of energy from one trophic level to the next is only about 10% efficient; about 90% of the useful energy is lost between each trophic level (Figure 8.2). Ecosystems are ultimately constrained by the first and second laws of thermodynamics. By extension, so are the insects that inhabit ecosystems.

8.2.2   Matter and ecosystems

The flow of matter through ecosystems differs fundamentally from the flow of energy. Energy flow through ecosystems is linear, and there is no recycling. It is dissipated in unusable forms into the environment during each stage of transformation until there is none remaining from the original pool. In contrast, materials such as nitrogen, carbon, phosphorus, potassium, and even water, flow through ecosystems in cycles. When matter cycles from the living world to the non-living physical environment and back again, we call this biogeochemical cycling. Other than sunlight and the occasional meteorite, the Earth is essentially a closed system. That means that matter cannot escape from the system and materials are reused and recycled both within and among ecosystems.

Insects are important components of several biogeochemical cycles as well as mediators of energy transformation. In the rest of this chapter, we describe

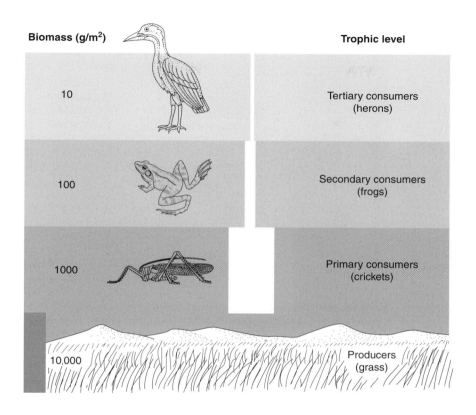

Figure 8.2 Losses in useful energy during transfer among trophic levels. It should be noted that $10,000\,\mathrm{g/m^2}$ of grass will support, on average, only $10\,\mathrm{g/m^2}$ of tertiary consumers, placing fundamental limitations on ecosystem processes. (From Raven et al. 1993.)

examples of the importance of insects to the cycling of materials and the transformation of energy in ecosystems. Specifically, we will examine the role of insect decomposers in the carbon cycle, the importance of leaf-shredding insects to ecosystem processes in streams, the role of insect defoliators as mediators of nutrient cycles, and the effects of insects on plant diversity and succession. We chose these examples from many possible systems because they are both topical and moderately well understood.

8.3 INSECTS AND THE TERRESTRIAL CARBON CYCLE

The carbon cycle describes how carbon flows from the physical environment to the living world, and back

again. Figure 8.3 is a simplified diagram of the global carbon cycle, including the "storage" of carbon in coal, oil, and natural gas that results from partial decomposition of plant and animal material. For our purposes, we concentrate here on the terrestrial components of the carbon cycle.

Carbon is incorporated from the physical environment into the living world by the process of photosynthesis; carbon dioxide (CO_2) from the atmosphere is captured by plants, and turned into organic compounds in living plant tissue using the energy of the sun:

$$6CO_2 + 12H_2O + \text{radiant energy} \rightarrow C_6H_{12}O_6 + 6H_2O + 6O_2$$

About 0.03% of the world's atmosphere is CO_2, and it acts as the major pool from which carbon enters the living world. Once the carbon is incorporated into

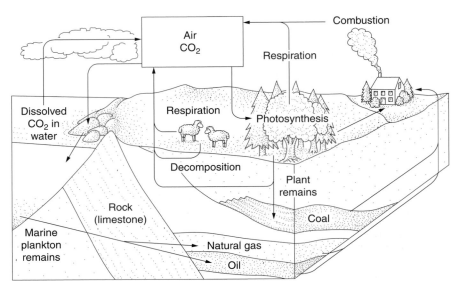

Figure 8.3 A simplified diagram of the global carbon cycle. The text refers only to terrestrial components of the cycle. It should be noted that incomplete decomposition of animal and plant remains can lead to carbon "storage" in the forms of coal, oil, and natural gas. (From Raven et al. 1993.)

plant tissue, it can follow several paths. First, some of it is used for the growth and reproduction of the plant itself. This costs energy, and the source of that energy is "respiration". We can think of this as the plant burning fuel, in this case carbon-containing molecules, to release the energy:

$$C_6H_{12}O_6 + 6H_2O + 6O_2 \rightarrow 6CO_2 + 12H_2O + \text{transformed energy}$$

It should be noted that respiration causes the release of CO_2, which then re-enters the atmosphere. This is one simple part of the carbon cycle: CO_2 is taken up by plants during photosynthesis, and so moves from the physical world to the living world; then, as plants respire and use the energy for growth and reproduction, carbon is returned to the physical world. Second, some of the carbon captured by plants is consumed by primary consumers such as insect herbivores and, in turn, by predators that eat herbivores. At each trophic level, carbon that was originally captured by plants is returned to the atmosphere by the respiration of organisms at that trophic level. For example, hawks that eat mice that eat grasshoppers that eat plants return a portion of the carbon originally captured by plants back to the atmosphere during respiration (Figure 8.4).

8.3.1 Insect decomposers

Insects play their major role in the carbon cycle during the decomposition process. Dead and decaying plant and animal tissues or waste products serve as sources of food for many kinds of decomposers, including insects, mites, fungi, and bacteria. As these tissues are processed by decomposers, CO_2 is released into the atmosphere, completing the major part of the carbon cycle. Blow flies and flesh flies (Diptera: Calliphoridae and Sarcophagidae, respectively) are well-known insect decomposers whose larvae often feed within carrion or excrement. Likewise, carrion beetles (Coleoptera: Silphidae) in the aptly named genus *Necrophorus* excavate chambers beneath the dead bodies of small mammals. The buried carcass provides food for their offspring. A dead mammal is indeed a "high-quality resource" (Carter et al. 2007), with a high water and organic nitrogen content. Carrion beetles are physically strong enough that a pair of beetles can move a large rat several feet before finding a suitable burial site. Although this feeding habitat may seem distasteful to some, these insects provide a valuable service by removing dead animals and animal waste from the environment, and completing the cycling of matter between the living and

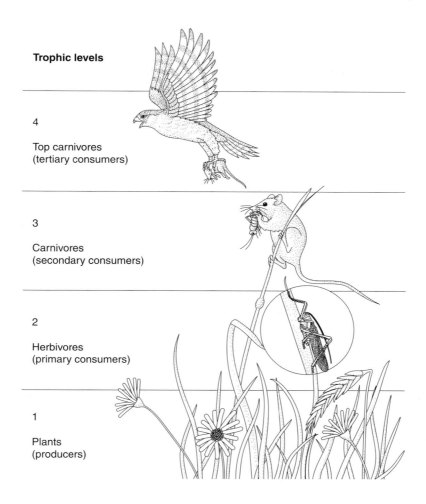

Trophic levels

4

Top carnivores
(tertiary consumers)

3

Carnivores
(secondary consumers)

2

Herbivores
(primary consumers)

1

Plants
(producers)

Figure 8.4 Some of the carbon trapped by plants during photosynthesis is ultimately released during respiration by top carnivores (tertiary consumers) such as hawks. (From Raven et al. 1993.)

non-living world. Similarly, many dung beetles chew off portions of animal feces, work them into balls, and can roll the dung considerable distances. They often work in pairs, with one adult beetle pulling from the front while the other pushes from behind. The dung ball is buried, and provides food and shelter for larvae of the next generation (Figure 8.5). The scarab, *Scarabaeus sacer* (Coleoptera: Scarabaeinae), which populates the perimeter of the Mediterranean, was sacred to the early Egyptians, who drew parallels between dung-rolling and the mysterious daily movement of the sun across the sky.

Dung beetles have actually been used as "decomposition control agents" in Australia. Although Australia has about 250 native species in the subfamily Scarabaeinae, their foraging habits appear much better adapted to the small, dry dung of marsupials than the large, wetter dung of cattle introduced by European settlers in the late 1700s. A single adult bovine drops an average of 12 dung pads per day and, if not removed, the dung can render inaccessible about 0.1 ha of pasture per bovine per year (Waterhouse 1977). With 3×10^7 cattle in Australia, dung pads can put about 2.5×10^6 ha of pasture land

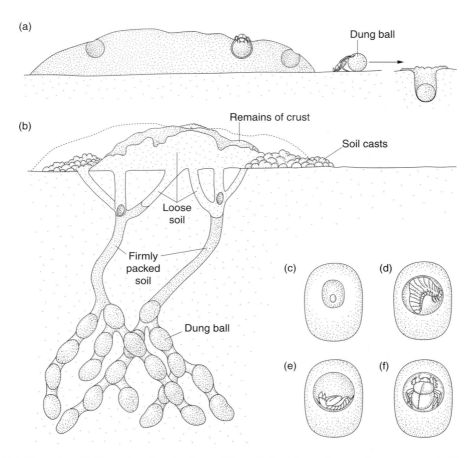

Figure 8.5 (a) Some dung beetles such as *Garreta nitens*, roll dung for burial some distance from its source. (b) Most form nests below the dung pat where (c) eggs, (d) larvae, (e) pupae, and (f) adults develop. (From Waterhousé 1977.)

out of service each year, constituting a truly enormous loss to the dairy and beef industries. In addition, dung pads act as resources for insect pests such as the bush fly, *Musca vetustissima*, and the buffalo fly, *Haematobia irritans exigua* (Diptera: Muscidae). In 1967, dung beetles from Africa, with its extensive large-mammal fauna, were introduced into Australia for the control of cattle dung. Dung beetles have spread widely in Australian pasture and have significantly reduced the buildup of cattle dung. Researchers continue to seek the most appropriate beetle species for both the removal of the dung itself and to control the insect pests that live within it (Kirk & Wallace 1990; Davis 1996). The battle is certainly not over. There is some evidence that modern grain feeds reduce

the palatability of cattle dung for dung beetles (Dadour & Cook 1996). None the less, the dung beetle introductions provide compelling evidence of the importance of certain insects to decomposition processes in ecosystems.

Perhaps the most impressive decomposers in the insect world are the termites (Isoptera) (Figure 8.6). They are essentially soil "engineers" (Jouquet et al. 2006) that, along with earthworms and ants, have fundamental effects on soil systems upon which other organisms such as microbes and plants rely. Although some termites feed on soil organic matter, most feed on dead and decaying plant material including leaf litter, roots, and woody debris (Whitford et al. 1988; Moorhead & Reynolds 1991). Dead plant material

(a)

(b)   (c)   (d)

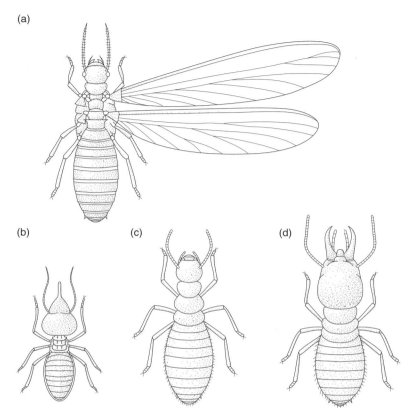

Figure 8.6 Castes of termites. (a) Sexual stage of *Amitermes tubiformans*. (b) Nasutus of *Tenuirostritermes tenuirostris*. (c) Worker, and (d) soldier of *Prorhinotermes simplex*. (From Borror et al. 1981.)

poses special problems to decomposers because it contains high concentrations of cellulose and lignin, neither of which is readily digested by animals. In fact, amongst the insects, some cockroaches and higher termites in the subfamily Nasutitermitinae are the only taxa known to synthesize enzymes capable of degrading cellulose (Martin 1991). None the less, termites are efficient decomposers of plant material in desert (Whitford et al. 1988), savanna (Wood & Sands 1978), and forest ecosystems (Bignell et al. 1997) because of their symbioses with microorganisms that live in the guts of termites and produce cellulolytic (cellulose-digesting) enzymes (see Chapter 6). The gut symbionts of various termite groups include both flagellate protozoans (Yoshimura et al. 1993) and bacteria (Basaglia et al. 1992). As an aside, it has been reported that some spirochetes that live

symbiotically in termite guts are able to fix atmospheric nitrogen, and may contribute this nitrogen to termite nutrition (Lilburn et al. 2001). Large amounts of ammonia (NH_3) build up in the nests of certain termite species, possibly to levels 300 times higher than in surrounding soil (Ji & Brune 2006). One crucial feature of termites relevant to the carbon cycle is the occurrence of anaerobic microsites in termite guts. When plant material decomposes in the absence of oxygen, the endproduct is methane (CH_4) instead of CO_2. Termites therefore have the potential to recycle significant amounts of carbon to the atmosphere in two gaseous forms. Methane is one of the principal greenhouse gases, contributing about 18% to the effects of such gases on climatic variation (Anon. 1992).

There is some debate on the relative importance of termites to global fluxes of CO_2 and CH_4. We might expect termites to be an important component of the terrestrial carbon cycle, if only because they are very abundant in many forest ecosystems. Forests contain more organic carbon than all other terrestrial systems and, at present, account for about 90% of the annual carbon flux between the atmosphere and the Earth's land surface (Groombridge 1992). The literature now contains estimates of termite abundance and biomass from each of the three global blocks of tropical forest, and these data suggest that termites may be an order of magnitude more abundant than the next most abundant arthropod group, the ants. The highest biomass reported for termites, 50–100 g/m^2 in southern Cameroon forest (Eggleton et al. 1996), is greater than any other component of the invertebrate (or vertebrate) biota and may constitute as much as 95% of all soil insect biomass (Bignell et al. 1997). If direct carbon fluxes by insects are significant components of ecosystem processes, by far the greatest contribution will be made by termites. In addition to direct carbon emissions during decomposition, the Macrotermitinae also cultivate a fungus on dead plant material. The metabolic rate of this cultivated fungus and the quantity of CO_2 that it releases exceed the corresponding values for its termite gardeners (Wood & Sands 1978).

Termites are likely of great ecological important in the global carbon cycle (Konig 2006). However, because of technical difficulties in estimating termite densities, their rates of respiration, and the effects of the environment on CO_2 and CH_4 emissions, there are probably no truly reliable estimates (Bignell et al. 1997). However, data suggest that 20% of all CO_2 produced in savanna ecosystems results from termite activity (Holt 1987): termites can consume up to 55% of all surface litter in such systems (Wood & Sands 1978). Globally, the number is likely to be much lower. Based on the most reliable estimates to date (Bignell et al. 1997), termites probably contribute up to 20% of the global production of CH_4 and 2% of the global production of CO_2. Although these numbers may not seem dramatic, we should remember that these are the contributions of a single order of insects that represent less that 0.01% of terrestrial species richness. In fact, the global CO_2 produced by termites each year probably exceeds that taken up by northern hemisphere regrowth forests (Houghton et al. 1990). Incidentally, termite densities drop dramatically in clear-cut forests (Watt et al. 1997c), and the practice of clear-cutting therefore has implications for global levels of CH_4 and CO_2 production.

8.4 LEAF-SHREDDING INSECTS AND STREAM ECOSYSTEMS

The role of insects in the flow of energy and matter through ecosystems is understood nowhere better than in stream systems. One of the reasons that ecosystem approaches have been so pervasive in stream entomology is the suggestion that many individual species may be "functionally redundant" (Lawton 1991), behaving in such similar ways that the presence or absence of a particular species matters less than the "functional group" to which it belongs (Wallace & Webster 1996). Insect functional groups in streams (Cummins & Klug 1979) are somewhat analogous to the "guilds" of insects described in Chapter 3, although there is even more emphasis on the mode of feeding by which resources are consumed. Functional groups include grazers, shredders, gatherers, filter-feeders, and predators, and are based on their morpho-behavioral mechanisms for gathering food rather than taxonomic relationships. Although all of these functional groups are relevant to the pathways taken by energy and materials through stream ecosystems (Wallace & Webster 1996), we focus here on shredding insects because of their profound effects on stream food chains, particularly in forested stream environments.

Upland streams in forested areas receive much of their energy base from outside the stream itself. Although primary production by algae and other plants occurs in such streams, low light levels under forest canopies reduce the importance of this energy pathway compared with terrestrial systems (above). Instead, much of the energy base in streams is derived from litter inputs (leaves, twigs, woody debris, etc.) from the terrestrial environment. Of course, this energy source is still the result of primary production, it just comes from outside the stream (allochthonous production) as opposed to inside the stream (autochthonous production). For example, in the eastern USA, litterfall into streams averages 600 g/m^2 per year, over half of which consists of leaves (Webster et al. 1995).

Shredders are the functional group of insect larvae (and sometimes other arthropods) that fragment (or comminute) the litter in streams into particle sizes that can be used by other invertebrate groups. Leaf-shredding insects are particularly common in temperate waters (Wantzen & Wagner 2006), and are dominated by three insect orders: the caddisflies

(Trichoptera) (Figure 8.7), stoneflies (Plecoptera) (Figure 8.8), and true flies (Diptera). It is important to point out, however, that many species in these three orders fall into other functional groups as well, such as grazers, filter-feeders, etc., and many related species seem to specialize on slightly different components of the fallen leaf resources in the ecosystem (Boyero et al. 2006), exhibiting some degree of niche partitioning. Most shredders appear to select leaf litter tissue that has been colonized and partially decomposed (or "conditioned") by fungi and bacteria (Cummins & Klug 1979). Indeed, shredders also ingest attached algae and bacteria along with litter tissue (Merritt & Cummins 1984), and it seems likely that they gain some of their energy and nutrient requirements from the microbes rather than the litter itself. In one experiment, caddisfly larvae fed on leaves from ponds that had been sterilized (to remove microbes and algae) before being eaten showed negative growth rates (Villanueva & Trochine 2005). Some shredding insects derive enzymes from microbial symbionts or ingested microbes that may be active in cellulose hydrolysis (Sinsabaugh et al. 1985), a similar strategy to that used by the termites in terrestrial systems described earlier in this chapter.

The major importance of shredding insects in stream ecosystems is their ability to turn the coarse particulate organic matter (CPOM) of litter into fine particulate organic matter (FPOM) and dissolved organic matter (DOM) (Wallace et al. 1982; Meyer & O'Hop 1983). Shredders are not especially efficient feeders, assimilating only 10–20% of the material that they ingest. Most litter passes through the gut, emerging as fine particles or dissolved fractions in the feces. FPOM and DOM are major sources of nutrition for gatherers, filter-feeders (e.g. blackfly larvae in the dipteran family Simulidae), and microbes in streams (Cummins et al. 1973; Short & Maslin 1977; Wotton 1994) (Figure 8.9). In other words, if allochthonous litter is a major energy source for streams, shredding insects are the catalyst that makes that energy available to a wide variety of stream biota. Moreover, as both FPOM and DOM are readily transported downstream (Cuffney et al. 1990; Cushing et al. 1993), shredders make the products of litter decomposition available to organisms at some distance from the major sites of litter input. In addition, defecation by shredding insects that burrow periodically in stream sediments increases the organic content of those sediments by 75–185% (Wagner 1991). Finally, insect shredders also promote wood decomposition by scraping, gouging, and tunneling into the woody debris (twigs, branches, stems) that falls into streams. Freshly gouged surfaces act as sites for microbial activity and subsequent decomposition (Anderson et al. 1984), and shredder activity has even been implicated in the collapse of highway bridges supported by untreated timber pilings.

In an experimental investigation of the role of insects in stream ecosystem processes, Wallace and colleagues from the University of Georgia eliminated more than 90% of the insect biomass in a southern

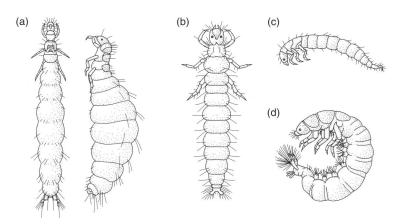

Figure 8.7 Aquatic larvae in the order Trichoptera (caddisflies): (a) *Hydroptila waubesiana*, (b) *Rhyacophila fenestra*, (c) *Polycentropus interruptus*, and (d) *Hydropsyche simulans*. (From Borror et al. 1981.)

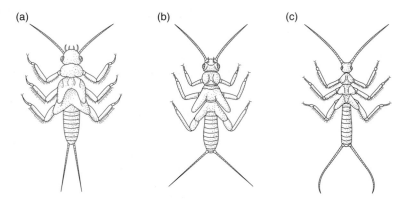

Figure 8.8 Aquatic larvae in the order Plecoptera (stoneflies): (a) *Isoperia transmarina*, (b) *Nemoura trispinosa*, and (c) *Taeniopteryx glacialis*. (From Borror et al. 1981.)

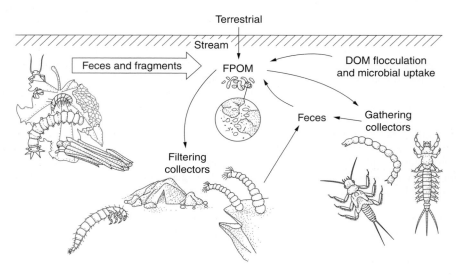

Figure 8.9 Sources and users of fine particulate organic matter (FPOM) in temperate streams. It should be noted that insect shredders (top left) are a major source of FPOM. They provide both fragments of leaf litter and feces that are necessary sources of food for filtering collectors and gathering collectors. DOM, dissolved organic matter. (After Cummins & Klug 1979, and redrawn from Allan 1995.)

Appalachian stream by applying insecticide. Their manipulation significantly reduced the rate of leaf litter break down and the export of FPOM in the stream (Wallace et al. 1982, 1991; Cuffney et al. 1990). Both litter decomposition and FPOM export recovered at the same time as the shredder community recovered from the insecticide treatment (Wallace et al. 1986) (Figure 8.10). Wallace's data suggest that insects accounted for 25–28% of annual leaf litter processing and 56% of FPOM export over a 3-year period in this stream. Larval insects including, but not exclusively, shredders are clearly critical to ecosystem processes in streams (Wallace & Webster 1996) and it is not surprising that they can be seriously affected by changes in water flow and quality. Natural events such as storm runoff, and alterations

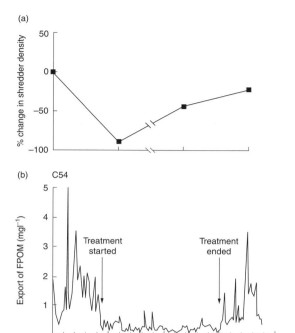

Figure 8.10 Changes in (a) shredder density (%), and (b) export of fine particulate organic matter (FPOM; mg/L) after insecticide treatment in a headwater stream in the southern Appalachians. Concentrations of FPOM return to normal when insect shredder density recovers from the treatment. (After Wallace et al. 1991; Whiles & Wallace 1995.)

in land use such as agriculture or urbanization, can change stream velocities, water chemistry, nutrient content, and so on, with serious consequences for the freshwater insect communities (Paul et al. 2006; Gresens et al. 2007).

8.5 INSECT DEFOLIATORS AND THE CYCLING OF NUTRIENTS

In Section 8.3 we pointed out that herbivores consume, and recycle, carbon that has been fixed by plants during the process of photosynthesis (see Figure 8.4). However, plant material contains much more than just carbon. Important nutrients such as nitrogen, phosphorus, and potassium are taken up by plants for their own growth and reproduction. As plant material is consumed by insect herbivores, some of those nutrients are assimilated in herbivore tissues, while others are recycled in herbivore feces. As herbivores die, or are consumed by predators, further nutrient cycling takes place. Given that insect herbivores are pervasive in terrestrial systems (see Chapter 3), they have the potential to influence the cycling of nutrients in ecosystems and the subsequent productivity of plants.

Here, we focus on the role of insect herbivores in the dynamics of nutrients, particularly nitrogen, in soils. The initial idea that herbivores can regulate nutrient cycling and primary production dates back to the 1960s and 1970s (Pitelka 1964; Chew 1974; Mattson & Addy 1975). Recent work has focused upon the detailed mechanisms by which such regulation might occur (Frost & Hunter 2004; Fonte & Schowalter 2005). There are seven broad mechanisms by which the activity of insect herbivores can cause changes in nutrient cycles and nutrient availability in soils (Hunter 2001d) (Table 8.1):

1 Insect herbivores can deposit significant quantities of fecal material (frass) onto litter and soil. Nitrogen returned to soils in insect frass can exceed that in leaf litter (Fogal & Slansky 1985; Grace 1986), and can double overall rates of nitrogen return from plants to soil (Hollinger 1986).

Table 8.1 Mechanisms by which insect herbivores can influence nutrient dynamics in soils. (From Hunter 2001d.)

1	Deposition of insect frass (feces)
2	Inputs of insect cadavers
3	Changes in throughfall chemistry
4	Changes in the quality/quantity of litter inputs (leaves and roots)
5	Changes in nutrient utilization by the plant community
6	Changes in root exudation and root–symbiont interactions
7	Alteration of canopy structure and soil microclimate

2 Nutrients returned to soils in insect cadavers are more easily decomposed than those in leaf litter (Schowalter et al. 1986) and can stimulate the decomposition of litter during defoliator outbreaks (Swank et al. 1981; Seastedt & Crossley 1984).

3 Insect defoliation changes the nutrient content of precipitation as it passes through plant canopies. Folivory influences the nutrient chemistry of this "throughfall", primarily through increased rates of nutrient leaching from damaged leaves (Tukey & Morgan 1963) and through the dissolution of frass from foliage.

4 Herbivory can change the quantity and quality of leaf litter that falls from plant canopies to the soil. Herbivore-mediated changes in litter inputs can occur through premature leaf abscission (Faeth et al. 1981), petiole clipping and foliar fragmentation (Risley 1986), wound-induced increases in foliar phenolics (Findlay et al. 1996), root mortality (Ruess et al., 1998), and community-wide changes in the relative abundance of plant species or genotypes that vary in their litter quality (Pastor et al. 1993; Uriarte 2000).

5 Herbivore-mediated changes in plant community composition not only influence litter quality, but may also affect the utilization of soil nutrients by the new community (Kielland et al. 1997).

6 Herbivory may influence root exudates or interactions between roots and their symbionts (Bardgett et al. 1998), both of which are known to influence nutrient dynamics.

7 Finally, herbivores can influence the structure of plant canopies and the cover that they provide, with concomitant changes in light availability, soil temperature, and moisture. Changes in soil microclimate that result from herbivore activity can alter the cycling of nutrients (Mulder 1999). Similarly, herbivore-induced changes in light availability may influence litter quality through effects on leaf chemistry (Hunter & Forkner 1999) or plant productivity and diversity (van der Wal et al. 2000).

These seven mechanisms are likely to vary in the speed with which they influence nutrient dynamics and primary production. Nutrient cycles are likely to respond rapidly to inputs of frass, cadavers, and modified throughfall because they do not require the decomposition of complex organic matter. These effects are analogous to McNaughton et al.'s (1988) "fast cycle" or the "acceleration hypothesis" (Ritchie et al. 1998). In contrast, herbivore-mediated changes in litter quality, canopy cover, and community composition will likely influence nutrient dynamics more slowly and are analogous to McNaughton et al.'s (1988) "slow cycle". In fact, wound-induced increases in recalcitrant compounds such as tannins and lignin, or herbivore-mediated selection of resistant plants, might act to decelerate nutrient cycling (Ritchie et al. 1998; Chapman et al. 2006). Both fast and slow cycle effects, and acceleration or deceleration of cycles, may occur in the same system but on different timescales. For example, herbivore activity may increase rates of nitrogen cycling and primary production in the short term (e.g. through frass inputs) while causing both to decline in the long term (e.g. through selection of resistant plants) (Uriarte 2000).

What is the empirical evidence to support the idea that insect herbivores mediate rates of nutrient cycling in soils? Some of the strongest evidence comes from studies of soil nutrient dynamics during insect outbreaks. Figure 8.11 describes the fate of nitrogen in the leaves of deciduous forest trees during periods of low and high insect densities (Lovett et al. 2002). When defoliation levels are low, about 70% of foliar nitrogen is resorbed by the trees before leaves senesce in the fall. Most of the rest falls to the forest floor in abscised leaves, with only a small proportion of total nitrogen lost to insect activity. In comparison, when defoliation levels are high, the fate of foliar nitrogen changes dramatically. Resorption can drop below 25%, and a significant proportion of foliar nitrogen is now added to the forest floor in frass and premature leaf abscission. Nitrogen levels in foliar leaching and insect biomass also increase. Overall, about 2.5 times as much nitrogen can be returned to the forest floor during defoliator outbreaks when compared with nitrogen return during endemic densities of insect herbivores. Though the term "defoliation" technically refers to the loss of leaf material by chewing, sap-feeders can also cause significant leaf loss (or its associated loss of photosynthetic production). Stadler et al. (2006) studied the effects of hemlock woolly adelgid, *Adelges tsugae* (Hemiptera: Adelgidae), on the throughfall of various chemicals in hemlock forests. They found that infested stands of trees had 24.6% higher dissolved organic carbon and 28.5% higher dissolved organic nitrogen in throughfall and litter when compared with uninfested stands.

What is the fate of this added nitrogen? The nitrogen in insect frass and prematurely abscised leaves is primarily in an organic form such as amino acids. Soil microbes (and some plants) may use organic nitrogen

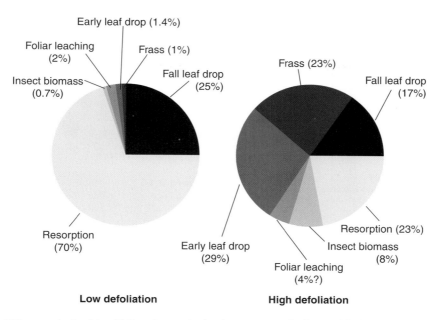

Figure 8.11 Differences in the fate of foliar nitrogen in deciduous trees under low and high defoliation conditions. (From Lovett et al. 2002.)

directly. However, organic nitrogen is also "mineralized" by some microbes, and reappears in mineral form (ammonium and nitrate) that can be used by microbes and taken up by plant roots. In a recent study, 40% defoliation of a northern hardwood forest resulted in a five-fold increase in soil nitrate availability (Reynolds et al. 2000) (Figure 8.12a). In addition, nitrate is readily leached from most soils, and stream nitrate concentrations doubled following the defoliation event (Figure 8.12b). Experiments with potted trees (Figure 8.13) have confirmed that insect frass can be a major source of mineral nitrogen in soils, and a source of nitrate lost in leachate (Frost & Hunter 2004). Additional experiments have shown that insect defoliators can significantly increase the rate of nitrogen cycling. Using a stable isotope of nitrogen (^{15}N), Frost and Hunter (2007) have traced nitrogen from insect frass into the soil, back into the trees, and into a second generation of insect herbivores, all within one growing season (Figure 8.14). In other words, nitrogen can cycle rapidly from one population of insects, through the soil, and into another population of insects within the same summer.

The acceleration of nutrient cycling by frass deposition, described above, can be further augmented by premature leaf abscission and the selection by insects of nutrient-rich, herbivore-tolerant plants (Ritchie et al. 1998; Chapman et al. 2003). There is evidence to suggest that both processes occur in nature. For example, pinyon pine trees, *Pinus edulis*, in Arizona are attacked by two insect herbivores: the stem-boring moth, *Dioryctria albovittella* (Lepidoptera: Pyralidae), and the scale insect, *Matsucoccus acalyptus* (Hemiptera/Homoptera: Margarodidae). Both herbivores induce premature leaf abscission, significantly increase the nitrogen concentration of needle litter, and decrease litter lignin : nitrogen and carbon : nitrogen ratios (Chapman et al. 2003). As a result, litter from attacked trees decomposes more rapidly than does litter from unattacked trees. Overall, the effect of insect herbivory is to increase the rate of nutrient cycling. Similarly, a 5-year experiment by Belovsky and Slade (2000) in a prairie ecosystem demonstrated that grasshopper feeding increased the rate of nitrogen cycling and increased productivity by 18%. In this case, grasshoppers appear to have

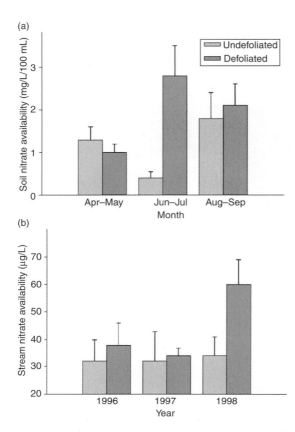

(a)

(b)

Figure 8.12 Increases in (a) soil nitrate and (b) stream nitrate following an outbreak of sawflies in 1998 in the southern Appalachians. (Data from Reynolds et al. 2000.)

Figure 8.13 An experimental array of potted oak trees used by Frost and Hunter (2004) to study the influence of insect frass on soil nutrient dynamics and nitrogen export. (Courtesy Chris Frost.)

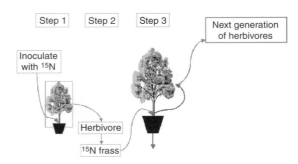

Figure 8.14 Experimental design used to trace nitrogen from foliage (step 1), into insect herbivores and their frass (step 2), back into the soil (step 3) and into a second generation of insect herbivores within the same growing season. (With permission of Chris Frost.)

selected for tolerant plants (see Chapter 3) that regrow rapidly after defoliation. The litter from these rapidly growing plants is high in nitrogen, and decomposes faster than litter from slow-growing plants.

Most of the data available at present suggest that insect herbivores increase rates of nutrient cycling. However, that need not always be the case. In the study by Belovsky and Slade (2000), herbivores selected for tolerant plants with high-quality litter. In contrast, selection for resistant plants (see Chapters 3 and 12), with high concentrations of defensive compounds, may have the opposite effect. Defenses such as lignin and tannin can reduce the rates of litter decomposition, and therefore reduce rates of nutrient cycling. This may be what occurred in the study by Uriarte (2000) during a 17-year manipulation of herbivore densities on goldenrod. In this case, herbivore attack decreased rates of nutrient cycling. Uriarte (2000) speculated that selection for resistant plants under herbivore attack resulted in goldenrod genotypes with low-quality foliage and litter. There is little doubt that genetic variation in foliar quality can influence rates of nutrient cycling (Madritch & Hunter 2002; Whitham et al. 2003), and long-term selection by insect herbivores has the potential to reduce rates of nutrient cycling. Some systems may therefore exhibit short-term acceleration of nutrient cycling by insect herbivores, but long-term deceleration. More long-term studies will be crucial in determining the relative importance of short-term and long-term effects of insects on nutrient dynamics in soils.

8.6 INSECTS, PLANT COMMUNITY STRUCTURE, AND SUCCESSION

We conclude this chapter on insects and ecosystems with a consideration of the effects of insects on plant community structure and succession. In Section 8.5, we learned that insects can influence the quality of foliage and litter in plant communities, with concomitant changes in ecosystem function. We might therefore expect that any temporal changes in plant community structure caused by insects would have significant impacts on ecosystem processes. In other words, if insect activity has an effect on ecosystem structure (the number and kinds of biota in an ecosystem), we might expect it to have simultaneous effects on ecosystem function. What is the evidence that insects influence the identity, diversity, and succession of plant species in communities?

The term succession describes temporal changes in community structure following disturbance or radical changes in resource availability. Patterns of species identity and diversity are often predictable as succession takes place (Figure 8.15), and ecologists have long been interested in studying the processes that determine the rate and direction of succession (Clements 1916; Gleason 1917). As well as predictable changes in community structure, succession also leads to predictable changes in ecosystem processes (Odum 1969) (Table 8.2). From this we can infer that, when insects alter the rate or direction of succession, they also influence ecosystem function.

The most dramatic examples of the effects of insects on succession come from studies of insect outbreaks.

Perhaps the best known example is one that we have encountered already, the spruce budworm, *Choristoneura fumiferana* (Lepidoptera: Tortricidae) (see Section 5.5.3). It is an important example because it illustrates a critical point – insects both respond to successional processes and modify successional processes. In the case of the spruce budworm, outbreaks become increasingly likely as the forest matures (Peterman et al. 1979) so that the successional stage of the forest determines the probability of outbreak. However, when outbreaks occur, the result is extensive tree mortality and the successional process is reset to a much earlier stage. Indeed, the spruce budworm has been called the "super silviculturalist" because of its ability to radically alter forest communities (Baskerville 1975). According to Baskerville's "cyclic model", virgin balsam fir stands in the wet boreal forest of northeastern America represent the end of the successional process. They are also the most susceptible to spruce budworm outbreaks. Extensive tree mortality resets the system to an earlier point in succession, and the forest community slowly recycles back to stands that are dominated by mature balsam fir (Bouchard et al. 2006). This successional process is responsible in part for the 40-year cycles of spruce budworm outbreak.

Other forest pests can also have dramatic effects upon forest stand structure and patterns of succession. Bark beetles, such as the southern pine beetle (*Dendroctonus frontalis*) and the mountain pine beetle (*Dendroctonus ponderosae*) (both Coleoptera: Scolytidae) cause extensive mortality of pines in eastern and western North America, respectively (Figure 8.16). One impact of these bark beetles is to reset the successional process in a manner similar to the spruce budworm (Fettig et al. 2007).

Of course, it is not just forest insects that exhibit outbreak dynamics. Fields of goldenrod in New York State are susceptible to outbreaks of the Chrysomelid beetle, *Microrhopala vittata* (Coleoptera: Chrysomelidae). Beetle outbreaks tend to open up the dense stands of goldenrod, increase light levels reaching the soil, and promote the establishment of other plant species (Carson & Root 2000). The overall effect of beetles is to increase plant diversity and increase the rate of succession.

The large outbreaks of insect pests described above are the most visually obvious examples of the impacts that insects can have on succession. However, some important effects of insects on succession are much

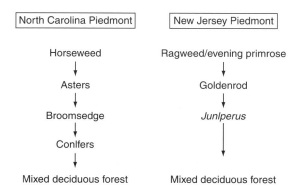

Figure 8.15 Typical patterns of secondary succession in abandoned fields as they succeed to mixed deciduous forest.

Table 8.2 Changes in a variety of ecosystem processes during ecological succession. (Modified and reprinted with permission from Odum, 1969. © 1969 American Association for the Advancement of Science.)

Variable	Early succession	Late succession
GPP/respiration	> or < 1	~1
GPP/biomass	High	Low
Biomass/energy	Low	High
Yield – NPP	High	Low
Food chains	Linear	Web-like
Total organic matter	Small	Large
Nutrients	Extrabiotic	Intrabiotic
Species richness	Low	High
Species evenness	Low	High
Biochemical diversity	Low	High
Stratification and pattern	Poorly organized	Well organized
Niche specialization	Broad	Narrow
Size	Small	Large
Life cycles	Short, simple	Long, complex
Mineral cycles	Open	Closed
Nutrient exchange	Rapid	Slow
Role of detritus	Unimportant	Important
Selection on growth form	r-Selection	k-Selection
Selection on production	Quantity	Quality
Symbiosis	Undeveloped	Developed
Nutrient conservation	Poor	Good
Stability	Low	High
Entropy	High	Low
Information content	Low	High

GPP, gross primary production = total photosynthesis; NPP, net primary production = total photosynthesis · respiration.

Figure 8.16 Extensive tree mortality caused by an outbreak of the mountain pine beetle, *Dendroctonus ponderosae*. (Photograph by Jerald E. Dewey, courtesy of Forestry Images.)

harder to observe without careful experimental studies. Chief among these are the impacts of below-ground herbivores on plant community structure and succession. Studies by Val Brown and colleagues have been instrumental in highlighting the effects of below-ground herbivores on plant diversity. Although root-feeding insects may be at lower densities than their foliar-feeding counterparts, the effects that they have on plant species richness can be more dramatic (Gange & Brown 2002). At the community level, insect herbivory below ground can decrease plant species richness and alter patterns of succession in European old fields (Brown & Gange 1989). Specifically, root-feeding insects can accelerate succession by reducing forb persistence and colonization, thereby promoting dominance by perennial grasses (Brown & Gange 1992). Experimental manipulation of artificial plant communities has confirmed that the general effect of soil fauna is to increase rates of succession (De Deyn et al. 2003), in part by their deleterious effects on dominant early-succession species. Applications of soil insecticide generally increase plant species richness, vegetative cover, and plant size (Masters & Brown 1997) although the effects of insects on succession depend in part upon the abundance of mycorrhizae in the soil (Gange & Brown 2002) (Figures 8.17 and 8.18). In some cases, removing below-ground insect herbivores has a greater impact on plant diversity than does the removal of above-ground insect herbivores (Schädler et al. 2004) (Figure 8.19). However, as we might expect, the effects of below- and above-ground herbivores on plant diversity and succession can sometimes interact. For example, in artificial plant communities,

wireworms (Coleoptera: Elateridae) acting alone have the effect of increasing plant diversity. In contrast, they act to decrease plant diversity in the presence of grasshoppers (van Ruijven et al. 2005) (Figure 8.20). Until relatively recently, the effects of soil insects on plant community dynamics and ecosystem processes have been "out of sight, out of mind" (Hunter 2001b). Thankfully, that is changing rapidly, and ecologists are starting to appreciate the importance of below-ground herbivory for ecological processes in a variety of terrestrial ecosystems.

The effects of insects on succession that we have discussed so far are examples from secondary succession – temporal community change that occurs following disturbances that do not completely eradicate biota. There is some, albeit limited, evidence to suggest that insect herbivores may also influence rates

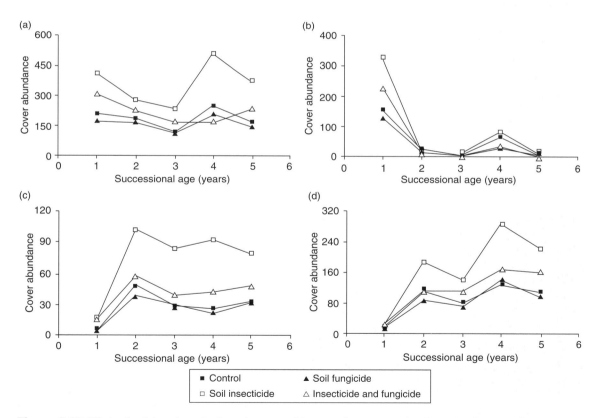

Figure 8.17 Effects of soil insects and arbuscular mycorrhizae on plant community dynamics during early succession: (a) total community cover abundance, (b) cover abundance of annual forbs, (c) cover abundance of perennial forbs, and (d) cover abundance of perennial grasses. Four treatments were used (From Gange & Brown 2002.)

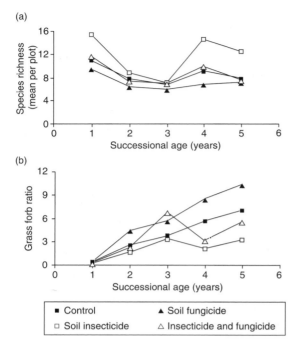

(a)

(b)

- ■ Control
- □ Soil insecticide
- ▲ Soil fungicide
- △ Insecticide and fungicide

Figure 8.18 Successional trends in (a) species richness, and (b) grass : forb ratio during early succession. Four treatments were used. (From Gange & Brown 2002.)

of primary succession. Primary succession occurs when disturbances are sufficiently severe to remove biota from a system. Volcanic activity and glaciation are forces that can remove biota (including soil and its associated organisms) from terrestrial systems. Sand accretion in some coastal and lake areas also generates landscapes upon which primary succession can then take place. What are the effects of insects on the rate and direction of primary succession in such habitats? Unfortunately, few ecologists have investigated the role of insects in such systems. Two of the best examples are provided below.

The 1980 eruption of Mount St. Helens, Washington, USA, created a $60\,km^2$ region of primary successional habitat and extirpated all plant and animal species from the area. In 1981, a species of lupine (*Lupinus lepidus* var. *lobbii* (Fabaceae)) colonized the otherwise barren north slope of Mount St. Helens. Lupines are a critical species during primary succession because they facilitate soil formation, fix nitrogen, and trap seeds and detritus. Initial rates of lupine spread were high and, by 1990, there existed a large

central core of high-density lupine. However, subsequent rates of spread were much lower, despite widespread habitat availability. Fagan and Bishop (2000) investigated the hypothesis that attack by insect herbivores was responsible for the decline in lupine spread. Lupines on Mount St. Helens are attacked by a range of insect herbivores including stem-boring, leaf-mining, and seed-eating lepidopterans (Bishop 2002). Fagan and Bishop used insecticide to eliminate insect herbivores from plots in the center of the core lupine population and from plots at the edge of the core region. They found that removing insect herbivores increased both the spreading of individual lupine plants and the production of new plants at the edge region (Figure 8.21). As a result, the intrinsic rate of increase of the lupine population was accelerated at the front of the lupine reinvasion. Interestingly, no such effects occurred in the center of the core region, suggesting that the effects of insect herbivores on well-established plots are minimal. However, the dramatic reduction in lupine population growth at the reinvasion edge should have significant consequences for the rate (and possibly direction) of primary succession. Specifically, rates of succession should decline because the lack of lupines will reduce rates of soil formation and nutrient accumulation.

At Grass Bay on Lake Huron (Michigan, USA), periodic changes in water levels create areas of bare sand upon which primary succession can take place. Bach (2001) has investigated the effects of sand accretion and insect herbivores on plant diversity during succession in this habitat. She began by excluding specialist flea beetles, *Altica subplicata* (Coleoptera: Chrysomelidae), for two consecutive years from the common dune willow, *Salix cordata*. She then monitored changes in the plant community for an additional 4 years. Herbivory on the willow decreased the proportional representation of herbaceous monocots in the plant community. In contrast, other grasses, *Aster* and *Solidago*, increased in response to previous damage to willow (Figure 8.22). Overall, willow herbivory caused significant changes in plant community structure and appeared to decrease the rate of succession.

It is dangerous to generalize from just two studies, but work on Mount St. Helens and at Grass Bay suggests that insect herbivores can act to reduce the rates of primary succession. This is in contrast to the effects of below-ground herbivores described above, in which herbivores increase rates of succession. The difference

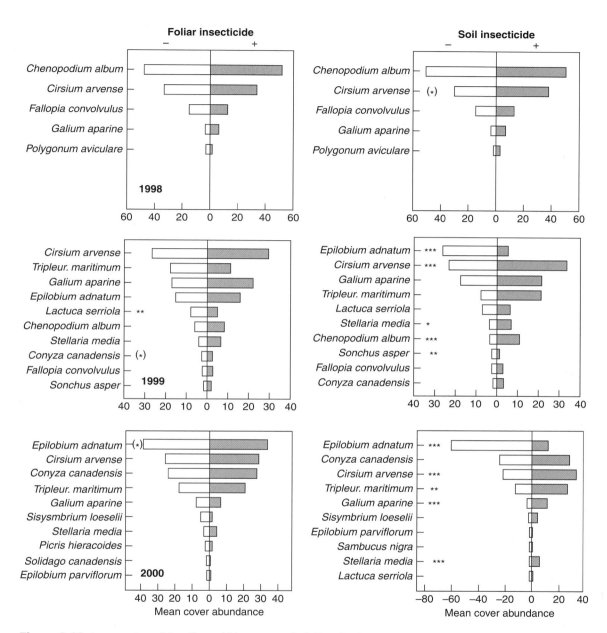

Figure 8.19 A comparison of the effects of foliar insecticide (left) and soil insecticide (right) on plant community dynamics during succession. Notice that the effects of removing soil insects become increasingly apparent over time (top to bottom, right-hand side). (From Schädler et al. 2004.)

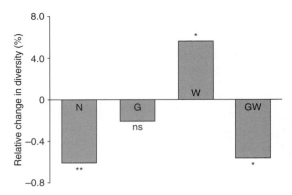

Figure 8.20 Interactive effects of below- and above-ground herbivores on plant diversity. Notice that wireworms (Coleoptera: Elateridae) increase plant diversity when acting alone, but decrease plant diversity in combination with grasshoppers. N, nematodes; G, grasshoppers; W, wireworms; GW, grasshoppers plus wireworms. (From van Ruijven et al. 2005.)

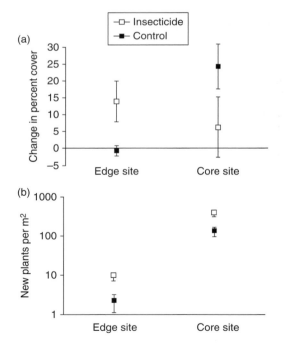

Figure 8.21 Effects of insect herbivores on lupine growth during primary succession on Mount St. Helens. (a) Change in percent cover between 1995 and 1996, and (b) number of new plants in 1996. (From Fagan & Bishop 2000.)

may lie in the relative roles of plant facilitation during primary and secondary succession. In primary succession, early plant colonists are critical for soil formation and nutrient accumulation, and likely facilitate colonization by additional plant species. When insects reduce the populations of early colonists, they retard the successional process. In contrast, during secondary succession, early-colonizing dominant plants may compete with potential new colonists during succession. Reductions in the fitness of early dominants caused by insect herbivores will therefore increase rates of succession.

Most of the evidence that insects affect plant community structure and succession come from studies of insect herbivores. In reality, we might expect that insect predators (see Chapter 5) and mutualists (see Chapter 6) are also capable of influencing the dynamics of plant communities. Unfortunately, the effects of these groups on plant community structure, and concomitant effects on ecosystem processes, are less well understood. For example, there is increasing concern that the loss of insects that act as plant mutualists may increase rates of plant extinction (Bond 1994). However, strong experimental support is generally lacking. One recent study suggests that the loss of seed-dispersing ants from South African shrublands has a significant impact on plant community structure. Areas of fynbos habitat in South Africa are being invaded by the Argentine ant, *Linepithema humile* (Hymenoptera: Formicidae). Argentine ants decimate native ant faunas in their introduced range and this is especially important in the fynbos where about 30% of the flora has seeds dispersed by native ants. Moreover, seed burial by ants is essential for seeds to survive the frequent natural fires. Argentine ants do not act as seed dispersers, and so the loss of native ants has the potential to reduce plant recruitment. Studies by Christian (2001) suggest that this is taking place. She has shown that Argentine ants have a particularly strong effect on those native ant species that disperse large seeds. As a consequence, there has been a disproportionate reduction in the densities of large-seeded plants (Figure 8.23) and an overall shift in plant community structure.

Insect predators and omnivores can also influence the dynamics of plant communities. Often, such effects are most readily observed during the invasion of exotic predator species. For example, the yellow crazy ant, *Anoplolepis gracilipes* (Hymenoptera: Formicidae), has

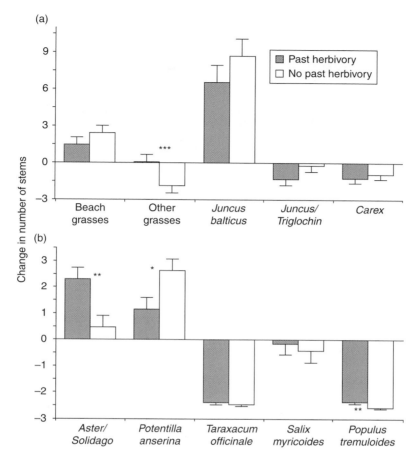

Figure 8.22 Effects of past willow herbivory on plant community dynamics during succession on sand dunes on Lake Huron, USA. (a) herbaceous monocots, and (b) herbaceous dicots. (From Bach 2001.)

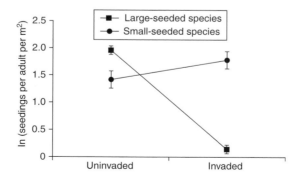

Figure 8.23 Consequences of Argentine ant invasion for post-fire seedling recruitment in South African fynbos. Ant invasion reduces the relative abundance of large-seeded species. (From Christian 2001.)

(a)

(b)

Figure 8.24 Effects of invasion by the yellow crazy ant on forest structure on Christmas Island: (a) an uninvaded site, and (b) a site after 1–2 years of ant invasion. (Courtesy Peter Green, from O'Dowd et al. 2003.)

invaded Christmas Island, a tropical island in the northeastern Indian Ocean. The crazy ant forms expansive and polygynous (multiple queened) super-colonies in which workers occur at extremely high densities (thousands of ants per square meter). In invaded areas, crazy ants extirpate red land crabs, the dominant native consumer on the forest floor. As a consequence, invasion of the ant releases seedling recruitment, increases the species richness of seedlings, and slows the break down of litter on the forest floor (O'Dowd et al. 2003). In addition, the invasive ants form mutualistic associations with another invasive species, honeydew-secreting scale insects in the forest canopy. High densities of scale insects, in combination with sooty mold that grows on their honeydew, causes canopy die-back and death of canopy trees. When two or more invasive species combine to cause dramatic changes in community structure and ecosystem processes, it is termed an "invasional meltdown". This certainly seems to be the case on Christmas Island, where crazy ants, in combination with scale insects, have caused drastic changes in the forest community (Figure 8.24). After only 2 years of ant activity, the forest floor becomes choked with understory vegetation, and the canopy begins to die back. Given what we know about the

effects of plant and litter quality on ecosystem processes (Section 8.5), it seems likely that the yellow ant is causing major changes in ecosystem function on Christmas Island.

8.7 CONCLUSION

In this chapter, we have presented examples of the roles that insects can play in ecosystem function. Of course, one of the most important functions that insects serve in ecosystems is the transformation of plant material into animal material during primary consumption. We learned in Chapters 1 and 3 that plant-feeding insects are abundant and more speciose than other kinds of insect. As such, they provide a crucial resource base for a diversity of secondary consumers including birds, mammals, reptiles, amphibians, fish, other insects and, in some parts of the world, humans. In other words, insects play a critical role as prey for many other kinds of organisms on our planet. In combination with their roles in decomposition, nutrient cycling, and plant community dynamics, it would seem wise to conserve as much insect biodiversity as possible. The diversity and conservation of insects are considered in detail in the next two chapters.

Chapter 9

BIODIVERSITY

9.1 INTRODUCTION

There is no doubt that insects make up most of the world's biodiversity. Of all the 1.7 million species described, approximately 45,000 are vertebrates, 250,000 are plants, and 950,000 (56%) are insects (World Conservation Monitoring Centre 1992; Purvis & Hector 2000). Of the estimated number of species in the world, some 64% are thought to be insects (Figure 9.1). The most diverse order of insects described is the Coleoptera (400,000 species; 24% of all species), followed by the Diptera (7%), Hymenoptera (8%), and Lepidoptera (9%) (May 1988; Stork 1997). Figure 9.1 shows estimates for the numbers of species in each of these orders, adjusted to take into account the fact that some insect groups are much better described than others. For example, 9% of all described species are Lepidoptera, but only 3% of all described and undescribed species are thought to be Lepidoptera. However, the adjusted estimate for Lepidoptera still leaves them eight times as speciose as vertebrates.

Insects are not only more numerically diverse than other organisms, they also perform a disproportionate number of functional roles: herbivory, predation, parasitism, decomposition, pollination, etc. The scale and breadth of insect diversity is well summed up by the title of E.O. Wilson's (1987) paper: 'The little things that run the world (the importance and conservation of invertebrates)'. This chapter discusses the measurement of biodiversity, particularly insect diversity, and the patterns in insect diversity that its measurement has revealed across the world.

9.2 MEASURING BIODIVERSITY

9.2.1 Defining biodiversity

Biodiversity may be defined as "the variability among living organisms from all sources including, *inter alia*, terrestrial, marine and other aquatic ecosystems and the ecological complexes of which they are part; this includes diversity within species, between species and of ecosystems" (CBD 1992). Most other definitions also include a hierarchical element, typically genes, species, and ecosystems. Some authors consider ecosystem processes as part of biodiversity but this is both confusing and unnecessary and is not followed here. However, the functional diversity of insects is such that they play many important roles in ecosystem processes, and the critical role of biodiversity in providing ecosystem services essential for life is considered in Section 9.3.5.

The fundamental level of biodiversity is the species, notwithstanding the other levels of biodiversity and the taxonomic problem of what constitutes a species. Species diversity tends to be expressed as species richness, i.e. the number of species in a given area or habitat (but see below). Although the terms "species richness" and "biodiversity" are often confused, species richness tends to be closely correlated with other measures of biodiversity (Gaston 1996; Brehm et al. 2007).

9.2.2 Biodiversity, species diversity, species richness and diversity indices

Species diversity has two components: the number of species present and the abundance of these species.

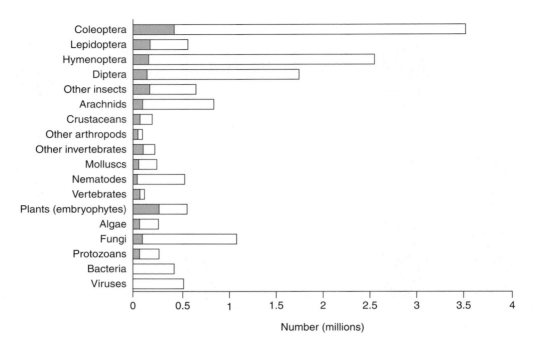

Figure 9.1 Estimated global numbers of species of insects and other groups of organisms. Shaded area is the number of described species. (After Stork 1997.)

Purvis and Hector (2000) illustrate this by comparing two insect communities, one with three species, the other with two (Figure 9.2). Although the second example has fewer species, both of the species present are equally abundant, whereas one species in the first example is much more abundant than the other two rare species present. It can be argued that the first community is more diverse because it has more species but it can also be argued that the second community is more diverse because there is less chance of encountering the same species twice (Purvis & Hector 2000). Many indices have been developed to measure species diversity, some placing greater emphasis on the number of species present, others emphasising the relative abundance of species present. Magurran (2004) has provided a critical review of the many diversity indices available and both her book and that of Krebs (1999) provide excellent guides to calculating diversity indices such as the log series index α, Simpson's index D, and Shannon index H'. Table 9.1 demonstrates the calculation of these and a few other indices for four ground-dwelling beetle communities.

Both Krebs and Magurran urge caution in the use of biodiversity indices because of the controversy over which indices of diversity are the "best". Magurran recommends the calculation of the Margalef and Berger–Parker indices because they are simple and easy to interpret, the log series α because of its wide use and theoretical robustness, and the Shannon index because of its widespread use (despite the widespread criticism it has attracted). Magurran also recommends the drawing of rank abundance graphs (Figure 9.3).

However, the problem with diversity indices, no matter how strongly based in ecological theory, is that they provide descriptors of diversity whose values mean very little to most people. Indeed, Pielou (1995) considered that diversity indices "are wholly unsuitable for measuring biodiversity". In contrast, the number of species in a particular area, community, or habitat is clearly a tangible entity (even if it is hard to estimate). It represents the number of species ecologically adapted to a habitat or the number of species we might be trying to conserve in a particular area. Moreover, information on abundance of the

Figure 9.2 Two samples of insects demonstrating two measures of diversity: species richness (sample A) and species evenness (sample B). (After Purvis & Hector 2000.)

species in a community is better interpreted graphically (Figure 9.3) than through a complex index.

9.2.3 Calculating species richness

Although conceptually a simple measure, species richness may not be simple to assess accurately. The total number of plant species in most temperate habitats may be easy to measure but the total number of species in most insect groups in most habitats must be estimated by sampling. Several books describe sampling methods for different insect groups (e.g. Southwood & Henderson 2000; Leather 2004). Sampling of insects (and other animals and plants) produces species accumulation curves such as those shown in Figure 9.4: the total number of species found by sampling rises rapidly at first, then the rate at which new species are recorded gradually declines, eventually reaching an asymptote, given sufficient sampling. There are several statistical techniques for estimating species richness from sampling data. These techniques are reviewed by Colwell and Coddington (1994) who conclude that there have been insufficient

studies on which to base a recommendation on which technique is best. One of the most frequently used techniques is the jack-knife method (Colwell & Coddington 1994; Borges & Brown 2004; King & Porter 2005; Hortal et al. 2006).

The species richness of different sites (or the same site at different times) can be compared by the total number of species sampled or the jack-knife (or other technique for extrapolation) estimate of species richness. The former approach is only valid when the same sampling intensity is used in each site or on each occasion. Statistical errors can be calculated for the jack-knife and similar methods (Colwell & Coddington 1994).

A widely used alternative to comparing the species richness of different sites by extrapolation is rarefaction. In essence, if there are differences in the total number of individuals sampled in different sites or the number of samples taken, rarefaction can be used to estimate the number of species present in samples of the same size. Gotelli and Colwell (2001) reviewed the problems associated with measuring and comparing species richness. They note, for example, that sample-based rarefaction methods can provide different

Table 9.1 Species diversity indices for ground beetle communities in four forest sites in the UK: Sitka spruce and birch in Moray (northeast Scotland) and Knapdale (southwest Scotland). (Data from Watt et al. 1998.)

	Moray spruce	Moray birch	Knapdale spruce	Knapdale birch
Philonthus decorus	0	246	0	39
Tachinus signatus	4	184	2	2
Abax parallelpipedus	0	0	6	167
Nebria brevicollis	2	149	0	0
Geotrupes stercorosus	0	0	2	91
Carabus problematicus	4	40	0	25
Leistus rufescens	20	22	5	6
Carabus violaceus	34	14	0	0
Calathus micropterus	4	38	0	0
Strophosomus melanogrammus	12	26	0	0
Barypeithes araneiformis	9	0	26	0
Catops nigrita	15	12	1	2
Olophrum piceum	7	17	1	0
Othius myrmecophilus	5	6	7	1
Pterostichus madidus	0	11	3	5
Catops tristis	3	13	1	1
Carabus glabratus	0	0	7	10
Tachinus lacticollis	3	12	2	0
Anthobium unicolor	2	10	0	0
Dalopius marginatus	3	9	0	0
Number of species	39	56	22	30
Number of individuals	187	905	78	394
α	1045	8766	396	3408
Margalef index	7.26	8.08	4.82	4.85
Berger-Parker index	0.182	0.272	0.333	0.424
$1 - D$ (Simpson's index)	0.931	0.850	0.854	0.751
Shannon index (of diversity)	3.142	2.538	2.478	1.953

results compared with individual-based methods, but conclude that it is not always clear which approach is more appropriate. Krebs (1999) provided examples of the use of rarefaction.

The enormous diversity of many insect communities makes them difficult to sample and forces the use of complex statistics such as those discussed here. This diversity also makes it hard to evaluate these statistics: there rarely is an inventory of species to compare the results against. Longino et al. (2002), however, were able to evaluate several methods of estimating the number of ant species in a tropical forest because they had collected data using a variety of methods over many years, producing a near-complete inventory of the area. A total of 437 species were recorded from a 1500 ha forest at La Selva, Costa Rica, using a variety of methods including canopy fogging, malaise traps, Berlese and Winkler samples of the litter fauna, baiting, and manual searching. They found that extrapolation based on limited sampling gave poor estimates of species richness and suggested that this was due to the large number of rare species found in such studies: 51 species (12% of the total) were only recorded once in the area. Longino et al. (2002) concluded that 20 of these species were possibly abundant in the area but difficult to sample; the rarity of the remaining 31 remained unexplained. Other studies comparing different approaches to estimating insect diversity include work on Costa Rican moths (Brehm et al. 2007) and research on methods for estimating arthropod species richness in Azorean forests (Hortal et al. 2006).

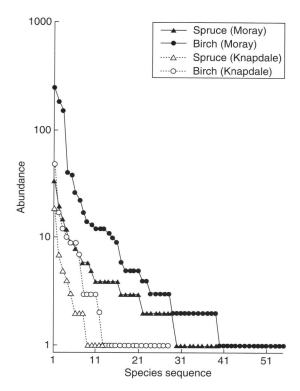

Figure 9.3 Species–abundance curves for ground-dwelling beetles in four Scottish forests: Sitka spruce and birch in Moray and Knapdale. (Abundance is the total number of individuals caught in 10 traps from May to September.) (Data from Watt et al. 1998.)

It is clearly difficult to estimate accurately the number of insects in most habitats. Nevertheless, the numbers of species present in different locations or in the same location at different times can be compared. The basis for success is the cautious application of methods for estimating species richness (Colwell 2000; Kery & Plattner 2007) based on rigorously standardized sampling effort.

9.2.4 Rarity and other attributes of species

A major problem with the use of species richness (or, indeed, any diversity index) is that it ignores any attribute or value that may be attached to a species. In particular, it fails to differentiate between common species and rare, potentially threatened species. A

possible solution is to calculate indices that weight species according to their degree of invasiveness (Robertson et al. 2003) or endemism (e.g. Kier & Barthlott 2001). Kessler et al. (2001), for example, developed a range size rarity index to measure the degree of endemism among plant and bird communities in Andean cloud forests. Most examples of the use of rarity-based indices involve plants or birds but Borges et al. (2005) produced a composite index based on both the diversity and rarity of soil surface-dwelling arthropods to assess the relative importance of potential protected areas in the Azores.

These approaches have not, as yet, been widely used and they run the risk of producing species richness indices as difficult to understand as diversity indices. There may, however, be much to be gained by separately presenting the species richness of clearly defined groups of species, such as native and alien species, or the numbers of species in different classes of rarity or risk from local, regional, or global extinction. Probably the best examples of classifying species according to rarity and risk of extinction are the Red List and Red Data Book (RDB) systems, considered later in relation to insect conservation (see Chapter 10). Unfortunately, although valuable for focusing attention on particular threatened species, particularly vertebrates, the problem with applying systems such as the RDB classification to insects is that, in most parts of the world for most insect groups, we do not know which insect species are endangered. Thus, although some measure of the status of insects is desirable, it is, apart from well-studied groups in a small number of countries, usually impossible to achieve.

The assessment of diversity in relation to other species attributes, such as the role they play in insect communities, is discussed in Section 9.3.4, and the various values attached to biodiversity are discussed in Section 9.3.5.

9.2.5 Comparing species composition

Insect ecologists frequently wish to compare the biodiversity of two or more sites with something more ecologically meaningful than species richness or any other measure of diversity. This usually involves a comparison of species composition. This can be done in various ways but the simplest way is to compare the presence of different species. Figure 9.5 shows the number and composition of four communities of

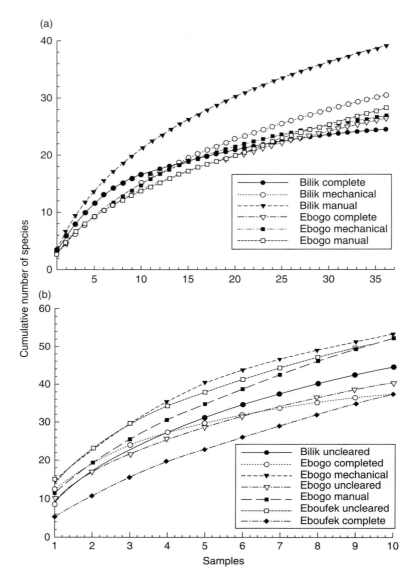

Figure 9.4 Species accumulation curves for ants in the Mbalmayo Forest Reserve, Cameroon: (a) forest canopy ants, and (b) leaf litter ants. (From Watt et al. 2002.)

orbatid mites in different sites in the coastal Sitka spruce forest of Vancouver Island (Winchester 1997). Winchester found that sites varied in their species richness and their species composition. Figure 9.5 summarizes variation in species composition in two ways: by the number of species shared between different (pairs of) sites and by the number of species unique to each site. This clearly shows how important the canopy and "forest interior" sites are in terms of the total number of species found there and, more importantly, in terms of the numbers of species only found in each of these sites.

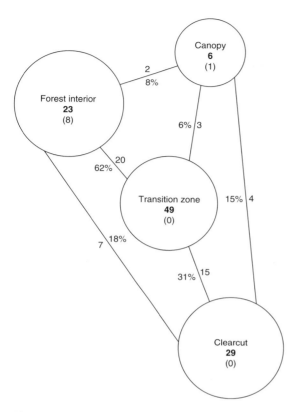

Figure 9.5 Staphylinid species in different habitats in the Upper Carmanah Valley, Vancouver Island, Canada. The number of species in each habitat is given in bold, the number of species occuring in common in different habitats is given along the lines joining the habitats, and the number of species unique to each habitat is given within parentheses within the circles. (From Winchester 1997.)

The approach illustrated above has two important limitations. First, its value is restricted to the comparison of a relatively small number of sites. A figure such as Figure 9.5 could not easily be interpreted with more than four sites. Second, it ignores the abundance of different species in each site. This is not a problem if we are sure that sampling tells us what species exist in each site but sampling, particularly of most insect groups, will often result in the collection of specimens that are simply dispersing through the site being sampled. The problem of these so-called "tourists" has long been recognized (Section 9.3.4). To a large extent it can be overcome by avoiding sampling techniques that tend to collect large numbers of tourists and by concentrating on insect groups that do not disperse widely. In addition, ecologists can use measures of the abundance of different species to compare the insect communities in different sites.

Various statistical techniques are needed to overcome the problems outlined above. The simplest approach is to use a similarity index to compare different sites. There are many similarity indices available, reviewed with worked examples by Krebs (1999). These indices vary from those that are based on the presence/absence of species, such as Jaccard and Sorensen coefficients, to those that are also based on the abundance of species. The latter include the percentage similarity (or Renkonen) index, which is notable for being easy to calculate, and the Morisita index, which is widely considered to be the best available. Table 9.2 shows an example of the use of the Morisita index. Major problems with indices of compositional similarity are that they are sensitive to sample size and, obviously, the unknown identity of species missed during sampling. Chao and colleagues have, therefore, developed a new approach to comparing the composition of different communities that attempts to take unseen shared species into account (Chao et al. 2005, 2006).

Similarity analysis results in the production of tables with similarity values comparing pairs of communities. For small numbers of sites no further analysis may be needed but these tables are often difficult to interpret. In most circumstances, therefore, cluster analysis on the index values is carried out. There are several cluster analysis techniques available (Krebs 1999) but, because of their general utility, they can be found within statistical packages such as SAS or SPSS. The latter package was used, for example, by Chey et al. (1998) in their comparisons of canopy arthropod communities in tree plantations in Sabah, Borneo. Figure 9.6 shows the results of average linkage cluster analysis on ant communities in West Africa.

Further analyses of the composition of insect communities can be carried out and, although complex, these analyses can be particularly useful for identifying the underlying factors causing changes in species composition. These multivariate analytical techniques include canonical correspondence analysis (CCA), detrended correspondence analysis (DCA or DECORANA) and two way indicator species analysis (TWINSPAN). It is beyond the scope of this book to describe these techniques but a brief description of a few examples should illustrate their value.

Table 9.2 Similarities between the canopy ant species composition of different plots in the Mbalmayo Forest Reserve as measured by the Morisita index. Sites are as in Figure 9.4. (From Watt et al. 2002.)

Treatments	Ebogo complete	Ebogo mechanical	Ebogo manual	Bilik complete	Bilik mechanical
Ebogo mechanical	0.9803				
Ebogo manual	0.9674	0.9293			
Bilik complete	0.0552	0.0537	0.0072		
Bilik mechanical	0.0473	0.0590	0.0046	0.3916	
Bilik manual	0.0149	0.0289	0.0131	0.0736	0.4650

Vanbergen et al. (2005) used CCA to show that distinct beetle assemblages were significantly associated with specific edaphic and botanical features of a land use gradient. Basset et al. (2001) also used this technique to assess the impact of logging on insects feeding on seedlings in forest gaps in Guyana. Letourneau and Goldstein (2001) used canonical discriminant analysis in quantifying the effect of organic farming on insect pests and their natural enemies on tomato. Chey et al. (1997) employed TWINSPAN to investigate variations in forest Lepidoptera collected by light trapping in relation to forest management and tree species composition, and Kaller and Kelso (2006) used DCA and CCA to assess the impact of feral pigs on water invertebrates. Collier (1995) used both DCA and TWINSPAN to identify the importance

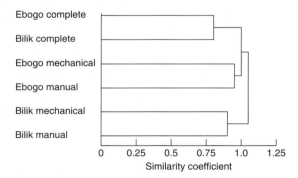

Figure 9.6 Ant communities from different forest canopies grouped according to the similarity of their species composition (using the Morisita index and average linkage cluster analysis). Sites are as in Figure 9.4 and Table 9.2. (From Watt et al. 2002.)

of factors such as the amount of native forest in the riparian zone of streams and rivers on the aquatic macroinvertebrate community downstream. The techniques used in these examples may be complex but they can have an important practical value by identifying factors that are critical for the maintenance of the biodiversity of particular environments. The last of the studies cited above, for example, identified the need to maintain native trees alongside streams and rivers.

9.2.6   The challenge of measuring biodiversity

Anyone wishing to measure the biodiversity, in the "complete" sense of the word, of a given area is faced with an enormous amount of work. Some measure of the amount of sampling and identification involved is given by studies of insects in forest canopies. In his classic study on estimating global species richness (Section 9.3.1), Erwin (1982) found 955 species of Coleoptera (excluding weevils, which were not sorted to species) in the canopy of only 19 trees in Peru. Lawton et al. (1998) recorded 132 species of butterflies, 342 canopy beetles, 96 canopy ants, 111 leaf litter ants, and 114 termites in a study of an area of about 10 ha of forest in southern Cameroon. They estimated that it took 3470 h to sample, sort, and identify the species.

Despite the enormity of the task, some ecologists have advocated the need to carry out a total inventory of all the plant and animal species in a given area. This approach, known as an all taxa biodiversity inventory (ATBI) has been championed by Janzen (1997). However, the purpose of an ATBI is much more than an inventory. Janzen (1997) outlines some of the

Plate 1.1 Lodgepole pine trees defoliated and killed by pine beauty moth, *Panolis flammea*, in northern Scotland.

Plate 1.2 Grassland stripped by armyworm (*Spodoptera* spp.) in Uganda. (Courtesy David Rogers.)

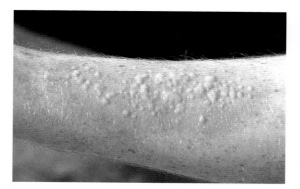

Plate 1.3 Skin rash caused by the urticating hairs of brown tail moth (*Euproctis chrysorrhoea*) larvae. (Courtesy Paul Embden.)

Plate 1.4 Hummingbird hawkmoth, *Macroglossum stelatarum*.

Plate 1.5 Male brown skimmer, *Orthetrum brunneum*.

Plate 1.6 Fried grasshoppers for sale on the streets of Bangkok.

Plate 2.1 Gypsy moth larva.

Plate 3.1 Defoliators. Lepidopteran larva on a tropical broadleafed tree, Borneo.

Plate 3.2 Defoliators. Sawfly (Hymenoptera: Symphyta) larvae on pine, UK.

Plate 3.4 Mines of the horse chestnut leaf-miner moth, *Cameraria ohridella*, with empty puparium.

Plate 3.3 Sap-feeder. Hemipteran shield bug, *Palomina prasina*, feeding on phloem sap.

Plate 3.5 Gall forming. Pineapple galls produced by conifer woolly aphids (Hemiptera: Adelgidae) in spruce. (Courtesy Matthew Mitchell.)

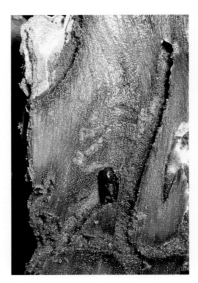

Plate 3.6 Bark boring. Adult female bark beetle (Coleoptera: Scolytidae) with maternal gallery, with larvae in separate tunnels.

Plate 3.7 Resin pitched out of spruce bark as a result of bark beetle tunneling.

Plate 5.1 Trees defoliated and killed by spruce budworm, *Choristoneura fumiferana* in Canada. (Courtesy George Varley.)

Plate 5.2 Lime aphid, *Eucallipterus tiliae*. (Courtesy Paul Embden.)

Plate 5.3 Spruce damaged by winter moth, *Operophtera brumata* in southern Scotland.

Plate 5.4 Ants tending aphids.

Plate 6.1 Cereal aphids on wheat ear.

Plate 6.2 Cossid moth larva in tunnel showing exit hole.

Plate 6.3 Water striders (Gerridae).

Plate 6.4   Adult antlion, *Myrmelion* spp., with full wings.

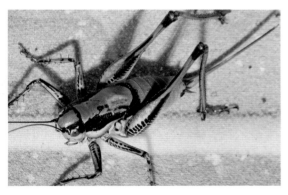

Plate 6.6   Adult female long-horned grasshopper, *Eupholidopterus chabrieri*, completely lacking wings.

Plate 6.5   Adult female leaf insect, *Phyllium* spp., with reduced wings. (Courtesy Paul Embden.)

Plate 6.7

Plate 6.8 Swallowtail butterfly, *Papilio* spp.

Plate 6.9 Heliconius butterfly. (Courtesy Paul Embden.)

Plate 6.10 Pine logs showing blue-stain fungus in cut ends after forest fire and subsequent attack by *Ips*, Queensland, Australia.

Plate 6.11 Termite mound in Turkana desert, Kenya.

Plate 6.12 Leaf-cutter ant with leaf fragment. (Courtesy Paul Embden.)

Plate 6.13 Leaf-cutter ant fungus garden. (Courtesy Paul Embden.)

Plate 6.14 *Zonocerus* grasshopper. (Courtesy Paul Embden.)

Plate 6.15   Larva of monarch butterfly, *Danaus plexippus*.

Plate 6.16   Aposomatic aphids, *Aphis nerii*, feeding on toxic milkweed.

Plate 6.17 Hornet, *Vespa crabro*, showing aposomatic (warning) coloration. (Courtesy Paul Embden.)

Plate 6.18 Hornet clearwing, *Sesia apiformis*, mimicking an aposomatic model. (Courtesy Paul Embden.)

Plate 6.19 Larva of *Oedoparena* spp., dipteran predator of barnacles. (Courtesy Chris Harley.)

Plate 9.1 A crane fly (family Tipulidae).

Plate 10.1 Silver-washed fritillary butterfly, *Argynnis paphia*.

Plate 10.2 Speckled wood butterfly, *Pararge aegeria*. (Courtesy Peter Henderson.)

Plate 11.1 Trees in English hedgerow with Dutch Elm disease. (Courtesy Don Barrett.)

Plate 11.2 Cereal plant infected with barley yellow dwarf virus (BYDV), (åKansas Department of Agriculture, USA.)

Plate 11.3 Pine tree in Japan showing classic wilt symptoms arising from attack by pine wilt nematode (PWN). (Courtesy Hugh Evans.)

Plate 12.1 Adult female cottony cushion scale, *Icerya purchasi*, with first instar crawler. (Courtesy Matthew Mitchell.)

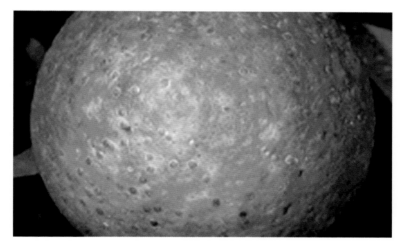

Plate 12.2 Citrus fruit infested with California red scale, *Aonidiella aurantii*. (Courtesy J.K. Clark, Regents, University of California.)

Plate 12.3   Brown tail moth, *Euproctis chryssorhoea*, larva killed by nuclear poly-hedrosis virus (NPV). (Courtesy Paul Embden.)

Plate 12.4   Rice paddy in north Vietnam.

Plate 12.5   Codling moth, *Cydia pomonella*, larva and damage. (Courtesy Doug Wilson, ARS, USDA.)

reasons for establishing ATBIs, including as a baseline for measuring the impact of global change (i.e. climate change, land use change, etc.) – "a gigantic canary in the mine" – and as a standard for calibrating and developing methods of measuring biodiversity.

The first ATBI, led by Daniel Janzen in Guanacaste in Costa Rica, was abandoned in 1997 (Kaiser 1997), although a national biodiversity survey continues. Another example of this approach is the ATBI of the Great Smoky Mountains National Park (Sharkey 2001; Gonzon et al. 2006). Petersen et al. (2005) report on the current status of knowledge from this study of just one group, the crane flies (Diptera: Ptychopteridae, Tipuloidea (see Plate 9.1, between pp. 310 and 311), Trichoceridae), where sampling resulted in the addition of 107 new records, bringing the total list to 250 species. Total richness was estimated to be between 450 and 500 species for the Great Smoky Mountains National Park. Bartlett and Bowman (2003) report research on another group of insects, the Fulgoroidea. Their preliminary inventory obtained 1290 specimens, representing eight families, 23 genera, and 37 species.

9.2.7 Biodiversity indicators

While ATBIs have a clear value, a complete inventory of all plants and animals is a time-consuming exercise. In most cases, however, the results of biodiversity assessments of given areas are needed rapidly as a basis for making decisions about, for example, where protected areas should be, how to manage particular areas such as nature reserves, and how to assess the success of management decisions. The often urgent need for biodiversity information has led to the development of biodiversity indicators and a range of rapid biodiversity assessment techniques.

A biodiversity indicator may be defined as a parameter that provides a measure of biodiversity, usually a relative measure of total biodiversity. This parameter is usually a biological measure of one or more taxonomic groups (or taxa) such as the abundance of one or more species, the species richness of a family or order of insects, or the number of genera. These indicators are often referred to as compositional indicators because they are direct measures of the composition of biodiversity.

There are many types of "bioindicator" and McGeoch (1998) provides a useful review, but in the context of this chapter we consider indicators that provide a quantitative measure of biodiversity rather than other types of bioindicators such as those that are used as surrogates for environmental quality, such as air quality or water pollution. One of the most widely used examples of the latter is the use of insects and other invertebrates to measure river quality in systems such as the River Invertebrate Prediction and Classification System (RIVPACS) (Clarke et al. 2003). RIVPACS is based on relationships between the macroinvertebrate fauna and the environmental characteristics of a large number of reference sites, selected to allow prediction of the macroinvertebrates likely to be present at any site in the absence of pollution or some other environmental pressure. The fauna observed at new test sites are then compared with the invertebrates present in the relevant reference sites to provide an indicator of ecological quality. Other examples of the use of insects as environmental bioindicators include the use of ants as indicators of restoration after, for example, mining (Andersen & Sparling 1997; Andersen & Majer 2004), and the use of ground beetles (carabids) to assess the hydrological condition of floodplain grasslands (Gerisch et al. 2006).

Returning to the key issue of using indicators to measure biodiversity, there are two main issues: (i) can we use insects in the general sense as indicators of other components of biodiversity; and (ii) can we use insects as indicators of particular habitats, or the condition of these habitats? The latter relates to the examples discussed above: researchers sometimes talk about the biological integrity of ecosystems and the use of insects, such as ants, to monitor this (e.g. Andersen et al. 2004). This is probably best applied to situations such as those that Andersen et al. (2004) concentrate on, where during a successful restoration project different taxonomic and functional groups return to make up the complex ecosystem that was there before being destroyed by mining. Another example might be the use of indicators based on the community structure of butterflies or other insects to detect tropical forest disturbance (e.g. Hill & Hamer 1998). However, our lack of understanding of how insects and other organisms react to the disturbance of tropical forests is so great that this type of indicator is unlikely to be useful (Watt 1998).

Nevertheless, some ecologists propose the use of specific indicator species to measure the degree of disturbance or fragmentation of a habitat. This approach is mostly related to the presence of particular bird species. For example, the resplendent quetzal (*Pharomachrus mocinno*), a large, frugivorous bird, has been used as an indicator of montane biodiversity in Central America (Powell & Bjork 1995). The use of individual species as indicators relates to the umbrella species concept, which argues that the conservation of such species confers protection on many other naturally co-occurring species (Sergio et al. 2006). Although most proposed umbrella species are mammals and birds, a number of studies consider insects and other invertebrates such as Lepidoptera (New 1997). In a review of over 100 papers on the subject, however, Roberge and Angelstam (2004) found that only some of the papers (18) critically evaluated the concept and concluded that the conservation of umbrella species could not ensure the conservation of all co-occurring species.

Although the idea of using single species as indicators of biodiversity is unsupported by research, many ecologists have explored the use of one or more groups of insects and other organisms as indicators of biodiversity. Birds, of course, are widely promoted as indicators of biodiversity and although the vastly greater abundance, diversity, and functional breadth of insects means that they must be considered as indicators of biodiversity, our greater knowledge of birds and the huge number of people engaged in monitoring their abundance mean that birds are the best available basis for national and international biodiversity schemes (Gregory et al. 2005). These facts should not stop us from critically assessing the validity of birds and other groups of organisms as indicators of biodiversity.

Amongst insects, a few groups stand out as being the most frequently recommended biodiversity indicators: tiger beetles (Cicindelidae), ground beetles (Carabidae), and butterflies. Pearson and Cassola (1992) argue that tiger beetles are particularly good indicators of biodiversity because of their stable taxonomy, well understood biology and life history, ease of sampling, worldwide distribution, presence in a wide range of habitat types, and specialization of individual species (within habitats). The value of using tiger beetles, rather than other taxa, Pearson and Cassola argue, is that the number of tiger beetle species can be reliably estimated within 50 h in a single site; it would take much longer to do the same for other taxa, even relatively apparent taxa such as birds or butterflies. Pearson and Carroll (1998) tested the use of tiger beetles as indicators of birds and butterflies using data from three grids of squares

(each 275×275 or 350×350 km) in North America, India, and Australia. They found that the number of tiger beetle species was a suitable indicator of the number of butterfly species in North America and the number of bird species in India but was not a useful indicator of bird or butterfly species in Australia or bird species in North America. Although tiger beetles may have some value as indicators of biodiversity at large spatial scales, there is little evidence to suggest that they are suitable indicators of biodiversity at local scales, which limits their effectiveness as indicators of the impact of local pressures on biodiversity. Nevertheless, the level of understanding of this group, which comprises almost 2600 known species, makes them an excellent group to study global distributions and trends (Pearson & Cassola 2005).

Another group of insects that has been promoted as an indicator taxon is the Carabidae or ground beetles. Carabids are more speciose than tiger beetles and much more research has been done on them (e.g. Stork 1990; Lovei & Sunderland 1996; Niemelä 2001). Research on this group has included work in forests (Koivula & Niemelä 2003; Rainio & Niemelä 2006), grassland (Grandchamp et al. 2005; Magura & Kodobocz 2007), agriculture (Cole et al. 2005), urban areas (Ishitani et al. 2003), and even golf courses (Tanner & Gange 2005). Their importance as natural enemies of insect pests has long been recognized (Kromp 1999), making them a key group in the evaluation of novel agricultural practices such as genetically modified crops (Dewar et al. 2003). Although the family is dominated by predacious species, including some that specialize on aquatic prey (Paetzold et al. 2005), the Carabidae are functionally diverse, including many seed-feeding species (Honek et al. 2006). Recent studies have included research on the influence on carabid communities of land use heterogeneity (Vanbergen et al. 2005), sheep and reindeer grazing (Suominen et al. 2003; Cole et al. 2006), and roads and forest fragmentation (Koivula & Vermeulen 2005). Indeed, so well studied are these insects that one paper was aptly titled "From systematics to conservation – carabidologists do it all" (Niemelä 1996).

Given the attention paid to ground beetles, it is no surprise that their potential role as indicators has often been discussed (Rainio & Niemelä 2003). Carabids have, for example, been proposed as indicators of grassland management practices (Eyre et al. 1989) and forest condition (Villa-Castillo & Wagner 2002). Rainio and Niemelä (2003), however, concluded that despite the large amount of research on carabids there is insufficient knowledge on how well they act as indicators of other taxa.

Many authors have also proposed butterflies as indicators (Thomas 2005). They are better known than any other group of insects, relatively easy to monitor (e.g. Collier et al. 2006), and represent the most well-known insect icon for biodiversity. Beccaloni and Gaston (1995), however, have gone further and suggested that because the proportion of different butterfly families and subfamilies in the tropical forests of Latin America is relatively constant, it may be possible to use a single butterfly group as an indicator of overall butterfly diversity. They suggest the use of ithomiine butterflies (Nymphalidae: Ithomiinae), which make up an average of 4.6% of all butterflies.

Each of these examples demonstrates different ways that insects can be used as indicators. Dung beetles (Vessby et al. 2002) and ants (Andersen et al. 2004) are among other insects that have been studied as potential indicators. We will return to this topic in the next chapter on conservation, but in the context of this chapter, the key issue is the degree to which the diversity of one taxonomic group can be used to measure biodiversity generally. In other words, does the diversity of tiger beetles, carabids, or butterflies, for example, provide an indicator of the diversity of other insect taxa or, indeed, for all animals and plants present in a particular area? Although other criteria are important for other types of indicators, the key attribute of biodiversity indicators is, clearly, the degree to which their diversity correlates with the diversity of other taxa. The evidence is mixed. Prendergast (1997; Prendergast et al. 1993) showed that there is a poor correlation between the richness of different taxa at a regional scale, and Lawton et al. (1998) showed the same for a range of insect and other taxa at a local scale. Some recent studies have provided support for single indicator taxa (e.g. Sauberer et al. 2004) but most show poor correlation between the diversity of different groups of species (e.g. Oerteli et al. 2005), supporting the conclusion that reliance on a single indicator would give a poor measure of overall biodiversity.

9.2.8 Rapid biodiversity assessment

Further critical evaluation of indicators is necessary but it is clear that the use of single taxa as indicators of biodiversity is misguided. What, then, is the best way to assess biodiversity? Although a detailed assessment of spatial and temporal patterns in biodiversity can only come from sampling many taxa, if we want to understand the impact of particular threats to biodiversity, plan conservation measures to combat these threats, or monitor the success of conservation measures, we need ways of measuring biodiversity rapidly. This has, of course, led to the development of indicators, as discussed above, but it has also led to the development of a variety of approaches that are often referred to as "rapid biodiversity assessments". To a large extent these approaches are, like the use of single indicator taxa, compromises: any reduction in the number of plant and animal groups sampled and any deviation away from statistically robust sampling will lead to a degree of uncertainty about the reliability of the results as a measure of total biodiversity.

The first step in rapid biodiversity assessment is the choice of groups or taxa to be sampled. To overcome the limitations discussed above that are inherent in choosing one indicator group, entomologists and others concerned with biodiversity have suggested that a range of taxa, sometimes referred to as "predictor sets" should be sampled to adequately measure overall biodiversity (Kitching et al. 2001; Basset et al. 2004a). The key criterion for choosing a set of taxa is that together they ensure that the insects sampled represent as wide a range of taxonomic and functional groups as possible. In tropical forests, for example, we might choose butterflies, termites, ants, and dung beetles. Together they represent four insect orders; butterflies are phytophagous, termites and dung beetles are decomposers, termites and ants are numerically the most dominant macroinvertebrate taxa in the soil and canopy, respectively (although neither group are restricted to these microhabitats), and ants are in themselves an extremely diverse group functionally including leaf-cutting ants, wood-eating ants, pollinators, Homoptera-tending ants, predators, and social parasites (Hölldobler & Wilson 1990). Other arthropod taxa have also been included in biodiversity assessments, including isopods (Hassall et al. 2006) and spiders (Oxbrough et al. 2005).

It is worth remembering that insects may form only part of a biodiversity assessment program. Indeed they are often excluded from rapid biodiversity assessment because they are considered to be too difficult to assess (Ward & Lariviere 2004), ignoring not only the fact that insects comprise the largest component of terrestrial biodiversity but also the possibility that insect sampling can be less expensive than conventional plant and vertebrate surveys (Oliver et al. 1998). Several studies have, however, combined sampling of insects and other taxa in biodiversity surveys. Sauberer et al. (2004), for example, included two groups of plants (bryophytes and vascular plants), five groups of invertebrates (gastropods, spiders, orthopterans, carabid beetles, and ants), and one vertebrate taxon (birds) in their study of biodiversity in agricultural areas in Austria.

Another important criterion for choosing taxa for biodiversity assessment is ease of identification. There are two aspects of this: availability of specialists and availability of user-friendly identification guides. The latter lag well behind the former and the only insects well served by identification guides for most of the world are butterflies (e.g. De Vries 1997; D'Abrera 2004; Larsen 2005). There is an urgent need to improve identification guides: Janzen (1997) wrote "No more turgid keys, please. Expert systems, picture keys ... are a major step forward." The web also presents new opportunities for making taxonomic information more widely available and improving its quality (Godfray 2002). Most insect groups require specialist taxonomists for identification to species, and in many cases, family. Moreover only a limited part of the world is reasonably well served by insect taxonomists (Gaston & May 1992) and globally taxonomy is in decline (Hopkins & Freckleton 2002; Kim & Byrne 2006).

There are, however, some promising signs. DNA barcoding also holds potential for rapid identification of species. This diagnostic technique uses short DNA sequence(s) for species identification and holds the prospect of handheld DNA-sequencing devices that can be applied in field biodiversity surveys (Savolainen et al. 2005). DNA barcoding is still a controversial technique: many insect taxonomists see no value in it (Lobl & Leschen 2005) but research on spiders has shown it to be rapid and accurate (Barrett & Hebert 2005). Barrett and Hebert (2005) used mitochondrial gene coding for cytochrome c oxidase I to successfully discriminate 168 spider species and 35 other arachnid species. Automated insect identification systems based on image analysis and pattern recognition has been developed for various insect groups such as leaf-hoppers (Dietrich & Pooley 1994) and spiders (Do et al. 1999) and, despite technical difficulties, it provides a potentially useful basis

for identifying insects (Gaston & O'Neill 2004; Ball & Armstrong 2006).

Despite the lack of identification guides for most insects, it would be a mistake to restrict studies on biodiversity to the most easily identifiable but least diverse groups, neglecting the difficult taxa that comprise most of insect, and total, biodiversity. While we wait for better identification guides and technologies, there are two alternative approaches that can be used: morphospecies and higher taxon measures of biodiversity as surrogates for species-based measures.

Morphospecies are specimens grouped into "recognizable taxonomic units" (RTUs) or "operational taxonomic units" (OTUs) (Oliver & Beattie 1993; Stork 1995; Ward & Lariviere 2004). Oliver and Beattie (1993, 1996a, 1996b) pioneered this approach for situations where, due to a lack of specialist taxonomists, technicians with basic training have to estimate the diversity of a sample of insects or other taxa. Examples of the use of morphospecies sorting include research on the impact of grazing in forests on spider communities in Australia (Harris et al. 2003), studies on tropical forest ants (Watt et al. 2002; Floren & Linsenmair 2005), and work on the movement of insects from habitat to habitat (Dangerfield et al. 2003). In the last example, 10,446 beetle, ant, wasp, fly, and collembolan specimens from 150 pitfall traps situated along the edge between riparian and saltbush habitats were sorted to 426 morphospecies. A large overlap in the composition of species present in the two habitats was recorded, with differences only becoming apparent gradually from the edge.

Although widely used, morphospecies sorting is a controversial approach (e.g. Goldstein 1997). In a review of nine studies, Krell (2004) reported a median error in the identification of 22% of the insects examined and identified several serious problems with this approach. Despite these concerns, it must not be forgotten that the taxonomic knowledge of insects in much of the world is very poor and few countries have a good supply of taxonomic specialists. In tropical forests, for example, there may be no alternative to using morphospecies sorting for the rapid assessment of the diversity of most insect groups. Table 9.3 shows the proportion of insects in different groups that had to be sorted to morphospecies in a study in Cameroon (Lawton et al. 1998). Only 1% of the butterflies could not be assigned to known species but over 80% of the canopy beetles had to be assigned to morphospecies.

Table 9.3 Numbers of species recorded in a study on forest biodiversity in Cameroon, the time taken to sample, sort and identify them and the percentage that could not be assigned to known species. (From Lawton et al. 1998.)

Group surveyed	Total species recorded on study plots during survey[*]	Scientist-hours to sample, sort and identify	Morphospecies that cannot be assigned to known species (%)
Birds (Aves)	78 (H)	50	0
Butterflies (Lepidoptera)	132 (I)	150	1
Flying beetles (Coleoptera)	358 (malaiso) (L) 467 (interception)	600 (both methods combined)	50–70
Canopy beetles (Coleoptera)	342 (L)	1,000	>80
Canopy ants (Hymenoptera: Formicidae)	96 (I)	160	40
Leaf litter ants (Hymenoptera: Formicidae)	111 (I)	160	40
Termites (Isoptera)	114 (H)	2000	30
Soil nematodes (Nematoda)	374 (I)	6000	>90

[*] H (high, >90%), I (intermediate, 50–90%) and L (low, <50%) are estimates of the degree of completeness in the species inventory for each group.

In another study, Brehm et al. (2005) assessed the diversity of geometrid moths in montane rainforest in southern Ecuador. From over 35,000 specimens, 1266 species were recorded but only 63% could be identified to known species. Brehm et al. (2005) concluded that the diversity of this family of moths is much higher in this area than anywhere else in the world.

The main problem with the use of morphospecies is that the identifications are, by definition, local. This is no obstacle to calculating species richness (providing the classification of individuals into morphospecies is accurate) but it prevents comparison of the species composition of different sites (or the same site at different times) unless the same set of "voucher specimens" is used. More generally, the use of morphospecies and the involvement of parataxonomists should be seen in context. The use of morphospecies is based on the concept of taxonomic sufficiency, that is "identifying organisms only to a level of taxonomic resolution sufficient to satisfy the objectives of a study" (Pik et al. 1999). Furthermore, sorting specimens into morphospecies should not always be seen as a final outcome of biodiversity studies but as a potential first step towards more detailed ecological research.

Such is the degree to which trained technicians, or parataxonomists, are associated with morphospecies sorting, that Krell (2004) refers to morphospecies as "parataxonomist units". However, parataxonomists play an increasingly important role in biodiversity studies, not simply limited to sorting specimens into morphospecies (Janzen 2004). Janzen described a parataxonomist as a "resident, field-based, biodiversity inventory specialist who is largely on-the-job trained out of the rural work force and makes a career of providing specimens and their natural history information to the taxasphere, and therefore to a multitude of users across society" and emphasized the importance of teams of parataxonomists, particularly to large inventories. Realizing the frequent criticism that parataxonomists lack taxonomic training, Janzen noted that their accuracy increases with experience and continuity of feedback from users of the specimens they collect and the information they provide.

Basset et al. (2004b) also discuss the role of para-taxonomists, focusing on their role in the biological monitoring of tropical forests. They argue that the urgent need for continuous data from tropical ecosystems can most efficiently be achieved through the involvement of parataxonomists.

Higher taxon measures of biodiversity provide a completely different approach to assessing biodiversity: instead of recording the number of species, the number of genera or other higher taxon measures such as the number of families or orders is recorded, although the latter is unsurprisingly less informative (Balmford et al. 2000). This approach has been applied to plants, vertebrates, and invertebrates for essentially the same reason as the use of morphospecies – identification to genera is easier and faster than identification to species. In an early example of this approach, Williams and Gaston (1994) recommended higher taxon measures on the basis of analyses of plants, vertebrates, and insect taxa, using data from North and Central America, Australia, and Britain. Like the use of morphospecies, however, this is a controversial approach. It has most frequently been applied to plants, and while some studies support its use (e.g. Balmford et al. 1996a, 1996b; Prinzing et al. 2003; Villasenor et al. 2005), others do not (e.g. Prance 1994).

Among studies on insects and other arthropods, support for higher taxon measures has come from research on butterflies (Cleary 2004), Coleoptera, Diptera, and Acari (Baldi 2003). Cardoso et al. (2004) tested the ability of family and genus richness to predict the number of spider (Araneae) species independently of sampling effort, location, and habitat type. They concluded that genus richness was a good surrogate of species richness but family richness was not. Brennan et al. (2006) compared the use of spider family richness with habitat structure as indicators of forest invertebrate diversity and found the former to be better.

In contrast, Andersen (1995) assessed the use of genus richness as a surrogate for ant species richness at 24 sites throughout Australia. He showed that, within a region, there was a strong relationship between the number of genera and the number of species, but overall the number of genera was a poor surrogate for the number of species (Figure 9.7). Other studies suggesting that this approach is not reliable include work on butterflies (Prendergast 1997) and hoverflies (Katzourakis et al. 2001).

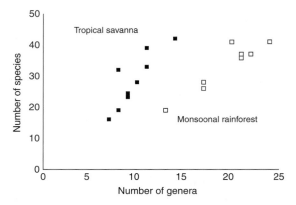

Figure 9.7 Relationship between numbers of ant genera and species in samples from tropical savanna and monsoonal rainforest in the Australian seasonal tropics. (From Andersen 1995.)

At the very least, the use of higher taxon measures should be cautiously applied in biodiversity assessment. It is unlikely, despite the results of the studies discussed here, that this approach is better suited to some insect groups than others, although it might be more successful for taxa in which a large proportion of species are found in a relatively small number of genera (Andersen 1995). The limitations discussed above about the lack of predictive power of single indicator taxa should also be borne in mind. Indeed, although Baldi (2003) found that higher taxon measures of beetles, flies, and mites were good surrogates of species richness in these orders, he noted that these measures were poor predictors of the species richness of other orders; e.g. the number of beetle genera did not correlate with the number of fly or mite species.

It is, nevertheless, important to put the use of this technique in context, noting in particular the purposes to which it may be put. Putting aside issues of practicality, there are two main reasons for using higher taxon measures of biodiversity: describing large-scale (including global) patterns in biodiversity and the selection of protected areas for conservation. Given the poor knowledge of most insect groups across the world, it seems sensible to consider patterns of diversity at higher taxon levels. This topic is discussed below in this section. Although many studies use higher taxon measures in these contexts (e.g. Cardoso et al. 2004), others use this approach in different contexts, for example evaluating the impact

Table 9.4 The proportion of different feeding guilds (as defined by Hammond (1990)) of beetles found in forest canopies in Sulawesi, Brunei, Australia, and the UK. (From Hammond et al. 1996.)

	Herbivores	Xylophages	Fungivores	Saprophages	Predators	Total of species
Sulawesi	25.1	16.1	27.5	13.8	17.4	1355
Brunei	34.6	7.7	18.5	15.5	23.7	875
Australia	19.8	21.1	23.1	8.4	27.3	454
UK	23.5	10.0	26.5	11.0	29.0	200

of forest management on forest arthropods (Simard & Fryxell 2003; Brennan et al. 2006). Such studies may be a useful step towards understanding the role of different influences on insect diversity but they are much less informative than studies that evaluate diversity at the species level.

Although most studies on biodiversity focus on taxonomic diversity, an alternative approach is to measure functional diversity. Such an approach depends, however, upon being able to place individual species in different functional groups. With most insect species still unidentified, let alone ecologically understood, it would appear to be a difficult task to categorize different species, particularly in functionally diverse groups such as ants, according to their functional roles. Nevertheless, the functional role of individual species in, for example, the mega-diverse Coleoptera can be assessed. In many beetle families the functional role is well known and stable across the family. Moreover, in beetle families where there is species to species variation in functional roles, the role of individual species can be assessed by examination of the mouthparts. Table 9.4 shows an example of the classification of beetles into fungivores, xylophages, decomposers, herbivores, and predators (Hammond et al. 1996).

Other useful related approaches to classifying insects in biodiversity surveys include classification as generalists or specialists (in terms of diet breadth) or classifying them in relation to their range. Andrew and Hughes (2004) used this approach to classify *Acacia*-feeding beetles in eastern Australia in order to investigate the potential effects of climate change. Beetles were classified as cosmopolitan species, generalist feeders, climate generalists, and specialists. Their subsequent analysis suggested that community structure would be fairly resilient to climate change but that community composition would alter.

Some biodiversity studies focus on the diversity of particular functional groups. Komonen et al. (2003), for example compared the insect communities associated with the wood-decaying *Fomitopsis* fungi in China and Finland. They compared diversity, uniqueness and food web structure of the two insect assemblages and found that species, genera, and family richness was greater in Finland than in China.

The notable aspect of examples such as these is that biodiversity studies can be extremely flexible and by combining taxonomic and ecological data, many important ecological questions can be addressed.

9.3 PATTERNS IN INSECT DIVERSITY

9.3.1 How many species are there in the world?

In a pioneering study, Erwin (1982) calculated that there were 30 million species of arthropods in the tropics alone. This was much higher than the then current estimate of 10 million species (Wilson 1992) and considerably higher than the estimate of 20,000 insect species in the world made by the British entomologist Ray in 1883 (Stork 1997). Although now regarded as an overestimate, Terry Erwin's estimate provided an important initial frame of reference for increasing concern about species extinctions and stimulated many others to address one of the major questions in the study of biodiversity: what is the number of species on the planet?

Erwin based his estimate on a study of insects in the canopy of tropical rainforest trees. Erwin's calculation was appealingly simple. He sampled the canopy of 19 trees of a single tree species (*Luehea seemannii*) in Peru. Erwin estimated that there were 1200 beetle species on these trees and that 20% of the herbivorous beetles, 5% of the predators, 10% of the fungivores, and 5% of the scavengers were specific to that tree species. That is, 163 (13.5% overall) of the 1200 beetles on *L. seemannii* were estimated to be host-specific. Erwin also estimated that there are 70 tree species or genera in each hectare of tropical forest, that beetles made up 40% of all arthropod species, and that there are twice as many arthropods in the canopy than the forest floor. Adding the non-specific beetles and other arthropods in the canopy and forest floor this gives an estimate of about 41,000 arthropod species per hectare of tropical forest. Erwin repeated his estimate for the tropics as a whole, based on 50,000 species of tropical trees and produced an estimate of almost 30,000,000 tropical arthropods rather than the 1.5 million species of arthropods estimated at the time.

Several ecologists have considered the assumptions made by Erwin. The most critical assumption is the estimate of host specificity. Erwin estimated 163 beetles per plant species. Gaston (1994), however, gives an estimate of 10 insects per plant species, which approximates to four specific beetle species, based on temperate countries. This figure may be less than that recorded in tropical areas. In Papua New Guinea, for example, Basset et al. (1996) recorded 391 species of phytophagous beetles (from 4696 individuals) on 10 species of tree. They estimated that there are between 23 and 37 monophagous leaf-feeding species on each tree species, higher than Gaston's estimate but much lower than Erwin's estimate of 136 species. Considering a different trophic group, Lucky et al. (2002) found 318 species of carabid beetles in nine sampling dates over approximately 3 years in a 100×1000 m terra firme forest plot in Ecuadorian Amazonia. They concluded that the large number of undescribed species (over 50%), the failure of the species accumulation curves to level off, and the temporal changes in species composition meant that the species diversity of carabids is much higher than previously thought.

However, other studies (e.g. Novotny et al. 2004) have cast doubt on high estimates of specificity. Novotny and Basset (2005), after reviewing studies of host specificity in tropical insect herbivores, concluded that few species are restricted to a single host plant and most are able to feed on whole plant genera. They also concluded that there is no major increase in host specificity from temperate to tropical communities. A lower figure for host specificity has a marked impact on global species estimates: on this basis Novotny et al. (2002) produced a global estimate of 4–6 million species.

Erwin's second assumption, that 40% of all canopy arthropods are beetles is probably an overestimate too (see below). Stork (1997) suggests a figure of about 20% may be more accurate and other estimates range from 18% to 33% (Ødegaard 2000a). Research in Borneo and Sulawesi (also see below) suggests that Erwin's third assumption, that 66% of beetle species are found in the canopy in comparison to the ground, probably underestimates the number of arthropods on the ground. Other estimates range from 25% to 50% (Ødegaard 2000a). An additional complication is that the proportion of insects found in the canopy probably varies according to insect taxon: Schulze et al. (2001) found a marked decrease in the abundance of fruit-feeding nymphalid butterflies during sampling of different strata from the ground to the forest canopy in Borneo.

Erwin's work prompted others to try and estimate the number of species in the world (e.g. Stork 1993; Ødegaard 2000a). The simplest approach is to base estimates on the ratio between different groups of insects. For example, Stork and Gaston (1990) used the ratio of butterfly species (67) to all species of insects (22,000) in the well-documented British fauna, and an estimate of 15,000–20,000 butterfly species worldwide, to give a global estimate of 4.9–6.6 million insect species.

Hodkinson and Casson (1991) estimated global insect species richness by using the ratios between: (i) the number of known and new Hemiptera in a sample of 1690 species collected in Sulawesi; and (ii) the number of species of Hemiptera and other insects. They estimated that 62.5% of the Hemiptera they collected were new to science and that there are 184,000–193,000 species in Hemiptera worldwide. Assuming that 7.5–10% of insect species are Hemiptera, they estimated that the number of insect species in the world is between 1.84 and 2.57 million. They also estimated global species richness by assuming that if the 500 tree species in their study area yielded 1056 new species of Hemiptera, the 50,000 species of tree in the tropics (estimated by Erwin) means that there are 105,600 undescribed Hemiptera species worldwide. Adding the 184,000–193,000 described species and using the ratio of Hemiptera to all insect species given above, Hodkinson and Casson again estimated that there are 1.84–2.57 million insect species worldwide. Hodkinson and Casson's estimates are flawed in several ways (Stork 1997): it is unlikely that they sampled all the Hemiptera in their study area, their ratio of described to new species is probably not representative, and the ratio of (described) Hemiptera to other insects is probably biased because of the large number of hemipteran pest species. Using data on beetles from the same study area, Stork (1997) produced an estimate of 5.0–6.7 million insects world-wide but warned that such estimates are very sensitive to the intensity of sampling.

Other methods have been used to estimate global species richness, such as extrapolation based on body size distribution (May 1988) and analysis of species turnover (Stork 1997). Hammond (1992) reviewed the different approaches to estimating the number of species in the world and, after consulting experts on different groups, produced a conservative "working estimate" of 12.5 million species of which 8 million are estimated to be insects. Dolphin and Quicke (2001) used two methods based on knowledge of the diversity of the parasitic wasp family Braconidae. The first method was based on extrapolation of the decreasing rate of species descriptions to the point at which this reaches zero. The second method was based on extrapolating their knowledge of braconid diversity in the Palaearctic using the global distribution of butterflies and mammals. Each method predicted that the actual number of Braconidae was 2–3 times that currently known. Applying their results to all insects produced a global estimate of 2.05–3.4 million species.

Ødegaard (2000a) carried out a thorough review of Erwin's research, which remains the most notable approach to estimating global biodiversity. Ødegaard compiled information on the various assumptions in Erwin's calculations and added some additional steps. One addition was based on the finding that for Heliconinae butterflies, a well-known group of insects, the number of insect associates does not increase proportionally with the number of host-plant species (Passifloraceae) recorded (Thomas 1990): relatively fewer insect species are associated with each plant species as the number of plant species recorded increases. This is due to relatively specialized insects using different species of plants in different regions of their range (May 1990). Ødegaard included a "between community correction factor" to account for this. He also added estimates of species found on lianas and epiphytes as well as trees. The full calculation is shown in Table 9.5. Ødegaard estimated that there are 4.8 million insect species globally but gave an estimated range of 2.4 to 10.2 million species and

Table 9.5 Revision of Erwin's (1982) estimate of the global number of insect species. (After Ødegaard 2000a.)

How many species are there?	Minimum	Working figure	Maximum
1. Number of phytophagous beetles in canopy			
(a) Effective number of species specialized on life forms of plants			
2.8–5.0 species effectively specialized on trees (Ødegaard 2000a)			
2.3–3.6 species effectively specialized on trees (Basset et al. 1996)			
3.5?–5.8 Lower quartile: 3.0; median: 3.9; upper quartile: 4.7 (based on mean value of each range) × 37,000 (total no. trees)	111,000	144,300	173,900
3.8–5.6 species effectively specialized on lianas (Ødegaard 2000a)			
Lower quartile: 3.8; median: 4.7; upper quartile: 5.6 × 18,500 (total no. lianas)	70,300	87,000	103,600
0.5 species effectively specialized on epiphytes (guess) × 20,000 (total no. epiphytes)	10,000	10,000	10,000
Total no. of species associated with trees, lianas, and epiphytes	191,300	241,300	287,500

(Continued)

Table 9.5 (*Continued*)

How many species are there?	Minimum	Working figure	Maximum
(b) Proportion of the total phytophagous beetle fauna associated with the plant species at a given site through the geographic range of these plant species			
50% (Cassidinae, Panama) 44% (Scolytinae, Norway) Lower quartile: 44%; median: 47%; upper quartile: 50%	383,000	513,300	652,000
(c) Between community correction factor (Thomas 1990) Working figure: 2.5 (May 1990)	153,000	205,300	261,000
2. Number of beetle species in the canopy The proportion of phytophagous beetles to total beetle fauna in the canopy 37% (Davies et al. 1997) 40% (Majer et al. 1994) 41% (Wagner 1997) 44% (Stork 1991) 47% (Basset 1991) 47% (Erwin & Scott 1980) 56% (Erwin 1983)			
Lower quartile: 40%; median: 44%; upper quartile: 47%	326,000	466,600	652,000
3. Number of beetles The proportion of canopy beetles to the total beetle fauna 25% (Hammond 1997) 33% (guess; Stork 1993) 50% (guess; May 1990) 66% (guess; Erwin 1982)			
Lower quartile: 29%; median: 42%; upper quartile: 58%	561,000	1,111,000	2,249,000
4. Number of arthropods The proportion of beetles to the total arthropod fauna 18% (Southwood et al. 1982) 22% (Majer et al. 1994) 23% (Stork 1988) 23% (Basset 1991) 33% (Hammond 1992)			
Lower quartile: 22%; median: 23%; upper quartile: 23%	2,441,000	4,830,000	10,224,000

concluded that the between community correction factor and the proportion of canopy (to total) species are major causes of uncertainty.

Perhaps this issue more than any other exposes our enormous ignorance of insect ecology. Every method used to estimate global species richness exposes several major gaps in our knowledge. Few estimates since Erwin's milestone study support his global

figure of about 30 million insect species. However, it is worth noting, for example, that Lewinsohn and Prado (2005) estimated that there are 1.4–2.4 million species in Brazil alone; new methods for sampling tropical forest canopies are demonstrating greater diversity than previously recorded (Dial et al. 2006); and the increasing discovery of cryptic species (e.g. Schonrogge et al. 2002; Schlick-Steiner et al. 2006) suggests that even more species remain to be discovered than we thought. The search for the answer to one of the most important questions in biodiversity continues.

9.3.2 Biogeographic patterns in biodiversity

Sixty-seven butterflies have been recorded in the UK, 321 in Europe, but over 700 have been recorded in one area (near Belem) in tropical Brazil (Robbins & Opler 1997). There are many similar examples of latitudinal variation in biodiversity, most demonstrating a general increase in biodiversity from polar to temperate to tropical latitudes (Gaston & Williams 1996; Willig et al. 2003). Latitudinal variation in biodiversity applies to vertebrates and plants as well as insects and other invertebrates, although there is evidence that genetic diversity shows the opposite trend, increasing from the equator to the poles (Molau 2004). Many underlying mechanisms for this variation have been suggested, for example the effects of environmental stability, environmental predictability, productivity, area, number of habitats, evolutionary time, and solar radiation (Rohde 1992). After reviewing the different hypotheses, Rohde concluded that the lower degree of variation in climate in tropical areas has permitted a greater degree of specialization, and therefore speciation, than temperate areas.

Although there is no consensus on what causes the latitudinal patterns in species richness, the fact remains that the diversity of most insects declines from tropical to temperate areas. Examples of research on latitudinal patterns within countries, within continents, and globally include ants (Gotelli & Ellison 2002), grasshoppers (Davidowitz & Rosenzweig 1998), assassin bugs (Rodriguero & Gorla 2004), termites (Eggleton et al. 1994), and swallowtail butterflies (Sutton & Collins 1991) (Figure 9.8). However, there are some interesting exceptions. Most aphids, for example, are found only in temperate regions. It has been argued that aphids are particularly adapted to temperate conditions, but some aphids are endemic to the tropics and are well adapted to tropical conditions (Dixon et al. 1987). Dixon et al. argue that the reason that there are so few species of aphids, particularly in the tropics, is because of: (i) their high degree of host specificity; (ii) their relatively poor efficiency in locating their host plants; and (iii) their inability to survive for long without feeding. The problems aphids face are not limited to

tropical regions: 90% of plant species worldwide are not used by aphids. However, the situation is worse in the more plant species-rich tropical regions where aphids' host plants are more difficult to locate. Thus, as the number of plant species in a particular geographic region increases, the number of aphid species actually declines (Figure 9.9). In contrast, phytophagous insects that are more efficient at locating their host plants, such as butterflies, are likely to become more species-rich as the number of plant species increases.

Sawfly diversity also increases from tropical to polar regions, and does so even more sharply than aphid diversity (Kouki et al. 1994) (Figure 9.10). Although sawflies, like aphids, are very host-specific, they are much better at locating their host plants. Thus Kouki et al. (1994) conclude that the latitudinal trends in sawflies and aphids are caused by different factors. They suggest that sawfly diversity parallels the diversity of their most widely used plants, the willows (*Salix* spp.).

Other exceptions to the rule that diversity increases from temperate to tropical latitudes include termites in Australia (Abensperg-Traun & Steven 1997) and ichneumonid wasps more generally (Janzen & Pond 1975). Several hypotheses have been put forward for the latter, including the "nasty host hypothesis" which proposes that the greater toxicity of butterflies in the tropics limits the diversity of these parasitic wasp species (Gauld et al. 1992; Sime & Brower 1998). Flea diversity also appears to be very weakly associated with latitude: host diversity and regional altitude are more important drivers of geographic variation in the Palaearctic (Krasnov et al. 2007).

Altitudinal gradients in biodiversity have also been studied by many ecologists. Because environmental variation along an altitudinal transect is similar to variation along a latitudinal transect, we would expect insect diversity, in general, to decrease from low to high altitudes. Several studies have considered trends in dung beetle diversity at a range of altitudes. For example, Escobar et al. (2005) studied the abundance, richness, and composition of Scarabaeinae dung beetles between 1000 and 2250 m above sea level in the Colombian Andes, finding a peak in species richness at middle elevations (Figure 9.11). Hanski and Niemelä (1990) studied the abundance and diversity of dung beetles along altitudinal transects in Sulawesi (from 300 to 1150 m above sea level) and Sarawak, Borneo (from 50 to 2400 m

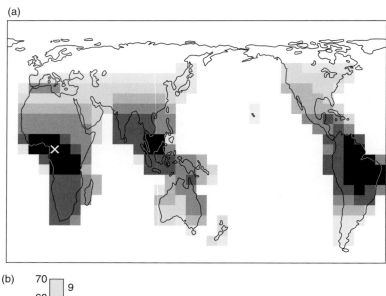

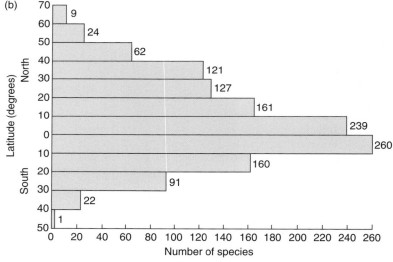

Figure 9.8 (a) Regional variation in the number of termite genera (represented by a logarithmic gray scale from a minimum (light gray) to a maximum (black with a cross); white, no data). (From Gaston &Williams 1996; data from Eggleton et al. 1994.) (b) Latitudinal gradients in the number of swallowtail butterfly species. (From Sutton & Collins 1991.)

above sea level). Along the latter transect the abundance of dung beetles fluctuated and did not show a marked drop in abundance until above 1700 m when abundance dropped sharply. In contrast, species richness declined steadily from 15–20 species at low altitudes to 5–10 species at around 800 m to less than five species above 1150 m (Figure 9.12). Species richness did not drop so sharply along the 300–1150 m transect in Sulawesi, even allowing for the fact that it reached a much lower altitude than the Sarawak transect: there were more than 10 species per site at 1150 m in Sulawesi. The difference was

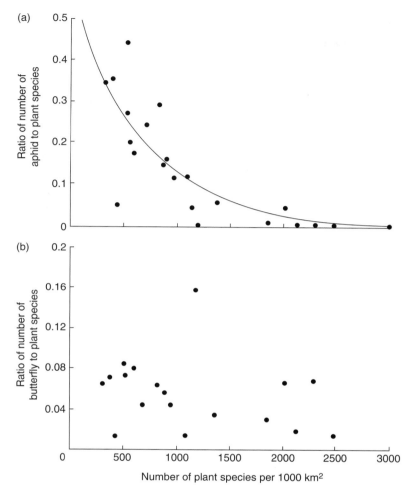

Figure 9.9 The relationship between the ratio of the numbers of (a) aphid and (b) butterfly species to plant species and the number of plant species in a range of countries. (From Dixon et al. 1987, with permission from University of Chicago Press.)

attributed to the effect of the size of the different study areas: the Sulawesi study was conducted in a much larger mountain range than the Sarawak study (see Section 9.3.3 below). Species composition as well as species richness changes along altitudinal gradients. Table 9.6 shows a few examples of the abundance of different dung beetle species at different altitudes. Research on another group of insects, the geometrid moths, also revealed a tendency toward maximum diversity at middle elevations (Brehm et al. 2007).

There are other large-scale patterns in biodiversity apart from the latitudinal variation discussed above. There is a surprising amount of variation, for example, from biogeographic region to region in addition to latitudinal variation. In general, biodiversity in tropical regions is greatest in the Neotropics, least in the Afrotropics and intermediate in the Indotropics (Gaston & Williams 1996). For example, the diversity of butterflies is particularly high in the Neotropics (Figure 9.13). Not all butterfly families, however, show this pattern. Swallowtails, for example, are most

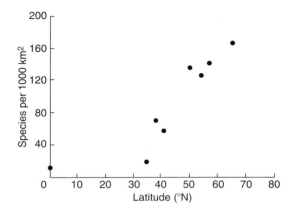

Figure 9.10 Latitudinal variation in the number of species of sawflies. (From Kouki et al. 1994.)

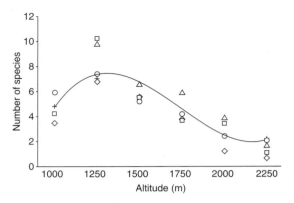

Figure 9.11 Influence of altitude on dung beetle diversity: number of species per trap at different elevations in five regions (each represented by a different symbol) in the Columbian Andes. The fitted line is based on data from all sites. (From Escobar et al. 2005.)

species-rich in East Asia and Australia (Sutton & Collins 1991) (Figure 9.14).

Another type of large-scale pattern is the presence of biodiversity "hotspots". The concept of hotspots or concentrations of biodiversity was developed by Myers (1988, 1990, 2003; Myers et al. 2000) who originally suggested that as many as 20% of plant species may be confined to 0.5% of the land surface of the Earth. He identified 18 biodiversity hotspots – concentrations of endemism as well as species richness, and called them hotspots because of the threats implicit in biodiversity being concentrated in relatively small areas. These areas included Hawaii, western Ecuador, southwestern Côte D'Ivoire, Madagascar, Sri Lanka, and northern Borneo.

Myers et al. (2000) revised the list of hotspots to 25, many of them much larger than the original list (Figure 9.15). Myers' original list was based on vascular plant diversity but the updated list is based on the endemism and degree of threat to plants, mammals, birds, reptiles, and amphibians. Unfortunately, the degree to which insect diversity fits the hotspot concept is unclear: unsurprisingly, suitable data are available only for some butterflies. Figure 9.16 compares areas of endemism of butterflies, birds, reptiles, and amphibians in Central America, showing a large degree of coincidence (Bibby et al. 1992).

Several ecologists have, however, questioned the concept that hotspots hold concentrations of diversity of different taxa (Prendergast et al. 1993; Williams et al. 1994; Prendergast 1997). The conclusion of these studies is that species-rich areas of different taxa tend not to coincide. Perhaps, as Lovejoy et al. (1997) points out, this is not very surprising given the diverse ecological requirements of different taxa. However, it may be that the hotspot hypothesis applies much better to the tropics and subtropics than to temperate and boreal areas (Myers 1997) and/or that concentrations of different taxa coincide at some scales but not at others. Coincident hotspots of different taxa are probably more apparent at larger spatial scales but less likely to occur at smaller scales, where natural and anthropogenic factors lead to different spatial patterns for different groups of plants and animals.

To a large extent, the hotspot concept is about the conservation of biodiversity and, therefore, we will return to it in the next chapter. The rigor of the concept has been questioned but it has served to focus minds on some major challenges in conserving biodiversity. The hotspot concept was developed from knowledge of plant diversity and has gained support from observations and from analysis of the causes of spatial patterns in plant diversity, particularly climate (Kleidon & Mooney 2000). Meanwhile research on organisms such as protozoans suggests that very small species have a cosmopolitan distribution (Fenchel & Finlay 2004, 2006). Once again, we have been shown how little we know about biodiversity, particularly insect diversity.

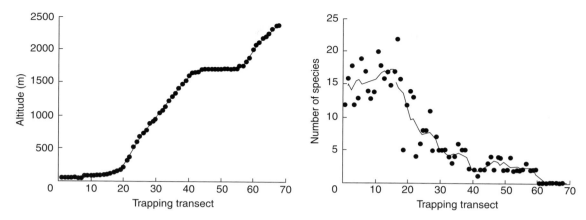

Figure 9.12 Altitudinal variation in the number of species of dung beetles in the Gunung Mulu National Park, Sarawak, Borneo, with the altitude of the trapping stations along the transect shown. (From Hanski & Niemelä 1990.)

Table 9.6 Altitudinal distribution of six common dung and carrion beetles in the Dumoga-Bone Reserve and on Gunung Muajat, Sulawesi, Indonesia. (After Hanski & Niemelä 1990.)

Altitude (m a.s.l.)	Number of individual beetles per trap					
	Onthophagus aereomaculatus	*Copris macacus*	*Onthophagus aper*	*Onthophagus* sp.n. (1)	*Copris* sp.n.	*Onthophagus* sp.n. (2)
200	2	7	22			
300	5	8	40			
350	8	11	43			
400	3	22	37			
450	8	16	60			
500	3	38	26			
550	1	27	10			
600		19	27	1		
650		16	11	1		
700		18	20			
750		7	11			
800		6	12			
850			5			
900			5			
950		1	2	4	1	
1000			3	1		
1050						
1100		1	1	2	1	
1150				3	3	
1300			1		1	
1600				2		2
1700				19		13
1750				26		30

sp.n., new species; m a.s.l., metres above sea level.

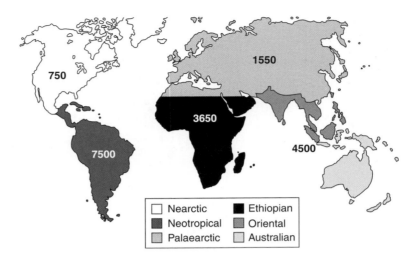

Figure 9.13 The number of butterfly species in different biogeographic realms. (From Robbins & Opler 1997.)

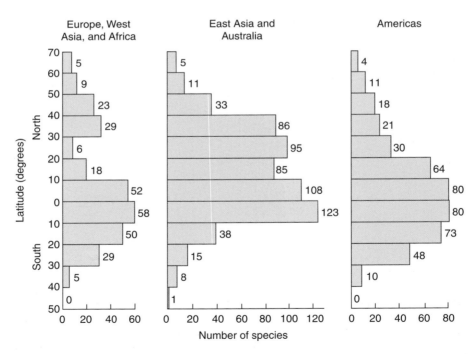

Figure 9.14 Latitudinal variation in the number of species of swallowtail butterflies in: Europe, West Asia, and Africa; East Asia and Australia; and the Americas. (From Sutton & Collins 1991.)

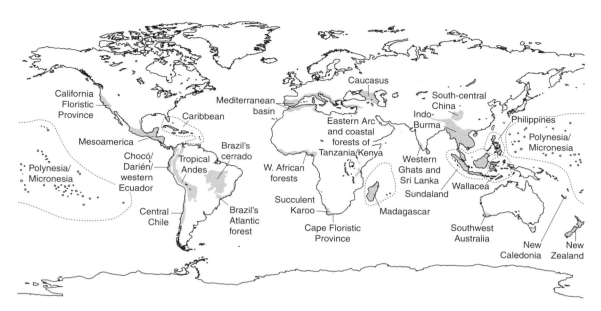

Figure 9.15 Twenty-five "biodiversity hotspots" selected on the basis of endemism and degree of threat. Each hotspot has at least 1500 endemic vascular plants (of the world total of 300,000) and has lost 70% or more of its primary vegetation. (From Myers et al. 2000.)

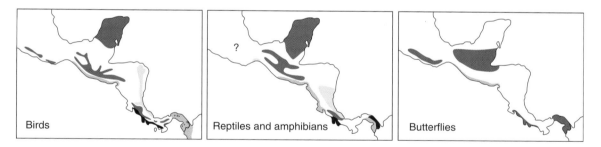

Figure 9.16 Areas of high endemism of birds, reptiles and amphibians, and butterflies in Central America. (From Bibby et al. 1992.)

9.3.3 Species–area relationships

One aspect of natural variation in biodiversity has long been of interest to ecologists – the relationship between the number of species in an area and the size of that area. So important is this relationship that it is often referred to as one of the few general rules, or laws, in ecology (Lawton 1999; Lomolino 2000; see

also Chapter 1). This rule, that the number of species inhabiting an area increases as the size of the area increases, was fundamental to the development of "island biogeography" (MacArthur & Wilson 1967) and has been applied to many "island" situations from the number of insect species on different plant species (Leather 1986) to the number of insects in urban roundabouts (Helden & Leather 2004). The

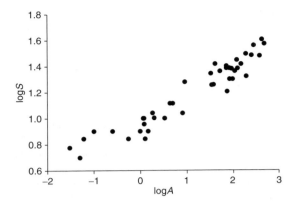

Figure 9.17 Relationship between the number of isopod species (*S*) in the Mediterranean Aegean islands and the size of these islands (*A*). (From Gentile & Argano 2005.)

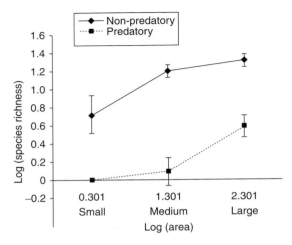

Figure 9.18 Numbers of non-predatory (oribatid mites and collembolans) and predatory (mesostigmatid mites) microarthropods in the humus soil of differently sized forest habitat fragments. (From Rantalainen et al. 2005.)

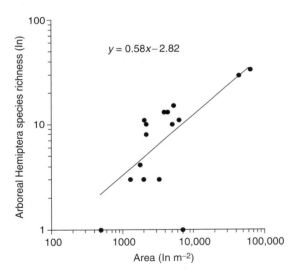

Figure 9.19 Numbers of tree-dwelling (arboreal) Hemiptera in urban road roundabouts of different sizes in Bracknell, England. (From Helden & Leather 2004.)

species–area relationship has also been important in estimating the impact of habitat loss and other pressures on biodiversity and is therefore a potentially key ecological tool for conservation (see Chapter 10).

Many studies show that the number of species on islands increases as island size increases, whether the study focuses on, for example, oceanic islands (Gentile & Argano 2005) (Figure 9.17), soil habitat fragments (Rantalainen et al. 2005) (Figure 9.18), or urban roundabouts (Helden & Leather 2004) (Figure 9.19).

There are two main theories to explain the species–area relationship: the equilibrium theory and the habitat diversity theory. The equilibrium theory (MacArthur & Wilson, 1976) is that the number of species on an island is a consequence of the balance between extinction and colonization by new species. Extinction and colonization rates are affected by the number of species on the island, the size of the island, and the distance of the island from other islands and the mainland. Although species are constantly colonizing the island and becoming extinct, an equilibrium is eventually reached which is directly related to island size and inversely related to the distance from other sources of species. Thus if the number of species on different islands is plotted, we see the types of examples shown in Figures 19.17–19.19. The species–area relationship can also be explained by the habitat diversity theory (Lack 1969, 1976): because larger islands tend to contain more habitats they will tend to contain more species.

These theories are not mutually exclusive and there is evidence to support both the effect of habitat diversity and the "colonization–extinction" equilibrium. Koh et al. (2002) showed that the number of springtail

(Collembola) and butterfly species on 17 Singapore islands increased as a function of island size (see Figure 9.17). In contrast, Torres and Snelling (1997), in a study on the numbers of ant species on Puerto Rico and 44 surrounding islands, found that island size had little influence on the numbers of ants. They described the situation as a ''non-equilibrium case'', with the number of colonizations exceeding the number of extinctions over a period of 18 years. They found that, of the factors studied (island size, isolation, etc.), habitat diversity had the greatest effect on the number of ant species on an island. One other situation where, unsurprisingly, the number of colonizations was shown to exceed the number of extinctions is the number of butterfly species on Krakatau (Figure 9.20). In the case of the ants of Puerto Rico, there is evidence that human occupation, disturbance, and movement between islands has led to an increase in both habitat diversity and colonization by ants. Torres and Snelling (1997) also suggest that there has been too much emphasis in island biogeography on the numbers of species rather than their identity. They point out that colonizations of islands by several ant species, such as the fire-ant *Solenopsis wagneri* (Hymenoptera: Formicidae), can have a disproportionate effect; these aggressive species cause the extinction of many other species.

Some studies have shown that both area per se (supporting the equilibrium theory) and habitat diversity have an influence on the numbers of insect species (and other taxa). Peck et al. (1999), for example, showed that the numbers of insects on different Hawaiian islands was related to island size and habitat diversity, using island elevation as a surrogate for the latter. Ricklefs and Lovette (1999) used a more direct measure of habitat diversity, measuring the total area of five vegetation types on each of 19 islands in the Lesser Antilles and applying Simpson's index to these data. The islands varied in size from 13 to 1510 km^2. Ricklefs and Lovette considered the species richness of birds, bats, butterflies, reptiles, and amphibians. In most cases simple correlations of species richness with island area or with habitat diversity were significant. However, multiple regressions to investigate the effects on species richness of area and habitat diversity together showed that area had a significant effect on birds and bats, and habitat diversity had a significant effect on birds, butterflies, reptiles, and amphibians. Thus Ricklefs and Lovette argue the relationship between the species richness of butterflies, reptiles, and amphibians and island area is probably a consequence of the relationship between habitat diversity and island area. In contrast, bird species richness appears to respond independently to both habitat diversity and area, and bat diversity is influenced only by island area. Interestingly, they also found that their measure of habitat diversity correlated with island elevation, providing support for the approach used by Peck et al. (1999).

Other studies have exposed differences between different taxonomic groups in their response to island size. Even within single taxa, different relationships between species richness and island area may exist. Steffan-Dewenter and Tscharntke (2000) studied the diversity of butterflies on calcareous grasslands ranging in size from 300 to 76000 m^2. They found that both plant and butterfly species richness increased with area (Figure 9.21). They also found when classified according to food plant specialization, the strongest relationship between diversity and habitat area occurred among monophagous species and the weakest among polyphagous species.

As some of the examples discussed above have shown, species–area relationships have been widely studied in habitat islands, such as grasslands, as well as ''true'' oceanic islands. Another example of the application of the species–area relationship has been its use in the study of biodiversity in forest fragments, notably in the Biological Dynamics of Forest Fragments Project in Brazil (Laurance et al. 2002). Figure 9.22 shows some results of this project: the numbers of species in different butterfly families in reserves ranging in size from 1 to 1000 ha. The numbers of species in some families respond to

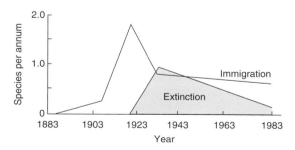

Figure 9.20 Annual immigration and extinction curves for butterflies on Krakatau (Krakatoa) 1883–1989. (From Bush & Whittaker 1991.)

(a)

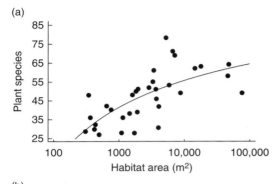

(b)

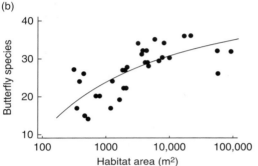

Figure 9.21 Relationships between the numbers of species of plants and butterflies in grassland habitats of different sizes. (From Steffan-Dewenter & Tscharntke 2000.)

fragment area, others do not. Lovei et al. (2006) studied the diversity of ground beetles in forest fragments. Their study is particularly important in separately considering species that are forest specialists, generalist species that can survive in both the forest and the surrounding landscape, and species that can be classified as forest edge specialists. The number of forest specialist species correlated positively with the size of the forest fragment. The number of generalist species, however, was marginally negatively related to forest patch size. Furthermore, the number of species that preferred the forest edge was correlated with the edge : area ratio. We will further discuss the application of species–area relationships to forests, and the general issue of using species–area relationships to predict the consequence of habitat loss, in the next chapter because of its importance in conservation.

For phytophagous insects, species–area relationships have been used to try to explain the number of insect species feeding on different plant species. Many studies have shown that the area occupied by a particular plant species is the most important determinant of the number of insect species that it supports (Janzen 1968; Strong et al. 1984) (Figure 9.23). Several of the studies on the influence of area on phytophagous insect species richness have focused on insects feeding on British trees, using data on the number of 10×10 km or 2×2 km squares occupied by each tree. Analysis based on the 2×2 km data explained 56% of the variation in the number of insect species on a tree (Kennedy & Southwood 1984). Interestingly, however, an analysis based on the most accurate data available on the area covered by each tree species, a thorough census (measuring the number of hectares occupied by each tree) carried out by the Britsh Forestry Commission, resulted in the explanation of only 17% of the variation in the number of insect species on a tree (Claridge & Evans 1990). This suggests that the number of insect species on a tree is influenced by the diversity of habitats occupied by, as well as total area covered by, a tree species: the 2×2 km data will reflect habitat diversity better than the census (hectare) data.

Claridge and Evans also obtained a better explanation of the number of species on a tree by separately considering the mainly native broad-leaved and the mainly non-native coniferous trees, particularly the latter (22% and 79%, respectively). It is worth recalling that Southwood (1961) concluded that the number of insects feeding on a tree species was determined by the present and past abundance of the tree. Data on the present abundance of trees may not accurately reflect the abundance of these species in the past, particularly in heavily deforested countries like the British Isles. However, data on recently introduced species, such as most of the conifers in the British Isles, provide a better test of Southwood's hypothesis, with notable success, despite the small number of tree species.

There have been other studies of the relationship between the number of insect species on plants, including several which focused attention on particular feeding guilds such as leaf-miners (e.g. Claridge & Wilson 1982). In these cases, and in general, the area occupied by a host plant was found to be a better predictor of the number of insect species when the studies were restricted to groups of taxonomically related host plants, such as Californian oaks (Opler 1974) and British Rosaceae (Leather 1986) than when the studies included taxonomically diverse plant species.

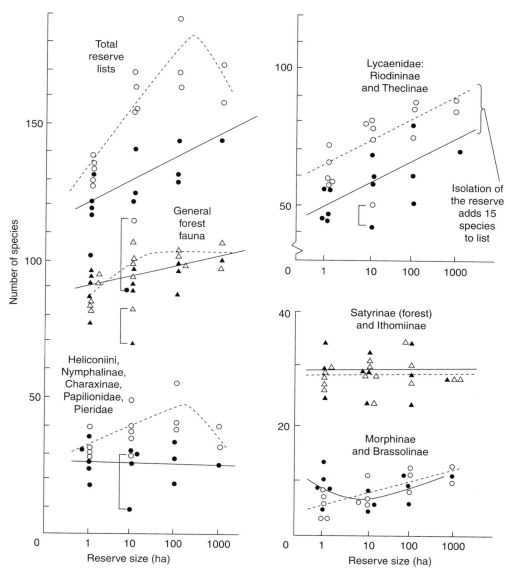

Figure 9.22 Relationships between number of butterfly species and reserve (forest fragment) size in the 'Biological Dynamics of Forest Fragments' project near Manaus in the Brazilian Amazon: the total number of butterfly species is given in the top left and separate groups are shown elsewhere. Isolated sites are shown as open symbols and broken lines; sites surrounded by forest are shown as solid symbols and lines. (Linked symbols in most graphs show the numbers of species in a 10 ha reserve before and after isolation.) Note (i) that isolation from surrounding forest results in a marked loss of species in some groups (e.g. Lycaenidae, top right); and (ii) the species richness of some groups increases with reserve area but other groups, notably the shade-loving Satyridae and Ithoniinae and general (widespread) species (center left) are insensitive to reserve size. (From Brown 1991.)

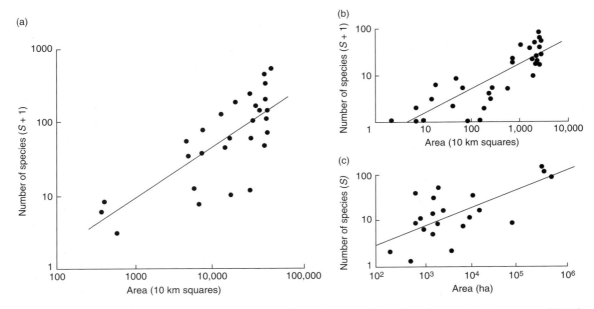

Figure 9.23 Relationships between numbers of species and host plant range for (a) phytophagous insects on genera of British trees; (b) phytophagous insects on British perennial herbs; and (c) insect pests on cacao (each point is a different country). (From Strong 1974; Lawton & Schroder 1977; Strong et al. 1984.)

The study on British Rosaceae mentioned above is a good example of the influence of plant architecture on the number of insect species feeding on it. Among the Rosaceae, trees (on average 53 insect species on native host species) have more species than shrubs (25 species), which have more species than herbaceous plants (eight species), even when the area occupied by each species is taken into account (Leather 1986). Thus, the more complex the structure of the plant, the more insect species feed on it, presumably because of the increasing number of niches available on more architecturally complex plants (see also Figure 9.23).

9.3.4 Insects and their habitats

In addition to the patterns of biodiversity discussed above there are, of course, marked differences in insect diversity between different habitats and ecosystems. Indeed, many of the patterns in biodiversity break down when we look at relatively small scales, often because of the influence of habitat on biodiversity.

Biodiversity is influenced by the successional stage of a habitat. Hilt and Fiedler (2005), for example, found that the abundance and diversity of arctiid moths were highest at advanced forest succession sites in southern Ecuador, followed by early succession sites, and were lowest in the mature forest (Figure 9.24a). The proportion of rare species showed the opposite pattern (Figure 9.24b). Ground beetle communities and those specializing on dead wood also respond to forest succession (Niemelä et al. 1993; Heyborne et al. 2003; Jacobs et al. 2007) (Figure 9.25). Changes in the richness and composition of insect communities during different successional stages can be found in many other situations including freshwater communities (e.g. Vieira et al. 2004) and wildlife carcasses (e.g. Watson & Carlton 2005; De Jong & Hoback 2006).

It would be impossible to discuss the diversity of insects in the enormous number of different habitats in the world. One area is worth particular mention if only because of its exceptional richness: the insect communities of tropical forest habitats. Unsurprisingly, our knowledge of the ecology of most tropical forest

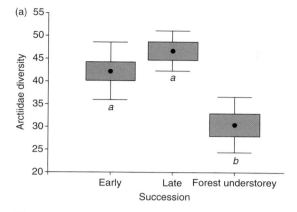

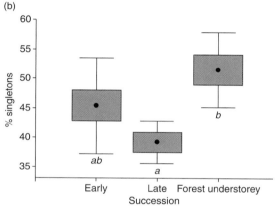

Figure 9.24 (a) Diversity of arctiid moths (measured by Fisher's alpha), and (b) proportion of singletons (species where only one individual was recorded locally, shown as a percentage of the species present) in forests in three different classes of succession habitat. Boxes with different letters indicate significant differences between habitats ($P < 0.05$, Scheffé test); means (black dots), $\pm$ SE (shaded boxes), and $\pm$ SD (bars) shown. (From Hilt & Fiedler 2005.)

insects is much poorer than that of temperate species. The ecology of a few tropical forest insect species is well known, notably that of ant species such as leaf-cutting ants (*Atta* spp.), army ants (e.g. *Eciton* spp), and the weaver ants (*Oecophylla* spp.) (all Hymenoptera: Formicidae) (Hölldobler & Wilson 1990). However, recent developments in forest canopy access methods have led to an increase in the study of tropical insect communities (Lowman & Nadkarni 1995; Stork et al. 1997; Basset et al. 2003; Dial et al. 2006; Grimbacher & Stork 2007).

Stork (1991), for example, surveyed the arthropod community of 10 forest trees in Borneo. Almost 24,000 individuals were sampled by insecticide knockdown fogging, a technique now widely used to sample the arthropod communities of the canopies of temperate and tropical forests (Stork et al. 1997). Stork (1991) found that the most abundant groups from the Borneo trees were Hymenoptera other than ants (27%), Diptera (22%), ants (18%), beetles (17%), and Hemiptera (11%) (Figure 9.26). Of the Hymenoptera, two-thirds were ants (18% of the total sample). Studies elsewhere in the tropics have shown that ants can be even more common. In Cameroon, for example, Watt et al. (1997a) found that ants made up 63% of the total canopy arthropods (Figure 9.26). Studies in Brazil, Indonesia, and elsewhere have shown similar levels of ant abundance (Stork et al. 1997).

The average number of arthropods in the Borneo study was 117 per square meter (in a range of 51 to 218 per m^2). (That is, the average numbers of arthropods caught by knockdown fogging in each $1\,m^2$ tray was 117.) Other studies of tropical canopy arthropods have produced similar results; for example, Marques et al. (2001) recorded 44 arthropods/m^2, but Watt et al. (2002) recorded 213 ants/m^2 in Cameroon. Surprisingly, results from studies of arthropods in temperate forests show even higher densities. Southwood et al. (1982) recorded 389 arthropods/m^2 on trees in England and South Africa, and Ozanne et al. (1997) recorded 1000–10,000 arthropods/m^2 in a transect across a Norway spruce forest. Floren et al. (2002) suggest that the high abundance of ants in tropical forest canopies is the cause of the low abundance of other arthropods.

What is most remarkable about tropical forest canopy arthropods is their species richness. Stork recorded over 3000 arthropod species in the canopy of 10 study trees in Borneo (excluding Psocoptera, Acari, and some other groups that were not identified to species). Hymenoptera were the most speciose order (23% of species), followed by Diptera (22%), beetles (21%), and Hemiptera (8%) (Figure 9.27). Ants made up only 3% of the total number of species. Studies elsewhere in the tropics also show that ants, in comparison to beetles for example, are numerically dominant but not proportionally rich in species. In contrast, the arthropod community of temperate trees is less species-rich. For example, Moran and Southwood (1982) recorded 465 arthropod species on *Quercus robur* in southern England (and fewer on

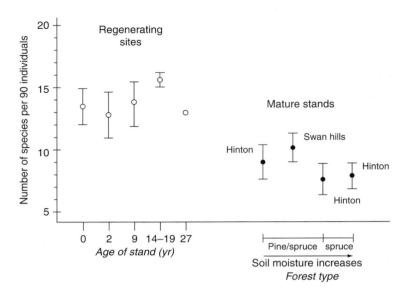

Figure 9.25 Number of species of ground beetle in Canadian forests estimated by rarefaction (per 90 individuals except for the total number of species recorded in the 27-year-old sites). Regenerating sites are shown as years since clear-felling. (From Niemelä et al. 1993.)

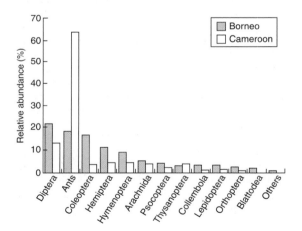
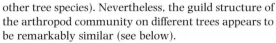

Figure 9.26 Relative abundance of different groups of canopy-dwelling arthropods in Borneo and Cameroon. (Data from Stork 1991; Watt et al. 1997b.)

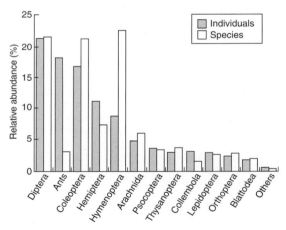

Figure 9.27 Relative abundance and number of species in different arthropod groups collected from tree canopies in Borneo. (From Stork 1991.)

other tree species). Nevertheless, the guild structure of the arthropod community on different trees appears to be remarkably similar (see below).

Until recently, most of the attention on the insects in tropical forest canopies focused on the trees themselves.

Ellwood and Foster (2004), however, showed that forest epiphytes also contain large numbers of insects. They studied the abundance of insects and other invertebrates in the bird's nest fern *Asplenium nidus* growing on dipterocarp rainforest in Borneo and found that the

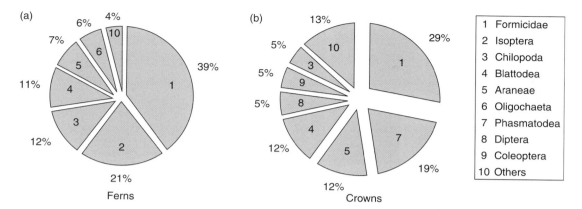

Figure 9.28 Percentage contribution of individual taxa to overall invertebrate biomass in (a) bird's nest fern (*Asplenium nidus*), and (b) the crowns of a lowland dipterocarp rainforest tree (*Parashorea tomentella*) in Borneo. (a) Percentage contributions of total invertebrate biomass in five ferns (22–26 kg dry weight; total invertebrate biomass, 439 g). (b) Percentage contributions of total invertebrate biomass in five tree crowns of (total invertebrate biomass, 21 g). (From Ellwood & Foster 2004.)

biomass of invertebrates in these ferns was as great as that found on the trees, doubling the estimate of invertebrate biomass: 2026 ± 373 g/ha on the trees crowns and 3776 ± 275 g/ha within the ferns. They also found that the ferns contained different proportions of the various invertebrate taxa present (Figure 9.28). These ferns contain large amounts of suspended soil and plant material and not only do they host large numbers of ants but they can also contain large numbers of termites: in 11 ferns studied, four contained a nest of *Hospitalitermes rufus* (Nasutitermitinae) and one contained a nest of an undescribed species of *Hospitalitermes* (Ellwood et al. 2002).

Another overlooked component of tropical forests is their lianas, despite the fact that their presence in tropical forests is one of the main features distinguishing them from temperate forests. Ødegaard (2000b), however, demonstrated how important they are, at least for phytophagous beetles. Using a crane to study these insects in dry tropical forest in Panama, he identified the hosts of 2561 individuals from 697 species. He found that lianas were as important as trees with a total of about 56 species associated with each tree species and 47 associated with each liana. However, estimates of specificity suggested that among foliage-feeders the lianas were more important (Figure 9.29).

Although the canopy of tropical forests, including its epiphytes and lianas, is undoubtedly rich in insect species, other parts of tropical forests hold large

numbers of species too. In a study of the Indonesian island of Seram, for example, Stork and Brendell (1993) measured the abundance of arthropods in the forest canopy, on tree trunks, on herb-layer vegetation, in the leaf litter, and in the soil. They collected over 32,000 individual insects and other arthropods and expressed the numbers caught as numbers per square meter (of ground). An average of 1201 arthropods/m^2 were recorded in the forest canopy, a much higher figure than recorded in the study discussed above on the nearby island of Borneo. However, almost twice that number of arthropods, 2372/m^2, were recorded in the soil. Only 12 arthropods/m^2 were sampled from herb-layer vegetation and 52 arthropods/m^2 from the tree trunks, but 602 arthropods/m^2 were found in the litter layer. Thus over twice as many arthropods were found in the soil and litter (70%) than in the canopy (28%). The canopy arthropods were dominated by ants (48%), but the most abundant arthropods in the soil and litter were Collembola (60% and 70%, respectively) and Acari (25% and 16%, respectively). Other studies of the soil and litter communities of tropical forests have shown them to be numerically dominated by termites (Eggleton et al. 1996, 2002; Jones et al. 2003).

Another aspect of the distribution of insects in tropical forests is the number and abundance of species associated with understory vegetation and different levels in the tree canopy. Schulze et al. (2001), for example, studied the vertical distribution

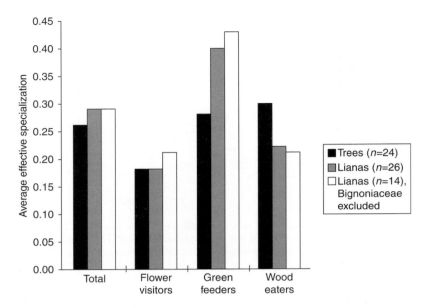

Figure 9.29 Host specificity of phytophagous beetle species associated with trees and lianas (and lianas excluding Bignonia-ceae) in a dry tropical forest in Panama as measured by the effective specialization index. Flower visitors, green leaf-feeders, and wood-eating species are shown separately, showing, for example, that green feeders of Bignoniaceae are less specialized than those feeding on other lianas. (From Ødegaard 2000b.)

of Lepidoptera in a forest in Borneo. They found that the abundance of fruit-feeding nymphalids decreased towards the canopy. However, hawkmoths and nectar-feeding butterflies and moths became more abundant in the upper canopy. Schulze and his coworkers suggested that these patterns were partly due to decreasing amounts of rotting fruit and increasing amounts of nectar resources in the upper canopy and partly because of the greater abundance of insectivorous birds in canopy. They also suggested that the latter might explain the tendency of stonger flying nymphalids to occur in the upper canopy. This study demonstrated that studies on Lepidoptera based only on sampling at ground level could provide misleading results, a conclusion confirmed by research on the impact of logging on forest butterflies (Dumbrell & Hill 2005; see also Chapter 10).

The biodiversity of tropical forests is remarkable but what are all these species doing? This question was also posed by Moran and Southwood (1982) for the insect arthropod communities of British and South African trees. Moran and Southwood were the first to analyze the feeding guild (or trophic)

structure of the arthropod fauna of trees. They classified the arthropods found in samples from six species of trees in each country into phytophages (herbivores: chewers and sap-suckers), epiphyte fauna, scavengers (of dead wood, etc.), predators, parasitoids, ants, and tourists. "Tourists" were defined as "non-predatory species which have no intimate or lasting association with the plant but which may be attracted to trees for shelter and sustenance (honey-dew and other substances), or as a site for sun-basking and sexual display." Moran and Southwood found that the ratio of different feeding guilds did not vary greatly either between tree species or between countries. Phytophages were the most abundant guild, making up about 68% of the individuals. They were also the most speciose guild: 25% of all species were phytophagous arthropods. Parasitoids, which comprised 3% of the total number of individuals, were the second most speciose guild (25%). The full guild structure is shown in Figure 9.30a.

Stork (1987) also carried out an analysis of the guild structure of the 3000 species recorded from the canopy of trees in Borneo (Figure 9.30b). As was

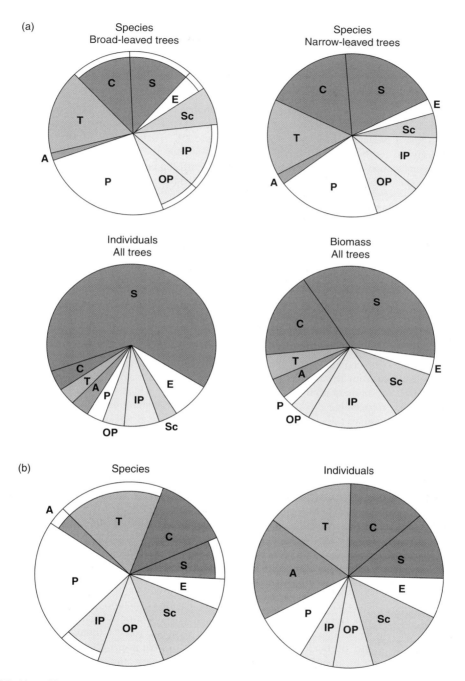

Figure 9.30 (a) Guild structure (in terms of numbers of species on broad-leaved and narrow-leaved trees, individuals, and biomass) of the arthropod fauna of South African and British trees. Segments with double external lines indicate guilds with constant proportions of species from tree to tree among broad-leaved species. (b) Guild structure (in terms of numbers of species and numbers of individuals) of the arthropod fauna of 10 Borneo trees. S, sap suckers; C, chewers; T. tourists; A, ants; P. parasitoids; IP, insect predators; OP, other predators; Sc, scavengers and fungivores; E, epiphyte fauna. Segments with double external lines indicate guilds with constant proportions of species from tree to tree. (From Moran & Southwood 1982.) (From Stork 1987.)

found in the studies on temperate trees, the most abundant group were phytophages (26%) but they were slightly less species-rich (22%) than parasitoids (25%). The abundance of parasitoids in the tropical forest study was similar to that of the temperate study. Indeed, overall, the relative abundance of different guilds in the temperate and tropical studies was surprisingly similar. There were relatively fewer phytophages, less insect predators, and more tourists in the tropical study. In terms of species, there were less phytophagous arthropods, more scavengers, and more ants in the tropical study. There were no other significant differences between the temperate and tropical studies (Stork 1987).

Didham et al. (1998) extended this approach to consider the impact of fragmentation on the guild structure of insect communities. Focusing on the beetle communities in an experimentally fragmented tropical forest landscape in central Amazonia, they identified 993 species from $920\,m^2$ of leaf litter at 46 sites. Beetle abundance increased significantly towards the edge of the forest fragments studied and although total species richness did not vary between the edge and the center of the fragments, the proportions of beetles in different feeding guilds changed significantly with distance from the forest fragment edge. When the impact of fragmentation on different guilds was considered separately, predators were found to be the most vulnerable.

Most of this section has focused on insects found in tropical forest habitats. The enormous diversity of insects found in the tropical forests and the fact that they are particularly threatened (see Chapter 10) means that tropical forest insects are especially interesting. However, insects are found throughout the world with adaptations that both ensure their survival in these places but also leave them vulnerable to habitat loss and degradation.

9.3.5 Biodiversity and its value

Before leaving the topic of biodiversity itself and, in the next chapter, turning to its conservation, it is worth considering the value of biodiversity and, thereby, the major arguments in support of its conservation. Biodiversity, of course, can be considered to have intrinsic value. Intrinsic value derives from ethical, esthetic, religious, philosophical, and cultural perspectives, many of which have been put forward in support of the conservation of biodiversity (Jepson & Canney 2003). Interestingly, quantitative studies have shown that 70–90% of the population in Western countries recognize the intrinsic right of biodiversity to exist (Van Den Born et al. 2001). Research in other countries suggests that both use (see below) and intrinsic values are important in defining people's attitudes, particularly locally perceived benefits (Bauer 2003).

With the notable exception of butterflies, however, esthetic arguments are rarely used in support of insect conservation, although the advance in microphotography has brought an appreciation of the esthetic aspects of many insects to a much wider audience (Thompson 2003). Other aspects of intrinsic value are even more rarely applied to insects, although it was the entomologist Edward O. Wilson who coined the word biophilia to describe the innate human appreciation of living things (Kellert & Wilson 1995).

Intrinsic values are often referred to as nonutilitarian values, contrasting with utilitarian, or economic, values of biodiversity. Much more attention has been placed on utilitarian values, which can be divided into non-use values and use values (to provide total economic value). Non-use values comprise existence values, which can be quantified as the amount people are willing to pay for the conservation of biodiversity (e.g. Wilson & Tisdell 2005; Christie et al. 2006). Use values are more tangible but, as with all values attributable to biodiversity, not necessarily easy to quantify (Turner et al. 2003).

A framework for considering the value of biodiversity is provided by the Millenium Ecosystem Assessment (2005). The Millenium Ecosystem Assessment, or MA, focused on the various ecosystem services provided by biodiversity and the roles of these services in human well-being. The MA framework (Figure 9.31) lists four types of ecosystem services: provisioning services such as food and fuel, supporting services such as primary production, regulating services such as pest and disease regulation; and cultural services. Clearly, biodiversity is responsible for these services and insects play a major role of them, even in providing food through edible insect species (Ramos-Elorduy 1997). The MA framework considers intrinsic values, classifying these as cultural ecosystem services, including esthetic, spiritual, educational, and recreational services. The strength of this framework is that it shows

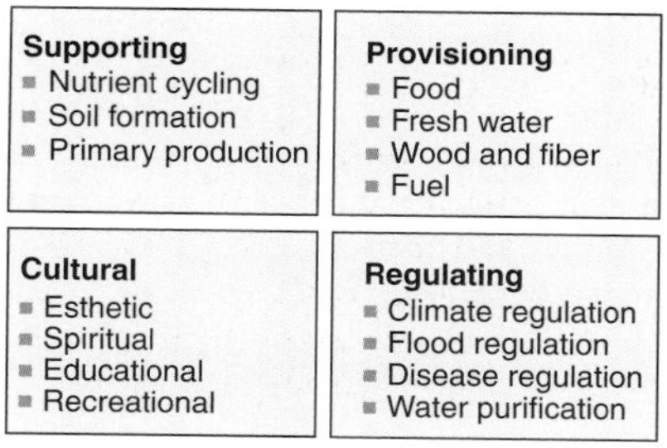

Supporting	Provisioning
■ Nutrient cycling ■ Soil formation ■ Primary production	■ Food ■ Fresh water ■ Wood and fiber ■ Fuel
Cultural	**Regulating**
■ Esthetic ■ Spiritual ■ Educational ■ Recreational	■ Climate regulation ■ Flood regulation ■ Disease regulation ■ Water purification

Figure 9.31 Categories of ecosystem services. (From Millenium Ecosystem Assessment 2005.)

the value of biodiversity, through ecosystem services, very clearly. The major constituents of human well-being – security, basic materials for life, health, and good social relations – are influenced, and in many cases provided, by biodiversity through ecosystem services. Indeed, such is the scale of the contribution of biodiversity to human societies that many people have attempted to put financial estimates on it. Pimentel et al. (1977), for example, estimated the economic benefit of biodiversity in the USA at \$300 billion. For further information on ecological and environmental economics see, for example, Pearce and Moran (1994), Constanza et al. (1997), Edwards-Jones et al. (2000), Hanley et al. (2001), and Common and Stagl (2005).

The question often arises, however: how much biodiversity is needed to provide these essential services? One important consideration is that biodiversity may provide novel products and other services in the future. These relate to what the economists call option values as distinct from currently provided direct and indirect use values. Option values most notably derive from genetic diversity, particularly the genetic material that might one day be incorporated into crops and domestic animals. Another well-known aspect of the option value of biodiversity derives from the enormous numbers of chemicals with medicinal and other valuable properties that are waiting to be discovered by "bioprospecting". Most bioprospecting focuses on plants, particularly where ethnobotanical information has identified candidate species (Martin 1995), although the current bioprospecting program of INBio (the Costa Rican Institute for Biodiversity) does include insects (R. Gamez, personal

communication). To date, however, bioprospecting appears not to have lived up to its promise (Firn 2003), and has sometimes been called "biopiracy" (Hamilton 2004) where, unlike the INBio example, potentially profitable plant material is collected in one country and processed in another.

From an entomological perspective, perhaps the most important direct option value derives from potential, future biological control agents. Amongst the many examples to date, where one or a few natural enemies have been introduced to control an insect pest, is control of the cassava mealybug *Phenacoccus manihoti* (Homoptera: Pseudococcidae) in Africa. After being free of significant pests for about 300 years after the introduction of cassava from South America, the cassava mealybug and the cassava green spider mite became serious pests in the 1970s, causing major economic problems (Norgaard 1988). The parasitoid *Apoanagyrus* (*Epidinocarsis*) *lopezi* (Hymenoptera: Encyrtidae) was successfully introduced from South America to 26 African countries, causing a satisfactory control on most farms affected by this pest (Neuenschwander 2001). Other examples are discussed in Chapter 12. Although the benefit of this particular natural enemy species has been realized, no doubt there are many more whose services to pest management will be required in the future.

The services currently delivered directly and indirectly by biodiversity, including insect decomposers, pollinators, and the natural enemies of pests is enormous. But do ecosystems services require so much diversity? Would ecosystems function in the same way with, for example, fewer decomposers, predators, and parasitic insects?

Many researchers have considered the effect of increasing biodiversity on ecosystem functioning (Lawton 1994; Johnson et al. 1996; Loreau et al. 2001; Cardinale et al. 2006). Different hypotheses are illustrated, shown in Figure 9.32, which show the effect of increasing species richness on the rate of an ecosystem process such as decomposition or primary productivity. First, the rivet hypothesis suggests that each species plays a significant role in affecting the ecosystem process: even a small decrease in biodiversity (similar to the loss of a single "rivet" from a complex machine; Lawton 1994) will result in a decrease in the rate of an ecosystem process (Figure 9.31b). Under this hypothesis, various forms of the

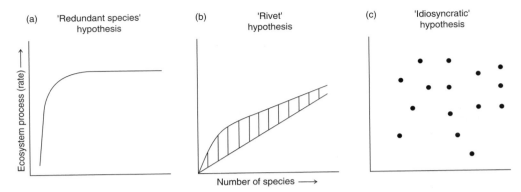

Figure 9.32 Three hypotheses for the relationship between ecosystem processes and the number of species present in an ecosystem. (From Lawton 1994.)

function between the ecosystem process and biodiversity are possible but all assume that each species has a unique contribution to that process. In contrast, the redundant species hypothesis suggests that ecosystem processes benefit from an increase in biodiversity up to a threshold level of biodiversity after which there is no increase in the rate of ecosystem processes: further species are redundant (Figure 9.31a). Thirdly, the idiosyncratic response hypothesis suggests that increasing biodiversity affects ecosystem function in an unpredictable way due to the complex and varied roles of individual species (Figure 9.31c).

Most research on the relationship between biodiversity and ecosystem functioning has been on plants. Several studies have demonstrated a positive impact of increasing species richness on plant productivity (e.g. Hector et al. 1999). It is, however, difficult to extrapolate these and other results to different trophic levels and ecosystem services: most studies have considered simplified ecosystems with single trophic levels, usually primary producers (Thebault & Loreau 2006). Furthermore, there are risks in applying the results of these studies in conservation (Srivastava & Vellend 2005). Nevertheless, research on the link between biodiversity and ecosystem functioning, despite showing that some ecosystem functions are insensitive to a degree of species loss, has highlighted the importance of individual, even rare, species (Hooper et al. 2005; Cardinale et al. 2006).

In one of the very few studies of insects, however, Klein (1989) examined the role of dung beetles

(Scarabaeinae) in the decomposition of dung in different habitats in central Amazonia – in continuous forest, forest fragments, and cleared pasture. He found that the rate of dung decomposition declined by about 60% from continuous forest to cleared pasture (Figure 9.33). The mean number of dung beetle species also declined about 80% from continuous forest to cleared pasture. However, the total abundance of dung beetles did not vary between different habitats. Taking this evidence together suggests that different dung beetle species play subtly different functional roles, and supports the rivet hypothesis (Figure 9.32b).

Perhaps because most of the research about the importance of biodiversity has focused on the effect of the number of species performing a particular function, such as the amount of plant biomass they produce, the unique role of individual species has often been overlooked. The unique ecological role of individual species is clearer for insects, however, particularly among the groups that we know most about such as the pollinators and insect natural enemies. The history of biological control, for example, gives ample evidence for the importance of individual species: there are many examples where the successful control of non-native pests requires the introduction of particular natural enemy species from the native ranges of these pests (see Chapter 12).

Clearly biodiversity matters and is worth conserving for its role in supplying essential ecosystem services (Balmford et al. 2002; Balvanera et al. 2006). In recent years the case for conserving biodiversity

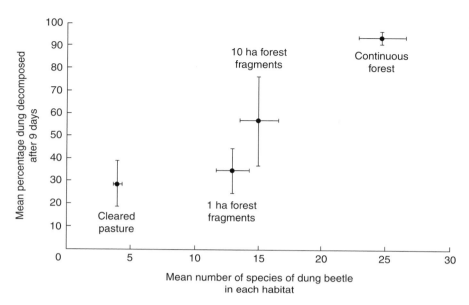

Figure 9.33 Impact of the number of dung beetle species on dung decomposition in Central Amazonia. (From Klein 1989; Didham et al. 1996.)

has increasingly focused on these and other utilitarian arguments: esthetic and ethical considerations, however, should not be ignored, particularly the contribution of biodiversity to the quality of life (Jepson & Canney 2003).

Whatever reasons may be put forward in support of insect conservation – and there are many arguments to consider – insects and other species are being threatened with extinction as never before. The next chapter considers the practice of insect conservation.

Chapter 10

INSECT
CONSERVATION

10.1 INTRODUCTION

To successfully conserve an insect, we need to know if it is threatened, why it is threatened and, based on this information, we need to develop ways of counteracting these threats. These are the three basic steps towards successful conservation of insects, and, indeed, biodiversity generally: assessment of status and trends in biodiversity, identification of the causes of loss of biodiversity, and the development and implementation of measures to conserve and restore biodiversity. Conservation is often considered as the last of these three steps but the first two are equally important, building a foundation for the last. Much of this book has already dealt, indirectly or indirectly, with all three of these steps. To put it another way, successful insect conservation depends upon sound insect ecology. In this chapter we will briefly consider key aspects the first two steps, loss of insect diversity and the causes of this loss, before we discuss conservation itself in more detail.

10.2 STATUS AND TRENDS IN INSECT DIVERSITY

10.2.1 Insect extinctions

Although global biodiversity loss is widely reported (Brook et al. 2003; Baillie et al. 2004), only 60–70 insect species are known to have become extinct globally since 1600 (Baillie et al. 2004; Dunn 2005), a small percentage (approximately 0.006–0.0006%) of the estimated total number of insect species (Groombridge 1992). In the same time, 313 other

invertebrate species, 22 reptiles, 93 fish, 77 mammals, 133 birds, and 110 higher plants have become extinct (Baillie et al. 2004), much higher percentages of the global species richness of these groups (Mawdsley & Stork 1995).

Clearly the figure for the number of insect species extinctions is an underestimate – for every plant, bird, and mammal that have become extinct, the insects that are solely dependent upon them as plant and animal hosts have become extinct too. For example, following the extinction of the passenger pigeon (*Ectopistes migratorius*), at least two insects, the lice *Columbicola extinctus* and *Campanulotes defectus* (Phthiraptera), were thought to have become extinct too (Stork & Lyal 1993). This example illustrates another point about extinction – it is difficult to know for certain that a species has become extinct. *Columbicola extinctus* is now known to live on the band-tailed pigeon, *Columba fasciata* (Clayton & Price 1999).

Nevertheless, the dependence of one species upon another makes it vulnerable to extinction if the species it is dependent upon is endangered. Koh et al. (2004) estimated the extent of this problem by considering several examples of coevolution: pollinating fig wasps (Chalcidoidea, Agaonidae) and figs (*Ficus*), parasites and their hosts, butterflies and their host plants, and (social parasite) butterflies and their host ants. (The last example is considered in more detail in Section 10.4.3.) Their analysis suggested that 6300 affiliate species are "coendangered" because of their association with host species that are already known to be endangered (Figure 10.1). Examples of species threatened because of the endangered status of their hosts are lice species recently discovered on the Iberian lynx, *Lynx pardinus*, in southern Spain (Perez & Palma

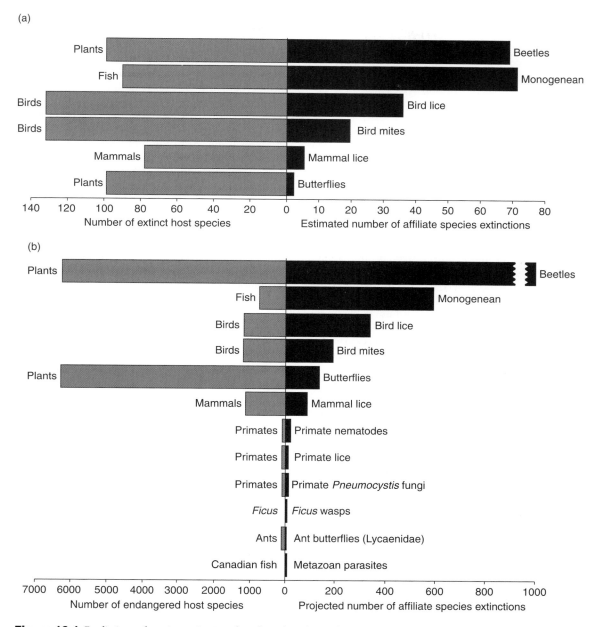

Figure 10.1 Predictions of species extinctions based on the relationships between different species, such as plants and the phytophagous insects dependent on them and mammals and their parasitic lice. These so-called affiliate extinctions are estimated (a) from the number of host species recorded as extinct, and (b) assuming all currently endangered hosts will go extinct. (From Koh et al. 2004.)

2001) and the Okarito brown kiwi in New Zealand (Palma & Price 2004) – two highly threatened hosts (Sales 2005).

Of the recorded extinctions, 54% are Lepidoptera and 16% Coleoptera (Groombridge 1992), probably because the former is the most well known order and the latter the most species-rich, not necessarily because most extinctions have taken place in these orders. The analysis by Koh et al. (2004) also suggests that recent extinctions amongst Hymenoptera and other orders are very likely.

Most insect extinctions (84%) recorded by Groombridge (1992) have been amongst island species: 69% on Hawaii alone. These figures partly reflect the fact that insects in places such as Hawaii and mainland USA (where most of the continental extinctions have been recorded) are much better known than in most other parts of the world. However, the inevitably smaller ranges and smaller population sizes of insects on islands than on the mainland probably means that insects restricted to islands have been and will continue to be more prone to extinctions than insects with mainland distributions. In addition, as discussed above, the smaller the island, the greater the chance of extinction. Furthermore, the adaptations of insects on remote islands may be such as to make them prone to extinction. This is certainly true of many island birds and mammals that evolved in the absence of predators. Tameness, flightlessness, and reduced reproductive rates are characteristic of many island birds and have probably contributed to the extinction of many of them by making them vulnerable to humans and introduced predators (Groombridge 1992). The evolution of flightlessness is also common among insect species endemic on islands. For example, 10 of the 11 orders of flighted insects that established in Hawaii have evolved flightless species (Howarth & Ramsay 1991). Flightless species have failed to evolve only in the Odonata in Hawaii. Insect species on islands may show adaptive radiation similar to the Galápagos finches studied by Darwin. The best-known example is probably the *Drosophila* of Hawaii (Carroll & Dingle 1996). Twenty percent of the world's *Drosophila* species are found in the Hawaiian Islands and most of these (484 of 509) are endemic.

Another example of the vulnerability of island insect is the Lord Howe Island stick-insect (*Dryococelus australis*; Phasmida). It was once abundant on Lord Howe Island, a small (about 11 × 2 km) island off the coast of Australia, but was thought to have been made extinct by black rats in the 1920s (Priddell et al. 2003). However, after freshly dead specimens were found in the 1960s, surveys in 2001–2 detected small populations on the island. Another insect species thought to be extinct until recently is the Pitt Island longhorn beetle, *Xylotoles costatus* (Coleoptera Cerambycidae), rediscovered in the Chatham Islands (Baillie et al. 2004).

These examples, like that of *Columbicola extinctus*, should not lead to complacency. Indirect evidence strongly suggests that many insects have become extinct and the size of the threats now facing them (Section 10.3) indicates that many more will become extinct unless action is taken to conserve them. As with many groups of species, some insects are becoming more widespread and abundant, while others are becoming rarer: the process of homogeneity (Samways 1996). So while most insects are not becoming globally extinct, many are becoming extinct locally or nationally.

10.2.2 Threatened insects

Recording insect extinction is important, but a precondition of successful conservation is identifying species that are vulnerable to extinction long before they become extinct. This is the rationale for the Red List developed by the IUCN (the World Conservation Union) for all animals and plants over more than 40 years (Baillie et al. 2004; Rodrigues et al. 2006). The Red List classifies species in different categories including critically endangered, endangered, and vulnerable (IUCN 2001) (Figure 10.2). Quantitative criteria are used to classify species in these categories (Table 10.1). Species classified as critically endangered, endangered, or vulnerable are frequently described together as "threatened".

The original IUCN Red List classification system operated for nearly 30 years and classified threatened species as extinct, endangered, vulnerable, or rare ("indeterminate" and "insufficiently known"; Mace & Stuart 1994). It was clearly valuable in producing databases of threatened species to provide a basis for setting conservation priorities and monitoring the success of conservation efforts (Miller et al. 1995). However, this system was frequently criticized for being subjective; classification of species by different authorities varied and did not always correspond with

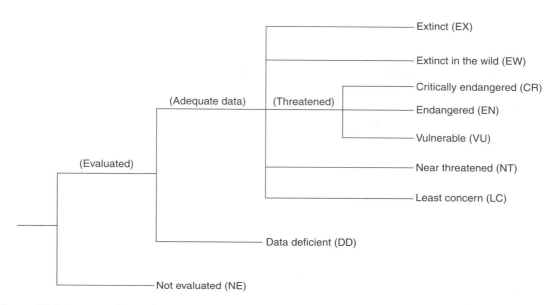

Figure 10.2 Structure of IUCN threatened categories. (From IUCN 2001.)

actual risks of extinction (Groombridge 1992). Thus the new IUCN classification system, described above, was adopted in 1994 (IUCN 1994).

The 2004 IUCN Red List named 15,589 species threatened with extinction (Baillie et al. 2004). Amongst insects, however, the identification of threatened species is made extremely difficult by our lack of knowledge of the ecology, distribution, and abundance of most species. The 2004 IUCN Red List names 559 insect species but this represents only 0.06% of all described insect species. The equivalent figures for plants and vertebrates were 3% and 9%, respectively. However, the detailed procedures adopted by the IUCN include an assessment of the amount of data available, and species are only evaluated if sufficient data exist. Only 771 insect species were, therefore, evaluated for the 2004 report. Thus 73% of all insects evaluated were classified in the threatened categories. Equivalent figures for plants and vertebrates were 70% and 23%, respectively. Although this suggests that insects are threatened to roughly the same amount as better known taxa, such is the degree of uncertainty about the status of insects that the true percentage of threatened species should be taken as lying somewhere between 0.06% and 73% (Baillie et al. 2004). It should be noted that the evaluation of insects was largely

based on swallowtail butterflies (Papilionidae) and the dragonflies and damselflies (Odonata). Although these groups are more vulnerable than many others, the 2004 Red List assessment could not include many insect groups likely to be vulnerable to extinction. For example, despite their vulnerability, Whiteman and Parker (2005) pointed out that only one of the 5000 known species of lice, and no fleas or other parasites are listed as threatened by the IUCN. As well as the intrinsic importance of lice and other parasitic species, these authors argue that knowledge of parasite ecology and evolution is important for the management of endangered vertebrates.

The parasite example demonstrates that for the majority of insects whose conservation status is poorly known, there is an alternative approach based on knowing the ecological requirements of species or communities. Thus, for example, if an insect species is restricted to feeding on a single host (plant, bird, or other insect species), threats to the host equate to threats to the insect species. This applies not only to the examples given above but also to many more. For example, the five species of *Maculinea* butterfly in Europe each depend upon a separate host *Myrmica* ant species and each has a rare specific parasitoid (Elmes & Thomas 1992). Thus not only is the

Table 10.1 IUCN criteria for critically endangered species. (From IUCN 2001, which lists criteria for all categories.)

A taxon is "critically endangered" when the best available evidence indicates that it meets any of the following criteria (A to E), and it is therefore considered to be facing an extremely high risk of extinction in the wild

A. *Reduction in population size based on any of the following*:
1. An observed, estimated, inferred, or suspected population size reduction of 90% over the last 10 years or three generations, whichever is the longer, where the causes of the reduction are clearly reversible *and* understood *and* ceased, based on (and specifying) any of the following:

 (a) direct observation,
 (b) an index of abundance appropriate to the taxon,
 (c) a decline in area of occupancy, extent of occurrence, and/or quality of habitat,
 (d) actual or potential levels of exploitation,
 (e) the effects of introduced taxa, hybridization, pathogens, pollutants, competitors or parasites

2. An observed, estimated, inferred, or suspected population size reduction of 80% over the last 10 years or three generations, whichever is the longer, where the reduction or its causes may not have ceased *or* may not be understood *or* may not be reversible, based on (and specifying) any of (a) to (e) under A1

3. A population size reduction of 80%, projected or suspected to be met within the next 10 years or three generations, whichever is the longer (up to a maximum of 100 years), based on (and specifying) any of (b) to (e) under A1

4. An observed, estimated, inferred, projected, or suspected population size reduction of 80% over any 10-year or three generation period, whichever is longer (up to a maximum of 100 years in the future), where the time period must include both the past and the future, and where the reduction or its causes may not have ceased *or* may not be understood *or* may not be reversible, based on (and specifying) any of (a) to (e) under A1

B. *Geographic range in the form of either B1 (extent of occurrence) or B2 (area of occupancy) or both*:
1. Extent of occurrence estimated to be less than $100\,km^2$, and estimates indicating at least two of (a) to (c):
 (a) severely fragmented or known to exist at only a single location,
 (b) continuing decline, observed, inferred, or projected, in any of the following:
  (i) extent of occurrence
  (ii) area of occupancy
  (iii) area, extent and/or quality of habitat
  (iv) number of locations or subpopulations
  (v) number of mature individuals,
 (c) extreme fluctuations in any of the following:
  (i) extent of occurrence
  (ii) area of occupancy
  (iii) number of locations or subpopulations
  (iv) number of mature individuals

2. Area of occupancy estimated to be less than $10\,km^2$, and estimates indicating at least two of (a) to (c):
 (a) severely fragmented or known to exist at only a single location
 (b) continuing decline, observed, inferred, or projected, in any of the following:
  (i) extent of occurrence
  (ii) area of occupancy
  (iii) area, extent and/or quality of habitat
  (iv) number of locations or subpopulations
  (v) number of mature individuals
 (c) extreme fluctuations in any of the following:
  (i) extent of occurrence
  (ii) area of occupancy
  (iii) number of locations or subpopulations
  (iv) number of mature individuals

(Continued)

Table 10.1 (*Continued*)

C. *Population size estimated to number fewer than 250 mature individuals and either:*

    1. An estimated continuing decline of at least 25% within 3 years or one generation, whichever is longer (up to a maximum of 100 years in the future) *or*

    2. A continuing decline, observed, projected, or inferred, in numbers of mature individuals *and* at least one of the following (a) to (b):

        (a) population structure in the form of one of the following:
            (i) no subpopulation estimated to contain more than 50 mature individuals, *or*
            (ii) at least 90% of mature individuals in one subpopulation
        (b) extreme fluctuations in number of mature individuals

D. *Population size estimated to number fewer than 50 mature individuals*

E. *Quantitative analysis showing the probability of extinction in the wild is at least 50% within 10 years or three generations, whichever is the longer* (up to a maximum of 100 years)

conservation of *Maculinea* butterflies dependent upon particular ant species, but also each parasitoid is dependent on a particular butterfly. For example, final instar larvae of *Maculinea rebeli* are social parasites of *Myrmica schencki*, and *Ichneumon eumerus* parasitizes *Maculinea rebeli* (Hochberg et al. 1996). In addition, young *Maculinea* larvae require specific host plants: for example, *M. alcon* is dependent upon its rare host plant *Gentiana pneumonanthe* (Mouquet et al. 2005a, 2005b). *Maculinea rebeli* is listed as vulnerable (IUCN 2004) but it and the others listed undoubtedly represent only a small proportion of the number of species threatened with extinction.

In addition to the global Red List classification system, national Red Data Books have been produced for plants, birds, mammals, amphibians, reptiles, and invertebrates such as the Red Data Book for British insects (Shirt 1987), updated as information on less well-known species becomes available (e.g. Parsons 1996). International Red Data Books are also being written including the *Red Data Book of European Butterflies* (Van Swaay & Warren 1999). Van Swaay et al. (2006) summarized the trends in Europe's 576 butterfly species using data from the *Red Data Book of European Butterflies*. They found that the distribution of butterflies had declined by an average of 11% over the last 25 years. Although the distributions of generalist species declined by 1%, specialist species declined more sharply: grassland butterflies by 19%, wetland species by 15%, and forest species by 14%.

10.2.3 Monitoring insects for conservation

Lists of threatened species are extremely valuable in prioritizing conservation measures but the only way of accurately deciding whether or not conservation measures for a particular insect species are needed (or to assess the success of conservation measures) is to monitor its abundance and distribution. For example, changes in the abundance and distribution of butterflies and day-flying moths are monitored through a variety of schemes, notably butterfly monitoring schemes in Europe (Pollard & Yates 1993; Roy et al. 2001) (Table 10.2). The UK scheme, for example, provides data on the abundance of British butterflies (Plate 10.1, between pp. 310 and 311) at a range of sites, potentially identifying species at risk locally and nationally (Figure 10.3).

The UK Butterfly Monitoring Scheme is linked to the UK Biological Records Centre, which collates data on the distribution of plants and animals, including dragonflies (Merritt et al. 1996), butterflies (Asher et al. 2001), and hoverflies (Ball & Morris 2000). Similar information is being collated elsewhere, for example for North American butterflies (Opler et al. 1995). Although the collection of such data may be seen as routine, there are important ecological issues associated with the mapping of species distributions, notably the issue of sampling bias. The likelihood of recording a species in a particular area will increase with the number of visits or samples, and, therefore, it is common for sampling effort to influence distribution maps of, for example, butterflies (Dennis et al. 1999)

Table 10.2 Butterfly monitoring schemes in Europe. (Source: Butterfly Conservation Europe.)

United Kingdom: all species since 1976, annually from hundreds of sites

Transcarpathia (Ukraine): field data collected for all species at 20–30 sites since 1983, but at present only analyzed for one species (*Erynnis tages*)

Germany: in the Pfalz region monitoring data on three habitat directive species (*Maculinea teleius*, *M. nausithous*, and *Lycaena dispar*) is available since 1989 from almost 100 sites

Germany: Nordrhein-Westfalen, all species since 2001. In 2005 data from over 100 sites available

Germany: In 2005 a nationwide monitoring scheme was launched. In the first year counts were made at a few hundred sites

The Netherlands: all species since 1990. In 2005 data available from 600 sites

Belgium (Flanders): all species since 1991 from 10–20 sites

Spain (Catalunya): all species since 1994 from 50–60 sites

Switzerland (Aargau): all species since 1998 from over 100 sites

Switzerland: In the rest of the country butterfly monitoring data have been collected since 2000, on at least 100 sites annually

Finland: all species since 1999 from around 100 sites

France (Doubs and Dordogne): all species since 2001 from 10 sites

Jersey (Channel Islands): all species since 2004 from 25 sites

Estonia: all species on seven transects since 2004

---

and hoverflies (Keil & Konvicka 2005) (Figure 10.4). This source of error, which is particularly likely to occur for rare species, may be minimized by programmed re-survey, collecting data on recording effort, collecting habitat and environmental data, and distinguishing between breeding individuals and vagrants (Dennis & Hardy 1999).

Maps of the changing distribution of insects and other species clearly show when they are becoming locally, regionally, or nationally extinct (Figure 10.5). Thomas et al. (2004c) used data on the changing distribution of butterflies, birds, and vascular plants in Britain to estimate global extinction rates. These data comprised surveys of the distributions of 254 native species of vascular plant, all 201 native breeding bird species, and all 58 native breeding butterfly species. Two survey periods were compared: early surveys from the 1950s to the 1970s and recent surveys in the late 1980s and 1990s. The surveys covered at least 98% of the 2861 10 × 10 km grid squares of England, Wales, and Scotland, and comprised over 15 million records from over 20,000 volunteer recorders – the most comprehensive data on the changing distribution of each taxon in the world. Thomas and his coworkers measured change in status by the difference between the total number of 10 km grid squares occupied in each survey. They found that 28% of the plant species had decreased

in Britain over 40 years, 54% of the bird species had decreased over 20 years, and 71% of the butterflies had decreased over about 20 years (Figure 10.6). This demonstration that butterflies are decreasing in status faster than plants and birds, which runs counter to reports of extinctions elsewhere, but clearly demonstrates the value of detailed, comprehensive surveys.

Although desirable, accurate monitoring of threatened species is not always possible – the rarity and inconspicuousness of threatened species usually prevent inexpensive, accurate assessment of abundance and distribution, and monitoring itself may be detrimental to the survival of insects. Martikainen and Kouki (2003), for example, estimated that at least 100,000 beetles would have to be collected in 10 boreal forest sites before their value in conserving rare species could be assessed. When accurate monitoring is not possible risk assessment may be based on "expert" judgment. That is, entomologists make a judgment of the threatened status of a species based on knowledge of its current abundance and distribution, any known changes in its abundance and distribution, and an assessment of threats to it including the vulnerability of its habitat. This, essentially, is the basis of classification systems for the threatened status of plants and animals, including the IUCN Red List classification, discussed above.

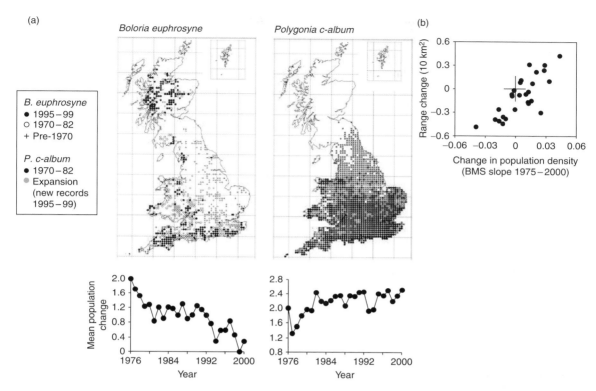

Figure 10.3 Monitoring change in the distribution and abundance of two butterflies in Britain, *Boloria euphrosyne* and *Polygonia c-album*. (a) Maps comparing change in distributions from 1970–82 to 1995–9 and mean population size of each species on sites monitored the Butterfly Monitoring Scheme (BMS). (b) Correlation between trends in the abundance of 27 butterfly species monitored by the BMS and changes in their ranges expressed as the difference in 10 × 10 km squares occupied by each species between the two survey periods. (From Warren et al. 2001; Thomas 2005.)

Another alternative to the monitoring of threatened insects is the use of indicators, a subject covered in detail in Chapter 9. In that chapter, the emphasis was on indicators of biodiversity generally rather than on threatened species, but clearly there are many situations where biodiversity as a whole is threatened and in those cases general indicators of biodiversity have an important role to play in the conservation of insects and other species. Thus all major initiatives on conservation include monitoring and indicators. The Convention on Biological Diversity (CBD), for example, has promoted the development of indicators in support of its global goal of significantly reducing the loss of biodiversity by 2010 (Dobson 2005). In Europe, a similar set of headline indicators is being developed in support of a target to halt biodiversity loss by 2010 (EASAC 2005) (Table 10.3). These

indicators include a few of the state of biodiversity, some that relate to the sustainable use of biodiversity, others that provide measures of threats to biodiversity, and some that relate to other aspects of biodiversity, including public awareness of biodiversity. A major problem with these indicators is the slow rate of development and implementation. For example, EASAC (2005) concluded that only two were ready for implementation in Europe: a European wild bird index, as a population trend measure (Gregory et al. 2005), and the coverage of protected areas. They also concluded that three other indicators were close to implementation: the extent of major habitats, the marine trophic index, which measures impacts of fishing on fish stocks (Pauly & Watson 2005), and the Red List index, which measures trends in threatened species. This index is based on the number of species in

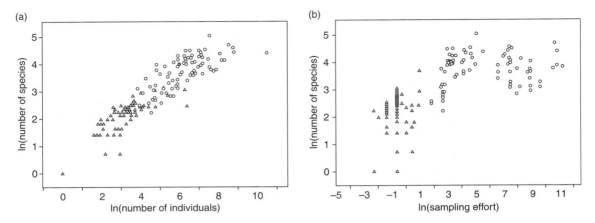

Figure 10.4 Influence of sampling intensity and the number of species recorded: the number of hoverfly (syrphid) species richness recorded with increasing number of individuals (a) and increasing sampling effort (b) as collated from published studies from Central Europe. (From Keil & Konvicka 2005.)

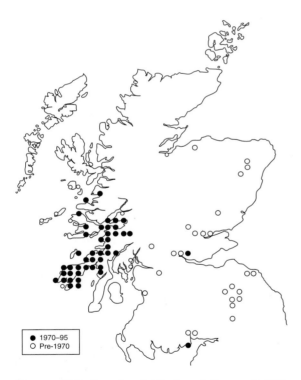

Figure 10.5 Changing distribution of the marsh fritillary butterfly *Eurodryas aurinia* in Scotland. (Data from the Biological Records Centre, Centre for Ecology and Hydrology UK.)

each IUCN Red List category (see above) and the number of species changing categories between assessments (Butchart et al. 2005). The index can be calculated for any set of species that has been fully assessed twice or more. Currently a Red List index is available only for the world's birds (Butchart et al. 2004).

These indicators provide very general information on biodiversity but are, nevertheless, important in influencing policy and funding for conservation. There is a lack of information on insects in these indicators but information on them could be incorporated in the Red List index. Population trend indicators could also be extended to include well-monitored groups such as butterflies (EASAC 2005). There is, however, a concern that these indicators do not provide information relevant to trends in the abundance and diversity of many taxa, including most insects. Information on trends in birds, butterflies, and some other groups is becoming increasingly available but research has shown poor correlation in the diversity of different taxa (see Chapter 9), limiting the relevance of this information.

At a more local and practical scale, indicators play an important role in the management of protected areas and in other circumstances where conservation of biodiversity is a management objective. The latter is increasingly true in forest management, where indicators have long been used for a range of purposes. Regarding biodiversity, the classification of indicators

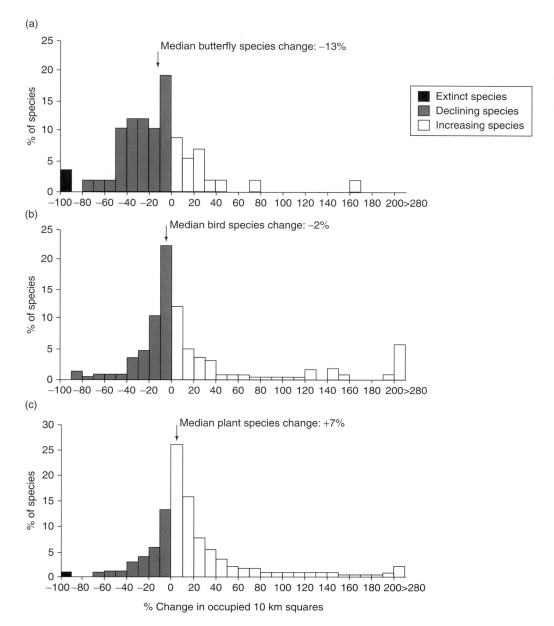

Figure 10.6 Changes in the number of 10×10 km squares in Britain occupied by (a) native butterfly, (b) bird, and (c) plant species between two censuses of each taxon. These censuses comprised: (a) all 58 native breeding butterfly species between 1970–82 and 1995–9; (b) all 201 native breeding bird species between 1968–72 and 1988–91; and (c) 254 native species of vascular plant between 1954–60 and 1987–99. (From Thomas et al. 2004c.)

Table 10.3 European biodiversity headline indicators. (from EASAC 2005.)

Trends in extent of selected biomes, ecosystems, and habitats
Trends in abundance and distribution of selected species
Coverage of protected areas
Change in status of threatened and/or protected species
Trends in genetic diversity of domesticated animals, cultivated plants, and fish species of major socioeconomic
    importance
Area of forest, agricultural, fishery and aquaculture ecosystems under sustainable management
Nitrogen deposition
Numbers and costs of invasive alien species
Impact of climate change on biodiversity
Marine trophic index
Water quality in aquatic ecosystems
Connectivity/fragmentation of ecosystems
Funding to biodiversity
Public awareness and participation

based on composition, structure, and function was developed for forests (Noss 1990). Ferris and Humphrey (1999), for example, recommended a combination of compositional and structural indicators to measure forest biodiversity. Based on their experience, mainly in conifer plantations, they recommended the use of two to three key compositional indicators, such as the species composition of broadleaved trees within conifer forests, and two or three key structural indicators, such as the quantity and quality of deadwood. Decaying wood provides a key resource for forest insects: Siitonen (2001), for examples, estimated that 20–25% (4000–5000) of all forest-dwelling species depend on deadwood habitats in Finland and that the diversity of some insect groups is correlated with the amount of deadwood present (Grove 2002; see also Section 9.2.7). Juutinen et al. (2006) examined the economics of using large-scale species inventories or a deadwood indicator for selecting sites for conservation and concluded that the latter was more cost-effective for saproxylic (deadwood-feeding) species but not for all species. They therefore identified the need for indicators for species not dependent on decaying wood. The diversity of these species may also depend on structural aspects of forests. Oxbrough et al. (2005), for example, studied the factors influencing spider diversity in Irish forests. They found that several aspects of forest structure affected spiders: the amount of ground vegetation, litter cover, litter depth, and twig cover. They also concluded that a mosaic of different aged stands was needed to sustain spiders with different habitat requirements.

Functional indicators, the third type of indicator proposed by Noss (1990), include measures of disturbance. They have been less well developed than compositional and structural indicators but are clearly relevant to forest biodiversity (Larsson 2001). One form of disturbance, whose measurement can form the basis of an indicator, is forest fire. Although forest fires can have a positive effect on biodiversity, some forest species are negatively affected by fire (Spies et al. 2006). Indeed, fire is a major focus of conflict between conservation and human activities in forest ecosystems (Niemelä et al 2005).

There is, of course, no fully satisfactory substitute for monitoring the distribution and abundance of insects of conservation concern. Indicators and expert judgment have important roles to play and both require development to make them more useful in insect conservation. But even when we do have detailed information on change in the distribution and abundance of insect species they must be put in context before action to conserve them can be adequately taken. Insect abundance fluctuates naturally (see Chapters 4 and 5), many insects are intrinsically rare, and local extinction and colonization are natural processes. Interpretation of change therefore requires ecological knowledge. It also requires knowledge of threats to insects, their habitats, and the species upon which they may depend (or which may threaten them).

10.3 THREATS TO INSECTS

Biodiversity faces many threats, often referred to as pressures or drivers of change. Sala et al. (2000) concluded that the most significant threats to terrestrial biodiversity over the next 100 years will be land use change, followed by climate change, nitrogen deposition, the spread of invasive species, and elevated carbon dioxide concentration. The Millenium Ecosystem Assessment (2005) defines drivers as "natural or human-induced factors that directly or indirectly cause a change in an ecosystem". Drivers are subdivided into direct drivers, which unequivocally influence ecosystems, and indirect drivers, which act more diffusely on one or more direct drivers. Direct drivers are referred to as pressures by many authors (e.g. Petit et al. 2001), often as part of the driver-pressure-state-impact-response model used for biodiversity, and environmental assessment generally (EEA 2005). Petit et al. (2001) concluded that land use change, particularly agricultural intensification,

was the most significant pressure on European biodiversity but that different parts of Europe experienced different pressures (Figure 10.7). At the global scale, the Millenium Ecosystem Assessment (2005) concluded that the major direct drivers of change in biodiversity and ecosystems were habitat change, climate change, invasive species, overexploitation, and pollution (from nitrogen and phosphorous), varying in importance from biome to biome (Figure 10.8).

Many entomologists have assessed the various threats to insects. Most conclude that the most serious class of threat to insects is land use change, particularly habitat destruction and fragmentation (e.g. Hafernik 1992), with global climate change posing an increasing, and synergistic, threat (Samways 2007). Threats to insects vary from region to region and are notably different between islands and continents. Gagne and Howarth (1985), for example, listed the following factors (in order of importance) for the extinction of 27 species of Macrolepidoptera in Hawaii: biological control introductions, habitat

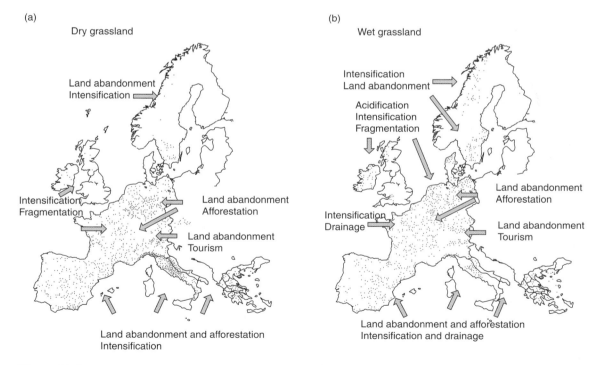

Figure 10.7 Examples of variation in major pressures on biodiversity in dry and wet grasslands in different parts of Europe. (From Petit et al. 2001.)

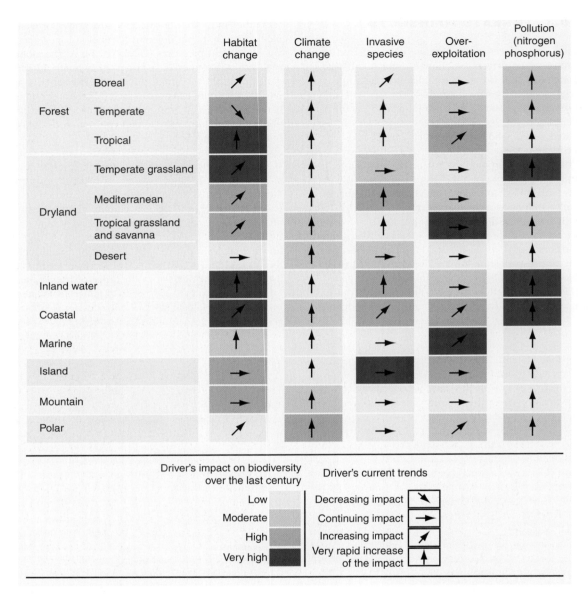

Figure 10.8 Impact of five major direct drivers (pressures) on biodiversity (habitat or land use change, climate change, invasive species, overexploitation, and pollution) on different biomes over the past 50–100 years. The cell color provides an estimate of the impact of each driver on biodiversity in each type of ecosystem: high impact means that the particular driver has significantly altered biodiversity in that biome; low impact indicates that it has had little influence on biodiversity in the biome. The arrows indicate the trend in the driver. (From Millenium Ecosystem Assessment 2005.)

loss, alien mammals, host loss, alien arthropods, and hybridization with a related invading alien species.

Such is the close association between insects and their environments that assessments by the Millenium Ecosystem Assessment and others of threats to biodiversity and ecosystems generally are equally relevant to insect conservation. The major threats are considered briefly below.

10.3.1 Land use change, habitat loss, and habitat fragmentation

The destruction, fragmentation, or modification of an insect's habitat poses an obvious threat to it (e.g. Jennings & Tallamy 2006). Probably the most serious example of this is deforestation, particularly the destruction of tropical forests (e.g. Benedick et al. 2006). As discussed in Chapter 9, what applies to "real" islands often applies to habitat islands. Thus insects, and other animals and plants, are likely to become more and more vulnerable to extinction as their habitats become smaller. Not surprisingly, therefore, several ecologists have used species–area relationships to predict the consequences of habitat loss, particularly deforestation, on global species richness (Reid 1992).

The relationship between the number of species (S) in an area and its size (A) (see Section 9.3.3) can be represented by the equation of a straight line:

$$\log S = \log c + z \log A$$

where c and z are constants. Studies on islands, mainland areas, and habitat islands have shown that the slope (z) generally lies in the range 0.10 to 0.50 (Lomolino 2000). Reid and Miller (1989) based their predictions of species extinctions on a slope of 0.15 to 0.40. These slopes predict that a 90% reduction in island or habitat size will result in the loss of 30–60% of the species present. Reid and Miller then took a deforestation rate of 0.5–1% loss of forest area per year, 1–2 times the FAO (1988c) estimate for 1980–5. The predicted rate of extinction based on these predictions was a loss of 2–5% of species per decade. Most predicted extinction rates largely based on habitat loss range from 2–6% to 8–11% of species per decade (Mawdsley & Stork 1995). Brook et al. (2003) used this method to infer local extinction rates in different taxa in Singapore from 1819 to 2002. During this time habitat loss exceeded 95%. Documented extinctions and inferred extinction rates for butterflies, freshwater fish, birds, and mammals ranged from 34% to 87%. A species–area model was also used to estimate that the current rate of habitat loss in Southeast Asia would lead to a loss of 13–42% species, of which about one-half would be global extinctions.

Individual threats to biodiversity do not, of course, act alone and, for example, habitat loss can interact with climate change. This issue, including the use of species–area relationships in predicting the consequences of

climate change on biodiversity, is discussed Section 10.3.2.

The main problem with predictions of species extinctions based on species–area relationships is that it assumes that as habitats are removed (e.g. deforestation), the species that formerly occupied them are unable to survive in the new habitat. This may be true when deforestation results in a change of land use to intensive agriculture, but when deforestation is followed by the growth of secondary forest or where forest plantations are established, the influence on biodiversity may not be so serious (Lugo 1995; Watt et al. 1997b, 1997c; Section 10.4.2). This and other problems with the application of species–area relationships to the prediction of species extinctions have resulted in extinction rates being overestimated (May et al. 1995). Kinzig and Harte (2000), for example, found that species–area relationships overestimated the extinction of endemic species due to habitat loss. However, they also found that beyond a certain threshold of habitat loss there was a rapid rise in the rate of extinctions, implying that extrapolations based on current observed rates of species loss may underestimate future extinctions due to continued habitat loss.

Although species–area relationships may not provide an accurate method for predicting species losses, deforestation remains a major threat to insects and other species. The total area of forest is currently about 4 billion hectares, but it is unevenly distributed with two-thirds of the total forest area found in the 10 most forest-rich countries (FAO 2005). Although its rate is reported to be slowing, deforestation – mainly the conversion of forests to agricultural land – is occurring at about 13 million hectares per year. The net rate of forest loss is slowed by forest plantations and natural expansion of forests: a net loss of 7.3 million hectares per year is estimated during 2000–5, compared with a loss of 8.9 million hectares per year from 1990 to 2000. The greatest rate of deforestation is in Africa and South America. Many ecologists have attempted to quantify the impact of deforestation on biodiversity (e.g. Watt et al. 2002; Brook et al. 2003; Stork et al. 2003; Section 10.4.2).

Deforestation is not the only threat to insects, but most detailed lists of threats to insects are dominated by habitat-related threats. In Britain, for example, the major threats to butterflies are considered to be (in decreasing order of importance): agricultural expansion, deforestation, (habitat) management practices, afforestation, and urbanization (Mawdsley & Stork 1995). The same authors concluded that the major threats to endangered beetles are: management practices, deforestation, agricultural expansion, natural succession, and removal of dead wood. Thus threats to different insect groups show a strong degree of overlap, at least within a single country. However, threats to different insect groups may be greatest in different habitats (Mawdsley & Stork 1995). In Britain, the habitats occupied by threatened butterflies are (in decreasing order of importance): chalk grassland, woodland edge, other grassland, hedgerow, heathland, and fen (measured by the number of threatened species found in each habitat). In contrast, the habitats occupied by threatened beetles are: deciduous woodland, other grassland, ancient woodland, coastal shingle, riparian habitats, and sand dunes.

Although habitat loss is an obvious pressure on biodiversity, the related threat of habitat fragmentation can have a serious impact on the diversity of insects and other species. Habitat fragmentation is common throughout the world, although rarely documented. Baguette et al. (2003), however, quantified the fragmentation of the habitat of the bog fritillary butterfly *Proclossiana eunomia* (Lepidoptera: Nymphalidae) in Belgium (Figure 10.9). The bog fritillary and many other species of butterflies and other insects continue to survive in habitat patches within fragmented landscapes. The study of butterflies in these situations, in particular, has given rise to the metapopulation concept (Levins 1970; Hanski & Gilpin 1997; Hanski 1999; Hanski et al. 2006). In a metapopulation, each local population has a separate probability of extinction and (re)colonization and occupied patches are connected by occasional migration. Mousson et al. (1999), for example, studied 18 populations of the cranberry fritillary *Boloria aquilonaris* (Lepidoptera: Nymphalidae) in peat bogs in southern Belgium. They found that the dynamics of the local populations was asynchronous and, in a mark–recapture experiment, they found a high degree of interchange between populations. They therefore concluded that these populations acted as a single metapopulation.

Studies on metapopulations are valuable in quantifying the viability of species in particular landscapes and recommending landscape management strategies for their conservation. Population viability analysis (PVA) can be used to assess whether metapopulations of species such as the marsh fritillary butterfly *Euphydryas aurinia* (Lepidoptera: Nymphalidae) will go extinct or not. Schtickzelle et al. (2005) for example, found that a patch system occupied by a metapopulation

of this species in Belgium could not survive under the present management of the area.

Metapopulation studies have been carried out on other insects too. Hein et al. (2003) studied movement behavior of the grey bush cricket *Platycleis albopunctata* (Orthoptera: Tettigoniidae) between patches, demonstrating that the amount of movement between patches of suitable habitat was affected by the kind of habitat encountered between patches, the so-called "matrix". Similar results have been observed in studies on butterflies: for example, Roland et al. (2000) showed that *Parnssius smintheus* (Papilionidae) moved twice as readily through open meadow than forest. They calculated that due to the reduction in alpine meadow habitat of 78% during the 43 years before their study, fragmentation had reduced butterfly movement by 41%. Ricketts (2001), in an aptly titled paper "The matrix matters: effective isolation in fragmented landscapes" studied the resistance to movement of willow thicket and conifer forest to a meadow-inhabiting butterfly community. Although the conifer forest was more resistant than willow for four of the species, there was no difference in the resistance of the two matrix habitats for the two remaining species.

Metapopulation theory is a useful framework for considering the impact of land use change on threatened insects, particularly in highly fragmented landscapes (Hanski 2004; Liu et al. 2006a). In other situations (Baguette 2004) and for particular species (Henle et al. 2004) it may not apply but, nevertheless, it provides strong theoretical support for insect conservation. Samways (2007) emphasized the importance of the metapopulation perspective in arguing that all insect conservation efforts should have the goal of achieving healthy insect populations, "which require the combined support of the metapopulation trio of large patch (habitat) size, good patch quality, and reduced patch isolation".

10.3.2 Climate change

Climate change is now seen as a major threat to biodiversity, second only to land use change (Sala et al. 2000; CBD 2003; Thomas et al. 2004a; Millenium Ecosystem Assessment 2005; Stork et al. 2007). This is partly because of the increasing evidence that the global climate is changing and the growing awareness that climate change has an enormous potential for affecting biodiversity.

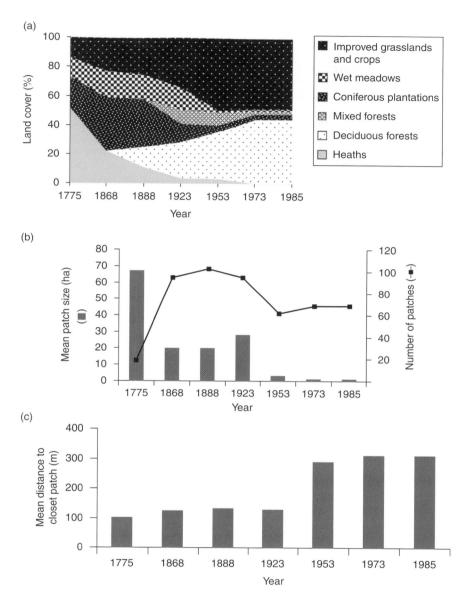

Figure 10.9 Fragmentation of the habitat of the bog fritillary butterfly in Lierneux, Belgium: (a) land cover change between 1775 and 1985, including wet meadows, the habitat of the butterfly; (b) changes in the number and size of suitable habitat and patch number in the landscape; and (c) mean distance to the nearest patch, a measure of spatial isolation of wet meadows patches. (From Baguette et al. 2003.)

The Earth's temperature has increased by an average of approximately 0.6–0.7°C during the past 100 years, with an increased rate of warming particularly since the mid-1970s (IPCC 2001; Brooker et al. 2007). Although the Earth's climate has always fluctuated, the current rate of climate change is greater than any experienced during the last 1000 years, and there is strong evidence that most of this increase is due to anthropogenic climate forcing as a result of increased release of carbon dioxide (CO_2) and other greenhouse gases into the atmosphere (IPCC 2001). The concentration of CO_2 is predicted to rise by 2100 to between 540 and 970 ppm, compared to about 368 ppm in 2000 and about 270 ppm in the pre-industrial era. Climate models predict a rise in temperature of 1.5–5.8°C between 1990 and 2100, which is 2–10 times greater than the rise in temperature observed in the last century. Annual precipitation is expected to rise on average by 5 to 20% over the same period, although the models suggest major regional and seasonal variations with precipitation increasing over high-latitude regions in summer and winter, in northern mid-latitudes, tropical Africa, and Antarctica in winter, and in southern and eastern Asia in summer. A decrease in winter precipitation is predicted for Australia, Central America, and southern Africa with consistent decreases in winter rainfall. The global mean sea level is expected to rise by 0.09–0.88 m between 1990 and 2100. An increase in the frequency, intensity, and duration of extreme events such as more hot days and heavy precipitation events is predicted, but the number of cold days is expected to decrease.

Such is the influence of climate on insects (see Chapter 2), that it is unsurprising that the changing climate is already reported to be having an effect on many species. In temperate countries, the bud burst and flowering of plant species is happening earlier, butterfly species are appearing earlier, amphibian and bird species are breeding earlier, and migrating bird species are arriving earlier (Walther et al. 2002; Parmesan & Yohe 2003; Root et al. 2003).

There is evidence that birds (Thomas & Lennon 1999) and butterflies (Warren et al. 2001) are extending their ranges polewards. There is evidence of shifts in the range of plants too, although not as rapid (Walther et al. 2005). Such shifts are not new: climate has had a dramatic influence on the distributions in the geological past of species such as beetles (Coope 1995), so much so that past temperatures can be estimated from the fossil beetle communities (Elias & Matthews 2002). The fossil record shows large-scale geographic changes in the distribution of Coleoptera and other insects in response to climate change during the latest glacial/interglacial cycle (Figure 10.10). Some species of beetles now found in northern Scandinavia occurred throughout the UK during previous "cold" episodes. Others, such as *Bembidion*

octomaculatum (Coleoptera: Carabidae), were widespread in the UK during a brief period of high temperatures, but are now found much further south and rarely in the UK (although, interestingly, *B. octomaculatum* became re-established in Sussex in the late 1980s). Overall, species extinctions of Coleoptera and other insects on a global scale during this cycle appear to have been uncommon despite the rapidity of climate change at the end of the glacial era (Coope 1995).

Although some movement in the distribution of insects and other species is already being detected (Thomas et al. 2006), it is likely that more significant changes in distribution will occur in the future. Models have been constructed to predict the impact of climate change of the distribution of insects, plants, birds, and other species (Berry et al. 2002). These models demonstrate how some species are likely to contract in range within a country, whereas others are likely to expand their range (Figure 10.11). Models predicting the impact of climate change on species distribution are based on analysis of species bioclimatic envelopes – the relationship between their observed distributions and climate. Although widely used, this approach does not work equally well for all species. Luoto et al. (2005) assessed the value of this approach for 98 butterfly species. They found that uncommon species or those at the edge of their range were better predicted by the models than widespread species, and species that had clumped distributions were better modeled than species with scattered distributions (Figure 10.12).

One of the major problems with developing bioclimatic models is that the observed distribution of species is determined not only by temperature but by other factors such as their interactions with other species. The distribution of insect herbivores, in particular, is likely to be largely determined by interactions with their host plants (Bale et al. 2002) (Figure 10.13). Interactions with natural enemies may limit the distribution of insects too. Randall (1982a, 1982b), for example, demonstrated that climate indirectly limits the distribution of insect species through a study on the population dynamics of the moth *Coleophora alticolella* (Lepidoptera: Coleophoridae) along an altitudinal transect in northern England. At the highest altitude, the abundance of *C. alticolella* is limited by low availability of larval food, seeds of the rush *Juncus squarrosus*, and at the lowest altitude by parasitism, which becomes more intense with decreasing altitude. That is, the altitudinal difference in climate does not

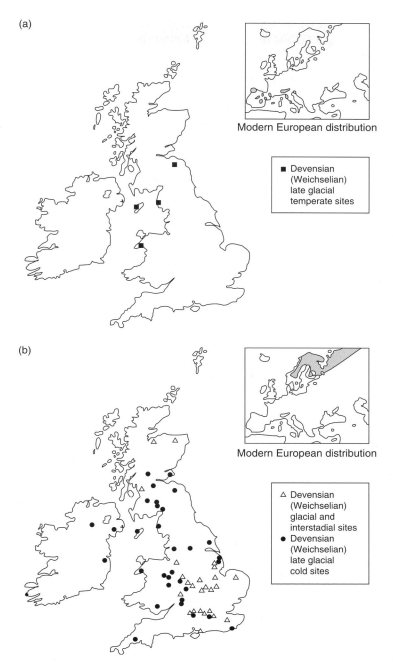

Figure 10.10 Comparison between the fossil record and present day distribution of two beetle species in the British Isles: (a) *Asaphidion cyanicorne* and (b) *Boreaphilus henningianus*. (From Coope 1995.)

(a) Trailing azalea

(i)   (ii)   (iii)

(c) Great burnet

(i)   (ii)   (iii)

(b) Hare's tail cotton grass

(i)   (ii)   (iii)

(d) Large skipper

(i)   (ii)   (iii)

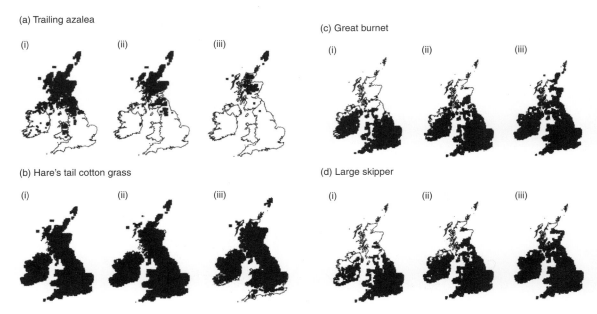

Figure 10.11 Contrasting examples of the impacts of climate change in Britain and Ireland using three simulated distributions. Two species that are likely to decrease in range in these countries by the 2050s because they are at the southern limit of their distribution: (a) trailing azalea *Loiseleuria procumbens*, and (b) hare's tail cotton grass *Eriophorum vaginatum*. Two species that are likely to increase their range because of an increase in favorable climatic conditions: (c) great burnet moth *Sanguisorba officinalis*, and (d) large skipper butterfly *Ochlodes venata*. The simulated distributions shown are: (i) current simulated distributions; (ii) simulated distributions in the 2050s under a low climate change scenario; and (iii) simulated distributions in the 2050s under a high climate change scenario. (From Berry et al. 2002.)

have a great influence directly on the moth but it has a profound influence through the insect-parasitoid relationship at low altitudes (and high temperatures) and the insect-plant relationship at high altitudes (and low temperatures). This study demonstrates the relevance of research along latitudinal gradients (Hodkinson 2005) and such studies (e.g. Andrew & Hughes 2005) are increasing our understanding of the future impact of climate change.

Despite concerns about bioclimate envelope models, Elith et al. (2006) – who reviewed 16 modeling methods applied to 226 species in six regions of the world – concluded that advances in this approach were providing improved results. However uncertain the results of these models may be, they have highlighted the scale of the impact of climate change on biodiversity and the need for conservation policies to take climate change into account (Harrison et al. 2006).

There is some evidence that species can adapt to a changing climate without changing their distribution. Partridge et al. (1994) showed that *Drosophila* adapts (in terms of survival, growth, and development) to changes in temperature within 5 years of continuous exposure. Species with much longer life cycles may not be able to adapt genetically to climate change. However, research on four butterfly species within the same family showed that those living in dry and open habitats had a maximum fecundity and survival rate at a higher temperature than the shade-dwelling species studied (Karlsson & Wiklund 2005) (Figure 10.14). Moreover, populations of the same species have been shown to be adapted to the different habitats they occupy. Populations of the woodland butterfly *Pararge aegeria* (Lepidotera: Nymphalidae) (Plate 10.2, between pp. 310 and 311) living in a shady woodland landscape have a higher fecundity at lower temperatures than those originating from open

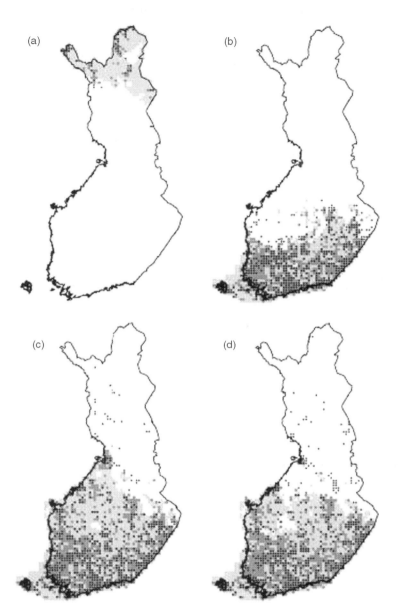

Figure 10.12 Variation in the accuracy of models for different species of butterfly in Finland: in each case points represent the sampling plots where the species was recorded; dark gray areas are where the model predicts the presence of the species and data are available; gray areas are where the model predicts the presence of the species but no butterfly data are available. (a) A highly accurate model for the dewy ringlet *Erebia pandrose*. (b) A good model for the high brown fritillary *Argynnis adippe*. (c) A fairly accurate model for the holly blue *Celastrina argiolus*. (d) A low accuracy model for the Idas blue *Plebeius idas*. (From Luoto et al. 2005.)

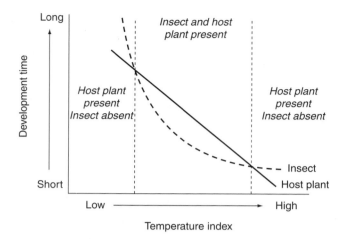

Figure 10.13 Model showing how the relative development rates of a host-specific insect and its host plant at different temperatures could set the distribution limits of such insect herbivore species. In that part of the range where temperatures are relatively low (low temperature index), the host plant grows too slowly to support insect development, but in that part of the range where temperatures are relatively high (high temperature index), the plant develops too quickly; only over the mid-part of the range is the insect herbivore able to match its phenology to that of its host plant. (From Bale et al. 2002.)

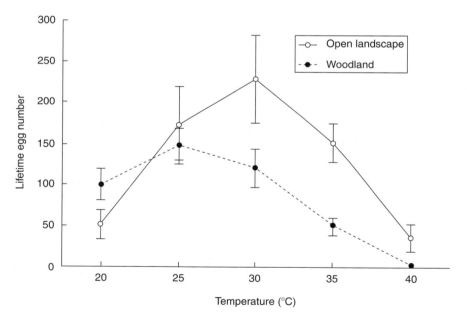

Figure 10.14 Average fecundity (total lifetime egg production) at five different temperatures of two butterfly species that are confined to dry and hot open habitats (grayling *Hipparchia semele* and the small heath *Coenonympha pamphilus*) and two shade-dwelling species (ringlet *Aphantopus hyperantus* and the speckled wood butterfly *Pararge aegeria*). The number of eggs are adjusted for female body size, and the mean and standard errors are shown. (From Karlsson & Wiklund 2005.)

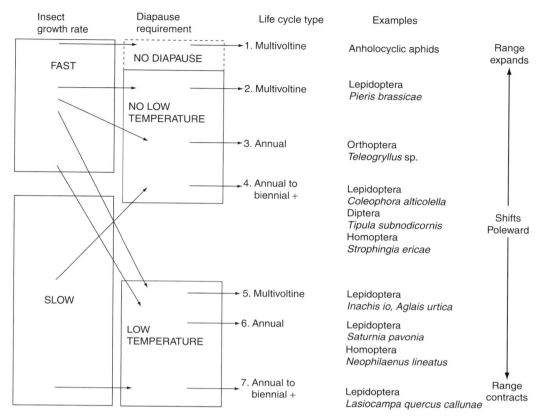

Figure 10.15 A model of insect response to climate warming where the growth period is assumed to be during the summer and the single diapause during the winter. The low temperature requirement for diapause indicates where temperatures lower than those favorable for insect development are needed for successful life cycle completion. Through combining fast/slow growth rates with temperature-related diapause, the life cycle of a species and its response to climate change through changes in its distribution (range) can be predicted. For example, fast-growing, non-diapausing species are likely to be multivoltine and will respond the most to climate warming by expanding their ranges. Fast-growing species that are not dependent on low temperatures to induce diapause are likely to be multivoltine or annual and will respond through range expansion. Fast-growing multivoltine or annual species that need low temperatures for diapause are likely to respond to climate warming with some range contraction. Slow-growing species that need low temperatures to induce diapause are likely to be unable to expand their ranges and may be detrimentally affected by climate change. (From Bale et al. 2002.)

agricultural landscapes and the opposite is true at higher temperatures (Karlsson & Van Dyck 2005).

As demonstrated throughout this book, insects have an enormous range of life histories and other characteristics. Some authors have attempted to predict how insects with different life histories will react to climate change (Bale et al 2002; Andrew & Hughes 2005) (Figure 10.15). It would be wrong, however, to consider the response of insects to climate change in isolation: its impact will depend upon interactions between other species of the same and different trophic levels (Bale et al. 2002; see also Chapter 2). Changing temperatures can also have an indirect influence on species that interact with other species which have a different capacity to change their distribution in response to climate change. Although insects appear to have an advantage over most other taxonomic groups because of their ability to disperse more rapidly than other groups of organisms, they are vulnerable when the species they depend upon are unable to change their distribution rapidly enough. Although the responses of particular species to changes in climate are poorly known, it is probable that different species will respond in different ways and to different degrees; this will lead to changes in the balance between competing species, plants, and herbivores, and insects and their predators (Lawton 1995).

A final consideration in assessing the threat of climate change to biodiversity is that it is only one component of global change; climate change and habitat loss, in particular, combine to produce what one writer referred to as "a deadly anthropogenic cocktail" (Travis 2003). Indeed, where we would expect that a species is likely to take advantage of climate change by extending its distribution, habitat loss or fragmentation might prevent such an expansion. Hill et al. (1999), for example, predicted that the speckled wood butterfly *P. aegeria* could increase its range substantially under predicted future climate change but that habitat availability would constrain its expansion. An analysis of 35 species of butterflies in the UK, all of which were predicted to have expanded their range in response to recent climate change, showed that all but five species had not done so because of a lack of suitable habitat (Hill et al. 2002).

Thomas et al. (2004a) used species–area models, the approach discussed above in relation to predicting the effect of habitat loss on biodiversity, to predict the impact of both climate change and habitat loss on global extinction rates of butterflies and other species. They argued that extinctions arising from reductions in area should apply not only to habitat loss per se but also to climatic unsuitability of that habitat. On this basis they predicted that for mid-range climate scenarios, 15–37% of the taxa they studied would become committed to extinction by 2050. Three different modeling methods were used for three different climate scenarios and they did the analyses twice, once assuming no capacity to disperse and one assuming the species could disperse. This produced a wide range of predicted extinctions (Table 10.4). They also predicted that habitat loss alone would result in extinction rates by 2050 of 1–29% in the areas they studied. Despite criticisms of their analysis, including concern about the use of species–area models (Buckley & Roughgarden 2004), Thomas et al. (2004a, 2004b) concluded that climate change represents the greatest threat to biodiversity in most if not all regions of the world.

10.3.3 Other factors

There are many other threats to biodiversity, many of which are associated with land use change (Young et al. 2005). These include pesticide application and other agricultural practices such as burning, atmospheric and aquatic pollution, the spread of invasive species, urbanization, and persecution of individual

Table 10.4 Projected extinction rates (%) for butterflies in different regions and for all taxa assessed globally (seven taxa including butterflies, mammals, birds, and plants) with three different climate change scenarios assuming species are fully able to disperse to take advantage of suitable climate elsewhere (with dispersal) or unable to disperse (no dispersal). In each case three different models are applied, providing three different extinction estimates. (From Thomas et al. 2004a.)

Taxa	Region	With dispersal			No dispersal		
		Minimum expected climate change	Mid-range climate change	Maximum expected climate change	Minimum expected climate change	Mid-range climate change	Maximum expected climate change
Butterflies	Mexico	1,3,4	3,4,5	–	6,9,11	9,12,15	–
	South Africa	–	13,7,8	–	–	35,45,70	–
	Australia	5,7,7	13,15,16	21,22,26	9,11,12	18,21,23	29,32,36
All taxa studied	Global estimate	9,10,13	15,15,20	21,23,32	22,25,31	26,29,37	38,42,52

species. Many of these threats do not apply as strongly to insects as to other species. Fire is clearly a major threat to plant diversity but many insects are relatively resistant to it (Parr et al. 2004). Roads endanger many vertebrates but roadside verges can promote insect diversity (Saarinen et al. 2005). In many cases, of course, we know very little about the threats facing insects, even in the urban environments that most of us live in (Niemelä et al. 2002).

Although the persecution of individual species is a major threat to birds and other vertebrates, particularly predatory species, it is rarely a serious threat to insects. Insect collectors, however, were probably responsible for the recorded extinction of the New Forest burnet moth in the UK in 1927 (Young & Barbour 2004; Section 10.5). Persecution or hunting of other species can also have knock-on effects on insects: mammal hunting in Panama, for example, affects the abundance and diversity of dung beetles (Andresen & Laurance 2007).

10.4 CONSERVATION AND RESTORATION

10.4.1 Threats to biodiversity and insect conservation

The conservation of insects cannot be viewed in isolation from other species, nor can it be effectively addressed unless we know as much as possible about the abundance and distribution of threatened species (Section 10.2) and unless, in particular, we know what the major threats to insects are (Section 10.3). Such is the dependency of insects on other species and on the habitats they occupy, that in considering the conservation of insects we should first consider ways of addressing threats to biodiversity generally.

Taking the first of the major pressures on biodiversity discussed above, land use change, habitat loss, and habitat fragmentation, the main responses have been to set up protected areas (Chape et al. 2005; Gaston et al. 2006) or to establish planning guidelines to minimize potentially damaging land uses (Dale et al. 2000). There are over 104,000 protected areas in the world, covering more than 12% of the land area, but only about 0.5% of the ocean surface area and about 1.4% of the marine coastal zone (Chape et al. 2005). Although areas have been set aside mainly for resource management reasons for hundreds of years,

the modern era of protected areas started when the Yellowstone National Park was established in 1872. The IUCN and the World Commission on Protected Areas (WPCA) define a protected area as "an area of land and/or sea especially dedicated to the protection and maintenance of biological diversity, and of natural and associated cultural resources, and managed through legal or other effective means." This wide definition covers many different types of protected areas; in response to these differences the IUCN proposed several categories (Table 10.5). Protected areas under the strictest protection belong to Category I but five other categories have been defined. Yellowstone, for example, is an IUCN Category II protected area, a national park. Major protected area programs include the UNESCO World Heritage Sites, UNESCO Man and the Biosphere (MAB) sites, ASEAN Heritage Parks and Reserves, sites established under the Ramsar Convention on Wetlands, and the European Natura 2000 network.

Despite the progress made in establishing protected areas, some ecosystems or biomes remain poorly protected, particularly those experiencing the greatest amount of conversion (Hoekstra et al. 2005) (Figure 10.16). For example approximately 16% of the tropical humid forests are protected to some degree. The value of protected areas for vertebrates, particularly birds, is better understood than most taxa. Round (1985), for example, found that 88% of Thailand's birds occurred in protected areas and Sayer and Stuart (1988) concluded that 90% of vertebrate species associated with tropical moist forests in Africa would be protected with a slight increase in the area of tropical moist forests currently under some form of protection. Other regions are less adequately protected and the presence of a species within a protected area does not guarantee its long-term survival (Groombridge 1992). Moreover, we do not know how many insect species these areas actually protect. Nevertheless, Samways (2007) argued that the establishment of protected areas is the first basic principle of insect conservation.

At the global level, the development of the hotspot concept (see Chapter 9) has attracted much attention. It has been argued that conservation effort should focus on these areas (see Figure 9.15). However, these areas are not necessarily the most rich in species and other important areas are excluded (Brummitt & Lughadha 2003). Nor are the proposed global diversity hotspots the areas where endemism or threat to

Table 10.5 IUCN protected area management categories and definitions. (Chape et al. 2005.)

Category Ia *Strict nature reserve: protected area managed mainly for science*
- Area of land and/or sea possessing some outstanding or representative ecosystems, geological or physiological features and/or species, available primarily for scientific research and/or environmental monitoring

Category Ib *Wilderness area: protected area managed mainly for wilderness protection*
- Large area of unmodified or slightly modified land, and/or sea, retaining its natural character and influence, without permanent or significant habitation, which is protected and managed so as to preserve its natural condition

Category II *National park: protected area managed mainly for ecosystem protection and recreation*
- Natural area of land and/or sea, designated to: (i) protect the ecological integrity of one or more ecosystems for present and future generations, (ii) exclude exploitation or occupation inimical to the purposes of designation of the area, and (iii) provide a foundation for spiritual, scientific, educational, recreational, and visitor opportunities, all of which must be environmentally and culturally compatible

Category III *Natural monument: protected area managed mainly for conservation of specific natural features*
- Area containing one or more, specific natural or natural/cultural feature which is of outstanding or unique value because of its inherent rarity, representative or esthetic qualities, or cultural significance

Category IV *Habitat/species management area: protected area managed mainly for conservation through management intervention*
- Area of land and/or sea subject to active intervention for management purposes so as to ensure the maintenance of habitats and/or to meet the requirements of specific species

Category V *Protected landscape/seascape: protected area managed mainly for landscape/seascape conservation and recreation*
- Area of land, with coast and sea as appropriate, where the interaction of people and nature over time has produced an area of distinct character with significant esthetic, ecological, and/or cultural value, and often with high biological diversity. Safeguarding the integrity of this traditional interaction is vital to the protection, maintenance, and evolution of such an area

Category VI *Managed resource protected area: protected area managed mainly for the sustainable use of natural ecosystems*
- Area containing predominantly unmodified natural systems, managed to ensure long-term protection and maintenance of biological diversity, while providing at the same time a sustainable flow of natural products and services to meet community needs

biodiversity is greatest (Orme et al. 2005). Criticisms of the hotspot strategy for conservation have been contested by Myers and Mittermeier (2003), but as Whittaker et al. (2005) point out there is a need for a critical analysis of the best way to focus effort on conservation. Table 10.6 compares the hotspot approach with two other global protected area planning frameworks.

Protected areas are often seen to be the most important strategy for conserving biodiversity (Pimm et al. 2001). There is often, however, local opposition to the establishment of protected areas, such as the opposition to the Natura 2000 network in Europe (Stoll-Kleemann 2001). Despite this and the many other conflicts between the conservation of biodiversity and human activities, approaches for managing these conflicts are being developed (Stoll-Kleemann 2005; Young et al. 2005).

Another serious problem with protected areas is that following climate change they may cease to be suitable environments for the species that they exist to protect. Even the biodiversity of Yellowstone National Park is threatened by climate change (Bartlein et al. 1997). We need to know where plants and animals would move, if free to do so in a future climate, so that species conservation efforts can be concentrated where they will be effective (Watt et al. 1997c). There is little point in trying to conserve species where their environment is no longer suitable and, therefore, there is a need to identify areas where new protected areas (or conservation efforts in general) should be

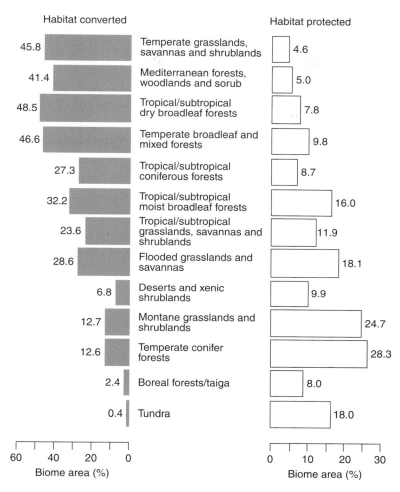

Habitat converted — Habitat protected

Value (converted)	Biome	Value (protected)
45.8	Temperate grasslands, savannas and shrublands	4.6
41.4	Mediterranean forests, woodlands and scrub	5.0
48.5	Tropical/subtropical dry broadleaf forests	7.8
46.6	Temperate broadleaf and mixed forests	9.8
27.3	Tropical/subtropical coniferous forests	8.7
32.2	Tropical/subtropical moist broadleaf forests	16.0
23.6	Tropical/subtropical grasslands, savannas and shrublands	11.9
28.6	Flooded grasslands and savannas	18.1
6.8	Deserts and xenic shrublands	9.9
12.7	Montane grasslands and shrublands	24.7
12.6	Temperate conifer forests	28.3
2.4	Boreal forests/taiga	8.0
0.4	Tundra	18.0

Biome area (%) — Biome area (%)

Figure 10.16 Amount of habitat conversion and area protected within each of the world's 13 terrestrial biomes. Biomes are ranked according to their relative degree of protection in relation to the least protected biome at the top. (From Hoekstra et al. 2005.)

targeted. Models are, however, now being developed to plan protected area networks suitable for future climates (Pyke & Fischer 2005). We can also take practical steps now by increasing the number of protected areas, particularly in areas of countries where they do not currently exist, particularly in areas with altitudinal variation.

The threat of climate change to biodiversity cannot, nevertheless, be overcome through the establishment of protected areas alone, although they should continue to form an important component of

a conservation strategy. The creation of "conservation corridors" (or "biological corridors") to link protected areas to promote the movement of species is seen as a useful approach to conserving biodiversity (Section 10.4.2) but it becomes particularly important as the climate changes (Williams et al. 2005). Past changes in climate occurred when the landscape was relatively intact. Movement of many species in response to future climate change will be severely restricted by the increasingly fragmented nature of our landscapes and suitable protected areas may be too far for the

Table 10.6 A typology of global protected area planning frameworks of global remit: biogeographic representation approaches, hotspot approaches, and the "important areas" approach. The first uses representative examples of ecosystem types identified using established biogeographic or ecological frameworks. Hotspot approaches target places rich in species and under threat. The "important areas" approach uses criteria such as the abundance of key species or the importance of a site for migration to identify priority areas. (From Whittaker et al. 2005.)

	Global protected area planning frameworks		
	Representation	**Hotspots**	**Important areas**
Basic idea	An example of each	Maximize number of species "saved" given available resources	Select a suite of sites that together protect key attributes of concern
Questions arising	What are the units of nature? What do we add next?	Where are places rich in species and under threat?	What are the key attributes and how can they be assessed?
Parameters	Vegetation formations Faunal regions Ecoregions	Species richness Species endemism Rate of habitat loss % Original habitat	Threatened species Endemic species Species assemblages Congregating species
Methods	% Dissimilarity Controlling factors	Species counts Threat/diversity indexes	Application of varied criteria
Schemes	Biogeographic provinces WWF ecoregions	Biodiversity Hotspots Endemic bird areas	Important bird areas Key biodiversity areas

threatened species to disperse to readily (Lawton 1995). This is one of several adaptation options. Another way of adapting to climate change includes the creation of landscape permeability to promote the spread of species (Kuefler & Haddad 2006), ultimately developing protocols to translocate species that cannot disperse fast enough.

Climate change is, of course, a threat to the environment generally, and so-called mitigation and adaptation methods are being developed and implemented (CBD 2003). Mitigation can be defined as an anthropogenic intervention to reduce the sources or enhance the sinks of greenhouse gases. There are two main approaches to mitigation: (i) reducing or limiting greenhouse gas emissions (e.g. the use of renewable energies); and (ii) the protection and enhancement of carbon sinks and reservoirs. The latter are mainly forestry-related activities to increase that amount of carbon sequestered from the atmosphere. Adaptation to climate change in its widest sense can be defined as adjustment in natural or human systems in response to actual or expected effects of climate change in order to moderate harmful effects or exploit beneficial opportunities. This may seem very far removed from

the conservation of insects but such is the threat of climate change that many ecologists identify the need for reducing greenhouse gas emissions and increasing carbon sequestration in order to reduce the rate of loss of global biodiversity (e.g. Thomas et al. 2004a). However, mitigation and adaptation activities can have detrimental effects on biodiversity (CBD 2003). The establishment of forest plantations is seen as a major approach to mitigation but when the primary concern is carbon sequestration, the resultant plantations may have poor biodiversity. Even some renewable energy methods may have an impact on biodiversity. There is concern, for example, that windfarms may have a negative impact on biodiversity, particularly for birds (Madders & Whitfield 2006). Climate change adaptation activities are more likely to have a negative impact on biodiversity. Sea walls, for example, will prevent flooding due to sea level rise but they are likely to have a negative impact on coastal habitats. Nevertheless, there is a growing awareness that an integrated approach is needed to address the problems of climate change and loss of biodiversity (CBD 2003) and practical measures are being developed (CBD 2005a, 2005b). These measures include ensuring that plantations established to sequester carbon are designed to maximize their biodiversity.

Clearly land use change and climate change, particularly working together, are the most serious threats to insect conservation and biodiversity generally. Ways of combating other threats, such as the introduction of alien species, need to be found. The Convention on Biological Diversity (CBD) is the main focus of effort on conserving biodiversity globally. It was signed by representatives of 150 countries at the Rio Earth Summit in 1992 and its objectives are (CBD 1992):

> the conservation of biological diversity, the sustainable use of its components and the fair and equitable sharing of the benefits arising out of the utilization of genetic resources, including by appropriate access to genetic resources and by appropriate transfer of relevant technologies, taking into account all rights over those resources and to technologies, and by appropriate funding. (CBD Article 1)

The CBD is therefore concerned with much more than the conservation of biodiversity and although some of its programs are focused on conservation (e.g. Mountain Biodiversity; Protected Areas; Climate Change and Biological Diversity), others relate much more to the use of biodiversity (e.g. Access to Genetic Resources and Benefit-sharing; Economics, Trade and Incentive Measures). A major goal of the CBD is to significantly halt the loss of biodiversity by 2010 (Balmford et al. 2005).

There are many other national and international efforts to conserve biodiversity and, in the last 20 years, non-governmental organizations have become a major force in conservation. Legislation and regulation play a role in conservation too, not always with positive outcomes (Smith et al. 2003). As well as legislation for protected areas (Dearden et al. 2005) and trade in endangered species (Reeve 2002), regulations also exists to limit the damage done by, for example, agricultural intensification. In many countries, agri-environment schemes exist to promote farming practices that are environmentally sensitive. These include, but are not restricted to, practices that are designed to conserve biodiversity. An individual scheme may have multiple objectives, making them difficult to evaluate (Carey et al. 2005). Agri-environment practices for biodiversity conservation include grass margins in cereal fields for butterflies (Field et al. 2005), planting pollen and nectar-rich plants for bumblebees (Pywell et al. 2006), mowing regimes for meadow butterflies (Johst et al. 2006), and the planting of seed-bearing crops for farmland birds (Stoate et al. 2004). The success of agri-environment schemes in conserving biodiversity has been questioned (Kleijn et al. 2001, 2006), although most studies on agri-environment schemes have been inadequate to evaluate their effectiveness for conservation (Kleijn & Sutherland 2003). However, there is evidence that specific measures can have benefits for insects and other species (e.g. Marshall et al. 2006; Pywell et al. 2006; Carvell et al. 2007) and that agri-environment schemes can have wider benefits at the landscape scale by being designed to increase habitat connectivity (Donald & Evans 2006).

10.4.2 Conserving insects

Although insect conservation should always remain closely linked to the conservation of biodiversity generally, there are limits to this approach, particularly for rare species of insects. Earlier in this book we discussed how the diversity of insects is generally poorly correlated with the diversity of other species. Panzer and Schwartz (1998), however, considered a "vegetation-based approach to insect conservation" but although they found that there was some degree of coincidence between plant family richness and insect species richness, they conclude that the conservation of rare insects required detailed knowledge of them.

Nevertheless, the knowledge of the habitat associations of threatened species is an important step towards their conservation. Although globally the habitat requirements of most insect species are not known, any information on the association between insect communities and particular habitats can help us to assess the risk posed by the loss of specific habitats to specific insect communities or insect species richness in general. For example, alpine meadows, pine forests, and freshwater lakes all have specific insect species associated with them and, therefore, any risk to these habitats represents a risk to these insect species.

There are two different strategies for conserving endangered species: "*in situ* conservation", the conservation of species within their natural or seminatural habitats, and "*ex situ* conservation", the conservation of species away from their natural habitats. *Ex situ* conservation is widely practiced for plants and vertebrates, particularly mammals, but only a few insects are held in zoos as part of an *ex situ* conservation strategy. Exceptions include the field cricket and the British wart-biter in London Zoo (Hodder & Bullock 1997).

The most effective approach to conserving insect species is to conserve them in their natural habitats. The full complexity of habitat conservation for insects cannot be covered here but there are four major ecological aspects to consider (Samways 2007).

1 All habitats important for insects should be retained. Applied at the landscape scale this means ensuring as much heterogeneity of suitable habitat as possible.

2 Habitat areas should be of sufficient size and shape to sustain insect populations.

3 There should be a sufficient number of areas of each habitat type, and these areas should be situated in such a way that the flow of individuals between different habitat areas ensures the long-term survival of the species unique to these habitats.

4 Appropriate habitat management strategies should be adopted. This may mean, for example, the simulation of natural disturbances.

The question of what comprises an important habitat is a crucial one. Because the ultimate aim of insect conservation is to ensure the survival of as many species as possible the most important habitats are those containing species that are solely dependent on those habitats for their survival. In some cases, the association between insects and habitats is well known from direct observation, particularly the habitat requirements of phytophagous insects. In other cases analyses of differences in insect species composition in different habitats (see Chapter 9) are needed to identify the species associated with particular habitats.

The question of habitat size was discussed in Section 10.3.1 and Chapter 1. The shape of habitat patches can be critical too. Usher (1995), for example, produced recommendations for the design of size and shape of "farm" woodlands for Macrolepidoptera (Figure 10.17). There has been a great deal of theoretical research on the relevance of the spatial design of habitat patches for insects and other species. This includes the study of metapopulation dynamics (Section 10.3.1). The metapopulation approach has led to specific recommendations on the conservation of particular species. Schtickzelle et al. (2005), for example, concluded that the metapopulation of the marsh fritillary butterfly *Euphydryas aurinia* in a Belgian landscape was destined for extinction (Section 10.3.1). They recommended a management regime of habitat restoration, enlargement of existing patches, and creation of new habitats in the landscape matrix.

In parallel to this research on metapopulations, although generally at a much greater spatial scale, there has been much practical consideration of the

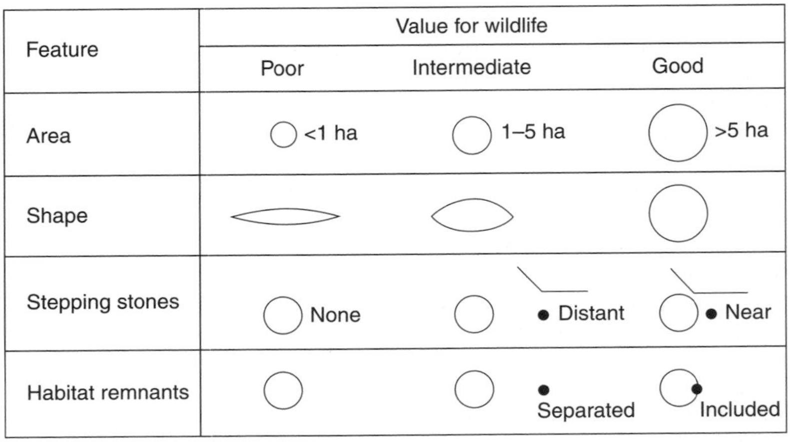

Feature	Value for wildlife		
	Poor	Intermediate	Good
Area	○ <1 ha	○ 1–5 ha	○ >5 ha
Shape	◁▷	◯	◯
Stepping stones	○ None	○ • Distant	○ • Near
Habitat remnants	○	○ • Separated	◯• Included

Figure 10.17 An assessment of the value of farm woodlands for Macrolepidoptera according to their area, shape the presence of stepping stones and habitat remnants. Woods are shown as open symbols, remnants as small black circles, and hedgerows as straight lines. (From Usher 1995.)

size, shape, and arrangement of protected areas. Much of this relates to general conservation objectives, notably whether protected areas should be single and large or several and small, the so-called SLOSS debate (e.g. Margules et al. 1982; Tscharntke et al. 2002; Whittaker et al. 2005; Van Teeffelen et al. 2006) and the idea of conservation corridors (e.g. Hobbs 1992; Rouget et al. 2006) linking protected areas.

Relatively few studies of the arrangement of, and connections between, protected areas have explicitly considered insects. However, several studies have focused on the design of farm woodlands, already mentioned above in relation to habitat patch size. Figure 10.17 also summarizes the value for Macrolepidoptera of placing farm woodlands near larger areas of woodland and of having other "stepping stone" patches of woodland. Much of this is ecological "common sense" but it is perhaps surprising how important habitat patch arrangement is for such mobile species as large moths.

Despite the importance of the biogeographic aspects of habitats discussed above, perhaps the most important aspect of conserving habitats for insects is habitat management. Returning to the example of farm woodland design, Dennis et al. (1995) considered a wide range of factors that might influence the number of arthropod species feeding on beech and other deciduous trees. He found that although factors such as woodland size and isolation had some effect, the most important factor was the amount of understory vegetation. Dennis concluded that the absence of an understory meant that the woods were more exposed and the overwintering stages of the insects associated with these tree species suffered greater mortality than insects in woodlands with understory vegetation. In this case, therefore, habitat island biogeographic effects are insignificant compared to habitat management.

One of the most interesting aspects of Dennis's work on farm woodland insects is that by focusing on phytophagous insects he knew what species were potentially present in his study areas and, therefore, which tree-feeding species could benefit from management. Habitat management often focuses on individual species and there are many cases where ecological research has identified the habitat requirements of individual species. The habitat requirements of temperate butterfly species have been particularly well studied. The heath fritillary (*Mellicta athalia*; Lepidoptera:

Nymphalidae), for example, requires sunny, sheltered woodland such as occurs in the first 3–4 years after coppicing and has become rare following the marked decline of this practice in England (Warren et al. 1984). The reintroduction of coppicing has been shown to benefit species such as the high brown fritillary (*Argynnis adippe*; Lepidoptera: Nymphalidae) (Usher 1995) (Figure 10.18). Most butterflies breeding in British woodland prefer very open, sunny rides or glades (i.e. less than 20% shade), but a few, such as *Pararge aegeria*, prefer shaded (40–90%) rides or glades (Warren & Key 1991). Although the habitat requirements of moths are poorly known in comparison to those of butterflies at least 60% of the 125 woodland Macrolepidoptera classified as "Nationally Notable" are thought to be associated with open woodland, rides, and clearings (Warren & Key 1991). The benefits of providing rides and glades in forests and woodland for Lepidoptera and other insects are such that precise recommendations for this form of management have been drawn up. For the least shade-tolerant species, Warren and Key cite minimum ride widths of 1–1.5 times the height of surrounding trees (Figure 10.19).

Apart from shade, the most important aspect of managing temperate woodland for insects is the provision of dead wood. Even modern conifer plantations can easily be managed to produce sunny conditions but the amount of dead wood in a forest takes many years to accumulate. The amount of dead

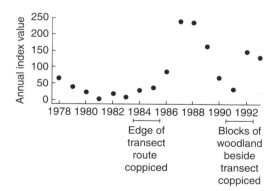

Figure 10.18 The effect of coppice management on the abundance of the high brown fritillary butterfly (*Argynnis adippe*) in a National Nature Reserve in England. The woodland was coppiced in about 1970 – other coppice dates marked. (From Usher 1995.)

Figure 10.19 Woodland ride in the UK cleared and widened to reduce shade.

Figure 10.20 Decaying stump of a rainforest tree.

wood in young plantations can be increased by cutting branches and felling trees but this tends to support only a limited number of deadwood species (Warren & Key 1991). Neither standing dead trees nor felled "healthy" trees have the diversity of niches for deadwood species that very old, naturally decaying trees have (Figure 10.20). Young felled trees rot from the outside (and rot develops inward) but old trees develop heart rot and rot holes. The size of decaying trees is important too: saproxylic insects tend to be sensitive to fluctuations in humidity but these are buffered in large trees and pieces of wood (Figure 10.21). In the absence of naturally decaying trees, there are a variety of ways of initiating premature decay: chainsaws to start the development of rot holes, inoculation of heart rot-inducing fungi, ring barking, fire, herbicides, and explosives (Warren & Key 1991).

The challenge for managing woodland for insects is to create habitat heterogeneity: variation in shade, for example, to benefit a range of butterflies and dead wood in a range of conditions to provide conditions for a range of saproxylic insects. The latter include species with contrasting requirements such as *Psilocephala melaleuca* (Diptera: Therevidae), which develops in very dry, red-rotted wood (Warren & Key 1991), to the many Diptera, Ephemeroptera, Trichoptera, and Coleoptera that live in dead wood in water (Dudley & Anderson 1982).

The above examples focused on insects in temperate woodlands. However, insect conservation is perhaps most urgently needed in tropical forests. These forests, despite covering only 8% of the land surface, probably contain 50–80% of the world's species (Stork 1988; Myers 1990). Few extinctions have been recorded in continental tropical forests but deforestation is a major threat to their biodiversity (Section 10.3.1).

Theoretical considerations of the impact of deforestation are no substitute for direct measurement of its impact on biodiversity. Accordingly, ecologists have tried to quantify the effects of deforestation by sampling insects in uncleared forest and areas cleared of forest for a range of purposes such as the establishment of forest plantations. In Cameroon, for example, several entomologists measured the effect of deforestation on ants, beetles, butterflies, and termites (Watt et al. 1997b, 1997c; Lawton et al. 1998; Watt et al. 2002; Stork et al. 2003). The species richness of each group of insects was assessed in: (i) forest with no evidence of previous clearance (primary forest); (ii) forest known to have been logged in the past (old secondary forest); (iii) plantations partially cleared (either manually or mechanically) and replanted with a West African tree species *Terminalia ivorensis*;

Figure 10.21 The variety of deadwood microhabitats, each with a distinctive fauna; 1, sun-baked wood; 2, fungus-infected bank; 3, fine branches and twigs; 4, bracket fungi; 5, birds' nests; 6, stumps; 7, hollow trees; 8, burnt wood; 9, large fallen timber; 10, dead outer branches; 11, rot-holes; 12, standing dead trees; 13, roots; 14, well-rotted timber; 15, wet fallen wood; 15, red-rotten heantwood. (From Kirby 1992.)

(iv) plantations completely cleared and replanted with *T. ivorensis*; and (v) fallow farmland. As might be expected the numbers of species of each group of insects are lowest in the fallow farmland (Figure 10.22). Although each group of insects shows a different response to forest disturbance, it is notable that, in terms of species richness, forest plantations can be comparable to uncleared forest for most of the insect groups studied. However, the type of plantation is crucial: plantations established after complete forest clearance are much poorer in insect diversity than plantations established after partial clearing of existing forest. In this study, the plantation tree was a native species: the contrast between the ecologically sensitive establishment of plantations of native trees and plantations of alien tree species established after complete forest clearance is likely to be even more marked than that observed in the Cameroon study. Sadly, reforestation projects are small in scale in comparison to deforestation. For example, during the 1980s, only $360 \, km^2$ of plantation forests were established each year in West Africa (Lawson 1994) in comparison to a rate of deforestation of an estimated $12,000 \, km^2$ annually (FAO 1991). In addition, these plantations often involved the use of alien tree species, have had a poor record of maintenance and survival, and were usually established in areas of existing forest (Lawson 1994).

Efforts to conserve insects frequently focus on individual species, particularly highly threatened ones. Hoyle and James (2005) considered the conservation of the world's smallest butterfly, the Sinai baton blue butterfly *Pseudophilotes sinaicus* (Lepidoptera: Lycaenidae). This species has a highly localized distribution because of its dependence on Sinai thyme, which only occurs in discrete patches in southern Sinai, Egypt, and is threatened both by climate change and by direct activities, mainly livestock grazing and the collection of medicinal plants. Hoyle and James concluded that, whatever its cause, the loss of habitat, particularly two specific patches that were considered very important for the persistence of this species, was a major risk to this species and that extinction was likely below a specified habitat area.

Another example is the silver-spotted skipper butterfly *Hesperia comma* (Lepidoptera: Hesperiidae) (Davies et al. 2005). This species became rare in Britain until, by 1982, less than 70 populations survived (Figure 10.23). *Hesperia comma* is restricted to a single host plant, sheep's fescue grass *Festuca*

ovina, and the decrease in the area of grazed calcareous grassland containing sheep's fescue grass caused the decline of this butterfly. However, a number of factors have led to the recovery of this species, including the recovery of rabbit populations after myxomatosis and the effect of climate warming on the butterfly's microhabitat. Most importantly, 698 management agreements were made under an agri-environment scheme, leading to an increase in traditional grassland grazing management. From 1982 to 2000 there was a 10-fold increase in the habitat area occupied by *H. comma* and a four-fold increase in the number of its populations. The agri-environment scheme was shown to be a major contributory factor (Figure 10.24). This example demonstrates the need to take a landscape approach to the conservation on individual species and illustrates the value of metapopulation models (Davies et al. 2005).

Although mainly perceived to be operating at a thematic level (see above), the CBD has been a major influence on the production of biodiversity action plans (BAPs) for the conservation of species and habitats. Many countries have BAPs, identifying national conservation priorities, and more detailed BAPs for species, habitats, and particular areas. The UK has individual action plans for 382 species and 45 habitats (www.ukbap.org.uk). Action on species and habitats is often coordinated under a local biodiversity action plan. In several cases species action plans are grouped because of the common policies and actions needed to conserve them. Thus, for example, six species are grouped together because of their association with exposed riverine sediments, usually shingle. These river shingle beetles comprise three species from the Carabidae, two Staphylinidae, and one Hydrophilidae (all from different genera). Although taxonomically different, their common habits mean that conservation action is more profitably considered for all six species. Another group is made up of 10 saproxylic beetles, all of which are associated with the dead wood of very old trees in deciduous woodlands. Inevitably, habitat action plans are relevant to particular species. Thus the specific action plan for lowland wood pasture and parkland takes this group of saproxylic beetles into account. Individual species action plans describe current status, factors causing decline, current conservation action, action plan objectives and targets, proposed actions, and links to other plans. The lead partner and organizations responsible for the proposed actions are also specified. Species covered by action

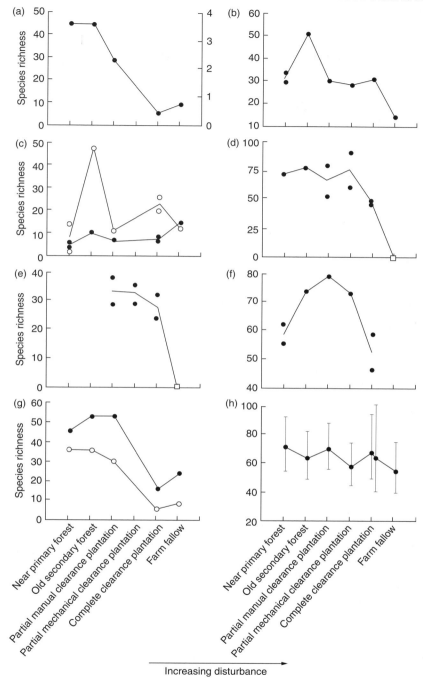

Figure 10.22 Trends in the number of insect and other species along a disturbance gradient in the Mbatmayo Forest Reserve in Southern Cameroon; (a) birds; (b) butterflies; (c) flying beetles (solid circles, malaise traps); (d) Canopy-dwelling beetles; (e) Canopy-dwelling ants; (f) leaf litter ants; (g) termites (two surveys shown); (h) nematodes (with 95% confidence limits). Note open symbols in (d) and (e) where the numbers of canopy-dwelling species are assumed to be zero in the absence of a canopy. (From Lawton et al. 1998. Reprinted by permission from *Nature*, © 1998 Macmillan Magazines Ltd.

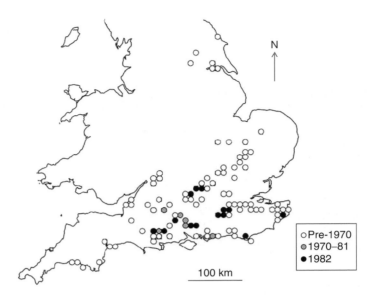

Figure 10.23 The decline of *Hesperia comma* in the UK using records from three time periods: pre-1970, 1970–81, and 1982. (From Davies et al. 2005.)

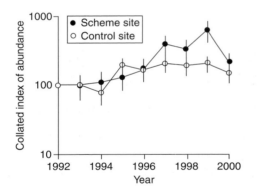

Figure 10.24 Abundance of *Hesperia comma* (mean population index and standard errors) between 1992 and 2000, on agri-environment scheme (Environmentally Sensitive Area and Countryside Stewardship Scheme) sites and control sites. The mean annual rate of increase was 22% and 9% on scheme and control sites, respectively. (From Davies et al. 2005.)

plans include the silver-spotted skipper butterfly (discussed above) and the cliff tiger beetle (*Cicindela germanica*; Coleoptera: Carabidae) (Table 10.7); these plans include details of future ecological research and monitoring requirements as well.

10.4.3 Restoring insect populations

When species have become locally extinct or are drastically reduced in number, the restoration of populations may be considered. More than 99% of the habitat of the Fender's blue butterfly *Icaricia icarioides fenderi* (Lepidoptera: Lycaenidae) has been lost in Oregon, USA, for example, prompting research to establish the best methods for restoring populations (Schultz 2001) in networks of connected patches (Schultz & Crone 2005).

Although the restoration of populations of threatened insects has been frequently done to prevent extinction, it has also been done following national extinction. In the UK, for example, the large blue butterfly *Maculinea arion* (Lepidoptera: Lycaenidae) became extinct in 1979 following a major loss of dry calcareous grassland habitat. Research on its complex ecology started many years before its extinction (Thomas 1995; Pullin 1996). The adult butterfly lays its eggs on thyme *Thymus praecox*, and the rapidly developing larvae feed for 2–3 weeks until, as final instar larvae, they fall to the ground. Here they are adopted by *Myrmica* ant workers who confuse them with their own larvae because the pheromones of the butterfly larvae mimic the pheromones of the *Myrmica*

Table 10.7 The UK Species Action Plan for the conservation of the cliff tiger beetle (*Cicindela germanica*). (From www. ukbap.org.uk with slight alterations. Plan coordinator: Roger Key.)

Current status
- *Cincindela germanica* occurs on, or near the base of, coastal cliffs or steep slopes, on bare or little-vegetated sand or silt near to freshwater seepages. It has, however, also been found on dry, gravelly but open situations. *Cicindela germanica* has an annual life cycle. It breeds in the summer and overwinters as larvae, sometimes at localized high densities, in vertical burrows in damp, open sand, or silt. Both larvae and adults are predatory on surface-active invertebrates, possibly chiefly on ants. Adults disperse by running rather than by flight
- This species used to occur along the English south coast from Hampshire to Devon, as well as in South Wales, and there are unconfirmed pre-1900 inland records from Berkshire and Kent. Since 1970 it has been found only on the coasts of Dorset and the Isle of Wight. Its wider distribution covers much of the Palearctic, from western Europe to eastern China, but it is generally local
- In Great Britain this species is now classified as "*Rare*"

Current factors causing loss or decline
- Cliff stabilization schemes
- Rapid coastal erosion exacerbated by coastal protection works
- Scrub encroachment on stabilized cliffs

Current action
Several sites are in current of proposed protected areas

Action plan objectives and targets
- Maintain populations at all known sites
- Restore populations to five suitable sites within the historic range by 2010

Proposed actions
1. Policy and legislation
- Address the requirements of this species in, for example, Shoreline Management Plans
2. Site safeguard and management
- Ensure that the species requirements are included in site objective and site management statements for all relevant protected areas
- Where possible, ensure that all occupied habitat is appropriately managed by 2008, including the maintenance of groundwater systems that supply natural freshwater seepages
- Ensure that the habitat requirements of this species are taken onto account in relevant coastal protection and development policies, plans, and proposals
- Consider establishing protected areas where it is necessary to secure the long-term protection and appropriate management of sites holding key populations of the species
3. Species management and protection
- Consider reintroducing *C. germanica* to a series of sites within the former range; if necessary to establish five new viable populations
4. Advisory
- Advise landowners and managers of the presence of the species and the importance of beneficial management for its conservation
5. Future research and monitoring
- Undertake surveys to determine the status of the species
- Conduct targeted ecological research to inform habitat management
- Establish an appropriate monitoring program for this species
- Pass information gathered during survey and monitoring of this species to a central database for incorporation into national and international databases
6. Communications and publicity
- Promote opportunities for the appreciation of the species and the conservation issues associated with soft-rock cliffs and coastal seepages

larvae. The workers take the butterfly larvae into their nest where they feed on the ant larvae. They remain in the ants' nest over winter, pupate and emerge as adult butterflies the following summer. As the butterfly moved closer to extinction in the UK, Jeremy Thomas and coworkers discovered that *Maculinea arion* can only survive in the nests of one ant species, *Myrmica sabuleti*; large blue larvae adopted by the workers of other ant species have very low survival rates (Thomas et al. 1989). Thus, although conditions were apparently suitable for the large blue butterfly with *Myrmica* nests present, it became extinct because the ant species present did not include *M. sabuleti*. The key to the successful restoration of the large blue butterfly, therefore, was to create conditions suitable for *M. sabuleti*, specifically closely grazed grassland. Grazing management programs, involving ponies in the winter and cattle in the summer, have been successful in creating conditions suitable both for the ant species specific to the large blue butterfly and also for its host plant, thyme. Thus only 4 years after its extinction in the UK, butterflies were successfully reintroduced from Sweden.

The key to the restoration of the large blue butterfly was the restoration of the habitat of the two species that it depends upon. Pullin (1996) gives other examples of restoration based on ecological research. Research on *Maculinea arion* has also been important in stimulating research on the ecological requirements of other rare *Maculinea* butterflies (Mouquet et al. 2005a, 2005b).

10.5 PROSPECTS FOR INSECT CONSERVATION

Although it had apparently become extinct in the UK in the early 1900s, a single population of the New Forest burnet moth *Zygaena viciae* (Lepidoptera: Zygaenidae) was found in Scotland in the 1960s (Young & Barbour 2004). The site, only about 1 ha in size, is situated between unstable basaltic cliffs and the sea. The moth was common in this site in the 1960s, then declined in number during the 1980s, until approximately 20 individuals were present in 1990 and the species was once again facing extinction

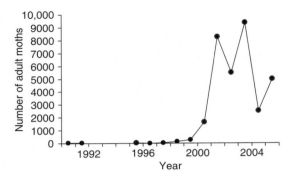

Figure 10.25 Abundance of the only known population of the New Forest burnet moth *Zygaena viciae* in the UK. (Data from Young & Barbour 2004; M.R. Young, personal communication; no estimates available for 1992–4.)

in the UK. Research on the New Forest burnet moth quickly led to the conclusion that overgrazing was causing its decline and a fence was erected to exclude sheep. Although storms have periodically damaged the fence around this exposed Atlantic site, the population steadily recovered to 2000–10,000 by 2000–5, the moth clearly benefiting from action taken to reduce grazing (Young & Barbour 2004) (Figure 10.25).

The conservation of biodiversity often seems to be an impossible task but, as this and other examples show, there is much that can be done practically to conserve insects and other species though the application of knowledge gained through research. Our knowledge of many endangered temperate insect species is remarkably good. Some of these insects are covered by protective legislation and conservation action plans have been prepared for many of them. In some cases, we know enough to be able to reintroduce species that have gone locally or nationally extinct. For example, when the large blue butterfly, *Maculinea arion*, went extinct in England, knowledge of its ecology was good enough to allow a successful reintroduction program (Section 10.4.3). Successful conservation of insects generally, particularly in the tropics and under the threat of increasing habitat loss and climate change, requires much more research on insect ecology.

Chapter 11

INSECTS AND DISEASES

11.1 INTRODUCTION

We discussed the ecology of insects that feed on plants (herbivores) in Chapter 3, and the ecology of insect-feeding insects (predators and parasitoids) in Chapter 5. Some insects such as mosquitoes and bedbugs also feed on vertebrates. In fact, we can draw parallels between this latest habit and that of sap feeding on plants, as both systems rely predominantly (though by no means exclusively) on piercing mouthparts, and utilize the circulatory fluids of the host, either sap or blood, as the primary food supply. During these feeding bouts, both plant- and animal-feeding insects may come into contact with pathogenic organisms present in the fluids of the host. As a result, evolution has forged some fascinating relationships between fluid-feeding insects and the pathogenic organisms that they spread. The insect, known as the vector, may or may not benefit from this association, but the host, be it a plant or animal, certainly does not. It becomes infected with pathogens, which cause disease.

11.2 DISEASES AND PATHOGENS

The dictionary definition of the term "disease" is loose and woolly; many forms of deterioration in the health of a plant or animal might be thought of as a disease, from heart problems in humans to drought stress in trees. In this chapter, we consider only those animal and plant diseases that are caused by the actions of pathogenic organisms with which insects have some sort of association. Even the term "pathogen" seems hard to define satisfactorily. Clearly, a pathogen is a type of parasite, but not all parasites are pathogens. Thus, fleas are well-known parasites of mammals and birds, whereas the disease-causing organisms (bacteria in this case) that they transmit to cause bubonic plague are pathogens. Perhaps the best way of defining pathogen is by description, without worrying too much about the reasons for the classification. In this context, a variety of pathogenic organisms commonly have associations with insects, including bacteria, fungi, nematodes, protozoans, spirochetes, phytoplasmas, and viruses.

Pathogens "in the wild" are thought to be powerful evolutionary forces that determine the structure and dynamics of plant and animal communities (Burdon et al. 2006). On the other hand, pathogens that infect our crops, our livestock, and indeed ourselves can cause enormous damage, hardship, economic loss, death, and debilitation. Most of the really prominent animal (including human) diseases are linked in some ways to arthropod (insect and tick) associations, including dengue, yellow fever, plague, malaria, leishmaniasis, encephalitis, chagas, sleeping sickness, and so on (Rathor 2000; Pugachev et al. 2003; Barrett & Higgs 2007). The picture is no better on the plant side, with cereal yellows, mosaic viruses, phytoplasms, and Dutch elm disease all global problems of trees and agricultural or horticultural crops (Weintraub & Beanland 2006). Some of these major diseases have been with us for centuries, if not thousands of years, (bubonic plague for example), whereas others are more recent (such as Japanese B encephalitis). Most worrisome are the so-called re-emergent diseases which were thought to have subsided or been eradicated, but now seem to be on the increase again; malaria is a case in point, as is dengue fever,

Table 11.1 The current burden of vector-borne human diseases. (Data from WHO Special Programme for Research and Training in Tropical Diseases; From Hemingway et al. 2006.)

Disease	Vector	Disease burden DALYs (thousands)	Deaths per year (thousands)
Malaria	*Anopheles* mosquitoes	42 280	1124
Dengue	*Aedes* mosquitoes	653	21
Lymphatic filariasis	*Anopheles* and *Culex* mosquitoes	5644	0
Leishmaniasis	Sandflies	2357	59
Chagas disease	Triatomid bugs	649	13

DALYs, disability-adjusted life years.

a mosquito-vectored disease that has increased globally four-fold since 1970, with 2.5 billion people now at risk from it (Derouich & Boutayeb 2006).

The effects on human populations of these diseases are often quoted but frankly are rather difficult to verify. None the less, the available statistics are extremely worrying. Yellow fever, for example, is still a major public health problem, especially in Africa and South America (Barrett & Higgs 2007). The World Health Organization (WHO) estimates that there are more than 200,000 cases of yellow fever annually, including 30,000 deaths. Malaria causes more than 300 million acute illnesses globally every year, of which at least a million die of the disease (WHO 2006). Worse yet, the number of cases of both yellow fever and malaria are increasing dramatically, despite the availability of prophylactics, vaccines, vector management tactics, and so on. Table 11.1 summarizes some of the worst (Hemingway et al. 2006). Note that a DALY (disability-adjusted life year) is a quantitative "burden of disease", reflecting the amount of healthy life lost to the particular cause or syndrome.

Monetary losses from pathogens can be extreme. For example, tick- and tsetse fly-borne diseases are said to cost Africa US$4–5 billion per year in livestock production losses (Eisler et al. 2003). The livestock disease nagana (related to sleeping sickness in humans) is thought to cost another US$4.5 billion in Africa (Gooding & Krafsur 2005) due to losses in production and heavy expenditure on drugs and vector management tactics. Many factors combine together to exacerbate vector pathogen and disease problems. These range from climate change (Hunter 2003b; Toussaint et al. 2006) to vector immunity (Dimopoulos 2003), to the underuse of real antipathogen drugs and the overuse of fake ones (Dondorp et al. 2004), or even to the effects of development and urbanization (Tauil 2006). Controversy, of course, is rife among experts apportioning blame for the continuation or re-emergence of diseases. In this chapter, we present a review of some major diseases and their insect associations. We also consider the potential for pest management as a system for improving human health (Baumgartner et al. 2003).

11.3 PATHOGEN SPREAD

Many pathogens move from one host individual to the next without specific or specialized associations with vector organisms. Take for example the fungus *Verticillium albo-atrum*, which infects the important crop alfalfa in Europe and North America, and causes the disease known as verticillium wilt (Huang 2003). The pathogen is able to spread by various methods. It can travel as spores on and in infected alfalfa seeds or other crop debris, via contact between the roots of adjacent plants, and even on agricultural equipment. The fungus can also blow on the wind or be carried by water. Clearly, the pathogen is a real opportunist, hitching a ride on any means readily available, which includes a variety of "innocent" insect species. In Canada, pests, pollinators, and predators are all able to carry the fungus on their bodies, so that aphids, weevils, grasshoppers, and bees are all implicated in the spread of the disease. Only in a general and very non-specific sense can they be classed as "pathogen vectors".

11.3.1 Insects as non-vectors of disease-causing pathogens

Of course, huge numbers of insect species feed on animals and plants without ever vectoring pathogens. Blood feeding, for example, is essential for the development of a great many insects and other arthropods (Sanders et al. 2003; Yuval 2006). In most cases, the process is relatively benign, if a trifle irritating. As an aside, fluid feeding is actually fairly risky for the feeder too (Hurd 2003) because most animals and plants have defensive responses that can cause harm if not death to their attackers.

Insects can also predispose animals and plants to pathogen infection without acting as vectors. Simply put, in the absence of attack by these non-vectors, the host would remain unavailable or resistant to infection. For example, larvae of screwworm flies, *Chrysomya bezziana* in the Old World and *Cochliomyia hominivorax* in the New World (Diptera: Calliphoridae), attack the flesh of a variety of mammals, causing a "disease" known as myiasis (Alexander 2006). In itself, no pathogen seems directly involved, with most damage to the host being caused by the feeding and burrowing activities of the fly maggots. However bacterial infections routinely and rapidly follow screwworm attack, so that treatment not only involves clearing up the insect infestation, but high doses of antibiotics to combat the secondary infections are also required.

An excellent example of a non-vector disease from plant pathology is beech bark disease. This serious malady infects European beech, *Fagus sylvatica*, in Europe and North America, and causes significant tree mortality (Munck & Manion 2006). The process begins when the beech scale insect, *Cryptococcus fagisuga* (Homoptera: Eriococcidae), uses its long piercing mouthparts to penetrate beech bark from the outside to feed on the living bark cambium inside (Houston 1994) (Figure 11.1). This feeding predisposes the bark to infection by airborne fungi, *Nectria coccinea* var. *coccinea* in Europe, and *N. coccinea* var. *faginata* and *N. galligena* in North America. Heavy infestations of

Figure 11.1 Beech scale on tree trunk.

the scale insect allow the weakly pathogenic *Nectria* spp. to spread rapidly within tree bark, unrestricted by defense reactions such as wound periderm (where special tissues develop around wounds and act as a barrier to infection) or callous formation. Without the scale insect, the fungus is usually unable to infect healthy trees on suitable soils. First instar scale "crawlers" move over the bark of host trees to establish their own feeding sites, and only a relatively small proportion move to adjacent trees on air currents within a stand. Only around 1% of crawlers spread greater distances, facilitated by wind above the beech canopies (Speight & Wainhouse 1989). New outbreaks are characterized by rapid population buildup on individual trees; established outbreaks, on the other hand, tend to fluctuate around lower infestation levels. Many factors are thought to contribute to variations in scale insect density, including host-plant resistance, weather conditions, and natural enemies (Wiggins et al. 2004). As the disease relies predominantly on the presence of the insect, *Nectria* epidemiology closely mimics that of *Cryptococcus*. Beeches free of the insect cannot be infected and subsequently killed by the fungus, so that though no pathogen transmission occurs, the insect is vital in the life of the fungus.

Table 11.2 Examples of insects that can act as vectors of pathogens, and the diseases with which they are associated (note that most genera consist of several species worldwide). (From Jeger et al. 2004.)

Insect genus or species	Order: family	Common name	Pathogen	Disease or syndrome
Animal diseases				
Aedes	Diptera: Culicidae	Mosquito	Virus	Rift Valley fever
Aedes	Diptera: Culicidae	Mosquito	Virus	Equine encephalitis
Aedes	Diptera: Culicidae	Mosquito	Virus	Yellow fever
Aedes	Diptera: Culicidae	Mosquito	Virus	Dengue fever
Anopheles	Diptera: Culicidae	Mosquito	Protozoan	Malaria
Chrysops	Diptera: Tabanidae	Horse fly	Nematode (filerial)	Loa loa (loiasis)
Culex	Diptera: Culicidae	Mosquito	Virus	Japanese encephalitis
Glossina	Diptera: Glossinidae	Tsetse fly	Protozoan (flagellate)	Sleeping sickness, nagana
Lutzomyia	Diptera: Psychodidae	Sandfly	Protozoan	Leishmaniasis
Simulium	Diptera: Simuliidae	Blackfly	Nematode (filarial)	Onchocerciasis
Triatoma	Hemiptera: Triatomidae	Kissing bug	Protozoan	Chagas disease
Xenopsylla cheopis	Siphonaptora: Pulicidae	Flea	Bacterium	Bubonic plague
Plant diseases				
Delphacodes	Hemiptera: Delphacidae	Plant-hopper	Virus	Maize dwarf virus
Monochamus alternatus	Coleoptera: Cerambycidae	Longhorn beetle	Nematode	Pine will
Myzus persicae	Hemiptera: Aphididae	Aphid	Virus	Sugar beet yellows
Scaphoidous titanus	Hemiptera: Cicadellidae	Leaf-hopper	Phytoplasma	Asters yellows
Scolytus	Coleoptera: Scolytidae	Bark beede	Fungus	Dutch elm disease
Sirex/Urocerus	Hymenoptera: Siricidae	Wood wasp	Fungus	Wet rot
Sitobion avenae	Hemiptera: Aphididae	Aphid	Virus	Barley yellow dwarf virus
Thrips	Thysanoptera: Thripidae	Thrips	Virus	Tomato spotted wilt

11.3.2   Insects as vectors of disease-causing pathogens

When insects carry pathogens on the outside or inside of their bodies, and transmit them to host organisms, we call them vectors. Undoubtedly, insects drive many disease epidemics (Jeger et al. 2004) and a great number of different insect groups act as vectors for an equally wide range of disease-causing pathogens. Table 11.2 describes just a few of these vectors and their associated diseases. Some of the most serious will be considered individually below, but we consider a few general features here. Taxonomically, pathogen transmission is fairly widespread within the Insecta. Two major vector orders, the Hemiptera and Diptera, are somewhat analogous as vectors of (mostly) plant and animal pathogens, respectively. Both have mouthparts derived from generalist mandibles, maxillae, and labiae, now adapted for the highly specialized function of piercing the epidermis of the host and transferring liquid contents to the insect's buccal cavity or mouth. Both transmit a whole range of pathogen types, though, with notable exceptions such as malaria and onchocerciasis, animal and plant viruses form the major group. However, there is one huge and important difference between the two orders. Juvenile stages of the exopterygote Hemiptera (nymphs) tend to exhibit much the same lifestyle as their parents, whereas larvae of the endopterygote Diptera have extremely different ecologies from the adults. Thus, only adult tsetse flies can act as vectors, whereas all stages and both sexes of plant-hoppers and aphids can, in theory, transmit pathogens. Bed bugs (Hemiptera: Cimicidae) and other blood-feeding Hemiptera are in principle much better vectors of pathogens than are mosquitoes or other dipterans, because young and old, nymph and adult, cimicids feed on mammalian hosts (Siva-Jothy 2006), whereas only adult female mosquitoes are blood-feeders.

Insects require certain morphological and behavioral traits to be effective pathogen vectors. In most cases (but see below), the ability to move is fundamental to the transmission of pathogens from one host individual to another. Typically, good disease vectors can move over long distances by flying and short distances by hopping or crawling, both within and between crops or animal hosts. Many disease-causing pathogens complete most of their life cycle inside the host organism (fungal fruiting bodies being one obvious exception), so insect vectors must have a means of penetrating the host's epidermis to collect pathogens from its circulatory or transport system. Vectors usually penetrate the host epidermis with piercing, chewing, or, in the case of sandflies, rasping mouthparts.

The majority of pathogens are fairly host-specific. Thus for example, barley yellow dwarf virus attacks only certain cereal species, whereas malaria is essentially a human disease (though non-cross-infective forms are also found in birds and monkeys). When pathogens have alternative or intermediate hosts, an efficient vector should restrict its own feeding habits to the same narrow host range, to avoid depositing pathogens in unsuitable hosts. When we consider all of the traits required to be an efficient pathogen vector, it is not at all surprising that insects are so well known in this capacity, with their excellent dispersal abilities, adaptable mouthpart structures, and complex host associations.

The majority of plant-feeding insects tend to transmit only one type of pathogen. For example, many aphids, whitefly, plant-hoppers, and thrips transmit primarily viruses. The plant viruses that lead to diseases such as leaf curl (Briddon 2003), spotted wilt (de Assis et al. 2004), citrus tristeza (McKenzie et al. 2004), papaya phytoplasma (Padovan & Gibb 2001), and sugar cane Fiji disease (Smith & Candy 2004) are largely both vector- and host-specific. Not only that, but specific viruses have specific relationships with their insect vectors, as detailed in Table 11.3 (Ng & Falk 2006). The epidemiology of non-persistent, semipersistent, etc. viruses differ markedly, as does their control (see later), and as Table 11.3 shows, the pathogen–vector relation can be relatively loose, as with non-persistent viruses, or extremely complex and intimate, as with persistent-propagative viruses. In the latter situation, not only does the pathogen multiply inside the insect, it can also be transmitted from mother to offspring in some cases, as with some phytoplasms (Weintraub & Beanland 2006).

In comparison, insect vectors that feed on animals can transmit a wide variety of pathogens. Hence, mosquitoes as a group are notorious vectors of every kind of animal pathogen from nematodes (Cancrini et al. 2003), through viruses (Russell 2002), to protozoa (Sachs 2002). The term arbovirus (arthropod-borne virus) is used to describe the wide range of animal viral diseases transmitted by arthropods – everything from yellow fever to Ross River virus. There are even some suggestions that human immunodeficiency virus (HIV) may be transmitted passively from apes to other apes (humans in this case) by vectors such as stable flies (Eigen et al. 2002). Supposedly immobile parasites such as human body lice (*Pediculus humanus*: Anoplura) are capable of carrying viral and bacterial pathogens that are the causative agents of diseases such as trench fever and typhus (Alsmark et al. 2004).

Pathogen transmission by insect vectors ranges from entirely accidental or facultative (as with the example of verticillium wilt in alfalfa mentioned above) to obligate. Reference to a few examples will illustrate this point. Cockroaches or houseflies walking over contaminated food can spread enterobacterial infections such as *Shigella* (Pellegrini et al. 1992) by passive transport on their bodies, legs, and mouthparts (Ugbogu et al. 2006). This sort of passive transmission also typifies certain plant diseases. Pine pitch canker is a common disease of conifer cones, caused by the fungus *Fusarium* spp. In California, pine pitch canker appears to be transmitted on the bodies of various species of small beetle that tunnel into cones and twigs of both native and exotic pines

Table 11.3 Retention and transmission characteristics of plant viruses (From Ng & Falk 2006.)

Transmission characteristic	Transmission relationship term			
	Nonpersistent	Semipersistent	Persistent-circulative	Persistent-propagative
Acquisition time	Sec-min	Min-hour	Hours	Hours
Retention time	Min-hours	Hours-days	Days-life	Days-life
Transtadial passage	No	No	Yes	Yes
Virus in hemolymph	No	No	Yes	Yes
Latent period	No	No	Hours	Days-weeks
Replication	No	No	No	Yes
Transovarial	No	No	No	Often
Example	TEV	BYV	BWYV	MMV

BWYV, Beet western yellows polerovirus; BYV, Beet yellows dosterovirus; MMV, Maize mosaic nuclearbabdovirus; TEV, Tobacco etch polyvirus

(Hoover et al. 1996). Simple artificial wounding of cones does not produce canker, showing that the beetles are required to vector the fungus into the cones, albeit accidentally on the part of the insect. In complete contrast, on the same genus of tree, the woodwasp *Sirex* spp. (Hymenoptera: Siricidae) is entirely dependent on the fungus *Amylostereum* spp. (Martinez et al. 2006). As the wasp vectors the fungus from one stressed pine tree to another, the fungus breaks down the host's heartwood to the point that woodwasp larvae can feed on otherwise unusable timber. The fungus, in return, is transported between ephemeral hosts, and in fact cannot live without its wasp symbiont (see also Chapter 6).

11.4 VECTOR ECOLOGY

Insect vectors are influenced by many abiotic and biotic factors, just as are other insects. For example, weather conditions may affect growth, reproduction, and dispersal (see Chapter 2), competition and natural enemies may limit their abundance (see Chapters 4 and 5), and the availability of both major and alternative hosts may influence their persistence in a habitat and their spread among habitats. We provide some general principles of vector ecology here, with further discussion during the detailed examples of specific insect-vectored diseases presented later.

11.4.1 Epidemiological theory

Let us consider briefly one or two aspects of epidemiological theory that are relevant to insect vectors of disease. We will need some basic understanding of vector–host–pathogen interactions in order to appreciate the case studies that are provided later in the chapter. For those interested in the ecology and mathematics of disease transmission, we strongly recommend reading Anderson and May's (1981) classic paper on the subject. That paper provides a framework from which more recent examples can be explored.

Here, we stick to a few basics. Fundamental to epidemiology is the basic case reproduction number (labeled R_0), which is defined as the average number of secondary cases of a disease arising from each primary one in a particular population of susceptible hosts (Dye 1992). A second important parameter is the vectorial capacity (labeled C), defined as the daily rate at which future inoculations arise from a currently infective population. Vector survival is most critical here; if, for example, a pathogen has to develop inside the body of the vector before being deposited in a new host (the time taken for this development is known as the extrinsic incubation period, which can be considered to be a latent period), then it is clearly vital that the vector lives longer than the pathogen incubation period. If the vector dies before its "on-board" pathogens are ready for inoculation, then of course transmission will never occur!

Two related equations, first derived in studies on malaria, describe the derivation of R_0 and C:

$$R_0 = \frac{ma^2bp^n}{-r \ln p}$$

$$C = \frac{ma^2bp^n}{-\ln p}$$

where m is the number of vectors per host, a is the number of times the vector bites the host per day, p is the daily survival rate of the vector, n is the extrinsic incubation period mentioned above, b is the proportion of vectors that actually generate a new infection in the host when it is bitten, and r is the daily rate of recovery of the infected host.

These somewhat cryptic equations are simply split into ma, the biting rate, and $pn/-\ln p$, the so-called longevity factor, and R_0 can be estimated as C/r (D.W. Kelly, unpublished data). So what does all this mean for vector ecology? Simply put, if R_0 is greater than unity (i.e. the recovery rate is less than the vectorial capacity), then the disease will spread through a host population, whereas if R_0 is less than unity (i.e. the recovery rate is greater than the vectorial capacity), the disease will become eradicated over time.

11.4.2   Host range

As we mentioned earlier, some vector species may be very specific to the hosts on which they feed, and hence are able to vector pathogens between only a narrow range of plants or animals. Elm bark beetles such as *Scolytus scolytus* and *S. multistriatus* (Coleoptera: Scolytidae) feed exclusively on trees in the genus *Ulmus*, and hence the fungi for which they act as vectors (Section 11.11.5) can never venture beyond this narrow host range. The transmission of mutualist fungi by bark beetles is discussed in more detail in Chapter 6.

Not all potential vectors have the ability to transmit pathogens between hosts, and pathogens in turn may be specific to certain species of vector. This is especially the case when pathogens replicate inside the vector which has acquired them before infecting a new host. The mosquitoes *Culex quinquefasciatus*

and *Aedes aegypti* (Diptera: Culicidae) are both renowned as vectors of a range of arboviruses (arthropod-borne viruses), though they do not perform this task equally well. Dengue fever is widespread over Southeast Asia and Central and South America, and although rarely fatal, causes serious aches, high fevers, extreme fatigue, and rashes (Schwartz et al. 1996). Dengue virus replication in the mosquito vector, however, does not occur in *Culex*, only in *Aedes*, so that only species in the latter genus can serve as vectors (Huang et al. 1992; Lourenco de Oliveira et al. 2004).

More often, though, insect vectors have a fairly wide range of hosts, many of which may act as reservoirs of pathogen infection. Reservoirs can maintain a disease within a locality even when the primary (or medically/economically important) host is absent. In contrast, a lack of reservoirs may reduce pathogen transmission. Blackflies (Diptera: Simuliidae), in the genus *Simulium*, are vectors of onchocerciasis (river-blindness) in humans (see Section 11.11.1). However, no species of blackfly feeds exclusively on humans (Service 2004), and some of those that do bite people may do so only when other large mammals such as donkeys or cattle are unavailable. In this particular case, the filarial nematode, *Onchocerca volvulus*, that causes river-blindness does not have any non-human reservoirs. As a consequence, village livestock may reduce disease transmission by keeping the vectors (blackfly) away from humans.

Bubonic plague (the Black Death of medieval Europe, see Section 11.11.4) is caused by the bacterium *Yersinia pestis* and is vectored by fleas (Siphonaptera). In urban areas, epidemics are primarily restricted to two hosts, rats and humans. However, outside urban areas, plague is primarily a disease of wild rodents and other mammals, including gerbils, voles, chipmunks, and ground squirrels (Service 2004). Plague provides a good example of how human activities can localize and focus a naturally occurring epizootic by forcing an otherwise wide host range into a narrow one, with extreme consequences. We term human diseases with animal reservoirs zoonoses, as opposed to anthroponoses, where only humans are involved.

Somewhat unlikely mammals may act as alternative hosts for insects that vector human diseases. One of the vectors of Chagas disease in South America, the bug *Triatoma sordida* (Heteroptera: Triatomidae), utilizes various food sources as well as humans.

Triatoma occur naturally in various habitats, including hardwood forests, where they can be found inhabiting areas dominated by trees, logs, cacti, or even bromeliads (Wisnivesky-Colli et al. 1997). In such diverse habitats, *Triatoma* encounter many mammal species, including humans, though most triatomes do not actually use human dwellings for habitats (dos Santos et al. 2005). Opossums for example provide an important alternative to humans, and a significant source of pathogen infection (Diotaiuti et al. 1993).

Many mosquitoes show wide host ranges too. As mentioned above, *Aedes albopictus*, which was accidentally introduced into North America in the 1980s, is a vector of dengue fever in Southeast Asia. Precipitin (antibody reaction) and ELISA (enzyme-linked immunosorbent assay) tests were carried out on wild-caught adult female mosquitoes to investigate the animals on which they had been feeding (Savage et al. 1993). Some of the results of these tests are shown in Figure 11.2. The mosquitoes not only fed on a wide range of mammalian species, but 17% of those tested had also fed on various birds. *Ae. albopictus* is clearly an opportunistic feeder with a wide range of hosts, and hence has the potential to become a vector of indigenous arboviruses. To make matters worse, many closely related vector species are able to successfully transmit the same pathogens. At the same time, related vector species often have subtle differences in their ecologies that allow them to exploit different environments. As a result, a single pathogen may occur over a huge geographic range

because it is vectored by a number of related species that are adapted to different habitats. There are, for example, more than 20 species of tsetse fly in the genus *Glossina* that are capable and indeed responsible for the spread of human trypanosomiasis (sleeping sickness) in Africa (Belete et al. 2004). Because each species is adapted to a different habitat, the disease remains endemic over a wide geographic range.

Of course, disease agents as well as vectors have close relatives. It is interesting (and a little worrying) to consider how many human diseases have close relatives in other animal species. Paul et al. (2003) review the literature on the genus *Plasmodium*, normally identified as the primary agent of human malaria. The genus is in fact estimated to include 172 species, 89 of which occur in reptiles, 32 in birds, and 51 in mammals. A mere two of the last group, *Plasmodium vivax* and *P. falciparum*, are known to cause serious disease in humans (Section 11.11.3). With this huge diversity of pathogens, it is not surprising to find that a large variety of vectors are implicated in their transmission from host to host, not just mosquitoes. It is sobering to realize that, as Paul et al. (2003) point out, the natural vectors of the majority of *Plasmodium* species remain unknown.

Some plant disease vectors may also show wide host ranges and hence are able to vector plant pathogens to and from large numbers of host species. The peach–potato aphid, *Myzus persicae* (Homoptera: Aphididae), is much less host-specific than its common name suggests. Studies in the UK showed that this aphid is able to transmit beet yellows virus (BYV) and/or beet mild yellowing virus (BMYV), both serious diseases of sugar beet, to nine species of common arable weed (Stevens et al. 1994). These include *Stellaria media* (chickweed), *Papaver rhoeas* (poppy), and *Capsella bursa-pastoris* (shepherd's purse). The weeds are likely to play an important role in the persistence of disease-causing pathogens in arable crops. Whitefly (Homoptera: Aleyrodidae) are also important vectors of plant viruses. In various parts of India, adults of *Bemisia tabaci* act as vectors of tobacco leaf curl virus (TbLCV). The virus causes a variety of debilitating symptoms in tobacco, from severe leaf curling and paling to irregular swellings and thickening of the veins (Valand & Muniyappa 1992). However, the whitefly does not confine its attentions to tobacco, and is able to transmit TbLCV to 35 other plant species, some of which are shown

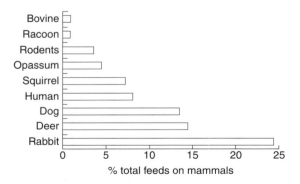

Figure 11.2 Mammalian host-feeding patterns of *Aedes albopictus* in North America as percentages of positive reactions in precipitin and ELISA tests. (From Savage et al. 1993.)

Table 11.4 Indication of the wide host plant range to which the whitefly *Bemisia tabaci* transmits tobacco leaf curl virus. (From Valand & Muniyappa 1992.)

Plant species	Common name
Beta vulgaris	Beetroot
Capsicum annuum	Sweet pepper
Carica papaya	Papaya
Cyamopsis tetragonoloba	Cluster bean or guar
Lycopersicon esculentum	Tomato
Phaseolus vulgaris	Bean
Petunia hybrida	Petunia
Sesamum indicum	Sesame

in Table 11.4. Once more, the role of non-crop reservoirs in the epidemiology and management of insect-vectored diseases needs very careful appraisal (Arli-Sokmen et al. 2005).

In reality, any specificity between vectors and the pathogens that they transmit seems to depend primarily on complex physiological interactions between vector and pathogen (see below). Gildow et al. (2000) showed that virus-vector cellular interactions regulated the transmission of a disease called soybean dwarf luteovirus. Though aphid vectors that were feeding on infected plants could acquire various type of plant virus, complex biochemical and cell membrane components, located in particular in the hindgut and salivary glands, resulted in highly specific virus transmission.

11.5 VECTOR DISPERSAL

From the pathogen's point of view, the prime importance of an insect vector is its ability to transport the pathogen from one host to another. As you might expect, the amount of vector movement can have a great influence on disease dynamics and spatial distribution (Ferriss & Berger 1993). Plant disease vectors may travel only short distances from one plant in a crop to an adjacent one, or over relatively long ones from distant reservoirs to crop fields. In this way, pathogens such as plant viruses can spread rapidly from mixed perennial pastures to commercial crops (Coutts & Jones 2002).

Insects that are able to fly, jump, or simply crawl may act as disease vectors. Even the least mobile of insects seem capable of some vector transmission. Mealybugs (Hemiptera: Pseudococcidae) are a very sedentary group of sap-feeders. Neither the nymphs nor the adults move very much, and the only really mobile stage is the first instar nymph, known as a crawler because of the way in which it moves around. The citrus mealybug, *Planococcus citri*, despite its name, is a polyphagous species found over most of Europe and North America. One of its claims to fame is as a vector of grapevine leafroll virus (GLR) in the Galician wine-growing region of northwestern Spain (Cabaleiro & Segura 1997), at least over short distances within a vineyard. The same vine disease is also found from Europe to America, and Australia to South Africa, and it causes various symptoms on commercial grape vines including uneven and retarded ripening of fruit, and small, pale-colored grapes that are difficult or impossible to export (Saayman & Lambrechts 1993). Scale insects (Hemiptera: Coccoidea) are even less mobile than mealybugs, though one species, *Pulvinaria vitis*, is also said to be able to vector GLR at least in experimental settings (Belli et al. 1994).

For small-scale vector movements, the type of crop and its planting system may help to dictate the success of the dispersal and hence spread of the pathogen. Leaf-hoppers (Homoptera: Cicadellidae) transmit a wide range of plant viruses and, as the name suggests, are able to move by jumping. In the case of the corn leaf-hopper, *Dalbulus maidis*, which spreads maize rayado fino virus in maize (*Zea mays*), plant-to-plant movement appears to depend on various planting characteristics and on the availability of alternative host plants (Summers et al. 2004). At its most basic, the shorter the distance between adjacent plants, the more likely a leaf-hopper is to move between them (Power 1992). However, things are never quite as simple as they seem, and several more complex observations have been made (Table 11.5). Thus, part of a vector–virus management system has to take into account the dispersal behavior of the insect, and the crop husbandry systems need to be altered accordingly.

Although dipteran vectors of animal pathogens can travel distances in the order of kilometers, evidence shows that short-range dispersal (displacement) is also important. This can occur, for example, in the transmission of malaria among members of a sleeping group of people, all of whom may be infected with a single strain of the parasite having been bitten by the

Table 11.5 Patterns of dispersal of the leaf-hopper *Dalbulus maidis* within a maize crop. (From Power 1992.)

1 Leaf-hoppers less abundant and spread of maize rayado fino virus slower in dense stands of maize than in sparse stands

2 With constant plant density, leaf-hoppers more abundant in stands with uniform spacing than in those with densely sown or clumped rows, though virus incidence did not differ

3 Leaf-hoppers were less likely to move to adjacent plants in uniform spacing patterns than in linear or clumped ones

4 Item 3 may explain the lack of higher virus incidence in uniform stands despite higher leaf-hopper abundance

same, single mosquito (D.W. Kelly, unpublished data). Displacement can be even more important for vectors such as bedbugs (Hemiptera: Cimicidae), which transmit pathogens mechanically (Jupp & Williamson 1990); mechanical transmission relies on the blood from one host being fresh.

Medium-range dispersal of insect vectors generally occurs during flight. An example is illustrated by phlebotomine sandflies (Diptera: Psychodidae), particularly notorious as vectors of leishmaniasis in both the Old and New Worlds. The leishmaniases are a spectrum of diseases caused by parasitic protozoans in the genus *Leishmania*, each with its own characteristic association of vectors and reservoir hosts. The parasites multiply in the gut of the sandfly once they have been acquired via a blood meal, and infected people may suffer in various ways. Skin infections can cause seriously debilitating sores and disfigurements, and visceral infections (known as kala-azar) are widespread through parts of Africa, Asia, and South America, which seem especially linked to human poverty (Alvar et al. 2006). Depending on the species, the parasite occurs in a variety of animal reservoirs such as dogs, foxes, and other primates apart from humans. Clearly, the dispersal of sandfly vectors from host to host will dictate the efficiency and spread of the disease. In a study in Kenya, Mutinga et al. (1992) used mark–recapture techniques to assess the flight range of 11 sandfly species, and found that 54% of their flies were recaptured

within a radius of a mere 10 m from the release site, and 95% within 50 m. Only two flies of a single species were found 1 km from the release point. In this example, it is apparent that hosts separated by some distance are not as likely to be infected as hosts close to each other, i.e. in high-density populations. In fact, the likelihood of vectors such as tsetse flies only moving short distances is one hope for the eradication of sleeping sickness from various parts of Africa. If the vector only tends to move over relatively short distances, isolated areas from where the vector has been eliminated using one method or another may remain fly-free for some time, if not permanently (Krafsur 2003).

Long-distance (or passive) dispersal of vectors is often more dependent on wind than on the flying abilities of the insects themselves. Biting midges such as *Culicoides brevitarsus* (Diptera: Ceratopogonidae) are mainly a physical nuisance – to quote Service (1996), one midge is an entomological curiosity, and a thousand are sheer hell! However, one or two arboviruses of livestock are also transmitted by midges, and as they are such small insects it is not surprising that their long-distance dispersal by wind is commonplace (Murray 1995). We described the whitefly *Bemisia tabaci* as a vector of TbLCV, but it is also responsible for the transmission of viral diseases of cassava and okra (Fargette et al. 1993). Field borders and edge effects influence the spatial spread of the insect and its associated plant viruses. Because field edges are differentially affected by prevailing wind strength and direction, pronounced gradients of insects and disease are generated through crops. Vector gradients are not limited to plant disease vectors. The gradients illustrated in Table 11.6 compare mosquito captures between forest and pasture sites, and between day and night (Weaver et al. 2004). Various species of *Culex* (otherwise known as *Melanoconion*) are vectors of a type of alphavirus that causes an exceedingly serious New World disease known as Venezuelan equine encephalitis (VEE). As the name suggests, VEE virus infects horses and donkeys as well as humans, wherein it causes, amongst other things, serious neurological disorders. As Table 11.6 illustrates, the flies are much more abundant in the forest than in open fields, but are able to move from one to the other, carrying viruses from forest equines to domestic versions, or indeed to humans living in the area. Note also that vector movement, as measured by trap captures, is much

Table 11.6 Culex (*Melanoconion*) spp. vector densities (numbers in flight traps) in and outside forests in Columbia. (From Weaver et al. 2004.)

Species	Nocturnal trap captures		Diurnal trap captures	
	Forest	Pasture	Forest	Pasture
Culex pedroi	21,727	69	26	1
Culex spissipes	21,115	587	53	2
Culex vomerifer	6492	8	9	0

greater at night than during the day. Presumably, blood meals are easier to obtain when most mammals, including humans, are asleep.

Similar movement patterns exist for another extremely serious human diseases, leishmaniasis. As mentioned earlier, this disease is caused by a protozoan pathogen which is vectored by sandflies (Diptera: Phleobotominae). These days, leishmaniasis occurs over a large proportion of the tropical and subtropical world. In Brazil, it seems to move between forested sites and cattle pastures with consummate ease (Massafera et al. 2005), presumably because of the vector's ability to migrate back and forth between adjacent habitats.

For vectors with limited dispersal abilities of their own, it is highly advantageous to hitch a ride with a more mobile animal. Hence plague fleas are well known to use their rat hosts not only as a source of food but also as a long-range dispersal mechanism (Pattison 2003). Woodwasps in the genus *Sirex* traveled from Europe to Australia and New Zealand during the early 1950s on ships carrying logs, thus carrying the rot fungus *Amylostereum areolatum* to the colonies. The woodwasp–fungus association has caused much more severe problems for foresters in Australia and New Zealand than those in Europe, because of the lack of natural biological control agents and the abundance of susceptible, particularly drought-stressed, trees (Speight & Wainhouse 1989).

11.6   PATHOGEN TRANSMISSION

Insect vectors transfer pathogens to new hosts in various ways, and the mechanism of transmission frequently dictates not only the type of pathogen to be vectored, but also its location within the body of the host. As most pathogens are relatively immobile

in their own right, part of the vector's role in disease transmission is to deliver the pathogen to the appropriate part of the host's tissue from where the pathogen can continue its life cycle without further dispersal. The host may even assist in this process; cockroaches provide an excellent example. The genus *Blatta* (Dictyoptera: Blattaria) is now established globally, and is known to frequent all manner of human habitation, including the supposed most hygienic hotels, restaurants, and even hospitals. Cockroaches are thought to transmit viruses, bacteria, protozoa, and nematodes (Service 2004), mainly by simple contamination of food over which they forage, disgorging their own partially digested meal, or depositing their excreta. All that is required for host infection beyond this point is that we eat the contaminated material, thus completing the job begun by the cockroach. It is therefore not surprising that most cockroach-vectored diseases of humans are intestinal. Many of the enterobacteria vectored by cockroaches are themselves considered to be opportunistic (Pellegrini et al. 1992), and the vector merely provides a fortuitous means of mechanical dispersal.

Sap-feeding insects transmit plant viruses in a variety of ways (Table 11.7). Simple mechanical transmission is again apparent, with viruses from the sap of an infected plant adhering to the mouthparts of the aphid or leaf-hopper, in a manner analogous to that of a contaminated hypodermic syringe. This is a similar system to that of biting and displacement seen in bloodsuckers. When the stylet bundle of an aphid, for example, penetrates the epidermis of an infected host plant, the virions (individual virus particles) of non-persistent, stylet-borne viruses bind to sites inside the tip of the bundle (Ng & Falk 2006). The binding of some types of viruses in fact requires

Table 11.7 Modes of transmission of plant viruses by sap feeding Hemiptera. (Partly from Agrios 1988.)

Non-persistent (stylet-borne) viruses
Mechanical transmission from sap or epidermal cell contents of infected host to a healthy one, e.g. turnip mosaic virus (TuMV)

Persistent (circulative) viruses
Vectors accumulate the virus internally, which passes through the tissues of the vector and is introduced to the new host again via the mouthparts, e.g. potato leaf roll virus (PLRV)

Propagative viruses
Persistent viruses, which multiply in the vector before transmission to the new host plant, e.g. tomato yellow leaf curl virus (TYLCV)

the presence of a "helper" proteinase which interacts with the protein coat of the virion – even a seemingly simple system can be complicated. Plant viruses are readily transmitted to new hosts during initial probings by sap-feeders, carried out in attempts to identify suitable host plants. In fact, the use of plant species or varieties that are resistant to sap-feeders may encourage the transmission of non-persistent plant viruses, as probings become more frequent if the insect cannot find suitable hosts. An example of this problem is shown with the aphid *Aphis craccivora* (Hemiptera/Homptera: Aphididae), where aphids change their behavior when confronted by resistant varieties of cowpeas. The number of probings increases significantly as the aphids sample numerous plants in an attempt to find a suitable host. This increased probing leads to a proliferation of plant virus incidence (Mesfin et al. 1992). Other plant pathogens may enter the foregut of the insect vector, but do not themselves replicate there. *Xylella fastidiosa* is a bacterium that causes a variety of plant diseases that is transmitted by leaf-hoppers (Hemiptera: Cicadellidae) and other hemipterans. There appears to be no latent period (Section 11.6.1), and the pathogen can be transmitted to a new host plant soon after feeding (Almeida & Purcell 2006). Furthermore, infectivity is lost from nymphs when they molt, indicating that the bacteria proceed no further than the vector's foregut.

Vectors of animal pathogens can also transmit diseases mechanically as described earlier for houseflies

and cockroaches. However, most vector–pathogen interactions are much more complex. Often, the pathogen is taken into the insect vector's body where it persists and frequently multiplies before being inoculated into a new animal host when the vector feeds again. As with plant–pathogen vectors, animal vectors possess mouthparts of various types to facilitate blood feeding, but the two major types are "cutters" and "piercers" (Brusca & Brusca 2001) (Figure 11.3). The basic components of an insect's mouthparts (from front to back: labrum, mandible, maxilla, and labium; Gullan & Cranston 2005) can still be identified, though with some difficulty. The result is one system that cuts holes in the skin of the vertebrate host whereupon the wound bleeds, and another that actually pierces the skin and reaches down into the surface tissues to extract blood in a similar way to aphids feeding on plants. Both systems usually employ the input of saliva from the vector, often as an anticoagulant or to increase the local pressure of fluid in the host to facilitate uptake. The input of saliva is the point at which pathogens are normally introduced to the new victim.

Cutters such as tsetse flies and blackflies have maxillae and mandibles that carry sharp teeth. These are used to cut through skin in a scissor-type action on an inward sweep. The resulting cut is deep and painful resulting in a welling pool of blood, which the fly's saliva stops from coagulating. There may also be a vasodilatory component to the saliva, which makes the wound bleed more copiously (Rajska et al. 2003). The labrum is then used like a sponge to ingest the pooled blood. Due to the cutting nature of the bite, the fly is frequently disturbed while feeding.

In mosquitoes, which are typical piercers, the outside of the proboscis is made up of the labium and labrum that house the mouthparts. The labellum at the tip of the proboscis is highly sensitive to host odors and is therefore important in finding a suitable blood meal (Kwon et al. 2006). Inside these are the hypopharynx, maxillae, and the mandibles. The hypopharynx is a canal that is connected to the salivary glands. During blood feeding, the hypopharynx pumps saliva into the wound to prevent the blood from coagulating, while the mandibles and maxillae cut the skin via a sawing action to penetrate the skin. Once that has been achieved, the labrum becomes a tube that enters the wound, and the pharyngeal pump then pumps blood up through it. The pharyngeal

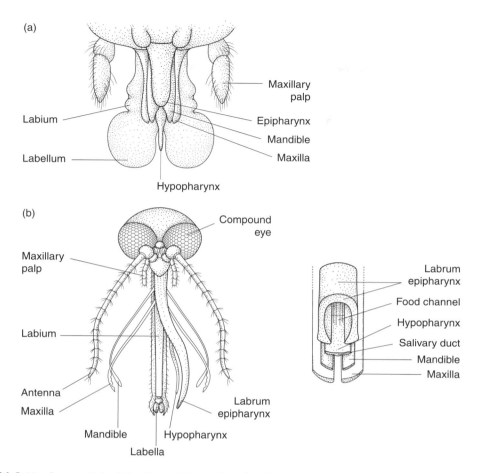

(a)

Maxillary palp

Labium

Epipharynx

Mandible

Maxilla

Labellum

Hypopharynx

(b)

Compound eye

Maxillary palp

Labium

Antenna

Maxilla

Mandible

Hypopharynx

Labella

Labrum epipharynx

Labrum epipharynx

Food channel

Hypopharynx

Salivary duct

Mandible

Maxilla

Figure 11.3 Mouthparts of (a) a biting fly, and (b) a sucking fly. (From Brusca & Brusca 2001.)

pump is an enlargement of the pharynx, which is located in the insect's esophagus. It pumps blood by expanding and contracting. Other piercers, such as bedbugs and kissing bugs (Hemiptera: Cimicidae and Triatomidae), are in the same insect order as the sap-feeding aphids, whitefly, and mealybugs, and their mouthparts are very similar in structure to their plant-feeding relatives.

Authors have categorized the vectors of animal pathogens by where the vector rests and feeds (Pates & Curtis 2005) (Table 11.8). Resting and feeding habits will influence potential strategies to control the vectors. Exophilic/exophagic mosquito species tend to be found in natural habitats such as forest, where they feed on primates or other mammals.

Humans become victims when they forage into the forest for food or new settlement. Control of exophilic mosquito vectors, even those species that come into buildings to feed (endophagic), is extremely difficult. Spraying pesticide inside houses is rarely effective because the insect is only present for a short time while feeding.

11.6.1 Pathogen multiplication

So far in this section we have considered simple mechanical transmission of pathogenic organisms by insect vectors. However, many associations among disease-causing agents and their vectors are more complex.

Table 11.8 Types of insect vector behavior. (From Pates & Curtis 2005.)

Term	Definition
Endophilic	Vector rests indoors
Exophilic	Vector rests outdoors
Endophagic	Feeds indoors
Exophagic	Feeds outdoors

In addition to relying upon the vector for dispersal, some pathogens also replicate or undergo part of their own life cycle within the insect. In a relatively simple form, pathogen transmission may include the passage of the disease-causing organism through the body of the vector, with the pathogen emerging at the other end of the alimentary tract in an infectious state, during which time the vector has moved on from one host to another. Triatomid bugs (Heteroptera: Triatomidae) that vector Chagas disease provide an example. The protozoan *Trypanosoma cruzi* is responsible for this human disease, which is widespread in Central and South America (Schofield et al. 2006). Parasites are ingested by the bugs via a blood meal on an infected host (humans or other mammals) and undergo their entire development within the gut of the vector (Service 2004). After a week or two, infective stages of the protozoan are present in the bug's excreta, which can enter a person's body via wounds, scratches, or even eyes and other mucous membranes. Transmission is therefore not through the bite of the insect, merely through its fecal deposits. Entry into the host's body may be enhanced when the person scratches his or her irritating bites. This type of transmission is termed stercorarian, as opposed to salivarian.

Vector–pathogen associations are more complex when the pathogen invades the vector's tissues and replicates before inoculation into a new host. Often, the vector's salivary glands are the site of this multiplication; ecologically, this is an excellent site, as the pathogen can build up large numbers immediately before the vector deposits it into the tissues of a new plant or animal host. As mentioned above, saliva is employed as an anticoagulant and vasodilator by insects that feed on animal hosts, whilst in plant-feeders, saliva may be used to increase sap pressure within host tissues to "force feed" the insect. In other words, the injection of saliva into the host is an extremely common part of feeding, and hence a very successful route for a pathogen to enter a new host. Figure 11.4 illustrates how plant pathogens, in this case viruses, can circulate through the tissues of an aphid (Schuman 1991; Gray & Gildow 2003). The illustration is typical of a persistent, circulatory virus such as barley yellow dwarf virus (Irwin & Thresh 1990). Because virus multiplication takes some time to complete, a latent period often occurs between the vector first acquiring the pathogen and its becoming infective for new hosts. This latent period may range from 12 h in some aphids to several weeks in some leaf-hoppers (Schuman 1991). Once the latent period is past, pathogens within vectors may remain infective for some time, possibly even up to 100 days (Chiykowski 1991).

A similar multiplication of pathogens occurs in insect vectors of animal diseases. The parasitic protozoans that are responsible for human malaria, *Plasmodium* spp., undergo several stages of their life cycle in the gut and hemocoel of the mosquito vector (Kettle 1995) (Figure 11.5). Development and replication within the mosquito may take up to 12 days depending on the temperature and species of *Plasmodium*. Finally, the infective stage, or sporozoite, is abundant in the vector's salivary glands and ready for inoculation into a fresh host at the next blood meal (Service 2004). Once infective, however, female mosquitoes usually remain so for the rest of their lives.

11.7 EFFECTS ON THE VECTOR

Insect-transmitted pathogens can have positive, negative, or neutral effects on the vector insect (Vega et al. 1995), and these effects may be direct, caused by the pathogen, or indirect via the host organism. Where simple mechanical transmission of a pathogen occurs, as in the case of cockroaches or houseflies walking over sources of contamination and infecting new sites, there will be no effect on the vector at all – the insect is merely a fortuitous carrier. However, as pathogen–vector associations become more intimate, the pathogen has to be able to survive and frequently replicate in the body of the insect vector. Anatomical and biochemical barriers to pathogens do exist in the insect body; key resistance factors include the toxicity of digestive fluids, impermeability of the peritrophic membrane, which surrounds the blood or

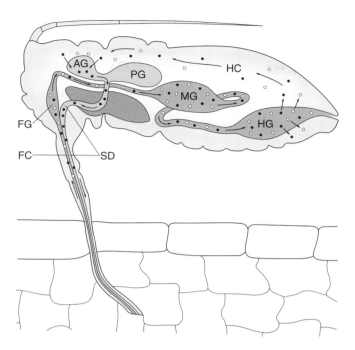

Figure 11.4. The circulative route of luteoviruses through an aphid. Ingested virus moves up the food canal (FC), through the foregut (FG) and accumulates in the midgut (MG) or hindgut (HG). Virus is then acquired into the hemocoel (HC). Virus may also accumulate in the hemocoel and remain viable for weeks. Transmissible virus (black dots) is transported into the accessory salivary gland (AG), but does not associate with the principal salivary gland (PG). Transmissible virus is then injected into the plant through the salivary duct (SD) when the aphid feeds on a plant. (From Gray & Gildow 2003).

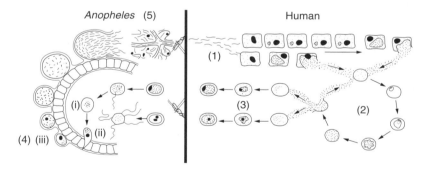

Figure 11.5. The life cycle of malaria parasites in the *Anopheles* mosquito and the human host, according to present views on exo-erythrocytic schizogony. 1, Sporozoites injected into a human by an *Anopheles* female develop in the liver either into latent hypnozoites (above), which sometime later undergo schizogony to cause relapses, or undergo immediate schizogony (below). Both release merozoites, which enter red blood cells. 2, Erythrocytic schizogony involving release of merozoites. 3, Some merozoites develop into male or female gametocytes, which develop further only when ingested by *Anopheles*. 4, Male gametocytes undergo exllagellation to produce male gametes, one of which will fuse with a female gamete to form a zygote (i). The zygote becomes a motile ookinete (ii), which passes between the cells of the midgut to form an oocyst (iii). 5, The oocyst enlarges; there is much nuclear division, ending in the formation of motile sporozoites, which invade the hemocoel and penetrate the salivary glands, from which they are passed into the host with the saliva when *Anopheles* next feeds. (From Kettle 1995.)

plant meal as it progresses down the vector's gut, and adverse intracellular environments of insect cells, which can be hostile to the multiplication of pathogens such as viruses (Glinski & Jarosz 1996). None the less, vector-borne pathogens are generally less virulent to the vector than to the host (Elliot et al. 2003).

However, certain pathogens have serious negative effects upon vector fitness. The fleas that carry bubonic plague between infected rats and humans provide a striking example (see also Section 11.11.4). Fleas in the genus *Xenopsylla* (Siphonaptera) are the most important vectors of so-called urban plague, and they acquire plague bacilli while feeding on the blood of diseased rats and other rodents (Gage & Kosoy 2005). The bacilli multiply enormously in the vector's stomach, and move forward into anterior sections of the gut, where they frequently block the alimentary tract. When a flea so afflicted feeds again, with increasing urgency because starvation will set in if the gut is blocked, some bacilli are regurgitated into the new, human, host. Eventually, the vector may starve to death, but not before bacilli have been transmitted onwards. The bacilli also kill the original source of infection, the rats. Consequently, three of the four organisms associated with infection are severely disadvantaged, with the obvious exception of the bacillus itself. How can natural selection favor a pathogen that kills its own vectors? As it turns out, plague can also spread through the air in high-density host populations and thus does not rely completely on a vector to move from one person to the next. As an aside, this ability of the plague to be transmitted through the air is responsible for the traditional nursery rhyme, "Ring-a-ring-roses".

Fleas and plague are an extreme example of deleterious effects of a pathogen on its vector. In many other cases, the abundance of the pathogenic organism within the vector insect determines the severity of effects in a density-dependent fashion. In other words, the higher the density of pathogens, the more seriously the vector is affected. Human onchocerciasis, discussed in detail below (Section 11.11.1), is widespread in parts of Africa and South America. It is caused by microfilarial nematodes in the genus *Onchocerca* that are vectored among human hosts by blackfly, *Simulium* spp. (Diptera: Simuliidae). In the laboratory, three species of blackfly showed decreasing survival times as the density of microfilariae in their blood meals increased (Basanez et al. 1996).

There is a latent period between the acquisition of nematodes by the insect and its ability to inoculate a new host, during which time the pathogen is developing in the vector's body. When pathogen loads are very high, flies may not live long enough to transmit the pathogen. Fly species vary in their life expectancy at high pathogen loads, with the result that only certain simuliids survive long enough to become infective and serve as suitable vectors.

Pathogens do not have things all their own way. Some vectors try to restrict the presence of pathogens in their bodies, and to prevent these pathogens from multiplying within their tissues. This suggests that acting as a vector is often disadvantageous for insects. Filariasis is a disease of humans and other animals that can take a variety of forms. The pathogenic agents are various nematode worms, including *Brugia malay* and *Wuchereria bancrofti* (Ichimori et al. 2007) that are vectored by culicid mosquitoes. One mosquito species, *Armigeres subalbatus* (Diptera: Culicidae), reacts to the presence of filarial nematodes in its tissues by encapsulating and killing up to 80% of the parasites within 36 h of acquiring them through a blood meal (Ferdig et al. 1993). This defense system is not without cost to the insect, however. Mosquitoes that encapsulate nematodes exhibit longer pre-oviposition periods, reduced ovary size, and reduced protein content within their eggs.

Insect vectors can also suffer indirectly from vectoring pathogens if those pathogens reduce the availability or quality of the hosts upon which the vectors rely. For example, insects that feed on plants might be negatively affected by pathogen-induced declines in plant health. Recall that in Chapter 3, we discussed "cross-talk" between insect and pathogen defense systems in plants (the jasmonate and salicylate pathways). In some cases, induction of defense against one type of attacker could act to suppress the induction of defense against another type of attacker. However, as we concluded in Chapter 3, there may be no simple relationship between disease infection and subsequent plant quality for insect herbivores. We might therefore expect to observe both positive and negative effects of plant disease transmission on vector herbivores. We consider some negative effects first. Larvae of the tortoise beetle *Cassida rubiginosa* (Coleoptera: Chrysomelidae) for example, feed on thistles (*Cirsium arvense*). Kluth et al. (2002) have demonstrated that survival to pupation was 89% when beetle larvae fed on healthy plants, but a mere

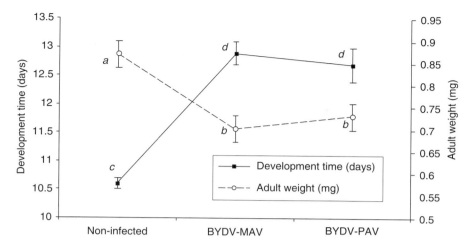

Figure 11.6. Development time of nymphs, and biomass of teneral adults, of the aphid *Sitobion avenae* feeding on wheat with or without an infection of two strains of barley yellow dwarf virus (BYDV) Means with the same letter not significantly different (From Fiebig et al. 2004.)

46% when they ate thistles infected with a systemic rust fungus, *Puccinia punctiformis*. *Cassida* is a mechanical, accidental vector of the fungus – spores are carried on its body, including mouthparts. In this case, *Cassida* suffers from the effects of the pathogen that it helps to transmit. Even phloem-feeding insects, often thought to thrive on stressed plants (see Chapters 3 and 12) may not benefit from disease infection of their host plants. The cereal aphid, *Sitobion avenae*, grows more slowly and achieves lower adult weights when feeding on wheat infected with barley yellow dwarf virus (BYDV) (Fiebig et al. 2004) (Figure 11.6). Interestingly, more winged aphids (alates) are produced from diseased plants. This is likely good news for the pathogen because higher numbers of alates means more dispersal to new plant hosts.

So far, we have considered the downside of being a pathogen vector. Does the process of vectoring pathogens ever increase the fitness of the vector? As we might expect, natural selection has favored the evolution of mutualism between some vectors and pathogens such that both parties benefit from the association. For example, Rivero and Ferguson (2003) have studied the effects of the malarial pathogen, *Plasmodium chaubaudi*, on its mosquito vector, *Anopheles stephensi* (Diptera: Culicidae). They originally predicted that mosquitoes infected with *Plasmodium* would have fewer energetic resources

and lower fecundity than uninfected mosquitoes. However, in experiments, they showed that flies infected with *Plasmodium* had higher levels of glucose in their bodies than did uninfected flies. Glucose, of course, is a vital nutritional and energetic resource for insects (and for all other animals) so the pathogen appears to have a beneficial effect on the vector.

A similar boost of energy levels has been reported in one vector of Rift Valley fever (RVF). RVF is a type of encephalitis that has been found in goats, cattle, horses, and camels as well as humans (Olaleye et al. 1996). RVF is present in at least a dozen African countries from Mauritania in the west to Zambia in the south and Egypt in the northeast. A broad range of arthropods can carry RVF including ticks and sandflies (Hogg & Hurd 1997), as well as many mosquitoes in the genera *Anopheles*, *Aedes*, and *Culex*. In *Culex pipiens* (Diptera: Culicidae), insects infected with RVF virus survive longer under conditions of carbohydrate deprivation than do their uninfected counterparts (Dohm et al. 1991). However, the reasons for this phenomenon are rather unclear.

In certain circumstances then, living on a diseased population of hosts, and carrying pathogens among them, may be beneficial to the vector. A blood-feeding mosquito or tsetse fly may be less likely to be disturbed during feeding if its host is debilitated by pathogens that cause fever or lassitude. It is also possible that

Table 11.9 Some of the advantages reported for aphids when vectoring barley yellow dwarf virus (BYDV). (From Irwin & Thresh 1990.)

Variable	Effect
Development rate	+
Longevity	+
Reproductive period	+
Number of offspring	+
Number of alatae (winged adults)	+

+, enhancement of the parameter in the presence of virus.

fevers open up the peripheral blood vessels in host skin, so providing extra food supplies for the insect. And, as we have seen, some pathogens may even promote vector survival while in the insect's tissues.

Mutualistic relationships between plant pathogens and their vectors appear to be more common than they are between animal pathogens and their vectors. In Figure 11.6, we illustrated the deleterious effects on the aphid *Sitobion avenae* of vectoring BYDV (Fiebig et al. 2004). As it turns out, this observation may be in the minority. For the cereal aphid, *Rhoplasiphum padi*, development rates and fecundities increase when aphids feed on hosts infected with BYDV (Jimenez-Martinez et al. 2004). In fact, many species of aphids (Homoptera: Aphididae) routinely act as vectors for BYDV, and Table 11.9 summarizes some of the benefits that accrue to the insects by acting as vectors. While not all species of aphid respond in the same fashion to BYDV infection of their host plants, one basic mechanism appears likely to fuel the benefits shown in the table. BYDV infection seems to increase the total amino acid content of cereal leaves, a commodity well known to favor aphid success (see Chapter 3). So, should insect vectors of plant pathogens actually choose to feed on infected host plants? One example concerns the bacterium *Xylella fastidiosa*, a pathogen that causes citrus variegated chlorosis (CVC). Unusually, this pathogen can only be vectored by xylem-feeding insects such as the so-called "sharpshooter" leaf-hoppers (Hemiptera: Cicadellidae) and some spittlebugs (Hemiptera: Cercopidae) (Marucci et al. 2005). In experiments with two leaf-hopper species, the insects preferred healthy plants to those with CVC symptoms (Figure 11.7), though it seems

that citrus plants infected by the bacterium but yet to show symptoms were equally acceptable to the vector as uninfected hosts. Clearly for the optimal success of transmission from plant to plant, the pathogen should be acquired by the vector soon after infection.

Plant viruses are transmitted by orders of insect in addition to Hemiptera/Homoptera, sometimes with similar benefits to the vector. The western flower thrips, *Frankliniella occidentalis* (Thysanoptera: Thripidae), is a serious pest of vegetable and ornamental crops worldwide, not only causing direct injury to the host plants, but also acting as a vector of tomato spotted wilt virus (TSWV). The vector undoubtedly reacts physiologically to the presence of the plant pathogen; the virus actually activates the thrips' immune system (Medeiros et al. 2004), but the insects produce more offspring when fed upon TSWV-infected pepper (capsicum) plants (Maris et al. 2004). Most significantly, thrips seem to select virus-infected plants when given a choice between diseased and disease-free individuals, providing strong support for a mutualistic interaction between pathogen and vector (Belliure et al. 2005).

We described some mutualistic associations between insect vectors and pathogenic fungi in Chapter 6. Perhaps the quintessence of obligate mutualism is seen in the so-called ambrosia beetles (Coleoptera: Platypodidae). Beetles such as *Trypodendron (Xyloterus) lineatum* can be serious pests of young trees in Western Europe, especially when the trees are temporarily stressed by heavy lepidopteran attack or storm damage (Bouget & Noblecourt 2005). Defoliation interrupts the normal transpiration stream of the tree and reduces sap pressure, a simple but usually effective physical defense mechanism. As a consequence, adult female beetles are able to bore through the bark and into the wood. So-called blue-stain fungi are intimately associated with ambrosia beetles, and wood tissues are inoculated with the fungi as the beetle tunnels along (see Chapter 6). Eggs are laid in chambers in the wood, the walls of which are infected with the fungi. The developing beetle larvae are in fact mycetophilic (fungus feeding), rather than direct wood-feeders, and harvest the fungal fruiting bodies that proliferate in their chambers. Without the fungus, beetle larvae cannot survive, and though the tree is not killed, severe wood degradation by the joint presence of beetle galleries and fungal-stained timber is the final result.

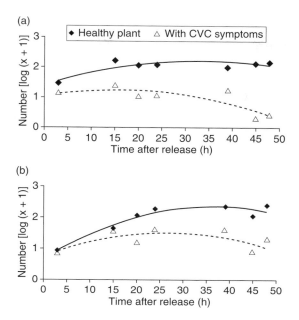

Figure 11.7. Regression curves of the relationship between observation time and mean number (x) of (a) *Dilobopterus costalimai* and (b) *Oncometopia facialis* per plant, in a choice test between healthy and CVC-symptomatic citrus plants. (From Marucci et al. 2005.)

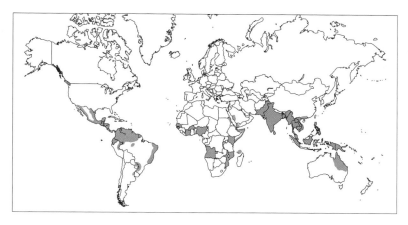

Figure 11.8. Distribution of dengue fever. (From Roberts 2002.)

11.8 EPIDEMIOLOGY: THE SPREAD OF DISEASE

If we examine the global distribution of some of the most important insect-vectored diseases, it is clear that disease spread can occur over grand scales. The map in Figure 11.8 illustrates the global distributions of dengue fever (Roberts 2002). Dengue or break-bone fever is caused by several different viruses, but is vectored mainly by the mosquito *Aedes aegyptii*

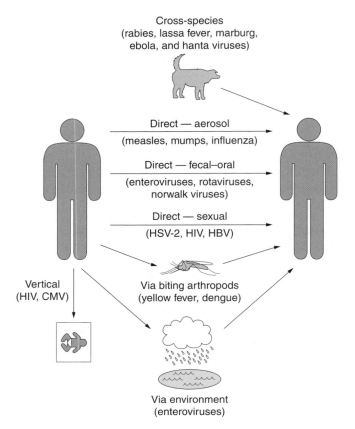

Figure 11.9. A schematic representation of the transmission routes of viral infections in humans, with some examples. Direct transmission can include sexual or fecal-oral contact. The sexually transmitted viruses, human immunodeficiency virus (HIV) and hepatitis B virus (HBV), may also be transmitted by other processes during which blood is exchanged. Some viruses that are transmitted by fecal-oral contact may also be released into the environment, which provides a source of infection. The cross-species virus transmission can be from rodents (the Hanta viruses and the arenaviruses such as Lassa fever), from monkeys (African hemorrhagic fever—the Marburg and Ebola viruses), or from domestic animals infected by wild animals (rabies). There is evidence of some transmission between humans for these infections, but the extent of this interperson transmission is limited. CMV, cytomegalovirus; HSV, herpes simplex virus. (From Garnett & Antia 1994.)

(Diptera: Culicidae) (Halstead 2008). Figure 11.8 shows the current patchy distribution of the disease but, as many travelers know, the problem is very much on the increase – a classic case of disease spread.

Of course, diseases are spread by a variety of mechanisms other than insect vectors (Garnett & Antia 1994) (Figure 11.9). The employment of a mobile vector organism, especially one in which the pathogen undergoes an essential part of its development, is a highly specialized form of dispersal, evolutionarily and ecologically complex. The alphaviruses provide

a case in point. They are typical RNA viruses that readily undergo mutations and are quick to exploit changes in their environment or the behavior of their hosts or vectors (Scott et al. 1994). Major alphavirus diseases include Ross River virus and three equine encephalitis viruses, eastern (EEE), western (WEE), and Venezuelan equine encephalitis (VEE) (Kramer et al. 2008). Ross River virus, far from being confined to the developing world, is present in urban areas such as Brisbane, Australia. It causes painful joints, rashes, and fevers, which can persist for several months, and it occurs in cattle, horses, and kangaroos as well as in humans. It is carried by various *Aedes* and *Culex* species that breed in fresh and brackish water in suburban areas (Ritchie et al. 1997). The equine encephalitis viruses, as the name suggests, are predominantly problematic in horses and donkeys, but humans do suffer as well. In 1995, a VEE outbreak in Colombia caused an estimated 75,000 human cases, of which 300 or so proved fatal (Rivas et al. 1997). Much further north and a year later, people living in Rhode Island, New York, Long Island, and Connecticut were forced to remain indoors in the evenings to avoid being bitten by mosquitoes carrying EEE; 1% of mosquitoes in the area were found to carry the potentially fatal virus (Anon. 1996). All three alphaviruses likely evolved from the same ancestor, but we can now recognize a number of different forms with different vectors and host mammals. This diversity may be due in part to the restricted dispersal of mammalian hosts, especially in South America, with cotton rats (*Sigmodon hispidus*) as likely non-primate/non-equine reservoirs of the disease (Carrara et al. 2005). With limited host dispersal comes the chance establishment of isolated, founder populations of virus, which then adapt and radiate under local selection pressures. Vector diversity in these restricted areas will also have a role to play in the evolutionary outcome. At its most sinister level, VEE may in fact emerge as an effective biological weapon (Weaver et al. 2004), where insect vectors may play a secondary but nevertheless crucial role in human disease epidemiology.

Disease-causing pathogens that are carried by insects rely, of course, on the dispersal activities of the vector. As we alluded to earlier, a vector can only spread a disease if the vector's lifespan exceeds the incubation period of the pathogen in the vector's tissues. Let us take, for example, the case of mosquitoes taking a blood meal that contains the malarial parasite *Plasmodium*. The latent period describes the time between pathogen uptake by the mosquito and the appearance of infectious forms of the pathogen in the mosquito salivary glands. If the latent period is close to, or even exceeds, the normal lifespan of the fly, then clearly there is little chance of the vector infecting humans (Anderson & May 1991). In fact, adult *Anopheles* mosquitoes have an expected lifespan under field conditions of only some 6–14 days, with *Plasmodium falciparum* latent periods of up to 12 days at 25–27°C (Anderson & May 1991). It seems very

likely that many flies die before they become able to infect humans. In the case of sandflies that transmit leishmania, the vector lifespan is particularly short. After the latent period, the average female might be able to bite only once or twice more before dying. In an apparent attempt to ameliorate this problem, the pathogen multiplies in the midgut, migrates forward, and blocks the pyloris (part of the foregut). As a consequence, the fly is forced to bite repeatedly and to regurgitate parasites as it does so (D.W. Kelly, unpublished data). Given that vector population dynamics, particularly infectious lifespan, are so crucial to pathogen transmission, we might expect that pathogens would evolve mechanisms to extend vector lifespan. As we described earlier, this appears to be the case for aphids that vector barley yellow dwarf virus.

11.9   HUMAN ACTIVITIES AND VECTORS

Many human activities favor the population growth of insect vectors. We provide them with abundant potential hosts (humans, livestock, crops), we aid in their dispersal, and we provide them with many available breeding sites, especially for their nymphs or larvae. Disease management strategies are often aimed at human activities as well as at the vectors or pathogens themselves.

A large number of animal and plant diseases occur in natural communities, and persist at endemic levels in wild plant and animal populations without affecting humans at all. Yellow fever provides a good example. Yellow fever and its mosquito vector *Aedes* spp. are said to have spread initially as a result of the African slave trade (Morse 1994). The zoonotic, so-called sylvatic, phase of African yellow fever is maintained in monkey populations in the natural forest. Monkeys seem to be little affected by the virus, with only a few individuals actually dying from yellow fever (Service 2004). When humans enter the forest in large numbers, or convert it to agricultural land, they are in the proximity of monkeys more frequently. The virus can then transfer via mosquitoes into human populations, where it is carried by other *Aedes* species, *Aedes aegypti* in particular. Figure 11.10 illustrates the complex interactions between natural and human host populations infected with African yellow fever. Note that the two separate systems, the urban versus jungle cycles,

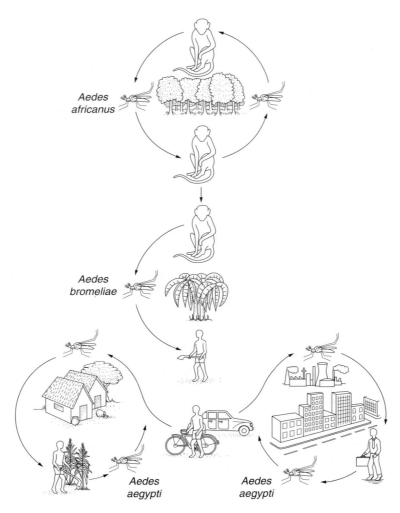

Figure 11.10 Diagrammatic representation of the sylvatic, rural, and urban transmission cycles of yellow fever in Africa. (From Service 1996.)

are self-sustaining (Barrett & Higgs 2007), and extra problems only really arise when mobile humans bring the jungle into town, as it were. Note that several vector species are involved, as different species of mosquito frequent different habitats. It is also important to realize that mammals including humans can also act as pathogen vectors. A villager bitten by a virus-infected mosquito while working on a farm may travel to the market in town to sell produce. If that individual is then bitten by another mosquito, which acquires the pathogen from its blood meal,

the mosquito may infect other people who have never left their urban environment. These days, urban forest (parks, gardens, semi-wild places in towns) can play the role of natural forest in harboring vectors (Lourenco de Oliveira et al. 2004), so even attempts to make urban areas better places to live may have their downside.

The spread of many other diseases is favored by human activities. Triatomid bugs that transmit the trypanosome responsible for Chagas disease live in the walls and rafters of village houses. If the houses provide refuges (such as non-plastered walls or untreated thatch) for the bugs, triatomids are able to conceal themselves until ready to feed on people (Cecere et al. 2003). In a similar way, malarial mosquitoes are encouraged in houses without proper ceilings, such as those found in rural Gambia (Lindsay et al. 2003). In Mexico, spatial analysis using geographic information systems (GIS) has been used to predict the abundance of mosquitoes that transmit malaria in rural villages (Beck et al. 1994). As we might expect, transitional swamps or unmanaged pastures favor mosquito populations because their aquatic larvae require slow moving or still water.

Irrigation ditches, water supplies, paddy fields, and water butts provide excellent sources of insect vectors (Sadanandane et al. 1991; Service 1991; Reisen et al. 1992). In fact, there is strong evidence that the mosquito *Ae. aegypti* has spread throughout the tropics by virtue of the international trade in second-hand truck and car tyres. Larvae can develop successfully in the small pools of water conveniently provided inside the rims of these artificial habitats. This type of provision of semipermanent breeding sites or larval habitats for vectors is potentially disastrous, but in many cases is so easily avoided. Let us not forget that mosquitoes have evolved to seek out and lay their eggs in many sorts of natural pools or ponds – epiphytic bromeliads are a classic example (Lounibos et al. 2003). Anything that humans produce and leave lying around that mimics a bromeliad or similar water source is going to attract ovipositing mosquitoes, and will soon contain their larvae. The longer these containers remain in place, and full of water, the more likely new adult vectors are to emerge. In western Kenya, larval densities of the mosquito *Anopheles gambiae* increase dramatically with the permanence of both natural pools and cement-lined ponds (Figure 11.11). A ready supply of breeding sites supports larval populations throughout the year, even during the dry season (Fillinger et al. 2004). It is a fact of life that in many tropical villages and towns, ideal containers for the survival of mosquito larvae are left lying around simply because no one clears them away. Everything from plastic buckets to Styrofoam cups may serve the purpose (Pena et al. 2003). A community-based scheme in Havana, Cuba has been established to remove bottles, pots, tires, and other mosquito breeding sites. A comparison using the so-called "house index" has shown that the number of receptacles containing

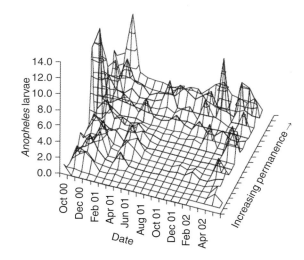

Figure 11.11 Densities of larvae of the mosquito *Anopheles gambiae* in western Kenya in relation to availability of artificial breeding sites. (From Fillinger et al. 2004.)

larvae of the dengue vector *Aedes aegypti* has declined as a result (Sanchez et al. 2005) (Figure 11.12). In other words, "social mobilization" can act as a form of disease control.

Human activities can also change the relative abundance and community structure of certain vector assemblages. In Sri Lanka, irrigation development for rice culture has provided newly abundant breeding sites for mosquitoes. The sudden availability of canals, reservoirs, and numerous paddy fields has caused long-term changes in the species composition of vector populations. Unfortunately, the vector community is now dominated by mosquito species with high potential to transmit human pathogens (Amerasinghe & Indrajith 1994).

So far in this section, we have discussed human influences on insect vectors on a fairly local scale. However, throughout recorded history, mankind has been traveling the world, carrying with him all manner of insect parasites and their associated pathogens (Lounibos 2002). Human disease such as malaria, yellow fever, typhus, and plague are now all over the world as we have seen. For centuries, global transport (largely by ship) of vectors, pathogens, and infected hosts has spread disease among countries. Outbreaks of plague as far back as the sixth and eighth centuries have been attributed to burgeoning trade. Starting in the 15th century, the

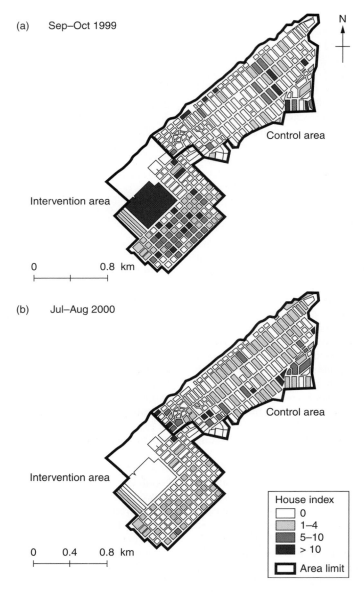

Figure 11.12 *Aedes aegypti* house indices (a) before and (b) after the intervention. (From Sanchez et al. 2005.)

yellow fever mosquito (and with it, the viral pathogen) migrated from West Africa to the New World, probably aboard slave ships. Even today, the importation of goods can result in the simultaneous importation of insect disease vectors (Table 11.10). A wide variety of modern goods, from tropical plants to aquarium fish, bring with them less than welcome insect vectors.

Perhaps the largest scale and most significant consequence of current human activity is global climate change, which may have very serious impacts on human health in the not too distant future

Table 11.10 Twentieth century invasions by container-inhabiting mosquitoes into the United States. (From Louinbos 2002.)

| Species | Regions | | Transport | Date |
	Donor	Recipient		
Aedes albopictus	Japan	Texas	Tyres	1985
Aedes atropalpus	Eastern USA	Illinois, Indiana, Nebraska, Ohio	Tyres	1970–80
Aedes bahamensis	Bahamas	South Florida	Tyres	1986
Aedes japonicus	Japan	Connecticut, New Jersey, New York	Tyres	1998
Aedes togoi	Asia	Pacific NW	Ships	1940–50
Toxorhynchites brevipalpis	E. Africa	Hawaii	Biocontrol	1950s
Toxorhynchites amboinensis	Pacific region	Hawaii	Biocontrol	1950s
Wyeomyia mitchellii	Caribbean or Florida	Hawaii	Bromeliads	1970s

(Kovats et al. 2005). As we described in Chapters 2–4, climate can drive the population dynamics of many insects, including crop pests and human disease agents. Many scientists have tried to attribute changes in vector and disease prevalence or distribution to changes in patterns of rainfall and temperature but a great deal of controversy still rages. For example, Zhou et al. (2005) are convinced that climate variability has a significant impact on malaria epidemics in the highlands of East Africa. They further suggest that anthropocentric climate change is likely to strongly influence epidemics in the future. Hay et al. (2005) disagree. They argue that the data presented by Zhou et al. (2005) do not warrant such conclusions. Indeed, in 2002, Hay et al. found no correlation between climate change and the resurgence of malaria in East Africa. Apparently, the experts disagree with one another, and the jury is still out. Plant diseases may also be influenced by climate change via a whole variety of interfaces. Plant growth rates are shown to vary under changed conditions of carbon dioxide, light, and water regimes, and undoubtedly insect vector populations may be affected by global warming, for example (see Chapter 2). Finally, the behavior of pathogens themselves may be influenced, to the detriment of the host plant (Garrett et al. 2006).

11.10 VECTOR CONTROL

Later in this chapter we discuss some case histories of major animal and plant diseases and their insect vectors. In this section, we cover some basic principles

in modern vector control that may be broadly applicable. To this end, we can recognize different types of control tactics, from simple mechanical control, to the uses of chemical and biological control, to population and genetic manipulations. The major aim of some of these tactics may be total vector eradication but, in most cases, this is impractical (and maybe not even desirable), and reducing vector–host interactions is a more realizable goal. It is important to note at this stage that whatever specific techniques (or combinations thereof) are employed to control disease vectors, education programs at local (village or community) levels are essential to ensure efficient and sustainable practices (Yasuoka et al. 2006).

11.10.1 Mechanical manipulations: removal of breeding sites

Both natural and artificial breeding sites of vectors that are allowed to persist close to areas frequented by humans and other animals can greatly exacerbate disease problems by significantly increasing the density of vectors. Such breeding sites include everything from bottles and tires (see above) to whole swamps, marshes, and other vegetated wild land. Removal, drainage, or destruction of these sites undoubtedly influences vector species and populations. Large-scale drainage of swamps and bogs appears to have been successful over the years in eradicating diseases such as malaria from very large areas including much of Western Europe, the UK, and the USA (Willott 2004). As we will see later, malaria has been present in Western Europe for thousands of years, and its eradication from places as far apart as England and Italy is thought to be at least a partial consequence of draining swamps or other marshy areas for agriculture, and of course, disease prevention (Kuhn et al. 2003). The snag is that, these days, conservation biologists have recognized the important role that wetlands can play in maintaining biodiversity and ecosystem processes. If we want the swamps and wetlands back again, as Willott (2004) points out, the return of mosquitoes has to be part of the deal. In addition, various other modern techniques for improving the quality of life such as irrigation and damming for water and power also provide new breeding sites for insect vectors. In Sri Lanka, five major dams were constructed and a big increase in the larval densities of anopheline mosquitoes was found in the resulting stagnant pools below the dams (Kusumawathie et al. 2006).

11.10.2   Mechanical manipulations: removal of refuges

Insect vectors rest or hide in many places when inactive. If vector refuges happen also to be in close proximity to buildings, livestock barns, and so on, then the chances of vectors finding hosts become very high. Removal of vector refuges can therefore be an effective tactic. Recall that Chagas disease is a trypanosome pathogen vectored by triatomid bugs (Hemiptera: Triatomidae). In central Argentina, Cecere et al. (2003) constructed three types of huts that mimicked rural houses. Huts with maximum refuges for the vector *Triatoma infestans* had roofs of tree leaves, providing a coarse texture, and unplastered walls of piled, mortarless bricks. Huts with intermediate refuges had the same roofs, but plastered walls; and those with minimum refuges had both plastered walls and roofs made of a locally available grass, giving the roof a fine texture. Each hut was artificially colonized with five female and three male *Triatoma* at the start of the experiment, and each night, for 2 years, a chicken was placed in each hut (as a blood source for the vectors). Figure 11.13 illustrates the average population densities of vectors in these "huts" at the end of the experiment. Although there is some variability, it is clear that vectors survived and reproduced more effectively in the huts with refuges. Vector control (in fact, prevention) in this case consists of using modern roofing materials and crack-free plaster walls.

Mechanical methods of vector elimination have worked extremely well in the control of Chagas disease, where removing the triatomid bug vectors has been a fundamental achievement of the Southern Cone Initiative (Dias et al. 2002). This international program began in 1991, and has now managed to eliminate Chagas transmission in Uruguay, Chile, and large parts of Brazil and Argentina. Continued vigilance and management will hopefully maintain and expand the scheme.

11.10.3   Mechanical manipulations: trapping

Many insect vectors locate their animal victims by sight, smell, or a combination of both. It is possible to reduce the density of some vectors by luring them to various types of trap where they can be killed by insecticides in or on the trap material. A trapping program has been established in Africa in attempts

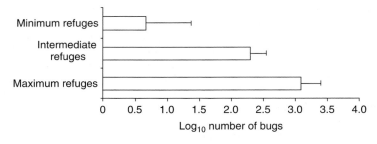

Figure 11.13 Mean number of *Triatoma infestans* in experimental huts 2 years after colonization, according to hut structure in terms of refuges for the bugs. (From Cecere et al. 2003.)

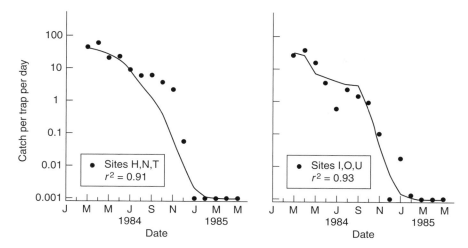

Figure 11.14 Mean daily catches of *Glossina pallidipes* pooled over monthly trapping periods and over three sites. The plotted lines indicate the predicted population changes resulting from the optimized simulation exercise. (From Hargrove 2003.)

to reduce populations of tsetse flies, the all-important vectors of sleeping sickness and nagana (Section 11.11.2). The structure (shape, size, color) of the traps is crucial to the success of these projects (Steverding & Troscianko 2004), and various designs are now in commercial use all over the tsetse belt of Africa. Some traps are designed to work without chemical bait or lure, whereas others may utilize a heady cocktail of aromas such as acetone, octenol, and cow urine (Belete et al. 2004). The spatial arrangement and density of traps is also important to trapping success. Hargrove (2003) has shown that baited traps in Zimbabwe facilitate a rapid and dramatic reduction in tsetse fly numbers (Figure 11.14). The combination of dry weather, ground spraying with insecticide, and the trapping effort itself were thought to result in the eradication of tsetse flies from closed (isolated) populations. Simulations run by Hargrove suggest that a trap density of four per square kilometer is appropriate.

11.10.4   Chemical control

Chapter 12 covers in detail the management by pesticides of insects that vector plant fungal and viral pathogens. For the vectors of pathogens that cause animal diseases, the use of residual insecticides applied to the walls and ceilings of dwellings and animal houses (abbreviated to RHS (residual household spray) and IRS (indoor residual spray)) has long been a stalwart of control. Chemicals such as DDT reduced malarial incidence in India from 75 million cases per year in the 1930s, to 110,000 per year in the 1960s (Pates & Curtis 2005). DDT was indeed the "wonder drug" of malaria control in years gone by, but it is now mainly unavailable and unusable due to problems of insect resistance, environmental degradation, and public health concerns. Ironically, a cheap, safe, and persistent DDT-like compound would be ideal to address the spread of disease that often follows large-scale disasters such as the Indian Ocean earthquake and tsunami in late 2004. Controversy now rages about whether we should reintroduce the careful use of DDT for IRS. Some see these proposals as a dreadful backward step to the bad old days of "Silent Spring"; others, more pragmatically, realize the lack of options available as malaria and its ilk rampage through developing countries (Schapira 2006).

Modern alternatives are mainly pyrethroid-based compounds such as cypermethrin, though with increasing incidence of resistance to these chemicals in vectors such as mosquitoes, older (much older) compounds such as carbamate insecticides may have to reappear (Conteh et al. 2004). Pyrethroids are more expensive and less persistent than DDT, but they are all we have. They can be effective in the control of mosquitoes that carry the virus responsible for dengue fever when sprayed in homes through ultra low volume (ULV) machines (Sulaiman et al. 2002; see also Chapter 12). Immediate (within 1 h) effects are limited, but as long as people wait a day after a house has been sprayed, virtually no mosquitoes survive. While house spraying may work for dengue vector control, it appears to have little impact on exophilic mosquito vectors of malaria that live outside and only venture indoors to feed (Pates & Curtis 2005). Another problem concerns the length of time that sprayed surfaces remain toxic to vectors. Sandflies vector leishmaniasis amongst other human diseases, and a major control tactic has been to use RHS techniques (Alexander & Maroli 2003). Unfortunately, but predictably, the insecticidal effect of sprayed compounds disappears after a while, with an effective period ranging from around 6 months to as little as a few days, depending on the type of compound used. The newest and most environmentally "friendly" insecticides are of course the least persistent. Of course, financial persistence is also required for vector control. Even if the spray programs kill vectors and reduce pathogen transmission, they will only persist if they are economically viable. Sometimes, the control costs seem trivial to those of us in wealthy nations. Conteh et al. (2004) analyzed the cost-effectiveness of malaria vector control using RHS/IRS in Mozambique. They estimated that the cost of vector control using insecticide spraying was in the region of US$3 per person covered per year. Simply put, malaria could be significantly reduced with long-term financial support, political will, and collaborative management coupled with community involvement and training.

Residual spraying has largely been replaced by the use of bednets impregnated with pyrethroid insecticides, and these are particularly effective against exophilic but not endophagic vectors (Pates & Curtis 2005). To quote Lindblade et al. (2004), insecticide-treated nets (ITN) are the most immediately available prevention tools to achieve substantial reductions in malaria-caused human mortality in sub-Saharan Africa. They are certainly at the forefront of the arsenal in the so-called "Roll Back Malaria" campaign. Not only does the routine use of ITNs reduce vector density and the incidence of malaria-infected bites on children, but there also appears to be a significant effect on untreated villages in the same area after a few years of treatment (Figure 11.15). As with spraying, costs must be taken into account. Wiseman et al. (2003) estimated that the net annual cost of ITN tactics per human life saved was around US$34 per year in western Kenya. At this monetary value for a human life, ITNs are a highly cost-effective use of scarce health care resources. Even better, as long as local people have access to the nets, and are willing to use them properly, the whole scheme can be run at a local community level. In Eritrea, for example, a steep decline in malaria morbidity was achieved between 2000 and 2004 (Nyarango et al. 2006), mainly through the combined use of ITNs and IRS. Finally, though, a word of caution. According to Hill et al. (2006), the use of ITNs is still unacceptably low. It seems that a mere 3% of African children are currently sleeping under an ITN, and only about 20% are sleeping under any kind of net, treated or untreated. A great deal of education and dissemination both of knowledge and hardware is clearly required.

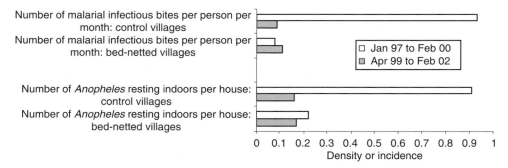

Figure 11.15 Results of trials on the effectiveness of insecticide-treated bednets on malaria and mosquitoes in Western Kenya. (From Lindblade et al. 2004.)

Finally of course there is the ever-present problem of resistance. Vector mosquitoes in China, for example, now show resistance to all the major insecticide groups (Cui et al. 2006). Enayati and Hemingway (2006) also point out that if mosquitoes develop resistance to pyrethroids via whatever mechanism, the crucial role of impregnated bednets may be compromised, or at worst, negated. As we will see in Chapter 12, if chemical control of vectors is somewhat half-hearted, resulting in a less than majority-kill (as might be expected in situations where users are trying to use dilute or ineffective chemicals to make them cover the maximum area for the minimum expense), then the selection for insecticide resistance will appear all the faster.

11.10.5 Biological control

We discuss the pros and cons of using biological control to manage insect vectors of plant diseases in Chapter 12. In truth, the biological control (i.e. by natural enemies) of animal disease vectors has not proven terribly effective. Adult flies have few if any significant predators or parasites, but their larvae, especially those that inhabit water bodies, may be vulnerable. The idea of using fish to remove mosquito larvae from ponds and streams is appealing (Floore 2006), but field studies suggest that ratios of predators to prey are generally too low for effective control.

However, toxins extracted from the soil bacterium *Bacillus thuringiensis* (*Bt*) show promise as agents of biological control. *Bt* in various guises is being used intensively and extensively for the control of crop pests (see Chapter 12). One variety of *Bt*, *Bt* var *israelensis*, has been used for the control of mosquito larvae in temporary and permanent water bodies ranging from discarded tires to small lakes (Sulaiman et al. 1997; Mulla et al. 2004). Mosquito-vectored diseases from dengue fever to malaria may be susceptible to *Bt* control, and the system is very simple. Aquatic mosquito and blackfly larvae are essentially filter-feeders, using their specialized mouthparts (totally different of course to those in the adult to come) to collect fine particles from the water around them. *Bt* var *israelensis* (trade name VectoBac) is normally sold in the form of fine granules, within which are micropellets containing the *Bt* toxin. These disperse rapidly in water, and are collected by first to fourth instar larvae of *Culex*, *Anopheles*, or *Aedes* mosquitoes, which subsequently die. VectoBac is extremely efficient in killing vector larvae (Figure 11.16). The downside is that the toxicity lasts only a short time, and treated water bodies can be recolonized by new mosquito larvae within a week or so of initial treatment (Russell et al. 2003). Nevertheless, this method is popular and commercial for small-scale vector management. In the USA, for example, mosquito "dunks" can be bought that release *Bt* var *israelensis* into tree crotches, birdbaths, flowerpots, and roof gutters. One snag with this method of control concerns the fact that the success of *Bt* (VectoBac) depends on the water conditions; turbidity, temperature, and even conductivity need to be taken into consideration when controlling blackfly (Wilson et al. 2005). *Bacillus sphaericus* (trade name

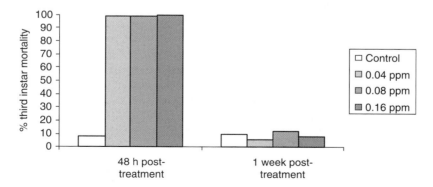

Figure 11.16 Field evaluation of *Bt* var *israelensis* (VectoBac) against the larvae of *Culex annulirostris* in Australia. (From Russell et al. 2003.)

Table 11.11 Larval control products: optimal application rates used for intervention, average yearly demand, and product price range. (From Fillinger & Lindsay 2006.)

Product	Formulation	ITU/mg	Application rate (kg/ha)	Average amount used per year (kg)	Product price* (US$ per kg)	Total costs per year (US$)
VectoBac®	WG	3000	0.2	5	20.00–30.00	100.00–150.00
	CG	200	5.0	17	2.16–3.00	36.72–51.00
VectoLex®	WDG	650	1.0	8	50.00–70.00	400.00–560.00
	CG	50	15.0	23	3.75–5.00	86.25–115.00

CG, corn granules; ITU, international toxic units; WDG, water dispersable granules; WG, wettable granules.
* Product pricing from factory (ex works) in 2005 not including transport and importation costs.

VectoLex) is a relative of *Bt* which has similar effects on filter-feeding aquatic dipteran larvae, with the added advantage of being more persistent than *Bt* in water, albeit more expensive. In a highly successful set of trials in Kenya, Fillinger and Lindsay (2006) used a combination of both microbial larvicides and achieved a reduction of over 95% in mosquito larvae populations from water courses, and a concomitant drop in the number of blood-fed adults per person of over 90%. Table 11.11 compares various parameters of the two biocontrol agents, including the all-important costs.

11.10.6 Genetic techniques

The term genetic control of vector insects covers methods whereby control is introduced into a wild insect population through mating (Pates & Curtis 2003). One system, the sterile insect technique (SIT), has been around for quite some time. Other more complex genetic modifications, such as the release of insects carrying a dominant lethal trait (RIDL), are on the increase (Alphey 2002). Most advanced of all are techniques that genetically modify vectors for resistance to the pathogen that they would normally vector (James 2002; Coleman & Alphey 2004). If successful, the pathogen–vector association is destroyed before the mammalian host can be infected. Development of transgenic mosquitoes and other vectors has yet to reach a commercial stage, but research continues apace.

SIT has already proven extremely successful in the control of screwworm fly, *Cochliomyia hominivorax*

(Diptera: Calliphoridae), in the southern USA, Mexico, and Central America (Benedict & Robinson 2003). Could SIT be used effectively against mosquitoes or blackfly? SIT is species-specific, environmentally benign, and, most importantly, increases in effectiveness as the population in which it acts declines in size. Put another way, SIT only works well in very small, preferably isolated, insect populations, a situation not always common in the cases of mosquitoes or blackfly. For SIT to work, large numbers of sexually active but genetically sterile males have to be released over large areas. Ideally, these sterile males should seriously outnumber naturally occurring, fertile males, so that females who mate only once "waste" their chance to produce fertile offspring. In practice, literally millions of male mosquitoes have to be reared in captivity, separated from the females, sterilized using chemicals or radiation, and then released into the wild. The whole procedure in fact sounds rather unlikely, and it still has to prove itself as a technique with wide applicability. There are clear problems, in particular the deleterious effects of the sterilization process on the insects before they are released. In the case of Mediterranean fruit fly, *Drosophila melanogaster* (Diptera: Drosophilidae), for example, irradiated males are less competitive with shorter lifespans and an estimated reduction in fitness of 4–10-fold (Alphey 2002). In reality, SIT is likely to work only in isolated populations where reinvasion is unlikely, and at the end of an integrated management program for vector control, when populations have been reduced significantly using other methods (Vinhaes & Schofield 2003). Thus one major but extremely localized success of SIT was the eradication of tsetse fly, *Glossina austeni* (Diptera: Glossinidae), from the island of Zanzibar (Reichard 2002). As Gooding and Krafsur (2005) point out however, such achievements are more "proofs of principle" than commercial programs.

11.11 CASE STUDIES

Many of the ecological principles discussed above are fundamental to the epidemiology and potential control of the world's most serious insect-vectored diseases. We now illustrate this further by presenting case studies of a few major animal and plant diseases, some of which we have encountered in previous sections. Readers should refer to the discussion of epidemiological theory to remind themselves that disease R_0 values below unity are required to achieve eradication, and that a vector's longevity is fundamental to its vectorial capacity.

11.11.1 Onchocerciasis

Onchocerciasis, or river-blindness, is a human disease caused by the filarial nematode parasite, *Onchocerca volvulus*. Though, strictly speaking, it is a non-fatal disease, the life expectancy of sufferers following the onset of blindness is much reduced. Adult filariae live in various human tissues, especially the skin, where they may produce unsightly nodules (Enk 2006). The dispersive or migratory stage of the parasite, microfilariae, cause much more serious skin irritation, and large populations may be found in the eyes of infected humans. Here, lesions are caused when the microfilariae die, which lead to increasing and finally complete opacity of the cornea, especially in older people.

Over the years, many millions of people in Africa, and a somewhat lesser number in Central America, have been infected. Statistics from WHO (1997, 2007) describe 120 million people at risk from onchocerciasis, and around 18 million actually infected. Of these, 99% are from 28 endemic countries in sub-Saharan Africa. We can get some idea of the scale of the problem by considering the situation in West Africa in the late 1980s. Here, a survey of over 600 villages in Guinea, Guinea-Bissau, western Mali, Senegal, and Sierra Leone found that 1,475,367 out of a

Figure 11.17 Blackfly feeding on human.

rural population of 4,464,183 people (33%) were infected by the parasite, and of those, 23,728 were blinded (de Sole et al. 1991). The only vectors of the parasite are blackflies in the genus *Simulium* (Diptera: Simuliidae) (Figure 11.17). The main species (or complex of species) in Africa is *Simulium damnosum* whereas, in Central and South America, *S. ochraceum* is the major vector.

When an adult fly cuts a hole or lesion in human skin, gaining access to superficial blood capillaries, secretions in the fly's saliva are inserted into the wound. The secretions not only inhibit the aggregation of blood platelets and slow down coagulation, but also increase vasodilation so that the wound bleeds more freely (Cupp & Cupp 1997). Some species of *Simulium* seem to have better vasodilatory abilities than others, and it is those that excel at this, such as *S. ochraceum*, that also happen to be amongst the most efficient vectors of the nematode. The saliva of one species of blackfly, *S. vittatum*, is even reported to affect immune cell responses in humans (Ribeiro & Francischetti 2003), indicating just how complex insect–host relationships can become. Of course, the salivary system has probably evolved to benefit the insect alone, but the prolonged bouts of undisturbed feeding will also benefit pathogen transmission. Blackfly use their rasping mouthparts to feed on blood, and the microfilariae contained within the blood then spend a developmental latent period inside the thoracic muscles of the vector. Nematodes then become concentrated again in the insect's proboscis before being introduced into a new host. There are no animal hosts except humans. However, *O. volvulus* can remain active in humans for 10–15 years, so that a high incidence of infection amongst people in rural communities maintains the parasite for a very long time, including during periods of low vector abundance.

The mainstay of the control of onchocerciasis is vector control based on insecticidal treatment (Davies 1994). Most insecticide programs target the larval stages of blackfly, which are spent in the fresh, flowing water of streams and rivers. Although larvae anchor themselves to streambeds, many can be swept downstream for long distances, dispersing vectors over wide areas. This phenomenon of stream transport, coupled with the ability of adult blackfly to disperse on the wind, has resulted in Ecuador in the formation of new foci of onchocerciasis in areas previously free of the disease (Guderian & Shelley 1992). However, various species of *Simulium* prefer different types of water, in a variety of environmental conditions. In Mexico and Guatemala, the main vector species of onchocerciasis predominate in shallow water at altitudes of between 800 and 1100 m above sea level (Ortega & Oliver 1990). Such specific observations are appropriate only for fairly local areas, but onchocerciasis infections in humans do seem in general to be positively correlated with altitude, and, not surprisingly, negatively correlated with distance from blackfly breeding sites (the nearest river) (Mendoza-Aldana et al. 1997). With this sort of ecological information, it is possible to restrict control operations to a subset of the total range of all *Simulium* species.

Undoubtedly, onchocerciasis vector control provides a very good example of situations where cheap and persistent insecticides, such as the otherwise abhorred DDT, are vitally important. Early attempts at larval control in the late 1940s and early 1950s using DDT in Kenya, Zaire, and Uganda resulted in virtually complete eradication of blackfly vectors, successes that remain up to the present day (Davies 1994). Admittedly, copious quantities of DDT were released into the environment (10 applications of the chemical were applied at 10-day intervals in Kenya in 1952 and 1953), but this is arguably a small price to pay for the cessation of a seriously debilitating human disease. DDT has now been replaced by pyrethroids as the insecticide of choice.

The WHO was responsible for the Onchocerciasis Control Programme (OCP), a huge and far-ranging undertaking employing aerial spraying of various larvicidal insecticides. This scheme began in 1974, and ended in 2002, with spectacular results. Figure 11.18 shows the extent of the project (WHO 2007). Over

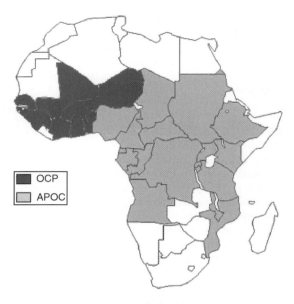

Figure 11.18 Extent of the Onchocercias Control Programme (OCP) and African Programme for Onchocerciasis Control Programme (APOC). (From WHO 2007.)

this period, insecticides were applied to vast stretches of rivers (more than 50,000 km at the peak of control activities, according to Resh et al. (2004)), covering 1,235,000 km^2. In the original central area of West Africa, in Burkina Faso and the Ivory Coast, transmission of the disease has been reduced to almost zero in 90% of the area, with the prevalence of onchocerciasis falling from 70% to a mere 3%. In central Sierra Leone, biting rates of potentially infective blackfly were reduced to 2% of their pretreatment levels after 4 years of efficient larviciding (Bissan et al. 1995). In many areas it has been possible to abandon treatment, and the niches of the blackfly vectors have been taken over, seemingly permanently, by other *Simulium* species that are unable to carry the parasite. Of course, it is dangerous to be complacent about the long-term effects of such a strategy. Though the OCP began to stop larviciding as early as 1989 in certain West African river basins, backup studies were required to ensure that vector flies did not re-establish themselves (Agoua et al. 1995). In one example, the precontrol infectivity levels of blackflies ranged from 2.5% to 8.9%, unacceptably high values. After the program, backup surveys revealed that these levels had been reduced to less than 1% in the main. In a few sites where infectivity remains significant, larviciding has had to be resumed.

On the whole, though, the OCP has been very successful, and over 90% of rivers in the original program areas are no longer in need of vector control. We might expect that the enormous quantities of insecticides applied to West Africa over 25 years would have exerted huge pressures on ecosystems, reducing biodiversity and selecting for resistance in the vector flies. However, it appears that little or no major effects have been detected, especially on fish populations (Resh et al. 2004), and now 25 million hectares of arable land in previously infected river valleys have been resettled.

The OCP in West Africa has now been extended into a second initiative, the African Programme for Onchocerciasis Control (APOC) (WHO 2007). Countries in Central and East Africa are now being targeted, including Burundi, Chad, Kenya, Malawi, Nigeria, Sudan, Tanzania, and Uganda (Figure 11.18). A filaricidal drug, ivermectin (a synthetic derivative of a product of the fungus *Streptomyces avermitilis*; Enk 2006), has become available, which drastically reduces the numbers of microfilariae in the skin. Ivermectin has become known as a "wonder drug" (Geary 2005), and since its introduction as a human medicine in 1985 it has undoubtedly alleviated the suffering of many thousands of people. As a result of reducing the microfilariae density in the skin, relatively few flies are able to pick up an infection during a blood meal. According to WHO (2007), community-directed treatment with ivermectin is the delivery strategy of APOC. It enables local communities to fight river blindness in their own villages, relieving suffering and slowing transmission. After just 8 years of operations, APOC has established 107 projects, which in 2003 treated 34 million people in 16 countries. The program intends over the following years to treat 90 million people annually in 19 countries, protecting an at-risk population of 109 million, and to prevent 43,000 cases of blindness every year. In addition to the standard techniques of drug distribution and vector control, APOC is also supporting the acquisition and use of REMO (Rapid Epidemiological Mapping of Onchocerciasis), a mapping system that delineates high-risk communities. In Cameroon, the levels of onchocerciasis endemicity were evaluated in 349 villages by examining the skin nodules caused by the parasite in men. REMO maps were then constructed,

which identified high-risk localities where ivermectin treatment was urgently required (Mace et al. 1997). In general, the success of onchocerciasis control is the result of secure financing, generous drug donation, and focused research and developmental research (Molyneux 2005), all of which may not last forever. At the moment, APOC is planned to close in 2010 (Amazigo & Boatin 2006), but onchocerciasis control must continue as long as there are blackflies to transmit the disease.

In the final analysis, the resounding success of the control of the insect vectors of onchocerciasis is at least partially attributable to the rather specific and identifiable ecology of blackfly. Insects that transmit other serious disease-causing pathogens, whose ecology is more complex and disparate, are much harder to deal with.

11.11.2 Trypanosomiasis

Trypanosomiases are a group of animal diseases caused by parasitic protozoa in the genus *Trypanosoma* (Barrett et al. 2003). Chagas disease in South and Central America is a trypanosomiasis vectored by triatomid bugs (Hemiptera: Triatomidae). The pathogenic agent of Chagas, *Trypanosoma cruzi*, has been found in 180 species belonging to 25 families of animals in the Americas (Ceballos et al. 2006), with marsupials such as possums, edentates such as anteaters, and carnivora such as skunks being most frequently infected. The major vector species, *Triatoma infestans* (Hemiptera/Heteroptera: Triatomidae), is common in wild (sylvatic) situations, despite having been targeted for eradication in more urban situations (Nicholson et al. 2006). The latter domestic infestations are much easier to deal with (see earlier). However, there are something like 30 different species of *Triatoma* in Mexico alone, at least 10 of which are implicated in the dissemination of Chagas disease (Cruz-Reyes & Pickering-Lopez 2006), so it is difficult if not impossible to make generalized control pronouncements that will work in all situations. None the less, remarkable successes have been achieved. Figure 11.19 shows estimates of the current (2006) distribution of *T. infestans* in South America, in relation to the maximum predicted distribution (Schofield et al. 2006). The Southern Cone Initiative (see earlier) has had a profound effect on Chagas disease, by virtue of eradicating its vector from very large areas of land

(a)

(b)

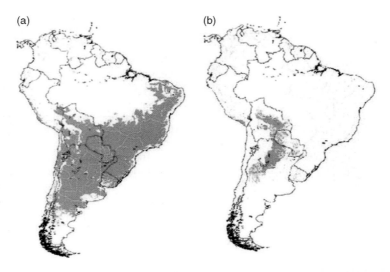

Figure 11.19 Apparent distribution of *Triatoma infestans*. (a) The maximum predicted distribution (in the absence of control interventions) from a geographic information system (GIS) thematic analysis, which reveals a few points outside the Southern Cone region (and in southern Chile and around the Peru–Bolivia–Brazil border) that might be suitable for *T. infestans* but from where this species has never been recorded. The predicted maximum distribution of *T. infestans* covers 6,278,081 km². (b) An estimate of current distribution based on reports from the Intergovernment Commission for the Southern Cone Initiative, from which the estimated distribution of *T. infestans* has been reduced to 913,485 km². (From Schofield et al. 2006; Courtesy of David E. Goria.)

(including, it is thought, the whole of Brazil). If only it were so simple in Africa.

Trypanosomiasis in Africa is vectored by tsetse flies (Figure 11.20). In sub-Saharan Africa, protozoan trypanosomes cause two related diseases, sleeping sickness in humans and nagana in livestock, both of which have enormous consequences for human health and livestock production. The vectors of trypanosomiasis are tsetse flies in the genus *Glossina* (Diptera: Glossinidae); these flies impose a major constraint on animal production in 8–10 million square kilometers (Krafsur 2003). Annual livestock losses in the 1990s were estimated to amount to US$5 billion (Kettle 1995). In the same vast area, some 50 million people were also at risk of the human form of trypanosomiasis (Chadenga 1994). According to WHO reports, men, women, and children in 36 African countries are at risk from the disease. The distribution of tsetse flies is roughly between the 15th parallel north and south of the equator (FAO 1998b) (Figure 11.21). Despite enormous research effort and investment by aid agencies in vector control for more than 70 years, the disease

Figure 11.20 Adult tsetse fly, *Glossina morsitans*, after a blood meal in Zambia. (Courtesy D.J.Rogers.)

still persists in vast areas of scrub savanna, forest edge, and riparian zones. Both sleeping sickness and its other-animal equivalent remain largely uncontrolled throughout much of sub-Saharan Africa (Welburn et al. 2006). Indeed, in some countries such as Angola, the Democratic Republic of Congo,

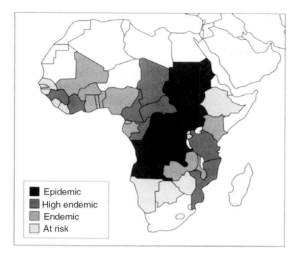

Figure 11.21 Distributions of sleeping sickness in Africa. (From FAO 1998b.)

In the case of sleeping sickness, two subspecies of the pathogen exist. *Trypanosoma brucei gambiense* occurs in western and Central Africa, whereas *T. brucei rhodesiense* is found in eastern and southern Africa. The latter tends to cause a more virulent and acute form of the disease than the former, although individual virulence can vary considerably (Dumas & Bouteille 1996). Once the pathogen has been transferred to a new host, he or she begins to exhibit fever, weakness, headache, joint pains, and itching. As the disease progresses, anemia, abortion, and kidney, cardiovascular and endocrine disorders occur. In the advanced stages, the victim becomes indifferent to events, exhibits lethargy and aggression in cycles, followed by extreme torpor and exhaustion. Death is preceded by a deep coma (WHO 1997). This final stage may only occur several years postinfection. With nagana, infected cattle may die within a few weeks; those that survive may persist with chronic infections for years, acting as reservoirs of the parasite within husbandry systems (Table 11.12). Trypanosomes also circulate in the host's bloodstream, where they can be acquired by both male and female tsetse flies while pool-feeding (lapping blood oozing from wounds cut by the insect's blade-like mouthparts) on the host. The protozoa multiply in the vector's gut and eventually migrate to its salivary glands, from where they can be inoculated into a new host.

Drugs do exist for combating sleeping sickness, but none are new, and one, melarsoprol, is more than 50 years old. Based on arsenic, this drug kills between 4% and 12% of people treated with it (Stich et al. 2003), hardly a good advertisement for its use. It is vital to achieve early detection of cases so that the use of these "hard" drugs can be minimized

and Sudan, it is rising to epidemic proportions again (Stich et al. 2003; WHO 2007) (Figure 11.22).

Tsetse flies are extremely unusual insects. Adult females do not lay eggs, but instead release single late-stage larvae (one larva roughly every 10 days) that have developed from eggs inside their mother's body. These larvae immediately pupate after being deposited in soil or litter. Hence, the insect's reproductive rate is very low, and tsetse flies are classic low-density pests (see Chapter 12). However, because of long lifespans and their efficiency at transmitting trypanosomes to mammalian hosts, they are exceedingly difficult to eradicate.

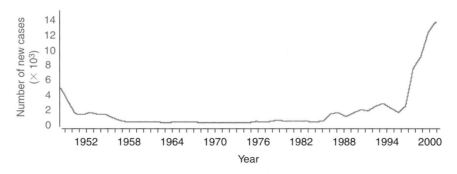

Figure 11.22 Sleeping sickness cases in Angola. (From Stich et al. 2003.)

Table 11.12 Feeding habits and host ranges of various species of tsetse fly (*Glossina*) that vector sleeping sickness to humans. (From Service 1996.)

Species	Location	Main hosts	Main habitat type
G. morsitans	East Africa	Wild pigs, wild and domesticated cattle	Savanna
G. morsitans	West Africa	Warthogs	Savanoa
G. pallipedies	East Africa	Wild cattle	Wooded savanna
G. palpalis	West Africa	Reptiles, humans	Riverine mangroves
G. swynnertoni	East Africa	Wild pigs	Savanna, dry thickets
G. tachinoides	South Nigeria	Domestic pigs	Riverine
G. tachinoides	West Africa	Cattle, humans	Riverine coastal

(Jannin 2005), but it seems highly unlikely that drug treatments will eliminate sleeping sickness on their own. Vector control is at the moment the only reasonable way to prevent the disease, and many techniques have been tried to this end. An effective solution might be to site villages and farms outside tsetse belts, but this of course prevents the habitation of large areas of Africa. The widespread use of insecticides is no longer possible or desirable; DDT was a useful tool for fly control in the old days, but its banning in the developed world has meant that DDT and related compounds are no longer an option for Africa (Langley 1994). However, if it is possible to predict where large numbers of flies are to be found at different times, then chemical control could be localized. In the Ivory Coast, for example, *Glossina palpalis* rests during the day on lianas, coffee bushes, and *Eupatorium* foliage (Seketeli & Kuzoe 1994), mostly between 10 cm and 2.5 m above the ground. Such sites are readily spot-treated with insecticides, thus minimizing wastage and pollution, and greatly reducing the cost of control. In Uganda, such bush resting sites were sprayed with the pyrethroid insecticide λ-cyhalothrin, which resulted in the numbers of human sleeping sickness cases dropping from between 4 and 12 per month to no more than one per month (Okoth et al. 1991).

As we described earlier, some countries have had success using specially designed tsetse fly traps impregnated with pyrethroid insecticides. In Uganda, a mere 10 of these traps per square kilometer reduced fly populations by more than 95% (Lancien 1991). The value of sterile male release, where artificially sterilized male tsetse flies are released into an area to outcompete natural, fertile ones, has also been

investigated (Moloo et al. 1988). A combination of traps and sterile male release led to the final eradication of *G. palpalis palpalis* in Nigeria in the 1980s. The traps were used to reduce fly populations along rivers and streams, with sterile males being released when the vector population was low enough to achieve a ratio of 10 sterile to one wild male (Myers et al. 1998). Certainly, the scattered and isolated populations of tsetse flies in parts of the range suggest that sterile insect techniques might usefully augment insecticide and trapping schemes (Krafsur 2003). In another example in Zimbabwe between 1984 and 1997, aerial and ground spraying, odor-baited traps, and insecticide treatment of cattle eliminated tsetse from around 35,000 km^2 (Torr et al. 2005), though getting rid of the vector does not always mean that the pathogen has also disappeared from the region (Cherenet et al. 2006), since it may remain in cattle and wild game for some time.

The Pan African Tsetse and Trypanosomiasis Initiative (PATTEC) was set up in 2000 to employ combinations of treatments (Kabayo 2002). The program aims to eradicate tsetse flies by attacking each pocket of infestation at a time, so creating tsetse-free zones that can ultimately be linked over large areas. However, it is dangerous to be too optimistic about tsetse fly control on such a scale. Rogers and Randolph (2002) have pointed out that eradication successes obtained in some countries are insignificant when compared with the continental scale of the problem. It is more than likely that flies from adjacent areas where control has been ineffective, or simply lacking, will reinvade areas in which tsetse flies have been controlled.

Finally, it is thought provoking to speculate about the effects of a really successful eradication program for nagana and sleeping sickness. If the non-human trypanosome can be controlled effectively, a great deal of already degraded land may become overstocked with grazing animals (FAO 1998b). Indeed, testes flies have even been called the "guardians of Africa" (D.J. Thompson, unpublished data). Human exploitation of dwindling resources may also increase dramatically if people are able to inhabit and develop previously dangerous places.

11.11.3   Malaria

The protozoan *Plasmodium* causes malaria; currently, *Plasmodium falciparum* is the most common (and most lethal) of the *Plasmodium* pathogens. Malaria is undoubtedly the world's most serious and widespread human disease transmitted by insects. It may have precipitated the decline of the Roman Empire and the fall of Greece (van Emden & Service 2004). Today, somewhere around 300 million people are infected (Collins & Paskewitz 1995), with 120 million clinical cases estimated globally per year (Coluzzi 1994). Between 1.5 and 2.7 million people die of the disease every year, and nearly 40% of the world's population live in regions where malaria is endemic. Guerra et al. (2006) have estimated the number of people at risk by region and according to parasite species (Table 11.13). Sub-Saharan Africa accounts for a large percentage (maybe up to 90%) of the reported malaria cases, mainly *P. falciparum*. In this region, more than 1 million children below the age of five are thought to die of the disease each year. Figure 11.23 illustrates the distribution of malaria in the world (WHO 2003). The disease is no longer endemic in large areas of Europe, Asia, North America, and Australia. It has been controlled in these regions by a combination of factors including the drainage of wetlands for farming, higher living standards, improved education, and the widespread use of insecticides such as DDT starting in the 1940s. However, malaria is still rife in most of Central and tropical South America, equatorial Africa, India, and the Far East.

The effects of malaria on people are far reaching (Snow & Gilles 2002), with young children and pregnant mothers the most seriously effected (Figure 11.24). The many direct consequences of infection vary from undernourishment to neurological disorders. Even the treatment, if available, can have serious side effects.

Table 11.13 Population at risk of malaria derived from extractions using the global spatial limits for *Plasmodium falciparum* and *P. vivax* in 2005. (From Guerra et al. 2006.)

WHO region	*P. falciparum* risk*	*P. vivax* risk*
SEARO	1.252	1.347
AFRO	0.525	0.050
WPRO	0.438	0.890
EMRO	0.245	0.211
AMRO	0.050	0.078
EURO	0.000	0.020
Total	2.510	2.596

WHO, World Health Organization; SEARO, South East Asian Regional Office; AFRO, African Regional Office; WPRO, Western Pacific Regional Office; EMRO. Eastern Mediterranean Regional Office; AMRO, American Regional Office; EURO, European Regional Office.
*The risk is given in billion (1,000,000,000) persons.

Indirect consequences include the economics of control. Regional and indeed international programs can of course cost a fortune, but local people also often have to bear some cost, where the poor are disproportionately affected (Chuma et al. 2007). In Kenya, the mean costs of malaria suppression were 5.9% and 7.1% of total household expenditure in rural areas, depending on season. Whilst wealthier households seemed able to cope, those starting out poor descended further into poverty.

Human malaria probably originated in tropical areas of the Old World, but Pleistocene glaciations delayed its spread in the northern hemisphere (de Zulueta 1994). During the last ice age, low temperatures in southern Europe precluded the activity of vector mosquitoes, and hence malaria was unable to spread. However, as temperatures rose and the glaciers retreated, about 10,000 years ago, some vector species were able to move northwards. The decline of malaria in Europe really began only during the 19th century, mainly because of new agricultural practices and changed social conditions such as better housing, drainage, and health care (Kuhn et al. 2003). Because, in this example, climate change appears to have been one of the driving forces in natural disease epidemiology, predicted rises in global

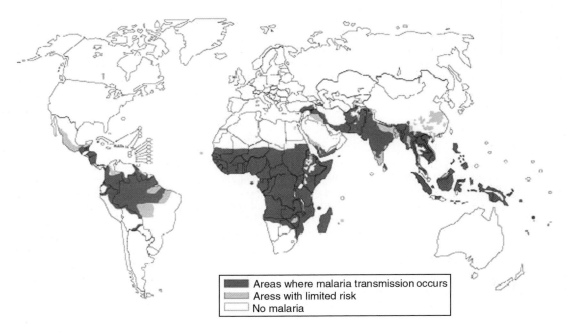

Figure 11.23 Distribution of malaria around the world. (From WHO 2003.)

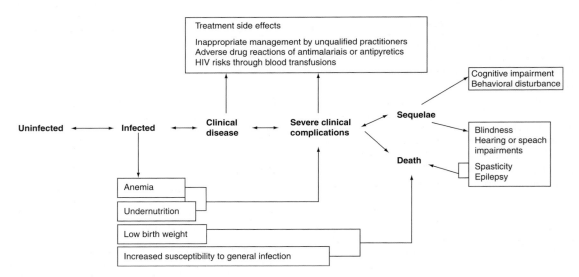

Figure 11.24 The direct, indirect, and consequential public health effects of *Plasmodium falciparum* malaria in Africa. (From Snow & Gilles 2002.)

temperatures may once again change the geographic distribution of insect vectors of malaria. Similar arguments can be applied to many other insect-vectored diseases such as yellow fever, dengue, and encephalitis (McMichael & Beers 1994). However, the links between climate variability and malaria incidence, on a local scale at least, are unclear, and, for now, controversial (Hay et al. 2005; Zhou et al. 2005). This means, sadly, that the development of reliable predictive systems for malaria outbreaks is still a long way off (Abeku et al. 2004; Thomas et al. 2004b).

Far from declining in the late 20th and early 21st centuries, malaria appears to be increasing rapidly in many parts of the world. Table 11.14 presents some examples from East Africa (Hay et al. 2002). Undoubtedly, even on a local scale, where you live can have a significant impact on the likelihood of catching malaria. Living in the forest as oppose to an open area dramatically increases the number of malaria cases in trials in India (Sharma et al. 2006) (Figure 11.25). In this case, living in the jungle between January to June was found to be relatively

Table 11.14 Resurgences of malaria caused by *Plasmodium falciparum* in the East Africa highlands. (Data from Hay et al. 2002.)

Region	Level of resurgence
Kericho tea estates, western Kenya	From 1988 to 1998, rise in severe malaria cases from 116 to 120 per 1000 people per year
Kabale, southwestern Uganda	Average monthly incidence in malaria risen from about 17 cases per 1000 (1992–6 average) to 24 cases per 1000 (1997–8 average)
Gkonko, southern Rwanda	Annual incidence of malaria risen from 160 to 260 cases per 1000 from 1976 to 1990
Muhanga, northern Burundi	Average of 18 deaths from malaria per 1000 during 1980s, risen to 25–35 deaths in 1991

safe since seasonal patterns of malaria are also clear, linked mainly to rainfall patterns. Even staying in one place can reduce risks if, for example, people live on higher ground. Githeko et al. (2006) found that malaria prevalence in children in Kenya was 68% for those living in valley bottoms, 40% at mid-hill, and only 27% at the hilltop, reflecting the behavior and microhabitat preferences of vector mosquitoes.

Resurgence of malaria is concurrent with the alarming appearance of vectors resistant to insecticides, and parasites resistant to drugs (Touré et al. 2004). Work in India in 1994, for instance, found that 95% of *P. falciparum* isolates from people in an epidemic in Rajasthan were showing resistance to chloroquine (Sharma et al. 1996), and there is now widespread resistance to chloroquine in many parts of the tropical world. This, together with a decrease in the efficacy of vector control and clinical services, especially in remote areas, is thought to be at the root of the resurgences (Hay et al. 2002).

The life cycle of the protozoan parasites that cause malaria is illustrated in Figure 11.5 (Kettle 1995). The infective stage for humans of *Plasmodium* is the sporozoite, and an important concept in malaria epidemiology and control is the sporozoite rate, i.e. the percentage of female mosquitoes with sporozoites in their salivary glands. The sporozoite rate is an indication of the likelihood of malaria being transmitted to a human after being bitten by one of the 40 or more species of *Anopheles* mosquito known to be important vectors of the disease worldwide.

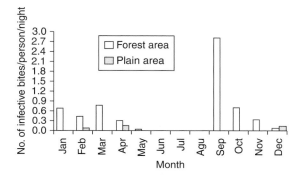

Figure 11.25 Number of infective bites per person per night (entomological inoculation rate (EIR)) in different months of the year (2003) in the forest and plain area study villages calculated from sporozoite index and human biting rates. (From Sharma et al. 2006.)

Table 11.15 Examples of sporozoite rates detected in anopheline mosquito vectors of malaria.

Country	Mosquito species	*Plasmodium* species	Sporozoite rate (%)	Reference
Cameroon	*An. gambiae*	*P. falciparum*	5.7	Robert et al. (1995)
China	*An. anthropophagus*	*P. falciparum*	10.9	Liu (1990)
Gambia	*An. gambiae*	n.a.	6.1–7.7	Thomson et al. (1995)
Kenya	*An. gambiae*	*P. falciparum*	5.4–13.6	Beier et al. (1990)
Madagascar	*An. gambiae*	n.a	1.7–3.2	Fontenille et al. (1992)
Papua New Guinea	*An. punctulatus*	*P. falciparum* or *P. vivax*	0–3.3	Burkot et al. (1988)
Sierra Leone	*An. gambiae*	n.a.	3.90	Bockarie et al. (1993)

n.a., not available.

Table 11.15 shows some published sporozoite rates from around the world.

Note that the high degree of variability in sporozoite rate is caused by ecological factors including the availability of breeding sites, the incidence of wet and dry seasons, and the species of vector and parasite involved. In theory, the average sporozoite rate in a mosquito species during a transmission season would reflect its transmission efficiency (Mendis et al. 1992). However, the detection of a high sporozoite rate does not necessarily point to high risk. If the population density of vectors is very low, the biting rate may also be very low, even if each bite has a fairly high probability of being an infective one. The combination of sporozoite rate and bite intensity is known as the entomological inoculation rate (EIR). One of the most important problems in malaria control is the fact that, in an increasing number of studies, a high incidence of the disease occurs even under conditions of very low levels of EIR. The consequence is that decreases in transmission by even the most efficient control systems may still not yield corresponding long-term reductions in the incidence of severe malaria (Mbogo et al. 1995).

The control and, ideally, eradication, of malaria is based on two approaches that can be combined, at least in theory. One tactic is management of the parasitic protozoan in humans, and the other is control of vector insects. The pharmacology and medical application of malarial prophylactic drugs such as chloroquine and maloprim are beyond the scope of this book, as are exciting but controversial developments in the immunization of people against the pathogen – although vaccines against malaria are still a long way off (Malkin et al. 2006). Suffice it to say that the use of drugs for malaria prevention is limited in scope, is expensive, and has undesirable side effects in the medium to long term. Moreover, the majority of people in the world at risk from malarial infections do not have access to drugs. Malarial vaccines are taking a long time to develop, and it is unlikely that the first-generation vaccines will be of much use to those who travel to malarial areas (Greenwood 1997; Tongren et al. 2004). Indeed, it is possible that these only partially effective new vaccines may actually upset the complex balance that exists between parasite and host in local populations, and may make matters worse rather than better. In reality, the development of new antimalarial drugs, an effective malaria vaccine, or new ways of killing mosquitoes, is still a long way from achieving the effective and dependable control of malaria (Beier 1998). The Roll Back Malaria (RBM) campaign was launched in 1998 by WHO, UNICEF, UNDP, and the World Bank. More than 40 countries have pledged to halve the so-called global burden of malaria by 2010, using a combination of best-practice methods. But, for most countries where humans live in permanent potential contact with vector mosquitoes, perhaps the only realistic management system is that of vector control. Details of potential control methods were provided earlier in this chapter.

11.11.4 Bubonic plague

Bubonic plague was certainly well known by the sixth century AD (Bari & Qazilbash 1995), but it has been traced as far back as 1700 BC in Pharaonic Egypt (Panagiotakopulu 2004). The Black Death, as it was known in the Middle Ages, killed somewhere in the region of 130 million people (or very approximately, 40% of humans in Europe) from the 14th right through until the 17th century. Population growth curves for Britain show a marked population "bump" spanning the 13th and 14th centuries when British populations were severely checked in their exponential growth (Pattison 2003). On a global scale, demographic models estimate that if such enormous mortality had not occurred, thus causing a severe setback in population growth, the world population might now exceed 9.8 billion, instead of the "mere" 6 billion we actually have. The rate of spread of the disease was stunning, especially for the Middle Ages. Epidemics of the Black Death spread from a small part of the Mediterranean coast of France and Italy in December 1347, reaching southern England by December 1348, and Scotland and Scandinavia by early 1350 (Scott & Duncan 2001). Plague is still very much with us today (Gage & Kosoy 2005) (Figure 11.26). WHO statistics (WHO 1994) show that during the 15 years from 1978 to 1992,

14,856 cases were reported from 21 countries in South America, Africa, and Asia. Of these, 1451 were fatal. Though not currently in the same league as malaria or sleeping sickness, plague has all the potential to cause serious human suffering in many countries where, for now, it is endemic.

Plague is essentially endemic in natural populations of wild rodents such as gerbils (Davis et al. 2004). This "enzootic cycle" in rodents is largely benign, although the pathogen appears able to infect virtually all mammals from rabbits to marsupials. When it infects highly susceptible hosts such as black rats (and, of course, humans), the cycle becomes epizootic and plague epidemics occur which may last for several years (Gage & Kosoy 2005) (Figure 11.27). The pathogenic bacterium responsible, *Yersinia pestis*, is in fact in the same family as *Escherichia coli* and *Salmonella typhi*, well-known human enteropathogens (Gage & Kosoy 2005). The major vector of *Y. pestis* from rodents to humans is the black rat flea, *Xenopsylla cheopis* (Siphonaptera). The reason that this species of flea is such a good pathogen vector is that the bacteria are able to multiply rapidly in the vector's midgut, forming bacterial colonies in just a

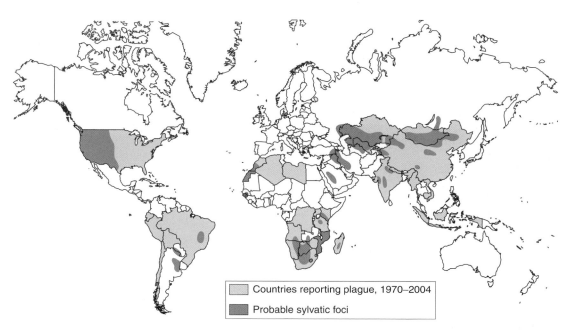

Figure 11.26 Distribution of plague foci and countries reporting plague. (From Gage & Kosoy 2005.)

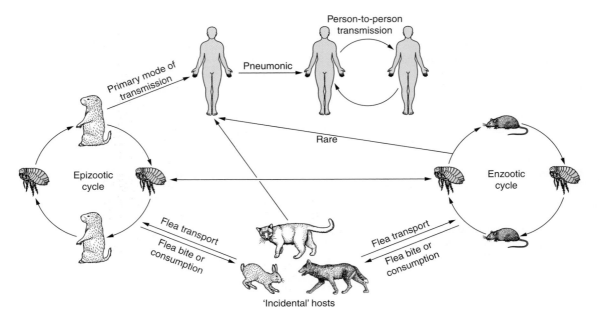

Figure 11.27 Natural cycles of plague. (From Gage & Kosoy 2005.)

few days. These colonies become large enough to block the insect gut, thus preventing it from feeding. The overall result is a starving flea that repeatedly tries to "clear its throat" of plague bacilli, resulting in frequent attempts to feed on and infect mammalian hosts. Black rats have been transported around the world for centuries if not thousands of years, and of course thrive in association with poor human habitation. Plague pandemics through the ages are thought to be a product of large rat populations in close association with humans, and a lack of vaccines and antibiotic medicines until recently.

So, it might be thought (or hoped) that plague is in decline in the world, and that most of us, whether we stay at home or travel overseas, are unlikely to run the risk of catching the disease. However, recent work by Stenseth et al. (2006) examined nearly 50 years of data from Central Asia where human plague is still reported regularly. It seems that *Y. pestis* in wild populations of gerbils increases with warmer springs and wetter summers, and it is a sobering thought that the very climatic conditions that may result from ongoing climate change in the 21st century are just those that prevailed at the time of the onset of the Black Death in the region.

11.11.5 Dutch elm disease

Dutch elm disease is caused by vascular-wilt fungi, *Ophiostoma* (= *Ceratocystis*), which are carried from infected to healthy trees by bark beetles, predominantly in the genus *Scolytus* (Coleoptera: Scolytidae). Fruiting bodies of the fungus proliferate in the larval galleries of the beetle under the bark of infested trees. When the young adults leave their pupal chambers in the bark to seek out healthy trees for maturation feeding (a pre-reproduction method for enhancing food reserves), fungal spores are carried on their bodies and mouthparts (Bevan 1987). There is no suggestion that this transmission is anything but mechanical. Maturation feeding occurs at the twig crotches of healthy trees, causing exposure of sapwood tissue by the chewing action of the beetles. Fungal spores are then introduced into these wounds, from where they are able to germinate and infect the tree's vascular system. Infected trees die rapidly (see Plate 11.1, between pp. 310 and 311), within 1 or 2 years of infection, providing highly suitable breeding sites for new generations of the vector. Figure 11.28 depicts this admirable cycle of infestation and infection. Undoubtedly, the association between beetle and

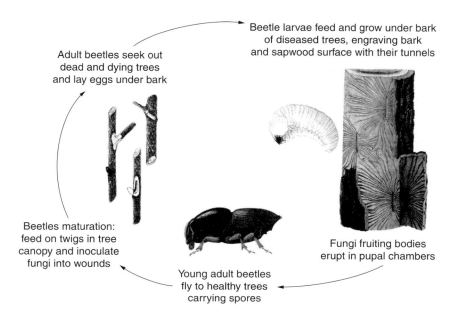

Adult beetles seek out
dead and dying trees
and lay eggs under bark

Beetle larvae feed and grow under bark
of diseased trees, engraving bark
and sapwood surface with their tunnels

Beetles maturation:
feed on twigs in tree
canopy and inoculate
fungi into wounds

Fungi fruiting bodies
erupt in pupal chambers

Young adult beetles
fly to healthy trees
carrying spores

Figure 11.28 Life cycle of scolytid bark beetles in association with elm disease fungi.

fungus is mutualistic, even though vectoring of the pathogen occurs simply via mechanical contamination of the beetles' external surfaces. Dutch elm disease has proved to be one of the most serious causes of tree decline and death in Western Europe and parts of North America. In Britain it is estimated that over 30 million elms have died, with over 1 billion worldwide. These huge mortalities were likely caused by a new, virulent species *Ophiostoma. novo-ulmi*, which has replaced the much less aggressive *O. ulmi* (Brasier & Buck 2001).

The spread of Dutch elm disease in Europe and North America during the 20th century exemplifies the complex interactions among vectors, pathogens, hosts, and human activity in the epidemiology of plant diseases (Brasier 1990). There were two epidemics during the 20th century. The first appeared in various parts of northwest Europe around 1920 and then spread eastwards across central Europe in the 1930s. It also spread in a westerly direction and reached Britain and North America during the 1920s. This first epidemic appears to have been caused by a weakly pathogenic, so-called non-aggressive, subgroup of *Ophiostoma*. The second epidemic, or more exactly, a set of epidemics, is still in progress, and is caused by two races of a much more aggressive

subgroup of the pathogen. One race, the Eurasian or EAN race, has been spreading across Europe, probably since the 1940s. The North American or NAN race has also been spreading across that continent for the same period of time. However, the NAN race crossed the Atlantic, presumably on logs and other wood products, to Britain and neighboring parts of Europe during the late 1960s. As a consequence, parts of Europe have both EAN and NAN epidemics (Brasier 1990) (Figure 11.29). These days, the EAN and NAN races of the fungus have been named as subspecies *novo-ulmi* and subspecies *americana* respectively (Temple et al. 2006).

The scolytid beetle vectors of Dutch elm disease are perhaps "innocent parties" in the tussle between pathogen and host tree. Both continents have indigenous elm bark beetle species, which can persist in the absence of any fungal pathogen by utilizing dead or dying trees that result from wind damage or old age. Beetles can only realistically transmit the fungi over relatively short distances, say from one tree to another in rather close proximity. However, when people start to move logs infested with beetle larvae or young adults either nationally by road transport or internationally by ship, the insect vector receives a considerable helping hand. A somewhat anecdotal example

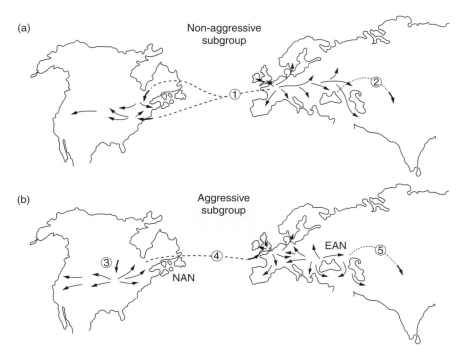

Figure 11.29 Proposed pattern of spread of (a) the non-aggressive, and (b) the EAN and NAN aggressive subgroups of *Ophiostoma ulmi* during the first and second epidemics of Dutch elm disease. Small arrows, overland spread; large arrows, major introductory events. (1) Introduction of the non-aggressive subgroup from northwest Europe to North America, *c.* 1920s. (2) Introduction of the non-aggressive subgroup from Krasnodor to Tashkent, *c.* late 1930s. (3) Introduction of a form close to the EAN aggressive subgroup into North America (Illinois area), *c.* 1940s, and its subsequent evolution into the NAN subgroup. (4) Introduction of the NAN subgroup from the Toronto area into the UK, *c.* 1960. (5) Introduction of the EAN subgroup into the Tashkent area, *c.* mid-1970s. (From Brasier 1990.)

of this comes from the first evidence of the second epidemic in central England in the 1970s, when dying trees were observed alongside major roads where trucks carrying elm logs would routinely stop. Beetles contaminated with fungal spores could easily hop off their truck and "high-tail it" to the nearest healthy elm tree growing by the roadside.

As with so many plant and animal pathogens discussed in this chapter, management of Dutch elm disease might be possible by controlling the vector. However, the prognosis is not particularly good. We know that stressed trees are much more likely to be attacked and colonized by various species of *Scolytus*, hence the success of the symbiosis. However, it also seems that environmental conditions influence a tree's susceptibility both to beetles and fungus (Solla & Gil 2002). Managing tree health

over large spatial scales is not particularly simple, and unlikely to be a viable method of control. Biological control (see Chapter 12) is possible in theory – large numbers of parasitic wasps and predatory beetles are known to attack scolytid larvae under the bark of infested trees (Manojlovic et al. 2003). None the less, biological control is unlikely to be efficient enough to prevent some young adult beetles surviving to carry the fungus to new elm trees. Similarly, attempts to use pheromone traps, while catching thousands of beetles, still do not reduce vector population densities far enough. The future would seem to be a countryside full of forever young elm trees which succumb to the beetles and the pathogens when they become old enough and large enough to be attacked, and are replaced by more new youngsters.

11.11.6   Barley yellow dwarf virus

If malaria is the current "champion" of human diseases, then barley yellow dwarf virus (BYDV) is a compelling candidate for the current champion of plant diseases (see Plate 11.2, between pp. 310 and 311). BYDV is a luteovirus that is the most economically important and widespread disease of cereals in the world (Irwin & Thresh 1990). It affects over 100 species of plants, including barley, wheat, oats, sorghum, rye, maize, rice, and many wild grasses. Wild grasses play an important role in the epidemiology of BYDV, acting as alternative hosts and vector–pathogen reservoirs within and around arable land. Over 20 species of aphid vector the virus, in a circulative, persistent manner. These include *Metopolophium dirhodum*, *Rhopalosiphum padi*, *Schizaphis graminum*, and *Sitobion avenae* (Hemiptera: Aphididae). All of these aphids are serious pests of cereals in the UK, continental Europe, and North America in their own right. Each aphid species tends to transmit rather specific strains of BYDV, which are designated under their own acronyms according to the vector species; the strain of BYDV vectored by *R. padi*, for example, is BYDV-RPV. The various strains of pathogen are to some extent regionally distributed, so that BYDV-RPV is more common in the south of the UK, whereas others predominate further north (Kendall et al. 1996). Transmission efficiency of the virus varies both with the strain of pathogen, and the genotype (and species) of aphid vector (Gray & Gildow 2003) (Figure 11.30). We suggested earlier that BYDV has some sort of mutualistic association with its aphid vectors. Little is known about potential mechanisms at the cellular level, although the aphids' own endosymbionts (see Chapter 6) may act to "chaperone" BYDV through the vector (Filichkin et al. 1997). In any event, the effects of BYDV on the host plant are definitely debilitating. Infection may contribute to crop death over the winter in colder regions, induce plant stunting, inhibit root growth, reduce or prevent flower production, and generally weaken the plant, rendering it more susceptible to other pathogens and environmental stress. Roots are particularly badly affected, with severely reduced lengths and general stunting observed only 4 days after inoculation with the virus (Hoffman & Kolb 1997; Riedell et al. 2003).

Some of the complexities of BYDV epidemiology are illustrated in Figure 11.31. The inevitable perturbation of agricultural systems, aphid dispersal patterns,

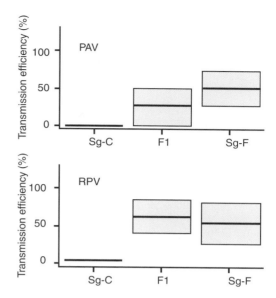

Figure 11.30 Mean transmission efficiency (horizontal line) of two luteoviruses that cause barley yellow dwarf virus, BYDV-RPV and BYDV-PAV, by two parental genotypes of *Schizaphis graminum*, Sg-C and Sg-F, and their F1 progeny ($n = 42$ and $n = 33$ for BYDV-PAV and BYDV-RPV, respectively). The shaded boxes indicate the 95% confidence interval around the mean and provide a measure of the transmission efficiency range. (From Gray & Gildow 2003.)

and the wide host range of vectors, all combine to make BYDV difficult to manage. There are three main avenues of control: (i) improve host-plant resistance to the virus and vector; (ii) the use of insecticides to control the vector; and (iii) manipulation of the crop environment or agricultural practices to minimize epidemics. Reliable host-plant resistance appears difficult to find. The effectiveness of resistance varies with type of cereal, species of vector, and even regional locality (Irwin & Thresh 1990), so the production of a coverall resistance mechanism is unlikely. However, some wild perennial grass species do exhibit real resistance, or at least a low rate of virus multiplication, and molecular techniques may allow such characteristics to be transferred to crop cereals. If this resistance to the pathogen can be achieved, then additional, heritable characteristics of cereals that deter aphid feeding may provide a dual-acting package (Gray et al. 1993).

Farmers can manipulate the crop environment to help manage BYDV epidemics. Most of these strategies

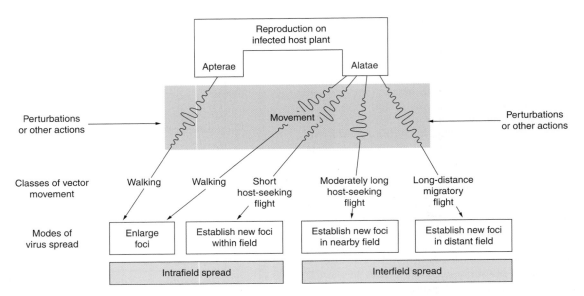

Figure 11.31 Conceptual diagram of aphid movement resulting from various perturbations to the system. These movements can result in spreading BYDV to other fields, or establishing new or enlarging existing BYDV foci within the field. (From Irwin & Thresh 1990, with permission, from the *Annual Review of Phytopathology*, ©1990 by Annual Reviews.)

target the aphid vectors rather than the pathogen directly, although eliminating perennial weeds and volunteer cereals adjacent to cereal crops may remove local reservoirs of BYDV infection. Strategies to deter aphids are many-fold. Tactics such as crop spacing and planting density are designed to reduce the crop's attractiveness to aphids. Some farmers may try to avoid peak aphid populations by sowing seed in the fall. Others may remove cereals from arable land for a time to break the otherwise permanent cycle of aphids and virus, using crop rotations. As it turns out, changing the sowing date has proved to be particularly successful. By delaying the sowing of wheat to late fall or early winter, workers in Australia were able to reduce BYDV levels and hence increase grain yield significantly (McKirdy & Jones 1997). The vectors were presumably unable to find the new plants effectively because of poor weather, though it must be said that these particular results were significant only in areas of high rainfall.

Cereal aphids that damage crops without vectoring viruses are routinely controlled with timely doses of chemical insecticides, the applications of which are based on monitoring and prediction strategies (see Chapter 12). When the pests are infected with BYDV, the situation becomes more complicated because virus spread is due to vector movement within and among the crops (Figure 11.31). Insecticidal compounds may increase rates of aphid movement so that they transmit the pathogen more effectively. Clearly, aphidicides that kill rapidly before the vector has time to move are the most desirable but, as only a few surviving aphids can still spread BYDV, very high percentage kills must be achieved. Another problem is the ability of aphids to move considerable distances from alternative wild host plants, or to move into a recently sprayed field from another, untreated one. If chemicals are to be used at all, appropriate timing of treatments is the key to success, so that the incidence of aphid migrations can be matched with susceptible stages in crop development (Mann et al. 1997). Another major difficulty with the economically viable management of BYDV is that epidemics tend to be sporadic, and hence chemical control of the aphids is not required every year or in every field (Fabre et al. 2003). Clearly, cereal growers would be

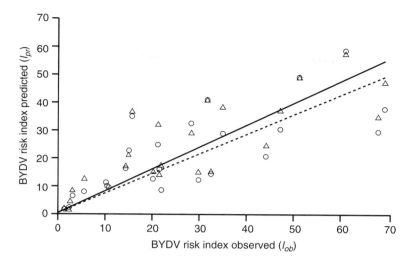

Figure 11.32 Relationship between the mean observed value of the BYDV risk index (I_{ob}) and its mean predicted value (I_{pr}) for simulations ran with actual observed temperatures (Δ, solid line: $n = 25$, $r^2 = 0.89$, $P < 10^{-3}$) and with 20-year average temperatures ($\bigcirc$, dotted line: $n = 25$, $r^2 = 0.85$, $P < 10^{-3}$). (From Fabre et al. 2006.)

able to reduce sprays if reliable detection and prediction systems were in place. Researchers frequently promise such "decision support systems" (DSS) but they less commonly reach widespread application. On the trail of such goals, Fabre et al. (2006) have shown that they can produce a risk index for BYDV based on temperature relationships for the aphid vector *Rhopalosiphum padi* in France, and based on a single early measurement of the proportion of cereal plants infested by the aphid in a season (Figure 11.32). Perhaps a user-friendly DSS is not so far away after all.

None of these strategies may as yet be commercially viable, especially in the areas of intensive, monocultural cereal production that occur in many regions of the world. In essence, the interactions among cereal aphids, strains of BYDV, wild plants, and commercial crops are exceedingly complex, and even now are not completely understood. This is another example of ecological complexity defeating human aspirations and economic targets.

11.11.7 Pine wilt

Previously, we discussed the ability of insects to carry nematodes responsible for human diseases, such as onchocerciasis. Insects can also transmit nematodes

that are pathogenic to plants; the potential global spread of pine wilt provides us with an example.

Pine wilt is a major killer of trees in the Far East, especially in China and Japan (see Plate 11.3, between pp. 310 and 311), where it was first described in 1905 (Fielding & Evans 1996). The pinewood nematode (PWN), *Bursaphelencus xylophilus*, is the causative pathogen, and is transmitted by beetles in the genus *Monochamus* (Coleoptera: Cerambycidae). This species is inferred to have been introduced into China, Korea, and Taiwan in the 1900s from North America (Togashi & Shigesada 2006).

Cerambycids, of which there are well over 20,000 species so far described in the world, are notorious pests of a huge range of tree species. The life cycle of *Monochamus* is fairly typical of the family. Adult females lay eggs in slits cut in the bark of trees. Individual trees selected for oviposition are usually stressed or lacking in vigor for one reason or another; healthy trees are not suitable for oviposition because their high resin flows prevent larval development. Tree stress may result from unsuitable soils, drought, or attack by primary pests such as defoliators. The hatching beetle larvae tunnel under the bark, feeding between the inner bark cambium and the sapwood, where they excavate broad, flattened tunnels filled with coarse frass consisting of wood fibers. This activity may kill trees by girdling them. Towards the end of the larval stage, cerambycid larvae "duck-dive" into the sapwood, where they excavate U-shaped galleries. They finally pupate, sometimes after several years as larvae, in chambers near the wood surface. Young adults then re-emerge through the bark of infested trees, leaving characteristic oval exit holes. Most species will then seek out new trees on which to lay eggs. However, some species, including *Monochamus*, undergo a period of maturation feeding, during which freshly emerged adults of both sexes fly to the tops of healthy trees to feed upon nutrient-rich tissues, before they are able to mature their eggs and sperm.

Pinewood nematodes live predominantly in the xylem tissues of host trees, where they feed on live wood cells, or on a variety of fungi (such as blue-stain fungi) that occur in moribund trees. This mycetophilic (fungus-feeding) habit explains in part

why *Monochamus* is such an important vector of PWN; *Monochamus* is associated with a particularly high abundance of fungi (Maehara & Futai 2002). Large number of nematodes in the xylem, combined with the phytotoxic effects of their feeding activities, cause blockages of the tree's transport vessels. As you might imagine, xylem blockage induces local drought effects, or wilt, in the plant's tissues. Seriously infected trees die rapidly from the top down. During winter and spring, dispersive PWN larvae move from the xylem to the beetle's pupal chambers. Here, the PWN molt to produce specialized forms (dauer larvae), which enter the body of the new adult beetles through their spiracles and into the tracheae (Aikawa et al. 2003). Up to 200,000 PWN may then accompany a single adult beetle on the next stage of its life cycle. There would appear to be a strong, perhaps coevolved (see Chapter 6), association between nematode and

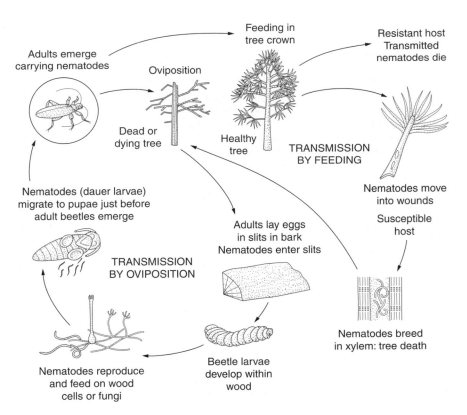

Figure 11.33 Life cycle of the pine wilt nematode and its cerambycid beetle vector. (From Fielding & Evans 1996.)

vector, especially during the dauer stage (Fielding & Evans 1996). Transmission of PWN to healthy trees occurs via the wounds caused by maturation feeding. The new nematode infection initiates a serious decline in tree health that, of course, provides suitable oviposition sites for a new generation of beetles. PWN can also be inoculated into trees at oviposition, but as these host trees are already likely to be infected in epidemic conditions, this form of transmission is considered secondary to the feeding process. Figure 11.33 summarizes the combined life cycles of vector and pathogen.

Despite its name, pine wilt is not completely restricted to pines. Fielding and Evans (1996) list 54 species of conifer susceptible to PWN, including firs, cedars, and larches. Perhaps most worrying for the UK and other parts of Western Europe is that the native Scots pine, *Pinus sylvestris*, beloved of conservationists and promoters of biodiversity, is highly susceptible to the beetle and its associated pathogen (Naves et al. 2006). Severe disease symptoms may not always develop, however. Experiments in Germany have shown that spruces (*Picea* spp.) largely tolerate PWN infestations without developing any wilt symptoms (Braasch 1996).

The spread of PWN in Japan has been rather pedestrian but undoubtedly resolute, indicative of the slow but sure march of the vector (Togashi & Shigesada 2006) (Figure 11.34). The extent of pine wilt damage to forests in Japan has been very considerable indeed, with many millions of *Pinus* spp. being killed over the years. For example, $2.4 \times 10^6 m^3$ of timber were reported lost to pine wilt in a single year (Mamiya 1984). Trees in forest stands often die back in clumps rather than as singleton trees (Togashi 1991), possibly indicating stress-related site interactions, or rapid transmission from one tree to the next. As with many other pathogens, human activities have made the outbreaks worse by artificially spreading the nematode in timber along railways and roads (Xu et al. 1996). Pine wilt disease has spread to many countries beyond Japan. In the USA, PWN outbreaks are associated with susceptible tree species, suitable vector species, and high summer temperatures (Fielding & Evans 1996). Nematodes have also been isolated (and quarantined) from imported wood material in Europe, arriving for example in Portugal in the late 1990s (Aikawa et al. 2003). Despite some near misses, the disease itself has yet to spread in Europe. Unfortunately, it seems highly likely that

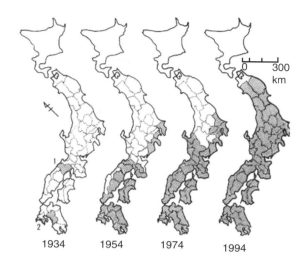

Figure 11.34 Range expansion of pine wilt in Japan. Shaded parts represent prefectures invaded by the pinewood nematode. The two numbers in the far left of the figure express the prefectures the nematode invaded early: 1 and 2 for Hyogo and Nagasaki Prefectures, respectively. (From Togashi & Shigesada 2006.)

PWN would survive, and that tree mortality would be severe in warmer southern countries. To avoid further introductions of PWN or its vector into Europe, strict legislation and quarantine practices have been developed. Legislation requires the heat treatment of coniferous material, such as chips, packaging, and pallets, from outside the European Community. In addition, round wood or sawn timber must be stripped of all bark and, if holes exist in the wood over 3 mm in diameter (indicating the likely presence of *Monochamus*), the material must be kiln-dried at the point of origin. Plant health risk and monitoring evaluation (PHRAME) systems are being developed at local and regional levels for pine wilt in Europe at the current time (H. Evans, personal communication).

11.12 CONCLUSION

Hopefully, you are now convinced that associations between insects and pathogens are common, complex, and fascinating. They range from coevolved mutualisms to incidental contact and transmission.

There are, of course, many serious animal and plant pathogens that are not vectored by insects. Influenza, SARS, HIV, and hepatitis are just a few of the current human health concerns that are unrelated to arthropods. Is there any potential for new associations to develop between insects and other human diseases? What, for example, is the likelihood of hepatitis or HIV being vectored by insects? This topic was vigorously debated in the 1980s, when it was concluded that potential vectors such as bedbugs (Hemiptera: Cimicidae) and horse flies (Diptera: Tabandidae) were unlikely to act as mechanical transmitters of HIV from one human to another (Jupp & Lyons 1987). However, by 1990, the same first author reported that the bedbug *Cimex lectularius* might be able to transmit hepatitis B virus mechanically, by multiple feeding on human hosts. There remains the potential for horse flies to transmit hepatitis C virus in a similar fashion (Silverman et al. 1996). With pathogens as mutable as RNA viruses, of which HIV is one, epidemiologists and insect ecologists must remain vigilant for the appearance of new insect vector–pathogen relationships.

Chapter 12

INSECT PEST MANAGEMENT

12.1 INTRODUCTION

Insects compete at many levels with humans for the crops that we grow and the livelihoods that we try to make from production systems, including agriculture, horticulture, and forestry. Without some form of control or management of these insects, enormous and unacceptable losses will regularly occur in all parts of the world. In previous chapters, we have described the complexity of interactions between insects and their environment. We are now able to consider how our understanding of insect–plant relationships, predator–prey dynamics, the abiotic environment, and evolutionary forces can be synthesized in modern approaches to insect pest management. No longer do we think of "pest control" as a strategy to reduce crop damage. This simplistic approach has been the cause of many basic problems and upsets, and has often been ineffective. Management these days is more concerned with the prevention of pest outbreaks whenever possible, so that more curative systems such as the use of chemical methods come into play only when prevention fails. The key to modern pest management is the goal of integrated pest management (IPM) in all appropriate cropping systems, and this chapter presents various important components that might function in complementary or additive ways to achieve this goal. The relative importance of management strategies within IPM has varied over time (Allen & Rajotte 1990) (Figure 12.1). Of the major components of IPM, the significance of biotechnology and plant resistance has increased markedly, while that of insecticides has declined. Biological and cultural control have seen some increase, and decision making using monitoring and pest thresholds remains as important now as it was 20 years ago.

12.2 THE CONCEPT OF "PEST"

The term "pest" is of course highly anthropocentric, but when seemingly small and innocent species such as the Russian wheat aphid *Diuraphis noxia* (Hemiptera: Aphididae), for example, ends up causing several hundred million US dollars worth of damage every year (Voothuluru et al. 2006), we must certainly take notice. Large numbers of insects in natural habitats are usually a source of wonder. Let us consider, for example, the enormous population densities of the monarch butterfly, *Danaus plexippus* (Lepidoptera: Nymphalidae and Danainae), which overwinter in Mexico. Adults from eastern North America migrate each fall to avoid climatic extremes, returning in the following spring when the larvae feed on milkweeds, *Asclepias* spp. (Gibbs et al. 2006). As we do not exploit milkweeds as a source of food or revenue, the dramatic densities of monarchs pose no problem. Only if *Danaus* in some way began eating a commercially important plant would it then earn the label "pest".

Of course, it is not just insects in large numbers that can be pests; in some cases, even one insect on its own can be too many to tolerate. These low-density pests include those species that act as vectors of many diseases of crop plants, domestic animals, and humans themselves (see Chapter 11). Clearly, strategies for the management of low-density pests must differ markedly from those appropriate for insects that become pests only by virtue of their high numbers (high-density pests) (Figure 12.2). In the former case, all insects

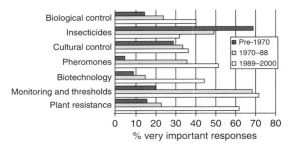

Figure 12.1 Influences of pest management technologies on the activities of extension workers; topics considered to be very important in responses from workers over 30 years. (From Allen & Rajotte 1990.)

must be prevented from attacking the host plant or animal, and prophylactic (insurance) measures are usually employed. In the latter case, however, it is desirable to wait until some threshold density is exceeded before any further action is taken. For example, not even one small larva of the codling moth, *Cydia pomonella* (Lepidoptera: Tortricidae), is tolerable in an apple to be sold in a supermarket, whereas fairly large numbers of defoliating Lepidoptera such as winter moth and green oak roller moth can be tolerated on oak trees. This tolerance of insect numbers usually relates to cost–benefit considerations, where the impact of the pest's damage in yield or monetary terms is weighed

against the cost of the management tactic to reduce it. The acquisition of the economic threshold based on sound and dependable impact data is a vital component of most IPM systems. Monitoring programs review the densities of pests on crops and provide data from which decisions about control tactics are made.

12.3 WHY PEST OUTBREAKS OCCUR

To prevent the numbers of insects on crops reaching levels where significant damage takes place, it is of fundamental importance to consider why such outbreaks occur (Hunter 2002). If we understand what has caused an upsurge in pest numbers, it may be possible to avoid pest increases, or at least reduce their intensity, in the future. Some factors that lead to pest outbreaks may be beyond our ability to manage. For example, it is well known that insect population densities can fluctuate tremendously over time (see Chapter 1). Cyclic dynamics are caused in the main by delayed density-dependent processes (see Chapters 4 and 5), but on occasion, stochastic events such as weather conditions may drive numbers higher and less predictably than in stable cycles (Zhang & Alfaro 2003). Even changes in the landscape or patterns of agriculture may lead to higher pest densities. Habitat fragmentation is thought to

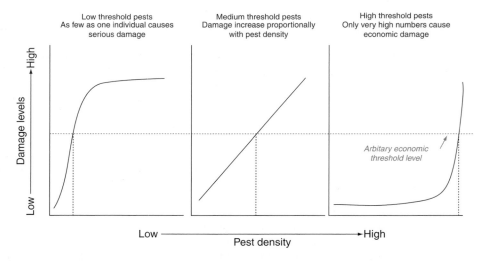

Figure 12.2 A classification of different types of insect pest according to their economic impact on crops.

Table 12.1 Some major reasons for insect pest outbreaks.

Natural disasters	Fire, windthrow, drought, waterlogging
Crop management	Species or varietal susceptibility
	Bad site choice and crop species matching
	Monocultures
	Proximity to pest reservoirs
	Bad nursery handling
	Bad crop handling
	Bad post-harvest handling
Pest Invasions	Accidental imports, lack of quarantine and inspection
	Intentional imports for other purposes
	Spread of a species
Misuse of control systems	Killing of natural enemies
	Resistance to pesticide in pest

Table 12.2 Summary of differences between natural habitats and managed cropland in relation to the increased liklihood of pest outbreaks.

	Natural habitats	Cropland
Alternative hosts for enemies	Many	Few
Artifical inputs such as agrochemicals	None	Many
Ease of finding specilaist host plants	Low	High
Habitat disturbance (soil or plant community)	Rare	Common, often annual
Numbers of natural enemies	High	Low
Plant species richness	High	Very low
Plant genetic diversity	High	Low
Plant species exotic	No	Yes
Plant species bred or domesticated	No	Yes
Plants chosen for high yields	No	Yes
Plant resistance to herbivory	Common	Rare

release some insect herbivores from regulation by their natural enemies (Kreuss & Tscharntke 1994; Speight & Wylie 2001). As with changing weather patterns, patterns of fragmentation may be beyond the control of any single farmer or pest manager.

Table 12.1 summarizes important factors that contribute to pest outbreaks (Speight 1997b), only some of which may be manipulated in a preventive pest management strategy. Many of them relate directly or indirectly to the provision of a large quantity of high-quality food and/or breeding sites for herbivorous insects. Intensive crop production systems, whether from agriculture, horticulture, or forestry, almost invariably represent gross departures from natural habitats within which these same insects have evolved with their enemies, host plants, and competitors. It is hardly surprising that the ecology of insects in these novel habitats also changes. Various important differences between natural habitats and cropland are summarized in Table 12.2; these, in most cases, are differences with which pest managers have to live. Most crop production in the world necessitates the modification and usually simplification of natural ecosystems to produce high-yielding, specific

crops. Even the plants themselves are frequently "unnatural", having been specially bred for certain desirable characteristics. Tongue somewhat in cheek, we might consider many examples of insect outbreak as caused mainly by the crop producer and not as the sole responsibility of the insects that are merely responding to an extra provision of resource. In reality, insect pest management cannot normally change crop husbandry techniques radically. The vast majority of pest managers throughout the world operate within intensive production systems and must attempt to manipulate them rather subtly to reduce the advantages provided by these systems to herbivorous insects. We now consider some of the major causes of insect pest outbreaks summarized in Table 12.1 in more detail.

12.3.1 Plant susceptibility

Whether or not a plant species, population, genotype, or individual is susceptible or resistant to an insect herbivore depends upon the resources (food and habitat) provided by the plant and the ability of the insect to utilize these resources. Two distinct ecological features of susceptibility are important in pest management. Susceptibility (or its opposite, resistance) can be environmentally derived and therefore not heritable. In contrast, resistance mechanisms may be genetically based and therefore passed on in selection and breeding programs to subsequent generations of the crop.

Environmental influences

As we described in Chapter 3, the conditions under which a plant grows are fundamental to determining its vigor. In turn, plant vigor can influence susceptibility to insect herbivores. Plants that are in some way stressed do, on many occasions, provide enhanced levels of organic nitrogen to insects feeding upon them, while showing reduced levels of physical and chemical defense (White 1993). Deterioration in the

health of plants tends to render them more susceptible to attack by certain groups of insects (Figure 12.3). Boring and sucking insects seem to perform better on stressed plants, whereas gall-formers and chewers are adversely affected (Koricheva et al. 1998). Coley et al. (2006) illustrate how this works for defoliators (Figure 12.4). In a survey of 85 species of Lepidoptera in Panama, they showed that larval growth rates were faster on younger than older (mature) leaves, reflecting the fact that higher nitrogen levels occur in younger leaves. The impact of borers on the other hand, such as those Lepidoptera that attack maize and sorghum, can be minimized by using nitrogenous fertilizers (Wale et al. 2006).

Adverse climatic conditions frequently influence the vigor or health of host plants, and in turn render the plants more or less susceptible to insect attack. Wood-borers, for example, are positively influenced by prolonged water stress in their host trees (Rouault et al. 2006). In Western Europe (excluding, for the moment at least, the UK) the bark beetle *Ips*

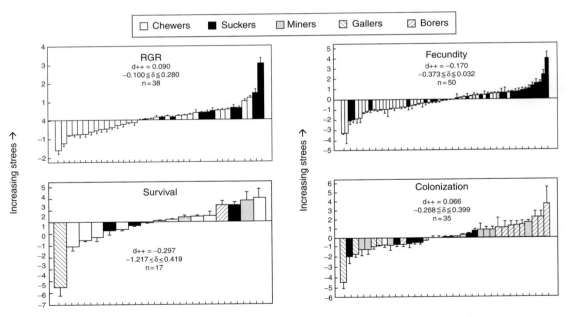

Figure 12.3 Responses of various insect herbivore guilds to plant stress as reported in the published literature (± 95% confidence intervals). Positive numbers on the *y*-axes indicate that increasing plant stress benefits insects. RGR, relative growth rate. (From Koricheva et al. 1998.)

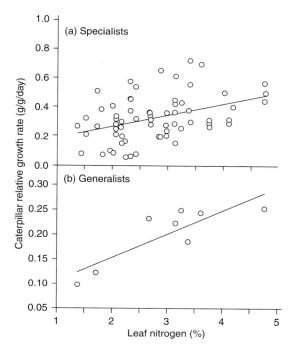

Figure 12.4 Effects of leaf nitrogen content on relative growth rates of lepidopteran caterpillars. (a) For specialist caterpillars, $r^2 = 0.18$, $P < 0.001$, $n = 69$ morphospecies, (b) For generalist caterpillars $r^2 = 0.68$, $P < 0.007$, $n = 8$ morphospecies. (From Coley et al. 2006.)

typographus (Coleoptera: Scolytidae) is a very serious pest of spruce, killing trees by the ring-barking action of larvae. However, this species is a so-called secondary pest, in that it is only able to attack, colonize, and breed in trees that have already suffered a primary stressor that has reduced their physical and secondary defenses (see below). Except in the instance of a mass outbreak of the beetle, when even healthy trees succumb to huge numbers of attacking pests, healthy trees remain immune. Figure 12.5 illustrates the results of two climatic effects, wind and lack of rain (drought) on tree mortality caused by *Ips*. In a managerial sense, it might seem difficult to control high winds or droughts. However, planting spruce on ridge tops or in heavily thinned stands will exacerbate wind problems. Likewise, spruce trees that are planted on dry, sandy soils are likely to suffer more from the effects of drought. In a study of *Dendroctonus micans* (Coleoptera: Scolytidae), a similar bark beetle to *Ips* in many ways (except for the fact that it seems

able to attack relatively healthy trees), Rolland and Lemperiere (2004) showed that spruce trees infested by the beetle in the Ardeche region of France had been weakened by a combination of cold winters and dry summers. In this example in fact, a combination of climatic stress and bark beetle outbreaks was thought to have led to forest die-back in a 15–20-year period. The simple message is that, while we might not be able to control the weather, a judicial choice of planting site can ameliorate deleterious effects of climate.

It is therefore vital, whenever possible, to grow plants under conditions that promote their health and reduce the likelihood of their becoming stressed if pests such as aphids, scales, and boring beetles are to be avoided. This is illustrated by outbreaks of the horse chestnut scale, *Pulvinaria regalis* (Hemiptera: Coccidae), on urban trees in the UK (Speight et al. 1998). The horse chestnut scale feeds as a nymph on leaves of lime, sycamore, and horse chestnut trees in urban areas in parts of Western Europe where adult

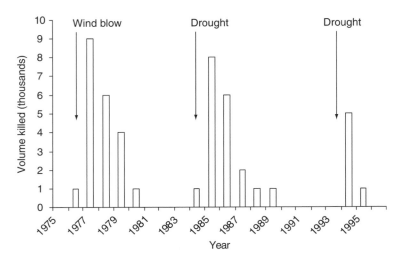

Figure 12.5 Spruce trees killed by *Ips typographus* in northern Europe.

females lay large, obvious egg batches on the main stems. In studies in Oxford, the density of the insect was found to rise dramatically on trees that were growing alongside streets or in car parks, where the permeability of the soil was seriously impaired (Figure 12.6). Furthermore, trees with morphological indications of low vigor such as die-back, rot, or wounds almost always showed higher levels of infestation no matter where they were growing.

Even manipulations of host plants during routine management practices can promote pest attack. Fertilizer applications are, of course, a mainstream input into agricultural and forest operations the world over, and Boege and Marquis (2006) found that fertilizer regimes on the tropical tree *Casearia nitida* promoted overall plant nitrogen, but depleted the levels of defensive phenolics (Figure 12.7). Environmental stress can promote pest problems too. In Kenya, pruning of *Cassia samia*, a leguminous fodder tree, to provide extra grazing for livestock, invokes significant increases in attacks by stem-boring moth larvae (Lepidoptera: Hepialidae), with a resultant decline and eventual death of the plant. Careless brashing of pole-stage *Acacia mangium* in parts of the Far East also increases the chance of attack by insect pests and fungi. Arable crops, too, can suffer varying rates of pest attack depending upon the growing environment. Mexican bean beetle, *Epilachna varivestis* (Coleoptera: Coccinellidae), is an unusual ladybeetle

in that its larvae are herbivores rather than predators, and can be serious pests of crops such as soybean in North America. Beanland et al. (2003) found that the development performance of *Epilachna* larvae was highest in plants grown in hydroponic solutions lacking trace minerals such as boron. High soil moisture also adversely affected soybean plants selected for resistance to the beetle. The expression of resistance decreased under conditions of high or low mineral proportions, overriding the genetically based resistance mechanisms.

Genetic resistance

The ability of insects to utilize a host plant can vary among genotypes within a species. For example, genotypes of cassava vary markedly in their susceptibility to the whitefly *Bemisia tabaci* (Hemiptera: Aleyrodidade). In this case, resistance to whitefly is expressed in oviposition preference, mortality, and development time (Bellotti & Arias 2001) (Figure 12.8), indicating a rather sophisticated, and above all, heritable, form of plant resistance. Van Lenteren et al. (1995) view host-plant resistance as a cornerstone of modern IPM. Some background to this topic is provided in Chapter 3, and readers should remind themselves of the difference between resistance and tolerance before proceeding. For pest management purposes, the trick is to isolate genetic mechanisms of

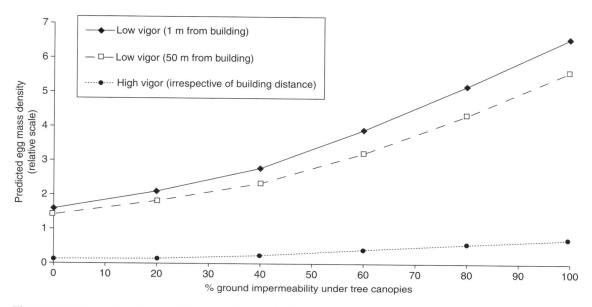

Figure 12.6 Regression lines describing the influence of soil impermeability and building distance of trees on the density of adult insects laying eggs. (From Speight et al. 1998.)

resistance or tolerance, and to incorporate these mechanisms into crops by breeding or genetic engineering.

These heritable defense mechanisms are normally considered under three separate headings: non-preference, antibiosis, and tolerance. The first two fall within the general category of resistance and include physical and/or chemical mechanisms for resistance. Tolerance, in contrast, describes the ability of the plant to compensate for attack without significant reduction in fitness or, in management terms, financial yield.

One problem in manipulating plant defense for pest management is the ability to distinguish between environmentally induced changes in susceptibility to insects (see above) and those that are genetically based. For example, antibiosis in the form of reduced weights of larvae and increased larval development periods were observed for the common cutworm, *Spodoptera litura* (Lepidoptera: Noctuidae), feeding on soybeans in Japan (Komatsu et al. 2004). However, these observations appear to have been mediated by environmental conditions such as local soil moisture and nutrient levels.

The various types of genetic defense mechanisms will be examined in turn, bearing in mind that they probably operate most of the time as complementary systems within the host plant–herbivore–enemy complex.

Non-preference
Non-preference, or antixenosis, occurs when insects avoid or reduce their feeding rate on one species or genotype of plant when compared with another on which they feed more readily. Most herbivorous insects in temperate regions tend towards host specialization, so that in a simple example, elm bark beetles (Coleoptera: Scolytidae) will not feed or breed in ash, nor will ash bark beetles utilize elm. There are also marked differences in insect preference among host plants within a plant genus. For example, the pine bark beetle, *Ips grandicollis* (Coleoptera: Scolytidae), imported from Europe into Australia some years ago, prefers to oviposit in logs of radiata pine (*Pinus radiata*) than in those of pinaster pine (*P. pinaster*) (Abbott 1993). Herbivores also discriminate among closely related crops, so that the African maize corn-borer, *Busseola fusca* (Lepidoptera: Noctuidae), for example, lays more eggs on maize than on sorghum (Rebe et al. 2004), an important consideration for mixed or trap cropping as pest management tactics (see later).

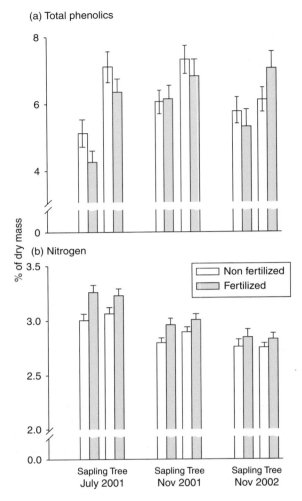

Figure 12.7 Plant quality of *Casearia nitida* saplings and reproductive trees under two fertilizer treatments in terms of (a) total phenolics, and (b) nitrogen. Means ± SE are shown. (From Boege & Marquis 2006.)

The most common way to investigate non-preference is to allow insects to choose between different crop genotypes under controlled conditions, so that if one genotype is repeatedly ignored or rejected by the pest, it provides evidence of antixenosis. The relative impact of insect feeding on the host plant can also provide an indication of preferences. The green spruce aphid, *Elatobium abietinum* (Hemiptera/Homoptera: Aphididae), responds to different species of spruce in

selection trials. Those species that originate in North America are much preferred to those of European provenance when equal choice is offered. Even within local areas of the USA where the pest has been introduced accidentally, such as Arizona, different species of spruce suffer significantly greater damage from the aphid (Lynch 2004). It is important to note that the most preferred species, *P. sitchensis* (Sitka spruce), grown in the UK is from North America, and it is the most widely planted of all spruce species in the UK. One obvious strategy for reducing aphid problems on spruce in Europe would be to plant European spruces, but forestry practices and market forces often prevail over sound pest management.

At a finer genetic scale, adults of the sorghum midge, *Contarinia sorghicola* (Diptera: Cecidomyiidae), prefer to lay their eggs on certain grain sorghum genotypes when compared to others (Sharma et al. 2002). Sometimes even slight differences in antixenosis can be magnified by antibiosis (below). For example, *Myzus persicae* and *Aphis craccivora* (Hemiptera/Homoptera: Aphididae), two widespread and internationally infamous species of aphid, exhibit only slight preference between susceptible and resistant varieties of lupin (Figure 12.9). However, experiments in Australia reveal a much more significant reduction in both aphid growth rate and survival after the choices have been made (Edwards et al. 2003). Always supposing that the crop husbandry system can utilize these resistant genotypes economically, choice of genotypes may provide a simple but effective method for managing the pest.

A common mechanism underlying non-preference is the hirsuteness (hairiness) of plant leaves. Hairs or trichomes on leaves make it more difficult for phloem-feeding insects to reach the leaf surface to penetrate its cuticle, or to move around. Clearly, the relative size of the hairs and the insect is crucial. Cultivars of maize with numerous trichomes on both upper and lower leaf surfaces are much less preferred as oviposition sites by the stem-borer *Chilo partellus* (Lepidoptera: Pyralidae) (Kumar 1997). Similarly, hirsute varieties of cotton support many fewer nymphs of the leaf-hopper *Empoasca* sp., whilst hairless varieties of alfalfa are much less able to deter the same insect pest as those that possess glandular hairs on their leaves (Ranger & Hower 2002) (Figure 12.10). A theme that we shall return to later describes the effects of genetically modified (transgenic) plants on insect pests. Bernal and Setamou (2003) discovered that

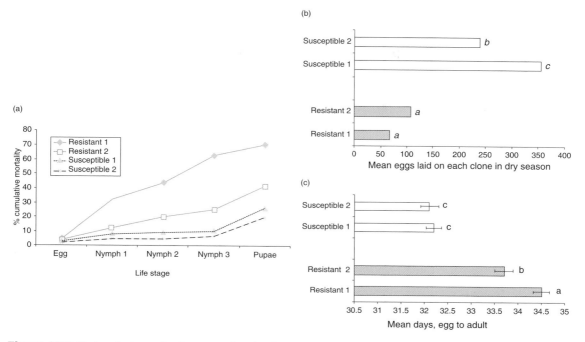

Figure 12.8 Various features of resistance to the whitefly *Bemisia tabaci* in different clones of cassava in Columbia. (From Bellotti & Arias 2001.)

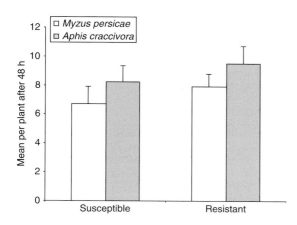

Figure 12.9 Performance of two species of aphid on lupins in Australia showing the number of alates choosing lupin varieties. (From Edwards et al. 2003.)

sugarcane which had been genetically modified to resist pests such as the stem-boring *Diatraea saccharalis* (Lepidoptera: Pyralidae) actually deterred egg laying by adult females, an unintended but additionally useful mechanism for crop resistance.

Antibiosis

Antibiosis acts on the performance of insects rather than their preference among host plants. Antibiosis mechanisms in plants can take several forms, such as variation in nutritional quality or chemical defenses (see Chapter 3). The effects on insects themselves are many and varied, and Table 12.3 provides just a few examples. Usually, antibiosis acts on the performance of the insect rather than on its preference for certain host plants. So, in trials on the resistance of sugarcane to whitegrubs (Coleoptera: Scarabaeidae), it was clear the antibiosis effects manifested themselves as deleterious effects on both grub survival and growth (Allsopp & Cox 2002). Resistant plants can also induce behavioral changes in insects. Figure 12.11 illustrates that the aphid *Aphis craccivora* (Hemiptera: Aphididae)

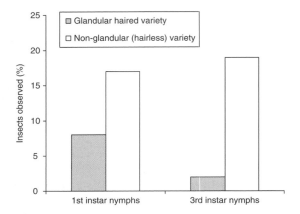

Figure 12.10 Percentage of *Empoasca fabae* nymphs jumping or falling off the surfaces of alfalfa plants of haired versus non-haired varieties (P = 0.013). (From Ranger & Hower 2002.)

reduces the time spent feeding on resistant varieties of cowpea, but increases the time spent probing (Mesfin et al. 1992). This system is of dubious advantage from the point of view of the transmission of plant pathogenic viruses by the pest, as increased probing may also increase the efficiency of the transmission of stylet-borne viruses (see Chapter 11).

Physical defense mechanisms can take the form of tough leaves, hairy leaves, or more complex mechanical systems within the host plant, but most resistance based on antibiosis is a function of host-plant chemistry,

either nutrients or defenses, or both. We have already seen how host-plant chemistry can fundamentally influence the ecology of herbivorous insects (see Chapter 3). Variation in primary metabolites (such as nutrients) may be particularly effective as defenses against highly oligophagous herbivores. Studies on black bean aphid, *Aphis fabae*, and peach–potato aphid, *Myzus persicae* (Hemiptera: Aphididae), in Poland found that resistant varieties of sugar beet had reduced concentrations of proteins when compared with susceptible ones. Moreover, the number of aphids per cultivar was directly proportional to the total content of free amino acids in the host plant (Bennewicz 1995). However, plant nitrogen levels do not always have a predictable effect on pest numbers (Fluegel & Johnson 2001). Somta et al. (2006) created a series of artificial seeds for the azuki bean weevil, *Callosobruchus chinensis* (Coleoptera: Bruchidae), to lay eggs in, and found that antibiosis effects varied considerably according to the protein and starch contents of the material.

Plant secondary compounds (allelochemicals) such as toxins vary considerably in their concentrations among plant genotypes. Given that plant chemical defenses may have evolved in part to deter insect herbivores, it is not surprising that we find them as the basis of many resistance mechanisms. Domestic potatoes (*Solanum tuberosum*) are derived from wild relatives and contain powerful glycoalkaloid antiherbivore compounds such as solanine and tomatine. Some genetic lines of potato are known to have antibiosis-like effects on Colorado potato beetle,

Table 12.3 Examples of the effects of antibiosis plant resistance to insects.

Plant	Insect	Effect of resistance
Alfalfa	*Spissistilus festinus* (Homoptera: Membracidae)	Reduction in male hopper weights
Chrysanthemum	*Spodoptera exigua* (Lepidoptera: Noctuidae)	Longer development time, reduced pupal weight
Cowpea	*Clavigralla tomentosicollis* (Hemiptera: Coreidae)	Increased nymph development time, high nymph mortality
Rice	*Nilaparvata lugens* (Homoptera: Delphacidae)	Decreased feeding rate, high nymphal duration and mortality
Sorghum	*Contarinia sorghicola* (Diptera: Cecidomyiidae)	Adult emergence significantly lower
Soybean	*Spilarctia casignata* (Lepidoptera: Arctiidae)	High larval and pupal mortalities, reduced pupal weight

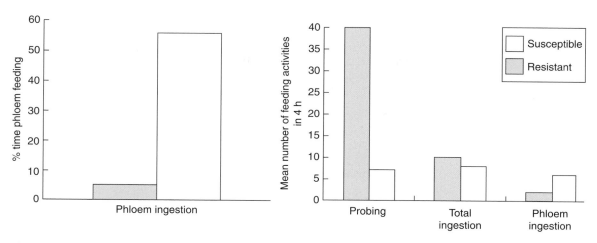

Figure 12.11 Variations in feeding activity of the aphid *Aphis craccivora* on resistant and susceptible cowpea cultivars in Nigeria. (From Mesfin et al. 1992.)

Leptinotarsa decemlineata (Coleoptera: Chrysomelidae), as measured by the weight gains of fourth instar larvae fed on the different lines (Horton et al. 1997). These effects result in very low defoliation levels in the field. Most of our common crops contain some kind of chemical associated with antibiosis. Cotton varieties, for example, vary in their production of the terpenoid aldehyde gossypol. Gossypol has important antibiosis activity against one of the most serious pests of cotton in Australia, the cotton bollworm *Helicoverpa armigera* (Lepidoptera: Noctuidae) (Du et al. 2004).

Often, chemical and physical defenses combine to produce resistance. The trichomes on the leaves and stems of tomatoes not only make it more difficult for small insects such as whitefly (Hemiptera/Homptera: Aleyrodidae) or even defoliating Coleoptera to reach the leaf surface, but they also produce poisonous exudates that can kill pests (Erb et al. 1994). Finally, we should note that some insect herbivores feeding on toxic plants have adapted to the chemicals from their hosts, and may even sequester toxins from plants to protect themselves against their natural enemies. As we might expect, sequestration of plant defensive chemicals by pests can reduce the efficacy of biological control (see Chapter 3).

It is sometimes quite difficult to pinpoint the causes of antibiosis, as physical and chemical features vary so much among plant genotypes, and some investigations may yield less than convincing results. For example, in experiments to evaluate rice cultivars

for resistance to the gall midge *Orseolia oryzae* (Diptera: Cecidomyiidae), eight resistant varieties had considerably higher concentrations of polyphenolic compounds in the basal stem where the insect larvae attack (Sain & Kalode 1994). Unfortunately, in the same study, seven other resistant cultivars showed significantly lower levels of the same compounds when compared with the susceptible varieties. The problem lies, in part, with two related issues. First, traits that appear to confer resistance against insects often covary. For example, the quality of oak foliage for insect herbivores declines seasonally, and is reflected in higher tannin levels, increased toughness, lower nitrogen content, and lower water content (see Chapter 3). If we observe antibiosis against an oak pest, it is extremely difficult to distinguish among the effects of the covarying resistance traits. Is it water or tannin concentration that is most important? As you might imagine, it is harder to select, engineer, or manage plants when the precise mechanism for resistance is unknown. In a related problem, resistance traits may actually represent suites of mechanisms that operate simultaneously. It may be that covarying traits are all necessary for the expression of resistance. In other words, it might be the combination of low water levels and high tannin concentrations that confer resistance in oak foliage. While genetic techniques such as quantitative trait loci (QTL) mapping and gene silencing may provide important insights into the genetic basis of resistance, it may not always be possible to select or engineer plants with the appropriate combination of resistance traits. That being said, recent advances in "gene pyramiding" and marker-assisted selection (see below) are providing plant breeders with remarkable tools for manipulating and managing genetic resistance traits.

Tolerance

Tolerance is described in some detail in Chapter 3, and we will not revisit it here in any depth. Suffice it to say that some plants are able to grow well and provide high yields despite being attacked by insect pests. If there is genetic variation for the ability to compensate for insect damage, that genetic variation can be exploited in plant breeding and engineering programs. In Chapter 3, we described tolerance in terms of plant fitness. In commercial production systems, tolerance and compensation are better described in terms of financial yield per unit area. Simply put, a tolerant crop genotype is one in which financial yields decline only minimally with herbivore damage. Apparent examples of genetic variation for tolerance do exist in cropping systems. The pod-borer *Helicoverpa armigera* (Lepidoptera: Noctuidae) attacks chickpeas, and highest yields are obtained with minimal damage (7%) to pods by the pest. None the less, one chickpea genotype can withstand three times this level of damage and still provide over 60% of maximum yield, indicating a degree of tolerance to the borer (Chauhan & Dahiya 1994).

12.3.2   Commercial use of genetic resistance

There is no doubt that plant selection, breeding, and engineering for resistance will constitute an increasing part of integrated insect pest management. Of particular importance in tropical countries is the development of crop cultivars that are resistant to several pest insects, pathogens, and even nematodes, all in the same genotype, as a whole battery of pests and diseases may attack a crop simultaneously (Nagadhara et al. 2003; Kranz 2005). Rice provides one of the best examples of multiple resistance of this type, and millions of hectares of multiple resistance varieties are now planted all over the world (see later in this chapter). Using multiple resistant varieties, growers can now expect to increase their yield up to 2 tons per hectare when compared with the older types of single resistance strains.

However, it is one thing to detect resistance, but it is quite another to decide what to do with it. The flowchart in Figure 12.12 illustrates a full trial system developed for the whitefly *Bemesia tabaci* mentioned earlier in this chapter. Notice the crucial importance in this type of scheme of considering the growers needs—new pest-resistant varieties may be entirely unsuitable in a culinary or commercial sense (Bellotti & Arias 2001). It is not unusual for different varieties of a crop to exhibit varying levels of resistance to a pest (Wearing et al. 2003) (Figure 12.13). In this example with apples, the cultivars vary in other characteristics such as yield, color, and taste. The apple grower's decision about which cultivars to use may be driven by market forces based on other apple traits so that the forms and intensity of pest management may have to be altered to fit.

It is rare that plant resistance on its own is sufficiently effective to reduce and maintain pest populations below economic thresholds all the time, but other control systems may be enhanced by the use of resistant crop cultivars. The actions of natural enemies may be a case in point. Gowling and van Emden (1994) established cage experiments wherein the cereal aphid *Metopolophium dirhodum* (Hemiptera: Aphididae) was grown on two wheat cultivars, one susceptible to aphids and the other partially resistant. The parasitic wasp *Aphidius rhopalosiphi* (Hymenoptera: Aphidiidae) was then introduced to both systems and the reductions in peak aphid numbers caused by parasitism were recorded. In the presence of this parasitoid, aphid populations were reduced by 30% on the susceptible host plant, but by 57% on the partially resistant one. Furthermore, the numbers of aphids that left the plants in the presence of the parasitoid nearly doubled on the resistant host. In other examples, this time from cucumbers (Van Lenteren et al. 1995) and soybeans (McAuslane et al. 1995), less hairy plants supported fewer whitefly (Homoptera: Aleyrodidae) than did hairy ones, because of higher parasitism by the parasitic wasp, *Encarsia* spp. (Hymenoptera: Aphelinidae) on the non-hairy varieties. The enemy was better able to forage for hosts when hairs or trichomes did not inhibit its progress and searching. Finally, combinations of host-plant resistance and predation by the mirid bug, *Cyrtorhinus lividipennis* (Hemiptera: Miridae), have a cumulative effect on population densities of the green leaf-hopper, *Nephotettix virescens* (Hemiptera: Cicadellidae), a serious pest of rice (see

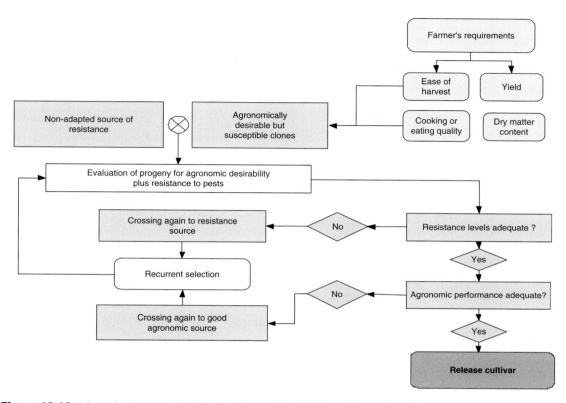

Figure 12.12 Scheme for incorperating whitefly resistance from "wild type" (non-adapted) cassava sources into commercial hybrids. (From Bellotti & Arias 2001.)

later). Again in cage studies, on a strain of rice (IR69) resistant to leaf-hoppers, the number of leaf-hoppers reached only six in the presence of the predator, and 31 in its absence (Figure 12.14). On a leaf-hopper-susceptible strain (IR22), however, there were 91 and 220 leafhoppers, respectively (Heinrichs 1996).

A whole range of plant characteristics influence the efficacy of insect predators and parasitoids, either adversely or synergistically. These include plant size, shape, hairs, waxes and colors, chemicals (both attractants and inhibitors), plant density, and vegetation diversity All of these factors impinge on the likely success of biological control programs (see Section 10.5).

A final problem remains. Insect populations generally express a wide range of genetic variability, and can develop resistance of their own to plant defenses (see Chapters 3 and 6). Nowhere is this more serious than in production systems where the widespread use

of just one variety of crop in monocultures favors the selection of resistance within insect populations. It should be no surprise that some pests have overcome the resistance of certain crop varieties and these new types of pest are known as biotypes. Various biotypes of brown plant-hopper *Nilaparvata lugens* (Hemiptera: Delphacidae) have become serious problems in rice cultivation in South and Southeast Asia. Rice variety IR26 was the first brown plant-hopper-resistant cultivar to be released by the International Rice Research Institute (IRRI) in 1974. Within the Philippines, brown plant-hopper outbreaks were observed in IR26 after only 2 or 3 years (roughly six crops) of commercial cultivation, because of the selection of a plant-hopper strain that could utilize the previously resistant rice. Within only a few years, the resistance becomes ineffective (Tanaka & Matsumura 2000). Other resistant crop varieties have suffered the same fate. For example, sorghum greenbug, *Schizaphis*

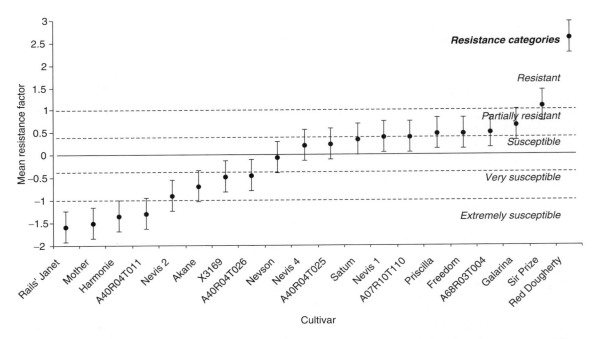

Figure 12.13 Ranked mean resistance (R_c) of apple cultivars to the lightbrown apple moth, *Epiphyas postvittana*. (From Wearing et al. 2003.)

graminum (Hemiptera: Aphididae), first became a pest on grain sorghum in parts of the USA in 1968 when a new population that was capable of surviving well on

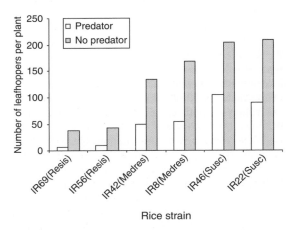

Figure 12.14 Numbers of the green leaf-hopper, *Nephotettix virescens*, on different varieties of rice plants with or without the mirid predator, *Cyrtorhinus lividipennis*. (From Heinrichs 1996.)

sorghum became abundant (Porter et al. 2000). This new population was designated "biotype C". In 1980, grain sorghum hybrids resistant to biotype C suffered damage from greenbugs in some areas of the central plains states. This damage was caused by a new green-bug biotype, later designated as "biotype E". These biotypes could not be identified by their appearance, since they looked the same. The different biotypes could only be identified by the feeding damage that they caused to different varieties of sorghum or other species of plants. A more recent biotype, "I", reported from Kansas could reproduce on biotype E-resistant sorghums; limited numbers of biotype I greenbugs were found in Nebraska (Table 12.4). Similar problems have arisen across the world where varieties of rice which at first showed resistance to the notorious brown plant-hopper, *Nilaparvata lugens* (Hemiptera: Delphacidae), became susceptible – in other words, resistance can break down, especially when new biotypes of the pest migrate from elsewhere (Tanaka & Matsumara 2000).

To cope with this biotype problem, so-called gene deployment strategies have been proposed, which

Table 12.4 Reaction of wheat resistance gene combinations to greenbug biotypes. Note that biotype F suceeds on all types of sorghum tested, despite some of them containing separate genotypes (pyramiding). (From Porter et al. 2000.)

Wheat entry	Resistant gene(s)	Greenbug biotype				
		E	F	G	H	I
Karl92	Nolle	S	S	S	S	S
Amigo	Gb2	S	S	S	S	S
Largo	Gb3	R	S	S	R	R
GRS 120 1	Gb6	R	S	R	S	R
Amigo/Largo F1	Gb2/Gb3	R	S	S	R	R
Amigo/G RS1201 F1	Gb2/Gb6	R	S	R	S	R
Largo/G RS1201 F1	Gb3/Gb6	R	S	R	R	R

Note that "pyramiding" resistant genes in one variety of plant does not confer additional protection against greenbugs.

include: (i) sequential release, where a crop variety with a single major resistant gene replaces a variety whose resistance has been overcome by a new pest biotype; (ii) gene pyramiding, where two or more major resistant genes are incorporated into the same crop variety to provide resistance to two or more biotypes of insect; (iii) horizontal resistance, where a type of resistance is expressed equally against all pest biotypes; (iv) gene rotation, where varieties with different major resistant genes are used in different cropping seasons to minimize selection pressure; and (v) geographic deployment, where crop varieties with different resistance genes are planted in adjacent cropping areas. It is very unlikely that the battle will ever be won completely, and continued research and development of plant resistance is vital to produce commercially viable crops using IPM.

12.4 ECOLOGICAL PEST MANAGEMENT

Many commercial practices carried out during crop production influence the potential for pest problems. From planting to harvest, and beyond, the ecology of the crop and its surroundings can be fundamental to the promotion of pest numbers by enhancing both biotic and abiotic factors under which the pests can thrive. There are of course some basic characteristics of farm and forest sites that are rather difficult to change without large amounts of money, or, indeed, a

wholesale removal to somewhere else. In Figure 12.15, for example, Coll et al. (2000) show how soil type and position in the field influence the intensity of attacks by potato tuber moth, *Phthorimaea operculella* (Lepidoptera: Gelechiidae), in potato fields. Small fields (with high edge to center ratios), on sandy soils, are probably not the best places to grow a crop of potatoes. Whatever the local conditions, every effort must be made to ensure that the crop plants are healthy, and that the habitat is optimized for the proliferation of natural enemies if the cropping system allows. All stages in crop husbandry may provide such enhancements, and each of these stages will be considered in turn.

12.4.1 Cultivation

Modern farming systems have a variety of cultivation or tillage tactics available to them. Traditional deep plowing may give way to minimum tillage, where seeds are directly drilled into the stubble or other remains of the previous crop. This system reduces the time and energy invested in reseeding, such that machinery is moved over the soil less frequently, with consequent savings on fuel, compaction, soil erosion, manpower, and so on. Table 12.5 provides some examples where minimum tillage for the most part favors the success of insect pests. In economic terms, there has to be a cost–benefit assessment to compare the advantages of minimum tillage with the promotion of insect (and, indeed, weed) problems. Declines in

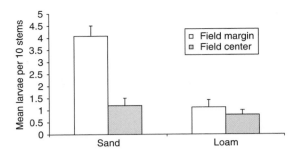

Figure 12.15 Mean seasonal number ($\pm$ SE) of potato tuber moth, *Phthorimaea operculella*, per 10 potato stems in crops in Israel. (Coll et al. 2000.)

Table 12.6 Mean ($\pm$ SE) of individuals of peach potato aphid, *Myzus persicæ*, recovered per plant after 2 weeks of growth on three species of winter weed in constant environment growth chambers. Initial infestation was 56 adult aphids per plant. (From Fernandez et al. 2002.)

Weed species	Aphid numbers
Brassica kaber (mustard)	207 $\pm$ 22
Malva parsiflora (mallow)	176 $\pm$ 8
Veronica hederiflora (speedwell)	132 $\pm$ 10

Table 12.5 Examples where minimum tillage increases the need for pest management.

Crop	Insect	Common name
Soybean	Hemiptera: Pentatomidae	Stinkbugs
Wheat	*Cephus cinctus* (Hymenoptera: Cephidae)	Wheat stem sawfly
Arable	Lepidoptera: Noctuidae	Cutworms
Corn	*Papaipema nebris* (Lepidoptera: Noctuidae)	Corn stalk-borer
Soybean	*Anticarsia gemmatalis* (Lepidoptera: Noctuidae)	Velvetbean caterpillar

pest populations that result from tillage may be desirable as a form of pest management (McLaughlin & Mineau 1995), though the reasons for this effect may be variable. Some pests in the soil, such as cutworms (Lepidoptera: Noctuidae) or wireworms (Coleoptera: Elateridae), may be exposed to desiccation or bird predation by whereas pests that feed on stubble after harvest may starve if the ground is tilled (Robertson 1993). It may also be possible that natural enemies such as predatory beetles vary in abundance according to soil moisture and vegetation cover (Yamazaki et al. 2003) but the links between such events and pest outbreaks have yet to be determined.

Additionally, cultivation of the fields should be linked with treatment of the field margins, or general non-crop vegetation adjacent to the fields. As we shall see later in this chapter, wild flowers and other native plants may support useful populations of natural enemies for biological control, but most benefits envisaged for a predator or parasitoid may also apply to pests. We know that weeds can make very good alternative host plants for crop pests. Table 12.6 shows the number of peach potato aphid, *Myzus persicae* (Hemiptera: Aphididae), after only 2 weeks of population growth on three common weed species (Fernandez et al. 2002). Weedy margins, or indeed weedy fields, can provide large ready-made populations of pests for a newly established crop.

12.4.2 Nursery management

For crops that are not sown directly into their final growing site, such as many horticultural and almost all forest species, careful management of the young plants in the nursery is crucial in giving them a vigorous start in life. In turn, healthy young plants are more likely to withstand pest attack following transplant. In tropical forestry, it is all too easy to damage little trees in the nursery by rough handling at the pricking out stage. In Sabah (northeast Borneo), a large number of nearly mature *Acacia mangium* were killed by secondary pests such as longhorn and roundhead borers (Coleoptera: Cerambycidae and Buprestidae), which were able to attack the sickly trees. These trees had reduced vigor caused by deformed root systems. The root damage was acquired in the nursery, 8 years previously, because of rough handling during potting, which produced severely coiled roots (Speight 1997a). Careful nursery management can take the place of conventional chemical control. Again in the tropics, nursery plants of hot pepper (*Capsicum* spp.) that are protected from the hot sun by screens have few problems with aphids, or the viruses that the aphids transmit, when compared with unshaded plants. In addition, these shaded plants establish much better in the field later on (Vos & Nurtika 1995).

12.4.3 Planting regime

Planting regimes for crops include many variables, including planting date, plant spacing, crop rotation, mixed or monoculture crops, intercropping, trap cropping, and so on, and most of these tactics can influence the potential for pest problems once the crop has established. Obviously, many of these tactics are inappropriate for intensive crop husbandry where high-yielding, monospecific or even monogenetic crops are grown year after year in the same place for sound and unavoidable economic reasons. However, in some situations, growers may be able to modify their tactics to incorporate some of these ecologically sound strategies.

Time of planting

Varying the time of planting may reduce the likelihood of pest problems. Where climates vary seasonally, many examples illustrate the fact that crops planted early in the season are less likely to suffer from pest outbreaks than those planted later. In crops such as cotton (Slosser et al. 1994) and rice (Thompson et al. 1994), insect pest problems are reduced or absent in early-planted crops. One reason for this may be that early-planted crops have a chance to become better established before insects appear. In this way, such plants may be less palatable to herbivores, or simply be able to tolerate higher pest densities before succumbing. In contrast, Charlet and Knodel (2003) found that populations of both adults and larvae of the sunflower beetle, *Zygogramma exclamationis* (Coleoptera: Chrysomelidae), decreased as crop planting date was delayed. Simple rules therefore rarely if ever hold true for all ecological situations, or they vary from year to year in efficacy (Sastawa et al. 2004). In South Korea, some pest problems on rice

are indeed reduced by early planting. However, other pests, such as striped rice-borer, *Chilo suppressalis* (Lepidoptera: Pyralidae), actually have higher population densities in early transplanted paddy fields (Ma & Lee 1996). The case study of barley yellow dwarf virus discussed in detail in Chapter 11 provides another example where early planting exacerbates the pest (and pathogen) problem.

Plant spacing

The effects of plant spacing on insect attack are complex, and need to be considered separately for each insect–crop interaction. Planting individual plants widely spaced provides a lot of bare soil which, as it turns out, is sometimes good and sometimes bad. Cabbage root fly, *Delia radicum* (Diptera: Anthomyiidae), adult females land about twice as often on *Brassica* plants growing in bare soil when compared with plants surrounded by other vegetation (Kostal & Finch 1994). However, as Figure 12.16 illustrates, *D. radicum* pupae survive better in plots where the cabbage plants are densely planted. In another example, Hamadain and Pitre (2002) showed that there were more eggs and larvae of the soybean looper, *Pseudoplusia includens* (Lepidoptera: Noctuidae), in narrow-row rather than wide-row planting regimes. Whether or not this initial greater infestation gives rise to significantly higher damage at harvest is less certain; many other factors such as predation or competition may come into play. It is rather hard therefore to make broad generalizations about plant spacing and its ability to reduce the likelihood of pest problems. The best strategy is probably to identify the most serious pest species in a crop or region of planting and follow specific planting guidelines.

Monocultures

The use, or some would say, misuse, of crop monocultures has been at the forefront of debate in pest management for a very long time. There can be problems with crop mixtures. They may produce decreased yields when compared with monocultures because of competition for resources between the two plant types. Because of different ripening times, or the need for different harvesting technologies, the use of mixtures may be commercially impractical. None the less, in terms of pest attack, extreme monocultures are generally more of a risk (Coyle et al. 2005).

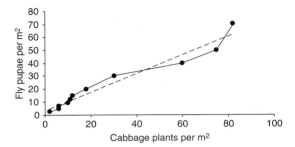

Figure 12.16 Relationship between plant density and number of cabbage rootfly pupae in cabbage crops in the UK. (From Kostal & Finch 1994.)

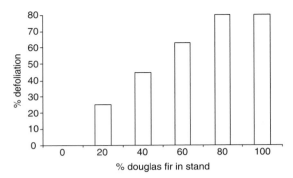

Figure 12.17 Defoliation of Douglas fir by the spruce budworm, *Choristoneura fumiferana*, according to the percentage of fir in a mixed stand of tree species. (From Fauss & Pierce 1969.)

Why do monocultures support higher densities of pests? There are several (not mutually exclusive) mechanisms that operate. First, monocultures make it easy for insect pests to locate suitable host plants within the cropping system. In some cases, a considerable proportion of insect mortality takes place during host location in early instars (Hunter et al. 1997). When every surrounding plant is of the same species, cultivar, or genotype, this source of mortality is greatly reduced. Second, monocultures present large targets to colonizing insects from outside the system. Even if colonization is through random dispersal, a target of many hectares will be "hit" more often than a smaller target. Third, large plant populations support large insect populations. This means that the probability of pest extinction is lower through what ecologists call "demographic stochasticity". Put simply, the probability that all individuals die or fail to breed in a given year declines as population size goes up. Finally, monocultures provide strong selection pressure for the development of resistance to plant defenses. If mutation or recombination generates an insect biotype capable of feeding on the crop, then a monoculture provides that biotype with a huge advantage in terms of food availability.

Early work demonstrated that the damage caused to Douglas fir by the spruce budworm, *Choristoneura fumiferana* (Lepidoptera: Tortricidae), rises dramatically as this preferred host constitutes more and more of the forest stand (Fauss & Pierce 1969) (Figure 12.17). In this case, it seems that the larvae, which disperse from high densities of their siblings to neighboring trees, are more likely to find an equally suitable host, free from competition from other larvae, in single-species

stands. Crop mixtures may interfere with the ability of insects to locate their hosts by chemical means. In line with the conventional wisdom of generations of gardeners, mixing carrots with onions reduces attack by carrot fly, *Psila rosae* (Diptera: Psilidae), compared with monoculture carrots. It is likely that the deterrent onion volatiles mask the subtler odor of carrots (Uvah & Coaker 1984). Additionally, pests may stay on plants within mixtures for less time than they do in monocultures.

A majority of trials investigating the importance of mixed cropping or intercropping (in other words, the avoidance of monocultures) illustrate the benefits of mixtures. Rukazambuga et al. (2002) looked at the pest status of banana weevil, *Cosmopolites sordidus* (Coleoptera: Curculionidae), in Uganda. The larvae of this beetle are found mainly in the rhizomes of their host, cooking bananas – a type of plantain – where they can cause up to 50% reduction in final fruit loss. When bananas were intercropped with millet, the numbers of weevil larvae were significantly reduced, and yield reductions were half those in monoculture bananas. In Kenya, Skovgard and Pats (1997) intercropped maize with cowpea (a cereal with a legume), and were able to reduce the number of maize stemboring Lepidoptera by 15–25%, with an increase in maize yield of 27–57%. In a similar system, Ogol et al. (1999) showed that an agro-forestry alley cropping of maize intercropped with the multipurpose tree *Leucaena leucocephala* significantly reduced maize-borer attacks. The width of the alley crop did not seem to be significant (Figure 12.18). Finally, intercropping

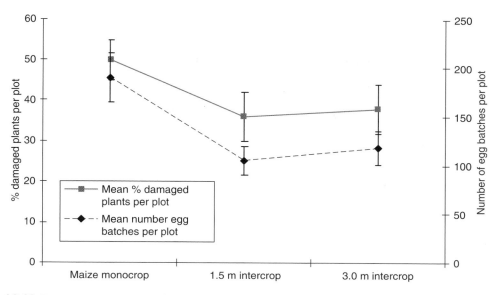

Figure 12.18 Damage to maize in Kenya by maize stem-borer, *Chilo partellus*, in monoculture maize as compared with alley crop mixtures with *Leucaena*. (From Ogol et al. 1999.)

cassava (manioch) with maize reduced egg and larval densities of the defoliating moth *Sesamia calamistis* (Lepidoptera: Noctuidae) by 67–83% on the latter crop (Schultness et al. 2004). Such tactics will, of course, succeed only if the grower is able to harvest both crops easily and independently – it is unlikely to work in large-scale arable systems.

Trap crops

So, two or more entirely different crop plants inter-cropped together may reduce the damage to one or both species. Conceptually, intercropping involves two (or more) crop species in close proximity to each other, within a plot or field (Hilje et al. 2001). In fact, the second crop species in a mixture may not in itself be of great commercial value; it may simply intercept dispersing pests which may otherwise descend upon a vulnerable crop (Tillman 2006). The undersowing of crops such as leeks with clover can produce drastic reductions in pests such as onion thrips, *Thrips tabaci* (Thysanoptera) (Theunisse & Schelling 1996), and these crops can be used to attract insect pests away from the more important target species. This may create a refuge for natural enemies, or, more likely, to manipulate the host-finding behavior of the insect

pest to protect the more susceptible, mainstream crop. The preferred plant species that keeps the pests away from the main crop is known as the trap crop. Trap crop plants may not only act as "sinks" for the insect pests themselves, but also in certain circumstances provide "soaks" for pathogens such as plant viruses which the insects vector (Shelton & Badenes-Perez 2006).

These tactics can be divided into several categories. Conventional trap cropping is where a trap crop planted next to a higher value crop is naturally more attractive to pests. As Figure 12.19 shows, cantaloupe (related to melons and squashes) has been used as a trap crop to reduce the numbers of whitefly, *Bemisia tabaci* (Hemiptera: Aleyrodidae), in cotton fields in central Arizona (Castle 2006). Somewhat unlikely species have been used as trap crops; tobacco has been shown to be an efficient trap to reduce the levels of tobacco budworm (hint in the name), *Heliothis virescens* (Lepidoptera: Noctuidae) in cotton (Tillman 2006), whilst chickpeas were highly attractive to ovipositing *Helicoverpa* spp (Lepidoptera: Noctuidae), again in cotton fields (Grundy et al. 2004).

Genetically engineered trap cropping is discussed later in this chapter, and it forms a special case of the more general dead-end trap cropping system (Shelton

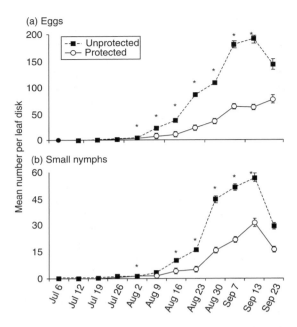

Figure 12.19 Differences in mean ($\pm$ SE) numbers of (a) eggs and (b) small nymphs on cotton protected by a cantaloupe trap crop or unprotected in the 1999 trial according to sampling date. Significant differences ($P < 0.05$) in comparisons made for each date are indicated by *. (From Castle 2006.)

& Badenes-Perez 2006). Dead-end trap crops are those which are highly attractive to insect pests, but on which they or their offspring cannot survive. Shelton and Nault (2004) describe how adult diamondback moths, *Plutela xylostella* (Lepidoptera: Plutellidae), were between 24 and 66 times more attracted to the non-crop plant *Barbarea vulgaris* than to cabbage. Unfortunately for the moths, their hatching larvae do not survive to adulthood on the trap crop.

Wild plants

Barbarea vulgaris (also known as golden rocket), mentioned above, is in fact a common biennial weed (Lu et al. 2004). Hence another type of crop mixture involves just one commercial plant species, but this time associated with non-crop, wild vegetation. In many instances, natural enemies such as parasitic wasps rely upon sources of pollen and nectar to provide energy for the adults. The establishment or

maintenance of wild flowers (so-called "companion" plants) in, and especially around, crops may enhance biological pest control (Bowie et al. 1995). It is unlikely that most growers will readily tolerate weeds and flowers within their intensively grown crops, but field boundaries can be suitable. So, sowing certain, but not all, species of wild flower in cabbage fields and adjacent habitats increases the fecundity and survival of parasitoids of the diamondback moth, *Plutella xylostella* (Lepidoptera: Plutellidae), in North America (Idris & Grafius 1995). A similar strategy increases the number of adult hover flies (Diptera: Syrphidae) in fields in New Zealand (White et al. 1995), the larvae of which prey on aphids. We should remember, however, that it is one thing to augment the number of natural enemies around a crop, but quite another to actually achieve significant reductions in crop losses caused by insect attack.

Shade

Growers have the choice of planting their crops in full sunshine or, alternatively, under the canopy of other plant species that provide shade. This system is one of the basic principles of agro-forestry. However, the likelihood of crop plants being more or less attacked by insect pests in the shade of other species is debatable. Shade may reduce the effects of direct sunlight, such as temperature and insolation, but it may also increase humidity and darkness, and hence increase the possibility of pathogenic infections (MacLean et al. 2003). Perhaps one of the most important commercial crops for third world countries, in terms of foreign exports at least, is coffee, and one of the most important pests of coffee is the coffee berry-borer (CBB), *Hypothenemus hampei* (Coleoptera: Curculionidae). Damage is caused by the female, which bores into green coffee berries to lay eggs. The hatching larvae feed on the growing coffee beans, causing two types of damage (CABI 2006). Premature fall of coffee berries and hence total loss of these to production may occur, or the damaged berries may be retained on the tree until harvest, making them of lower commercial value by reducing the weight of the bean and downgrading the quality and affecting the flavor of the coffee. Losses due to CBB were estimated in 1994–5 to amount to: 2.18 million bags in Central America, then worth approximately US$328 million, and the infestation of 650,000 ha of coffee plantations in Colombia, reducing the estimated national crop production from

13 million to 11.5 million bags, at an estimated cost in excess of US$100 million (CABI 2006). The virtues of growing coffee under shade have been much extolled, and many third world countries now plant coffee under tree canopies. Whether or not this system does anything for biodiversity is questionable (and irrelevant), but the effects of this shade on insect pests suggest that the relationships are complex (Soto-Pinto et al. 2002). More research is needed on whether CBB can be controlled using planting techniques.

12.4.4 Growing crops

Once an annual crop has established, there is little opportunity for ecological manipulation, and further pest management takes the form of biological or chemical control, or a combination of both. However, in a perennial system, such as plantation forestry, further manipulations can be carried out, such as stand thinning. This is a common enough silvicultural practice, where a number of trees are removed from a growing stand so that those that remain are subject to significantly less competition for light and other resources, and hence grow more vigorously. Many, though not all, judicious thinning programs reduce the problems from forest pests such as bark beetles or ambrosia beetles (Coleoptera: Scolytidae and Platypodidae). There is no doubt that thinning can dramatically enhance the performance of trees. In ponderosa pine stands in California, the growth of trees in a thinned stand 5 years after thinning is five times greater than in control plots where no thinning has taken place

(Fiddler et al. 1989). This increase in vigor should reduce the susceptibility of trees to insect attack, as long as sanitation removes pest-infested timber to prevent population buildups (Hayes & Daterman 2001). Sure enough, pine beetle, *Dendroctonus ponderosae* (Coleoptera: Scolytidae), attacks are significantly lower in thinned stands of ponderosa pine (Brown et al. 1987) (Figure 12.20).

12.4.5 Senescing crops

The quality of plants for insect herbivores varies with the age of the plant and its modules (stems, leaves, and so on). For example, organic nitrogen is mobilized in senescing foliage, providing resources that some insects use to enhance their performance. This is rarely a problem that reaches economic proportions. In contrast, as trees become older, their growth rates decline, until they reach a stage called overmaturity. Once more, low vigor now renders them susceptible to secondary insect attack, and silvicultural practices need to ensure that commercial crops are harvested before this stage ensues. A good example comes from Sabah in northeast Borneo, where longhorn beetles (Coleoptera: Cerambycidae) and ambrosia beetles (Coleoptera: Platypodidae) attack *Acacia mangium* trees (M.R. Speight, unpublished data). The age at which such trees should be harvested to avoid overmaturity and susceptibility to attack is as little as 8 years. However, in some cases trees are left until at least 14 years of age, and should have been felled and processed years earlier.

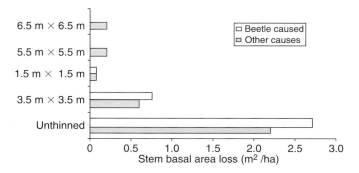

Figure 12.20 Losses to ponderosa pine stands caused by attacks of *Dendroctonus ponderosae* according to regime. (From Brown *et al.* 1987).

12.4.6   Post-harvest

Insect pest problems do not necessarily cease once the crop has been harvested. For example, felled trees must be removed from commercial forest stands within a few weeks of spring cutting in the UK to prevent bark beetle populations escalating in the plantations. Treatments such as bark peeling or constant water spraying may have to be employed to protect logs from beetle attack before they are processed. This last strategy may continue for several years if large quantities of timber are suddenly made available by large-scale sanitation felling after disasters such as forest fire (F.R. Wylie, unpublished data). In agriculture, storing products such as grain can be very problematic. A variety of pests are notorious destroyers of stored products all over the world. In the UK, the grain beetle, *Oryzaephilus surinamensis* (Coleoptera: Cucujidae), can proliferate very quickly in grain silos where the wheat or barley has been inadequately dried before storage (Figure 12.21) (Beckett & Evans 1994). At moisture contents of over 30% or so, the beetles aggregate and breed in "hotspots" in the silos, where their larvae tunnel into the grain, destroying it. Usually, some form of insecticidal control has to be administered, although controlled atmospheres such as elevated carbon dioxide levels may be established in the storage containers (Arbocast 2003).

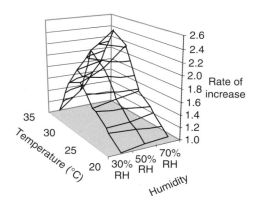

Figure 12.21 Rate of increase of the weevil. *Oryzaephilus surinamenis*, in stored wheat at different temperatures and humidities. (From Beckett & Evans 1994.)

12.5   BIOLOGICAL CONTROL

To distinguish it from the multifarious techniques of crop ecosystem management, we define biological control as the use of natural enemies of insect pests to reduce the latter's population densities. In most cases, the goal is to maintain low densities (i.e. regulate the pest population) through time without further manipulation (see Chapter 5). Thus the success (or otherwise) of top-down control of a pest by a natural enemy is the basis of the system (Mills 2001). In fact, perhaps pedantically, the term biological "control" is a misnomer, as what we are in fact looking for is biological "regulation", a fundamentally density-dependent process. The first insect parasites were mentioned in Chinese literature around 300 AD (Cai et al. 2005) and, over the last century or so, biological control has had some rousing successes and suffered many failures. It is vital to appreciate the limitations of the various types of biological control, to place them in the perspective of other pest management strategies, as part of IPM (Speight & Wylie 2001).

12.5.1   Limitations of biological control

Many types of natural enemies of insect pests exist, each with its own advantages and disadvantages as a potential agent of biological control; Table 12.7 summarizes the most important ones. We should point out that the third column in the table, which indicates the likely success of the various enemy types, is a very general view, and that for nearly all of the categories there are examples where biological control has been successful (see below). However, one fundamental facet of biological control is the ability of a particular natural enemy, or group of enemies, to respond to differing densities of the pest's population. Specifically, the rate (or per cent) mortality caused by a natural enemy has to increase as pest density increases if the biological control agent is to regulate the pest population.

A parasitoid is a special type of parasitic insect in which the larva of a wasp or fly consumes the body of its host so that the host dies within one generation of the parasitoid. We have discussed the population dynamics of predator–prey and parasitoid–host interactions in detail in Chapter 5. Reference to this will help to explain why parasitoids and particularly predators often fail to regulate pest populations when

Table 12.7 Major types of natural enemies of insect pests which may have a role in biological control.

Type of enemy	Commonly attacked pests	Likely efficiency in biological control
Vertebrates		
Mammals (mice, voles)	Soil larvae and pupae of defoliators	Low
Birds	Defoliating larvae of moths and sawflies	Low
Amphibia (frogs)	Soil larvae	Low
Fish	Aquatic insect larvae	Low
Invertebrates		
Spiders	Small flying or crawling pests, e.g. aphids	Low to medium
Predatory insects (beetles, bugs, lacewings, hoverflies)	Aphids, lepidopteran, dipteran, and sawfly, soil larvae and pupae	Low to medium
Parasitic insects (wasps and flies)	All pest types, all life stages	Medium to high
Pathogens		
Bacteria		
Fungi		
Nematodes		
Protozoa	Most pest types, except those permanently	Medium to very high
Viruses	concealed within the host plant	

pests have already reached high, epidemic levels. When pest numbers exceed some regulation threshold of their enemies, food limitation usually brings about pest population decline; such pests are known as resource limited. Only pathogens such as viruses seem to have no upper limit to their ability to reduce very high insect pest numbers (see Chapter 1). A further problem is the fact that many predators, parasitoids, and even pathogens have their own natural enemies, so-called higher order predators or hyperparasitoids. Natural enemies of herbivorous insects are often not top of their food chain, and it is likely that in some cases their ability to regulate pest densities effectively is constrained by the action of trophic levels above them (Harvey et al. 2003).

As food limitation is such an important influence on pest population dynamics, any consideration of a biological control program must also include a careful appraisal of the ways in which the pest responds to its host plant. This is especially important where that plant is presented in an intensively grown crop system, or when the host plant is growing poorly. The failure of many biological control attempts may at least in part be due to a lack of understanding of these basic relationships.

12.5.2 Biological control manipulations

Biological control centers on a number of techniques that are employed to maximize the efficiency of natural enemies. These techniques vary according to the type of pest, its origin, and the nature of the crop ecosystem or husbandry system.

The control of an introduced pest by its introduced natural enemy is known as *classical* biological control, mainly because of early success stories such as control of the cottony cushion scale (see below). Natural enemies may be introduced from elsewhere to control native pests whose own complex of predators, parasitoids, and pathogens fail to keep numbers low. This may occur by chance, and is called *fortuitous* control, or by design where, for example, a virus disease is moved from one region or country to another to establish it in an area where the pest is active but the pathogen is absent. Native or exotic natural enemies may be encouraged in crop systems by the provision of suitable habitats or alternative hosts. This is *conservation*, and the background populations of enemies may be topped up by limited introductions of more of the same species in *augmentation*. Finally, large numbers of usually exotic enemies may be released into a pest-ridden crop; this is *inundation*, a good example of which takes place routinely in some glasshouse systems.

12.5.3   Biological control using predators and parasitoids

Predators and parasitoids can exert heavy impacts on pest insects and classical biological control generally consists of large-scale releases of enemies into crops to regulate future pest populations. A good example of this comes from Taiwan, where the coconut leaf beetle, *Brontispa longissima* (Coleoptera: Hispidae), causes severe yield losses to coconuts (Chang 1991). The larvae and adults of the beetle feed on developing leaflets in the central leaf bud of coconut palms, gnawing long incisions in the tissues, thus causing severe die-back when the leaves expand. In 1983, a parasitic wasp, *Tetrastichus brontispae* (Hymenoptera: Eulophidae), was introduced into Taiwan from Guam, and Figure 12.22

illustrates the suppression of coconut leaf beetle populations following successive releases of the parasitoid. By 1991, about 980,000 wasps had been released, preventing the potential destruction of around 80,000 coconut palms.

One of the first really good examples of a successful biological control program, and a blueprint for many others to follow, is the oft-quoted story of the cottony cushion scale, *Icerya purchasi* (Homoptera: Margarodidae) (see Plate 12.1, between pp. 310 and 311), in citrus orchards in California in the 19th century. The pest was accidentally introduced into California in 1868 from Australia on acacia plants, and, within 20 years, the scale was seriously threatening the citrus industry, with no obvious way of controlling it. A search was undertaken for natural enemies in the pest's native home, where eventually two species were discovered that seemed to hold the scale in check. The vedalia ladybird, *Rodolia cardinalis* (Coleoptera: Coccinellidae), and a parasitic fly, *Cryptochaetum iceryae* (Diptera: Tachinidae), were

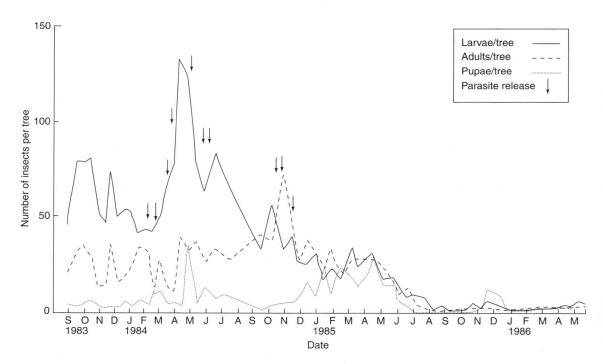

Figure 12.22 Population densities of coconut leaf beetle in Taiwan, before and after release of the parasitoid *Tetrastichus brontispae*. (From Chang 1991.)

released into Californian citrus groves. Eighteen months later, the cottony cushion scale was reduced to a non-economically threatening level over the entire region (Huffaker 1971). In this case, the pest's biology acted in favor of a successful program. Scale insects such as *Icerya*, although having serious impacts on their host plant, do not reproduce quickly, have relatively low fecundities, and tend to disperse from tree to tree rather poorly, giving their predators opportunities to restrict population growth. Their track record for biological control is much better when compared with, for example, closely related aphids (Hirose 2006).

Since those days, biological control programs have been established all over the temperate and tropical world. One modern example mimics that of the cottony cushion scale from a century earlier rather well, and concerns a pest called the cassava mealybug, *Phenacoccus manihoti* (Homoptera: Pseudococcidae). Table 12.8 describes the program. The native home of the mealybug is South America, and like so many other exotic pests, it was introduced accidentally into Africa in the late 1960s or early 1970s (Herren

Table 12.8 Biological control of the cassava mealybug, *Phenacoccus manihoti* (Homoptera: Pseudococcidae) by *Epidinocarsis lopezi* (Hymenoptera: Encrytidae) in Africa. (From Herren et al. 1987.)

Pest spread

1973	Congo/Zaire river
1976	Gambia
1979	Nigeria and Benin
1985	Sierra Leone and Malawi
1986	25 countries (70% of African cassava belt)

Parasitoid spread

1981	Introduced to Nigeria from Paraguay
1983	Recovered from all samples within 100 km of release site
1985	Over 50 releases in 12 Africa countries
1986	Established in 16 countries, covering 750,000 km^2

Result

Pest numbers reduced by 20–30 times, saving around 2.5 tonnes / ha in savanna region

1990). The sap-feeding mealybug causes stunting of the growing shoots of cassava. Peak densities of the pest vary a great deal, from 600 to 37,000 bugs per plant (Schulthess et al. 1991), an enormous infestation at the upper end of the scale. Yield losses to the crop itself, the cassava tubers, caused by the mealybug were of the order of 52–58% when compared with non-infested plants (Sculthess et al. 1991). Between 1981 and 1995, more than 2 million exotic parasitoids and predators were released against the cassava mealybug in Africa (Neuenschwander 2001), but of all these imports, the South American parasitic wasp, *Epidinocarsis lopezi* (Hymenoptera: Encrytidae), was most successful. This species was first imported into Nigeria, where it was reared in an insectary to bulk up numbers before being released into cassava crops. Further releases were carried out, so that by 1985, the parasitoid was established over 420,000 km^2 in West Africa and 210,000 km^2 in Central Africa (Herren et al. 1987). Note that no more releases have been made since 1994 (Neuenschwander 2001). After this establishment, densities of cassava mealybug were very much reduced. In Chad, for example, average pest numbers were around 1.6 per shoot (Neuenschwander et al. 1990), and the biological control program resulted in yield increases of around 2.5 tonnes/ha in savanna regions of West Africa (Neuenschwander et al. 1989).

It has to be said that some workers feel that the introduction of a single exotic parasitoid species into a huge region of ecological diversity typified by the African cassava growing region may not solve the pest problem entirely (Fabres & Nenon 1997), and 25 years or so after the first parasitoid introductions, this biological control is not as efficient as had been expected. To make matters worse, the spread of the enemies seems to be sporadic and unpredictable. In a similar biological control program for the cassava mealybug, this time in the Bahia region of Brazil, released parasitoids dispersed over a disappointing few hundred kilometers only (Bento et al. 1999). The best solution may be the integration of cultural practices such as intercropping and the use of cassava varieties tolerant to the mealybug, with classical biological control (Neuenschwander 2001; Schulthess et al. 2004).

Another successful introduction of an exotic parasitoid is the biological control of the California red scale, *Aonidiella aurantii* (Hemiptera: Diaspididae) (see Plate 12.2, between pp. 310 and 311). This sap-feeder is a worldwide pest of citrus (Murdoch et al. 1996), including oranges, lemons, and grapefruits. It is also a serious pest of many other tree species, such as neem (*Azidirachta indica*) (Speight & Wylie 2001), indicating its enormous host-plant range. The scale insect attacks all aerial parts of the tree, including twigs, leaves, branches, and fruit. Severe infestations cause leaf yellowing and twig die-back, and infested fruit becomes seriously devalued (University of California 1998). One parasitoid wasp in particular has successfully suppressed red scale in many regions including California for years. *Aphytis melinus* (Hymenoptera: Aphelinidae) was originally found in India and Pakistan (De Bach 1964), but is now firmly established in the USA. The adult female parasitoid lays eggs under the armored cover of the scale insect, where the resulting larvae eat the pest. In addition, *Aphytis* frequently kills scales without parasitizing them, by probing their bodies with its ovipositor. This coupled with the fact that the parasitoid has about three generations for every one of the scale's makes it a particularly efficient biological control agent. In fact, in southern California at least, *Aphytis* appears capable of maintaining *Aonidiella* populations at something like 1/200th of the density reached by the pest in the absence of control (Murdoch et al. 2006).

Curiously, though the interaction between parasitoid and host appears to be extremely stable, it has been hard to detect the mechanism underlying density dependence in this system (Murdoch et al. 1995). Indeed, the parasitoid–host interaction is so stable that there is almost no variation in density with which to measure density dependence. Clearly, the suppression of the pest is efficient and dependable. When the scale is unable to hide from attacks by the parasitoid by sheltering in bark crevice "refuges", the mean percentage parasitism reaches over 19% (though it can peak at over 30% on second instar male scales). Pest densities are drastically and rapidly reduced (Figure 12.23), and the numbers of first instar nymphs that mature to reproduce (the recruitment rate) is reduced by more than four times in regions of the tree exposed to parasitism (Murdoch et al. 1995, 2006). For a number of years, it was thought that the refuges played the dominant role in the stability of the scale–parasitoid system, but models based upon these parameters were never fully satisfactory. Recent studies may finally have uncovered the secret behind the stable regulation and high suppression of

the scale by the parasitoid (Murdoch et al. 2005). The combination of invulnerable adult scales (maintaining stability) and rapid parasitoid development (high suppression) appear to be responsible for the long-term success of this biological control program.

Apart from the very first example of commercial biological control – the cottony cushion scale story described above – there are fewer examples of success using a single species of predator than there are using parasitoids as natural enemies. Predator release has helped to control the great spruce bark beetle, *Dendroctonus micans* (Coleoptera: Scolytidae), in the UK. Unlike most species of bark beetle, whose larvae excavate solitary tunnels away from their maternal gallery in the bark cambium of infested trees, and who generally attack non-vigorous or moribund trees and logs, *D. micans* larvae exhibit aggregative behavior. All of the larvae from one brood live together in a chamber under the bark of spruce trees. They take part in communal feeding, which enables them to overcome the tree's resin defenses, and hence fairly vigorous trees can be attacked and killed (Evans & Fielding 1994). In continental Europe, *D. micans* has been an extremely serious pest of spruce for many years (Legrand 1993), and it was first discovered in the UK in 1982 (Gilbert et al. 2003), having been accidentally introduced on imported timber, it is assumed, some years previously. In 1983, a predatory beetle, *Rhizophagus grandis* (Coleoptera: Rhizophagidae), was introduced into infested spruce stands in Wales and the west of England from Belgium (Fielding et al. 1991; King et al. 1991). The predators are extremely efficient at locating prey eggs and larvae under the bark of attacked trees, using the volatile compounds in the pests' frass as an attractant, though there appears to be a lag of several years before the predator "catches up" with its prey (Gilbert & Grégoire 2003). Because *D. micans* larvae aggregate in one brood chamber, once a female *R. grandis* has located one prey larva, she is able to lay sufficient eggs to produce larvae to destroy the whole pest brood. *D. micans* has a relatively long generation time when compared with *R. grandis*, and the pest disperses very slowly by virtue of the adults' high flight threshold temperature (see Chapter 2). There is now strong evidence that *R. grandis* has established well in the UK, and is an important factor in the regulation of *D. micans* (Evans & Fielding 1994) (Figure 12.24).

A final example of classical biological control involves the island of St. Helena in the south

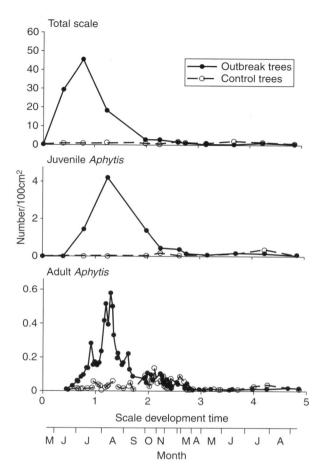

Figure 12.23 Mean densities in four outbreak trees and 10 control trees over five scale development times, approximately 16 months. Development is determined by temperature, and red scale takes about three times longer (in degree-days) to develop from crawler to reproductive adult than does *Aphytis*. (From Murdoch et al. 2006.)

Atlantic. Gumwood, *Commidendrum robustum*, is one of about 14 threatened and endemic species of tree growing on St. Helena in the subtropical southern Atlantic Ocean. Originally threatened by overexploitation for timber and firewood, the remaining trees began to be attacked by a scale insect, *Orthezia insignis* (Homoptera: Ortheziidae), in 1991. The pest originates from South and Central America, but is now widespread. The predatory ladybird, *Hyperaspis pantherina* (Coleoptera: Coccinellidae), has been viewed as a significant control agent for some time in other countries. In the 1990s, the International Institute of Biological Control reared and then released the beetle on St. Helena with great success (Fowler 2004) (Figure 12.25). This, according to the author and his colleagues, is probably the first case of biological control used against an insect pest to save a plant species from extinction.

As defined, classical biological control describes the introduction of non-native enemies to control non-native pests. As ecologists, we should not be surprised that introducing non-native enemies, even

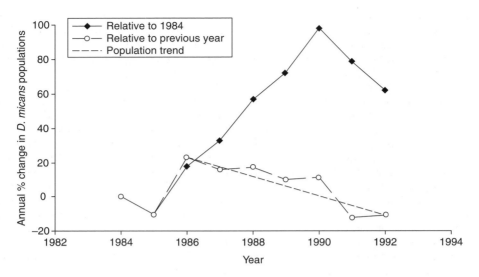

Figure 12.24 The decline in the numbers of spruce bark beetlee, *Dendroctonus micans*, over a 5-year period after the predator, *Rhizophagus grandis*, was released in the UK. (From Evans & Fielding 1994.)

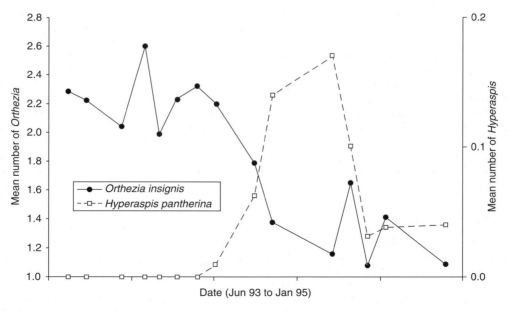

Figure 12.25 Mean numbers of the scale insect *Orthezia insignis* feeding on the endemic gumwood tree, *Commidendrum robustum*, on St. Helena, in relation to mean numbers of its released predator, *Hyperaspis pantherina*. (From Fowler 2004.)

with the best of intentions, can have unexpected and deleterious effects on native species and ecosystems. This inherent danger of classical biological control is receiving increasing attention (Simberloff & Stiling 1996, 1998; Frank 1998; Louda & Stiling 2004) and there are certainly cases where introduced agents of control have caused problems. In two noteworthy cases, the biological control agents are herbivorous insects and the target pests are weeds. The moth, *Cactoblastis cactorum* (Lepidoptera: Pyralidae), is a "poster child" of biological control in Australia, where it has largely contained populations of the invasive cactus, *Opuntia*. However, *C. cactorum* has recently invaded the United States and threatens native cacti, including the endangered semaphore cactus, *Opuntia corallicola*, in Florida (Stiling 2004). *Cactoblastis* is moving rapidly northward and westward from Florida, threatening cacti throughout the southern United States. In a second example, the weevil, *Rhinocyllus conicus* (Coleoptera: Curculionidae), was introduced into North America to control non-native thistle species. Unfortunately, the weevil has expanded its host range (and geographic range) since introduction and now attacks native and endangered thistle species (Louda et al. 1997, 2005). Looking beyond weeds, there are many examples where introduced parasites and predators of insect pests have had deleterious effects on native species (Simberloff & Stiling 1996; Stiling 2004) and the fields of biological control and conservation are somewhat at odds. Of course, one of the touted benefits of biological control is reduced reliance on insecticides with their deleterious effects on natural systems. If biological control itself is a conservation threat, then its role must be (dispassionately) reassessed. We clearly need high-quality screening of potential control agents accompanied by adequate risk assessment before introduction. Wherever possible, augmentation of native natural enemies may be less of a risk. This strategy is described next.

Natural biological control

Classical biological control describes the introduction of enemies from elsewhere – a very active management approach. However, a great deal of interest has been expressed over many years in the ability of naturally occurring predators and parasitoids to regulate native pest species. As we have seen, natural enemies from exotic, far-flung countries exert successful biological control on crop pests, and native, even endemic, parasitoids and predators tend to be ignored. This may be unfair; Neuenschwander (2001) for example shows that exotic predators of the cassava mealybug were no more useful at biological control of the pest than local African species.

Cereal aphids epitomize the nature of a pest, with their highly specialized lifestyles, high fecundities, and high rates of dispersal. It is against this background that the abilities of predators and parasitoids have to be assessed. Figure 12.26 illustrates a typical example of pest enemy dynamics in a field of winter wheat in southern UK in early summer. If the weather is warm and dry at the start of the milk-ripe stage in wheat, aphids can undergo explosive population increases. Winged forms (alatae) appear a little after the main peak, responding both to reductions in food availability as the grain matures, and of course increasing competition for food and space from daughters and siblings. Parasitic wasps and predatory beetles are incapable of reacting fast enough or efficiently enough to keep the aphid populations below economic thresholds (see later in this chapter). However, not everyone agrees that generalist predators are such weak agents of control, and some workers feel that generalist natural enemies, occurring in and around cereal fields may exert significant pressures on aphids and their ilk. In Figure 12.27, Schmidt et al. (2003) illustrate that if all natural enemies are kept away from cereal aphids, their populations build to very high levels. However, when full access for all manner of aphid enemies is allowed, the aphid populations drop very significantly. Finally, Collins et al. (2002) have shown that even very generalist predators such as ground beetles, rove beetles, and orb spiders can reduce cereal aphid populations (Figure 12.28). Of course the only real validation of these findings is to compare the pest population reductions with the minimum economic thresholds for cereal aphids (see later). If natural biological control cannot reduce pest populations below these thresholds, then other tactics, such as plant resistance or, more realistically in winter wheat, insecticidal control, will have to be integrated.

Drawbacks and problems with predators and parasitoids as agents of control

Unfortunately, not all biological programs are successful. To make things worse, it can take a rather long time to come to these conclusions. Gillespie et al.

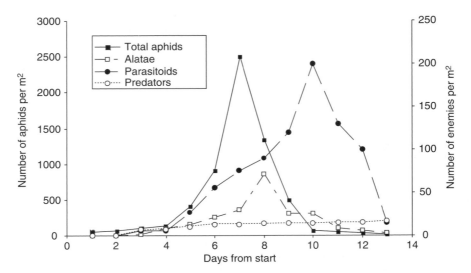

Figure 12.26 Densities of cereal aphids and their natural enemies on winter wheat in southern England (hypothetical data).

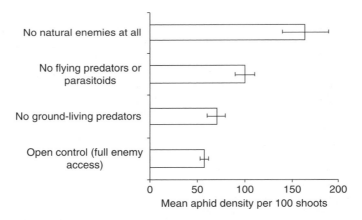

Figure 12.27 Densites of cereal aphids on winter wheat plants at the end of the milk-ripe stage in relation to natural enemy access to the pests. (From Schmidt et al. 2003.)

(2006) discuss the attempts to control the cabbage seedpod weevil, *Ceutorhynchus obstrictus* (Coleoptera: Curculionidae), by introducing European parasitoids into North America. The first multiple-species introductions took place in 1949, but by 2005 only one species of parasitoid is still to be found, and that in low numbers. It seems that in some cases the actions of predators and parasitoids make little or no difference to the course of insect pest outbreaks. Some insect pest populations simply do not respond to the actions of their natural enemies in a manner useful for biological control. We discussed the green spruce aphid, *Elatobium abietinum* (Hemiptera: Aphididae), earlier in this chapter. *E. abietinum* can cause serious defoliation and increment loss to spruce trees in Western Europe and, as Chapter 2 describes, one of the most important mortality factors acting upon green spruce aphid is late winter temperature. Declines in spring

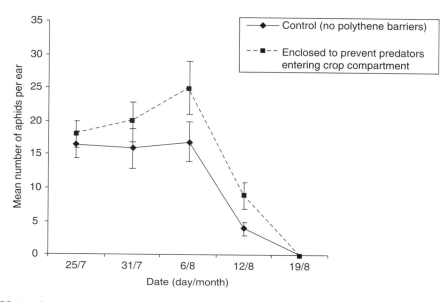

Figure 12.28 Number of aphids per ear of winter wheat in predator-free and control plots. (From Collins et al. 2002.)

populations coincide with the production of alatae (winged adults) in response to changes in host quality and photoperiod (Leather & Owuor 1996). There are species of predator within the system. The two-spot ladybird, *Adalia bipunctata* (Coleoptera: Coccinellidae), has both larvae and adults that consume spruce aphids, but studies have shown that they fail to make an appreciable impact on pest population density because of their slow numerical response (see Chapter 5).

Even when high rates of parasitism are recorded within a pest population, the pest may still be a serious problem if those individuals that survive parasitism are capable of causing serious damage. Moreover, if alternative sources of mortality, such as weather or competition, drive population change, natural enemies may merely track the population fluctuations of the pest, and not regulate it. This is sometimes referred to as "donor control". The basic idea is simple – predators track changes in the supply of their pest resource but do not influence the rate at which that resource is renewed (Hawkins 1992). Let us take, for example, the case of the horse chestnut scale, *Pulvinaria regalis* (Homoptera: Coccidae), in the UK. Horse chestnut scales feed as young nymphs on leaves of a variety of urban trees, causing considerable growth losses (Speight 1992). Since *P. regalis* arrived in the UK in the 1970s, at least one parasitoid species,

Coccophagus obscurus (Hymenoptera: Chalcidae), has broadened its host range to include *P. regalis*. The parasitoid is now frequently found attacking young *P. regalis* nymphs, but its response to the density of the herbivore is inversely density dependent (Figure 12.29), and thus non-regulatory. Rates of parasitism actually

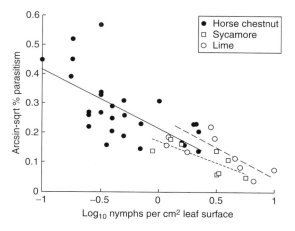

Figure 12.29 Parasitism of horse chestnut scale *Pulvinaria regalis* by *Coccophagus obscurus*, showing inverse density dependence. All regression lines significant at $P < 0.05$ or below. (M.R. Speight, unpublished data.)

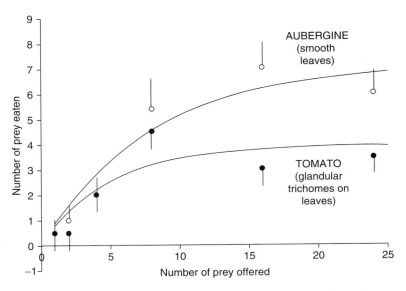

Figure 12.30 Functional responses of the predatory bug, *Podisus* spp, on fourth instar *Spodoptera* larvae, on two different host plants. (From de Clerq et al. 2000.)

decline as pest density increases. It seems that the most important mortality factor in the life of the pest is the often huge losses of first instar crawlers as they migrate on the wind from tree to tree down urban streets.

Readers may wish to familiarize themselves with the definitions of predator functional and numerical responses provided in Chapter 5. The functional response of predators and parasitoids places a fundamental limitation upon their ability to regulate pest populations. In Figure 12.30, the number of armyworm caterpillars eaten by the predatory bug, *Podisus* (Hemiptera: Lygaeidae), increases at a declining rate as the number of caterpillars increases (de Clerq et al. 2000). Eventually, the predators become satiated with prey, and their consumption reaches an asymptote. As an aside, notice again how food plant traits can interfere with predator foraging; hairy leaves significantly handicap *Podisus*. The take-home message is that constraints on food consumption and handling (functional responses) and constraints on population growth and recruitment (numerical responses) can limit the efficacy of natural enemies as agents of biological control.

Even when inundative releases are carried out, biological control may not work adequately. One

example of this is the attempt to control the sap-feeding bug, *Lygus hesperus* (Hemiptera: Miridae), a pest in commercial strawberry fields in the USA, using repeated releases of the egg parasitoid *Anaphes iole* (Hymenoptera: Mymaridae). Following the release of a staggering 37,000 adult parasitoids per hectare of strawberries, 50% of *Lygus* eggs were found to be parasitized. This resulted in a 43% reduction in the numbers of pest nymphs, and a 22% reduction in the amount of fruit damaged (Norton & Welter 1996). Whether or not the huge inundation program for parasitoid release merits the rather small reduction in damage will only be revealed by detailed cost–benefit analysis. Cereal aphid control has suffered similar disappointments. Levie et al. (2005) released 21,000 individuals per hectare of the aphid parasitoid *Aphidius rhopalosiphi* (Hymenoptera: Aphidiinae) in Belgium winter wheat fields. There was only a small increase in aphid parasitism within that growing season, and it had disappeared by the following season. In fact, inundative or augmentative biological control does not have a particularly stellar track record. Collier and van Steenwyk (2004) reviewed a large number of attempts at this type of pest management, and found that only 15% of them achieved the reduced target pest density, while a depressing 64% of attempts failed completely.

The habitat of the pest as well as its relationships with its host plant may influence the potential success of a biological control program. Many pests are in some way concealed, by living in shoots, stems, galls, mines, roots, or bark (see Table 12.17, below), and although there are examples of such pests being regulated successfully by natural enemies (see *Dendroctonus micans* example above), there are many occasions where parasitoids and predators simply are unable to attack enough pests in a given time to affect the latter's density significantly. In general terms, the success of biological control appears inversely related to the proportion of insect pests that are in some way protected from enemy attack, i.e. they inhabit some sort of refuge (Mills 2001). This is not to say that borers can never be controlled biologically, but there is a conventional wisdom suggesting that when ecological factors such as food, climate, and refuges promote the proliferation of the pest, then we should perhaps look elsewhere for primary pest management systems. We can then retain biological control by predators and parasitoids as a backup measure within an IPM system (see below).

One final consideration hinges on whether or not it is better to introduce more than one natural enemy at a time to improve the chances of efficient biological control. Two issues arise. First, natural enemies can compete with one another for prey (see Chapter 4) and may prove less effective together than either would on its own. Second, many predators participate in intraguild predation (see Chapter 5) in which predators consume other predators as well as herbivorous prey (Denno et al. 2003). Clearly, if one agent of control is consuming another agent of control, suppression of the pest may be compromised. So what is the end result of adding multiple predators or parasitoids to a system? The answer, of course, is variable. Figure 12.31 illustrates the additional impact of a predator on the biological control of an aphid (Snyder et al. 2004). Here, the predator seems to provide a backup to the parasitoid, assisting in preventing even a brief peak in pest population density. Specialist parasitoids, such as an egg parasitoid and a later larval parasitoid, might well complement each other, but a problem certainly arises when predators such as ladybirds are introduced into a system already utilizing parasitoids. Experiments have shown that this type of predator will readily prey on parasitized aphids as well as healthy ones (in fact, the former may be easier to catch), and the effects of multiple natural enemies in this scenario are

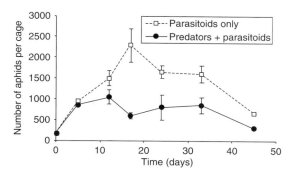

Figure 12.31 Densities of the aphid *Macrosiphum euphorbia* through time related to natural enemy prescence. (From Snyder et al. 2004.)

clearly non-additive (Ferguson & Stiling 1996). In addition, large generalist predators can reduce the success of biological control by eating smaller predators in preference to the pest (Prasad & Snyder 2004).

12.5.4   Biological control using pathogens

For many years now, pathogenic organisms have been growing in popularity as alternative biological control agents to parasitoids and predators, and also as alternatives to synthetic chemical insecticides (Lacey & Shapiro-Ilan 2008). There are a variety of organisms that fall under the general heading of "pathogens" (Table 12.9). The use of these so-called microbiological insecticides, both natural and genetically engineered forms, is at the forefront of insect pest management in many parts of the world. Fungi are now used routinely and commercially in glasshouses all over the world, mainly to control aphids, and they have real potential to replace synthetic insecticides for locust control, except in emergencies (Lomer et al. 2001). They may also be highly effective in damp tropical situations (Gopal et al. 2006) where they can be a much better solution than traditional insecticides. Nematodes are now widely available to combat concealed pests such as weevil larvae in soil and under bark, and fly larvae in mushroom houses (Georgis et al. 2005). For the sake of brevity, however, we consider only two major groups of pathogens here, bacteria and viruses.

Bacteria

Bacteria used in pest management belong to three main species: *Bacillus popilliae*, *B. thuringiensis*, and

Table 12.9 Summary of types of pathogenic organism used in the biological control of insect pests.

| Pathogen | Commercial/experimental example | | | |
	Pathogen	Insect	Common name	System
Bacterium	*Bacillus thuringiensis*	*Plutella xylostella*	Diamondback moth	Arable: brassicas
Fungus	*Verticillium lecanii*	*Myzus persicae*	Peach potato aphid	Glasshouses
Nematode	*Nosema locustae*	*Locusta migratoria*	Locust	Grassland
Protozoan	*Vairimorpha necatrix*	Lepidoptera: Noctuidae	Leather jackets	Various
Virus	*H$_p$NPV*	*Hyblaea puera*	Teak defoliator	Teak plantations

B. sphaericus. *B. popilliae* is a bacterium that causes "milky disease" in numerous species of scarab beetles all over the world (Cherry & Klein 1997), and is used, not always successfully, to control turf pests such as Japanese chafer beetle larvae (Coleoptera: Scarabaeidae) (Redmond & Potter 1995). Unlike *B. popilliae*, *B. thuringiensis* and *B. sphaericus* produce toxins that have insecticidal activity. *B. sphaericus* is able to grow saprophytically in polluted water (Payne 1988), and it is now widely used against mosquito larvae in many parts of the world (Zahiri & Mulla 2003). *B. thuringiensis* is the most widespread bacterial pesticide, and is available in a wide variety of strains and commercial formulations all over the world. *B. thuringiensis*, or *Bt* for short, is a spore-forming pathogen (Crickmore 2006) that is easy to grow on many media, including cheap waste products of the fish and food processing industries (Payne 1988). *Bt* was discovered early in the 20th century, and we now know that the spores produce a proteinaceous toxic crystal with strong insecticidal properties (Burges 1993). This so-called δ-endotoxin (labeled a Cry protein) is the product of a single bacterial gene, which has great potential for breeding and selection for specific targets, and for genetic engineering. Normally, it is the δ-endotoxin itself that forms the active ingredient in commercial formulations of *Bt*, so that the pathogen is unable to replicate in the environment and is instead a true biological insecticide. In all cases, the toxin has to be ingested by the host. Unlike fungi, *Bt* cannot enter the body through the cuticle or spiracles. When an insect consumes the protein, protease enzymes in the insect's digestive system cut the normally non-toxic protein into a smaller piece that is highly toxic to insects. The smaller, activated form of the δ-endotoxin binds to a specific receptor on the surface of cells lining the insect's gut, causing a

disruption of electrolyte balance, leading to death (Figure 12.32). *Bt* toxins are considered very safe for human consumption because the intestinal walls of mammals do not have the endotoxin receptor necessary for the toxic effect, and the proteins are degraded quickly in the stomach.

Bt can be targeted at pests in the form of sprays or genetically engineered plants. Although you might not know it from recent controversies surrounding genetically modified (GM) crops, *Bt* sprays have been used in agriculture and forestry for several decades. Since *Bt* has to be ingested to kill insects, boring, galling, and mining insects are essentially immune once concealed within their host plants. The only way to control a lepidopteran larva such as a shoot- or fruit-borer with *Bt* sprays is to ensure that the very young larvae hatching from eggs on the outside of the plant consume the toxin before entering the plant tissues. *Bt* is very inefficient at killing shoot-boring pests (Figure 12.33), and it is on these occasions that applications of pathogens are regularly put to shame by the synthetic insecticides. Timing of spray applications for concealed pests is therefore critical (see Section 12.6.1).

For pests exposed on the surface of plants or soil, there are now several distinct varieties or subspecies of *Bt* available, each with a variety of serotypes. *Bt* varieties have different properties and are used against different insect groups (Table 12.10). *Bt kurstaki* is still the most widely used, although *Bt israelensis* is being employed increasingly in vector control. Even though *Bt* sprays rarely rival the performance of synthetic chemical insecticides, they are the preferred method of control in many systems. Aerial spraying is a routine annual treatment for pests over very wide areas. With the changing attitudes to insecticides and the environment, *Bt* has replaced synthetic

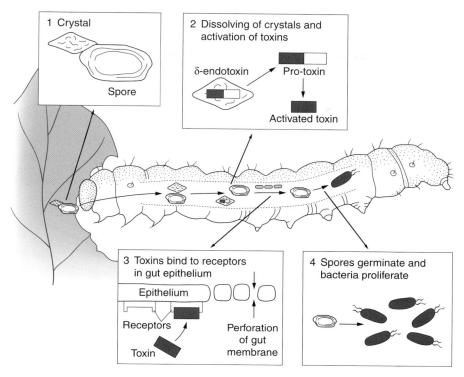

Figure 12.32 The mechanism of toxicity of *Bt*. (Courtesy www.inchem.org.)

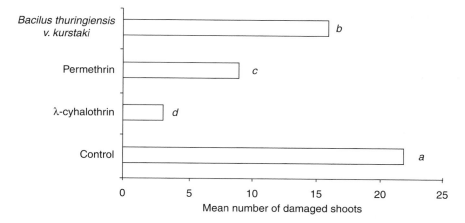

Figure 12.33 Mean percentage of loblolly pine shoots damaged by larvae of pine tip moth, comparing the efficiency of four types of insecticide applied externally in Southeast USA. Bars with the same letter are not significantly different at $P < 0.05$. (From Nowak et al. 2000.)

Table 12.10 Varieties of *Bacillus thuringiensis* and their susceptible hosts.

Variety	Susceptible host
Bt kurstaki	Many lepidopteran larvae
Bt israelensis	Dipteran larvae, especially mosquitoes
Bt tenebrionis (also known as *Bt san diego*)	Coleopteran larvae, including mealworms and colorado beetles
Bt aizawa	Wax moth larvae

compounds in the fight against defoliating insects in many countries. In New Zealand, for example, it has been used to eradicate the tussock moth, *Orgyia thyellina* (Lepidoptera: Lymantriidae), over 4000 ha of Auckland suburbs (Hosking et al. 2003).

One problem with *Bt* is its rather broad spectrum of activity in comparison to species-specific pathogens (see below). Clearly, growers would prefer to use a chemical that will kill a whole range of pests, rather than having to select a specific compound for each insect species they wish to control. For the grower, this wide host range is good news, though for the conservationist, this may be less attractive. A formulation targeted at one species of Lepidoptera will be largely effective against all Lepidoptera. It would indeed be folly to use *Bt* against crop pests in close proximity to a silk farm; silkworm larvae (Lepidoptera: Saturnidae) are eminently susceptible to *Bt kurstaki*. Likewise, insects of conservation importance might be susceptible, so *Bt* has to be used with great care and foresight. However, all strains of *Bt* are entirely specific to arthropods, and hence have no environmental impact on vertebrates, except perhaps indirectly by removing the food of certain insectivorous birds and mammals. Whatever the concerns with *Bt* sprays, they are generally much more benign than many synthetic chemical insecticides. *Bt* is now used throughout the world, and is even available in some countries, including the UK and Australia, as a garden insecticide for the domestic market.

Of course, *Bt* sprays are not problem-free. Any mortality factor to which organisms are exposed will, if not immediately fatal to all genotypes, select for resistance down the generations. However, unlike with synthetic insecticides (below), resistance to

Bt sprays is not commonly encountered in field populations. One of the only insects so far to show resistance to *Bt* formulations such as Dipel is the diamondback moth, *Plutella xylostella* (Lepidoptera: Plutellidae) (Janmaat & Myers 2003), whose global ubiquity on *Brassica* crops means that it is routinely exposed to almost continuous *Bt* toxins. Greenhouse populations of other Lepidoptera such as the cabbage looper, *Trichoplusia ni* (Lepidoptera: Noctuidae), and even borers such as the European corn-borer, *Ostrinia nubialis* (Lepidoptera: Pyralidae), can exhibit even higher levels of resistance to *Bt* sprays. The rate of resistance development is a function of selection pressure, which, in this case, is the proportion of larvae killed by an application. The greater the mortality, the greater the selection pressure for resistance development, and the faster that resistance will be observed (Huang et al. 1999) (Figure 12.34). Of course, if mortality reaches 100%, there are no pests remaining upon which selection can act. One control strategy might be to use high enough doses to kill nearly all individuals within a given area, yet leave some areas entirely unsprayed so that genes for susceptibility remain in the pest population. One source for optimism is that resistance to *Bt* sprays such as Dipel appears to be very short lived. If the selection pressure is removed for only a few generations, then resistance breaks down again (Huang et al. 1999) (Figure 12.35).

Today, *Bt* is used as an insecticide primarily by incorporating *Bt* genes into plant tissues by genetic modification. According to Christou and Twyman (2004) nearly 900 million people have to live with chronic hunger, with 3 billion suffering from nutrient deficiencies. Figure 12.36 shows the global adoption of "biotech" crops – note that only insect and herbicide resistance are routinely available for major global crop types (Castle et al. 2006). These modified plants, so-called "transgenics", are engineered to "express" parts of the *Bt* genome such that, when an insect eats the plant, the food is already toxic. The insects die in just the same manner as those exposed to *Bt* sprays, but the benefits of internal expression are many-fold:

(i) Internal plant tissues can be rendered toxic, whereas sprays normally act only on plant surfaces where they can be deactivated by weather and ultraviolet radiation. Therefore borers as well as defoliators can be killed as soon as they feed on small quantities of the crop, whether they are internally or externally located.

(ii) Every pest individual that feeds on the transgenic crop is exposed to the toxin. Even the most careful of

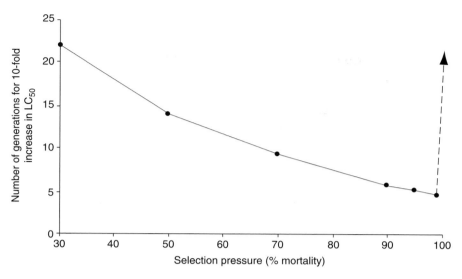

Figure 12.34 Number of generations of European corn-borer, *Ostrinia nubilalis*, required to produce a 10-fold increase in LC_{50} according to selection pressure of exposure to varying concentrations of Dipel. (From Huang et al. 1999.)

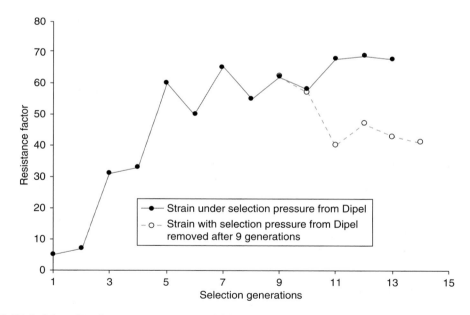

Figure 12.35 Stability of Dipel resistance in *Ostrinia nubilalis*. (From Huang et al. 1999.)

spray programs will miss a proportion of individuals within the pest population, but GM crops target all individuals as soon as they begin to eat.

(iii) Non-target plants within and beside the target crops are not exposed to *Bt* toxins. Sprays, in contrast, do not distinguish crops from weeds and wildflowers.

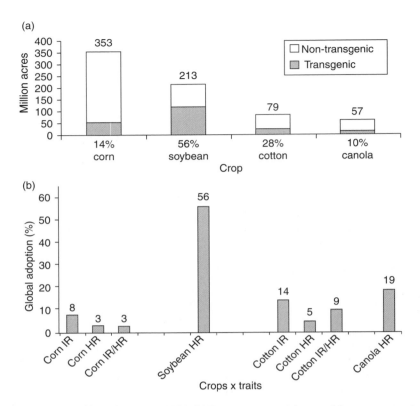

Figure 12.36 Global adoption of biotech crops in 2004. (a) Bars represent total acres of four crops grown globally with the percentage acreage devoted to biotech crops. (b) Bars represent the major crops and traits with the percentage global crop acreage devoted to that trait. HR, herbicide resistance; IR, insect resistance. (From Castle et al. 2006.)

(iv) Sprays are notorious for drifting beyond the target crop and entering adjacent ecosystems. In GM crops, the plant residues and exudates present the only potential avenues of contamination beyond the crop system.

(v) In theory at least, only those herbivores that consume GM plants should be affected.

(vi) Finally, costs of aerial application are avoided because the seed already contains *Bt* genes before it is sown.

Figure 12.37 illustrates successful control of Lepidoptera on GM tomatoes in India (Kumar & Kumar 2004), whilst Christou et al. (2006) review the benefits of transgenic crops in the USA. Here, crops such as canola, corn, cotton, papaya, squash and soybean planted in 2003 produced an additional 2.4 million tonnes and increased income by US$1.9 billion. Pesticide use was reduced by 21,000 tonnes

(Cattaneo et al. 2006). As with all good things, there are potential drawbacks to the use of GM plants. None the less, transgenics have been the target of unusually fierce debate and it is well known that the use of such crops has met with extreme resistance from some members of the public. Concerns have been expressed about potential risks to non-target insects, beneficial organisms, the whole agro-ecosystem, and human health. Political groups and the media have cast an aura of danger around the deployment of transgenic crops by using the term "Frankenstein crops" or "Franken-foods". In parts of Europe, transgenics are met with an unhelpful mixture of hyperbole and hysteria.

It is essential to investigate possible side effects of any new technology, and GM crops are no exception. There now exist a wealth of scientific data with which to assess more accurately and (we might hope)

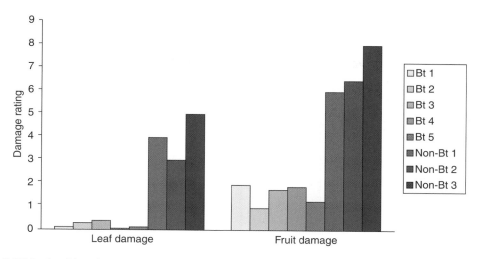

Figure 12.37 Leaf and fruit damage to tomato plants caused by *Helicoverpa armigera* on tomato hybrids in the field in India. (From Kumar & Kumar 2004.)

dispassionately the potential risks of GM crops. Most importantly, it seems that average yields between wild-type (normal) and transgenic crops are not significantly different (Cattaneo et al. 2006). In addition, early concerns that pollen from *Bt* crops might pose a significant risk to non-target Lepidoptera on wild-flowers near agricultural fields appear to be unfounded.

The larvae of the North American black swallowtail butterfly, *Papilio polyxenes* (Lepidoptera: Papilionidae), feed on umbelliferous plants at field edges or along roadsides, and so are prime targets for any influence of GM crops nearby. Figure 12.38 illustrates that the pollen of *Bt* maize, which can contain *Bt* endotoxin, penetrates only a small distance into field margins, with no apparent effects on either the growth or survival of swallowtail larvae (Wraight et al. 2000). O'Callaghan et al. (2005) have provided a comprehensive review of many recent studies of potential risks posed by *Bt* crops. The authors conclude "the extensive testing on non-target plant-feeding insects and beneficial species that has accompanied the long-term and wide-scale use of *Bt* plants has not detected significant adverse effects." Furthermore, the reviewed studies show that *Bt* plants have little or no impact on soil organisms, including earthworms and microflora. Table 12.11 summarizes the major tests and results described by O'Callaghan and her colleagues. As we pointed out earlier in this chapter, *Bt* strains such as *Bt kurstaki* are very specific to insect

groups (in this case, Lepidoptera), so it is hardly surprising that crops engineered with *Bt kurstaki* have little impact upon insects such as lacewings, ladybirds, parasitic wasps, or bees. We should also recognize that the alternative to *Bt* crops is not simply *Bt*-free crops. Realistically, it is likely to be *Bt*-free crops sprayed regularly with synthetic insecticides, the majority of which have more adverse effects on the environment than does *Bt*.

Despite bad press, transgenics expressing *Bt* have proven commercially successful. In 2002, some 14.5 million hectares of transgenic crops were grown worldwide (Zhao et al. 2003a), with corn (maize) and cotton the most commonly planted. In the USA, commercial benefits from the use of *Bt* corn were estimated at US$190 million in 1997 (Babu et al. 2003). *Bt* cotton has proven extremely beneficial in a number of countries (Pemsl et al. 2004) (Table 12.12). Note that although the cost of transgenic seed can be very high, net profits are favorable and reductions in the use of conventional pesticides likely provide additional health and environmental benefits.

None the less, there is one significant problem posed by the widespread deployment of *Bt* crops. The use of *Bt* in transgenic plants may significantly increase the risk and rate of resistance development within insect populations (Sanchis et al. 1995; Tabashnik et al. 2006). Clearly, large-scale plantings of *Bt* cotton and maize must exert a huge selection

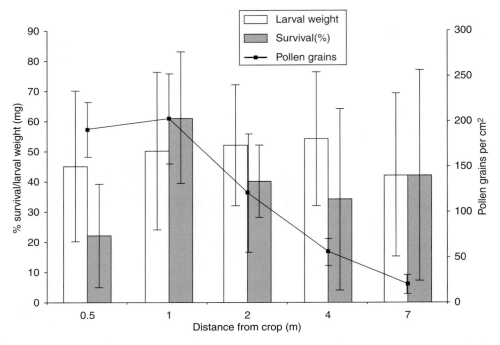

Figure 12.38 Pollen loads, larval mass, and survivorship of early instar black swallowtails as a function of distance from the edge of field of *Bt* corn. (From Wraight et al. 2000.)

pressure for resistance development, and in the long term, the very success of GM crops could also be their downfall. At least four basic strategies have been proposed to delay the adaptation of pests to GM crops. These are: (i) the mixture of toxic and non-toxic cultivars in a field (the so-called "refuge approach"); (ii) the use (or stacking) of two or more distinctly different toxin genes in one cultivar; (iii) the use of low doses of toxins that act in conjunction with natural enemies; and (iv) the differential expression of toxin genes in certain plant parts or at certain times of year (Gould 1998).

Refuges are probably the most widespread tactic used to combat *Bt* resistance in crop systems, whereby certain sections of a crop or field are planted with non-transgenics. This may be in strips, or more usually as a bounding margin with the inside section transgenic. The theory is simple enough. Pest insects will be able to survive to adulthood in the refuges, free of the selection pressure for resistance development. The refuges support a source of susceptible genotypes within the insect population that will mate with those

individuals from the *Bt* crop, diluting the proportion of resistance genes in the population. The strategy may be particularly effective if there exists some fitness cost to resistance. In the USA, where resistance management tactics based upon refuges are widespread, at least 20% of every *Bt* corn crop must be a *Bt*-free refuge. Moreover, if the *Bt* corn is grown in a cotton-growing region, the refuge must be 50% of the crop. The combined effects of refuges and gene pyramiding (see below) on predicted rates of resistance development are illustrated in Figure 12.39 (Roush 1998).

It is also possible to plant mosaics of the same crop but with different *Bt* strains in the same field, on the fair assumption that insect pests tend to develop resistance to only one of them at a time. Mosaics should reduce the selection pressure for resistance on both strains.

Finally, pyramiding is a technique whereby two or more strains of the *Bt* toxin gene are used in the same crop plant. In fact, *Bt* strains are known to produce over 100 toxic proteins (Gould 2003), but only a very few of these are commercially developed at the

Table 12.11 Effects of insect-resistant genetically modified plants on non-target plant-feeding insects and their enemies: summary of review findings. (From O'Callaghan et al. 2005.)

Types of insect	Crop	Quoted effect
Pollinators	Maize, cotton	No effect
Natural enemies (ladybirds, lacewings, parasitoids)	Cotton	No effect
Natural enemies (lacewings)	Maize	Little impact
Earthworms	Maize	No deleterious effects
Woodlice	Maize	No adverse effects
Collembola	Cotton, potato	No adverse effects
Mites	Cotton, potato	No effect
Nematodes	Maize	No effect

Viruses

Insects, like humans, suffer from a wide variety of virus diseases (see Plate 12.3, between pp. 310 and 311). By the end of the 1980s, over 1600 virus isolates had been discovered attacking insects (Payne 1988). However, only one group, the Baculoviridae, is exclusively found in the arthropods (Cory & Myers 2003), with no known vertebrate associations (Szewczyk et al. 2006). Baculoviruses are large rod-shaped DNA viruses and, though modern nomenclature changes rapidly, they have been split in the past into three basic types. Both the nuclear polyhedrosis viruses (nucleopolyhedroviruses or NPVs) and the granulosis viruses (granuloviruses or GVs) have their virus rods, called virions, embedded in a proteinaceous sheath, the polyhedral inclusion body (PIB) (Figure 12.41). In the NPVs, each PIB is relatively large (0.8–15 μm in diameter), certainly visible under a compound microscope, and contains many virions; whereas GV PIBs are much smaller (0.3–0.5 μm) and usually contain but one virion. *Oryctes*-type viruses (recently renamed "unclassified") are a special type of baculovirus, which have so far been found only commonly in beetles such as the palm rhinoceros beetle, *Oryctes* spp. (Coleoptera: Scarabaeidae) (Huger 2005; Ramle et al. 2005).

The development of baculoviruses for the biological control of insect pests is a slow process, hampered not so much by biology, ecology, or technology, but more by public and even scientific distrust of the perceived danger of "germ warfare" in our crops. None the less, insect pest populations can be suppressed very efficiently by baculoviruses, and many defoliating pests of farm and forest crops are now controllable with

moment. In Figure 12.40, Zhao et al. (2003a) illustrate the effects of pyramiding on the mortality of diamondback moth larvae already resistant to one or other of the Cry1C or Cry1Ac *Bt* strains. Note that insect densities are lower on the pyramided, two-gene plants, than on the single-gene plants in mosaics or sequential plantings.

Table 12.12 Performance differences (% change) between *Bt* and conventional cotton varieties. (From Pemsl et al. 2004.)

	Argentina	China	India	Mexico	South Africa
Bt cotton as % of total cotton area	5	40	2	71	10
Lint yield (% change)	33	19	80	11	65
No. of chemical sprays	−2.4	−13.2	−3	−2.2	n.a.
Pest control costs (% change)	−47	−67	−39	−77	−58
Seed costs (% change)	530	95	82	165	89
Profit (% change)	31	340	83	12	299

n.a., data not available

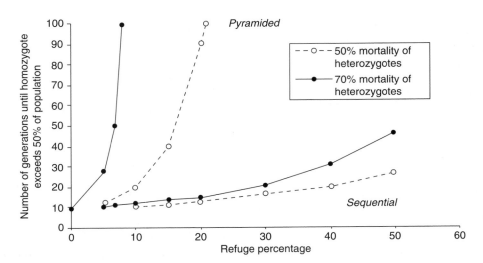

Figure 12.39 Simulated evolution of resistance in insects to *Bt* in transgenic crops comparing the use of two sequential toxins with the use of the toxins jointly in a pyramidal variety, for a range of percentages of the population in non-toxic refuges. Thus, with a refuge per cent exceeding around 20%, insect resistance to *Bt* in transgenic crops should not evolve if two toxins are used in a pyramided fashion, on the understanding that resistance only develops to one toxin at a time, not both. (From Roush 1998.)

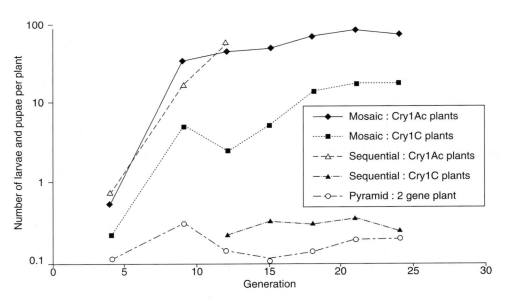

Figure 12.40 Populations of Cry1Ac/Cry1C-resistant diamondback moths, *Plutella xylostella*, in cages with different *Bt* broccoli treatments. (From Zhao et al. 2003a.)

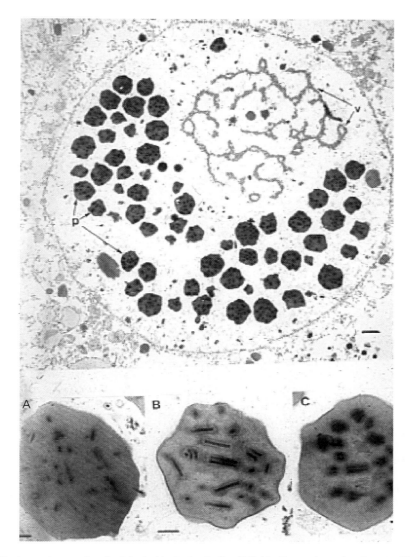

Figure 12.41 Electron micrographs of polyhedral inclusion bodies (PIBs) in the nucleus of an insect mid-gut cell. Rod-like structures are the virus particles (virions) themselves. (B) and (C) show single PIBs containing virus rods.

viruses (Cory & Myers 2003). Of course, viral control agents are not perfect and even heavy mortality of the pest does not guarantee satisfactory pest suppression.

All baculoviruses so far employed are essentially the natural result of insect–pathogen coevolution. Pest managers are merely in the business of moving the viruses from one place to another, where for some reason the pathogen has not established itself. Natural

epizootics of baculoviruses are regularly recorded, and various pest species seem to suffer natural declines as a result. For example, the European spruce sawfly, *Gilpinia hercyniae* (Hymentoptera: Diprionidae), in Canada; the teak defoliator, *Hyblaea puera* (Lepidoptera: Hyblaeidae), in India; and the yellow butterfly, *Eurema blanda* (Lepidoptera: Pieridae), in Borneo, are all known to exhibit regular and spectacular population declines

Table 12.13 Characterstics of nuclear polyhedrosis viruses. (From Ivory & Speight 1994.)

Specificity	Complete within the phylum Arthropoda; usually within families, genera, even species No significant effects on vertebrates ever detected
Occurrence	Natural within insect populations, especially in indigenous areas
Life stage	Infects larvae; eggs, pupae, and adults may rarely carry inactive virus
Mode of action	Infection from contaminated substrate by ingestion only Virus attacks nuclei of gut cells in sawfly larvae and of other cells (e.g. fat body) in Lepidopters
Transfer between hosts	In feces of infected but living larvae, from disintegrating dead bodies. In guts of predators (e.g. birds and beetles), maybe on parasitoid ovipositors
Persistence	Environmentally very stable except in UV light Persist outside host on leaves or bark from generation to generation of pest, or in soil for years Secondary epizootics may negate need for repeated applications
Efficiency	Very high in pest epidemics (ease of transfer of virus between hosts greatest in high pest densities) Enormous replication potential
Cost	Cheap to produce and bulk-up by rearing host insect, or from field collections Complex technology not usually required beyond semipurification
Drawbacks	Must be ingested; not plant systemic, limited or no use for sap-feeders or borers High host specificity: often not cross-infective from one insect species to another May not be available from nature Do not kill immediately; may be weeks until infected pests die Must be applied in the same way (timing and technology) as an insecticide Public and government distrust; unfounded worries about safety; overtight rules

because of NPV epizootics (Entwistle et al. 1983; Ahmed 1995; M.R. Speight, unpublished data). Whether naturally erupting, or purposely applied, NPVs have a variety of characteristics that render them very suitable indeed for certain types of biological control of insect pests (Table 12.13). Their major advantages over other types of natural enemy hinge upon their ability to remain potentially infective for considerable periods of time outside the host's body, and their enormous replication rate once ingested by the usually very species-specific target (Cory et al. 2000; Biji et al. 2006) (Figure 12.42). A control program in Colombia to manage defoliating Lepidoptera on cultivated palms illustrates this replication rate. The moth *Sibine fusca* (Lepidoptera: Limacodidae) was successfully eradicated by aerial spraying of a viral solution containing the equivalent of a mere 10 dead caterpillars per hectare (Philippe et al. 1997). It must be noted that though high levels of host species specificity may seem desirable, cost-effective control agents such as

viruses should ideally be pathogenic for multiple pest species (Bourner & Cory 2004).

Figure 12.43 illustrates the various pathways by which baculoviruses – NPVs in particular – are able to spread rapidly and independently from an epicenter of infection into the surrounding crops or habitat, where new hosts may encounter them, either immediately or during subsequent generations of the pest (Richards et al. 1998; Raymond et al. 2005).

First, infected larvae themselves may transmit NPV to clean substrates such as leaves and stems as they move around. Larvae of the cabbage moth, *Mamestra brassicae* (Lepidoptera: Noctuidae), when infected with a sublethal dose of NPV, live longer (Vasconcelos et al. 2005) and hence move further than non-infected ones. Their feces and other body secretions may contaminate new leaf surfaces where more susceptible young larvae can pick up a fatal dose. Other moth larvae move over much greater distances. Gypsy moth, *Lymantria dispar* (Lepidoptera:

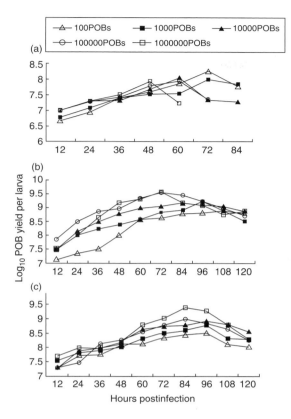

Figure 12.42 Growth of nuclear polyhedrosis viruses (NPVs) through different periods of time after infection in terms of $\log_{10}$ polyhedral occlusion body (POB) yield per larva against five HpNPV doses in three larval instars. (From Biji et al. 2006.)

Lymantriidae), larvae balloon away from high-density patches of their peers. Larval dispersal is a good predictor of early virus spread through the forest (Dwyer & Elkinton 1995). Other organisms may also help to disperse NPVs. During an outbreak of soybean looper, *Anticarsia gemmatalis* (Lepidoptera: Noctuidae), in southern USA into which a pest-specific NPV had been introduced, bird droppings in the area contained viable NPVs (Fuxa & Richter 1994). Because there is no vertebrate toxicity, NPVs are able to pass right through a predator's body, causing no ill effect to the animal. They emerge in the feces still infective, but at some distance from where the original larva was eaten. Even invertebrate enemies have

been shown to be effective vectors of NPVs. Again in the case of *A. gemmatalis*, NPV from the pest population was found in at least six species of predatory hemipterans, one coleopteran, nine species of spider, plus one hymenopteran and two dipteran parasitoids (Fuxa et al. 1993). Rain is well known to redistribute NPVs from soil reservoirs where they are protected from ultraviolet radiation, back onto plants. Rainsplash may also redistribute virus from the upper to lower branches of trees (D'Amico & Elkinton 1995) (Figure 12.44).

Optimism about the future of NPVs in crop protection must be slightly tempered with experience. For example, aerial application trials with a new NPV of the spruce budworm, *Choristoneura fumiferana* (Lepidoptera: Tortricidae), in Ontario, Canada, did not reduce tree defoliation, even after double, high-dose applications of virus (Cadogan et al. 2004).

Though public and professional resistance to the commercial use of insect viruses continues in some regions, including parts of Europe and America, NPVs are used regularly, commercially and over large areas of cropland in some parts of the world. We alluded earlier to one of the best examples, soybean looper, *Anticarsia gemmatalis*, the control of which in Brazil is carried out with NPVs over several million hectares of commercial soybean crops (Moscardi 1999). Figure 12.45 illustrates the successful use of NPV in Peru (Zeddam et al. 2003) against an oil-palm defoliator, *Euprosterna elaeasa* (Lepidoptera: Limacodidae). NPVs are undoubtedly a better solution both economically and environmentally. Clearly, these systems work very efficiently, especially when compared with conventional insecticidal spraying. They could be considered as models for the management of other crop pests, especially in Europe and North America where legislation requires environmentally benign control systems as alternatives to the large-scale use of chemical insecticides. Of course, like any other pesticide, stringent safety tests are necessary for all commercial use of baculoviruses. At a minimum, tests on mammals should include acute oral toxicity, dermal toxicity, eye irritation, and carcinogenicity. In practice, even detailed tests with NPVs such as the reaction of vertebrate embryos or the mutagenicity of human body cells have shown no responses (Liu et al. 1992). A final public concern might be that an NPV will mutate to become infective to vertebrates, including humans. However, DNA viruses do not mutate as readily as RNA viruses such as influenza or HIV.

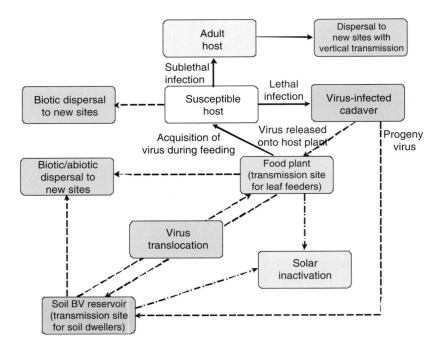

Figure 12.43 The fate of lepidopteran baculoviruses (BVs) in the environment. Solid arrows, BV transmission routes; dashed arrows, BV dispersal routes; dash-dot arrows, BV inactivation. (From Richards et al. 1998.)

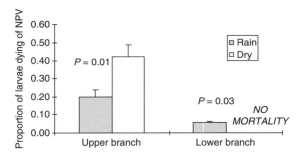

Figure 12.44 Mean proportion of *Lymantria dispar* larvae killed by nuclear polyhedrosis virus (NPV) when branches were exposed to rain. (From D'Amico & Elkinton 1995.)

They appear unlikely to make the leap from arthropods to mammals.

The next stages in the commercial development of baculoviruses center on techniques to increase virus productivity *in vivo* (i.e. in dead insect larvae; Subramanian et al. 2006) and then on the genetic engineering of natural NPVs to diminish some of their less desirable properties. The former systems may be fairly acceptable to consumers and environmentalists, the latter is less likely to be immediately popular. Nevertheless, research continues to decrease the time taken for NPVs to kill hosts (Cory & Myers 2003). Unlike chemical insecticides, which normally kill pests upon application, it may take many days or even several weeks before a host dies from NPV. This also generates a time lag before insect corpses break down to release PIBs back onto the crop for further infection. During that time, unacceptable feeding damage may still occur, especially on high-value crops. In an effort to speed up mortality after an NPV treatment, researchers have tried to genetically engineer baculoviruses. For example, the NPV of the alfalfa looper, *Autographa californica* (Lepidoptera: Noctuidae), has been modified to express an insect-selective toxin gene derived from scorpion venom (Cory et al. 1994). In field trials with the engineered (recombinant) NPV used against the cabbage looper, *Trichoplusia ni* (Lepidoptera: Noctuidae), larvae died

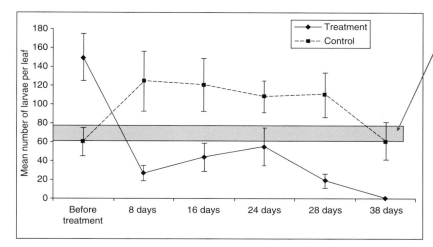

Note

Above the CRITICAL THRESHOLD up to 80% of leaf surfaces can be destroyed, resulting in 50% loss in production

NPV-killed larvae homogenized in distilled water and filtered through cheesecloth. Aerial treatments carried out through hydraulic nozzles attached to Cessna wings, at rate of 3×10^{13} PIBs/ha and flow rate of 50

Cost estimates of different biological and chemical treatments against oil palm caterpillar				
Treatment	Agent	Activity	Cost (US$/ha) product only	Product + application
Root absorption	Monocrotophos	Organophosphate insecticide	28	41
Ground application (fogger)	Alpha-cypermethirn	Pyrethroid insecticide	7	13
Aerial application	Flufenoxuran	Insect moulting inhibitor	10	14
Aerial application	EuclNPV	Virus	2	5

(Note :- Aerial application costs approx. US$3.3 per ha, irrespective of product)

Figure 12.45 Density of oil-palm leaf-eater caterpillars alter aerial treatment of nuclear polyhedrosis virus (NPV) in Peru. (From Zeddam et al. 2003.)

significantly faster than when treated with the natural (wild-type) virus. The virus yield from larval cadavers infected with the recombinant virus was significantly lower than that from the wild type, indicating that the engineered virus was inherently less virulent. Using a less virulent virus with an additional toxin may be beneficial, reducing the probability of transmission to other hosts. Similarly promising results have been achieved using genetically enhanced isolates of *A. californica* NPV that express insect-specific neurotoxin genes from spiders (Hughes et al. 1997). One species of spider utilized in this way belongs to the same genus, *Tegenaria*, as the benign European house spider. Engineered viruses are not yet deemed acceptable for widespread commercial use. The levels of public anxiety over the use of *Bt* suggest that recombinant viruses will be perceived with considerable

distrust. None the less, baculoviruses may prove to be a valuable weapon in the future arsenal of pest managers (Richards et al. 1998) and we should improve our knowledge of their genetics, molecular biology, and ecology so that we can use them safely and efficiently. Indeed, it is hoped that scientists and layman alike will realize that biopesticides such as NPVs and *Bt* will have a bright future (Inceoglu et al. 2006).

12.6 CHEMICAL CONTROL

The use of chemicals in insect pest management has come a long way since the indiscriminate use of organochlorine insecticides in the 1940s and 1950s. It comprises three distinct sets of tactics, one involving

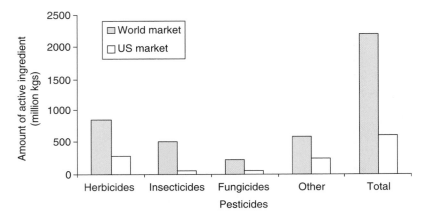

Figure 12.46 Amounts of active ingredient of pesticides used (2001 estimates). (From US EPA 2004.)

the use of (mainly) synthetic toxins, another that of insect growth regulators, and a third that of insect pheromones.

12.6.1   Insecticides

The use of pesticides (herbicides and fungicides as well as insecticides) has grown enormously. India, for example, produced somewhere in the region of 5000 metric tonnes in 1958, which has now risen to 85,000 metric tones (Gupta 2004). This is, however, still a low figure – India consumes a mere 0.5 kg/ha of pesticides, compared with Korea and Japan who use 6.6 and 12.0 kg/ha, respectively. This high use of pesticides is hardly surprising. Farmers, foresters, horticulturalists, and the communities that they support cannot tolerate the huge reductions in crop yields caused by the depredations of insect pests. In the developing world, gross yield is often the overriding priority, in attempts to generate sufficient quantities of food and fiber for growing populations. In the developed world, net yield and hence profit is often more important than gross yield. At the same time, most consumers in developed countries are intolerant of apples with grubs in them, cut flowers with leaf-miners, or strawberries full of holes. Organic farming strives to produce high-quality crops without resorting to chemical inputs, and organic fields have been shown to support significantly higher numbers of beneficial arthropods when compared with intensively managed ones (Berry et al. 1996). But with the

best will in the world, the pragmatist must accept that farmers will continue to need pesticides of various types to meet global food requirements. Indeed many developing countries in Africa and Asia, for example, still to this day have no real alternatives to pesticides, and a large number may not yet be aware of the existence of things like the natural enemies of insect pests (Badenes-Perez & Shelton 2006).

Figure 12.46 illustrates the magnitude of pesticide use in the world in 2001 (US EPA 2004). Insecticides made up over a quarter of the total market value of around US$22 billion, with the developed world using the lion's share.

History

The fact that we still use so much pesticide may come as a shock when you consider how long ago Rachel Carson wrote the book *Silent Spring* (Carson 1962). Carson's book is still often quoted, although seldom actually read, and has for all intents and purposes entered the realms of race memory. The basic arguments are plain enough. In 1944, *Scientific American* published a glowing testimonial to the new insecticide, dichloro-diphenyl-trichloroethane, otherwise known as DDT, saying that "painstaking investigations have shown it [DDT] to be signally effective against many of the most destructive insects that feed on our crops." Indeed, DDT appeared to be a godsend. Figure 12.47 illustrates how effective it was in virtually wiping out malaria in Italy (Casida & Quistad 1998).

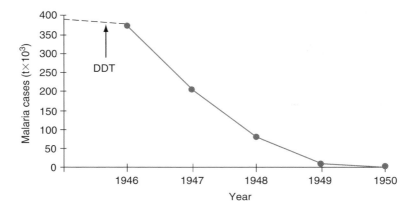

Figure 12.47 Impact of DDT on malaria cases in Italy. (From Casida & Quistad 1998.)

Toxicity

However, by the late 1950s, Carson and others were arguing that cyclodeine organochlorines (of which DDT is a member) were killing large numbers of insects, fish, birds, and mammals. The compounds have broad-range toxicity and, above all, persist for a long time in the environment, becoming concentrated in the tissues of organisms towards the top of food webs. Cyclodeine organochlorines are found in a whole range of localities, including human adipose tissue (Barquero & Constenla 1986), and dolphins (Tanabe et al. 1993), both scavenging and fish-eating birds (Ramesh et al. 1992), and in mayflies, bark, and leaves flowing in streams from tropical forest watersheds (Standley & Sweeney 1995). DDT is highly toxic to a myriad of organisms, including estuarine bacteria (Rajendran et al. 1990), zebrafish (Njiwa et al. 2004), and peregrine falcons (Thomas et al. 1992). Deposits of DDT and other organochlorine insecticides have been found in whale blubber (Stern et al. 2005), shellfish (Sankar et al. 2006), human mammary gland tissue (Munoz de Toro et al. 2006), and human breast milk (Bouwman et al. 2006). We should note, however, that DDT is much less toxic to humans in the short term than many of the newer and supposedly more desirable insecticides on the market today, though it is implicated in medical conditions such as memory loss (Ribas-Fito et al. 2006) and cancer (Morisawa et al. 2002) in humans.

For the sake of balance, it is worth noting that the deleterious effects of DDT and its relatives are variable depending upon climatic conditions. One fundamental criticism of organochlorines has been their long half-life – the time taken for 50% of the compound to degrade. In reality, organochlorines degrade at very different rates, depending upon the prevailing climate (Table 12.14). In Egypt for example, where DDT and its relatives have been banned for 25 years or more, residues can still be found in water and sediments (Mansour 2004). However, DDT does disappear eventually, and work in Californian salt marshes has shown that DDE and DDD accounted for more than 90% of DDT-derived compounds, indicating that DDT had degraded very significantly (Hwang et al. 2006). The combined effects of soil pH, photon fluxes, humic break down, and ultraviolet light can cause significant degradation of DDT (Quan et al. 2005), and there can be little doubt that the turnover of DDT in the tropics is significantly faster than that in temperate areas (Berg 1995), with a half-life measured in a few months rather than years. Therefore, in wet tropical countries, DDT is a completely different chemical from the notorious destroyer of ecosystems that we hear about in northern temperate regions (Samuel & Pillai 1989). It may be that organochlorines are relatively safe for certain types of pest management in the tropics where cheapness and human safety are paramount, such as for vector control (see Chapter 11). Whether or not the old organochlorine insecticides might be of some practical use in the 21st century, the fact remains that most countries in the world have or are in the process of banning their use.

Table 12.14 Published half-life of various organochlorine insecticides.

Compound	Country/region	Climatic regime	Half-life	Reference
BHC (HCH)	USA (assumed)	Temperate	5–6 years	Carson (1962)
Heptachlor	USA (assumed)	Temperate	4–5 years	Carson (1962)
DDT	Norway	Temperate (marine)	5 years $\pm$ 2.3 years	Skare et al. (1985)
DDT and HCH	Not specified	Temperate	6 months to 3 years	Rajukkannu et al. (1985)
DDT	Highland Indonesia	Tropical	159 days	Sjoeib et al. (1994)
DDT	Lowland Indonesia	Tropical	236 days	Sjoeib et al. (1994)
DDT and HCH	Southern India	Tropical	35–45 days	Rajukkannu et al. (1985)

With the demise of the organochlorines as safe insecticides, new compounds took their place. The organophosphates, a group of chemicals that includes highly effective chemical warfare agents, are nearly as old as the organochlorines. They now are used in enormous quantities worldwide, accounting for an estimated 34% of worldwide insecticide sales (Singh & Walker 2006). One such compound is chlorpyrifos, commercialized in 1965 and now thought to be one of the top five insecticides used in the world (Mori et al. 2006). Chlopyrifos poisoning in humans affects many parts of the body, including the central and peripheral nervous system, the eyes, the digestive tract, and the respiratory system.

The carbamate group of chemicals followed, accompanied by the pyrethroid insecticides, derived originally from naturally occurring compounds in flowers such as chrysanthemum. These are on the whole much less toxic to humans, but do have serious side effects on animals such as fish (Velisek et al. 2006).

Make no mistake, pesticides in general are harmful to human health. It has been reported that an estimated 1 million to 5 million cases of pesticide poisonings occur every year, resulting in 20,000 fatalities among agricultural workers. Most of these poisonings take place in developing countries, and although developing countries use 25% of the world's production of pesticides, they experience 99% of the deaths (UNEP 2004). In addition to mortality, most of the people who do not die (the majority of them children) suffer sublethal effects that can be serious and long lasting. Exposure to pyrethroid insecticides can result in skin irritations, while exposure to organophosphates can cause complex systemic illnesses resulting from cholinesterase inhibition (Rafai et al.

2007). People who apply the chemicals tend to suffer most, because of their use of inappropriate (and often unknown) compounds, by the use of incorrect or outdated technology, and by not using protective clothing. There is also the significant problem of pesticide companies in some countries who exert an unethical but tight hold on growers, where aggressive marketing and unscrupulous dealers are commonplace (Rengam 1992).

Resistance

Irrespective of the hazards to people, the insect pests themselves are becoming resistant to many insecticides, especially where repeated applications are required during one generation of a particular crop, such as cotton. Ironically, high pesticide use and subsequent resistance promote higher pest densities (Figure 12.48) with a vicious cycle of positive feedback demanding ever more pesticide to attempt control of ever-larger pest populations (Way & Heong 1994). For some pest species, insecticides have simply ceased to be a viable control tactic because of resurgence and resistance. For example, the tobacco budworm, *Heliothis virescens* (Lepidoptera: Noctuidae), a defoliator of numerous crops in the USA, is now partially or totally resistant to compounds in all the major groups of insecticides, including the organophosphates, the carbamates, and the pyrethroids (Elzen 1997). Table 12.15 summarizes the colossal incidence of resistance now found in many species of crop pest (Whalon et al. 2004). Incipient resistance in only a small percentage of the pest population can become widespread very quickly over a few generations if the efficiency of pest control is fairly low (Figure 12.49; and see below).

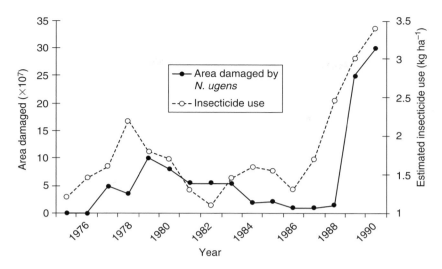

Figure 12.48 Relationship between increase in insecticide use and the area of infestation from the brown plant-hopper in Thailand. (From Way & Heong 1994.)

Table 12.15 Number of times various species of insect pest have been reported to show resistance to pesticides since the year 2000. (From Whalon et al. 2004.)

Pest species	Common name	No. of cases of resistance
Aedes aegypti	Yellow fever mosquito	21
Aphis gossypii	Cotton aphid	39
Bemisia tabaci	Tobacco whitefly	36
Helicoverpa armigera	Cotton bollworm/tobacco budworm	32
Leptinotarsa decemlineata	Colorado beetle	41
Musca domestica	Common house fly	46
Myzus persicae	Peach–potato aphid	68
Plutela xylostella	Diamondback moth	73

To make matters worse, disease vectors in the vicinity of crops that are regularly treated with pesticide may also become resistant, purely by association. We have already mentioned similar problems occurring in the insect vectors of human diseases such as malaria in Chapter 11. In China, for example, large quantities of the major classes of insecticides (organochlorines, organophosphates, carbamates, and pyrethroids) are applied both indoors and outdoors in attempts to control mosquito vectors in the genera *Anopheles*,

Aedes, and *Culex* (Cui et al. 2006). All have evolved resistance to most compounds – only some of the carbamates have yet to be affected. In an even more extreme situation, *Culex* larvae in Israel are now so resistant to the organophosphate chlorpyrifos that it has been withdrawn from use (Orshan et al. 2006). Finally, Oliveira et al. (2005) suggest a bleak future, because resistance to insecticides may take a long time to disappear once developed in an insect pest population. They feel that resistance is likely to be

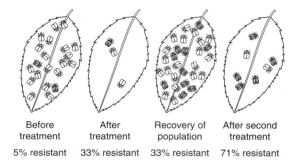

Before treatment | After treatment | Recovery of population | After second treatment
5% resistant | 33% resistant | 33% resistant | 71% resistant

Figure 12.49 Development of resistance to an insecticide, assuming a 50% kill of resistant phenotypes and 90% kill of non-resistant ones per treatment.

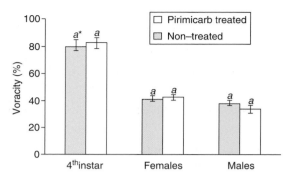

Figure 12.50 Voracity of *Coccinella undecimpunctata* fourth instars and adults (females and males) treated and not treated with pirimicarb, using *Aphis fabae* as prey. Means in each column for each developmental instars followed by different letters are significantly different at $P < 0.05$ (LSD test). (From Moura et al. 2006.)

maintained for long periods, since there are no obvious adaptation disadvantages with possessing it.

Effects on natural enemies

If and when crop producers are able to rely on natural enemies such as predatory or parasitic insects to assist with pest management, it is clearly important not to destroy these enemies using insecticides targeted at the pests themselves. The then shocking, but these days predictable, effect of insecticides on the natural enemies of cottony cushion scale are extremely well documented (see De Bach 1964). In more modern times, the search is on for insecticidal compounds that can act in harmony with predators and parasitoids so that chemical control can reduce or prevent epidemics, down to levels where biological control can "mop up" the rest. This of course is no easy task, since insect nerve poisons (which is what most of insecticides are) affect pests and enemies alike. Pirimicarb for example is a carbamate insecticide widely used for the control of pests such as aphids, and as gardeners and the eco-friendly profess, ladybirds (Coleoptera: Coccinellidae) are important aphid predators. Figure 12.50 shows that there is no difference between the impact of predation by the laybird *Coccinella undecimpunctata* on the black bean aphid, *Aphis fabae* (Hemiptera: Aphididae) (Moura et al. 2006), with or without sprays of pirimicarb. It has to be said that this is an unusual situation – broad-scale sprays of toxic chemicals are rarely able to kill pests whilst keeping enemies alive and well. Perhaps one light on the horizon is that some parasitic wasps may also be developing insecticide resistance (Baker et al. 1997).

Modern compounds

We should therefore be looking to more modern groups of insecticides to alleviate some of the problems with the old ones. Ideally, novel toxins should be under continual development and commercialization, but this is no longer a viable expectation. Because of insecticide resistance, and toxicological and environmental considerations, coupled with the cost of development and registration, the number of insecticides available for use has declined, and the number of new insecticides submitted for laboratory and field trials has dwindled even more (Gratz & Jany 1994). Essentially, while legislation and safety testing are indispensable when new chemicals are to be released into the environment (we do not want *another* "silent spring"), these mean that the cost of producing desirable but safe compounds becomes prohibitive, and agrochemical companies see little profit in the venture. Some new compounds do appear from time to time (Palumbo et al. 2001) (Table 12.16). A relatively recent addition to the insecticide armory comes from the chloronicotinyl (nictinoid) group of compounds. Imidacloprid is one of several new-generation systemic insecticides available to crop producers (Casida & Quistad 1998; Wu et al. 2006), with widespread efficacy against many pests from aphids to sugarcane white grubs (McGill et al. 2003). These compounds appear to be particularly effective in the control of plant viruses such as beet severe curly top by controlling their vectors, in this case sugarbeet leaf-hopper, *Circulifer tenellus*

Table 12.16 Relatively new chemical compounds with insecticidal properties. (From Palumbo et al. 2001.)

Chemical group	Active compound	Trade name(s)	Activity	Target insects
Chloronicotinyls/ neonicotinoids	Imidacloprid Acetomiprid Thiomethoxam	Admire, Provado Assail, Rescate Platinum, Actara	Neurotoxins, plant-systemic	Aphids, whiteflies, hoppers, thrips, beetle larvae, etc.
Pyridazinone	Pyridaben	Pyramite	Metabolic toxin, non-systemic, good residuality	Aphids, whiteflies hoppers, mealybugs, thrips
Phenylpyrazoles	Fipronil	Regent	Neurotoxin, systemic and contact good residuality	Coleoptera, Lepidoptera, Isoptera, Diptera, Homoptera
Amino triazinones	Pymetrozine	Fulfill	Antifeedant, slow acting contact and systemic	Aphids, whitelfly
Oxidiazines	Indoxocarb	Avaunt	Neurotoxin	Lepidoptera, Homoptera, thrips
Macrocyclic lactones	Spinosad	Success	Neurotoxin	Coleoptera, Diptera, Hymenoptera, Isoptera, Thysanoptera
Moulting disruptants	Buprofezin	Applaud	Chitin sysnthesis inhibitor	Hoppers, scales, whitefly, lepidopteran larvae
	Methoxyyfenozide	Intrepid	Insect hormone disruptors	
Juvenoids	Pyriproxyfen	Knack	Embryogenesis and metamorphosis suppressant	Scales, thrips, whitefly

(Hemiptera: Cicadellidae) (Strausbaugh et al. 2006). However, there were some signs that pests such as aphids may have been developing tolerance or resistance to these chemicals more than a decade ago (Nauen & Elbert 1997).

All the compounds considered so far have been synthetic. It is well known that plants produce a huge range of so-called secondary chemicals that act as defenses against herbivory (see Chapter 3), and one or two of these compounds are being considered as commercial insecticides. The best known is neem oil, a chemical produced from seed kernels of the neem tree *Azadirachta indica*, available commercially under the trade name Margosan-O. Neem oil has significant effects on a variety of pest insects. Larvae of the gypsy moth, *Lymantria dispar* (Lepidoptera: Lymantriidae), grow much more slowly when exposed to neem oil than do non-treated controls. In one particular experiment, most untreated larvae were in their last instar, ready to pupate, when the treated ones were still only

in the second and third instar (Shapiro et al. 1994). Neem was shown to be effective against citrus psyllid *Diaphorina citri* (Homoptera: Psyllidae) in Florida (Weathersbee & McKenzie 2005), and maize spotted stem-borer *Chilo partellus* (Lepidoptera: Pyralidae) in Kenya (Tekie et al. 2006). Trials of neem have also proved encouraging for the control of insects attacking brassicas, potato, and aubergine (eggplant) (el Shafie & Basedow 2003; Thacker et al. 2003). However, it seems unlikely that neem oil will replace the more toxic and effective synthetics in the near future.

A final group of insecticides includes the compound spinosad, essentially another natural insecticide, this time derived from an actinomycete fungus, *Saccharopolyspora spinosa*, found in soil. This product kills insects by contact or ingestion only, and has proved successful against a wide variety of crop pests such as flour beetle, *Tribolium castaneum* (Coleoptera: Tenebrionidae), in stored products (Toews & Subramanyam 2004), beet armyworm, *Spodoptera*

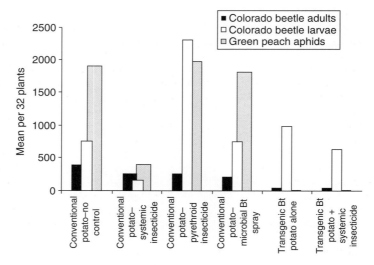

Figure 12.51 Abundances of Colorado beetle. *Leptinotarsa decemlineata*, and green peach aphid, *Myzus persicae*, on various types of potato crop in Oregon in 1993. Significant differences are not shown. (From Reed et al. 2001.)

exigua (Lepidoptera: Noctuidae) (Wang et al. 2006), and oriental fruit fly, *Bactrocera dorsalis* (Diptera: Tephritidae) (Hsu & Feng 2006). Spinosad is in additional thought to be beneficial because it has some degree of selectivity within insect groups. So, Galvan et al. (2006) found that adult ladybird *Harmonia axyridis* (Coleoptera: Coccinellidae), an important generalist predator of crop aphids in the USA, was tolerant to the chemical. Note, of course, that this effect will only be of use if the aphids are highly susceptible, and if the predator exerts a heavy impact on the prey.

Overall, we have to make do with the relatively few insecticides that we are still content to use, on the understanding that they remain our most powerful tool in pest management. It is therefore vital that we use them carefully and efficiently, to extend their useful lives for as long as possible.

Efficiency of insecticide use

Historically, the efficiency of insecticide use, measured as the percentage of total chemical applied to a crop that actually kills insects, has been woefully poor. On occasion, less than 1% of the chemical sprayed actually kills any insects. This incredible waste is expensive, and is compounded by the certain knowledge that the vast percentage of insecticide that does not kill the target will go somewhere else in runoff and

pollution. Obviously, modern pest management must address the problem of maximizing insecticide efficiency while minimizing waste. Integrating several tactics, namely choice of compound and its concentration, application timing, and application technology may achieve this ideal.

Choice of compound

Figure 12.51 illustrates how different types of insecticide vary in their efficiency in killing one well-known pest, the Colorado potato beetle, *Leptinotarsa decemlineata* (Coleoptera: Chrysomelidae) (Reed et al. 2001). Notice the differences in efficiency and methods of delivery of contact and systemic insecticides, as well as the comparisons with biological agents, in this case, *Bacillus thuringiensis*. Insecticides are predominantly nerve poisons, and they gain access to the nervous tissues of pests in various ways. Contact chemicals may be applied directly to the insect's body, or the target may acquire them as it roams over the plant surface. The poison is then absorbed through the cuticle. Fumigants may penetrate the pest's body through the spiracles, and compounds that cannot penetrate the epidermis can be ingested when the insect eats contaminated foliage or drinks the droplets from leaf surfaces. All of these types of compound are effective only on pests that easily come into contact with externally applied insecticides. To kill insects feeding on internal tissues of plants, such as sap-feeders, or pests contained within the plant such as borers, a systemic insecticide is required. Systemics are absorbed by the roots, stems, or foliage of the plant and are translocated in the sap vessels to wherever the insect is feeding. Because different insecticides have different characteristics, it is vital to choose one that is appropriate for the target pest's ecology and habitat. Using a contact-only insecticide against a boring beetle or moth larva will be doomed to failure.

Although we considered vector insect control in detail in Chapter 11, one point is worth emphasizing here. The choice of insecticide may influence not just the pest population, but also the spread of any pathogen with which it is associated. For example, Figure 12.52 illustrates that a seed dressing of the systemically active compound imidacloprid decreases the incidence of two pathogenic viruses vectored by bean aphids, with very significant increases in yield of field beans (Makkouk & Kumari 2000). Notice that the concentration of the insecticide used can be very important as well. Yield losses generally decline as concentration increases but there is no value in exceeding a certain dose. The increased effects on yield are negligible above, in this case, about 0.5 g active ingredient per kilogram of bean seeds.

Concentration of compound

You might think that a modern, environmentally sound, pest control program would use as little insecticide as possible, introducing minimum amounts of toxic compounds into the ecosystem. Indeed it can be seen that on some cases, the "recommended" dose of insecticide is unnecessarily high (Figure 12.52), in that a much lower dose can have the same effect, as Barcic et al. (2006) found when investigating the chemical and biological control of Colorado beetle, *Leptinotarsa decemlineata* (Coleoptera: Chrysomelidae) in Croatia. However, it is better to be safe than sorry; if the insecticide is applied in too low a concentration, too few pests may be killed. Let us take, for example, a population of aphids feeding on a cereal crop at the height of the plants' nutrient availability (the so-called milk-ripe stage), and during optimally warm, settled weather. Such an aphid might have a generation time of around 12–30 days (Zhou & Carter 1992), and an average fecundity per female (the entire population will likely be female), of approximately 40–80 nymphs (Kocourek et al. 1994). If a chemical spray kills 90% of the pest population, the 10% that survive will be largely freed from intraspecific competition. Within 2 weeks or so, the population might easily climb back to its prespray density. An added problem is the high

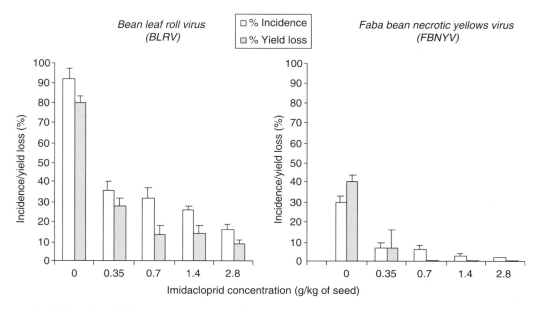

Figure 12.52 The effect of different concentrations of the insecticide imidacloprid applied to beans as a seed dressing on two aphid-vectored plant viruses in Syria. (From Makkouk & Kumari 2000.)

likelihood that any incipient resistance mechanism in the pest population would be concentrated in these survivors, so that the resulting generations would quickly and efficiently build up resistance to the insecticide. As it turns out, many examples of insecticide resistance in pests are likely caused by the failure to kill a large enough percentage of a pest population at an early stage in the selection for resistance. In the absence of any expected backup from natural enemies, a pest manager to should strive to kill a very high percentage of the pest population, up to and beyond 99% (see Figure 12.4a).

Timing of application

To kill as many pests as possible, and to reduce the damage done by them to a crop, it is critical to apply an insecticide at the right time. In basic toxicological terms, young insects are much more susceptible to poisons of a given concentration than are older ones. Take as an example the southern corn rootworm, *Diabrotica undecimpunctata howardi* (Coleoptera: Chrysomelidae). The adult of this pest is also known as the spotted cucumber beetle (Davidson & Lyon 1979), but it is the larvae that are the real problem

on a variety of crops from maize to peanuts. The tiny white larvae burrow into the roots and lower stems of crop plants, frequently causing the crop to lodge (fall over). Injury can be particularly serious to young plants, which cannot withstand much damage. In peanut culture in the USA, there is a choice of times to apply insecticides against this pest (Brandenburg & Herbert 1991): (i) a pre-planting treatment; (ii) during the flowering stage; or (iii) later as the crop matures. The actual yield of peanuts is not strikingly different among the timings, but the advantages of an earlier season treatment include: (i) less damage to the crop, because it is smaller and thus less affected by machinery; (ii) early-season control of other insect pests, not just rootworm; and (iii) fewer problems with non-target species because small plants allow soil incorporation of granular insecticides, so that predatory birds are less likely to pick up the poisons.

It is obviously important to target pests when they are most exposed to sprays. Given that many insect species spend most of their lives feeding in concealed locations, it is useless to attempt control when the likelihood of their coming into contact with poisons or pathogens is remote. Table 12.17 provides examples

Table 12.17 Examples of insect pests concealed for most of their active lives, for whom the accurate timing of spray application is very important.

Insect	Common name	Habit	Crop	Vulnerable life stage
Hylobius abietis	Great spruce weevil	Larvae under bark of stump roots. Adults girdle young transplants	Pine, spruce	Adult on young trees
Cydia pomonella	Codling moth	Larvae in fruit	Apples	Young larvae before bud entry
Rhyacionia buoliana	European pine shoot moth	Larvae in leading shoots	Pine	Young larvae before shoot boring
Euproctis chrysorrhoea	Browntail moth	Larvae partially in silk tents	Many shrubs and trees. Public health pest	Young larvae on foliage
Laspeyresia nigricana	Pea moth	Larvae in pea pods	Peas	Very young larvae on outside of plant
Phyllonorycter crataegella	Apple blotch leaf-miner moth	Leaf-mining larvae	Apple foliage	Adult before oviposition
Panolis flammea	Pine beauty moth	Young larvae feed head-down in needle bases	Pine	Older larvae feeding externally (note: applies to *Bt* control only)

of pests that are normally difficult to reach with chemicals, and for which application timing is critical. The key to accurate timing of an application often lies with accurate monitoring and prediction techniques that tell the grower when the vulnerable life stage of the pest is about to appear, or is at a peak. Many commercial monitoring systems employ pheromones to predict the best time to apply an insecticide (see below), but simple counts of appropriate life stages may also be employed. The tomato fruit worm, *Helicoverpa zea* (Lepidoptera: Noctuidae), feeds on the foliage and especially green fruits of tomatoes, rendering them unmarketable. The eggs of the pest are fairly easy to see on crops, and egg-scouting guarantees that insecticide treatment is carried out only when fruit worm eggs are found on the foliage (Zehnder et al. 1995). An application just as these eggs hatch ensures that the very young larvae are killed before any damage is done. This system is much preferable to routine spraying, and requires around 50% fewer insecticide treatments, saving about US$100 per hectare.

Application technology

Having chosen the appropriate insecticide and the concentration and time at which to apply it, the fourth vital decision concerns the method by which it is applied to the crop. Figure 12.53 shows just how complex spray application can be in terms of influencing the effectiveness of not only killing pests but also causing side effects such as environmental pollution and, most importantly to crop producers, wasting money (Al-Sarar et al. 2006). Nowhere else in applied ecology does the need to combine widely differing sciences come so much to the fore: physics, engineering, chemistry, and biology are integrated into a package that optimizes the delivery of the insecticide to the target while minimizing waste and off-target pollution. There are many ways of delivering a pesticide to a target, including the use of dusts, smokes, granules, and liquid-based sprays. Perhaps the most widespread method is the last system, where the insecticide itself, the active ingredient, is dissolved or suspended in an oil- or water-based formulation, and delivered to the target insects through some sort of machine.

The type of machine is critical to the success of the operation, and old-style technologies based on a hydraulic nozzle principle are still commonly used. Hydraulic nozzles force liquid under pressure through a constriction, producing a supposedly fine spray or

mist of droplets. However, a typical hydraulic nozzle produces droplets in a wide variety of sizes, whereas a spinning disk system produces drops in a very much narrower size spectrum. There is an intimate relationship between the size of droplets in a spray cloud, the volume of spray used per unit area of target, and the efficiency of the operation (Matthews 1992). As can be seen from Figure 12.54, smaller droplets cover a much larger area of substrate (Al-Sarar et al. 2006); in this example it was clearly shown that larvae of the fall armyworm, *Spodoptera armigera* (Lepidoptera: Noctuidae) were less susceptible to cypermethrin and developed resistance to it faster when sprayed with larger (volume median diameter 519 μm) drops than with smaller ones (volume median diameter 163 μm). Thus, the size of drops is of paramount importance (Table 12.18).

For a variety of reasons, which include the fluid dynamics of differing particle sizes in turbulent or linear airflow, only a small size range of insecticide droplets actually come into contact with, and therefore kill, insect pests. The optimal droplet size for insects on foliage can be as low as 30–50 μm. Anything much larger does not penetrate crop canopies well and, if the drops do manage to contact foliage, they do not readily adhere because of their size. Very tiny drops, on the other hand, will tend to stay in air suspension even at very low wind velocities (Hiscox et al. 2006). Moreover, they contain such small amounts of active ingredient because of their small volume that they may be unable to kill adequately.

The majority of drops from a normal hydraulic nozzle are well outside the size range for killing insects efficiently (Matthews 1992), and 5% or less may be in the desired range. Hence, most of a conventional application of insecticide is very unlikely to kill pests. When adequate control is achieved by hydraulic spraying, we can probably infer that the same result could have been achieved with much lower quantities of insecticide, using droplets only in the 30–50 μm size range. A technology called controlled droplet application (CDA) does just this, and allows the use of ultralow volumes (ULV) of pesticide, often as little as 1 L/ha. ULV/CDA applications also tend to be much more efficient (Table 12.19). CDA technology is based on two types of machine, one of which relies on electrostatic charging of spray droplets, and the other on rapidly spinning disks, the so-called centrifugal nozzle. The latter system is much more widely used than the former. Oil formulations of the active ingredient are

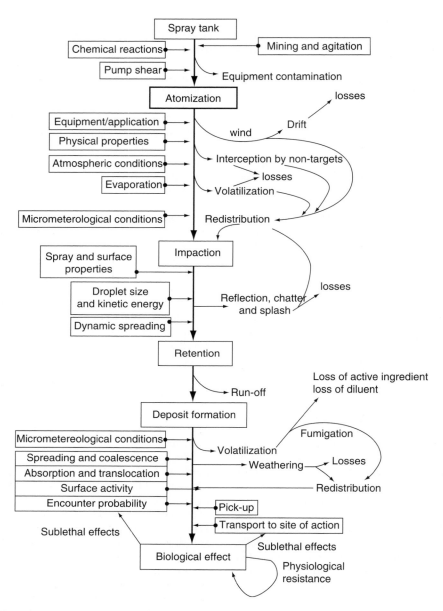

Figure 12.53 Dose transfer process: a roadmap for the delivery of toxicants to target organisms in agro-ecosystems. (From Al-Sarar et al. 2006.)

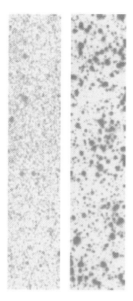

Figure 12.54 Small droplet coverage (163 μm diameter, left) and large droplet coverage (519 μm diameter, right). (From Al-Sarar et al. 2006.)

Table 12.18 Optimal droplet size for pesticides used on a range of targets. (From Matthews 1992.)

Target	Droplet diameter (μm)
Flying insects	10–50
Insects on foliage	30–75
Foliage	40–100
Soil (avoiding drift)	>200

Table 12.19 Aerial spraying against pine beauty moth showing the fate of fenitrothion.

	Ultralow volume (g/ha)	Low volume (g/ha)
Insecticide applied	300	300
Amount lost outside target area	3	60
Amount lost to ground	13.5	115
Amount collected by target	283.5	125
% Insecticide collected by target	94.5%	41.7%

gravity or pump fed through flow-rate constrictors onto metal or plastic disks spinning at high velocity. The liquid is flung to the edges of the disks by centrifugal force, where it streams off the disks in filaments that fragment into droplets. The size of these droplets depends mainly on the flow rate of the liquid, and the velocity of the spinning disk. In reality, such machines are not actually sprayers at all; they are merely droplet producers. Some rely on air currents to disperse the drops within the target crop, whereas others provide a gasoline- or battery-driven fan to blow the drops onto the target. These air-assisted ULV machines permit the application of insecticides at considerably lower volumes than conventional technologies (Mulrooney et al. 1997).

A variety of ULV/CDA machines are now available, from simple and cheap handheld devices to complex and powerful fixed-wing- or helicopter-mounted rigs for large-scale applications. Electrostatic machines which produce tiny charged particles of chemical, whilst very efficient in certain circumstances (Kirk et al. 2001), have not met with much commercial success. An advantage of using size-controlled drops

in a spray program is the reduction in waste and lower economic and environmental impact. However, small droplets can be subject to spray drift. Figure 12.55 illustrates the mortality of an aphid parasitoid, *Aphidius colemani* (Hemiptera: Aphididae), in kohlrabi crops sprayed with the carbamate insecticide lannate (methomyl) (Langhof et al. 2003). In this trial, modern hydraulic nozzles (jet ID 120–025) were mounted on a standard tractor boom rig, and the chemical was applied at a rate of 0.45 kg/ha in 400 L water (medium rather than low volume). Wind speed averaged 3.6 m/s, with the experimental drift areas downwind of the treated plots. Natural enemy mortality was still significant within 1 m of the sprayed area. However, by 2 m distance, the mortality on the day of spraying was approximately five times less. In general, the smaller the droplet, the further we might expect it to drift downwind. However, it is worth considering two

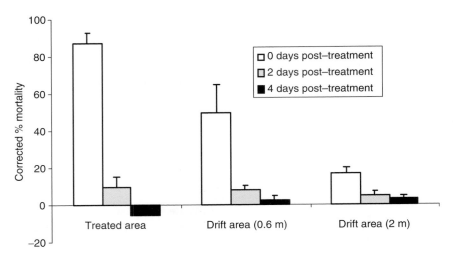

Figure 12.55 Mean corrected mortality of the parasitoid *Aphidius colemani* exposed to residues of the insecticide Lannate in treated and adjacent drift areas of a kohlrabi crop. (From Langhof et al. 2003.)

additional points. First, the behavior of tiny particles in turbulent rather than laminar airflow suggests that very small drops may drift less downwind than do larger ones. Second, even if very small drops drift off target, each individual one contains very small quantities of toxin. Modern designs of spray nozzles can reduce drift but sometimes at the cost of killing fewer insects (Lesnik et al. 2005). A final solution might be to change spray nozzles or other items of equipment close to field margins to minimize contamination (Matthews & Friedrich 2004).

We do not mean to suggest that older spray technologies are of no value. In some developing countries, managers still use classic hydraulic nozzles, usually from leaky and unreliable, but nevertheless cheap and familiar, knapsack sprayers (Matthews et al. 2003). They may achieve the same level of pest control, but often at much greater cost in terms of chemicals, time, labor, and environmental impact. As Figure 12.56 shows, the use of a CDA spinning disk machine for the control of various groundnut pests in India does not produce strikingly different results when compared with a conventional knapsack sprayer. However, the crucial point is that there is a very much reduced demand on labor and costs with the CDA machine. Spinning disks are particularly appropriate when large volumes of water are in short supply, or the crop to be treated is isolated.

Clearly, if 1 L/ha will do the same job as 100 L/ha, then the former has to be preferable, everything else being equal.

As with all systems, CDA/ULV does have its drawbacks. For example, the insecticides used in the newer machinery are more complex and expensive, often oil-based formulations, which may not be easily available in many countries. They are an alien technology to many growers, from British farmers to subsistence growers in the third world, who may well be happier relying on their well-tried (though probably much less efficient) traditional machines. Furthermore, the use of CDA technology requires more care and thought. Wind speed and direction have to be monitored carefully, and as the droplets are so small, they can frequently be hard to see, thus making attempts to cover all the target crop uniformly rather haphazard. On the plus side, most application systems for pathogenic organisms such as bacteria, and especially viruses, now also use CDA spinning disk technology. Clearly, spray programs in different crop systems against different pests have specific requirements, and hence the future for pesticide application may lie with variable rate application (Liu et al. 2006a), where computer-controlled and laser-guided machines produce just the right compound and introduce it to the optimum place in the crop to do maximum pest damage.

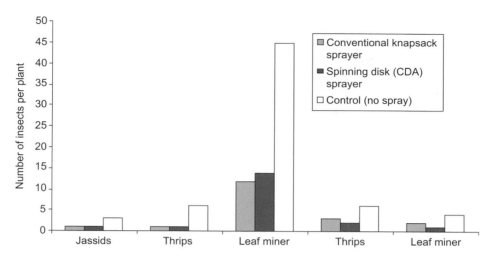

Figure 12.56 Number of insect pests per groundnut plant in India 36 h after spraying insecticides with two types of spray machine.

Conclusions

It should be clear by now that there are many situations where the use of insecticides in pest management is necessary and appropriate. Current pest management in tomato crops in Virginia, USA, illustrates this rather well. A whole host of insects feed on mono-crop tomatoes in Virginia, including cutworms, Colorado potato beetle, tobacco hornworm, cabbage looper, psyllids, aphids, stinkbugs, flea beetles, whitefly, and wireworms (Nault & Speese 2002). The grower's reaction to such a barrage of pests is to utilize management tactics that minimize effort and pest damage, while maximizing economic return. Insecticides remain the economically sensible and preferred method of pest management, despite the availability of other, more environmentally friendly techniques. The percentage of insect-damaged fruit in insecticide-treated plots is no more than 0.5%, whereas fruit yield in untreated plots drops by 33% in both spring and fall tomato crops. The loss in return per hectare in an untreated tomato field would range from US$3015 to $17,883 in the spring crop, and US$2555 to $11,074 in the fall crop. This provides an estimate of the impact of insect pests on tomato yield in the absence of pest control. In fact, tomato growers in Virginia spend less than 5% of their overall production costs on insecticides, despite the high frequency of insecticide use. For example, the cost of

treating a 1 ha field 15 times (i.e. one application per week from transplanting to harvest) with a broad-spectrum insecticide would be US$569, including application costs, whereas the overall production costs for fresh-market staked tomatoes are more than US$15,000 (production, harvest, and marketing costs). Although there are potentially large risks of insecticide resistance, no resistance had developed by 2002. Given the ongoing risks of fruit loss to insects, and the relatively low cost of using insecticides, growers are unlikely to deviate from calendar-based spray programs. Perhaps the only concessions to environmental risk are the increased use of biologically based pesticides and pest monitoring (scouting) (Nault & Speese 2002).

12.6.2   Insect growth regulators

In relatively recent years, companies have developed much more subtle types of insecticide that interfere in various ways with the growth and development of insects. Insect growth regulators (IGRs) can be split into two groups: the juvenile hormones which regulate the control of larval development and pupation, and the molting hormones (ecdysones) which control skin shedding in nymphs and larvae (see Chapter 7). Hydroprene is a juvenile hormone analog used commonly in controlling stored product pests

(Mohandass et al. 2006). Like others in the group, this compound has very low mammalian toxicity, and disappears quickly from the environment. Ecdysones, such as diflubenzuron, inhibit chitin production, which prevents larval and nymphal ecdysis. They also have ovicidal properties, where treated eggs fail to hatch (Cloyd et al. 2004). They have been used on a wide range of pests and cropping systems, from orchards to forests, and mushroom growing to vector control. IGRs normally have to be ingested by target insects, although even sap-feeding scales and aphids have been controlled. Indeed, Grafton-Cardwell et al. (2006) suggest that IGRs such as pyriproxfen control the notorious California red scale, *Aonidiella aurantii* (Hemiptera: Diaspididae) extremely well, though the same authors found that cottony cushion scale, *Icerya purchasi* (Hemiptera: Margarodidae), was only incompletely controlled by the same chemical. Another class of growth regulators, the non-steroidal ecdysones, induces premature and lethal molting in lepidopteran, dipteran, and coleopteran larvae (Zhao et al. 2003b). Field trials with ecdysones in cotton crops have decreased populations of bollworm larvae by 68–73%, and damage to apical cotton buds by 57–77%. As can be seen in Figure 12.57, spraying low volumes of the ecdysone tebufenozide (Mimic®) against larvae of the autumn gum moth, *Mnesampela privata* (Lepidoptera: Geometridae), on eucalyptus trees in southern Australia resulted in 100% mortality, though only after 3 weeks post-spraying (Elek et al. 2003). This delayed effect might be cause for concern were it not for the fact that pest insects coming into contact with

the chemical ceased feeding almost immediately. Aerial spraying using spinning disk technology was found to control 90% of the larvae, with a much reduced impact on natural enemies.

As usual, there may be side effects. IGRs such as diflubenzuron and its relatives can readily affect the chitin synthesis of any insect with which they come into contact, so economically important species such as silk worms, *Bombyx mori* (Lepidoptera: Saturnidae), and beneficials such as bumblebees, *Bombus terrestris* (Hymenoptera: Apidae), can be severely affected (Mommaerts et al. 2006). One final point: even with ecdysones, prolonged exposure of insect populations has resulted in some species, such as the Australian sheep blowfly, *Lucilia cuprina* (Diptera: Calliphoridae), becoming tolerant (Kotze et al. 1997).

12.6.3   Pheromones

Many animal phyla, both terrestrial and aquatic, communicate with members of their own species using complex chemical signals, known as pheromones. Aggregation pheromones may be employed to concentrate pests in areas where they can be dealt with by other control methods, whilst anti-aggregation pheromones are used to protect vulnerable or susceptible crops such as stressed trees in a forest (Borden et al. 2006). Chemical communication for the purposes of finding a mate probably reaches its quintessence in the insects (Wyatt 2003). Pheromones are species-specific cocktails of chemicals (Symonds & Elgar 2004) and, in most cases, work by air currents delivering volatile "plumes" downwind from a point source, which is very often the female, or less often the male, who is ready to mate. These sex-attractant pheromones are widely used in the management of insect pests, either by luring adults to artificial pheromone sources, or to broadcast sex attractants through a crop so that the point source system is overwhelmed.

Sex pheromones have now been identified for a very large number of insect species. A publication on the internet, "Pherolist" (Witzgall et al. 2004) (Figure 12.58) cites over 670 genera from nearly 50 families of Lepidoptera for whom the pheromones are known. Most pheromone compounds are of a rather low molecular weight, highly volatile, and species specific. They may convey complex signals about the location of a mate or host, but in general are easy to synthesize in a laboratory, and an enormous range

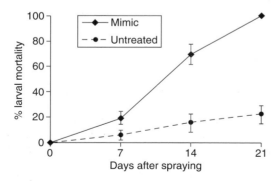

Figure 12.57 Field mortality of *Mnesampela privata* larvae after hand-spraying 172g Mimic® in approximately 10 L water/ha.(From Elek et al. 2003.)

HOME PRODUCTS PHEROLIST RESEARCH CONTACT

Phero Net

CERTIFIED LURES FOR INSECT MONITORING

Pherolist

Search

Family
Genus
Species
Common name

Compound

Source
Author

Help
Contribute

Taxa > Lepidoptera > Lymantriidae > Lymantria > < >

Lymantria dispar Linnaeus

gypsy moth - lövskogsnunna - Schwammspinner - bombyx disparate

© Ernst Priesner

Displaying reference **1** to **3** out of **3**

cis7,8epo-2me-18Hy	P	Bierl, 1970, Sc, 170:87
cis7,8epo-2me-18Hy	A	Beroza, 1973, EnvEn, 2:966
7R8Sepo-2me-18Hy	P	Cardé, 1977, EnvEn, 6:768

Figure 12.58 Extract from Pherolist (www-pherolist.slu.se). (From Witzgall et al. 2004.)

of synthetic compounds are now available on the market for a whole host of insect pests. Cocktails of pheromone compounds and their allies have enormous effects on insect behavior (Kawazu et al. 2004) (Figure 12.59). To be useful in pest management, the design of the lure or trap is almost as important as the correct pheromone chemistry, and many trap designs are now available. The chemicals themselves are usually placed in slow-release formulations, such as plastic or polythene capsules, microfibers, or rubber bungs, so that the active ingredient is slowly released into the habitat. The traps in which these lures are located have to be of a design whereby air currents flow readily through them to take the attractants away, and many also have some form of capture device such as a sticky insert or bottle, to retain the insects that are lured to the point source. In many cases, the design of the trap is equally important as the chemicals placed inside (Mulder et al. 2003). Crucially for management, there has to be a strong, dependable

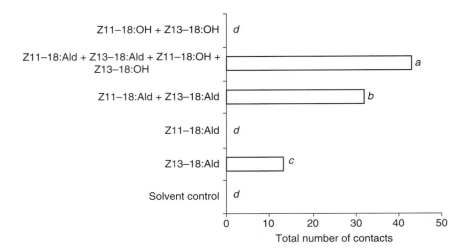

Figure 12.59 Effects of the composition of synthetic pheromone components on the responses of adult male *Cnaphalocrocis medinalis* in the lab. Means with the same letter are not significantly different at $P < 0.05$ by the Tukey–Kramer test. (From Kawazu et al. 2004.)

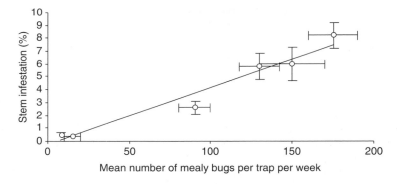

Figure 12.60 Seasonal average stem infestation of *Planococcus ficus* in relation to the number of males caught in pheromone-baited traps in vineyards. (From Walton et al. 2004.)

and above all predictable link between insects caught in pheromone traps, and the infestation or damage estimates caused by the pests. Figures 12.60 and 12.61 illustrate such relationships for two distinctly different insect pests, phloem-feeding mealybugs (Walton et al. 2004) and defoliating moth larvae (Morewood et al. 2000). Notice from the data for *Lymantria monacha* (Figure 12.61) that both trap design and pheromone concentration influence the reliability of the prediction system.

Sex pheromones can be used in several ways in pest management, namely mass trapping, monitoring, and mating disruption.

Mass trapping

The concept of mass trapping using sex-attractant pheromones is simple enough. Sufficient adult insects (normally male) are lured into traps where they are collected and killed, so that the likelihood of females in

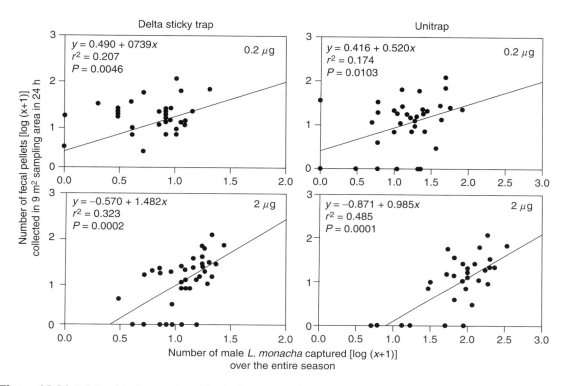

Figure 12.61 Relationships between larval fecal pellet counts and numbers of male *Lymantria monacha* caught in two types of pheromone trap in Germany and the Czeck Republic. (From Morewood et al. 2000.)

the crop mating to produce the next generation of pests is very much reduced, or, ideally, removed altogether. A fundamental problem with this idea is that most male insects are polygamous, so that even if only a few survive, they may be able to fertilize a large number of females. Furthermore, the immigration of new males from adjacent habitats to fill the vacuum left by mass trapping may result in failure, unless the crop or stand is isolated (el Sayed et al. 2006).

None the less, successes have been achieved with mass trapping. In oil palm plantations in Costa Rica, nearly 250 pheromone traps were located in a 30 ha area, and the numbers of the weevil pest, *Rhynchophorus palmarum* (Coleoptera: Curculionidae), were counted in the traps for 17 consecutive months (Oehlschlager et al. 1995). The weevil larvae burrow in the crowns of growing palms, feeding on the young tissue and causing the destruction of the growing point. By the end of the mass trapping period, over

60,000 adult weevils had been removed from the crop, and expected outbreaks of the pest during the following dry season did not occur. For mass trapping to have some chance of success, the number of traps per unit area must be considerable. In China, for example, Wang et al. (2005) used 25 pheromone traps per hectare of tea plantation in attempts to mass trap males of the tea tussock moth, *Euproctis pseudoconspersa* (Lepidoptera: Lymantriidae). In 2 years, they caught nearly 150,000 male moths, and managed to reduce larval densities in the crop by 39–51%. Though this level of pest mortality might not be sufficient on its own, mass trapping could certainly be considered as part of a management scheme where insecticide use could be significantly reduced.

Monitoring

The most widespread uses of insect sex pheromones and aggregation pheromones in the crop protection

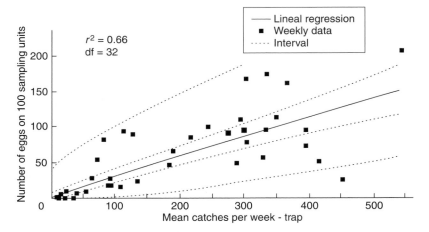

Figure 12.62 Relationship between the mean weekly catches of *Helicoverpa armigera* in pheromone traps and egg density in carnation fields. (From Izquierdo, 1996).

world are as tools for monitoring pest densities (Jactel et al. 2006). With increasing concerns over resistance development and pollution, most growers now restrict spraying to periods when pest densities are economically high. Traps are used to estimate the numbers of adult insects on a day-by-day or week-by-week basis. These estimates are then related to dependable economic threshold figures (see later), and decisions made about if and when to treat. As mentioned above, of crucial importance to the success of any monitoring program is establishing the link between the numbers of adult insects caught in a trap and the expected density of the pest population in the next generation. Figure 12.62 illustrates one example of this relationship for the bollworm, *Helicoverpa armigera* (Lepidoptera: Noctuidae), in southern Spain, where it is a serious pest of tomatoes and carnations (Izquierdo 1996). The relationship between mean weekly captures of males in pheromone traps and the numbers of eggs laid in carnation fields is significant, though the data are rather variable. Plant phenological and weather data can also be added to the prediction system to improve the regression fit shown in the figure, but there is clear potential for this monitoring tool. For example, an average of 300 males per trap per week indicates that the grower should expect between 50 and 100 eggs per 100 sampling units later in the season. The next step, of course, is to relate such a density to a prediction of economic damage. In a rather similar monitoring

program in Finland, this time for European pine sawfly, *Neodiprion sertifer* (Hymenoptera: Diprionidae), Lyytikainen-Saarenmaa et al. (2006) found that pheromone trap captures of 800–1000 or higher during a season indicated a risk of moderate defoliation the following year. Though both examples appear to make predictions of pest damage based on somewhat imprecise data, pest managers are provided with sufficient early warning of the likelihood of damage, and also its potential severity.

The practice of relying on pheromone trap captures to predict pest problems to come is commercially viable in a number of other crops too, from timber yards to apple orchards and pea fields. Timber storage yards, by their very nature, provide massive quantities of highly suitable breeding material for bark- and wood-boring beetles, in the form of logs waiting to be processed. Although felled logs can no longer be damaged economically by bark beetles (Coleoptera: Scolytidae), they can still produce enormous numbers of beetles that can mass attack standing trees in the vicinity of the timber yard. Furthermore, wood-borers such as some of the ambrosia beetles, *Trypodendron* spp. (Coleoptera: Platypodidae), cause extensive degradation of stored logs, not only by their tunneling in the timber itself, but also by their inoculations of blue-stain fungi into the wood (see Chapter 6). Pheromone traps baited with synthetic aggregation pheromones for *Trypodendron lineatum* and *Ips typographus* are used routinely in wood yards all over Europe

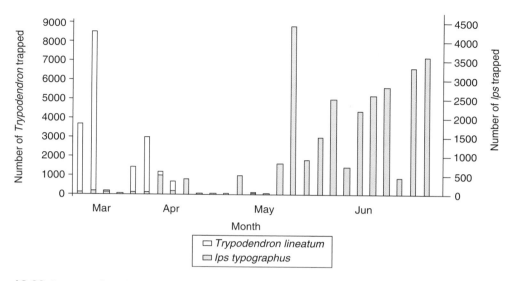

Figure 12.63 Captures of two beetle species in aggregation pheromone traps in a woodyard in Slovenia. (From Babuder et al. 1996.)

to monitor densities of these pests in the locality. One example from Slovenia is illustrated in Figure 12.63 (Babuder et al. 1996). If large numbers of beetles are detected during the monitoring procedure, action is taken to protect the timber in the yard, such as debarking, water storage, or insecticidal sprays.

Mating disruption

As male moths usually find their mates by following the female's pheromone plume upwind to its source, it is possible to reduce matings by broadcasting synthetic sex pheromone through a crop. Disruptants work in various ways. Males may become habituated to the sex attractant, which then loses its "appeal"; synthetic lures may "compete" with calling females, so that males are attracted to the wrong place; or the synthetic may simply smother and camouflage the females' less powerful scent. Many different formulations of synthetic sex pheromone are now available for disruption, which include polyvinyl chloride (PVC) beads, twisted-tie or wire "ropes", flakes, wafers, or polythene microtubules all saturated with the sex pheromone. These can either be applied from the air to broadcast the disrupting chemicals, or attached to trees or stakes within a crop to provide point sources (de Lame & Gut 2006).

As with most pest management techniques, we might expect both failures and successes. In the case of codling moth, *Cydia pomonella* (Lepidoptera: Tortricidae), attacking organically grown apples in Ontario, Canada, mating disruption using pheromones reduced the number of adult male moths found in the crop by between 75% and 96% when compared with control plots (Trimble 1995). However, the pheromone treatment did not prevent an increase in codling moth damage and, in fact, pest larval densities rose considerably. A very similar trial in British Columbia, Canada, is in stark contrast to this. The synthetic pheromone of codling moth, codlemone, was released at varying concentrations into orchards over a 6-month period. Fruit damage ranged from 0% to 1.5% overall in the treated orchards, but reached as high as 43.5% in untreated controls (Judd et al. 1996). Why the results of these two trials on the same insect using very similar tactics produced conflicting results is rather difficult to explain. It is, of course, important to remember the pest's ecology. A low-density pest such as codling moth can still cause economic damage by larval tunneling in apples and other pome fruits, even when adults are at fairly low densities. (Later in this chapter, we provide a more detailed discussion of codling moth IPM.)

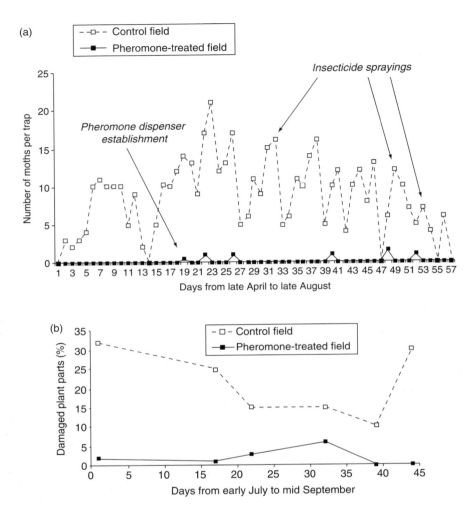

Figure 12.64 (a) Pheromone trap captures of a adult male *Pectinophera gossypiella* in cotton fields with and without mating disruption treatment. (b) Damage in cotton fields caused by *P. gossypiella* with and without mating disruption pheromone dispensers. (From Lykouressis et al. 2005.)

None the less, pheromone mating disruption can be commercially successful. One of the most notorious cotton pests is the pink bollworm, *Pectinophera gossypiella* (Lepidoptera: Gelechiidae). The larvae of the moth are well protected from conventional insecticides because they feed within flower buds and cotton bolls, and thus the pest is particularly difficult to control. Figure 12.64 illustrates results from a mating disruption program against pink bollworm in Greece, showing that disruption really can work on a large

commercial scale (Lykouressis et al. 2005). Pheromone rope dispensers, 1000 per hectare, reduced the number of male pink bollworm moths to virtually zero, accompanied by a huge reduction in damaged plants without the need for copious chemical sprays. Along with reduced pesticide input come the benefits of less resistance in the pest population, less harm to neutral and beneficial insects, and even a decrease in direct phytotoxic effects of the insecticides on the crop itself.

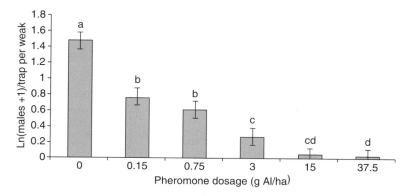

Figure 12.65 Male gypsy moths ($\pm$ SE) recaptured in plots treated with various doses of pheromone in Virginia in 2002. Bars with the same letter are not significantly different. (From Tcheslavskaia et al. 2005.)

Even more notorious, to foresters at least, is the gypsy moth, *Lymantria dispar* (Lepidoptera: Lymantriidae), probably the most serious defoliator of broadleaved trees in North America. Tcheslavskaia et al. (2005) were able to reduce the mating success of female adult moths by over 99% using the synthetic pheromone disparlure in Virginia, USA. The formulation of pheromone consisted of plastic flakes (3 mm $\times$ 1 mm $\times$ 0.5 mm) composed of PVC outer layers and an inner polymer layer containing disparlure. The flakes were mixed with diatomaceous earth to reduce clogging and were applied from a fixed-wing plane at various doses. As Figure 12.65 shows, the number of male moths caught in pheromone traps was virtually zero at and above a dose of 15 g pheromone active ingredient per hectare, a very good indication that disruption as working well. Above a certain dose, no further effect can be expected, though it is possible that higher concentration of pheromone may prolong the efficacy of mating disruption (Gillette et al. 2006).

12.7 INTEGRATED PEST MANAGEMENT

We are now ready to bring together all of the individual techniques and tactics for insect pest management into a coherent plan. We can hopefully combine the most appropriate characteristics of each technique to develop an ecologically sound, economically viable package, called integrated pest management (IPM). Various definitions of IPM are available. One that we

favor suggests that IPM is a decision support system for the selection and use of pest control tactics, singly or harmoniously coordinated into a management strategy, based on cost–benefit analyses that take into account the interests of and impacts on producers, consumers, and society, which focuses on the pest's biology, and how it interacts with the environment, natural enemies, and the crop itself. All these factors must also be judged in the context of economic injury levels.

Figure 12.66 illustrates a framework for IPM (Speight 1997a) including the various management options and their relationships to one another. The management options can be considered as tools within a hypothetical toolbox, each tool representing one of the management tactics that we have considered in sequence in this chapter. Clearly, not all tactics are available or even desirable for every IPM program, just as not all tools in a toolbox are required for every home repair project. However, elements of each will be apparent in the examples that we provide later in this chapter. First and foremost, IPM is a preventative strategy, which relies on control measures only if and when prevention fails for some reason. As Figure 12.66 suggests, if the tactics in stages A and B are adopted efficiently, then stage D may never be required. Complacency must be avoided, however, so that stage C links prevention (A and B) with action (D) via monitoring and predictive systems that depend on a sound knowledge of economic thresholds (see below). IPM can also be viewed as a flowchart (Figure 12.67). This scheme, designed for tropical

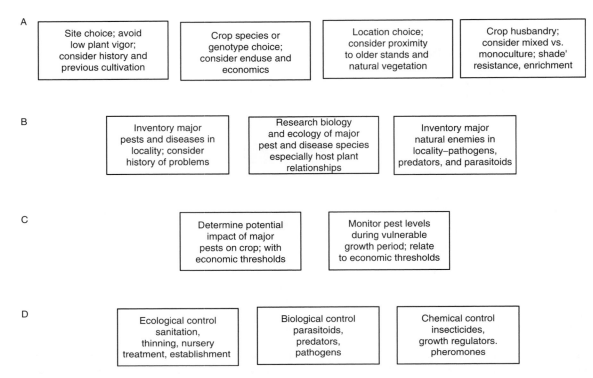

Figure 12.66 Theoretical components of a generalized IPM system. Stages A and B are entirely preventative, stage C involves monitoring and prediction, whilst stage D covers control strategies which are available should prevention fail, or monitoring reveal high risk.

forestry, provides a sequence of events, or management decisions, in a complete package that prevents, or failing that, controls, a pest problem. Note that for a management program to fit the definition of IPM, it would include, at bare minimum, a monitoring and prediction system that might entail no further action.

Pragmatically, IPM is usually adopted with enthusiasm only when other, essentially easier tactics have failed, or proved to be too expensive. IPM is particularly attractive when it can be shown to save money. For example, Trumble et al. (1997) analyzed the profits to be made using an IPM program in celery crops in California. They compared the standard system of chemical pesticide use, which consisted of nine applications through the crop's life, with an IPM strategy based on sampling and pest threshold assessments followed by three or four treatments with *Bacillus thuringienis* when required. In a commercial operation, the IPM program generated a net profit of over US$410 per hectare compared to standard chemical use, with the added bonus of approximately 40% reduction in pesticides.

Efficient IPM is rarely achieved without availability to the growers of support systems from well-informed experts, via advisory or extension schemes. The absence of such "luxuries" in many developing countries, especially in subsistence crops, is a very serious drawback to the use of much needed IPM. Even when an IPM program has reached the stage where it should be extended to commercial growers in the thick of pest problems, merely sending students on IPM training courses may not equip them to set up and execute an IPM system for real. The bottom line is that IPM often requires a more complex and sophisticated infrastructure than that of "traditional" methods. Basic research provides the vital ecological and economic data for a particular pest–crop interaction. Support services including database reference,

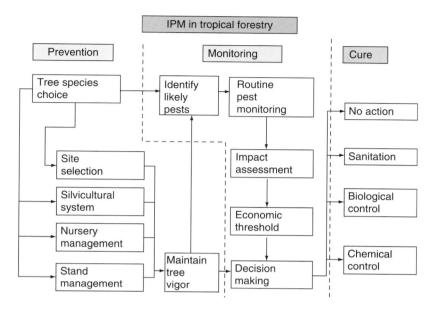

Figure 12.67 Flowchart of an IPM framework in tropical forestry.

extension services, and funding bodies then take this information and pass it on to the industry, wherein pest management becomes a crucial and integral part of crop husbandry. Ideals are infrequently met, but some examples of IPM systems from different crop systems should illustrate how some of these ideals can become reality.

12.7.1   Monitoring and economic thresholds

For efficient management, we must be able to rely on dependable data to tell us when pest densities in our crops reach levels at which financial crop losses are greater than the cost of control. Rice IPM is described below, but let us take at this stage the example of rice stem-boring Lepidoptera from India (Muralidharan & Pasalu 2006). The larvae of these moths are from various species (Lepidoptera: Pyralidae and Noctuidae), but all of them feed either in the stems or the seed heads of the crop, causing destructive syndromes of dead heart or white earhead. Data from over 20 years' study were combined, and it was found that only 1% crop damage from both types could result in 6.4% overall yield loss, equivalent to around 280 kg/ha. Armed with such information,

growers (or their advisors) can make decisions about control tactics. In the rice example, if effective control of whatever sort costs significantly less per hectare than the market value of rice lost, then intervention will result in net profit. A snag with this plan, of course, occurs when rice growers are not selling their produce, merely consuming it directly.

Economic thresholds have been established for a whole range of crop–pest interactions, and IPM cannot function properly without them. Table 12.20 lists some of these thresholds. To collect these data, growers need some sort of capture or counting system. We described the use of sex pheromone trapping earlier in this chapter, and other approaches require growers to examine their crops at crucial times of growth according to prescribed protocols (a set of procedures known as scouting). Such tactics do, of course, demand more care and attention from growers, but successful IPM relies upon the intimate involvement of agriculturists, horticulturists, and silviculturists to promote the health and hygiene of their crops. Perhaps one of the greatest strengths of monitoring and prediction based on cost–benefit analyses is the luxurious ability to decide confidently to do nothing, even when some pests can be observed in the crops.

Table 12.20 Examples of economic thresholds (critical densities) for various crop pests in the UK (from ADAS/MAFF).

Crop	Pest	Common name	Threshold density/infestation
Apples	*Rhopalosiphum insertum*	Apple grass aphid	Aphids on 50% of trusses at budburst
	Dysaphis plataginea	Rosy apple aphid	Any aphids present on trusses at budburst
	Operophtera brumata	Winter moth	Larvae on 10% of trusses at budburst
Field beans	*Aphis fabae*	Black bean aphid	5% or more plants on southwest headland infested (spring-sown crops only)
Peas	*Cydia nigicana*	Peamoth	10 or more moths per pheromone trap on two consecutive 2-day periods
Potato		Various aphids	Average 3–5 aphids per true leaf in a sample of 30 each of top, middle, and lower leaves taken across the field
Rape	*Meligethes* spp.	Blossom beetle	15–20 adults per plant at fowering bud stage
Wheat	*Metapolophium dirhodum*	Rose grain aphid	30 or more aphids per flag leaf at flowering
	Sitobion avenae	Cereal grain aphid	5 aphids per ear at the start of flowering, with weather fine and settled

Figure 12.68 illustrates how economic thresholds and scouting are used in New Zealand to inform control of the so-called tomato grub, *Helicoverpa armigera* (Cameron et al. 2001). As an aside, this same insect is also known as the cotton bollworm and the tobacco budworm – a fearsome reputation. Commercial IPM programs established against this pest include the release and widespread establishment of two species of hymenopteran parasitoid. The parasitoids attack and kill the pest larvae, and parasitism rates reach 60–80%. However, as is frequently the case with biological control, parasitism on its own cannot reduce overall tomato fruit damage to less than 10% at harvest, the economically tolerable level. As a result, applications of insecticide are still required on occasion. The problem is that about half of the fields are sprayed needlessly when control is carried out as a routine operation. So, growers assess economic thresholds using one pheromone trap per field, which is inspected twice a week. If adult moth captures exceed five per day, hand scouting begins, which comprises searching the top half of a tomato plant, looking for the eggs and larvae of *Helicoverpa*. Ten random plants in each of four parts of the field are inspected, and spraying is only carried out if growers discover more than one larva per examined plant. Using economic threshold assessments, tomato growers are able to maintain damage levels between 2% and 3%, reducing overall pesticide use by around 50%.

Economic damage threshold models have now been developed which incorporate all manner of abiotic and biotic factors into predictive systems for decision-making in pest management. One example comes from the control of pollen beetle, *Meligethes aeneus* (Coleoptera: Nitidulidae), in oilseed rape in Denmark. Hansen (2004) produced the following predictive model:

$$YL = 0.253BD + 0.340DD_{7b} - 1.721P_{7b} - 0.296DD_{7a}$$

where the variables were beetle-days (BD), day-degrees (DD_{7b}) and precipitation (P_{7b}) during the 7 days before the start of immigration, and day-degrees (DD_{7a}) during the 7 days after the start of immigration. The pollen beetle infestation was expressed as beetle-days, calculated as number of pollen beetles per plant per day and summarized from immigration until end of flowering. Using this rather complex procedure with measurements of rainfall, accumulated temperature, and pest density, yield losses (YL) could be predicted and then related to tolerable (below economic threshold) levels.

On some occasions, economic thresholds may be too general to be of much help. In the case of tropical irrigated rice in Asia (see next section), Matteson (2000) feels that general economic (action) thresholds are not sufficiently specific to local farms or areas, and therefore should be discarded. He believes that such

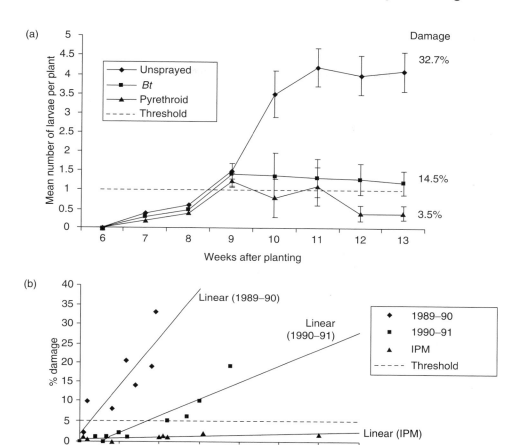

Figure 12.68 (a) Efficiency of permethrin (pyrethroid) or *Bt* when applied above a threshold of one *Helicoverpa armigera* larva per plant, and the impact on tomato damage at harvest, in New Zealand. (b) Relationship between peak numbers of *H. armigera* larvae per tomato plant estimated by scouting, and per cent fruit damage at harvest, unsprayed in 2 years and sprayed at the economic threshold (IPM). (From Cameron et al. 2001.)

thresholds alarm farmers and pressurize them to apply insecticides unnecessarily.

12.7.2 Examples of IPM in practice

There are, of course, numerous examples of IPM in practice available in scientific publications and on websites. It can be difficult to distinguish between the research of scientists and ongoing commercial practice, and so the following few examples have been chosen in an attempt to synthesize both elements.

Insect pests of rice in South and Southeast Asia

Rice is the most important food crop in the world (Bajaj & Mohanty 2005), but unlike most other major cereal crops such as wheat and maize, rice is devoted almost completely to feeding humans directly (Frei & Becker 2005). The annual world production of rice is at an all-time high (IRRI 2007) (Table 12.21), but demand will undoubtedly increase as human populations continue to expand. Most of this production is intensive, where the crop is newly established every year by transplantation into standing water

Table 12.21 Paddy rice cultivation in 2005 (FAO statistics).

	Area harvested (ha)	Production (millions of metric tonnes)
All world	153,511,755	614,654,895
Developing world	149,671,913	589,390,422
China	29,900,000	184,254,000
India	43,000,000	129,000,000

(see Plate 12.4, between pp. 310 and 311). This system favors the rice, but it also benefits certain pests and diseases by producing suitable microclimatic conditions (Lee 1992). Willocquet et al. (2004) estimate that between 120 and 200 million tonnes of rice are lost to pests annually in Asia. Changes in rice cultivation such as double cropping (two crops per year), growing high-yielding but susceptible varieties that require heavy fertilizer treatment, earlier transplanting, close spacing, and intensive use of pesticides have all contributed to very substantial increases in pest problems. An additional problem is the enormous diversity of insects that can damage rice. In Taiwan, for example, more than 40 diseases and 135 species of insect attack rice (Chen & Yeh 1992). Many of these

pests are by no means restricted to one country, and are distributed across very large areas of the world. Among the most serious in South and Southeast Asia are the brown plant-hopper, *Nilaparvata lugens* (Hemiptera: Delphacidae), green rice leaf-hopper, *Nephotettix* spp. (Hemiptera: Cicadellidae), whitebacked plant-hopper, *Sogatella furcifera* (Hemiptera: Delphacidae), rice stem-borer, *Chilo suppressalis* (Lepidoptera: Pyralidae), rice leaf-folder, *Cnaphalocrocis medinalis* (Lepidoptera: Pyralidae), and armyworm, *Spodoptera* spp. (Lepidoptera: Noctuidae). Different pests (and diseases) dominate according to season (Thuy & Thieu 1992) (Figure 12.69), but the major culprits seem omnipresent. Humans have definitely made matters worse in some cases. For example, brown plant-hopper densities increased between 100 and 1000 times after the use of large quantities of pesticides in Indonesia (FAO 1998a). The pest resurgence was likely due to deleterious effects of insecticide on natural enemies. The scale of this sort of problem is quantified in Table 12.22 (IRRI 2004).

It is quite understandable that rice farmers want to avoid risks from pest damage and the plant pathogens which they often transmit (Saha et al. 2006). However, risk-averse tactics only make sense if they are based on reality, and many rice farmers in Asia at least tend to seriously overestimate crop losses from pests (Escalada & Heong 2004). IPM tactics for rice farmers must therefore not only succeed in managing pests and diseases, but must also instill sufficient

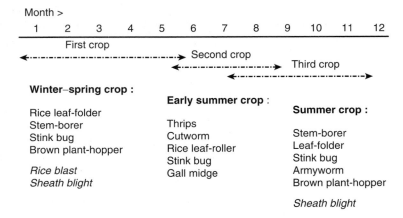

Figure 12.69 Cropping seasons of rice in Vietnam, showing the variation in major pest species according to season (plant diseases in italics). (From Thuy & Thieu 1992.)

Table 12.22 Differential effects of insecticide sprays on arthropod abundances over a whole season in a rice field. (From IRRI 2004.)

Functional group	Unsprayed plot	Sprayed plot	% Difference
All pests	14,402	14,221	−1.2
Leaf-hoppers	8,191	7,229	−11.7
Plant-hoppers	3,879	4,776	+23.1
All predators	46,248	26,967	−41.7
Mirids	4,255	4,211	+1.0
Spiders	1,063	417	−60.8
Velids	39,845	18,012	−54.8
All parasitoids	2,359	1,477	−37.4

confidence to gain the trust of farmers. Figure 12.70 illustrates a complete IPM package for rice (IRRI 2004), where the insect problems appear in the middle of the flowchart, but are only one facet of the overall "paradigm shift" (Matteson 2000) in Asian rice agriculture.

Modern rice IPM is based on a farmer-first approach where participatory but informal education takes place in farmer field schools (FFSs). In FFSs, or community IPM, the goal is to make farmers experts in their own fields and help them to make their own decisions with confidence. The FFS approach is guided by learning activities that aim to promote farmers' abilities to "gather, analyze and interpret information,

take action based on the information and evaluate the results in ways that would influence the next action by focusing on specific conditions and problems that farmers face in the field" (IRRI 2004). Matteson (2000) illustrates how these systems operate (Figure 12.71).

Huan et al. (2005) show how farmers in the Mekong delta of Vietnam were able to maintain if not improve their net profits by reducing insecticide treatments (and, less importantly, use of fertilizers and fungicides) (Table 12.23). Reduced reliance on insecticides for pest control in rice, especially during the first 40 days or so of a new crop, is made possible by virtue of an increase in other forms of management. These

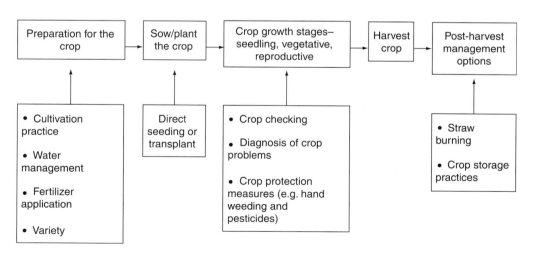

Figure 12.70 Stages in rice production with IPM. (From IRRI 2004.)

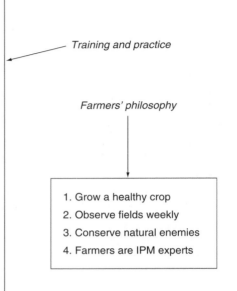

1. Group training so that farmers can learn from each other, with frequent discussions and group reinforcement of decisions

2. A curriculum pared down to essentials and simplified, having the most important points repeated often

3. Twenty to 40 hours of good-quality instruction in the rice paddy, distributed so that farmers can practice skills and crop protection decision-making each week during an entire growing season

4. Class experiments and demonstrations that engage farmers'curiosity and encourage imaginative inquiry and self-reliance

5. Periodic follow-up as farmers gain confidence in their independent decision-making

Training and practice

Farmers' philosophy

1. Grow a healthy crop
2. Observe fields weekly
3. Conserve natural enemies
4. Farmers are IPM experts

Figure 12.71 Rice IPM in Southeast Asia: training and instructions for farmers. (From Matteson 2000.)

Table 12.23 Mean number of insecticide sprays and net profit margins for rice farmers taking part in farmer's participatory evaluation schemes in the Mekong Delta of Vietnam. (From Huan et al. 2005.)

	Normal pesticide routines	**Reduced insecticide routine***
Mean no. of insecticide sprays applied, winter/spring	1.61	0.36
Mean no. of insecticide sprays applied, summer/fall	1.31	0.3
Net profit margins (US$/ha), winter/spring	608.5	666.54
Net profit margins (US$/ha), summer/fall	437.18	481.28

* No insecticide used for the first 40 days for leaf-roller control; all other sprays as required.

include the screening, breeding, and use of multiple-resistant rice varieties, the enhancement and protection of natural enemies including microbials such as fungi, bacteria, and viruses, and the adoption of new pesticide formulations and technologies such as ULV to apply them. Farmers also modify their rice husbandry systems to grow the crop in as healthy and vigorous a manner as possible. Two of the

many benefits of these IPM systems are illustrated in Figure 12.72 (Berg 2001). Here, the number of pesticide applications declines dramatically under IPM, and spraying does not begin until several weeks later than is typical under normal pest management systems. Even if yields do not dramatically increase, the sheer luxury of not having to go out and spray poisonous chemicals is often well worth having

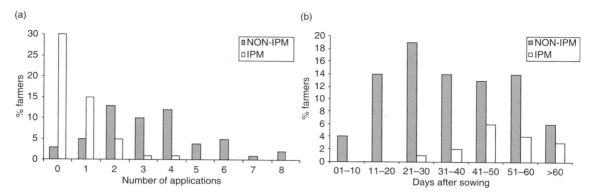

Figure 12.72 (a) Number of pesticide applications on the first rice crop in the Mekong delta in 1999 according to management technique. (b) Number of days between saving of rice and first insecticide application in the Mekong delta in 1999 according to management technique. (From Berg 2001.)

(Huan et al. 2005). In Indonesia, the Food and Agriculture Organization (FAO 1998a) has helped to train 200,000 farmers in IPM techniques for rice production, which has saved the Indonesian government around US$120 million in pesticide subsidies to farmers. In China, the adoption of IPM for rice pests and diseases has resulted not only in an increase in crop yield, but also in a reduction in the use of insecticides and a substantial decrease in the workload of farmers (Zhaohui et al. 1992). There is to this day, however, a concern that FFSs and their ilk will not reach enough people and villages to make a big difference (Tripp et al. 2005).

Some rice pests are amenable to their own bespoke IPM systems that, whilst ignoring other pest species, work well for a particular target. Thus the rice water weevil, *Lissorhoptrus oryzophilus* (Coleoptera: Curculionidae), is a very serious invasive pest in mainland China (Chen et al. 2005). This species was introduced into Asia from the USA in the mid-1970s, and since then it has spread into most areas where climatic conditions allow, with devastating effects. By 2003, it was found infesting 400,000 ha of mainland China. The insect has a very wide host range, in China feeding on 64 plant species from 10 families, mainly grasses and their relatives, and the major damage is caused, as almost always, by the larvae that feed on the roots causing serious yield losses. Despite its sudden and rather recent appearance, Chinese rice farmers are able to control the pest fairly well using an integration of tolerant rice cultivars, insecticide

applications, water management (weevil larvae do not like dry soils), late planting, light trapping, and the use of entomopathogenic fungi such as *Beauveria bassiana* and *Metarhizium anisopliae*.

The development of appropriate IPM tactics for a particular combination of pests, crop type, country, and socioeconomic factors is under the central guidance of institutions such as the IRRI and FAO's intercountry program on rice IPM (Teng 1994). The participation of these large institutions illustrates the need for coordinated, multidisciplinary approaches to the widespread adoption of IPM tactics. A truly international collaboration will involve the development of prediction systems for the migration of pests such as the rice plant-hoppers, *Sogatella furcifera* and *Nilaparvata lugens*, mentioned above. Otuka et al. (2005) suggest a "when, whence and whither" approach to predict the occurrence of pest migrations, arrivals, and damage likelihoods, based on insect phenology and weather patterns over regional scales (see also Chapter 2). Clearly, such predictive systems are only of value to growers if they can use the information to reduce pest damage. These IPM systems are complex and large scale and are necessarily under constant development and fine-tuning. Given changes in abiotic, biotic, and social conditions, no pest management package can ever be considered permanent.

Perhaps the most satisfying type of IPM is where insect pest management is almost incidental, and is a spinoff benefit from other more fundamental management systems. One such example is rice–fish farming integrations now being expanded in China and other Asian countries. Of course, there is nothing new about growing fish in paddy fields – the Chinese have been doing it for probably 2000 years – but in recent years, UNESCO and FAO have listed the practice as one of their "globally important ingenious agricultural heritage systems" (GIAHS) (Lu & Li 2006). Rice–fish culture can produce huge quantities of vital carp and other fish protein (nearly 850,000 tonnes in China in 2001, for example) (Frei & Becker 2005); as a side effect, the biological control of pests such as weevils, caterpillars, and some hoppers is enhanced by the simple predatory influence of the growing fish so that rice yields in these integrated systems are significantly enhanced. An added bonus is that farmers who also harvest the fish are rightly reluctant to use harmful pesticides in their paddy complexes for fear of poisoning themselves via the food chain.

Undoubtedly the most controversial technique to be included as a component of rice IPM in Asia is that of GM crops. We have discussed transgenic plants modified to contain and express *Bacillus thuringiensis* (*Bt*) toxicity early in this chapter, and as might be expected, rice is one of the staple crops at the forefront of this development. Most research is targeted at lepidopteran stem-borer or stem-folder species (High et al. 2004; Tang et al. 2006) – sap-feeders such as leafhoppers are less affected by *Bt* (Chen et al. 2006). Undoubtedly, transgenic *Bt* rice works. Table 12.24 shows the results of field trials in China using two insect-resistant varieties of rice, GM Xianyou 63 and GM II-Youming 86, compared to the use of non-GM rice (Huang et al. 2005). There is clear evidence that using GM transgenic rice varieties significantly reduces the use of insecticides in rice agriculture, and results such as these naturally encourage Asian countries such as China to commercialize the production of GM rice; undoubtedly GM rice will play a crucial role in providing the future's rice harvest. Sadly, but unsurprisingly, not everyone agrees (Cyranoski 2005). There is a great deal of debate and controversy about the long-term benefits of planting GM rice over a huge scale, and concerns continue about public health, outbreeding into wild plant communities, pest resistance, and so on, and it remains to be seen whether or not in the life of this book, countries such as China and India do indeed commercialize GM rice. Realistically, in fact, it may be that insect-resistant GM varieties of rice are already available on the black market, should one know where to go shopping.

Table 12.24 Pesticide use and yields of insect-resistant GM rice in China ($\pm$ SE). (From Huang et al. 2005.)

	GM rice	Non-GM rice
No. of pesticide sprays	0.50 ± 0.074	3.70 ± 0.13
Expenditure on pesticide (yuan/ha)	31 ± 4.5	243 ± 12.3
Pesticide use (kg/ha)	2.0 ± 0.25	21.2 ± 1.04
Pesticide spray labor (days/ha)	0.73 ± 0.14	9.10 ± 0.51
Rice yields (kg/ha)	6364 ± 118.0	6151 ± 101.1

IPM in orchards in the USA

Apple orchards are plagued by a wide range of insect and mite pests, and the relative importance of the various species depends very much on the geographic location, the type of management, the crops being cultivated, and the legislation pertaining to approved pesticides, which vary from country to country. Non-insect arthropods such as mites are also notorious pests in apple orchards, and their management often centers on natural control by a predatory mite released periodically into the orchard. As many pesticides that might be useful against insect pests are also toxic to the predatory mites, IPM employs insecticide-resistant predators and microbial pathogens wherever possible, which do no harm to other arthropods. Many insects cause commercial damage to pome fruits (apples, pears, etc), including aphids such as *Myzus persicae* and *Aphis pomi* (Hemiptera: Aphididae), leaf-folders or leaf-rollers such as the tufted apple bud moth, *Platynota idaeusalis*, and the obliquebanded leaf-roller, *Choristoneura rosacana*, and fruit-borers such as the Oriental fruit moth, *Grapholita molesta*, and codling moth, *Cydia pomonella* (see Plate 12.1, between pp. 310 and 311) (all Lepidoptera: Tortricidae) (Brunner et al. 2005; Robertson et al. 2005; Whitaker et al. 2006; Myers et al. 2007).

In the past, as might be expected, pest control has relied on routine and multiple applications of

Table 12.25 Objectives of IPM programes for coding moth *Cydia pomonelia*, in the USA. (Partly from Calkins & Faust 2003.)

A	To enhance the efficiency of non-insecticidal systems and eliminate post-harvest residues, by reducing non-essential poisons
B	To increase the use of pheromone mating disruption over large areas
C	To improve the use and efficiency of biological control tactics using natural enemies such as parasitoids
D	To increase the use of insect-specific pathogens such as bacteria (*Bt*) and viruses (GV)
E	To develop sterile insect release techniques to reduce fertile adult populations
F	To encourage the use of physical and cultural techniques such as orchard sanitation, hand pruning, and band trapping
G	To develop monitoring programs for pest densities and fruit damage, in order to use poisons only when strictly necessary
H	To aid fruit producers to make the transition from high reliance on insecticides to more environmentally benign IPM
I	To improve the perception amongst consumers that fruit production is based on environmentally friendly techniques

synthetic insecticides, especially the organophosphates and pyrethroids, and, as a result, some chemical residues can undoubtedly be detected in harvest apples and other fruit (Rawn et al. 2006). There is therefore a trend to reduce pesticide inputs in orchards, and, at an extreme, to go "fully organic" where no poisons are used at all (Judd & Gardiner 2005), and where some aspects of fruit quality and marketability are allegedly maximized (Peck et al. 2006). Realistically, IPM for fruit pests is the economically sensible and environmentally reasonable way to go, and Table 12.25 summarizes the aims and objectives for codling moth. As can be seen the overall set of tactics is complex and requires a great deal of research and development, as well as good will and a certain degree of faith on the part of growers and managers. Some techniques have more potential or general use than others. Sterile insect release (SIR), for example, is not widespread at the moment, though it may supplement mating disruption and band trapping. It requires a great deal of support and investment, and seems relatively ineffective in colder regions (Judd & Gardiner 2005). Biological control that relies on natural enemies such as parasitic wasps and even generalist predators within the orchard and also in edge vegetation, although environmentally desirable, is not efficient enough in most cases (Miliczky & Horton 2005); the use of pathogens such as *Bt*, and granulosis viruses in particular, are more effective (Steineke & Jehle 2004). Undoubtedly, the use of insecticides will

not go away entirely, even with the deployment of pest-resistant fruit cultivars where market pressures allow (Hogmire & Miller 2005), so sensible IPM will still rely on certain toxic compounds, applied to maximize efficiency and minimize waste (Walklate et al. 2006), with decisions about if and when to spray based on reliable action thresholds determined by monitoring and prediction using pheromone traps (Knight & Light 2005). In general, IPM tactics in orchards are efficient and sophisticated, as one detailed example will demonstrate.

One of the most serious single insect pests of apples in Europe and North America is the codling moth, *Cydia pomonella* (Lepidoptera: Tortricidae). Codling moth larvae are the archetypal "grub in the apple" and as such are classic low-density pests (see Figure 12.2). Estimates from the USA suggest that, in the absence of any control, codling moth could infest up to 95% of all apples in an orchard. Codling moth is also a concealed pest (see Table 12.17), which is very difficult to control once it has settled inside the developing fruit. There is a very narrow time-window for control between egg hatch and larval tunneling in the fruit. For an IPM program to be successful, it is essential that growers can follow the date of egg hatch and know when young larvae are susceptible to conventional insecticides, or a granulosis virus such as Cyd-X.

The life cycle of codling moth in the USA is illustrated in Figure 12.73. Mature larvae overwinter in leaf litter or under bark cracks, and new adults of the

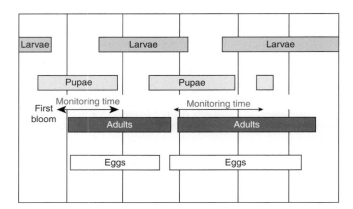

Figure 12.73 Codling moth life history. The arrows indicate when adults should be monitored with pheromone traps. To determine biofix, traps should be set out prior to moth flight. (From Alston et al. 2006.)

first generation appear around apple blossom time in May. Eggs are laid on leaves close to new fruit and, crucially, according to ambient temperature, hatch within 8–14 days. The young larvae quickly bore into the very young apples, wherein they are safe from most external mortality factors. Pheromone traps are used routinely in orchards to establish the dates of peak male flight, which can readily be used to predict subsequent egg laying and larval hatching. Figure 12.74 illustrates the system used in Utah (Alston et al. 2006), which uses the first day of consistent moth flight (at least two moths caught in pheromone traps on two or more consecutive nights) as the "biofix". From then on, the accumulated day-degrees (DDs) for egg hatch are measured. A day-degree (see Chapter 2) is essentially a measure of the metabolic processes that take a poikilothermic organism from one stage of development to the next. For example, if development from egg laying to egg hatch requires 100 DD, then this could mean 10 days at 10°, or 5 days at 20°, and so on. In practice, it is a little more complicated, but apple growers in the USA have access to websites on which they can calculate DD accumulation for codling moth. Growers can then apply insecticidal sprays of one sort or another at, for example, 250 DD (above a base of 50 and below a ceiling of 80) after the biofix. (Note that in all of this section, temperature is measured in degrees Fahrenheit – for those under 30 years old, not living in America, deg C = 5(deg F – 32)/9). Commercial IPM

in both conventional and especially organic orchards is also regularly based on mating disruption (Witzgall et al. 2008), where pheromone dispensers or aerosol canisters (known as puffers) are placed in tree canopies, again using biofix decisions to get the timing right.

IPM in European forestry

Of all the types of crop production in the world, plantation forestry is undoubtedly the least amenable to control once pest outbreaks have begun. As might be expect, forestry production has to rely most heavily on prevention as the mainstay of its IPM tactics (Speight & Wylie 2001; Speight & Evans 2004). Figure 12.75 illustrates typical management systems (if they exist at all) for four different types of forest pest. It is pretty clear that once things go wrong in forestry, it is highly unlikely that they can be fixed. In many parts of the tropical world, IPM of forest insects just does not work once preventative techniques have failed. For example, the neem scale, *Aonidiella orientalis* (Hemiptera: Diaspididae), in Chad (Lale 1998) and mahogany shoot-borer, *Hypsipyla* spp. (Lepidoptera: Pyralidae), on a global scale (Speight & Cory 2000) are essentially uncontrollable, even with IPM, at least for now. However, IPM in forestry in certain parts of the world has been successful, especially in "high tech" countries such as the USA, UK, and Australia. The spruce bark beetle, *Ips typographus* (Coleoptera: Scolytidae), provides an example (Evans & Speight 2004).

Degree days (DD)	Adults emerged %	Eggs hatched %	Management event
100*	0	0	• Place traps in orchards
150–200	First moths expected	0	• Check traps every 1–2 days until biofix is determined
First generation			
0 (biofix)**	First consistent moth catch	0	• Reset degree Days to 0
50–79	5–9	0	• First eggs are laid • Apply insectides that need to be present before egg-laying
100–200	15–40	0	• Early egg-laying period • Apply insecticides that target early egg-laying period
200–250	45–50	1–3	• Beginning of egg hatch • Apply insecticides that target newly hatched larvae
340–640	67–98	12–80	• Critical period for control, high rate of egg hatch
920	100	99	• End of egg hatch for 1st generation
Second generation			
1000–1050	5–8	0	• First eggs of 2nd generation are laid • Apply insecticides to target early egg-laying
1100	13	1	• Beginning of egg hatch
1320–1720	46–93	11–71	• Apply insecticides that target newly hatched larvae • Critical period for control, high rate of egg hatch
2100	100	99	• End of egg hatch for 2st generation
Third generation			
2160	1	15	• Beginning of egg hatch • Keep fruit protected through September 15 • Check pre-harvest interval of material used to ensure that final spray is not too near harvest

*Begin accumulating degree days after temperatures begin to exceed 50°F, typically on January 1 for southern Utah or March 1 for northern Utah
**Biofix is when at least 2 moths caught on consecutive nights.

Figure 12.74 Major events in the codling moth management program, based on accumulated degree-days. (From Alston et al. 2006.)

We have mentioned *I. typographus* on various occasions in this book – it is a classic "secondary pest" that is theoretically unable to attack and colonize trees unless they are stressed in some way (windblow, drought, competitive suppression, and chainsaws are all examples of stressors). However, some stressors are hard to avoid during typical silviculture, and even harder to predict once the stand is established. A complicating factor is a mass outbreak, where huge numbers of adult beetles inundate the defenses of even healthy trees. Disturbance events such as major tree killing by wind can convert *Ips* from non-epidemic to epidemic behavior (Okland & Bjornstad 2006), causing huge losses to growing timber, and, indeed, exerting profound changes to entire landscapes (Aukema et al. 2006). Hence, *Ips* is an entomological horseman of the apocalypse, rearing its head in huge abundance after natural disasters. Undoubtedly, the

<table>
<tr>
<td>

Management tactics–borers

PREVENTION

      Match tree species to site
      Avoid soil aridity or waterlogging
      Avoid overmaturity
      Avoid root damage in nursery
      Avoid log piles etc near stands
      Avoid pruning/brashing damage

CURE

      None

</td>
<td>

Management tactics–defoliators

PREVENTION

      None (NB prophylactic spraying in
      nurseries should not be carried out)

CURE

      Nurseries–local spray of insecticide only
      when serious leaf loss is observed
      Plantations–none

</td>
</tr>
<tr>
<td>

Management tactics–sap-feeders

PREVENTION

      Avoid non-vigorous trees (see Borers)
      (NB prophylactic spraying in
      nurseries should not be carried out)

CURE

      Nurseries–local spray of insecticide only
      when serious leaf damage is observed
      Plantations–none

</td>
<td>

Management tactics–root-feeders

PREVENTION

      Nurseries–avoid root damage
      Plantations–incorporation of insecticide
      granules in planting hole if experience
      shows serious losses from termites

CURE

      Nurseries–soil drench of insecticide for
      white grubs if losses serious
      Plantation–none

</td>
</tr>
</table>

Figure 12.75 General management tactics for four categories of insect pest in forests.

abundance of dead and fallen timber in a spruce forest provides highly suitable and copious breeding sites for *Ips* larvae, which will eventually give rise to huge densities of fresh young adults to attack standing trees. Eriksson et al. (2006) show how the amount of freshly dead wood (DWR, down wood retention) coupled with its size (diameter) is closely correlated with the number of egg galleries (and resulting new adults) (Figure 12.76). Bark thickness is especially important (Wermelinger 2004) since beetle larvae and pupae are unable to reach maturity in thin bark because of their own physical size. Unsurprisingly then, in Western Europe in the late 1990s, the enormous buildup of *Ips* in large-diameter, wind-damaged timber killed large numbers of spruce trees (Wermlinger 2004), sparking an unprecedented surge of research into *I. typographus* IPM development. To make matters worse, it is suggested that as global

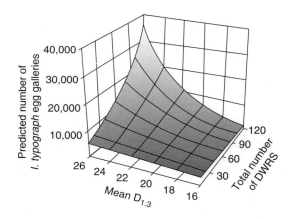

Figure 12.76 The predicted number of *Ips typographus* egg galleries in relation to the total number and $D_{1.3}$ (diameter at breast height) of DWRS. (From Eriksson et al. 2006.)

Table 12.26 Major site and stand features predisposing spruce forests to attack by *Ips typographus*. (Mainly from Netherer & Nopp-Mayr 2005.)

Parameter	Criterion	Predispostion rating (high liklihood of attacks)
Site level features		
Terrain	Solar irradiation	High
	Slope position	Ridge (upper slope)
	Aspect	South or west facing
Climate	Mean temperature	Warm (two generations/year)
	Summer rainfall	< 360 mm
Soil	Hydrology	Excessively to well drained
Predisposition	Storm (wind) damage	High
Stand level features		
Tree species composition	% spruce	> 70% high
		50–70% medium/high
Structure	Stand age	> 100 years high
		80–100 years medium/high
	Edge trees	Medium/high
	Tree health	Poor

warming produces longer summers in Western Europe, *Ips* will be able to increase its number of generations per year and become even more of a threat (Schlyter et al. 2006). Biological control from predators and parasitoids has little impact (Feicht 2004).

Natural spruce forests in Europe have coevolved with *I. typographus*, making it relatively easy to suggest features of forests and the trees within them that predispose to insect attack and subsequent damage and death of trees. Netherer and Nopp-Meyer (2005) provide a series of criteria that comprise both site and stand features known to predispose spruce forest to *Ips* (Table 12.26). In other words, these provide risk or hazard ratings for pest problems – simply put, stands on dry soils at the top of ridges with old and infirm trees containing lots of dead timber are most at risk. Using these predictions, Figure 12.77 summarizes the methods available for *Ips* IPM. The summary contains a wide variety of tactics that may not all be equally useful in all parts of the region. Countries such as the UK that, sadly for forest entomologists, have not yet acquired a significant population of *Ips*, rely on stringent plant health and quarantine inspections to prevent imports from continental Europe. Other countries such as Slovakia, where *I. typographus* is endemic, have set up community-wide monitoring, detection, and mass-trapping schemes

(Novotny 2004). On-site inspections are still the mainstay of detecting spruce stands already infested with *I. typographus*, an expensive and labor-intensive process especially in rather isolated mountainous regions. Franklin et al. (2004) suggest that sales of damaged trees in timber yards might be a useful proxy of attack levels. More work is required to estimate the reliability of this form of "remote sensing".

Ironically, the increasing trend of maintaining large quantities of dead wood in European forests to enhance biodiversity (see Chapter 10) may increase problems with *Ips*. Dead wood provides breeding and overwintering material, allowing sizable populations of bark beetles to build up even when climatic conditions do not stress standing timber (Hedgren 2003). It may be possible to protect fallen logs and trees of conservation value using antiattractant chemicals, at least on a small scale. Jakus et al. (2003) were able to decrease *Ips* attacks on spruce by up to 80% using various mixtures of non-host volatiles. Whether or not such treatment would defeat the object of encouraging rare insect species in the logs is unknown. Recourse to toxic synthetic insecticides is of course always possible, and DeGomez et al. (2006), for example, have shown that spraying trees and logs with pyrethroids protects trees from beetle attack for a whole season. This is, however, unlikely to be a popular tactic in most cases. In all probability,

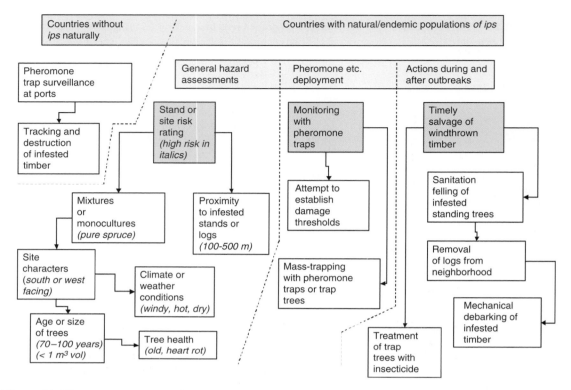

Figure 12.77 Generalized summary of components in the IPM of *Ips typographus* in Europe. (Mainly from Wermelinger 2004.)

chemical control of bark beetles such as *I. typographus* will rely on pheromones to provide either barriers to attack, or attractants (and lethal) point sources (Dahlsten et al. 2004). Either way, susceptible trees and stands should be expected to gain a degree of protection.

A major problem in developing a comprehensive IPM program against *I. typographus* has been obtaining dependable risk forecasts for any one year in a particular area, before substantial damage to standing trees has already occurred. Wermelinger (2004) feels that monitoring using pheromone trapping is not sufficiently dependable although, in local areas, it may have potential. In the southern Alps, adult beetles caught in pheromone traps by the end of June provide a useful prediction of future attacks and timber losses later in the summer. Detection by June leaves enough time to plan and apply specific control measures (Faccoli & Stergulc 2004). In

Figure 12.78, it can be seen that reliance on prediction models for trap captures can be used to dictate management tactics (Faccoli & Stergulc 2006), assuming that short-term population densities of the beetle are elated one to another. Once the go ahead has been decided, control methods include increasing the numbers of pheromone traps for mass trapping, using trap trees, or cutting recently attacked trees in localized sanitation programs. Note that the threshold level for management in this example is around $100 \, m^3$ per 1200 ha. Efforts to reduce the damage further would probably cost more than the value of the saved timber. The spring monitoring in this program reduces the costs and labor required to continue monitoring for the whole season. Even better, Faccoli and Stergulc (2004) suggest that, since *Ips* breeds mainly in trees with thick bark, monitoring would only be required in a subset of stands of susceptible stage.

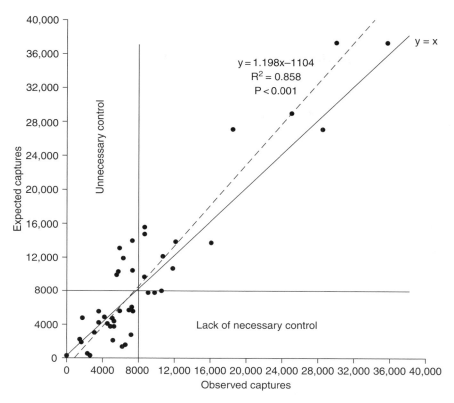

Figure 12.78 Accuracy of the model showing correlation between observed and expected captures of *Ips typographus* adult beetles. Using a mean of 8000 insects per trap as threshold risk for outbreaks, in 12% of the cases the model leads to unjustified sanitary measures, whereas in 6% of the cases no control is applied although it would be necessary. (From Faccoli & Stergulc 2006.)

12.7.3 Conclusions

Ips typographus provides an excellent example of how a close relationship between an insect and its host plant can become a serious economic problem, once humans decide that they also want to exploit the plant in question. Outbreaks of insects on trees are completely natural events in the ecology of forests, often predictable in the natural course of succession. Insect herbivores that we happen to call pests are, as we said at the beginning, simply in the wrong place at the wrong time for our own economic or esthetic goals.

REFERENCES

Abbott, I. (1993) Review of the ecology and control of the introduced bark beetle, *Ips grandicollis* (Eichhoff) (Coleoptera: Scolytidae) in Western Australia, 1952–90. *CalmScience*, **1** (1), 35–46.

Abeku T-A., Hay S-I., Ochola S. et al. (2004) Malaria epidemic early warning and detection in African highlands. *Trends in Parasitology*, **20** (9), 400–5.

Abensperg-Traun, M. & Steven, D. (1997) Latitudinal gradients in the species richness of Australian termites (Isoptera). *Australian Journal of Ecology*, **22**, 471–6.

Abrahamson, W.G. & Weis, A.E. (1997) *Evolutionary Ecology across Three Trophic Levels*. Princeton University Press, Princeton, NJ.

Abrams, P. (1983) Theory of limiting similarity. *Annual Review of Ecology and Systematics*, **14**, 359–76.

Abril, A.B. & Bucher, E.H. (2002). Evidence that the fungus cultured by leaf-cutting ants does not metabolize cellulose. *Ecology Letters*, **5** (3), 325–8.

Ackonor, J.B. & Vajime, C.K. (1995) Factors affecting *Locusta migratoria migratorioides* egg development in the Lake Chad basin outbreak area. *International Journal of Pest Management*, **41** (2), 87–96.

Adams, E.S. & Tschinkel, W.R. (1995) Effects of foundress number on brood raids and queen survival in the fire ant, *Solenopsis invicta*. *Behavioral Ecology Sociobiology*, **37**, 233–42.

Addo-Bediako, A., Chown, S.L. & Gaston, K.J. (2000). Thermal tolerance. *Proceedings of the Royal Society of London Series B, Biological Sciences*, **267** (1445), 739–45.

Adler, F.R. & Karban, R. (1994) Defended fortresses or moving targets? another model of ducible defenses inspired by military metaphors. *American Naturalist*, **144**, 813–32.

Adler, L.S., Schmitt, J. & Bowers, M.D. (1995) Genetic variation in defensive chemistry in *Plantago lanceolata* (Plantaginaceae) and its effect on the specialist herbivore *Junonia coenia* (Nymphalidae). *Oecologia (Berlin)*, **101**, 75–85.

Agelopoulos, N.G. & Keller, M.A. (1994a) Plant–natural enemy association in the tritrophic system *Cotesia rubecula–Pieris rapae*–Brassicaceae (Cruciferae). II. Preference of *C. rubecula* for landing and searching. *Journal of Chemical Ecology*, **20**, 1735–48.

Agelopoulos, N.G. & Keller, M.A. (1994b) Plant–natural enemy association in the tritrophic system *Cotesia rubecula–Pieris rapae*–Brassicaceae (Cruciferae). III. Collection and identification of plant and frass volatiles. *Journal of Chemical Ecology*, **20**, 1955–68.

Agoua, H., Alley, E.S., Hougard, J.M., Akpoboua, K.L.B., Boatin, B. & Seketeli, A. (1995) Procedure of definitive cessation of larviciding in the Onchocerciasis Control Programme in West Africa: entomological post-control studies. *Parasite*, **2** (3), 281–8.

Agrawal, A.A. (2000) Specificity of induced resistance in wild radish: causes and consequences for two specialist and two generalist caterpillars. *Oikos*, **89**, 493–500.

Agrawal, A.A. (2004) Plant defense and density dependence in the population growth of herbivores. *American Naturalist*, **164**, 113–20.

Agrawal, A.A. & Fishbein, M. (2006) Plant defense syndromes. *Ecology*, **87**, S132–S149.

Agrawal, A.A., Strauss, S.Y. & Stout, M.J. (1999) Costs of induced responses and tolerance to herbivory in male and female fitness components of wild radish. *Evolution*, **53**, 1093–104.

Agrawal, A.A., Underwood, N. & Stinchcombe, J.R. (2004) Intraspecific variation in the strength of density dependence in aphid populations. *Ecological Entomology*, **29**, 521–6.

Agrell, J., McDonald, E.P. & Lindroth, R.L. (2000) Effects of CO_2 and light on tree phytochemistry and insect performance. *Oikos*, **88**, 259–72.

Agrios G.N. (1988) *Plant Pathology*, 3rd edn, p. 803. Academic Press, London.

Aguilar, J.M. & Boecklen, W.J. (1992) Patterns of herbivory in the *Quercus grisea–Quercus gambelli* species complex. *Oikos*, **64**, 498–504.

Ahmad, S., Govindarajan, S., Funk, C.R. & Johnson-Cicalese, J.M. (1985) Fatality of house crickets on perennial ryegrasses (*Lolium perenne*) infected with a fungal endophyte. *Entomologia Experimentalis et Applicata*, **39**, 183–90.

Ahmad, S., Govindajaran, S., Johnson-Cicalese, J.M. & Funk, C.R. (1987) Association of a fungal endophyte in perennial ryegrass with antibiosis to larvae of the southern armyworm, *Spodoptera eridania*. *Entomologia Experimentalis et Applicata*, **43**, 287–94.

Ahmed, S.I. (1995) Investigations on the nuclear polyhedrosis of teak defoliator, *Hyblaea puera* (Cram) (Lepidoptera: Hyblaeidae). *Journal of Applied Entomology*, **119** (5), 351–4.

Ahn, K.J., Maruyama, M. & Jeon, M.J. (2003) Redescription of the intertidal genus Brachypronomaea Sawada (Coleoptera: Staphylinidae: Aleocharinae) from the Ryukyu Islands, Japan with a discussion of its phylogenetic relationships. *Journal of the Kansas Entomological Society* **76** (4), 622–9.

Aide, T.M. (1992) Dry season leaf production an escape from herbivory. *Biotropica*, **24**, 532–7.

Aikawa, T., Togashi, K. & Kosaka, H. (2003) Different developmental responses of virulent and avirulent isolates of the pinewood nematode, *Bursaphelenchus xylophilus* (Nematoda: Aphelenchoididae), to the insect vector, *Monochamus alternatus* (Coleoptera: Cerambycidae). *Environmental Entomology*, **32**, 96–102.

Akimoto, S. & Yamaguchi, Y. (1994) Phenotype selection on the process of gall formation of a *Tetranuera* aphid (Pemphigidae). *Journal of Animal Ecology*, **63**, 727–38.

Akinlosotu, T.A. (1973) *The role of* Diaeretiella rapae *(McIntosh) in the control of the cabbage aphid*. PhD Thesis, University of London.

Alborn, H.T., Turlings, T.C.J., Jones, T.H., Stenhagen, G., Loughrin, J.H. & Tumlinson, J.H. (1997) An elicitor of plant volatiles from beet armyworm oral secretion. *Science*, **276**, 945–9.

Aleksic, I., Gliksman, I., Milanovic, D. & Tucic, N. (1993) On *r*- and *K*-selection: evidence from the bean weevil (*Acanthoscelides obtectus*). *Zeitschrift fuer Zoologische Systematik und Evolutionsforschung*, **31** (4), 259–68.

Alexander, B. & Maroli, M. (2003) Control of phlebotomine sandflies. *Medical and Veterinary Entomology*, **17**, 1–18.

Alexander, J.L. (2006) Zoonosis update – screwworms. *Journal of the American Veterinary Medical Association*, **228** (3), 357–67.

Allan, J.D. (1995) Stream ecology. *Structure and Function of Running Waters*. Chapman & Hall, London.

Allee, J.P., Pelletier, C.L., Fergusson, E.K. & Champlin, D.T. (2006) Early events in adult eye development of the moth, *Manduca sexta*. *Journal of Insect Physiology*, **52** (5), 450–60.

Allen, W.A. & Rajotte, E.G. (1990) The changing role of extension entomology in the IPM era. *Annual Review of Entomology*, **35**, 379–97.

Allsopp, P.G. & Cox, M.C. (2002) Sugarcane clones vary in their resistance to sugarcane whitegrubs. *Australian Journal of Agricultural Research*, **53** (10), 1111–36.

Almeida, R.P.P. & Purcell, A.H. (2006) Patterns of *Xylella fastidiosa* colonization on the precibarium of sharpshooter

vectors relative to transmission to plants. *Annals of the Entomological Society of America*, **99** (5), 884–90.

Alonso-Amelot, M.E. & Oliveros-Bastidas, A. (2005) Kinetics of the natural evolution of hydrogen cyanide in plants in neotropical *Pteridium arachnoideum* and its ecological significance. *Journal of Chemical Ecology*, **31** (2), 315–31.

Alphey, L. (2002) Re-engineering the sterile insect technique. *Insect Biochemistry and Molecular Biology*, **32**, 1243–7.

Al-Sarar, A., Hall, F.R. & Downer, R.A. (2006) Impact of spray application methodology on the development of resistance to cypermethrin and spinosad by fall armyworm *Spodoptera frugiperda* (JE Smith). *Pest Management Science*, **62** (11), 1023–31.

Alsmark, C-M., Frank, A-C., Karlberg, E-O. et al. (2004) The louse-borne human pathogen *Bartonella quintana* is a genomic derivative of the zoonotic agent *Bartonella henselae*. *Proceedings of the National Academy of Sciences of the USA*, **101** (26), 9716–21.

Alstad, D.N. (1998) Population structure and the conundrum of local adaptation. In *Genetic Structure and Local Adaptation in Natural Insect Populations. Effects of Ecology, Life History, and Behavior* (eds S. Mopper & S.Y. Strauss), pp. 3–21. Chapman & Hall, New York.

Alstad, D.N. & Corbin, K.W. (1990) Scale insect allozyme differentiation within and between host trees. *Ecological Ecology*, **4**, 43–56.

Alston, D., Murray, M. & Reding, M. (2006) Codling moth – Utah pests fact sheet. http://extension.usu.edu/files/publications/factsheet/ENT-13–06.pdf.

Alvar, J., Yactayo, S. & Bern, C. (2006) Leishmaniasis and poverty. *Trends In Parasitology*, **22** (12), 552–7.

Alvarez, N., Mercier, L., Hossaert-McKey, M. et al. (2006) Ecological distribution and niche segregation of sibling species: the case of bean beetles, *Acanthoscelides obtectus* Say and *A. obvelatus* Bridwell. *Ecological Entomology*, **31** (6), 582–90.

Alward, R.D. & Joern, A. (1993) Plasticity and overcompensation in grass responses to herbivory. *Oecologia*, **95**, 358–64.

Amazigo, U. & Boatin, B. (2006) The future of onchocerciasis control in Africa. *Lancet*, **368** (9551), 1946–7.

Amerasinghe, F.P. & Indrajith, N.G. (1994) Post-irrigation breeding patterns of surface water mosquitoes in the Mahaweli Project, Sri Lanka, and comparisons with preceding developmental phases. *Journal of Medical Entomology*, **31** (4), 516–23.

Andersen, A.N. (1995) Measuring more of biodiversity–genus richness as a surrogate for species richness in Australian ant faunas. *Biological Conservation*, **73**, 39–43.

Andersen, A.N. (1997) Using ants as bioindicators: multiscale issues in ant community ecology. *Conservation Ecology [online]*, **1**, 8.

Andersen, A.N., Fisher, A., Hoffmann, B.D., Read, J.L. & Richards, R. (2004) Use of terrestrial invertebrates for biodiversity monitoring in Australian rangelands, with particular reference to ants. *Austral Ecology*, **29**, 87–92.

Andersen, A.N. & Majer, J.D. (2004) Ants show the way Down Under: invertebrates as bioindicators in land management. *Frontiers in Ecology and the Environment*, **2**, 291–8.

Andersen, N.M. (1991) Marine insects: genital morphology, phylogeny and evolution of sea skaters, genus *Halobates* (Hemiptera. Gerridae). *Zoological Journal of the Linnean Society*, **103** (1), 21–60.

Andersen, N.M. (1999) The evolution of marine insects: phylogenetic, ecological and geographical aspects of species diversity in marine water striders. *Ecography*, **22** (1): 98–111.

Andersen, N.M., Cheng, L., Damgaard, J. & Sperling, F.A.H. (2000) Mitochondrial DNA sequence variation and phylogeography of oceanic insects (Hemiptera: Gerridae: Halobates spp.). *Marine Biology*, **136**, 421–30.

Andersen, N.M., Farma, A., Minelli, A. & Piccoli, G. (1994) A fossil *Halobates* from the Mediterranean and the origin of sea skaters (Hemiptera: Gerridae). *Zoological Journal of the Linnean Society*, **112** (4), 479–89.

Andersen, N.M. & Grimaldi, D. (2001) A fossil water measurer from the mid-Cretaceous Burmese amber (Hemiptera: Gerromorpha: Hydrometridae). *Insect Systematics and Evolution* **32** (4), 381–92.

Andersen, A.N. & Sparling, G.P. (1997) Ants as indicators of restoration success: relationship with soil microbial biomass in the Australian seasonal tropics. *Restoration Ecology*, **5**, 109–14.

Anderson, J.B. & Brower, L.P. (1996) Freeze-protection of overwintering monarch butterflies in Mexico: critical role of the forest as a blanket and an umbrella. *Ecological Entomology*, **21** (2), 107–16.

Anderson, N.H., Steedman, R.J. & Dudley, T. (1984) Patterns of exploitation by stream invertebrates of wood debris (xylophagy). *Proceedings of the International Association of Theoretical and Applied Limnology*, **22**, 1847–52.

Anderson, R.M. & May, R.M. (1981) The population-dynamics of micro-parasites and their invertebrate hosts. *Philosophical Transactions of the Royal Society of London Series B, Biological Sciences*, **291** (1054), 451–524.

Anderson, R.M. & May, R.M. (1991) *Infectious Diseases of Humans: Dynamics and Control.* Oxford University Press, Oxford.

Andreadis, S.S., Milonas, P.G. & Savopoulou-Soultani, M. (2005) Cold hardiness of diapausing and non-diapausing pupae of the European grapevine moth, *Lobesia botrana*. *Entomologia Experimentalis et Applicata*, **117**, 113–18.

Andresen, E. & Laurance, S.G.W. (2007) Possible indirect effects of mammal hunting on dung beetle assemblages in Panama. *Biotropica*, **39**, 141–6.

Andrew, N.R. & Hughes, L. (2004) Species diversity and structure of phytophagous beetle assemblages along a latitudinal gradient: predicting the potential impacts of climate change. *Ecological Entomology*, **29**, 527–42.

Andrew, N.R. & Hughes, L. (2005) Diversity and assemblage structure of phytophagous Hemiptera along a latitudinal gradient: predicting the potential impacts of climate change. *Global Ecology and Biogeography*, **14**, 249–62.

Andrewartha, H.G. & Birch, L.C. (1954) *The Distribution and Abundance of Animals*, p. 782. University of Chicago Press, Chicago, IL.

Anon. (1992) *Climate Change Supplementary Report to the Intergovernmental Panel on Climate Change Scientific Assessment.* Cambridge University Press, New York.

Anon. (1996) No-go verandahs. *The Guardian*, 19 Sept 1996.

Antoine, P.O., De Franceschi, D., Flynn, J.J. et al. (2006) Amber from western Amazonia reveals Neotropical diversity during the middle Miocene. *Proceedings of the National Academy of Sciences of the USA*, **103** (37), 13595–600.

Appel, H.M. (1993) Phenolics in ecological interactions: the importance of oxidation. *Journal of Chemical Ecology*, **19**, 1521–52.

Appel, H.M. & Maines, L.W. (1995) The influence of host plant on gut conditions of gypsy moth (*Lymantria dispar*) caterpillars. *Journal of Insect Physiology*, **41**, 241–6.

Arbocast, R.T. (2003) Humidity response of adult male *Oryzaephilus surinamensis* (Coleoptera: Cucujidae) with special reference to the effect of carbon dioxide. *Environmental Entomology*, **32** (2), 264–9.

Arditi, R., Callois, J.M., Tyutyunov, Y. & Jost, C. (2004) Does mutual interference always stabilize predator–prey dynamics? A comparison of models. *Comptes Rendus Biologies*, **327** (11), 1037–57.

Arillo, A. & Engel, M.S. (2006) Rock crawlers in Baltic amber (Notoptera: Mantophasmatodea). *American Museum Novitates*, **3539**, 1–10.

Arimura, G., Ozawa, R., Shimoda, T., Nishioka, T., Boland, W. & Takabyashi, J. (2000) Herbivory-induced volatiles elicit defence genes in lima bean leaves. *Nature*, **406**, 512–15.

Arli-Sokmen, M., Mennan, H., Sevik, M.A. & Ecevit, O. (2005) Occurrence of viruses in field-grown pepper crops and some of their reservoir weed hosts in Samsun, Turkey. *Phytoparasitica*, **33** (4), 347–58.

Armbruster, P. & Hutchinson, R.A. (2002) Pupal mass and wing length as indicators of fecundity in *Aedes albopictus* and *Aedes geniculatus* (Diptera: culicidae). *Journal of Medical Entomology*, **39**, 699–704.

Arnold, T.M. & Schultz, J.C. (2002) Induced sink strength as a prerequisite for induced tannin biosynthesis in developing leaves of Populus. *Oecologia*, **130**, 585–93.

Arocha-Pinango, C.I., de Bosch, N.B., Torres, A. et al. (1992) Six new cases of a caterpillar-induced bleeding syndrome. *Thrombosis and Haemostasis*, **67** (4), 402–7.

Ash, S. (1996) Evidence of arthropod–plant interactions in the Upper Triassic of the southwestern United States. *Lethaia*, **29** (3), 237–48.

Asher, J., Warren, M., Fox, R., Harding, P., Jeffcoate, G. & Jeffcoate, S. (2001) *The Millennium Atlas of Butterflies in Britain and Ireland.* Oxford University Press, Oxford.

Ashworth, A.C. & Kuschel, G. (2003) Fossil weevils (Coleoptera: Curculionidae) from latitude 85degrees Antarctica. *Palaeogeography Palaeoclimatology Palaeoecology*, **191** (2), 191–202.

Asiimwe, P., Ecaat, J.S., Otim, M., Gerling, D., Kyamanywa, S. & Legg, J.P. (2007) Life-table analysis of mortality factors affecting populations of *Bemisia tabaci* on cassava in Uganda. *Entomologia Experimentalis et Applicata* **122** (1), 37–44.

Askew, R.R. (1971) *Parasitic Insects*. Heinemann, London.

Auerbach, M. (1991) Relative impact of interactions within and between trophic levels during an insect outbreak. *Ecology*, **72**, 1599–608.

Aukema, B.H., Carroll, A.L., Zhu, J., Raffa, K.F., Sickley, T.A. & Taylor, S.W. (2006) Landscape level analysis of mountain pine beetle in British Columbia, Canada: spatiotemporal development and spatial synchrony within the present outbreak. *Ecography*, **29** (3), 427–41.

Averof, M. & Cohen, S.M. (1997) Evolutionary origin of insect wings from ancestral gills. *Nature*, **385** (6617), 627–30.

Awmack, C.S., Harrington, R. & Leather, S.R. (1997) Host plant effects on the performance of the aphid *Aulcorthum solani* (Kalt.) (Homoptera: Aphididae) at ambient and elevated CO_2. *Global Change Biology*, **3**, 545–9.

Ayal, Y. (2007) Trophic structure and the role of predation in shaping hot desert communities. *Journal of Arid Environments*, **68** (2), 171–87.

Ayala, F.J. (1970) Competition, coexistence, and evolution. In *Essays in Evolution and Genetics* (eds M.K. Hecht & W.C. Steere), pp. 121–58. Appleton-Century-Crofts, New York.

Ayala, F.J. (1971) Frequency-dependent competition. *Science*, **171**, 820–4.

Ayres, M.P., Suomela, J. & McLean, S.F. Jr. (1987) Growth performance of *Epirrita autumnata* (Lepidoptera: Geometridae) on mountain birch: trees, broods, and tree × brood interactions. *Oecologia*, **74**, 450–7.

Azerefegne, F., Solbreck, C. & Ives, A.R. (2001) Environmental forcing and high amplitude fluctuations in the population dynamics of the tropical butterfly *Acraea acerata* (Lepidoptera: Nymphalidae). *Journal of Animal Ecology*, **70** (6), 1032–45.

Babu, R-M., Sajeena, A., Seetharam, K. & Reddy, M.S. (2003) Advances in genetically engineered (transgenic) plants in pest management: an over view. *Crop Protection*, **22** (9), 1071–86.

Babuder, G., Pohleven, F. & Brelih, S. (1996) Selectivity of synthetic aggregation pheromones Linoprax© and Pheroprax© in the control of the bark beetles (Coleoptera: Scolytidae) in a timber storage yard. *Journal of Applied Entomology*, **120**, 131–6.

Bach, C.E. (2001) Long-term effects of insect herbivory and sand accretion on plant succession on sand dunes. *Ecology*, **82**, 1401–16.

Badenes-Perez, F.R. & Shelton, A.M. (2006) Pest management and other agricultural practices among farmers growing cruciferous vegetables in the Central and Western highlands of Kenya and the Western Himalayas of India. *International Journal of Pest Management*, **52** (4), 303–15.

Baguette, M. (2004) The classical metapopulation theory and the real, natural world: a critical appraisal. *Basic and Applied Ecology*, **5**, 213–24.

Baguette, M., Mennechez, G., Petit, S. & Schtickzelle, N. (2003) Effect of habitat fragmentation on dispersal in the butterfly *Proclossiana eunomia*. *Comptes Rendus Biologies*, **326**, S200–S209.

Baillie, J.E.M., Hilton-Taylor, C. & Stuart, S.N. (2004) *2004 IUCN Red List of Threatened Species. A Global Species Assessment*. IUCN (The World Conservation Union), Gland, Switzerland and Cambridge, UK.

Bajaj, S. & Mohanty, A. (2005) Recent advances in rice biotechnology – towards genetically superior transgenic rice. *Plant Biotechnology Journal*, **3** (3), 275–307.

Baker, J.E., Perez-Mendoza, J. & Beeman, R.W. (1997) Inheritance of malathion resistance in the parasitoid *Anisopteromalus calandrae* (Hymenoptera: Pteromalidae). *Journal of Economic Entomology*, **90** (2), 304–8.

Bakhvalov, S.A. & Bakhvalova, V.N. (1990) Ecology of a baculovirus in *Ocneria monacha* L. (Lepidoptera: Lymantriidae): virus persistence in insect populations. *Ekologiya (Sverdlovsk)*, **6**, 53–9.

Bakker, K., Bagchee, S.N., Van Zwet, W.R. & Meelis, E. (1967) Host discrimination in *Pseudeucoila bochei* (Hymenoptera: Cynipidae). *Entomologia Experimentalis et Applicata*, **10**, 295–311.

Balas, M.T. & Adams, E.S. (1996) Nestmate discrimination and competition in incipient colonies of fire ants. *Animal Behavior*, **51**, 49–59.

Baldi, A. (2003) Using higher taxa as surrogates of species richness: a study based on 3700 Coleoptera, Diptera, and *Acari* species in Central-Hungarian reserves. *Basic and Applied Ecology*, **4**, 589–93.

Baldwin, I.T. (1998) Jasmonate-induced responses are costly but benefit plants under attack in native population. *Proceedings of the National Academy of Sciences of the USA*, **95**, 8113–18.

Baldwin, I.T. & Schultz, J.C. (1983) Rapid changes in tree leaf chemistry induced by damage: Evidence for communication between plants. *Science*, **221**, 277–9.

Baldwin, I.T. & Schultz, J.C. (1988) Phylogeny and the patterns of leaf phenolics in gap and forest-adapted *Piper* and *Miconia* understory shrubs. *Oecologia (Heidelberg)*, **75**, 105–9.

Bale, J.S., Masters, G.J., Hodkinson, I.D., et al. (2002) Herbivory in global climate change research: direct effects of rising temperature on insect herbivores. *Global Change Biology*, **8**, 1–16.

Ball, S.G. & Morris, R.K.A. (2000) *Provisional Atlas of British Hoverflies (Diptera, Syrphidae)*. Biological Records Centre, Huntingdon, UK.

Ball, S.L. & Armstrong, K.F. (2006) DNA barcodes for insect pest identification: a test case with tussock moths

(Lepidoptera: Lymantriidae). *Canadian Journal of Forest Research (Revue Canadienne De Recherche Forestiere)*, **36**, 337–50.

Balmford, A., Bennun, L., ten Brink, B. et al. (2005) The convention on biological diversity's 2010 target. *Science*, **307**, 212–13.

Balmford, A., Bruner, A., Cooper, P. et al. (2002) Ecology – economic reasons for conserving wild nature. *Science*, **297**, 950–3.

Balmford, A., Green, M.J.B. & Murray, M.G. (1996a) Using higher-taxon richness as a surrogate for species richness. 1. Regional tests. *Proceedings of the Royal Society of London Series B, Biological Sciences*, **263**, 1267–74.

Balmford, A., Jayasuriya, A.H.M. & Green, M.J.B. (1996b) Using higher-taxon richness as a surrogate for species richness. 2. Local applications. *Proceedings of the Royal Society of London Series B, Biological Sciences*, **263**, 1571–5.

Balmford, A., Lyon, A.J.E. & Lang, R.M. (2000) Testing the higher-taxon approach to conservation planning in a megadiverse group: the macrofungi. *Biological Conservation*, **93**, 209–17.

Baltensweiler, W. (1984) The role of environment and reproduction in the population dynamics of the larch budmoth *Zieraphera diniana* Gn. (Lep.: Torticidae). In *Advances in Invertebrate Reproduction*, Vol. 3 (eds W. Engels, W.H. Clark, A. Fischer, P.J.W. Olive & F.F. Went), pp. 291–301. Elsevier, Amsterdam.

Baltensweiler, W. & Fischlin, A. (1988) The larch budmoth in the Alps. In *Dynamics of Forest Insect Populations: Patterns, Causes, Implications* (ed. A.A. Berryman), pp. 331–51. Plenum, New York.

Balvanera, P., Pfisterer, A.B., Buchmann, N., et al. (2006) Quantifying the evidence for biodiversity effects on ecosystem functioning and services. *Ecology Letters*, **9**, 1146–56.

Barbour, D.A. (1985) Patterns of population fluctuation in the pine looper moth *Bupalus piniaria* L. in Britain. In *Site Characteristics and Population Dynamics of Lepidotern and Hymenopteran Forest Pests* (eds D. Bevan & J.T. Stoakley), pp. 8–20. HMSO, London.

Barcic, J.I., Bazok, R., Bezjak, S., Culjak, T.G. & Barcic, J. (2006) Combinations of several insecticides used for integrated control of Colorado potato beetle (*Leptinotarsa decemlineata*, Say., Coleoptera: Chrysomelidae). *Journal of Pest Science*, **79** (4), 223–32.

Bardgett, R.D., Wardle, D.A. & Yeates, G.W. (1998) Linking above-ground and below-ground interactions: how plant responses to foliar herbivory influence soil organisms. *Soil Biology and Biochemistry*, **30** (14), 1867–78.

Bari, A. & Qazilbash, A.A. (1995) Plague: a medieval killer in the present world: a review. *Punjab University Journal of Zoology*, **10**, 125–35.

Barlow, N.D. & Dixon, A.F.G. (1980) *Simulation of Lime Aphid Population Dynamics*. Pudoc, Wageningen.

Barnby, M.A. & Resh, V.H. (1988) Factors affecting the distribution of an endemic and a widespread species of

brine fly (Diptera: Ephydridae) in a northern California (USA) thermal saline spring. *Annals of the Entomological Society of America*, **81** (3), 437–46.

Barnes, R.K.S. (2001) *The Invertebrates: a Synthesis*. Blackwell Publications, Oxford.

Barquero, M. & Constenla, M.A. (1986) Remnants of organochlorine pesticides in human fatty tissue in Costa Rica. *Revista de Biologia Tropical*, **34** (1), 7–12.

Barrett, A.D.T. & Higgs, S. (2007) Yellow fever: a disease that has yet to be conquered. *Annual Review of Entomology*, **52**, 209–29.

Barrett, M.P., Burchmore, R.J.S., Stich, A. et al. (2003) The trypanosomiases. *Lancet (North American Edition)*, **362**, 1469–80.

Barrett, R.D.H. & Hebert, P.D.N. (2005) Identifying spiders through DNA barcodes. *Canadian Journal of Zoology- (Revue Canadienne de Zoologie)*, **83**, 481–91.

Bartlein, P.J., Whitlock, C. & Shafter, S.L. (1997) Future climate in the Yellowstone National Park region and its potential impact on vegetation. *Conservation Biology*, **11**, 782–92.

Bartlet, E., Parson, D., Williams, I.H. & Clark, S.J. (1994) The influence of glucosinolates and sugars on feeding by the cabbage stem flea beetle, *Psylliodes chrysocephala*. *Entomologia Experimentalis et Applicata*, **73**, 77–83.

Bartlett, C.R. & Bowman, J.L. (2003) Preliminary inventory of the planthoppers (Hemiptera: Fulgoroidea) of the Great Smoky Mountains National Park, North Carolina and Tennessee, USA. *Entomological News*, **114**, 246–54.

Barto, E.K. & Cipollini, D. (2005) Testing the optimal defense theory and the growth-differentiation balance hypothesis in Arabidopsis thaliana. *Oecologia*, **146**, 169–78.

Barton, K.E. (2007) Early ontogenetic patterns in chemical defense in Plantago (Plantaginaceae): genetic variation and trade-offs. *American Journal of Botany*, **94**, 56–66.

Basaglia, M., Concheri, G., Cardinali, S., Pasti-Grigsby, M.B. & Nuti, M.P. (1992) Enhanced degradation of ammonium-pretreated wheat straw by lignocellulolytic *Streptomyces* spp. *Canadian Journal of Microbiology*, **38**, 1022–5.

Basanez, M.G., Townson, H., Williams, J.R., Frontado, H., Villamizar, N.J. & Anderson, R.M. (1996) Density-dependent processes in the transmission of human onchocerciasis: relationship between microfilarial intake and mortality of the simuliid vector. *Parasitology*, **113** (4), 331–55.

Baskerville, G.L. (1975) Spruce budworm – super silviculturist. *Forestry Chronicle*, **51**, 138–40.

Bass, M. & Cherett, J.M. (1995) Fungal hyphae as a source of nutrients for the leaf-cutting ant *Atta sexdens*. *Physiological Entomology*, **20** (1), 1–6.

Basset, Y. (1991) The taxonomic composition of the arthropod fauna associated with an Australian rain-forest tree. *Australian Journal of Zoology*, **39**, 171–90.

Basset, Y. (1992) Host specificity of arboreal and free-living insect herbivores in rain forests. *Biological Journal of the Linnean Society*, **47** (2), 115–33.

Basset, Y., Charles, E., Hammond, D.S. & Brown, V.K. (2001) Short-term effects of canopy openness on insect herbivores in a rain forest in Guyana. *Journal of Applied Ecology*, **38**, 1045–58.

Basset, Y., Mavoungou, J.F., Mikissa, J.B. et al. (2004a) Discriminatory power of different arthropod data sets for the biological monitoring of anthropogenic disturbance in tropical forests. *Biodiversity and Conservation*, **13**, 709–32.

Basset, Y., Novotny, V., Miller, S.E. & Kitching, R.L. (eds.) (2003) *Arthropods of Tropical Forests. Spatio-temporal Dynamics and Resource Use in the Canopy*. Cambridge University Press, Cambridge, UK.

Basset, Y., Novotny, V., Miller, S.E., Weiblen, G.D., Missa, O. & Stewart, A.J.A. (2004b) Conservation and biological monitoring of tropical forests: the role of parataxonomists. *Journal of Applied Ecology*, **41**, 163–74.

Basset, Y., Samuelson, G.A., Allison, A. & Miller, S.E. (1996) How many species of host-specific insects feed on a species of tropical tree? *Biological Journal of the Linnean Society*, **59**, 201–16.

Bassuk, N.L., Hunter, L.D. & Howard, B.H. (1981) The apparent involvement of polyphenol oxidase and phloridzin in the production of apple rooting cofactors. *Journal of Horticultural Science*, **56**, 313–22.

Batchelor, T.P., Hardy, I.C.W., Barrera, J.F. & Perez-Lachaud, G. (2005) Insect gladiators II: competitive interactions within and between bethylid parasitoid species of the coffee berry borer, *Hypothenemus hampei* (Coleoptera: Scolytidae). *Biological Control*, **33** (2), 194–202.

Bauer, H. (2003) Local perceptions of Waza National Park, northern Cameroon. *Environmental Conservation*, **30**, 175–81.

Bauerfeind, S.S. & Fischer, K. (2005) Effects of food stress and density in different life stages on reproduction in a butterfly. *Oikos*, **111** (3), 514–24.

Baumgartner, J., Schulthess, F. & Xia, Y. (2003) Integrated arthropod pest management systems for human health improvement in Africa. *Insect Science and its Application*, **23** (2), 85–98.

Bayoh, M.N. & Lindsay, S.W. (2003) Effect of temperature on the development of the aquatic stages of *Anopheles gambiae* sensu stricto (Diptera: Culicidae). *Bulletin of Entomological Research*, **93**, 375–81.

Bazely, D.R., Da Myers, J.H. & Silva, K.B. (1991) The response of numbers of bramble prickles to herbivory and depressed resource availability. *Oikos*, **61**, 327–36.

Beanland, L., Phelan, P.L. & Salminen, S. (2003) Micronutrient interactions on soybean growth and the developmental performance of three insect herbivores. *Environmental Entomology*, **32** (3), 641–51.

Beaver, R.A. (1989) Insect–fungus relations in the bark and ambrosia beetles. In *Insect–Fungus Interactions. 14th Symposium of the Royal Entomological Society of London in Collaboration with the British Mycological Society* (eds N. Wilding, N.M. Collins, P.M. Hammond & J.F. Webber), pp. 121–43. Academic Press, London.

Beccaloni, G.W. & Gaston, K.J. (1995) Predicting the species richness of neotropical forest butterflies – Ithomiinae (Lepidoptera, Nymphalidae) as indicators. *Biological Conservation*, **71**, 77–86.

Becerra, J.X. (1994) Squirt-gun defense in *Bursera* and the Chrysomelid counterploy. *Ecology*, **75**, 1991–6.

Becerra, J.X. (1997) Insects on plants: macroevolutionary chemical trends in host use. *Science*, **276**, 253–6.

Becerra, J.X. (2003) Synchronous coadaptation in an ancient case of herbivory. *Proceedings of the National Academy of Sciences of the USA*, **100**, 12804–7.

Becerra, J.X. (2004) Molec.ular systematics of *Blepharida* beetles (Chrysomelidae: Alticinae) and relatives. *Molecular Phylogenetics and Evolution*, **30** (1), 107–17.

Becerra, J.X. & Venable D.L. (1999) Macroevolution of insect–plant associations: the relevance of host biogeography to host affiliation. *Proceedings of the National Academy of Sciences of the USA*, **96**, 12626–31.

Beck, J., Kitching, I.J. & Linsenmair, K.E. (2006) Wallace's line revisited: has vicariance or dispersal shaped the distribution of Malesian hawkmoths (Lepidoptera: Sphingidae)? *Biological Journal of the Linnean Society*, **89** (3), 455–68.

Beck, L.R., Rodriguez, M.H., Dister, S.W. et al. (1994) Remote sensing as a landscape epidemiological tool to identify villages at high risk for malaria transmission. *American Journal of Tropical Medicine and Hygiene*, **51** (3), 271–80.

Beckers, G.J.M. & Spoel, S.H. (2006) Fine-tuning plant defence signalling: salicylate versus jasmonate. *Plant Biology*, **8**, 1–10.

Beckett, S.J. & Evans, D.E. (1994) The demography of *Oryzaephilus surinamensis* (L) (Coleoptera, Silvanidae) on kibbled wheat. *Journal of Stored Products Research*, **30** (2), 121–37.

Beddington, J.R., Free, C.A. & Lawton, J.H. (1975) Dynamic complexity in predator–prey models framed in difference equations. *Nature*, **225**, 58–60.

Begon, M., Harper, J.L. & Townsend C.R. (1986) *Ecology – Individuals, Populations and Communities*. Blackwell Scientific Publications, Oxford.

Begon, M., Harper, J.L. & Townsend, C.R. (1990) *Ecology – Individuals Populations and Communities*. Blackwell Scientific Publications, Cambridge, MA.

Beier, J.C. (1998) Malaria parasite development in mosquitoes. *Annual Review of Entomology*, **43**, 519–43.

Beier, J.C., Perkins, P.V., Koros, J.K., et al. (1990) Malaria Sporozoite Detection by Dissection and ELISA to assess infectivity of Afrotropical *Anopheles* (Diptera, Culicidae). *Journal of Medical Entomology*, **27** (3), 377–84.

Beingolea, G.O.D. (1985) The locust *Schistocerca interrita* in northern coast of Peru during 1983. *Revista Peruana de Entomologia*, **28**, 35–40.

Bejer-Peterson, B. (1972) The nun moth, *Lymantria monacha* L., in Denmark (Lep., Lymantriidae). *Entomologiske Meddelelser*, **40**, 129–39.

Belete, H., Tikubet, G., Petros, B., Oyibo, W-A. & Otigbuo, I-N. (2004) Control of human African trypanosomiasis: trap and odour preference of tsetse flies (*Glossina morsitans submorsitans*) in the upper Didessa river valley of Ethiopia. *Tropical Medicine and International Health*, **9** (6), 710–14.

Bell, G., Lechowicz, M.J., Appenzeller, A. et al. (1993) The spatial structure of the physical environment. *Oecologia (Heidelberg)*, **96** (1), 114–21.

Belli, G., Fortusini, A., Casati, P., Belli, L., Bianco, P.A. & Prati, S. (1994) Transmission of a grapevine leafroll associated closterovirus by the scale insect *Pulvinaria vitis* L. *Rivista di Patologia Vegetale*, **4** (3), 105–8.

Belliure, B., Janssen, A., Maris, P.C., Peters, D. & Sabelis, M.W. (2005) Herbivore arthropods benefit from vectoring plant viruses. *Ecology Letters*, **8** (1), 70–9.

Bellotti, A.C. & Arias, B. (2001) Host plant resistance to whiteflies with emphasis on cassava as a case study. *Crop Protection*, **20**, 813–23.

Belovsky, G.E. & Joern, A. (1995) Regulation of rangeland grasshoppers: differing dominant mechanisms in space and time. In *Population Dynamics: New Approaches and Synthesis* (eds N. Cappuccino & P.W. Price), pp. 359–86. Academic Press, London.

Belovsky, G.E. & Slade, J.B. (2000) Insect herbivory accelerates nutrient cycling and increases plant production. *Proceedings of the National Academy of Sciences of the USA*, **97**, 14414–17.

Belsky, A.J., Carson, W.P., Jensen, C.L. & Fox, G.A. (1993) Overcompensation by plants, herbivore optimization or red herring? *Evolutionary Ecology*, **7**, 109–21.

Benedick, S., Hill, J.K., Mustaffa, N. et al. (2006) Impacts of rain forest fragmentation on butterflies in northern Borneo: species richness, turnover and the value of small fragments. *Journal of Applied Ecology*, **43**, 967–77.

Benedict, M.Q. & Robinson, A.S. (2003). The first releases of transgenic mosquitoes: an argument for the sterile insect technique. *Trends in Parasitology*, **19**, 349–55.

Bengtsson, M., Karpati, Z., Szocs, G., Reuveny, H., Yang, Z.H. & Witzgall, P. (2006) Flight tunnel responses of Z strain European corn borer females to corn and hemp plants. *Environmental Entomology*, **35** (5), 1238–43.

Bennett, S. & Mill, P.J. (1995) Lifetime egg production and egg mortality in the damselfly *Pyrrhosoma nymphula* (Sulzer) (Zygoptera: Coenagrionidae). *Hydrobiologia*, **310** (1), 71–8.

Bennewicz, J. (1995) Assessment of susceptibility of varieties and lines of sugar beet to feeding of aphids: black bean aphid (*Aphis fabae* Scop.) and green peach aphid (*Myzus persicae* Sulz.). *Hodowla Roslin Aklimatyzacja i Nasiennictwo*, **39** (3), 41–72.

Benrey, B. & Denno, R.F. (1997) The slow-growth–high-mortality hypothesis: a test using the cabbage butterfly. *Ecology*, **78**, 987–99.

Bento, J.M.S., de Moraes, G.J.C., Bellotti A., Castillo, J.A., Warumby, J.F. & Lapointe, S.L. (1999) Introduction of parasitoids for the control of the cassava mealybug *Phenacoccus herreni* (Hemiptera: Pseudococcidae) in north-eastern Brazil. *Bulletin of Entomological Research*, **89** (5), 403–10.

Benz, G. (1974) Negative feedback by competition for food and space, and by cyclic induced changes in the nutritional base as regulatory principles in the population dynamics of the larch budmoth, *Zeiraphera diniana* (Guenee) (Lep., Tortricidae). *Zeitschrift fuer Angewandte Entomologie*, **76**, 196–228.

Berenbaum, M. (1980) Adaptive significance of midgut pH in larval lepidoptera. *American Naturalist*, **115**, 138–46.

Berenbaum, M.R. (1983) Coumarins and caterpillars: a case for coevolution. *Evolution*, **37**, 163–79.

Berenbaum, M.R. (1995) Turnabout is fair play: secondary roles for primary compounds. *Journal of Chemical Ecology*, **21** (7), 925–40.

Berg, H. (1995) Modelling of DDT dynamics in Lake Kariba, a tropical man-made lake, and its implications for the control of tsetse flies. *Annales Zoologici Fennici*, **32** (3), 331–53.

Berg, H. (2001) Pesticide use in rice and rice-fish farms in the Mekong Delta, Vietnam. *Crop Protection*, **20**, 897–905.

Bergelson, J. & Crawley, M.J. (1992) The effects of grazers on the performance of individuals and populations of scarlet gilia, *Ipomopsis aggregata*. *Oecologia (Heidelberg)*, **90**, 435–44.

Bergivinson, D.J., Hamilton, R.I. & Arnason, J.T. (1995a) Leaf profile of maize resistance factors to European corn borer, *Ostrinia nubilalis*. *Journal of Chemical Ecology*, **21**, 343–54.

Bergivinson, D.J., Larsen, J.S. & Arnason, J.T. (1995b) Effect of light on changes in maize resistance against the European corn borer, *Ostrinia nubilalis* (Hubner). *Canadian Entomologist*, **127**, 111–22.

Bernal, J.S., Luck, R.F. & Morse, J.G. (1999) Host influences on sex ratio, longevity, and egg load of two *Metaphycus* species parasitic on soft scales: implications for insectary rearing. *Entomologia Experimentalis et Applicata*, **92**, 191–204.

Bernal, J.S. & Setamou, M. (2003) Fortuitous antixenosis in transgenic sugarcane: antibiosis-expressing cultivar is refractory to ovipositing herbivore pests. *Environmental Entomology*, **32** (4), 886–94.

Bernays, E.A. & Graham, M. (1988) On the evolution of host specificity in phytophagous arthropods. *Ecology*, **69**, 886–92.

Berry, N.A., Wratten, S.D., Mcerlich, A. & Frampton, C. (1996) Abundance and diversity of beneficial arthropods in conventional and organic carrot crops in New Zealand. *New Zealand Journal of Crop and Horticultural Science*, **24** (4), 307–13.

Berry, P.M., Dawson, T.P., Harrison, P.A. & Pearson, R.G. (2002) Modelling potential impacts of climate change on the bioclimatic envelope of species in Britain and Ireland. *Global Ecology and Biogeography*, **11**, 453–62.

Berryman, A.A. (1973) Population dynamics of the fir engraver, *Scolytus ventralis* (Coleoptera: Scolytidae). I. Analysis of population behavior and survival from 1964 to 1971. *Canadian Entomologist*, **105**, 1465–88.

Berryman, A.A. (1987) The theory and classification of outbreaks. In *Insect Outbreaks* (eds P. Barbosa & J.C. Schultz), pp. 3–29. Academic Press, San Diego.

Berryman, A.A. (1999) *Principles of Population Dynamics and their Application*. Samuel Thornes, London.

Beshers, S.N. & Traniello, J.F.A. (1994) The adaptiveness of worker demography in the attine ant *Trachymyrmex septentrionalis*. *Ecology (Tempe)*, **75** (3), 763–75.

Béthoux, O. (2005) Cnemidolestodea (Insecta): an ancient order reinstated. *Journal of Systematic Palaeontology*, **3**, 403–8.

Béthoux, O., J. McBride, J. & Maul C. (2004). Surface laser scanning of fossil insect wings. *Palaeontology (Oxford)*, **47** (1), 13–19.

Bevan, D. (1987) *Forest Insects. Forestry Commission Handbook 1*. HMSO, London.

Bezemer, T.M. & Jones, T.H. (1998) Plant–insect herbivore interactions in elevated atmospheric CO_2: quantitative analyses and guild effects. *Oikos*, **82**, 212–22.

Bibby, C.J., Crosby, M.J., Heath, M.F., et al. (1992) *Putting Biodiversity on the Map: Global Priorities for Conservation*. International Council for Bird Preservation, Cambridge.

Bignell, D.E., Eggleton, P., Nunes, L. & Thomas, K.L. (1997) Termites as mediators of carbon fluxes in tropical forests: budgets for carbon dioxide and methane emissions. In *Forests and Insects* (eds A.D. Watt, N.E. Stork & M.D. Hunter), pp. 109–34. Chapman & Hall, London.

Bignell, D.E., Slaytor, M., Veivers, P.C., Muhlemann, R. & Leuthold, R.H. (1994) Functions of symbiotic fungus gardens in higher termites of the genus *Macrotermes*: evidence against the acquired enzyme hypothesis. *Acta Microbiologica et Immunologica Hungarica*, **41** (4), 391–401.

Biji, C.P., Sudheendrakumar, V.V. & Sajeev, T.V. (2006) Quantitative estimation of *Hyblaea puera* NPV production in three larval stages of the teak defoliator, *Hyblaea puera* (Cramer). *Journal of Virological Methods*, **136**, 78–82.

Biller, A., Boppre, M., Witte, L. & Hartmann, T. (1994) Pyrrolizidine alkaloids in *Chromolaena odorata*: chemical and chemoecological aspects. *Phytochemistry (Oxford)*, **35**, 615–19.

Birch, L.C. (1953) Experimental background to the study of the distribution and abundance of insects. I. The influence of temperature, moisture, and food on the innate capacity for increase of three grain beetles. *Ecology*, **34**, 698–711.

Birch, M.C. (1978) Chemical communication in pine bark beetles. *American Scientist*, **66**, 409–19.

Birch, M.C., Svihra, P., Paine, T.D. & Miller, J.C. (1980) Influence of chemically mediated behavior on host tree colonization by four cohabitating species of bark beetles. *Journal of Chemical Ecology*, **6**, 395–414.

Birkett, M.A., Campbell, C.A.M., Chamberlain, K. et al. (2000) New roles for cis-jasmone as an insect semiochemical and in plant defense. *Proceedings of the National Academy of Sciences of the USA*, **97**, 9329–34.

Birkle, L.M., Minto, L.B. & Douglas, A.E. (2002). Relating genotype and phenotype for tryptophan synthesis in an aphid–bacterial symbiosis. *Physiological Entomology*, **27** (4), 302–6.

Bishop, J.G. (2002) Early primary succession on Mount St. Helens: impact of insect herbivores on colonizing lupines. *Ecology*, **83**, 191–202.

Bissan, Y., Hougard, J.M., Doucoure, K., et al. (1995) Drastic reduction of populations of *Simulium sirbanum* (Diptera: Simuliidae) in central Sierra Leone after 5 years of larviciding operations by the Onchocerciasis Control Programme. *Annals of Tropical Medicine and Parasitology*, **89** (1), 63–72.

Bitsch, C. & Bitsch, J. (2004) Phylogenetic relationships of basal hexapods among the mandibulate arthropods: a cladistic analysis based on comparative morphological characters. *Zoologica Scripta*, **33** (6), 511–50.

Bjornstad, O.N., Peltonen, M., Liebold, A.M. & Baltensweiler, W. (2002). Waves of larch budmoth outbreaks in European Alps. *Science*, **298**, 1020–3.

Blackman, R.L. & Eastop, V.F. (1984) *Aphids on the World's Crops: an Identification and Information Guide*. Wiley, Chichester, UK.

Blackman, M.W. (1924) The effect of deficiency and excess in rainfall upon the hickory bark beetle. *Journal of Economic Entomology*, **17**, 460–70.

Blakely, N. & Dingle, H. (1978) Butterflies eliminate milkweed bugs from a Carribean island. *Oecologia*, **37**, 133–6.

Blank, R.H., Gill, G.S.C., Oslon, M.H. & Upsdell, M.P. (1995) Greedy scale (Homoptera: Diaspididae) phenology on taraire based on Julian day and degree-day accumulations. *Environmental Entomology*, **24** (6), 1569–75.

Blankenhorn, W.U. (2000) The evolution of body size: what keeps organisms small? *Quarterly Review of Biology*, **75**, 385–40.

Blattner, F.R., Weising, K., Banfer, G., Maschwitz, U. & Fiala, B. (2001) Molecular analysis of phylogenetic relationships among myrmecophytic *Macaranga* species (Euphorbiaceae). *Molecular Phylogenetics and Evolution*, **19**, 331–44.

Blossey, B. & Hunt-Joshi, T.R. (2003) Belowground herbivory by insects: influence on plants and aboveground herbivores. *Annual Review of Entomology*, **48**, 521–47.

Bluthgen, N. & Wesenberg, J. (2001) Ants induce domatia in a rain forest tree (*Vochysia vismiaefolia*). *Biotropica*, **33**, 637–42.

Bockarie, M.J., Service, M.W., Toure, Y.T., Traore, S., Barnish, G. & Greenwood, B.M. (1993) The ecology and behaviour of the forest form of *Anopheles gambiae s.s. Parassitologia*, **35** (Suppl), 5–8.

Boecklen, W.J. & Spellenberg, R. (1990) Structure of herbivore communities in two oak (*Quercus* spp.) hybrid zones. *Oecologia (Heidelberg)*, **85**, 92–100.

Boege, K. & Marquis, R.J. (2006) Plant quality and predation risk mediated by plant ontogeny: consequences for herbivores and plants. *Oikos*, **115**, 559–72.

Bokononganta, A.H., Vanalphen, J.J.M. & Neuenschwander, P. (1996) Competition between *Gyranusodea tebygi* and *Anagyrus mangicola*, parasitoids of the mango mealybug, *Rastrococcus invadens*. Interspecific host discrimination and larval competition. *Entomologia Experimentalis et Applicata*, **79**, 179–85.

Bombick, D.W., Arlian, L.G. & Livingston, J.M. (1987) Toxicity of jet fuels to several terrestrial insects. *Archives of Environmental Contamination and Toxicology*, **16** (1), 111–18.

Bond W.J. (1994) Do mutualisms matter – assessing the impact of pollinator and disperser disruption on plant extinction. *Philosophical Transactions of the Royal Society of London Series B, Biological Sciences*, **344**, 83–90.

Bond, W.J. & Stock, W.D. (1989) The costs of leaving home: ants disperse myrmecochorous seeds to low nutrient sites. *Oecologia*, **81**, 412–17.

Bonsall, M.B., Jones, T.H. & Perry, J.N. (1998) Determinants of dynamics: population size, stability and persistence. *Trends in Ecology and Evolution*, **13**, 174–6.

Bonsall, M.B., Van der Meijden, E. & Crawley, M.J. (2003) Contrasting dynamics in the same plant–herbivore interaction. *Proceedings of the National Academy of Sciences of the USA*, **100**, 14932–6.

Borchert, R. (1994) Water status and development of tropical trees during seasonal drought. *Trees*, **8**, 115–25.

Borden, J.H. (1984) Semiochemical-mediated aggregation and dispersion. In *Insect Communication. 12th Symposium of the Royal Entomological Society, London* (ed. T. Lewis), pp. 123–49. Academic Press, San Diego, CA.

Borden, J.H., Birmingham, A.L. & Burleigh, J.S. (2006) Evaluation of the push–pull tactic against the mountain pine beetle using verbenone and non-host volatiles in combination with pheromone-baited trees. *Forestry Chronicle*, **82** (4), 579–90.

Borges, P.A.V., Aguiar, C., Amaral, J. et al. (2005) Ranking protected areas in the Azores using standardised sampling of soil epigean arthropods. *Biodiversity and Conservation*, **14**, 2029–60.

Borges, P.A.V. & Brown, V.K. (2004) Arthropod community structure in pastures of an island archipelago (Azores): looking for local–regional species richness patterns at fine-scales. *Bulletin of Entomological Research*, **94**, 111–21.

Borkent, A. (1996) Biting midges from Upper Cretaceous New Jersey amber (Ceratopogonidae: Diptera). *American Museum Novitates*, **3159**, 1–29.

Borror, D.J., De Long, D.M. & Triplehorn, C.A. (1981) *An Introduction to the Study of Insects*, 5th edn. Saunders College Publishing, Philadelphia, PA.

Bostock, R.M. (1999) Signal conflicts and synergies in induced resistances to multiple attackers. *Physiological and Molecular Plant Pathology*, **55**, 99–109.

Bouchard, M., Kneeshaw, D. & Bergeron, Y. (2006) Forest dynamics after successive spruce budworm outbreaks in mixedwood forests. *Ecology*, **87** (9): 2319–29.

Bouchard, R.W., Carrillo, M.A. & Ferrington, L.C. (2006) Lower lethal temperature for adult male *Diamesa mendotae* Muttkowski (Diptera: Chironomidae), a winter-emerging aquatic insect. *Aquatic Insects*, **28** (1), 57–66.

Boudreaux, H.B. (1987) *Arthropod Phylogeny with Special Reference to Insects*. Krieger, Melbourne.

Bouget, C. & Noblecourt, T. (2005) Short-term development of ambrosia and bark beetle assemblages following a windstorm in French broadleaved temperate forests. *Journal of Applied Entomology*, **129** (6), 300–10.

Bourchier, R.S. & Smith, S.M. (1996) Influence of environmental conditions and parasitoid quality on field performance of *Trichogramma minutum*. *Entomologia Experimentalis et Applicata*, **80** (3), 461–8.

Bourner, T.C. & Cory, J.S. (2004) Host range of an NPV and a GV isolated from the common cutworm, *Agrotis segetum*: pathogenicity within the cutworm complex. *Biological Control*, **31** (3), 372–9.

Bouwman, H., Sereda, B. & Meinhardt, H.M. (2006) Simultaneous presence of DDT and pyrethroid residues in human breast milk from a malaria endemic area in South Africa. *Environmental Pollution*, **144** (3), 902–17.

Bowie, M.H., Wratten, S.D. & White, A.J. (1995) Agronomy and phenology of 'companion plants' of potential for enhancement of biological control. *New Zealand Journal of Crop and Horticultural Science*, **23** (4), 423–7.

Box, G.E.P. & Jenkins, G.M. (1976) *Time Series Analysis: Forecasting and Control*. Holden-Day, San Francisco, CA.

Boyero, L., Pearson, R.G. & Camacho, R. (2006) Leaf breakdown in tropical streams: the role of different species in ecosystem functioning. *Archiv fur Hydrobiologie*, **166** (4), 453–66.

Braasch, H. (1996) Pathogenicity tests with *Bursaphelenchus mucronatus* on pine and spruce seedlings in Germany. *European Journal of Forest Pathology*, **26** (4), 205–16.

Braby, M.F. & Trueman, J.W.H. (2006) Evolution of larval host plant associations and adaptive radiation in pierid butterflies. *Journal of Evolutionary Biology*, **19**, 1677–90.

Braddy, S.J. & Briggs, D.E.G. (2002) New Lower Permian nonmarine arthropod trace fossils from New Mexico and South Africa. *Journal of Paleontology*, **76** (3), 546–57.

Bradshaw, W.E. (1983) Estimating biomass of mosquito populations. *Environmental Entomology*, **12** (3), 779–81.

Brady, S.G., Schultz, T.R., Fisher, B.L. & Ward, P.S. (2006) Evaluating alternative hypotheses for the early evolution and diversification of ants. *Proceedings of the National Academy of Sciences of the USA*, **103** (48), 18172–7.

Braendle, C., Davis, G.K., Brisson, J.A. et al. (2006) Wing dimorphism in aphids. *Heredity*, **97** (3), 192–9.

Braendle, C., Miura, T., Bickel, R., Shingleton, A-W., Kambhampati, S. & Stern, D-L. (2003) Developmental origin and evolution of bacteriocytes in the aphid–*Buchnera* symbiosis. *PLoS Biology*, **1**, 70–6.

Brandao, C.R.F., Martins-Neto, R.G. & Vulcano, M.A. (1989) The earliest known fossil ant (first southern hemisphere Mesozoic record) (Hymenoptera: Formicidae: Myrmeciinae). *Psyche*, **96** (3/4), 195–208.

Brandenburg, R.L. & Herbert, D.A. (1991) Effect of timing on prophylactic treatments for southern corn rootworm (Coleoptera: Chrysomelidae) in peanut. *Journal of Economic Entomology*, **84** (6), 1894–8.

Brandle, M., Knoll, S., Eber, S., Stadler, J. & Brandl, R. (2005) Flies on thistles: support for synchronous speciation? *Biological Journal of the Linnean Society*, **84**, 775–83.

Brasier, C.M. (1990) The unexpected element: mycovirus involvement in the outcome of two recent pandemics, Dutch elm disease and chestnut blight. In *Pests, Pathogens and Plant Communities* (eds J.J. Burdon & S.R. Leather), pp. 289–308. Blackwell Scientific Publications, Oxford.

Brasier, C.M. & Buck, K.W. (2001) Rapid evolutionary changes in a globally invading fungal pathogen (Dutch elm disease). *Biological Invasions*, **3**, 223–33.

Braun, S. & Fluckiger, W. (1984) Increased population of the aphid *Aphis pomi* at a motorway. Part 2. The effect of drought and deicing salt. *Environmental Pollution Series*, **36 A**, 261–70.

Brehm, G., Colwell, R.K. & Kluge, J. (2007) The role of environment and mid-domain effect on moth species richness along a tropical elevational gradient. *Global Ecology and Biogeography*, **16**, 205–19.

Brehm, G., Homeier, J. & Fiedler, K. (2003) Beta diversity of geometrid moths (Lepidoptera: Geometridae) in an Andean montane rainforest. *Diversity and Distributions*, **9** (5), 351–66.

Brehm, G., Pitkin, L.M., Hilt, N. & Fiedler, K. (2005) Montane Andean rain forests are a global diversity hotspot of geometrid moths. *Journal of Biogeography*, **32**, 1621–7.

Brennan, K.E.C., Ashby, L., Majer, J.D., Moir, M.L. & Koch, J.M. (2006) Simplifying assessment of forest management practices for invertebrates: how effective are higher taxon and habitat surrogates for spiders following prescribed burning? *Forest Ecology and Management*, **231**, 138–54.

Briddon, R.W. (2003) Cotton leaf curl disease, a multicomponent begomovirus complex. *Molecular Plant Pathology*, **4** (6), 427–34.

Briere, J.F., Pracros, P., Le Roux, A.Y. & Pierre, J.S. (1999) A novel rate model of temperature-dependent development for arthropods. *Environmental Entomology*, **28** (1), 22–9.

Briers, R.A., Cariss, H.M. & Gee, J.H.R. (2003) Flight activity of adult stoneflies in relation to weather. *Ecological Entomology*, **28** (1), 31–40.

Brodmann, P.A., Wilcox, C.V. & Harrison, S. (1997) Mobile parasitoids may restrict the spatial spread of an insect outbreak. *Journal of Animal Ecology*, **66**, 65–72.

Bronstein, J.L., Alarcon, R. & Geber, M. (2006) The evolution of plant–insect mutualisms. *New Phytologist*, **172** (3), 412–28.

Brook, B.W., Sodhi, N.S. & Ng, P.K.L. (2003) Catastrophic extinctions follow deforestation in Singapore. *Nature*, **424**, 420–3.

Brooker, R., Young, J. & Watt, A.D. (2007) Climate change and biodiversity: impacts and policy development challenges – a European case study. *International Journal of Biodiversity Science and Management*, **3**, 12–30.

Brower, L.P. & Brower, J.V.Z. (1964) Birds, butterflies and plant poisons: a study in ecological biochemistry. *Zoologica*, **49**, 137–59.

Brower, L.P., Nelson, C.J., Seiber, J.N., Fink, L.S. & Bond, C. (1988) Exaptation as an alternative to coevolution in the cardenolide-based chemical defense of monarch butterflies (*Danaus plexippus* L.) against avian predators. In *Chemical Mediation of Coevolution* (ed. K.C. Spencer), pp. 447–75. Academic Press, San Diego, CA.

Brown, J.M., Abrahamson, W.G., Packer, R.A. & Way, P.A. (1995) The role of natural enemy escape in a gallmaker host-plant shift. *Oecologia*, **104**, 52–60.

Brown, K.S. (1991) Conservation of neotropical environments: insects as indicators. In *The Conservation of Insects and their Habitats* (eds N.M. Collins & J.A. Thomas), pp. 350–404. Academic Press, London.

Brown, M.W. (1993) Resilience of the natural arthropod community on apple to external disturbance. *Ecological Entomology*, **18** (3), 169–83.

Brown, M.W., Nebeker, T.E. & Honea, C.R. (1987) Thinning increases loblolly pine vigour and resistance to bark beetles. *Southern Journal of Applied Forestry*, **11** (1), 28–31.

Brown, V.K. & Gange, A.C. (1989) Herbivory by soil-dwelling insects depresses plant species richness. *Functional Ecology*, **3**, 667–71.

Brown, V.K. & Gange, A.C. (1992) Secondary plant succession: how is it modified by insect herbivory? *Vegetatio*, **101**, 3–13.

Browne, J. & Peck, S.B. (1996) The long-horned beetles of south Florida (Cerambycidae: Coleoptera): biogeography and relationships with the Bahama Islands and Cuba. *Canadian Journal of Zoology*, **74** (12), 2154–69.

Bruchim, Y., Ranen, E., Saragusty, J. & Aroch, I. (2005) Severe tongue necrosis associated with pine processionary moth (*Thaumetopoea wilkinsoni*) ingestion in three dogs. *Toxicon*, **45** (4), 443–7.

Bruhl, C.A., Eltz, T. & Linsenmair, K.E. (2003) Size does matter – effects of tropical rainforest fragmentation on the leaf litter ant community in Sabah, Malaysia. *Biodiversity and Conservation*, **12** (7), 1371–89.

Bruin, J., Dicke, M. & Sabelis, M.W. (1992) Plants are better protected against spider-mites after exposure to volatiles from infested conspecifics. *Experienta (Basel)*, **48**, 525–9.

Bruin, J., Sabelis, M.W. & Dicke, M. (1995) Do plants tap SOS signals from their infested neighbors? *Trends in Ecology and Evolution*, **10**, 167–70.

Brummitt, N. & Lughadha, E.N. (2003) Biodiversity: where's hot and where's not. *Conservation Biology*, **17**, 1442–8.

Bruna, E.M., Lapola, D.M. & Vasconcelos, H.L. (2004) Inter-specific variation in the defensive responses of obligate plant-ants: experimental tests and consequences for herbivory. *Oecologia*, **138**, 558–65.

Brunner, J.F., Beers, E.H., Dunley, J.E., Doerr, M. & Granger, K. (2005) Role of neonicotinyl insecticides in Washington apple integrated pest management. Part I. Control of lepidopteran pests. *Journal of Insect Science*, **5**, Article number 14.

Brusca, R.C. & Brusca, G.J. (2001) *Invertebrates*. Sinauer Associates, Sunderland, MA.

Bryant, J.P., Chapin, F.S. & Klein, D.R. (1983) Carbon/nutrient balance of boreal plants in relation to vertebrate herbivory. *Oikos*, **40**, 357–68.

Bryant, J.P., Reichardt, P.B., Clausen, T.P. & Werner, R.A. (1993) Effects of mineral nutrition on delayed inducible resistance in Alaska paper birch. *Ecology*, **74**, 2072–84.

Bryceson, K.P. (1989) The use of Landsat MSS data to determine the distribution of locust eggbeds in the Riverina region of New South Wales, Australia. *International Journal of Remote Sensing*, **10** (11), 1749–62.

Bryceson, K.P. (1990) Digitally processed satellite data as a tool in detecting potential Australian plague locust outbreak areas. *Journal of Environmental Management*, **30** (3), 191–208.

Buchner, P. (1965) *Endosymbiosis of Animals with Plant Microorganisms*. Wiley, New York.

Buckley, L.B. & Roughgarden, J. (2004) Biodiversity conservation – effects of changes in climate and land use. *Nature*, **430**.

Bulmer, M. (1994) *Theoretical Evolutionary Ecology*. Sinauer Associates, Sunderland, MA.

Bultman, T.L., Borowicz, K.L., Schneble, R.M., Coudron, T.A. & Bush, L.P. (1997) Effect of a fungal endophyte on the growth and survival of two *Euplectrus* parasitoids. *Oikos*, **78**, 170–6.

Bultman, T.L. & Faeth, S.H. (1985) Patterns of intra- and inter-specific association in leaf-mining insects on three oak species. *Ecological Entomology*, **10**, 121–9.

Bultman, T.L. & Faeth, S.H. (1987) Impact of irrigation and experimental drought stress on leaf-mining insects of Emory oak. *Oikos*, **48**, 5–10.

Burdon, J.J., Thrall, P.H. & Ericson, L. (2006) The current and future dynamics of disease in plant communities. *Annual Review of Phytopathology*, **44**, 19–39.

Burges, H.D. (1993) Foreword. In Bacillus thuringiensis, an Environmental Biopesticide: Theory and Practice (eds P.F. Entwistle, J.S. Cory, M.J. Bailey & S. Higgs), pp. xv–xvii. John Wiley, Chichester, UK.

Burghardt, F., Proksch, P. & Fiedler, K. (2001) Flavonoid sequestration by the common blue butterfly *Polyommatus icarus*: quantitative intraspecific variation in relation to larval hostplant, sex and body size. *Biochemical Systematics and Ecology*, **29**, 875–89.

Burkot, T.R., Graves, P.M., Paru, R., Wirtz, R.A. & Heywood, P.F. (1988) Human malaria transmission studies in the *Anopheles punctulatus* complex in Papua New Guinea: sporozoite rates, inoculation rates, and sporozoite densities. *American Journal of Tropical Medicine and Hygiene*, **39** (2), 135–44.

Burnett, T. (1958) Dispersal of an insect parasite over a small plot. *Canadian Entomologist*, **90**, 279–83.

Buse, A. & Good, J.E.G. (1996) Synchronization of larval emergence in winter moth (*Operophtera brumata* L.) and budburst in pedunculate oak (*Quercus robur* L.) under simulated climate change. *Ecological Entomology*, **21** (4), 335–43.

Bush, M.B. & Whittaker, R.J. (1991) Krakatau: colonization patterns and hierarchies. *Journal of Biogeography*, **18**, 341–56.

Bustamante-Sanchez, M.A., Grez, A.A. & Simonetti, J.A. (2004) Dung decomposition and associated beetles in a fragmented temperate forest. *Revista Chilena de Historia Natural*, **77** (1), 107–20.

Butchart, S.H.M., Stattersfield, A.J., Baillie, J. et al. (2005) Using Red List indices to measure progress towards the 2010 target and beyond. *Philosophical Transactions of the Royal Society of London Series B, Biological Sciences*, **360**, 255–68.

Butchart, S.H.M., Stattersfield, A.J., Bennun, L.A. et al. (2004) Measuring global trends in the status of biodiversity: Red List indices for birds. *Plos Biology*, **2**, 2294–304.

Byers, J.A. (1993) Avoidance of competition by spruce bark beetles, *Ips typographus* and *Pityogenes chalcographus*. *Experientia*, **49**, 272–5.

Bylund, H. & Tenow, O. (1994) Long-term dynamics of leaf miners, *Eriocrania* spp, on mountain birch – alternate year fluctuations and interaction with *Epirrita autumnata*. *Ecological Entomology*, **19**, 310–18.

Byrne, D.N., Rathman, R.J., Orum, T.V. & Palumbo, J.C. (1996) Localized migration and dispersal by the sweet potato whitefly, *Bemisia tabaci*. *Oecologia (Berlin)*, **105** (3), 320–8.

Cabaleiro, C. & Segura, A. (1997) Field transmission of grapevine leafroll associated virus 3 (GLRaV-3) by the mealybug *Planococcus citri*. *Plant Disease*, **81** (3), 283–7.

CABI (2006) www.cabi.org.

Cabrero-Sanudo, F.J. & Lobo, J.M. (2003) Estimating the number of species not yet described and their characteristics: the case of Western Palaearctic dung beetle species (Coleoptera, Scarabaeoidea). *Biodiversity and Conservation*, **12**, 147–66.

Cadogan, B-L., Scharbach, R-D., Brown, K.W., Ebling, P.M., Payne, N.J. & Krause, R.E. (2004). Experimental aerial application of a new isolate of the nucleopolyhedrovirus, CfMNPV against Choristoneura fumiferana (Lepidoptera: Tortricidae). *Crop Protection*, **23** (1), 1–9.

Cai, W., Yan, Y. & Li, L.Y. (2005) The earliest records of insect parasitoids in China. *Biological Control*, **32** (1), 8–11.

Calkins, C.O. & Faust, R.J. (2003) Overview of areawide programs and the program for suppression of codling moth in the western USA directed by the United States Department of Agriculture – Agricultural Research Service. *Pest Management Science*, **59** (6/7), 601–4.

Cambell, R.W. (1975) The gypsy moth and its natural enemies. *USDA Forest Service, Agricultural Information Bulletin*, **381**.

Cameron, P.J., Walker, G.P., Herman, T.J.B. & Wallace, A.R. (2001) Development of economic thresholds and monitoring systems for *Helicoverpa armigera* (Lepidoptera: Noctuidae) in tomatoes. *Journal of Economic Entomology*, **94**, 1104–12.

Cameron, T.C., Wearing, H.J., Rohani, P. & Sait, S.M. (2007) Two-species asymmetric competition: effects of age structure on intra- and interspecific interactions. *Journal of Animal Ecology*, **76** (1), 83–93.

Campbell, B.C., Bragg, T.S. & Turner, C.E. (1992) Phylogeny of symbiotic bacteria of four weevil species (Coleoptera: Curculionidae) based on analysis of 16S ribosomal DNA. *Insect Biochemistry and Molecular Biology*, **22** (5), 415–21.

Campbell, B.C. & Duffy, S.S. (1979) Tomatine and parasitic wasps: potential incompatability of plant antibiosis with biological control. *Science*, **205**, 700–2.

Campbell, C.A.M. & Muir, R.C. (2005) Flight activity of the damson-hop aphid, *Phorodon humuli*. *Annals of Applied Biology*, **147**, 109–18.

Camuffo, D. & Enzi, S. (1991) Locust invasions and climatic factors from the Middle Ages to 1800. *Theoretical and Applied Climatology*, **43** (1/2), 43–74.

Camus, P.A. & Barahona, R.M. (2002) Insectos del intermareal de Concepcion, Chile: perspectivas para la investigacion ecologica. *Revista Chilena de Historia Natural*, **75** (4), 793–803.

Cancrini, G., di Regalbono, A.F., Ricci, I., Tessarin, C., Gabrielli, S. & Pietrobelli, M. (2003) *Aedes albopictus* is a natural vector of *Dirofilaria immitis* in Italy. *Veterinary Parasitology*, **118** (3/4), 195–202.

Cardinale, B.J., Srivastava, D.S., Duffy, J.E. et al. (2006) Effects of biodiversity on the functioning of trophic groups and ecosystems. *Nature*, **443**, 989–92.

Cardosa, M.Z. & Gilbert, L.E. (2007) A male gift to its partner? Cyanogenic glycosides in the spermatophore of longwing butterflies (*Heliconius*). Naturwissenschaften, **94** (1), 39–42.

Cardoso, P., Silva, I., de Oliveira, N.G. & Serrano, A.R.M. (2004) Higher taxa surrogates of spider (Araneae) diversity and their efficiency in conservation. *Biological Conservation*, **117**, 453–9.

Carey, P.D., Manchester, S.J. & Firbank, L.G. (2005) Performance of two agri-environment schemes in England: a comparison of ecological and multi-disciplinary evaluations. *Agriculture Ecosystems and Environment*, **108**, 178–88.

Carney, S.E., Byerley, M.B. & Holway, D.A. (2003) Invasive Argentine ants (*Linepithema humile*) do not replace native ants as seed dispersers of *Dendromecon rigida* (Papaveraceae) in California, USA. *Oecologia*, **135**, 576–82.

Carpaneto, G.M., Piattella, E. & Spampinato, M.F. (1995) Analysis of a scarab dung beetle community from an Apenninic grassland (Coleoptera, Scarabaeoidea). *Bollettino dell'Associazione Romana di Entomologia*, **50** (1–4), 45–60.

Carrara, A.S., Gonzales, M., Ferro, C. et al. (2005) Venezuelan equine encephalitis virus infection of spiny rats. *Emerging Infectious Diseases* **11** (5), 663–9.

Carroll, A.L. & Quiring, D.T. (1994) Intratree variation in foliar development influences the foraging strategy of a caterpillar. *Ecology*, **75**, 1978–90.

Carroll, M.J., Schmelz, E.A., Meagher, R.L. & Teal, P.E.A. (2006) Attraction of *Spodoptera frugiperda* larvae to volatiles from herbivore-damaged maize seedlings. *Journal of Chemical Ecology*, **32**, 1911–24.

Carroll, S.P. & Dingle, H. (1996) The biology of post-invasion events. *Biological Conservation*, **78**, 207–14.

Carroll, S.P., Marler, M., Winchell, R. & Dingle, H. (2003) Evolution of cryptic flight morph and life history differences during host race radiation in the soapberry bug, *Jadera haematoloma* Herrich-Schaeffer (Hemiptera: Rhopalidae). *Annals of the Entomological Society of America*, **96** (2), 135–43.

Carroll, W.J. (1956) History of the hemlock looper, *Lambdina fiscellaria* (Guen.) (Lepidoptera: Geometridae), in Newfoundland, and notes on its biology. *Canadian Entomologist*, **88**, 587–99.

Carson, W.P. & Root, R.B. (2000) Herbivory and plant species coexistence: community regulation by an outbreaking phytophagous insect. *Ecological Monographs*, **70**, 73–99.

Carson, R. (1962) *Silent Spring*. Hamish Hamilton, London.

Carter, D.O., Yellowlees, D. & Tibbett, M. (2007) Cadaver decomposition in terrestrial ecosystems. *Naturwissenschaften*, **94** (1), 12–24.

Carvell, C., Meek, W.R., Pywell, R.F., Goulson, D. & Nowakowski, M. (2007) Comparing the efficacy of agri-environment schemes to enhance bumble bee abundance and diversity on arable field margins. *Journal of Applied Ecology*, **44**, 29–40.

Casher, L.E. (1996) Leaf toughness in *Quercus agrifolia* and its effects on tissue selection by first instars of *Phryganidia californica* (Lepidoptera: Dioptidae) and *Bucculatrix albertiella* (Lepidoptera. Lyonetiidae). *Annals of the Entomological Society of America*, **89**, 109–21.

Casida, J.E. & Quistad, G.B. (1998) Golden age of insecticide research: past, present or future? *Annual Review of Entomology*, **43**, 1–16.

Cassier, P., Levieux, J., Morelet, M. & Rougon, D. (1996) The mycangia of *Platypus cylindrus* Fab. & *P. oxyurus* Dufour (Coleoptera: Platypodidae): structure and associated fungi. *Journal of Insect Physiology*, **42** (2), 171–9.

Casson, D.S. & Hodkinson, I.D. (1991) The Hemiptera (Insecta) communities of tropical rain forest in Sulawesi (Indonesia). *Zoological Journal of the Linnean Society*, **102** (3), 253–76.

Castle, L.A, Wu, G.S. & McElroy, D. (2006) Agricultural input traits: past, present and future. *Current Opinion in Biotechnology*, **17** (2), 105–12.

Castle, S.J. (2006) Concentration and management of *Bemisia tabaci* in cantaloupe as a trap crop for cotton. *Crop Protection*, **25**, 574–84.

Caterino, M.S., Shull, V.L., Hammond, P.M. & Vogier, A.P. (2002) Basal relationships of Coleoptera inferred from 18S rDNA sequences. *Zoologica Scripta*, **31** (1), 41–9.

Cates, R.G. (1996) The role of mixtures and variation in the production of terpenoids in conifer–insect–pathogen interactions. In *Recent Advances in Phytochemistry*, Vol. 30 (eds J.T. Romeo, J.A. Saunders & P. Barbosa), pp. 179–216. Plenum Press, New York.

Cattaneo, M.G., Yafuso, C., Schmidt, C. et al. (2006) Farm-scale evaluation of the impacts of transgenic cotton on biodiversity, pesticide use, and yield. *Proceedings of the National Academy of Sciences of the USA*, **103** (20), 7571–6.

CBD (1992) *The Convention on Biological Diversity*. UNEP, Montreal.

CBD (2003) *Interlinkages between Biological Diversity and Climate Change. Advice on the Integration of Biodiversity Considerations into the Implementation of the United Nations Framework Convention on Climate Change and its Kyoto Protocol*. Secretariat of the Convention on Biological Diversity, Montreal.

CBD (2005a) *Integration of Biodiversity Considerations in the Implementation of Adaptation Activities to Climate Change at the Local, Subnational, National, Subregional and International Levels*. Report No. UNEP/CBD/AHTEG-BDACC/1/2. Secretariat of the Convention on Biological Diversity, Montreal.

CBD (2005b) *Proposed Framework: Risk Assessment and Management Guidance from Adaptation Projects on Biodiversity – Practical Guidance*. Report No. UNEP/CBD/AHTEG-BDACC/1/3. Secretariat of the Convention on Biological Diversity, Montreal.

Ceballos, L.A., Cardinal, M.V., Vazquez-Prokopec, G.M. et al. (2006) Long-term reduction of *Trypanosoma cruzi* infection in sylvatic mammals following deforestation and sustained vector surveillance in northwestern Argentina. *Acta Tropica*, **98** (3), 286–96.

Cecere, M.C., Canale, D.M. & Gurtler, R.E. (2003) Effects of refuge availability on the population dynamics of *Triatoma infestans* in central Argentina. *Journal of Applied Ecology*, **40**, 742–56.

Chadenga, V. (1994) Epidemiology and control of trypanosomiasis. *Onderstepoort Journal of Veterinary Research*, **61** (4), 385–90.

Challice, J.S. & Williams, A.H. (1970) A comparative biochemical study of phenolase specificity in *Malus*, *Pyrus* and other plants. *Phytochemistry*, **9**, 1261–9.

Chang, Y.C. (1991) Integrated pest management of several forest defoliators in Taiwan. *Forest Ecology and Management*, **39**, 65–72.

Chao, A., Chazdon, R.L., Colwell, R.K. & Shen, T.J. (2005) A new statistical approach for assessing similarity of species composition with incidence and abundance data. *Ecology Letters*, **8**, 148–59.

Chao, A., Chazdon, R.L., Colwell, R.K. & Shen, T.J. (2006) Abundance-based similarity indices and their estimation when there are unseen species in samples. *Biometrics*, **62**, 361–71.

Chape, S., Harrison, J., Spalding, M. & Lysenko, I. (2005) Measuring the extent and effectiveness of protected areas as an indicator for meeting global biodiversity targets. *Philosophical Transactions of the Royal Society of London Series B, Biological Sciences*, **360**, 443–55.

Chapman, J.W., Reynolds, D.R., Smith, A.D., Riley, J.R., Pedgley, D.E. & Woiwod, I.P. (2002) High-altitude migration of the diamondback moth *Plutella xylostella* to the UK: a study using radar, aerial netting, and ground trapping. *Ecological Entomology*, **27**, 641–50.

Chapman, S.K., Hart S.C., Cobb N.S., Whitham, T.G. & Koch G.W. (2003) Insect herbivory increases litter quality and decomposition: an extension of the acceleration hypothesis. *Ecology*, **84**, 2867–76.

Chapman, S.K., Schweitzer, J.A. & Whitham, T.G. (2006). Herbivory differentially alters plant litter dynamics of evergreen and deciduous trees. *Oikos*, **114**, 566–74.

Chappell, M.A. & Rogowitz, G.L. (2000) Mass, temperature and metabolic effects on discontinuous gas exchange cycles in eucalyptus-boring beetles (Coleoptera: Cerambycidae). *Journal of Experimental Biology*, **203** (24), 3809–20.

Chararas, C. (1979) *Ecophysiologie des Insectes Parasites des Forêts*. Printed by the author, Paris.

Charlet, L.D. & Knodel, J.J (2003) Impact of planting date on sunflower beetle (Coleoptera: Chrysomelidae) infestation, damage, and parasitism in cultivated sunflower. *Journal of Economic Entomology*, **96** (3), 706–13.

Charnov, E.L., Los-Den Hartogh, R.L., Jones, W.T. & van den Assem, J. (1981) Sex ratio evolution in variable environments. *Nature*, **289**, 27–33.

Chauhan, R. & Dahiya, B. (1994) Responses to different chickpea genotypes by *Heliocoverpa armigera* at Hisar. *Indian Journal of Plant Protection*, **22** (2), 170–2.

Chemengich, B.T. (1993) Ecological relationships between Argomyzidae feeding on leguminous plants and species–area effects in Kenya. *Insect Science and its Application*, **14**, 603–9.

Chen, C.N. & Yeh, Y. (1992) Integrated pest management in Taiwan. In *Integrated Pest Management in the Asia–Pacific Region* (eds P.A.C. Ooi, G.S. Lim, T.H. Ho, P.L. Manalo & J. Waage), pp. 283–302. CAB International, Wallingford, UK.

Chen, H., Chen, Z.M. & Zhou, Y.S. (2005) Rice water weevil (Coleoptera: Curculionidae) in mainland China: invasion, spread and control. *Crop Protection*, **24** (8), 695–702.

Chen, M., Ye, G.Y., Liu, Z.C. et al. (2006) Field assessment of the effects of transgenic rice expressing a fused gene of cry1Ab and cry1Ac from *Bacillus thuringiensis* Berliner on nontarget planthopper and leafhopper populations. *Environmental Entomology*, **35** (1), 127–34.

Chen, W.L., Leopold, R.A. & Harris, M.O. (2006) Parasitism of the glassy-winged sharpshooter, *Homalodisca coagulata* (Homoptera: Cicadellidae): functional response and superparasitism by *Gonatocerus ashmeadi* (Hymenoptera: Mymaridae). *Biological Control*, **37** (1), 119–29.

Cheng, L., Baars, M.A. & Oosterhuis, S.S. (1990) *Halobates* in the Banda Sea (Indonesia): monsoonal differences in abundance and species composition. *Bulletin of Marine Science*, **47** (2), 421–30.

Cheng, L. & Pitman, R.L. (2002) Mass oviposition and egg development of the ocean-skater *Halobates sobrinus* (Heteroptera: Gerridae). *Pacific Science*, **56** (4), 441–5.

Chenuil, A. & McKey, D.B. (1996) Molecular phylogenetic study of a myrmecophyte symbiosis: did *Leonardoxa*–ant associations diversify via conspecification? *Molecular Phylogenetics and Evolution*, **6**, 270–86.

Cherenet, T., Sani, R.A., Speybroeck, N. et al. (2006) A comparative longitudinal study of bovine trypanosomiasis in tsetse-free and tsetse-infested zones of the Amhara Region, northwest Ethiopia. *Veterinary Parasitology*, **140** (3/4), 251–8.

Cherrett, J.M., Powell, R.J. & Stradling, D.J. (1989) The mutualism between leaf-cutting ants and their fungus. In *Insect–Fungus Interactions. 14th Symposium of the Royal Entomological Society of London in Collaboration with the British Mycological Society* (eds N. Wilding, N.M. Collins, P.M. Hammond & J.F. Webber), pp. 93–120. Academic Press, London.

Cherry, R.H. & Klein, M.G. (1997) Mortality induced by *Bacillus popilliae* in *Cyclocephala parallela* (Coleoptera: Scarabaeidae) held under simulated field temperatures. *Florida Entomologist*, **80** (2), 261–5.

Chesson, P.L. (1985) Coexistence of competitors in spatially and temporally varying environments: a look at the combined effect of different sorts of variability. *Theoretical Population Biology*, **28**, 263–87.

Chesson, P.L. & Warner, R.R. (1981) Environmental variability promotes coexistence in lottery competitive systems. *American Naturalist*, **117**, 923–43.

Chew, R.M. (1974) Consumers as regulators of ecosystems: an alternative to energetics. *Ohio Journal of Science*, **74**, 359–70.

Chey, V.K., Holloway, J.D. & Speight, M.R. (1997) Diversity of moths in forest plantations and natural forests in Sabah. *Bulletin of Entomological Research*, **87**, 371–85.

Chey, V.K., Holloway, J.D. & Speight, M.R. (1998) Canopy knockdown of arthropods in exotic plantations and natural forest in Sabah, north-east Borneo, using insecticidal mist-blowing. *Bulletin of Entomological Research*, **88**, 15–24.

Chilcutt, C.F. & Tabashnik, B.E. (1997) Host-mediated competition between the pathogen *Bacillus thuringiensis* and the parasitoid *Cotesia plutellae* of the diamondback moth (Lepidoptera, Plutellidae). *Environmental Entomology*, **26**, 38–45.

Chiykowski, L.N. (1991) Vector–pathogen–host plant relationships of clover phyllody mycoplasma-like organism and the vector leafhopper *Paraphlepsius irroratus*. *Canadian Journal of Plant Pathology*, **13** (1), 11–18.

Chong, J.H. & Oetting, R.D. (2006) Functional response and progeny production of the Madeira mealybug parasitoid, *Anagyrus* sp nov nr. sinope: the effects of host and parasitoid densities. *Biological Control*, **39** (3), 320–8.

Choong, M.F. (1996) What makes a leaf tough and how this affects the patterns of *Castanopsis fissa* leaf consumption by caterpillars. *Functional Ecology*, **10**, 668–74.

Chown, S.L. (2002) Respiratory water loss in insects. *Comparative Biochemistry and Physiology Part A, Molecular and Integrative Physiology*, **133A** (3), 791–804.

Christian, C.E. (2001) Consequences of a biological invasion reveal the importance of mutualism for plant communities. *Nature*, **413**, 635–9.

Christian, C.E. & Stanton, M.L. (2004) Cryptic consequences of a dispersal mutualism: seed burial, elaiosome removal, and seed-bank dynamics. *Ecology*, **85**, 1101–10.

Christianini, A.V. & Machado, G. (2004) Induced biotic responses to herbivory and associated cues in the Amazonian ant-plant *Maieta poeppigii*. *Entomologia Experimentalis et Applicata*, **112**, 81–8.

Christie, M., Hanley, N., Warren, J., Murphy, K., Wright, R. & Hyde, T. (2006) Valuing the diversity of biodiversity. *Ecological Economics*, **58**, 304–17.

Christou, P., Capell, T., Kohli, A., Gatehouse, J.A. & Gatehouse, A.M.R. (2006) Recent developments and future prospects in insect pest control in transgenic crops. *Trends in Plant Science*, **11**, 302–8.

Christou, P. & Twyman, R.M. (2004) The potential of genetically enhanced plants to address food insecurity. *Nutrition Research Reviews*, **17** (1), 23–42.

Chuma, J., Gilson, L. & Molyneux, C. (2007) Treatment-seeking behaviour, cost burdens and coping strategies among rural and urban households in Coastal Kenya: an equity analysis. *Tropical Medicine and International Health*, **12** (5), 673–86.

Chung, A.Y.C., Khen, C.V., Unchi, S. & Momin, B. (2002) Edible insects and entomophagy in Sabah, Malaysia. *Malayan Nature Journal*, **56** (2), 131–44.

Cipollini, D.F. Jr. (1997) Wind-induced mechanical stimulation increases pest resistance in common bean. *Oecologia*, **111**, 84–90.

Cipollini, M.L., Paulk, E., Mink, K., Vaughn, K. & Fischer, T. (2004) Defense tradeoffs in fleshy fruits: effects of resource variation on growth, reproduction, and fruit secondary chemistry in *Solanum carolinense*. *Journal of Chemical Ecology*, **30**, 1–17.

Cizek, L. (2005) Diet composition and body size in insect herbivores: why do small species prefer young leaves? *European Journal of Entomology*, **102**, 675–81.

Clancy, K.M. & Price, P.W. (1987) Rapid herbivore growth enhances enemy attack: sublethal plant defenses remain a paradox. *Ecology*, **68**, 736–8.

Claridge, M.F. & Evans, H.F. (1990) Species–area relationships: relevance to pest problems of British trees? In *Population Dynamics of Forest Insects* (eds A.D. Watt, S.R. Leather, M.D. Hunter & N.A.C. Kidd), pp. 59–69. Intercept, Andover, UK.

Claridge, M.F. & Wilson, M.R. (1978) British insects and trees: a study in island biogeography or insect/plant coevolution? *American Naturalist*, **112** (984), 451–6.

Claridge, M.F. & Wilson, M.R. (1982) Insect herbivore guilds and species–area relationships – leafminers on British trees. *Ecological Entomology*, **7**, 19–30.

Clarke, A. (2003) Costs and consequences of evolutionary temperature adaptation. *Trends in Ecology and Evolution*, **18** (11), 573–81.

Clarke, R.T., Wright, J.F. & Furse, M.T. (2003) RIVPACS models for predicting the expected macroinvertebrate fauna and assessing the ecological quality of rivers. *Ecological Modelling*, **160**, 219–33.

Clay, K. & Cheplick, G.P. (1989) Effect of ergot alkaloids from fungal endophyte-infected grasses on fall armyworm (*Spodoptera frugiperda*). *Journal of Chemical Ecology*, **15**, 169–82.

Clay, K., Hardy, T.N. & Hammond, A.M. (1985) Fungal endophytes of grasses and their effects on an insect herbivore. *Oecologia*, **66**, 1–5.

Clay, K. & Holah, J. (1999) Fungal endophyte symbiosis and plant diversity in successional fields. *Science*, **285**, 1742–4.

Clayton, D.H. & Price, R.D. (1999) Taxonomy of New World Columbicola (Phthiraptera: Philopteridae) from the Columbiformes (Aves), with descriptions of five new species. *Annals of the Entomological Society of America*, **92**, 675–85.

Cleary, D.F.R. (2004) Assessing the use of butterflies as indicators of logging in Borneo at three taxonomic levels. *Journal of Economic Entomology*, **97**, 429–35.

Clements, F.E. (1916) *Plant Succession: an Analysis of the Development of Vegetation*. Publication No. 242. Carnegie Institute of Washington, Washington, DC.

Close, D., McArthur, C., Paterson, S., Fitzgerald, H., Walsh, A. & Kincade, T. (2003) Photoinhibition: a link between effects of the environment on eucalypt leaf chemistry and herbivory. *Ecology*, **84**, 2952–66.

Cloutier, C. & Douglas, A-E. (2003) Impact of a parasitoid on the bacterial symbiosis of its aphid host. *Entomologia Experimentalis et Applicata*, **109** (1), 13–19.

Cloyd, R.A., Keith, S.R. & Galle, C.L. (2004) Effect of the insect growth regulator novaluron (Pedestal) on silverleaf whitefly reproduction. *Horttechnology*, **14** (4), 551–4.

Cobb, N.S. & Whitham, T.G. (1993) Herbivore deme formation on individual trees: a test case. *Oecologia*, **94**, 496–502.

Cognato, A.I., Harlin, A.D. & Fisher, M.L. (2003) Genetic structure among pinyon pine beetle populations (Scolytinae: *Ips confusus*). *Environmental Entomology*, **32**, 1262–70.

Cole, L.J., McCracken, D.I., Downie, I.S. et al. (2005) Comparing the effects of farming practices on ground beetle (Coleoptera: Carabidae) and spider (Araneae) assemblages of Scottish farmland. *Biodiversity and Conservation*, **14**, 441–60.

Cole, L.J., Pollock, M.L., Robertson, D., Holland, J.P. & McCracken, D.I. (2006) Carabid (Coleoptera) assemblages in the Scottish uplands: the influence of sheep grazing on ecological structure. *Entomologica Fennica*, **17**, 229–40.

Coleman, P-G. & Alphey, L. (2004) Editorial: genetic control of vector populations: an imminent prospect. *Tropical Medicine and International Health*, **9** (4), 433–7.

Coley, P.D. (1986) Costs and benefits of defense by tannins in a neotropical tree. *Oecologia (Heidelberg)*, **70**, 238–41.

Coley, P.D. & Aide, T.M. (1991) Comparison of herbivory and plant defenses in temperate and tropical broad-leaved forests. In *Plant–animal Interactions. Evolutionary Ecology in Tropical and Temperate Regions* (eds P.W. Price, T.M. Lewinshon, G.W. Fernandes & W.W. Benson), pp. 25–49. John Wiley & Sons, Chichester, UK.

Coley, P.D., Bateman, M.L. & Kursar, T.A. (2006) The effects of plant quality on caterpillar growth and defense against natural enemies. *Oikos*, **115**, 219–28.

Coley, P.D., Bryant, J.P. & Chapin, F.S. (1985) Resource availability and plant antiherbivore defense. *Science*, **230**, 895–9.

Colfer, R.G. & Rosenheim, J.A. (2001) Predation on immature parasitoids and its impact on aphid suppression. *Oecologia*, **126**, 292–304.

Coll, M., Gavish, S. & Dori, I. (2000) Population biology of the potato tuber moth, *Phthorimaea operculella* (Lepidoptera: Gelechiidae), in two potato cropping systems in Israel. *Bulletin of Entomological Research*, **90**, 309–15.

Collier, K.J. (1995) Environmental factors affecting the taxonomic composition of aquatic macroinvertebrate communities in lowland waterways of Northland, New Zealand. *New Zealand Journal of Marine and Freshwater Research*, **29**, 453–65.

Collier, N., Mackay, D.A., Benkendorff, K., Austin, A.D. & Carthew, S.M. (2006) Butterfly communities in South Australian urban reserves: estimating abundance and diversity using the Pollard walk. *Austral Ecology*, **31**, 282–90.

Collier, T. & van Steenwyk, R. (2004) A critical evaluation of augmentative biological control. *Biological Control*, **31** (2), 245–56.

Collins, F.H. & Paskewitz, S.M. (1995) Malaria: current and future prospects for control. *Annual Review of Entomology*, **40**, 195–219.

Collins, K.L., Boatman, N.D., Wilcox, A., Holland, J.M. & Chaney, K. (2002) Influence of beetle banks on cereal, aphid predation in winter wheat. *Agriculture Ecosystems and Environment*, **93**, 337–50.

Coluzzi, M. (1994) Malaria and the afrotropical ecosystems: impact of man-made environmental changes. *Parassitologia (Rome)*, **36** (1/2), 223–7.

Colwell, R. (2000) EstimateS. *http://viceroy.eeb.uconn.edu/EstimateS*.

Colwell, R.K. & Coddington, J.A. (1994) Estimating terrestrial biodiversity through extrapolation. *Philosophical Transactions of the Royal Society of London Series B, Biological Sciences*, **345**, 101–18.

Comins, H.N. & Hassell, M.P. (1979) The dynamics of optimally foraging predators and parasitoids. *Journal of Animal Ecology*, **48**, 335–51.

Comins, H.N., Hassell, M.P. & May, R.M. (1992) The spatial dynamics of host–parasitoid systems. *Journal of Animal Ecology*, **61**, 735–48.

Commins, H.N. & Noble, I.R. (1985) Dispersal, variability and transient niches: species coexistence in a uniformly varying environment. *American Naturalist*, **126**, 706–23.

Common, M. & Stagl, S. (2005) *Ecological Economics: an Introduction*. Cambridge University Press, Cambridge, UK.

Compton, S.G., Lawton, J.H. & Rashbrook, V.K. (1989) Regional diversity, local community structure and vacant niches: the herbivorous arthropods of bracken in South Africa. *Ecological Entomology*, **14** (4), 365–73.

Connell, J.H. (1980) Diversity and the coevolution of competitors, or the ghost of competition past. *Oikos*, **35**, 131–8.

Connell, J.H. (1983) On the prevalence and importance of interspecific competition: evidence from field experiments. *American Naturalist*, **122**, 661–96.

Connell, J.H. & Slatyer, R.O. (1977) Mechanisms of succession in natural communities and their role in community stability and organisation. *American Naturalist*, **111**, 1119–44.

Conteh, L., Sharp, B-L., Streat, E., Barreto, A. & Konar, S. (2004) The cost and cost-effectiveness of malaria vector control by residual insecticide house-spraying in southern Mozambique: a rural and urban analysis. *Tropical Medicine and International Health*, **9** (1), 124–32.

Cooke, B.J. & Lorenzetti, F. (2006) The dynamics of forest tent caterpillar outbreaks in Quebec, Canada. *Forest Ecology and Management*, **226** (1–3), 110–21.

Coope, G.R. (1994) The response of insect faunas to glacial–interglacial climatic fluctuations. *Philosophical Transactions of the Royal Society of London Series B, Biological Sciences*, **344** (1307), 19–26.

Coope, G.R. (1995) The effects of Quaternary climatic changes on insect populations: lessons from the past. In *Insects in a Changing Environment* (eds R. Harrington & N.E. Stork), pp. 29–48. Academic Press, London.

Coppedge, B.R., Stephen, F.M. & Felton, G.W. (1995) Variation in female southern pine beetle size and lipid content in relation to fungal associates. *Canadian Entomologist*, **127** (2), 145–54.

Cornelissen, T. & Stiling, A. (2006) Does low nutritional quality act as a plant defence? An experimental test of the slow-growth, high-mortality hypothesis. *Ecological Entomology*, **31**, 32–40.

Cornell, H.V. & Hawkins, B.A. (1995) Survival patterns and mortality sources of herbivorous insects – some demographic trends. *American Naturalist*, **145**, 563–93.

Corrêa, M.M., Fernandes, W.D. & Leal, I.R. (2006) Ant diversity (Hymenoptera: Formicidae) from Capoes in Brazilian pantanal: relationship between species richness and structural complexity. *Neotropical Entomology*, **35**, 724–30.

Cory, J.S., Hirst, M.L., Sterling, P.A. & Speight, M.R. (2000) Narrow host range nucleopolyhedrovirus for control of the browntail moth (Lepidoptera: Lymantriidae). *Environmental Entomology*, **29** (3), 661–7.

Cory, J.S., Hirst, M.L., Williams, T. et al. (1994) Field trial of a genetically improved baculovirus insecticide. *Nature*, **370** (6485), 138–40.

Cory, J.S. & Myers, J.H. (2003) The ecology and evolution of insect baculoviruses. *Annual Review of Ecology, Evolution and Systematics*, **34**, 239–72.

Costa-Neto, E.M. (2003) Insects as sources of proteins for man: valorization of disgusting resources. *Interciencia*, **28** (3), 136.

Costanza, R., dArge, R., deGroot, R. et al. (1997) The value of the world's ecosystem services and natural capital. *Nature*, **387**, 253–60.

Coulson, R.N., Pope, D.N., Gagne, J.A., Fargo, W.S. & Pulley, P.E. (1980) Impact of foraging by *Monochamus titilator* (Col. Cerambycidae) on within-tree populations of *Dendroctonus frontalis* (Col. Scolyditae). *Entomophaga*, **25**, 155–70.

Coulson, S.J., Hodkinson, I.D., Strathdee, A.T. et al. (1995) Thermal environments of Arctic soil organisms during winter. *Arctic and Alpine Research*, **27** (4), 364–70.

Coulson, S.J., Hodkinson, I.D., Webb, N.R., Mikkola, K., Harrison, J.A. & Pedgeley, D.E. (2002) Aerial colonization of high Arctic islands by invertebrates: the diamondback moth *Plutella xylostella* (Lepidoptera: Yponomeutidae) as a potential indicator species. *Diversity and Distributions*, **8** (6), 327–34.

Coutts, B.A. & Jones, R.A.C. (2002) Temporal dynamics of spread of four viruses within mixed species perennial pastures. *Annals of Applied Biology*, **140** (1), 37–52.

Coyle, D.R., Nebeker, T.E., Hart, E.R. & Mattson, W.J. (2005) Biology and management of insect pests in North American intensively managed hardwood forest systems. *Annual Review of Entomology*, **50**, 1–29.

Craig, T.P., Itami, J.K. & Price, P.W. (1990) Intraspecific competition and facilitation by a shoot-galling sawfly. *Journal of Animal Ecology*, **59**, 147–59.

Craighead, F.C. (1925) Bark-beetle epidemics and rainfall deficiency. *Journal of Economic Entomology*, **18**, 577–86.

Crickmore, N. (2006) Beyond the spore – past and future developments of *Bacillus thuringiensis* as a biopesticide. *Journal of Applied Microbiology*, **101** (3), 616–19.

Cronin, J.T. & Strong, D.R. (1993) Superparasitism and mutual interference in the egg parasitoid *Anagrus delicatus*

(Hymenoptera: Mymaridae). *Ecological Entomology*, **18**, 293–302.

Cross, W.F., Benstead, J.P., Rosemond, A.D. & Wallace, J.B. (2003) Consumer-resource stoichiometry in detritus-based streams. *Ecology Letters*, **6**, 721–32.

Crowe, M.L. & Bourchier, R.S. (2006) Interspecific interactions between the gall-fly *Urophora affinis* Frfld. (Diptera: Tephritidae) and the weevil *Larinus minutus* Gyll. (Coleoptera: Curculionidae), two biological control agents released against spotted knapweed, *Centaurea stobe* L. ssp *micranthos*. *Biocontrol Science and Technology*, **16** (4), 417–30.

Crozier, L. (2003) Winter warming facilitates range expansion: cold tolerance of the butterfly *Atalopedes campestris*. *Oecologia*, **135**, 648–56.

Cruz-Reyes, A. & Pickering-Lopez, J.M. (2006) Chagas disease in Mexico: an analysis of geographical distribution during the past 76 years – a review. *Memorias do Instituto Oswaldo Cruz*, **101** (4), 345–54.

CSIRO (1979) *The Insects of Australia*. Melbourne University Press, Melbourne.

Cuffney, T.F., Wallace, J.B. & Lugthart, G.J. (1990) Experimental evidence quantifying the role of benthic invertebrates in organic matter dynamics of headwater streams. *Freshwater Biology*, **23**, 281–99.

Cugala, D., Schulthess, F., Ogol, C.P.O., Omwega, C.O. (2006) Assessment of the impact of natural enemies on stemborer infestations and yield loss in maize using selected insecticides in Mozambique. *Annales de la Societe Entomologique de France*, **42** (3/4), 503–10.

Cui, F., Raymond, M. & Qiao, C.L. (2006) Insecticide resistance in vector mosquitoes in China. *Pest Management Science*, **62** (11), 1013–22.

Cummins, K.W. & Klug, M.J. (1979) Feeding ecology of stream invertebrates. *Annual Review of Ecology and Systematics*, **10**, 147–72.

Cummins, K.W., Petersen, R.C., Howard, F.O., Wuycheck, J.C. & Holt, V.I. (1973) The utilization of leaf litter by stream detritivores. *Ecology*, **54**, 336–45.

Cupp, E.W. & Cupp, M.S. (1997) Black fly (Diptera: Simuliidae) salivary secretions: importance in vector competence and disease. *Journal of Medical Entomology*, **34** (2), 87–94.

Currie, C.R. & Stuart, A.E. (2001) Weeding and grooming of pathogens in agriculture by ants. *Proceedings of the Royal Society of London Series B, Biological Sciences*, **268**, 1033–9.

Currie, C.R., Wong, B., Stuart, A.E. et al. (2003) Ancient tripartite coevolution in the attine ant–microbe symbiosis. *Science (Washington DC)*, **299** (5605), 386–8.

Curtis, A.D. & Waller, D.A. (1995) Changes in nitrogen fixation rates in termites (Isoptera: Rhinotermitidae) maintained in the laboratory. *Annals of the Entomological Society of America*, **88** (6), 764–7.

Cushing, C.E., Minshall, G.W. & Newbold, J.D. (1993) Transport dynamics of fine particulate organic matter in two Idaho streams. *Limnology and Oceanography*, **38**, 1101–15.

Cyranoski, D. (2005) Pesticide results help China edge transgenic rice towards market. *Nature*, **435** (7038), 3.

D'Abrera, B. (2004) *Butterflies of the Afrotropical Region, Part 2. Nymphalidae (complete), Libytheidae*. Hill House Publishers, London.

D'Amico, V. & Elkinton, J.S. (1995) Rainfall effects on transmission of gypsy moth (Lepidoptera: Lymantriidae) nuclear polyhedrosis virus. *Environmental Entomology*, **24** (5), 1144–9.

Dadour, I.R. & Cook, D.F. (1996) Survival and reproduction in the scarabaeine dung beetle *Onthophagus binodis* (Coleoptera: Scarabaeidae) on dung produced by cattle on grain diets in feedlots. *Environmental Entomology*, **25**, 1026–31.

Dahlsten, D.L., Six, D.L., Rowney, D.L., Lawson, A.B., Erbilgin, N. & Raffa, K.F. (2004) Attraction of *Ips pini* (Coleoptera: Scolytinae) and its predators to natural attractants and synthetic semiochemicals in northern California: implications for population monitoring. *Environmental Entomology*, **33** (6), 1554–61.

Dale, V.H., Brown, S., Haeuber, R.A. et al. (2000) Ecological principles and guidelines for managing the use of land. *Ecological Applications*, **10**, 639–70.

Damgaard, J., Andersen, N.M., Cheng, L. & Sperling, F.A.H. (2000) Phylogeny of sea skaters, *Halobates eschscholtz* (Hemiptera, Gerridae), based on mtDNA sequence and morphology. *Zoological Journal of the Linnean Society*, **130** (4), 511–26.

Damman, H. (1987) Leaf quality and enemy avoidance by the larvae of a pyralid moth. *Ecology*, **68**, 88–97.

Damman, H. (1993) Patterns of herbivore interaction among herbivore species. In *Caterpillars: Ecological and Evolutionary Constraints on Foraging* (eds N.E. Stamp & T.M. Casey), pp. 132–69. Chapman & Hall, New York.

Danforth, B-N., Brady, S-G., Sipes, S.D. & Pearson, A. (2004) Single-copy nuclear genes recover Cretaceous-age divergences in bees. *Systematic Biology*, **53** (2), 309–26.

Dangerfield, J.M., Pik, A.J., Britton, D. et al. (2003) Patterns of invertebrate biodiversity across a natural edge. *Austral Ecology*, **28**, 227–36.

Danks, H.V. (1994) Regional diversity of insects in North America. *American Entomologist*, **40** (1), 50–5.

Danyk, T.P. & Mackauer, M. (1996) An extraserosal envelope in eggs of *Proan pequodorum* (Hymenoptera, aphidae), a parasitoid of pea aphid. *Biological Contrology*, **7**, 67–70.

Darwin, C. (1859) *On the Origin of Species by Means of Natural Selection*. John Murrary, London.

Davidowitz, G. & Rosenzweig, M.L. (1998) The latitudinal gradient of species diversity among North American grasshoppers (Acrididae) within a single habitat: a test of the spatial heterogeneity hypothesis. *Journal of Biogeography*, **25**, 553–60.

Davidson, D.W., Cook, S.C. & Snelling, R.R. (2004) Liquid-feeding performances of ants (Formicidae): ecological and evolutionary implications. *Oecologia*, **139** (2), 255–66.

Davidson, D.W., Cook, S.C., Snelling, R.R. & Chua, T. H. (2003) Explaining the abundance of ants in lowland tropical rainforest canopies. *Science*, **300**, 969–72.

Davidson, D.W., Longino, J.T. & Snelling, R.R. (1988) Pruning of host plant neighbors by ants: an experimental approach. *Ecology*, **69**, 801–8.

Davidson, J. & Andrewartha, H.G. (1948a) Annual trends in a natural population of *Thrips imaginis* (Thysanoptera). *Journal of Animal Ecology*, **17**, 193–9.

Davidson, J. & Andrewartha, H.G. (1948b) The influence of rainfall, evaporation and atmospheric temperature on fluctuations in the size of a natural population of *Thrips imaginis* (Thysanoptera). *Journal of Animal Ecology*, **17**, 200–22.

Davidson, R.H. & Lyon, W.F. (1979) *Insect Pests of Farm, Garden and Orchard*. John Wiley & Sons, New York.

Davies, J., Stork, N., Brendell, M. & Hine, S. (1997). Beetle species diversity and faunal similarity in Venezuelan rainforest tree canopies. In *Canopy Arthropods* (eds N. Stork, J. Adis & R. Didham), pp. 85–103. Chapman & Hall, London.

Davies, J.B. (1994) Sixty years of onchocerciasis vector control: a chronological summary with comments on eradication, reinvasion and insecticide resistance. *Annual Review of Entomology*, **39**, 23–46.

Davies, S.J., Lum, S.K.Y., Chan, R. & Wang, L.K. (2001) Evolution of myrmecophytism in western Malesian *Macaranga* (Euphorbiaceae). *Evolution*, **55**, 1542–59.

Davies, Z.G., Wilson, R.J., Brereton, T.M. & Thomas, C.D. (2005) The re-expansion and improving status of the silver-spotted skipper butterfly (*Hesperia comma*) in Britain: a metapopulation success story. *Biological Conservation*, **124**, 189–98.

Davis, A.L.V. (1996) Habitat associations in a South African, summer rainfall, dung beetle community (Coleoptera. Carabaeidae, Aphodiidae, Staphylinidae, Histeridae, Hydrophilidae). *Pedobiologia*, **40**, 260–80.

Davis, H.G. (2005) *r*-Selected traits in an invasive population. *Evolutionary Ecology*, **19** (3), 255–74.

Davis, S., Begon, M., De-Bruyn, L. et al. (2004) Predictive thresholds for plague in Kazakhstan. *Science* (Washington DC), **304** (5671), 736–8.

Day, K.R., Docherty, M., Leather, S.R. & Kidd, N.A.C. (2006) The role of generalist insect predators and pathogens in suppressing green spruce aphid populations through direct mortality and mediation of aphid dropping behaviour. *Biological Control*, **38** (2), 233–46.

Day, W.H. (1994) Estimating mortality caused by parasites and diseases of insects – comparisons of the dissection and rearing methods. *Environmental Entomology*, **23**, 543–50.

de Assis, F.A.D.A., Deom, C.A. & Sherwood, J.L. (2004) Acquisition of tomato spotted wilt virus by adults of two thrips species. *Phytopathology*, **94** (4), 333–6.

De Bach, P. (ed.) (1964) *Biological Control of Insect Pests*. Chapman & Hall, London.

de Clercq, P., Mohaghegh, J. & Tirry, L. (2000) Effect of host plant on the functional response of the predator *Podisus nigrispinus* (Heteroptera: Pentatomidae). *Biological Control*, **18**, 65–70.

De Deyn, G.B., Raaijmakers, C.E., Zoomer, H.R. et al. (2003) Soil invertebrate fauna enhances grassland succession and diversity. *Nature*, **422**, 711–13.

De Jong, G. (1981) The evolution of dispersal pattern on the evolution of fecundity. *Netherlands Journal of Zoology*, **32**, 1–30.

De Jong, G.D. & Hoback, W.W. (2006) Effect of investigator disturbance in experimental forensic entomology: succession and community composition. *Medical and Veterinary Entomology*, **20**, 248–58.

de Mazancourt, C. & Loreau, M. (2000) Grazing optimization, nutrient cycling, and spatial heterogeneity of plant–herbivore interactions: should a palatable plant evolve? *Evolution*, **54**, 81–92.

De Moraes, C.M., Lewis, W.J., Paré, P.W., Alborn, H.T. & Tumlinson, J.H. (1998) Herbivore-infested plants selectively attract parasitoids. *Nature*, **393**, 570–3.

De Moraes, C.M., Mescher, M.C. & Tumlinson, J.H. (2001) Caterpillar-induced nocturnal plant volatiles repel conspecific females. *Nature*, **410**, 577–80.

De Sole, G., Baker, R., Dadzie, K.Y. et al. (1991) Onchocerciasis distribution and severity in five West African countries. *Bulletin of the World Health Organization*, **69** (6), 689–98.

De Vries, P.J. (1997) *The Butterflies of Costa Rica and their Natural History*. Princeton University Press, Princeton, NJ.

de Zulueta, J. (1994) Malaria and ecosystems: from prehistory to posteradication. *Parassitologia (Rome)*, **36** (1/2), 7–15.

Dearden, P., Bennett, M. & Johnston, J. (2005) Trends in global protected area governance, 1992–2002. *Environmental Management*, **36**, 89–100.

Debouzie, D., Desouhant, E., Oberli, F. & Menu, F. (2002) Resource limitation in natural populations of phytophagous insects. A long-term study case with the chestnut weevil. *Acta Oecologica, International Journal of Ecology*, **23**, 31–9.

Dedryver, C.A., Hulle, M., Le, G.J.F., Caillaud, M.C. & Simon, J.C. (2001) Coexistence in space and time of sexual and asexual populations of the cereal aphid *Sitobion avenae*. *Oecologia Berlin*, **128** (3), 379–88.

DeGomez, T.E., Hayes, C.J., Anhold, J.A., McMillin, J.D., Clancy, K.M. & Bosu, P.P. (2006) Evaluation of insecticides for protecting southwestern ponderosa pines from attack by engraver beetles (Coleoptera: Curculionidae: Scolytinae). *Journal of Economic Entomology*, **99** (2), 393–400.

Dejean, A., Delabie, J.H.C., Cerdan, P., Gibernau, M. & Corbara, B. (2006) Are myrmecophytes always better protected against herbivores than other plants? *Biological Journal of the Linnean Society* **89** (1), 91–8.

del Campo, M.L., Miles, C.I., Schroeder, F.C., Mueller, C., Booker, R. & Renwick, J.A. (2001) Host recognition by the tobacco hornworm is mediated by a host plant compound. *Nature*, **411**, 186–9.

Delledonne, M., Xia, Y., Dixon, R.A. & Lamc, C. (1998) Nitric oxide functions as a signal in plant disease resistance. *Nature*, **394**, 585–8.

Delmotte, F., Leterme, N., Bonhomme, J., Rispe, C. & Simon, J.C. (2001) Multiple routes to asexuality in an aphid species. *Proceedings of the Royal Society of London Series B, Biological Sciences*, **268**, 2291–9.

Delucchi, V. (1982) Parasitoids and hyperparasitoids of *Zeiraphera diniana* (Lep., Tortricidae) and their role in population control in outbreak areas. *Entomophaga*, **27**, 77–92.

Dempster, J.P. (1983) The natural control of populations of butterflies and moths. *Biological Reviews of the Cambridge Philosophical Society*, **58**, 461–81.

Dempster, J.P. & McLean, I.F.G. (1998) *Insect Populations*. Chapman & Hall, London.

Dempster, J.P. & Pollard, E. (1981) Fluctuations in resource availability and insect populations. *Oecologia*, **50**, 412–16.

Den Boer, P.J. (1991) Seeing the trees for the wood – random-walks or bounded fluctuations of population-size. *Oecologia*, **86**, 484–91.

Dennis, P., Usher, G.B. & Watt, A.D. (1995) Lowland woodland structure and pattern and the distribution of arboreal, phytophagous arthropods. *Biodiversity and Conservation*, **4**, 728–44.

Dennis, P. & Wratten, S.D. (1991) Field manipulation of populations of individual staphylinid species in cereals and their impact on aphid populations. *Ecological Entomology*, **16**, 17–24.

Dennis, R.L.H. & Hardy, P.B. (1999) Targeting squares far survey: predicting species richness and incidence of species for a butterfly atlas. *Global Ecology and Biogeography*, **8**, 443–54.

Dennis, R.L.H., Sparks, T.H. & Hardy, P.B. (1999) Bias in butterfly distribution maps: the effects of sampling effort. *Journal of Insect Conservation*, **3**, 33–42.

Denno, R.F. & Fagan, W.F. (2003) Might nitrogen limitation promote omnivory among carnivorous arthropods? *Ecology*, **84**, 2522–31.

Denno, R.F., Gratton, C., Dobel, H. & Finke, D.L. (2003) Predation risk affects relative strength of top-down and bottom-up impacts on insect herbivores. *Ecology*, **84** (4), 1032–44.

Denno, R.F. & McClure, M.S. (1983) *Variable Plants and Herbivores in Natural and Managed Systems*. Academic Press, New York.

Denno, R.F., McClure, M.S. & Ott, J.R. (1995) Interspecific interactions in phytophagous insects: competition reexamined and resurrected. *Annual Review of Entomology*, **40**, 297–331.

Denno, R.F., Peterson, M.A., Gratton, C. et al. (2000) Feeding-induced changes in plant quality mediate interspecific competition between sap-feeding herbivores. *Ecology*, **81**, 1814–27.

Denno, R.F. & Roderick, G.K. (1992) Density-related dispersal in planthoppers: effects of interspecific crowding. *Ecology*, **73**, 1323–34.

Derouich, M. & Boutayeb, A. (2006) Dengue fever: mathematical modelling and computer simulation. *Applied Mathematics and Computation*, **177** (2), 528–44.

Despland, E. & Noseworthy, M. (2006) How well do specialist feeders regulate nutrient intake? Evidence from a gregarious tree-feeding caterpillar. *Journal of Experimental Biology*, **209** (7), 1301–9.

Despland, E., Rosenberg, J. & Simpson, S.J. (2004) Landscape structure and locust swarming: a satellite's eye view. *Ecography*, **27** (3), 381–91.

Dethier, V.G. (1941) Chemical factors determining the choice of food plants by *Papilio* larvae. *American Naturalist*, **75**, 61–73.

Dethier, V.G. (1954) Evolution of feeding preferences in phytophagous insects. *Evolution*, **8**, 33–54.

Deuve, T. (2002) Deux remarquables *Trechinae anophtalmes* des cavites souterraines du Guangxi nord-occidental, Chine (Coleoptera, Trechidae). *Bulletin de la Societe Entomologique de France*, **107** (5), 515–23.

Dewar, A.M., May, M.J., Woiwod, I.P. et al. (2003) A novel approach to the use of genetically modified herbicide tolerant crops for environmental benefit. *Proceedings of the Royal Society of London Series B, Biological Sciences*, **270**, 335–40.

Dhileepan, K. (2004) The applicability of the plant vigor and resource regulation hypotheses in explaining *Epiblema* gall moth–*Parthenium* weed interactions. *Entomologia Experimentalis et Applicata*, **113**, 63–70.

Dial, R.J., Ellwood, M.D.F., Turner, E.C. & Foster, W.A. (2006) Arthropod abundance, canopy structure, and microclimate in a Bornean lowland tropical rain forest. *Biotropica*, **38**, 643–52.

Dias, J.C.P., Silveira, A.C. & Schofield, C.J. (2002) The impact of Chagas disease control in Latin America: a review. *Memorias do Instituto Oswaldo Cruz*, **97**, 603–12.

Diaz, B.M. & Fereres, A. (2005) Life table and population parameters of *Nasonovia ribisnigri* (Homoptera: Aphididae) at different constant temperatures. *Environmental Entomology*, **34** (3), 527–34.

Didham, R.K., Ghazoul, J., Stork, N.E. & Davis, A.J. (1996) Insects in fragmented forests: a functional approach. *Trends in Ecology and Evolution*, **11**, 255–60.

Didham, R.K., Lawton, J.H., Hammond, P.M. & Eggleton, P. (1998) Trophic structure stability and extinction dynamics of beetles (Coleoptera) in tropical forest fragments. *Philosophical Transactions of the Royal Society of London Series B, Biological Sciences*, **353**, 437–51.

Dietrich, C.H. & Pooley, C.D. (1994) Automated identification of leafhoppers (Homoptera, Cicadellidae, Draeculacephala Ball). *Annals of the Entomological Society of America*, **87**, 412–23.

Dietrich, C.H. & Vega, F.E. (1995) Leafhoppers (Homoptera: Cicadellidae) from Dominican amber. *Annals of the Entomological Society of America*, **88** (3), 263–70.

Digweed, S.C. (2006) Oviposition preference and larval performance in the exotic birch-leafmining sawfly *Profenusa thomsoni*. *Entomologia Experimentalis et Applicata*, **120** (1), 41–9.

Dimopoulos, G. (2003) Insect immunity and its implication in mosquito–malaria interactions. *Cellular Microbiology*, **5**, 3–14.

Diotaiuti, L., Loiola, C.F., Falcao, P.L. & Dias, J.C.P.O. (1993) The ecology of *Triatoma sordida* in natural environments in two different regions of the state of Minas Gerais, Brazil. *Revista do Instituto de Medicina Tropical de São Paulo*, **35** (3), 237–45.

Diss, A.L., Kunkel, J.G., Montgomery, M.E. & Leonard, D.E. (1996) Effects of maternal nutrition and egg provisioning on parameters of larval hatch, survival and dispersal in the gypsy moth, *Lymantria dispar* L. *Oecologia (Berlin)*, **106** (4), 470–7.

Dixon, A.F.G. (2007) Body size and resource partitioning in ladybirds. *Population Ecology*, **49** (1), 45–50.

Dixon, A.F.G., Kindlmann, P., Leps, J. & Holman, J. (1987) Why there are so few species of aphids, especially in the tropics. *American Naturalist*, **129**, 580–92.

Djieto-Lordon, C., Dejean, A., Gibernau, M., Hossaert-Mckey, M. & McKey, D. (2004) Symbiotic mutualism with a community of opportunistic ants: protection, competition, and ant occupancy of the Myrmecophyte *Barteria nigritana* (Passifloraceae). *Acta Oecologica, International Journal of Ecology*, **26**, 109–16.

Do, M.T., Harp, J.M. & Norris, K.C. (1999) A test of a pattern recognition system for identification of spiders. *Bulletin of Entomological Research*, **89**, 217–24.

Dobson, A. (2005) Monitoring global rates of biodiversity change: challenges that arise in meeting the Convention on Biological Diversity (CBD) 2010 goals. *Philosophical Transactions of the Royal Society of London series B, Biological Sciences*, **360**, 229–41.

Dohm, D.J., Romoser, W.S., Turell, M.J. & Linthicum, K.J. (1991) Impact of stressful conditions on the survival of *Culex pipiens* exposed to rift valley fever virus. *Journal of the American Mosquito Control Association*, **7** (4), 621–3.

Dolch, R. & Tscharntke, T. (2000) Defoliation of alders (*Alnus glutinosa*) affects herbivory by leaf beetles on undamaged neighbours. *Oecologia*, **125**, 504–11.

Dolphin, K. & Quicke, D.L.J. (2001) Estimating the global species richness of an incompletely described taxon: an example using parasitoid wasps (Hymenoptera: Braconidae). *Biological Journal of the Linnean Society*, **73**, 279–86.

Donald, P.F. & Evans, A.D. (2006) Habitat connectivity and matrix restoration: the wider implications of agri-environment schemes. *Journal of Applied Ecology*, **43**, 209–18.

Dondorp, A-M., Newton, P-N., Mayxay, N. et al. (2004) Fake antimalarials in Southeast Asia are a major impediment to malaria control: multinational cross-sectional survey on the prevalence of fake antimalarials. *Tropical Medicine and International Health*, **9** (12), 1241–6.

Dormont, L., Baltensweiler, W., Choquet, R. & Roques, A. (2006) Larch- and pine-feeding host races of the larch bud moth (*Zeiraphera diniana*) have cyclic and synchronous population fluctuations. *Oikos*, **115**, 299–307.

dos Santos, C.B., Ferreira, A.L., Leite, G.R., Ferreira, G.E.M., Rodrigues, A.A.F. & Falqueto, A. (2005) Peridomiciliary colonies of *Triatoma vitticeps* (Stal, 1859) (Hemiptera, Reduviidae, Triatominae) infected with *Trypanosoma cruzi* in rural areas of the state of Espirito Santo, Brazil. *Memorias do Instituto Oswaldo Cruz*, **100** (5), 471–3.

Dostal, P. (2005) Effect of three mound-building ant species on the formation of soil seed bank in mountain grassland. *Flora*, **200**, 148–58.

Douglas, A.E. (1996) Reproductive failure and the free amino acid pools in pea aphids (*Acyrthosiphon pisum*) lacking symbiotic bacteria. *Journal of Insect Physiology*, **42** (3), 247–55.

Douglas, A.E. (1998) Nutritional interactions in insect–microbial symbioses: aphids and their symbiotic bacteria *Buchnera*. *Annual Review of Entomology*, **43**, 17–37.

Douglas, A.E. (2006) Phloem-sap feeding by animals: problems and solutions. *Journal of Experimental Botany*, **57**, 747–54.

Douglas, A.E., Francois, C.L.M.J. & Minto, L.B. (2006) Facultative 'secondary' bacterial symbionts and the nutrition of the pea aphid, *Acyrthosiphon pisum*. *Physiological Entomology*, **31** (3), 262–9.

Doutre, M.S. (2006) Urticaria – introduction. *Clinical Reviews in Allergy and Immunology*, **30**, 1–2.

Dowd, P.F. & Shen, S.K. (1990) The contribution of symbiotic yeast to toxin resistance of the cigarette beetle (*Lasioderma serricorne*). *Entomologia Experimentalis et Applicata*, **56** (3), 241–8.

Drake, B., Gonzalez-Meler, M. & Long, S.P. (1997) More efficient plants: a consequence of rising atmospheric CO_2. *Annual Review of Plant Physiology and Plant Molecular Biology*, **48**, 607–37.

Drake, V.A. & Farrow, R.A. (1989) The 'aerial plankton' and atmospheric convergence. *Trends in Ecology and Evolution*, **4** (12), 381–5.

Dransfield, R. (1975) *The ecology of grassland and cereal aphids*. PhD Thesis, University of London.

Drew, A.E. & Roderick, G.K. (2005) Insect biodiversity on plant hybrids within the Hawaiian silversword alliance (Asteraceae: Heliantheae-Madiinae). *Environmental Entomology*, **34**, 1095–108.

Drukker, B., Bruin, J., Jacobs, G., Kroon, A. & Sabelis, M.W. (2000a) How predatory mites learn to cope with variability in volatile plant signals in the environment of their herbivorous prey. *Experimental and Applied Acarology*, **24**, 881–895.

Drukker, B., Bruin, J. & Sabelis, M.W. (2000b) Anthocorid predators learn to associate herbivore-induced plant volatiles with presence or absence of prey. *Physiological Entomology*, **25**, 260–5.

Drukker, B., Scutareanu, P. & Sabelis, M.W. (1995) Do anthocorid predators respond to synomones from *Psylla*-infested pear trees under field conditions? *Entomologia Experimentalis et Applicata*, **77**, 193–203.

Du, L., Ge, F., Zhu, S.R. & Parajulee, M.N. (2004) Effect of cotton cultivar on development and reproduction of *Aphis gossypii* (Homoptera: Aphididae) and its predator *Propylaea japonica* (Coleoptera: Coccinellidae). *Journal of Economic Entomology*, **97** (4), 1278–83.

Du, Y.J., Poppy, G.M. & Powell, W. (1996) Relative importance of semiochemicals from first and second trophic levels in host foraging behavior of *Aphidius ervil*. *Journal of Chemical Ecology*, **22**, 1591–605.

Dudgeon, D. & Chan, I.K.K. (1992) An experimental-study of the influence of periphytic algae on invertebrate abundance in a Hong Kong stream. *Freshwater Biology*, **27**, 53–63.

Dudley, T. & Anderson, N.H. (1982) A survey of invertebrates associated with wood debris in aquatic habitats. *Melanderia*, **39**, 1–22.

Duelli, P., Studer, M., Marchand, I. & Jakob, S. (1990) Population movements of arthropods between natural and cultivated areas. *Biological Conservation*, **54** (3), 193–208.

Duffels, J.P. & van Mastrigt, H.J.G. (1991) Recognition of cicadas (Hemiptera: Cicadidae) by the Ekagi people of Irian Jaya (Indonesia), with a description of a new species of *Cosmopsaltria*. *Journal of Natural History*, **25** (1), 173–82.

Duffey, S.S. & Stout, M.J. (1996) Antinutritive and toxic components of plant defense against insects. *Archives of Insect Biochemistry and Physiology*, **32**, 3–37.

Dufour, D.L. (1987) Insects as food: a case study from the northwest Amazon. *American Anthropologist*, **89** (2), 383–97.

Dumas, M. & Bouteille, B. (1996) Human African trypanosomiasis. *Comptes Rendus des Séances de la Société de Biologie et de ses Filiales*, **190** (4), 395–408.

Dumbrell, A.J. & Hill, J.K. (2005) Impacts of selective logging on canopy and ground assemblages of tropical forest butterflies: implications for sampling. *Biological Conservation*, **125**, 123–31.

Duncan, F.D. & Byrne, M.J. (2000) Discontinuous gas exchange in dung beetles: patterns and ecological implications. *Oecologia Berlin*, **122** (4), 452–8.

Duncan, F.D., Krasnov, B. & McMaster, M. (2002) Metabolic rate and respiratory gas-exchange patterns in tenebrionid beetles from the Negev Highlands, Israel. *Journal of Experimental Biology*, **205** (6), 791–8.

Dungey, H.S., Potts, B.M., Whitham, T.G. & Li, H.F. (2000) Plant genetics affects arthropod community richness and composition: evidence from a synthetic eucalypt hybrid population. *Evolution*, **54**, 1938–46.

Dunn, D.W., Crean, C.S. & Gilburn, A.S. (2002) The effects of exposure to seaweed on willingness to mate, oviposition, and longevity in seaweed flies. *Ecological Entomology*, **27** (5), 554–64.

Dunn, R.R. (2005) Modern insect extinctions, the neglected majority. *Conservation Biology*, **19**, 1030–6.

Dupont, A., Belanger, L. & Bousquet, J. (1991) Relationships between balsam fir vulnerability to spruce budworm and ecological site conditions of fir stands in central Quebec. *Canadian Journal of Forest Research*, **21**, 1752–9.

Dupuy, C., Huguet, E. & Drezen, J.M. (2006) Unfolding the evolutionary story of polydnaviruses. *Virus Research*, **117**, 81–9.

Duringer, P., Schuster, M., Genise, J.F. et al. (2006) The first fossil fungus gardens of Isoptera: oldest evidence of symbiotic termite fungiculture (Miocene, Chad basin). *Naturwissenschaften*, **93** (12), 610–15.

Durner, J., Wendehenne, D. & Klessig, S.F. (1998) Defense gene induction in tobacco by nitric oxide, cyclic GMP, and cyclic ADP-ribose. *Proceedings of the National Academy of Science of the USA*, **95**, 10328–33.

Durrett, R. & Levin, S. (1996) Spatial models for species–area curves. *Journal of Theoretical Biology*, **179** (2), 119–27.

Dutra, H.P., Freitas, A.V.L. & Oliveira, P.S. (2006) Dual ant attraction in the neotropical shrub *Urera baccifera* (Urticaceae): the role of ant visitation to pearl bodies and fruits in herbivore deterrence and leaf longevity. *Functional Ecology*, **20** (2), 252–60.

Dwyer, G. & Elkinton, J.S. (1995) Host dispersal and the spatial spread of insect pathogens. *Ecology (Washington DC)*, **76** (4), 1262–75.

Dye, C. (1992) The analysis of parasite transmission by bloodsucking insects. *Annual Review of Entomology*, **37**, 1–20.

Dyer, L.A. & Bowers, M.D. (1996) The importance of sequestered iridoid glycosides as a defense against an ant predator. *Journal of Chemical Ecology*, **22**, 1527–39.

Dyer, M.I. & Bokhari, U.G. (1976) Plant–animal interactions: studies of the effects of grasshopper grazing on blue grama grass. *Ecology*, **57**, 762–72.

Dyer, M.I., Moon, A.M., Brown, M.R. & Crossley, D.A. Jr (1995) Grasshopper crop and midgut extract effects on plants: an example of reward feedback. *Proceedings of the National Academy of Sciences of the USA*, **92**, 5475–8.

EASAC (2005) *A Users Guide to Biodiversity Indicators*. The Royal Society, London.

Eastop, V.F. (1973) Deductions from the present day host plants of aphids and related species. *Symposium of the Royal Entomological Society of London*, **6**, 157–77.

Eck, G., Fiala, B., Linsenmair, K.E., Bin Hashim, R. & Proksch, P. (2001) Trade-off between chemical and biotic antiherbivore defense in the South East Asian plant genus *Macaranga*. *Journal of Chemical Ecology*, **27**, 1979–96.

Edenius, L., Danell, K. & Bergstrom, R. (1993) Impact of herbivory and competition on compensatory growth in woody plants winter browsing by moose on scots pine. *Oikos*, **66**, 286–92.

Edmunds, G.F. Jr (1973) Ecology of black pineleaf scale (Homoptera. *Diaspididae*). *Environmental Entomology*, **2**, 765–77.

Edmunds, G.F. Jr & Alstad, D.N. (1978) Coevolution in insect herbivores and conifers. *Science*, **199**, 941–5.

Edson, J.L. (1985) The influences of predation and resource subdivision on the coexistence of goldenrod aphids. *Ecology*, **66**, 1736–43.

Edwards, C.A., Sunderland, K.D. & George, K.S. (1979) Studies on polyphagous predators of cereal aphids. *Journal of Applied Ecology*, **16**, 811–23.

Edwards, O.R., Ridsdill-Smith, T.J. & Berlandier, F.A. (2003) Aphids do not avoid resistance in Australian lupin (*Lupinus angustifolius*, L-luteus) varieties. *Bulletin of Entomological Research*, **93**, 403–11.

Edwards, P.J. & Wratten, S.D. (1985) Induced plant defenses against insect grazing: fact or artefact? *Oikos*, **44**, 70–4.

Edwards, W., Dunlop, M. & Rodgerson, L. (2006) The evolution of rewards: seed dispersal, seed size and elaiosome size. *Journal of Ecology*, **94**, 687–94.

Edwards-Jones, G., Hussain, S.S. & Davies, B. (2000) *Ecological Economics*. University of Cambridge, Cambridge, UK.

EEA (2005) *The European Environment – State and outlook 2005*. European Environment Agency, Copenhagen.

Eggleton, P. & Belshaw, R. (1992) Insect parasitoids – an evolutionary overview. *Philosophical Transactions of the Royal Society of London Series B, Biological Sciences*, **337**, 1–20.

Eggleton, P., Bignell, D.E., Hauser, S., Dibog, L., Norgrove, L. & Madong, B. (2002) Termite diversity across an anthropogenic disturbance gradient in the humid forest zone of West Africa. *Agriculture Ecosystems and Environment*, **90**, 189–202.

Eggleton, P., Bignell, D.E., Sands, W.A. et al. (1996) The diversity, abundance and biomass of termites under differing levels of disturbance in the Mbalmayo Forest Reserve, southern Cameroon. *Philosophical Transactions of the Royal Society of London Series B, Biological Sciences*, **351**, 51–68.

Eggleton, P., Williams, P.H. & Gaston, K.J. (1994) Explaining global termite diversity – productivity or history. *Biodiversity and Conservation*, **3**, 318–30.

Ehrlich, P.R. & Raven, P.H. (1964) Butterflies and plants: a study in coevolution. *Evolution*, **18**, 586–608.

Eigen, M., Kloft, W.J. & Brandner, G. (2002) Correction of previews 200200338961. Transferability of HIV by arthropods supports the hypothesis about transmission of the virus from apes to man. Correction of abstract. *Naturwissenschaften*, **89**, 280.

Eigenbrode, S.D., Moodie, S. & Castagnola, T. (1995) Predators mediate host plant resistance to a phytophagous pest in cabbage with glossy leaf wax. *Entomologia Experimentalis et Applicata*, **77**, 335–42.

Eigenbrode, S.D., Trumble, J.T., Millar, J.G. & White, K.K. (1994) Topical toxicity of tomato sesquiterpenes to the beet armyworm and the role of these comounds in resistance derived form an accession of *Lycopersicon hirsitum* f. *typicum*. *Journal of Agricultural and Food Chemistry*, **42**, 807–10.

Eisler, M.C., Torr, S.J., Coleman, P.G., Machila, N. & Morton, J.F. (2003) Integrated control of vector-borne diseases of livestock: pyrethroids: panacea or poison? *Trends in Parasitology*, **19**, 341–5.

Eisner, T., Eisner, M., Rossini, C. et al. (2000) Chemical defense against predation in an insect egg. *Proceedings of the National Academy of Sciences of the USA*, **97**, 1634–9.

El Sayed, A.M., Suckling, D.M., Wearing, C.H. & Byers, J.A. (2006) Potential of mass trapping for long-term pest management and eradication of invasive species. *Journal of Economic Entomology*, **99** (5), 1550–64.

el Shafie, H.A.F. & Basedow, T. (2003) The efficacy of different neem preparations for the control of insects damaging potatoes and eggplants in the Sudan. *Crop Protection*, **22** (8), 1015–21.

Elek, J.A., Steinbauer, M.J., Beveridge, N. & Ebner, P. (2003) The efficacy of high and low volume spray applications of Mimic ® (tebufenozide) for managing autumn gum moth larvae *Mnesampela privata* (Lepidoptera: Geometridae) in eucalypt plantations. *Agricultural and Forest Entomology*, **5**, 325–32.

Elias, S.A. & Matthews, J.V. (2002) Arctic North American seasonal temperatures from the latest Miocene to the Early Pleistocene, based on mutual climatic range analysis of fossil beetle assemblages. *Canadian Journal of Earth Sciences*, **39**, 911–20.

Elith, J., Graham, C.H., Anderson, R.P. et al. (2006) Novel methods improve prediction of species' distributions from occurrence data. *Ecography*, **29**, 129–51.

Elkinton, J.S., Parry, D. & Boettner, G.H. (2006) Implicating an introduced generalist parasitoid in the invasive brown-tail moth's enigmatic demise. *Ecology*, **87**, 2664–72.

Elliot, S.L., Adler, F.R. & Sabelis, M.W. (2003) How virulent should a parasite be to its vector? *Ecology*, **84** (10), 2568–74.

Elliott, N.C., Tao, F.L., Giles, K.L., Royer, T.A., Greenstone, M.H. & Shufran, K.A. (2006) First quantitative study of rove beetles in Oklahoma winter wheat fields. *Biocontrol*, **51** (1), 79–87.

Ellwood, M.D.F. & Foster, W.A. (2004) Doubling the estimate of invertebrate biomass in a rainforest canopy. *Nature*, **429**, 549–51.

Ellwood, M.D.F., Jones, D.T. & Foster, W.A. (2002) Canopy ferns in lowland dipterocarp forest support a prolific abundance of ants, termites, and other invertebrates. *Biotropica*, **34**, 575–83.

Elmes, G.W. & Thomas, J.A. (1992) Complexity of species conservation in managed habitats – interaction between Maculinea butterflies and their ant hosts. *Biodiversity and Conservation*, **1**, 155–69.

Elser, J.J., Acharya, K., Kyle, M. et al. (2003) Growth rate-stoichiometry couplings in diverse biota. *Ecology Letters*, **6**, 936–43.

Elser, J.J., Dobberfuhl, D.R., Mackay, N.A. & Schampel, J.H. (1996) Organism size, life history, and N:P stoichiometry. *Bioscience*, **46**, 674–84.

Elser, J.J., Sterner, R.W., Gorokhova, E. et al. (2000) Biological stoichiometry from genes to ecosystems. *Ecology Letters*, **3**, 540–50.

Elton, C. (1927) *Animal Ecology*. Sidwick & Jackson, London.

Elton, C. (1933) *The Ecology of Animals. Methuen's Monographs on Biological Subjects*. Methuen, London.

Elzen, G.W. (1997) Changes in resistance to insecticides in tobacco budworm populations in Mississippi, 1993–5. *Southwestern Entomologist*, **22** (1), 61–72.

Embree, D.G. (1966) The role of introduced parasites in the control of the winter moth in Nova Scotia. *Canadian Entomologist*, **98**, 1159–68.

Emecki, M., Navarro, S., Donahaye, E., Rindner, M. & Azrieli, A. (2004) Respiration of *Rhyzopertha dominica* (F.) at reduced oxygen concentrations. *Journal of Stored Products Research*, **40**, 27–38.

Emmerson, M., Bezemer, H., Hunter, M.D. & Jones, T.H. (2005) Global change alters the stability of food webs. *Global Change Biology*, **11** (3), 490–501.

Emmerson, M., Bezemer, T.M., Hunter, M.D., Jones, T.H., Masters, G.J. & Van Dam, N.M. (2004) How does global change affect the strength of trophic interactions? *Basic and Applied Ecology*, **5** (6), 505–14.

Enayati, A.A. & Hemingway, J. (2006) Pyrethroid insecticide resistance and treated bednets efficacy in malaria control. *Pesticide Biochemistry and Physiology*, **84** (2), 116–26.

Endo, C. (2006) Seasonal wing dimorphism and life cycle of the mole cricket *Gryllotalpa orientalis* (Orthoptera: Gryllotalpidae). *European Journal of Entomology*, **103** (4), 743–50.

Engel, M-S. & Grimaldi, D-A. (2004) New light shed on the oldest insect. *Nature (London)*, **427** (6975), 627–30.

English-Loeb, G.M. (1989) Nonlinear responses of spider mites to drought-stressed host plants. *Ecological Entomology*, **14**, 45–55.

Enk, CD. (2006) Onchocerciasis – river blindness. *Clinics in Dermatology*, **24** (3), 176–80.

Entwistle, P.F., Adams, P.H.W., Evans, H.F., Rivers, C.F., Bird, F.T. & Burk, J.M. (1983) Epizootiology of a nuclear polyhedrosis virus (Baculoviridae) in European spruce sawfly (*Gilpinia hercyniae*): spread of disease from small epicentres in comparison with spread of baculovirus diseases in other hosts. *Journal of Applied Entomology*, **2**, 473–87.

Erb, W.A., Lindquist, R.K., Flickinger, N.J. & Casey, M.L. (1994) Resistance of selected *Lycopersicon* hybrids to greenhouse whitefly (Homoptera: Aleyrodidae). *Florida Entomologist*, **77** (1), 104–16.

Eriksson, M., Lilja, S. & Roininen, H. (2006) Dead wood creation and restoration burning: implications for bark beetles and beetle induced tree deaths. *Forest Ecology and Management*, **231**, 205–13.

Eriksson, M., Pouttu, A. & Roininen, H. (2005) The influence of windthrow area and timber characteristics on colonization of wind-felled spruces by Ips typographus (L.). *Forest Ecology and Management*, **216**, 105–16.

Erwin, T.L. (1982) Tropical forests: their richness in Coleoptera and other arthropod species. *Coleopterists Bulletin*, **36**, 74–5.

Erwin, T.L. (1983) Tropical forest canopies: the last biotic frontier. *Bulletin of the Entomological Society of America*, **30**, 14–19.

Erwin, T.L. & Scott, J. (1980) Seasonal and size patterns, trophic structure, and richness of Coleoptera in tropical arboreal ecosystem: The fauna of the tree *Luehea seemannii* Triana and Planch in the canal zone of Panama. *Coleopterists Bulletin*, **34**, 305–22.

Erzinclioglu, Y.Z., Baker, J.M. & Howell, S.E. (1990) Cyclorrhaphous maggots from a hypersaline oil spill site. *Entomologist*, **109** (4), 250–5.

Escalada, M-M. & Heong, K-L. (2004) A participatory exercise for modifying rice farmers' beliefs and practices in stem borer loss assessment. *Crop Protection*, **23** (1), 11–17.

Escarre, J., Lepart, J. & Sentuc, J.J. (1996) Effects of simulated herbivory in three old field Compositae with different inflorescence architectures. *Oecologia (Berlin)*, **105**, 501–8.

Escobar, F., Lobo, J.M. & Halffter, G. (2005) Altitudinal variation of dung beetle (Scarabaeidae: Scarabaeinae) assemblages in the Colombian Andes. *Global Ecology and Biogeography*, **14**, 327–37.

Esper, J., Buntgen, U., Frank, D.C., Nievergelt, D. & Liebhold, A. (2007) 1200 years of regular outbreaks in alpine insects. *Proceedings of the Royal Society of London Series B, Biological Sciences*, **274**, 671–9.

Eubanks, M.D. & Denno, R.F. (2000) Host plants mediate omnivore–herbivore interactions and influence prey suppression. *Ecology*, **81**, 936–47.

Eubanks, M.D., Nesci, K.A., Petersen, M.K., Liu, Z. & Sanchez, H.B. (1997) The exploitation of an ant-defended host plant by a shelter-building herbivore. *Oecologia (Berlin)*, **109**, 454–60.

Evans, E.W. & England, S. (1996) Indirect interactions in biological-control of insects – pests and natural enemies in alfalfa. *Ecological Applications*, **6**, 920–30.

Evans, G. (1977) *The Life of Beetles*, p. 232. Allen & Unwin, London.

Evans, H.F. & Fielding, N.J. (1994) Integrated management of *Dendroctonus micans* in the UK. *Forest Ecology and Management*, **65**, 17–30.

Evans, H.F. & Speight, M.R. (2004) Health and protection: integrated pest management practices. In *Encyclopedia of Forest Sciences*, Vol. 1: A–L (eds J. Burley, J. Evans & J.A. Youngquist), pp. 1045–52. Elsevier, Amsterdam.

Eyre, M.D., Luff, M.L., Rushton, S.P. & Topping, C.J. (1989) Ground beetles and weevils (Carabidae and Curculionoidea) as indicators of grassland management-practices. *Journal of Applied Entomology (Zeitschrift fur Angewandte Entomologie)*, **107**, 508–17.

Fabre, F., Dedryver, C.A., Leterrier, J.L. & Plantegenest, M. (2003) Aphid abundance on cereals in autumn predicts yield losses caused by Barley yellow dwarf virus. *Phytopathology*, **93**, 1217–22.

Fabre, F., Pierre, J.S., Dedryver, C.A. & Plantegenest, M. (2006) Barley yellow dwarf disease risk assessment based on Bayesian modelling of aphid population dynamics. *Ecological Modelling*, **193**, 457–66.

Fabre, J.H. (1882) *Nouveaux Souverirs Entomologiques: Etudes sur L'instincte et les Moeurs des Insectes*. Librairie Delagrave, Paris.

Fabres, G. & Nenon, J.P. (1997) Biodiversity and biological control: the case of the cassava mealybug in Africa. *Journal of African Zoology*, **111** (1), 7–15.

Faccoli, M. & Stergulc, F. (2004) *Ips typographus* (L.) pheromone trapping in south Alps: spring catches determine damage thresholds. *Journal of Applied Entomology*, **128** (4), 307–11.

Faccoli, M. & Stergulc, F. (2006) A practical method for predicting the short-time trend of bivoltine populations of *Ips typographus* (L.) (Col., Scolytidae). *Journal of Applied Entomology*, **130**, 61–6.

Faeth, S. (1985) Host leaf selection by leaf-miners: interaction among three trophic levels. *Ecology*, **66**, 479–94.

Faeth, S. (1986) Indirect interactions between temporally separated herbivores mediated by the host plant. *Ecology*, **67**, 479–94.

Faeth, S. (1987) Community structure and folivorous insect outbreaks: the role of vertical and horizontal interactions. In *Insect Outbreaks* (eds P. Barbosa & J.C. Schultz), pp. 135–71. Academic Press, New York.

Faeth, S.H. (1992) Interspecific and intraspecific interactions via plant responses to folivory – an experimental field-test. *Ecology*, **73**, 1802–13.

Faeth, S.H. (2002) Are endophytic fungi defensive plant mutualists? *Oikos*, **98**, 25–36.

Faeth, S.H., Connor, E.F. & Simberloff, D.S. (1981) Early leaf abscission: a neglected source of mortality for folivores. *American Naturalist*, **117**, 409–15.

Faeth, S.H. & Fagan, W.F. (2002) Fungal endophytes: common host plant symbionts but uncommon mutualists. *Integrative and Comparative Biology*, **42**, 360–8.

Fagan, W.F. & Bishop, J.G. (2000) Trophic interactions during primary succession: herbivores slow a plant reinvasion at Mount St. Helens. *American Naturalist*, **155**, 238–51.

Fagan, W.F., Siemann, E., Mitter, C.M., Denno, R.F., Huberty, A.F. & Woods, H.A. (2002) Nitrogen in insects: implications for trophic complexity and species diversification. *American Naturalist*, **160**, 784–802.

Fain, H.D. (1995) Genetic foraging for variability. *Evolutionary Theory*, **11** (1), 15–29.

Fajer, E.D. (1989) The effects of enriched CO_2 atmospheres on plant–insect herbivore interactions: growth responses of larvae of the specialist butterfly, *Junonia coenia* (Lepidoptera: Nymphalidae). *Oecologia*, **81**, 514–20.

Fajer, E.D., Bowers, M.D. & Bazzaz, F.A. (1991) The effects of enriched CO_2 atmospheres on the buckeye butterfly, *Junonia coenia*. *Ecology*, **72**, 751–4.

Fan, Y.H. & Li, W.T. (2007) Permanence in delayed ratio-dependent predator–prey models with monotonic functional responses. *Nonlinear Analysis, Real World Applications*, **8** (2), 424–34.

FAO (1991) *Second Interim Report on the State of Tropical Forests by Forest Resources Assessment 1990 Project*. Tenth World Forestry Congress, Paris. Food and Agriculture Organization, Rome.

FAO (1998a) *FAOSTAT* database on rice production. http://apps.fao.org/lim500/nph-wrap.pl?Production.Crops. Primary&Domain=SUA.

FAO (1998b) The Programme against African trypanosomiasis. http: //www.fao.org/waicent/faoinfo/agricult/aga/agah/pd/vector.htm.

FAO (1998c) *An Interim Report on the State of the Forest Resources in the Developing Countries*. Food and Agriculture Organization, Rome.

FAO (2005) *Forest Resources Assessment 2005 Progress towards Sustainable Forest Management*. Food and Agriculture Organization, Rome.

FAO (2007) http://faostat.fao.org/site/526/default.aspx.

Fargette, D., Muniyappa, V., Fauquet, C., N'guessan, P. & Thouvenel, J.C. (1993) Comparative epidemiology of three tropical whitefly-transmitted gemininviruses. *Biochimie (Paris)*, **75** (7), 547–54.

Farrar, R.R., Kennedy, G.G. & Kashyap, R.K. (1992) Influence of life-history differences of 2 tachinid parasitoids of *Helicoverpa zea* (Boddie) (Lepidoptera, Noctuidae) on their interactions with glandular trichome methyl ketone-based insect resistance in tomato. *Journal of Chemical Ecology*, **18**, 499–515.

Fauss, D.L. & Pierce, W.R. (1969) Stand conditions and spruce budworm damage in a western Montana forest. *Journal of Forestry*, **67**, 322–9.

Faveri, S.B. & Vasconcelos, H.L. (2004) The *Azteca–Cecropia* association: are ants always necessary for their host plants? *Biotropica*, **36**, 641–6.

Fayers, S.R. & Trewin, N.H. (2005) A hexapod from the Early Devonian Windyfield chert, Rhynie, Scotland. *Palaeontology*, **48**, 1117–30.

Feeny, P. (1970) Seasonal changes in oak leaf tannins and nutrients as a cause of spring feeding by winter moth caterpillars. *Ecology*, **51**, 565–81.

Feeny, P. (1976) Plant apparency and chemical defense. Rec. Adv. Phytochem. 10: 1–40. In *Coevolution* (ed. J.B. Harborne), pp. 163–206. Academic Press, London.

Feicht, E. (2004) Parasitoids of *Ips typographus* (Col., Scolytidae), their frequency and composition in uncontrolled and controlled infested spruce forest in Bavaria. *Journal of Pest Science*, **77** (3), 165–72.

Feinsinger, P. (1983) Coevolution and pollination. In *Coevolution* (eds D.J. Futuyma & M. Slatkin), pp. 282–310. Sinauer Associates, Sunderland, MA.

Feller, I.C. (1995) Effects of nutrient enrichment on growth and herbivory of dwarf red mangrove (*Rhizophora mangle*). *Ecological Monographs*, **65**, 477–505.

Fenchel, T. & Finlay, B.J. (2004) The ubiquity of small species: patterns of local and global diversity. *Bioscience*, **54**, 777–84.

Fenchel, T. & Finlay, B.J. (2006) The diversity of microbes: resurgence of the phenotype. *Philosophical Transactions of the Royal Society of London Series B, Biological Sciences*, **361**, 1965–73.

Feng, H.Q., Wu, K.M., Ni, Y.X., Cheng, D.F. & Guo, Y.Y. (2006) Nocturnal migration of dragonflies over the Bohai Sea in northern China. *Ecological Entomology*, **31**, 511–20.

Feng, Y., Chen, X.M. & Ye, S. (2001). The common edible species of wasps in Yunnan and their value as food. *Forest Research*, **14** (5), 578–81.

Fensham, R.J. (1994) Phytophagous insect–woody sprout interactions in tropical eucalypt forest: I. Insect herbivory. *Australian Journal of Ecology*, **19** (2), 178–88.

Ferdig, M.T., Beernsten, B.T., Spray, F.J., Li, J. & Christensen, B.M. (1993) Reproductive costs with resistance in a mosquito–filarial worm system. *American Journal of Tropical Medicine and Hygiene*, **49** (6), 756–62.

Ferguson, K.I. & Stiling, P. (1996) Non-additive effects of multiple natural enemies on aphid populations. *Oecologia (Berlin)*, **108** (2), 375–9.

Fernandes, G.W. (1994) Plant mechanical defenses against insect herbivory. *Revista Brasileira de Entomologia*, **38** (2), 421–33.

Fernandez, Q.C., Fereres, A., Godfrey, L. & Norris, R.F. (2002) Development and reproduction of *Myzus persicae* and *Aphis fabae* (Hom., Aphididae) on selected weed species surrounding sugar beet fields. *Journal of Applied Entomology*, **126** (4), 198–202.

Fernandez-Marin, H., Zimmerman, J.K. & Wcislo, W.T. (2004) Ecological traits and evolutionary sequence of nest establishment in fungus-growing ants (Hymenoptera, Formicidae, Attini). *Biological Journal of the Linnean Society*, **81**, 39–48.

Ferrell, G.T. & Hall, R.C. (1975) Weather and tree growth associated with white fir mortality caused by fir engraver and roundheaded fir borer. *US Forest Service Research Paper, PSW*, **109**.

Ferrenberg, S.M. & Denno, R.F. (2003) Competition as a factor underlying the abundance of an uncommon phytophagous insect, the salt-marsh planthopper *Delphacodes penedetecta*. *Ecological Entomology*, **28**, 58–66.

Ferrier, S.M. & Price, P.W. (2004) Oviposition preference and larval performance of a rare bud-galling sawfly (Hymenoptera: Tenthredinidae) on willow in northern Arizona. *Environmental Entomology*, **33** (3), 700–8.

Ferris, R. & Humphrey, J.W. (1999) A review of potential biodiversity indicators for application in British forests. *Forestry*, **72**, 313–28.

Ferriss, R.S. & Berger, H. (1993) A stochastic simulation model of epidemics of arthropod-vectored plant virus. *Phytopathology*, **83** (12), 1269–78.

Fettig, C.J., Klepzig, K.D., Billings, R.F. et al. (2007) The effectiveness of vegetation management practices for prevention and control of bark beetle infestations in coniferous forests of the western and southern United States. *Forest Ecology and Management*, **238** (1–3), 24–53.

Fiala, B. & Linsenmair, K.E. (1995) Distribution and abundance of plants with extrafloral nectaries in the woody flora of a lowland primary forest in Malaysia. *Biodiversity and Conservation*, **4**, 165–82.

Fiala, B. & Maschwitz, U. (1991) Extrafloral nectaries in the genus *Macaranga* (Euphorbiaceae) in Malaysia: comparitive studies of their possible significance as predispositions for myrmecophytism. *Biological Journal of the Linnean Society*, **44**, 287–306.

Fiala, B. & Maschwitz, U. (1992) Domatia as most important adaptations in the evolution of myrmecophyes in the paleotropical tree genus *Macaranga* (Euphorbiaceae). *Plant Systematics and Evolution*, **180**, 53–64.

Fiddler, G.O., Hart, D.R., Fiddler, T.A. & Mcdonald, P.M. (1989) Thinning decreases mortality and increases growth of ponderosa pine in northeastern California (USA). *US Forest Service Research Paper, Pacific South West*, **194** (I/II), 1–7.

Fiebig, M., Poehling, H. M. & Borgemeister, C. (2004) Barley yellow dwarf virus, wheat, and Sitobion avenae: a case of trilateral interactions. *Entomologia Experimentalis et Applicata*, **110**, 11–21.

Fiedler, K. (1996) Host-plant relationships of lycaenid butterflies: large-scale patterns, interactions with plant chemistry, and mutualism with ants. *Entomologia Experimentalis et Applicata*, **80**, 259–67.

Field, R.G., Gardiner, T., Mason, C.F. & Hill, J. (2005) Agri-environment schemes and butterflies: the utilisation of 6 m grass margins. *Biodiversity and Conservation*, **14**, 1969–76.

Fielding, N.J. & Evans, H.F. (1996) The pine wood nematode, *Bursaphelenchus xylophilus* (Steiner and Buhrer) Nickle (= *B. lignicolis* Mamimya and Kiyohara): an assessment of the current position. *Forestry (Oxford)*, **69** (1), 35–46.

Fielding, N.J., O'Keefe, T. & King, C.J. (1991) Dispersal and host-finding capability of the predatory beetle, *Rhizophagus grandis* (Coleoptera: Rhizophagidae). *Journal of Applied Entomology*, **112** (1), 89–98.

Filichkin, S.A., Brumfield, S., Filichkin, T.P. & Young, M.J. (1997) *In vitro* interactions of the aphid endosymbiotic symL chaperonin with barley yellow dwarf virus. *Journal of Virology*, **71** (1), 569–77.

Fillinger, U. & Lindsay, S.W. (2006) Suppression of exposure to malaria vectors by an order of magnitude using microbial larvicides in rural Kenya. *Tropical Medicine and International Health*, **11** (11), 1629–42.

Fillinger, U., Sonye, G., Killeen, G.F., Knols, B.G.J. & Becker, N. (2004) The practical importance of permanent and semipermanent habitats for controlling aquatic stages of *Anopheles gambiae* sensu lato mosquitoes: operational observations from a rural town in western Kenya. *Tropical Medicine and International Health*, **9**, 1274–89.

Fincke, O.M. (1994) Population regulation of a tropical damselfly in the larval stage by food limitation, cannibalism, intraguild predation and habitat drying. *Oecologia (Berlin)*, **100** (1/2), 118–27.

Findlay, J.A., Li, G., Penner, P. & Miller, J.D. (1995) Novel diterpenoid insect toxins from a conifer endophyte. *Journal of Natural Products (Lloydia)*, **58**, 197–200.

Findlay, S., Carreiro, M., Krischik, V. & Jones, C.G. (1996) Effects of damage to living plants on leaf litter quality. *Ecological Applications*, **6**, 269–75.

Fineblum, W.L. & Rausher, M.D. (1995) Tradeoff between resistance and tolerance to herbivore damage in a morning glory. *Nature*, **377**, 517–20.

Fink, L.S. & Brower, L.P. (1981) Birds can overcome the cardenolide defence of monarch butterflies in Mexico. *Nature*, **291**, 67–70.

Finke, D.L. & Denno, R.F. (2002) Intraguild predation diminished in complex-structured vegetation: implications for prey suppression. *Ecology*, **83**, 643–52.

Finlay, B.J., Thomas, J.A., McGavin, G.C., Fenchel, T. & Clarke, R.T. (2006) Self-similar patterns of nature: insect diversity at local to global scales. *Proceedings of the Royal Society of London Series B, Biological Sciences*, **273**, 1935–41.

Firn, R.D. (2003) Bioprospecting – why is it so unrewarding? *Biodiversity and Conservation*, **12**, 207–16.

Fischer, K. & Fiedler, K. (2000) Sex-related differences in reaction norms in the butterfly *Lycaena tityrus* (Lepidoptera: Lycaenidae). *Oikos*, **90**, 372–80.

Fischer R.C., Olzant S.M., Wanek W. & Mayer V. (2005) The fate of *Corydalis cava* elaiosomes within an ant colony of *Myrmica rubra*: elaiosomes are preferentially fed to larvae. *Insectes Sociaux*, **52**, 55–62.

Flamm, R.O., Wagner, T.L., Cook, S.P., Pulley, P.E., Coulson, R.N. & Mcardle, T.M. (1987) Host colonization by cohabitating *Dendroctonus frontalis*, *Ips avulsus*, and *I. calligraphus* (Coleoptera. *Scolytidae*). *Environmental Entomology*, **16**, 390–9.

Flatt, T. & Scheuring, I. (2004) Stabilizing factors interact in promoting host–parasite coexistence. *Journal of Theoretical Biology*, **228** (2), 241–9.

Floate, K.D. & Whitman, T.G. (1995) Insects as traits in plant systematics: their use in discriminating between hybrid cottonwoods. *Canadian Journal of Botany*, **73**, 1–13.

Floate, K.D., Kearsley, M.J.C. & Whitham, T.G. (1993) Elevated herbivory in plant hybrid zones: *Chrysomela confluens*, *Populus* and phenological sinks. *Ecology (Tempe)*, **74**, 2056–65.

Floate, K.D., Martinsen, G.D. & Whitham, T.G. (1997) Cottonwood hybrid zones as centres of abundance for gall aphids in western North America: importance of relative habitat size. *Journal of Animal Ecology*, **66** (2), 179–88.

Floore, T.G. (2006) Mosquito larval control practices: past and present. *Journal of the American Mosquito Control Association*, **22** (3), 527–33.

Floren, A., Biun, A. & Linsenmair, K.E. (2002) Arboreal ants as key predators in tropical lowland rainforest trees. *Oecologia*, **131**, 137–44.

Floren, A. & Linsenmair, K.E. (2005) The importance of primary tropical rain forest for species diversity: an investigation using arboreal ants as an example. *Ecosystems*, **8**, 559–67.

Floyd, R.B., Farrow, R.A. & Neumann, F.G. (1994). Inter- and intra-provenance variation in resistance of red gum foliage to insect feeding. *Australian Forestry*, **57**, 45–8.

Fluegel, S.M. & Johnson, J.B. (2001) The effect of soil nitrogen levels and wheat resistance on the Russian wheat aphid, *Diuraphis noxia* (Homoptera: Aphididae). *Journal of the Kansas Entomological Society*, **74** (1), 49–55.

Fogal, W.H. & Slansky, F. Jr. (1985) Contribution of feeding by European sawfly larvae to litter production and element flux on Scots pine plantations. *Canadian Journal of Forest Research*, **15**, 484–7.

Foggo, A., Ozanne, C.M.P., Speight, M.R. & Hambler, C. (2001) Edge effects and tropical forest canopy invertebrates. *Plant Ecology*, **153** (1/2), 347–59.

Foggo, A. & Speight, M.R. (1993) Root damage and water stress: treatments affecting the exploitation of the buds of common ash *Fraxinus excelsior* L., by larvae of the ash bud moth *Prays fraxinella* Bjerk. (Lep., Yponomeutidae). *Oecologia*, **96**, 134–8.

Foggo, A., Speight, M.R. & Gregoire, J.C. (1994) Root disturbance of common ash, *Fraxinus excelsior* (Oleaceae), leads to reduced foliar toughness and increased feeding by a folivorous weevil, *Stereonychus fraxini* (Coleoptera, Curculionidae). *Ecological Entomology*, **19**, 344–8.

Fonseca, C.R. (1994) Herbivory and the long-lived leaves of an Amazonian ant-tree. *Journal of Ecology*, **82**, 833–42.

Fonte, S.J. & Schowalter, T.D. (2005) The influence of a neotropical herbivore (*Lamponius portoricensis*) on nutrient cycling and soil processes. *Oecologia*, **146** (3), 423–31.

Fontenille, D., Lepers, J.P., Coluzzi, M., Campbell, G.H., Rakotoarivony, I. & Coulanges, P. (1992) Malaria transmission and vector biology on Sainte Marie Island, Madagascar. *Journal of Medical Entomology*, **29** (2), 197–202.

Forgie, S.A., Philips, T.K. & Scholtz, C.H. (2005) Evolution of the Scarabaeini (Scarabaeidae: Scarabaeinae). *Systematic Entomology*, **30**, 60–96.

Forup, M.L. & Memmott, J. (2005) The. relationship between the abundances of bumblebees and honeybees in a native habitat. *Ecological Entomology*, **30**, 47–57.

Foster, M.A., Schultz, J.C. & Hunter, M.D. (1992) Modeling gypsy-moth virus leaf chemistry interactions – implications of plant-quality for pest and pathogen dynamics. *Journal of Animal Ecology*, **61**, 509–20.

Foster, W.A. & Treherne, J.E. (1986) The ecology and behavior of a marine insect, *Halobates fijiensis* (Hemiptera

(Heteroptera): Gerridae). *Zoological Journal of the Linnean Society*, **86** (4), 391–412.

Fowler, S.V. (2004) Biological control of an exotic scale, *Orthezia insignis* Browne (Homoptera: Ortheziidae), saves the endemic gumwood tree, *Commidendrum robustum* (Roxb.) DC. (Asteraceae) on the island of St. Helena. *Biological Control*, **29**, 367–74.

Fowler, S.V. & Lawton, J.H. (1985) Rapidly induced defenses and talking trees: the devil's advocate position. *American Naturalist*, **126**, 181–95.

Fox, J.W., Wood, D.L., Akers, R.P. & Parmeter, J.R.J.R. (1992) Survival and development of *Ips paraconfusus* Lanier (Coleoptera: Scolytidae) reared axenically and with tree-pathogenic fungi vectored by cohabiting *Dendroctonus* species. *Canadian Entomologist*, **124** (6), 1157–67.

Fraenkel, G.S. (1959) Raison d'etre of secondary plant substances. *Science*, **129**, 1466–70.

Franceschi, V.R., Krokene, P., Christiansen, E. & Krekling, T. (2005) Anatomical and chemical defenses of conifer bark against bark beetles and other pests. *New Phytologist*, **167**, 353–75.

Francis, J.E. & Harland, B.M. (2006) Termite borings in Early Cretaceous fossil wood, Isle of Wight, UK. *Cretaceous Research*, **27**, 773–7.

Frank, J.H. (1998) How risky is biological control? Comment. *Ecology*, **79** (5), 1829–34.

Franklin, A., De-Canniere, C. & Gregoire, J.C. (2004) Can sales of infested timber be used to quantify attacks by *Ips typographus* (Coleoptera, Scolytidae)? A pilot study from Belgium. *Annals of Forest Science*, **61** (5), 477–80.

Frederickson, M.E. (2005) Ant species confer different partner benefits on two neotropical Myrmecophytes. *Oecologia*, **143**, 387–95.

Free, C.A., Beddington, J.R. & Lawton, J.H. (1977) On the inadequacy of simple models of mutual interference for parasitism and predation. *Journal of Animal Ecology*, **46**, 543–4.

Frei, M. & Becker, K. (2005) A greenhouse experiment on growth and yield effects in integrated rice–fish culture. *Aquaculture*, **244** (1–4), 119–28.

Freitas, A.V.L. & Oliveira, P.S. (1996) Ants as selective agents on herbivore biology: effects on the behaviour of a non-myrmecophilous butterfly. *Journal of Animal Ecology*, **65**, 205–10.

Frey, M., Stettner, C., Pare, P.W., Schmelz, E.A., Tumlinson, J.H. & Gierl, A. (2000) An herbivore elicitor activates the gene for indole emission in maize. *Proceedings of the National Academy of Sciences of the USA* **97**, 14801–6.

Fritz, R.S., Nichols-Orians, C.M. & Brunsfield, S.J. (1994) Interspecific hybridization of plants and resistance to herbivores: hypothesis, genetics, and variable responses in a diverse herbivore community. *Oecologia (Berlin)*, **97**, 106–17.

Fritz, R.S., Sacchi, C.F. & Price, P.W. (1986) Competition versus host plant phenomenon in species composition: willow sawflies. *Ecology*, **67**, 1608–18.

Fritz, R.S. & Simms, E.L. (eds) (1992) *Plant Resistance to Herbivores and Pathogens. Ecology, Evolution, and Genetics.* University of Chicago Press, Chicago.

Frost, C.J. & Hunter, M.D. (2004) Insect canopy herbivory and frass deposition affect soil nutrient dynamics and export in oak mesocosms. *Ecology*, **85**, 3335–47.

Frost, C.J. & Hunter, M.D. (2007) Recycling of nitrogen in herbivore feces: plant recovery, herbivore assimilation, soil retention, and leaching losses. *Oecologia*, **151**, 42–53.

Fu, S., Kiselle, K.W., Coleman, D.C., Hendrix, P.F. & Crossley, D.A. (2001) Short-term impacts of aboveground herbivory (grasshopper) on the abundance and 14C activity of soil nematodes in conventional tillage and no-till agroecosystems. *Soil Biology and Biochemistry*, **33**, 1253–8.

Fukatsu, T. (1994) Endosymbiosis of aphids with microorganisms: a model case of dynamic endosymbiotic evolution. *Plant Species Biology*, **9** (3), 145–54.

Fuxa, J.R. & Richter, A.R. (1994) Distance and rate of spread of *Anticarsia gemmatalis* (Lepidoptera: Noctuidae) nuclear polyhedrosis virus released into soybean. *Environmental Entomology*, **23** (5), 1308–16.

Fuxa, J.R., Richter, A.R. & Strother, M.S. (1993) Detection of *Anticarsia gemmatalis* nuclear polyhedrosis virus in predatory arthropods and parasitoids after viral release in Louisiana soybean. *Journal of Entomological Science*, **28** (1), 51–60.

Gade, G. (2002) Sexual dimorphism in the pyrgomorphid grasshopper *Phymateus morbillosus*: from wing morphometry and flight behaviour to flight physiology and endocrinology. *Physiological Entomology*, **27**, 51–7.

Gage, K.L. & Kosoy, M.Y. (2005) Natural history of plague: perspectives from more than a century of research. *Annual Review of Entomology*, **50**, 505–28.

Gagne, W.C. & Howarth, F.G. (1985) Conservation status of endemic Hawaiian Lepidoptera. In *Proceedings of the 3rd Congress European Lepidopterologists, Cambridge 1982*, pp. 74–84. Societus Europaea Lepidopterologica, Karlsruhe.

Galil, J. & Eisikowitch, D. (1968) On the pollination ecology of *Ficus sycomorus* in East Africa. *Ecology*, **49**, 259–69.

Galvan, T.L., Koch, R.L. & Hutchison, W.D. (2006) Toxicity of indoxacarb and spinosad to the multicolored Asian lady beetle, *Harmonia axyridis* (Coleoptera: Coccinellidae), via three routes of exposure. *Pest Management Science*, **62** (9), 797–804.

Gammans, N., Bullock, J.M. & Schonrogge, K. (2005) Ant benefits in a seed dispersal mutualism. *Oecologia*, **146** (1), 43–9.

Gan, J.B. (2004) Risk and damage of southern pine beetle outbreaks under global climate change. *Forest Ecology and Management*, **191** (1–3), 61–71.

Gange, A.C. (1995) Aphid performance in an alder (*Alnus*) hybrid zone. *Ecology*, **76**, 2074–83.

Gange, A.C. & Brown, V.K. (1989) Effects of root herbivory by an insect on a foliar-feeding species, meditated through changes in the host plant. *Oecologia*, **81**, 38–42.

Gange, A.C. & Brown, V.K. (2002) Soil food web components affect plant community structure during early succession. *Ecological Research*, **17**, 217–27.

Garcia-Barros, E. (1992) Evidence for geographic variation of egg size and fecundity in a satyrine butterfly, *Hipparchia semele* (L.) (Lepidoptera, Nymphalidae–Satyrinae). *Graellsia*, **48**, 45–52.

Garnett, G.P. & Antia, R. (1994) Population biology of virus–host interactions. In *The Evolutionary Biology of Viruses* (ed. S.S. Morse), pp. 51–73. Raven Press, New York.

Garrett, K.A., Dendy, S.P., Frank, E.E., Rouse, M.N. & Travers, S.E. (2006) Climate change effects on plant disease: genomes to ecosystems. *Annual Review of Phytopathology*, **44**, 489–509.

Gassmann, A.J. & Hare, J.D. (2005) Indirect cost of a defensive trait: variation in trichome type affects the natural enemies of herbivorous insects on *Datura wrightii*. *Oecologia*, **144**, 62–71.

Gaston, K.J. (1991) The magnitude of global insect species richness. *Conservation Biology*, **5** (3), 283–96.

Gaston, K.J. (1994) Spatial patterns of species description – how is our knowledge of the global insect fauna growing. *Biological Conservation*, **67**, 37–40.

Gaston, K.J. (1996) What is biodiversity? In *Biodiversity: a Biology of Numbers and Difference* (ed. K.J. Gaston), pp. 1–9. Blackwell Science, Oxford.

Gaston, K.J., Charman, K., Jackson, S.F. et al. (2006) The ecological effectiveness of protected areas: the United Kingdom. *Biological Conservation*, **132**, 76–87.

Gaston, K.J., Gauld, I.D. & Hanson, P. (1996) The size and composition of the hymenopteran fauna of Costa Rica. *Journal of Biogeography*, **23** (1), 105–13.

Gaston, K.J. & Hudson, E. (1994) Regional patterns of diversity and estimates of global insect species richness. *Biodiversity and Conservation*, **3** (6), 493–500.

Gaston, K.J. & May, R.M. (1992) Taxonomy of taxonomists. *Nature*, **356** (6367), 281–2.

Gaston, K.J. & O'Neill, M.A. (2004) Automated species identification: why not? *Philosophical Transactions of the Royal Society of London Series B, Biological Sciences*, **359**, 655–67.

Gaston, K.J. & Williams, P.H. (1996) Spatial patterns in taxonomic diversity. In *Biodiversity: a Biology of Numbers and Difference* (ed. K.J. Gaston), pp. 202–29. Blackwell Science, Oxford.

Gathmann, A., Greiler, H.J. & Tscharntke, T. (1994) Trap-nesting bees and wasps colonising set-aside fields: succession and body size, management by cutting and sowing. *Oecologia*, **98** (1), 8–14.

Gauld, I.D., Gaston, K.J. & Janzen, D.H. (1992) Plant allelochemicals, tritrophic interactions and the anomalous diversity of tropical parasitoids – the nasty host hypothesis. *Oikos*, **65**, 353–7.

Gaume, L., Zacharias, M. & Borges, R.M. (2005) Ant-plant conflicts and a novel case of castration parasitism in a Myrmecophyte. *Evolutionary Ecology Research*, **7**, 435–52.

Gaunt, M.W. & Miles, M.A. (2002) An insect molecular clock dates the origin of the insects and accords with palaeontological and biogeographic landmarks. *Molecular Biology and Evolution*, **19** (5), 748–61.

Gaylord, E.S., Preszler, R.W. & BoeckleN, W.J. (1996) Interactions between host plants, endophytic fungi, and a phytophagous insect in an oak (*Quercus grisea* × *Quercus gambelii*) hybrid zone. *Oecologia*, **105**, 336–42.

Ge, S.Q., Yang, X.K., Cui, J. & Wen, H.L. (2003). Disscussion on the study of phylogeny of Strepsiptera. *Acta Zootaxonomica Sinica*, **27** (3), 417–27.

Geary, T.G. (2005) Ivermectin 20 years on: maturation of a wonder drug. *Trends in Parasitology*, **21** (11), 530–2.

Geervliet, J.B.F., Posthumus, M.A., Vet, L.E.M. & Dicke, M. (1997) Comparative analysis of headspace volatiles from different caterpillar-infested or uninfested food plants of *Pieris* species. *Journal of Chemical Ecology*, **23**, 2935–54.

Gefen, E. & Ar, A. (2006) Temperature dependence of water loss rates in scorpions and its effect on the distribution of *Buthotus judaicus* (Buthidae) in Israel. *Comparative Biochemistry and Physiology A, Molecular and Integrative Physiology*, **144** (1), 58–62.

Gehring, W.J. & Wehner, R. (1995) Heat shock protein synthesis and thermotolerance in *Cataglyphis*, an ant from the Sahara desert. *Proceedings of the National Academy of Sciences of the USA*, **92** (7), 2994–8.

Gemperli, A., Sogoba, N., Fondjo, E. et al. (2006) Mapping malaria transmission in West and Central Africa. *Tropical Medicine and International Health*, **11**, 1032–46.

Genise, J.F., Sciutto, J.C., Laza, J.H., Gonzalez, M.G. & Bellosi, E.S. (2002). Fossil bee nests, coleopteran pupal chambers and tuffaceous paleosols from the Late Cretaceous Laguna Palacios Formation, Central Patagonia (Argentina). *Palaeogeography Palaeoclimatology Palaeoecology*, **177** (34), 215–35.

Gentile, G. & Argano, R. (2005) Island biogeography of the Mediterranean sea: the species–area relationship for terrestrial isopods. *Journal of Biogeography*, **32**, 1715–26.

Gerardo, N.M., Mueller, U.G. & Currie, C.R. (2006) Complex host–pathogen coevolution in the Apterostigma fungus-growing ant–microbe symbiosis. *BMC Evolutionary Biology*, **6**, Article Number 88.

Gering, J.C., DeRennaux, K.A. & Crist, T.O. (2007) Scale dependence of effective specialization: its analysis and implications for estimates of global insect species richness. *Diversity and Distributions*, **13**, 115–25.

Gerisch, M., Schanowski, A., Figura, W., Gerken, B., Dziock, F. & Henle, K. (2006) Carabid beetles (Coleoptera, carabidae) as indicators of hydrological site conditions in floodplain grasslands. *International Review of Hydrobiology*, **91**, 326–40.

Gershenzon, J. (1994) Metabolic costs of terpenoid accumulation in higher plants. *Journal of Chemical Ecology*, **20**, 1281–328.

Gibbs, D., Walton, R., Brower, L. & Davis, A.K. (2006) Monarch butterfly (Lepidoptera: Nymphalidae) migration monitoring at Chincoteague, Virginia and Cape May, New Jersey: a comparison of long-term trends. *Journal of the Kansas Entomological Society*, **79** (2), 156–64.

Gibson, C.W.D. (1980) Niche use patterns among some Stenodemini (Heteroptera: Miridae) of limestone grassland, and an investigation of the possibility of interspecific competition between *Notostira elongata* Geoffroy and *Megaloceraea recticornis* Geoffroy. *Oecologia*, **47**, 352–64.

Gilbert, L.E. (1975) Ecological consequences of a coevolved mutualism between butterflies and plants. In *Coevolution of Animals and Plants* (eds L.E. Gilbert & P.R. Raven), pp. 210–40. University of Texas Press, Austin.

Gilbert, L.E. (1991) Biodiversity of a Central American *Heliconius* community: pattern, process, and problems. In *Plant–Animal Interactions: Evolutionary Ecology in Tropical and Temperate Regions* (eds P.W. Price, T.M. Lewinshon, G.W. Fernandes & W.W. Benson), pp. 403–27. John Wiley & Sons, New York.

Gilbert, L.E. & Raven, P.H. (eds) (1975) *Coevolution of Animals and Plants*. University of Texas Press, Austin.

Gilbert, L-I. (2004). Halloween genes encode P450 enzymes that mediate steroid hormone biosynthesis in *Drosophila melanogaster*. *Molecular and Cellular Endocrinology*, **215**, 1–10.

Gilbert, M., Fielding, N., Evans, H.F. & Gregoire, J.C. (2003) Spatial pattern of invading Dendroctonus micans (Coleoptera: Scolytidae) populations in the United Kingdom. *Canadian Journal of Forest Research*, **33** (4), 712–25.

Gilbert, M. & Grégoire, J-C. (2003) Site condition and predation influence a bark beetle's success: a spatially realistic approach. *Agricultural and Forest Entomology*, **5** (2), 87–96.

Gildow, F.E., Reavy, B., Mayo, M.A. et al. (2000) Aphid acquisition and cellular transport of potato leafroll virus-like particles lacking P5 readthrough protein. *Phytopathology*, **90**, 1153–61.

Gillespie, D.R., Mason, P.G., Dosdall, L.M., Bouchard, P. & Gibson, G.A.P. (2006) Importance of long-term research in classical biological control: an analytical review of a release against the cabbage seedpod weevil in North America. *Journal of Applied Entomology*, **130** (8), 401–9.

Gillette, N.E., Stein, J.D., Owen, D.A. et al. (2006) Verbenone-releasing flakes protect individual *Pinus contorta* trees from attack by *Dendroctonus ponderosae* and *Dendroctonus valens* (Coleoptera: Curculionidae, Scolytinae). *Agricultural and Forest Entomology*, **8** (3), 243–51.

Gillman, M. & Hails, R. (1997) *An Introduction to Ecological Modelling: Putting Practice into Theory*. Blackwell Science, Oxford.

Gilpin, M.E. & Ayala, F.J. (1973) Global models of growth and competition. *Proceedings of the National Academy of Science of the USA*, **70**, 3590–3.

Gish, M. & Inbar, M. (2006) Host location by apterous aphids after escape dropping from the plant. *Journal of Insect Behavior*, **19** (1), 143–53.

Githeko, A.K., Ayisi, J.M., Odada, P.K. et al. (2006) Topography and malaria transmission heterogeneity in western Kenya highlands: prospects for focal vector control. *Malaria Journal*, **5**, Article number 107.

Gleason, H.A. (1917) The structure and development of the plant association. *Bulletin of the Torrey Botanical Club*, **43**, 463–81.

Glinski, Z. & Jarosz, J. (1996) Immune mechanisms of vector insects in parasite destruction. *Central European Journal of Immunology*, **21** (1), 61–70.

Glynn, C. & Herms, D.A. (2004) Local adaptation in pine needle scale (*Chionaspis pinifoliae*): natal and novel host quality as tests for specialization within and among red and Scots pine. *Environmental Entomology*, **33**, 748–55.

Godfray, H.C.J. (2002) Challenges for taxonomy: the discipline will have to reinvent itself if it is to survive and flourish. *Nature*, **417**, 17–19.

Godfrey, L.D., Godfrey, K.E., Hunt, T.E. & Spomer, S.M. (1991) Natural enemies of European corn-borer *Ostrinia nubilalis* (Hübner) (Lepidoptera, Pyralidae) larvae in irrigated and drought-stressed corn. *Journal of the Kansas Entomological Society*, **64**, 279–86.

Goehring, L. & Oberhauser, K.S. (2002) Effects of photoperiod, temperature, and host plant age on induction of reproductive diapause and development time in *Danaus plexippus*. *Ecological Entomology*, **27**, 674–85.

Goldstein, P.Z. (1997) How many things are there? A reply to Oliver and Beattie, Beattie and Oliver, Oliver and Beattie, and Oliver and Beattie. *Conservation Biology*, **11**, 571–4.

Gonzalez-Megias, A. & Gomez, J.M. (2003) Consequences of removing a keystone herbivore for the abundance and diversity of arthropods associated with a cruciferous shrub. *Ecological Entomology*, **28**, 299–308.

Gonzon, A.T., Bartlett, C.R. & Bowman, J.L. (2006) Planthopper (Hemiptera: Fulgoroidea) diversity in the great Smoky Mountains National Park. *Transactions of the American Entomological Society*, **132**, 243–60.

Gooding, R.H. & Krafsur, E.S. (2005) TSETSE genetics: contributions to biology, systematics, and control of tsetse flies. *Annual Review of Entomology*, **50**, 101–23.

Gordeev, M.I. & Stegniy, V.N. (1987) Inversion Polymorphism in Malaria Mosquito *Anopheles Messeae*. 7. Fecundity and Genetic-Populational Structure of the Species. *Genetika*, **23** (12), 2169–74.

Gordo, O. & Sanz, J.J. (2006) Temporal trends in phenology of the honey bee *Apis mellifera* (L.) and the small white *Pieris rapae* (L.) in the Iberian Peninsula (1952–2004). *Ecological Entomology*, **31**, 261–8.

Gotelli, N.J. & Colwell, R.K. (2001) Quantifying biodiversity: procedures and pitfalls in the measurement and comparison of species richness. *Ecology Letters*, **4**, 379–91.

Gotelli, N.J. & Ellison, A.M. (2002) Biogeography at a regional scale: determinants of ant species density in new england bogs and forests. *Ecology*, **83**, 1604–9.

Gould, F. (1998) Sustainability of transgenic insecticidal cultivars – integrating pest genetics and ecology. *Annual Review of Entomology*, **43**, 701–26.

Gould, F. (2003) *Bt*-resistance management – theory meets data. *Nature Biotechnology*, **21** (12), 1450–1.

Gould, J.R., Elkinton, J.S. & Wallner, W.E. (1990) Density-dependent suppression of experimentally created gypsy-moth, *Lymantria dispar* (Lepidoptera, Lymantriidae), populations by natural enemies. *Journal of Animal Ecology*, **59**, 213–33.

Gowling, G.R. & van Emden, H.F. (1994) Falling aphids enhance impact of biological control by parasitoids on partially aphid-resistant plant varieties. *Annals of Applied Biology*, **125** (2), 233–42.

Grace, J.R. (1986) The influence of gypsy moth on the composition and nutrient content of litter fall in a Pennsylvania oak forest. *Forest Science*, **32**, 855–70.

Grafton-Cardwell, E.E., Lee, J.E., Stewart, J.R. & Olsen, K.D. (2006) Role of two insect growth regulators in integrated pest management of citrus scales. *Journal of Economic Entomology*, **99** (3), 733–44.

Grammatikopoulous, G. & Manetas, Y. (1994) Direct absorption of water by hairy leaves of *Phlomis fruticosa* and its contribution to drought avoidance. *Canadian Journal of Botany*, **72**, 1805–11.

Grandchamp, A.C., Bergamini, A., Stofer, S., Niemelä, J., Duelli, P. & Scheidegger, C. (2005) The influence of grassland management on ground beetles (Carabidae, Coleoptera) in Swiss montane meadows. *Agriculture Ecosystems and Environment*, **110**, 307–17.

Grant, P.R. (2005) The priming of periodical cicada life cycles. *Trends in Ecology and Evolution*, **20**, 169–74.

Gratz, N.G. & Jany, W.C. (1994) What role for insecticides in vector control programs? *American Journal of Tropical Medicine and Hygiene*, **50** (Suppl 6), 11–20.

Graur, D. & Martin, W. (2004) Reading the entrails of chickens: molecular timescales of evolution and the illusion of precision. *Trends in Genetics*, **20** (5), 242–7.

Gray, S. & Gildow, F.E. (2003) Luteovirus–aphid interactions. *Annual Review of Phytopathology*, **41**, 539–66.

Gray, S.M., Smith, D. & Altman, N. (1993) Barley yellow dwarf virus isolate-specific resistance in spring oats reduced virus accumulation and aphid transmission. *Phytopathology*, **83** (7), 716–20.

Grayson, J., Edmunds, M., Evans, E.H. & Britton, G. (1991) Carotenoids and coloration of popular hawkmoth caterpillars *Laothoe populi*. *Biological Journal of the Linnean Society*, **42**, 457–66.

Green, T.R. & Ryan, C.A. (1972) Wound-induced proteinase inhibitor in plant leaves: a possible defense mechanism against insects. *Science*, **175**, 776–7.

Greenblatt, J.A. & Barbosa, P. (1981) Effects of host diet on two pupal parasitoids of the gypsy moth: *Brachymeria intermedia* (Nees.) and *Coccygominus turionellae* (L.). *Journal of Applied Entomology*, **18**, 1–10.

Greenwood, B.M. (1997) What can be expected from malaria vaccines? *Annals of Tropical Medicine and Parasitology*, **91** (Suppl 1), S9–S13.

Gregory, R.D., van Strien, A., Vorisek, P. et al. (2005) Developing indicators for European birds. *Philosophical Transactions of the Royal Society of London Series B, Biological Sciences*, **360**, 269–88.

Grenier, A.M. & Nardon, P. (1994) The genetic control of ovariole number in *Sitophilus oryzae* L. (Coleoptera, Curculionidae) is temperature sensitive. *Genetics Selection Evolution (Paris)*, **26** (5), 413–30.

Gresens, S.E., Belt, K.T., Tang, J.A., Gwinn, D.C. & Banks, P.A. (2007) Temporal and spatial responses of Chironomidae (Diptera) and other benthic invertebrates to urban stormwater runoff. *Hydrobiologia*, **575**, 173–90.

Grez, A.A. (1992) Species richness of herbivorous insects versus patch size of host plant: an experimental test. *Revista Chilena de Historia Natural*, **65** (1), 115–20.

Grimaldi, D. (1997) The birdflies, genus *Carnus*: species revision, generic relationships, and a fossil Meoneura in amber (Diptera: Carnidae). *American Museum Novitates*, **0** (3190), 1–30.

Grimaldi, D. (2003) A revision of Cretaceous mantises and their relationships, including new taxa (Insecta: Dictyoptera: Mantodea). *American Museum Novitates*, **3412**, 1–47.

Grimaldi, D., Bonwich, E., Delannoy, M. & Doberstein, S. (1994) Electron microscopic studies of mummified tissues in amber fossils. *American Museum Novitates*, **0** (3097), 1–31.

Grimbacher, P.S. & Stork, N.E. (2007) Vertical stratification of feeding guilds and body size in beetle assemblages from an Australian tropical rainforest. *Austral Ecology*, **32**, 77–85.

Grinnell, J. (1904) The origin and distribution of the chestnut-backed chickadee. *Auk*, **21**, 364–82.

Gripenberg, S. & Roslin, T. (2007) Up or down in space? Uniting the bottom-up versus top-down paradigm and spatial ecology. *Oikos*, **116** (2), 181–8.

Groombridge, B. (ed.) (1992) *Global Biodiversity. Status of the Earth's Living Resources*. Chapman & Hall, London.

Grove, S.J. (2002) Tree basal area and dead wood as surrogate indicators of saproxylic insect faunal integrity: a case from the Australian lowland tropics. *Ecological Indicators*, **1**, 171–88.

Grover, P.B. Jr. (1995) Hypersensitive response of wheat to the Hessian fly. *Entomologia Experimentalis et Applicata*, **74**, 283–94.

Growchowska, M.J. & Ciurzynska, W. (1979) Differences between fruit-bearing and non-bearing apple spurs in activity of an enzyme system decomposing phloridzin. *Biological Plant*, **21**, 201–5.

Grundy, P.R., Sequeira, R.V. & Short, K.S. (2004) Evaluating legume species as alternative trap crops to chickpea for management of *Helicoverpa* spp. (Lepidoptera: Noctuidae) in central Queensland cotton cropping systems. *Bulletin of Entomological Research*, **94** (6), 481–6.

Gu, H.N., Hughes, J. & Dorn, S. (2006) Trade-off between mobility and fitness in *Cydia pomonella* L. (Lepidoptera: Tortricidae). *Ecological Entomology*, **31** (1), 68–74.

Guderian, R.H. & Shelley, A.J. (1992) Onchocerciasis in Ecuador: the situation in 1989. *Memorias do Instituto Oswaldo Cruz, Rio de Janeiro*, **87** (3), 405–15.

Guerenstein, P.G. & Hildebrand, J.G. (2008) Roles and effects of environmental carbon dioxide in insect life. *Annual Review of Entomology*, **53**, 161–78.

Guerold, F., Vein, D., Jacquemin, G. & Pihan, J.C. (1995) The macroinvertebrate communities of streams draining a small granitic catchment exposed to acidic precipitations (Vosges Mountains, northeastern France). *Hydrobiologia*, **300–1**, 141–8.

Guerra, C.A., Snow, R.W. & Hay, S.I. (2006) Mapping the global extent of malaria in 2005. *Trends in Parasitology* **22** (8), 353–8.

Gullan, P.J. & Cranston, P.S. (1996) *The Insects, An Outline of Entomology* (reprint). Chapman & Hall, London.

Gullan, P.J. & Cranston, P.S. (2005) *The Insects – an Outline of Entomology*, 3rd edn. Blackwell Publishing, Oxford.

Gupta, P.K. (2004) Pesticide exposure - Indian scene. *Toxicology*, **198** (1–3), 83–90.

Gur'yanova, T.M. (2006) Fecundity of *Neodiprion sertifer* (Hymenoptera, Diprionidae) related to cyclic outbreaks: invariance effects. *Zoologichesky Zhurnal*, **85** (9), 1085–95.

Haack, R.A. & Benjamin, D.M. (1982) The biology and ecology of the two-lined chestnut borer, *Agrilus biliniatus* (Coleoptera: Buprestidae), on oaks, (*Quercus* spp.) in Wisconsin. *Canadian Entomologist*, **114**, 385–96.

Hackstein, J.H.P. & Stumm, C.K. (1994) Methane production in terrestrial arthropods. *Proceedings of the National Academy of Sciences of the USA*, **91** (12), 5441–5.

Hafernik, J.E.J. (1992) Threats to invertebrate biodiversity: implications for conservation strategies. In *The Theory and Practice of Nature Conservation, Preservation and Management* (eds P.L. Fielder & S.K. Jain), pp. 172–95. Chapman & Hall, London.

Haggis, M.J. (1996) Forecasting the severity of seasonal outbreaks of African armyworm, *Spodoptera exempta* (Lepidoptera: Noctuidae) in Kenya from the previous year's rainfall. *Bulletin of Entomological Research*, **86** (2), 129–36.

Haggstrom, H. & Larsson, S. (1995) Slow larval growth on a suboptimal willow results in high predation mortality in the leaf beetle *Galerucella lineola*. *Oecologia*, **104**, 308–15.

Hagler, J.R. (2006) Foraging behavior and prey interactions by a guild of predators on various lifestages of *Bemisia tabaci*. *Annals of Applied Biology*, **149** (2), 153–65.

Hagler, J.R. & Naranjo, S.E. (2004) A multiple ELISA system for simultaneously monitoring intercrop movement and feeding activity of mass-released insect predators. *International Journal of Pest Management*, **50** (3), 199–207.

Hairston, N.G., Smith, F.E. & Slobodkin, L.B. (1960) Community structure, population control, and competition. *American Naturalist*, **44**, 421–5.

Hale, B.K., Bale, J.S., Pritchard, J., Masters, G.J. & Brown, V.K. (2003) Effects of host plant drought stress on the performance of the bird cherry-oat aphid, *Rhopalosiphum padi* (L.): a mechanistic analysis. *Ecological Entomology*, **28**, 666–77.

Halkett, F., Plantegenest, M., Prunier-Leterme, N., Mieuzet, L., Delmotte, F. & Simon, J.C. (2005) Admixed sexual and facultatively asexual aphid lineages at mating sites. *Molecular Ecology*, **14**, 325–36.

Hall, M.C., Stiling, P., Moon, D.C., Drake, B.G. & Hunter, M.D. (2005) Effects of elevated CO_2 on foliar quality and herbivore damage in a scrub oak ecosystem. *Journal of Chemical Ecology*, **31**, 267–86.

Halstead, S.B. (2008) Dengue virus–mosquito interactions. *Annual Review of Entomology*, **53**, 273–91.

Hamadain, E.I & Pitre, H.N (2002) Oviposition and larval behavior of soybean looper, *Pseudoplusia includens* (Lepidoptera: Noctuidae), on soybean with different row spacings and plant growth stages. *Journal of Agricultural and Urban Entomology*, **19** (1), 29–44.

Hamer, K.C., Hill, J.K., Benedick, S. et al. (2003) Ecology of butterflies in natural and selectively logged forests of northern Borneo: the importance of habitat heterogeneity. *Journal of Applied Ecology*, **40** (1), 150–62.

Hamilton, A.C. (2004) Medicinal plants, conservation and livelihoods. *Biodiversity and Conservation*, **13**, 1477–517.

Hamilton, J.G., Zangerl, A.R., DeLucia, E.H. & Berenbaum, M.R. (2001) The carbon–nutrient balance hypothesis: its rise and fall. *Ecology Letters*, **4**, 86–95.

Hamilton, W.D. (1964) The genetical evolution of social behaviour. *Journal of Theoretical Biology*, **7**, 1–52.

Hamilton, W.D. (1967) Extraordinary sex ratios. *Science*, **156**, 477–88.

Hamilton, W.D., Axelrod, R. & Tanese, R. (1990) Sexual reproduction as an adaptation to resist parasites (a review). *Proceedings of the National Academy of Sciences of the USA*, **87** (9), 3566–73.

Hammond, P.M. (1990) Insect abundance and diversity in the Dumoga-Bone National Park, North Sulawesi, with special reference to the beetle fauna of lowland rainforest in the Toraut region. In *Insects and the Rain Forests of South East Asia (Wallacea)* (eds W.J. Knight & J.D. Holloway), pp. 197–254. Royal Entomological Society, London.

Hammond, P.M. (1992) Species inventory. In *Global Biodiversity, Status of the Earth's Living Resources* (ed. B. Groombridge), pp. 17–39. Chapman & Hall, London.

Hammond, P.M., Stork, N.E. & Brendell, M.J.D. (1996) Tree-crown beetles in context: a comparison of canopy and other ecotone assemblages in a lowland tropical forest in Sulawesi. In *Canopy Arthropods* (eds N.E. Stork, J. Adis & R.K. Didham), pp. 184–223. Chapman & Hall, London.

Han, E.N. & Bauce, E. (1995) Glycerol synthesis by diapausing larvae in response to the timing of low temperature exposure, and implications for overwintering survival of the spruce budworm, *Choristoneura fumiferana*. *Journal of Insect Physiology*, **41** (11), 981–5.

Han, R., Ge, F., Yardim, E.N. & He, Z. (2005) The effect of low temperatures on diapause and non-diapause larvae of the pine caterpillar, *Dendrolimus tabulaeformis* Tsai et Liu (Lepidoptera: Lasiocampidae). *Applied Entomology and Zoology*, **40**, 429–35.

Han, S.S., Lee, M.H., Lim, C.Y. et al. (1996) Cabbage worm, *Artogeia rapae*, immunodeficiency syndrome caused by virus-like particles in the parasitoid wasp. *Korean Journal of Entomology*, **26**, 279–87.

Hanhimaeki, S., Senn, J. & Haukioja, E. (1995) The convergence in growth of foliage-chewing insect species on individual mountain birch trees. *Journal of Animal Ecology*, **64**, 543–52.

Hanks, L.M. & Denno, R.F. (1993a) Natural enemies and plant water relations influence the distribution of an armored scale insect. *Ecology*, **74**, 1081–91.

Hanks, L.M. & Denno, R.F. (1993b) The role of demic adaptation in colonization and spread of scale insect populations. In *Evolution of Insect Pests: Patterns and Variation* (eds K.C. Kim & B.A. McPheron), pp. 393–411. John Wiley, New York.

Hanks, L.M., Paine, T.D., Millar, J.G. & Hom, J.L. (1995) Variation among *Eucalyptus* species in resistance to eucalyptus long-horned borer in Southern California. *Entomologia Experimentalis et Applicata*, **74**, 185–94.

Hanley, N., Shogren, J.F. & White, B. (2001) *An Introduction to Environmental Economics*. Oxford University Press, Oxford.

Hansen, L-M. (2004) Economic damage threshold model for pollen beetles (*Meligethes aeneus* F.) in spring oilseed rape (*Brassica napus* L.) crops. *Crop Protection*, **23** (1), 43–6.

Hansen, L.O. & Somme, L. (1994) Cold hardiness of the elm bark beetle *Scolytus laevis* Chapuis, 1873 (Col., Scolytidae) and its potential as Dutch elm disease vector in the northernmost elm forests of Europe. *Journal of Applied Entomology*, **117** (5), 444–50.

Hanski, I. (1999) *Metapopulation Ecology*. Oxford University Press, Oxford.

Hanski, I. & Gilpin, M. (eds). (1997) *Metapopulation Biology. Ecology, Genetics and Evolution*. Academic Press, London.

Hanski, I. & Gyllenberg, M. (1997) Uniting two general patterns in the distribution of species. *Science*, **275** (5298), 397–400.

Hanski, I. & Niemelä, J. (1990) Elevational distributions of dung and carrion beetles in northern Sulawesi. In *Insects and the Rain Forests of South East Asia (Wallacea)* (eds W.J. Knight & J.D. Holloway), pp. 145–52. Royal Entomological Society, London.

Hanski, I., Saastamoinen, M. & Ovaskainen, O. (2006) Dispersal-related life-history trade-offs in a butterfly metapopulation. *Journal of Animal Ecology*, **75**, 91–100.

Hanski, L. (2004) Metapopulation theory, its use and misuse. *Basic and Applied Ecology*, **5**, 225–229.

Harada, H. & Ishikawa, H. (1993) Gut microbe of aphid closely related to its intracellular symbiont. *Biosystems*, **31** (2/3), 185–91.

Harborne, J.B. (1982) *Introduction to Ecological Biochemistry*. Academic Press, London.

Hardy, I.C.W., Griffiths, N.T. & Godfray, H.C.J. (1992) Clutch size in a parasitoid wasp – a manipulation experiment. *Journal of Animal Ecology*, **61**, 121–9.

Hargrove, J.W. (2003) Optimized simulation of the control of tsetse flies *Gloissina pallidipes* and *G-m. morsitans* (Diptera: Glossinidae) using odour-baited targets in Zimbabwe. *Bulletin of Entomological Research*, **93**, 19–29.

Hariyama, T., Takaku, Y., Hironaka, M., Horiguchi, H., Komiya, Y. & Kurachi, M. (2002) The origin of the iridescent colors in Coleopteran elytron. *Forma*, **17**, 123–32.

Harley, C.D.G. & Lopez, J.P. (2003) The natural history, thermal physiology, and ecological impacts of intertidal mesopredators, *Oedoparena* spp. (Diptera: Dryomyzidae). *Invertebrate Biology*, **122**, 61–73.

Harris, M.K., Chung, C.S. & Jackman, J.A. (1996) Masting and pecan interaction with insectan predehiscient nut feeders. *Environmental Entomology*, **25**, 1068–76.

Harris, R., York, A. & Beattie, A.J. (2003) Impacts of grazing and burning on spider assemblages in dry eucalypt forests of north-eastern New South Wales, Australia. *Austral Ecology*, **28**, 526–38.

Harrison, J.F., Lafreniere, J.J. & Greenlee, K.J. (2005) Ontogeny of tracheal dimensions and gas exchange capacities in the grasshopper, *Schistocerca americana*. *Comparative Biochemistry and Physiology a, Molecular and Integrative Physiology*, **141**, 372–80.

Harrison, P.A., Berry, P.M., Butt, N. & New, M. (2006) Modelling climate change impacts on species' distributions at the European scale: implications for conservation policy. *Environmental Science and Policy*, **9**, 116–28.

Harrison, R.D. & Rasplus, J.Y. (2006) Dispersal of fig pollinators in Asian tropical rain forests. *Journal of Tropical Ecology*, **22**, 631–9.

Hartley, S.E. & Lawton, J.H. (1987) The effects of different types of damage on the chemistry of birch foliage and the responses of birch feeding insects. *Oecologia*, **74**, 432–7.

Hartley, S.E. & Lawton, J.H. (1990) Damage-induced changes in birch foliage: mechanisms and effects on insect herbivores. In *Population Dynamics of Forest Insects* (eds A.D. Watt, S.R. Leather, M.D. Hunter & N.A.C. Kidd), pp. 147–55. Intercept, Andover.

Hartmann, S., Nason, J.D. & Bhattacharya, D. (2002) Phylogenetic origins of *Lophocereus* (Cactaceae) and the *Senita* cactus–*Senita* moth pollination mutualism. *American Journal of Botany*, **89**, 1085–92.

Hartmann, T., Theuring, C., Beuerle, T. & Bernays, E.A. (2004) Phenological fate of plant-acquired pyrrolizidine alkaloids in the polyphagous Arctiid *Estigmene acrea*. *Chemoecology*, **14**, 207–16.

Hartmann, T., Theuring, C., Witte, L., Schulz, S. & Pasteels, J.M. (2003) Biochemical processing of plant acquired pyrrolizidine alkaloids by the neotropical leaf-beetle *Platyphora boucardi*. *Insect Biochemistry and Molecular Biology*, **33**, 515–23.

Harvey, J.A. (2000) Dynamic effects of parasitism by an endoparasitoid wasp on the development of two host species: implications for host quality and parasitoid fitness. *Ecological Entomology*, **25**, 267–78.

Harvey, J.A., Harvey, I.F. & Thompson, D.J. (1995) The effect of host nutrition on growth and development of the parasitoid wasp *Venturia canescens*. *Entomologia Experimentalis et Applicata*, **75**, 213–20.

Hassall, M., Jones, D.T., Taiti, S., Latipi, Z., Sutton, S.L. & Mohammed, M. (2006) Biodiversity and abundance of terrestrial isopods along a gradient of disturbance in Sabah, East Malaysia. *European Journal of Soil Biology*, **42**, S197–S207.

Hassell, M.P. (1971) Mutual interference between searching insect parasites. *Journal of Animal Ecology*, **40**, 473–86.

Hassell, M.P. (1978) *The Dynamics of Arthropod Predator–Prey Systems*. Princeton University Press, Princeton, NJ.

Hassell, M.P. (1981) Arthropod predator–prey systems. In *Theoretical Ecology: Principles and Applications* (ed. R.M. May), pp. 105–31. Blackwell Scientific Publications, Oxford.

Hassell, M.P. (1985a) Insect natural enemies as regulating factors. *Journal of Animal Ecology*, **54**, 323–34.

Hassell, M.P. (1985b) Parasitism in patchy environments: inverse density dependence can be stabilizing. *IMA Journal of Math Applied in Medicine and Biology*, **1**, 123–33.

Hassell, M.P. (2000) *The Spatial and Temporal Dynamics of Host–Parasitoid Interactions*. Oxford University Press, Oxford.

Hassell, M.P. & May, R.M. (1973) Stability in insect host–parasite models. *Journal of Animal Ecology*, **42**, 693–736.

Hassell, M.P. & May, R.M. (1974) Aggregation in predators and insect parasites and its effect on stability. *Journal of Animal Ecology*, **43**, 567–94.

Hassell, M.P., Lawton, J.H. & Beddington, J.R. (1977) Sigmoid functional responses by invertebrate predators and parasitoids. *Journal of Animal Ecology*, **46**, 249–62.

Hassell, M.P., Lawton, J.H. & May, R.M. (1976) Patterns of dynamical behaviour in single species populations. *Journal of Animal Ecology*, **45**, 471–86.

Hassell, M.P., Southwood, T.R.E. & Reader, P.M. (1987) The dynamics of the viburnum whitefly, *Aleurotrachelus jelinekii* Fraunf. A case study on population regulation. *Journal of Animal Ecology*, **56**, 283–300.

Hassell, M.P. & Varley, G.C. (1969) New inductive population model for insect parasites and its bearing on biological control. *Nature*, **223**, 1113–37.

Hastings, A., Byers, J.E., Crooks, J.A. et al. (2007) Ecosystem engineering in space and time. *Ecology Letters*, **10** (2), 153–64.

Hatcher, P.E. & Paul, N.D. (2000) Beetle grazing reduces natural infection of *Rumex obtusifolius* by fungal pathogens. *New Phytologist*, **146**, 325–33.

Hatcher, P.E., Paul, N.D., Ayres, P.G. & Whitaker, J.B. (1994a) The effect of a foliar disease on the development of *Gastrophysa viridula* (Coleoptera, Chrysomelidae). *Ecological Entomology*, **19**, 349–60.

Hatcher, P.E., Paul, N.D., Ayres, P.G. & Whittaker, J.B. (1994b) Interactions between *Rumex* spp., herbivores, and a rust fungus: *Gastrophysa viridula* grazing reduces subsequent infection by *Uromyces rumicis*. *Functional Ecology*, **8**, 265–72.

Haukioja, E. (1980) On the role of plant defences in the fluctuations of herbivore populations. *Oikos*, **35**, 202–13.

Haukioja, E. (1991) Cyclic fluctuations in density: interactions between a defoliator and its host tree. *Acta Oecologica*, **12** (1), 77–88.

Haukioja, E. & Niemelä, P. (1977) Retarded growth of a geometrid larva after mechanical damage to leaves of its host tree. *Annales Zoologici Fennici*, **14**, 48–52.

Hawkeswood,T.J. (1988) A survey of the leaf beetles (Coleoptera: Chrysomelidae) from the Townsville district, northern Queensland, Australia. *Giornale Italiano di Entomologia*, **4** (20), 93–112.

Hawkins, B.A. (1992) Parasitoid–host food webs and donor control. *Oikos*, **65**, 159–62.

Hawkins, B.A., Cornell, H.V. & Hochberg, M.E. (1997) Predators, parasitoids, and pathogens as mortality agents in phytophagous insect populations. *Ecology*, **78**, 2145–52.

Hawkins, B.A., Mills, N.J., Jervis, M.A. & Price, P.W. (1999) Is the biological control of insects a natural phenomenon? *Oikos*, **86**, 493–506.

Hay, S.I., Rogers, D.J., Randolph, S.E. et al. (2002) Hot topic or hot air? Climate change and malaria resurgence in East African highlands. *Trends in Parasitology*, **18** (12), 530–4.

Hay, S.I., Shanks, G.D., Stern, D.I., Snow, R.W., Randolph, S.E. & Rogers, D.J. (2005) Climate variability and malaria epidemics in the highlands of East Africa. *Trends in Parasitology*, **21** (2), 52–3.

Hayes, J.L. & Daterman, G.E. (2001) Bark beetles (Scolytidae) in eastern Oregon and Washington. *Northwest Science*, **75**, 21–30.

Haynes, S., Darby, A-C., Daniell, T-J. et al. (2003). Diversity of bacteria associated with natural aphid populations. *Applied and Environmental Microbiology*, **69** (12), 7216–23.

Heard, S.B., Stireman, J.O., Nason, J.D. et al. (2006) On the elusiveness of enemy-free space: spatial, temporal, and host-plant-related variation in parasitoid attack rates on three gallmakers of goldenrods. *Oecologia*, **150** (3), 421–34.

Hector, A., Schmid, B., Beierkuhnlein, C. et al. (1999) Plant diversity and productivity experiments in European grasslands. *Science*, **286**, 1123–7.

Hedgren, P-O. (2003) Granbarkborren (*Ips typographus*) och naturvarden. *Entomologisk Tidskrift*, **124**, 159–73.

Heie, O.E., Pettersson, J., Fuentes-Contreras, E. & Niemeyer, H.M. (1996) New records of aphids (Hemiptera: Aphidoidea) and their host plants from northern Chile. *Revista Chilena de Entomologia*, **23**, 83–7.

Hein, S., Gombert, J., Hovestadt, T. & Poethke, H.J. (2003) Movement patterns of the bush cricket *Platycleis albopunctata* in different types of habitat: matrix is not always matrix. *Ecological Entomology*, **28**, 432–8.

Heinrichs, E.A. (1996) Management of rice insect pests. University of Minnesota, http: //www.ent.agri.umn.edu/academics/classes/ipm/chapters/heinrich.htm.

Heinz, K.M. & Nelson, J.M. (1996) Interspecific interactions among natural enemies of *Bemisia* in an inundative biological-control program. *Biological Contrology*, **6**, 384–93.

Helden, A.J. & Leather, S.R. (2004) Biodiversity on urban roundabouts – Hemiptera, management and the species–area relationship. *Basic and Applied Ecology*, **5**, 367–77.

Helms, S.E., Connelly S.J. & Hunter M.D. (2004) Effects of variation among plant species on the interaction between a herbivore and its parasitoid. *Ecological Entomology*, **29**, 44–51.

Helms, S.E. & Hunter, M.D. (2005) Variation in plant quality and the population dynamics of herbivores: there is nothing average about aphids. *Oecologia*, **145**, 197–204.

Hemingway, J., Beaty, B.J., Rowland, M., Scott, T.W. & Sharp, B.L. (2006) The innovative vector control consortium: improved control of mosquito-borne diseases. *Trends in Parasitology*, **22** (7), 308–12.

Henle, K., Davies, K.F., Kleyer, M., Margules, C. & Settele, J. (2004) Predictors of species sensitivity to fragmentation. *Biodiversity and Conservation*, **13**, 207–51.

Herms, D.A. & Mattson, W.J. (1992) The dilemma of plants: to grow or to defend. *Quarterly Review of Biology*, **67**, 283–335.

Herren, H.R. (1990) Biological control as the primary option in sustainable pest management: the cassava pest project. *Mitteilungen der Schweizerischen Entomologischen Gesellschaft*, **63** (3/4), 405–14.

Herren, H.R., Neuenschwander, P., Hennessey, R.D. & Hammond, W.N.O. (1987) Introduction and dispersal of *Epidinocarsus lopezi* (Hymenoptera: Encyrtidae), an exotic parasitoid of the cassava mealybug, *Phenacoccus manihoti*, in Africa. *Agriculture, Ecosystems and Environment*, **19** (2), 131–44.

Herrera, C.M. (2005) Plant generalization on pollinators: species property or local phenomenon? *American Journal of Botany*, **92**, 13–20.

Hertel, W. & Pass, G. (2002) An evolutionary treatment of the morphology and physiology of circulatory organs in insects. *Comparative Biochemistry and Physiology a, Molecular and Integrative Physiology*, **133A(3)**, 555–75.

Heyborne, W.H., Miller, J.C. & Parsons, G.L. (2003) Ground dwelling beetles and forest vegetation change over a 17-year-period, in western Oregon, USA. *Forest Ecology and Management*, **179**, 123–34.

Higashi, M., Abe, T. & Burns, T.P. (1992) Carbon–nitrogen balance and termite ecology. *Proceedings of the Royal Society of London Series B, Biological Sciences*, **249** (1326), 303–8.

High, S.M., Cohen, M.B., Shu, Q.Y. & Altosaar, I. (2004) Achieving successful deployment of *Bt* rice. *Trends in Plant Science*, **9** (6), 286–92.

Hilje, L., Costa, H.S. & Stansly, P.A. (2001) Cultural practices for managing *Bemisia tabaci* and associated viral diseases. *Crop Protection*, **20** (9), 801–12.

Hill, J., Lines, J. & Rowland, M. (2006) Insecticide-treated nets. *Advances in Parasitology*, **61**, 77–28.

Hill, J.K., Collingham, Y.C., Thomas, C.D. et al. (2001) Impacts of landscape structure on butterfly range expansion. *Ecology Letters*, **4**, 313–21.

Hill, J.K. & Hamer, K.C. (1998) Using species abundance models as indicators of habitat disturbance in tropical forests. *Journal of Applied Ecology*, **35**, 458–60.

Hill, J.K., Thomas, C.D., Fox, R. et al. (2002) Responses of butterflies to twentieth century climate warming: implications for future ranges. *Proceedings of the Royal Society of London Series B, Biological Sciences*, **269**, 2163–71.

Hill, J.K., Thomas, C.D. & Huntley, B. (1999) Climate and habitat availability determine 20th century changes in a butterfly's range margin. *Proceedings of the Royal Society of London Series B, Biological Sciences*, **266**, 1197–206.

Hill, J.K., Thomas, C.D. & Lewis, O.T. (1996) Effects of habitat patch size and isolation on dispersal by *Hesperia comma* butterflies: implications for metapopulation structure. *Journal of Animal Ecology*, **65** (6), 725–35.

Hilt, N. & Fiedler, K. (2005) Diversity and composition of Arctiidae moth ensembles along a successional gradient in the Ecuadorian Andes. *Diversity and Distributions*, **11**, 387–98.

Hinkle, G., Wetter, J.K., Schultz, T.R. & Sogin, M.L. (1994) Phylogeny of the attine ant fungi based on analysis of small subunit ribosomal RNA gene sequences. *Science*, **266** (5191), 1695–7.

Hirose, Y. (2006) Biological control of aphids and coccids: a comparative analysis. *Population Ecology*, **48** (4), 307–15.

Hiscox, A.L., Miller, D.R., Nappo, C.J. & Ross, J. (2006) Dispersion of fine spray from aerial applications in stable atmospheric conditions. *Transactions of the Asabe*, **49** (5), 1513–20.

Hjalten, J., Danell, K. & Ericson, L. (1993) Effects of simulated herbivory and intraspecific competition on the compensatory ability of birches. *Ecology*, **74**, 1136–42.

Hoback, W.W., Golick, S.A., Svatos, T.M., Spomer, S.M. & Higley, L.G. (2000) Salinity and shade preferences result in ovipositional differences between sympatric tiger beetle species. *Ecological Entomology*, **25**, 180–7.

Hobbs, R. (1992) The role of corridors in conservation: solution or bandwagon? *Trends in Ecology and Evolution*, **7**, 389–92.

Hobson, K.R., Wood, D.L., Cool, L.G. et al. (1993) Chiral specificity in responses by the bark beetle *Dendroctonus valens* to host kairomones. *Journal of Chemical Ecology*, **19**, 1837–46.

Hochberg, M.E., Elmes, G.W., Thomas, J.A. & Clarke, R.T. (1996) Mechanisms of local persistence in coupled host–parasitoid associations: the case model of *Maculinea rebeli* and *Ichneumon eumerus*. *Philosophical Transactions of the Royal Society of London Series B, Biological Sciences*, **351**, 1713–24.

Hochberg, M.E. & Ives, A.R. (2000) *Parasitoid Population Biology*. Princeton University Press, Princeton, NJ.

Hochberg, M.E. & Waage, J.K. (1991) A model for the biological control of *Oryctes rhinoceros* (Coleoptera: Scarabaeidae) by means of pathogens. *Journal of Applied Ecology*, **28** (2), 514–31.

Hodder, K.H. & Bullock, J.M. (1997) Translocations of native species in the UK: Implications for biodiversity. *Journal of Applied Ecology*, **34**, 547–5.

Hodges, R.J., Addo, S. & Birkinshaw, L. (2003) Can observation of climatic variables be used to predict the flight dispersal rates of Prostephanus truncatus? *Agricultural and Forest Entomology*, **5**, 123–35.

Hodkinson, I.D. (2005) Terrestrial insects along elevation gradients: species and community responses to altitude. *Biological Reviews*, **80**, 489–513.

Hodkinson, I.D. & Casson, D. (1991) A lesser prediction for bugs: Hemiptera (Insecta) diversity in tropical forests. *Biological Journal of the Linnean Society*, **43**, 101–9.

Hoekstra, J.M., Boucher, T.M., Ricketts, T.H. & Roberts, C. (2005) Confronting a biome crisis: global disparities of habitat loss and protection. *Ecology Letters*, **8**, 23–9.

Hoffman, T.K. & Kolb, F.L. (1997) Effects of barley yellow dwarf virus on root and shoot growth of winter wheat seedlings grown in aeroponic culture. *Plant Disease*, **81** (5), 497–500.

Hoganson, J.W. & Ashworth, A.C. (1992) Fossil beetle evidence for climatic change 18000–10000 years BP in south–central Chile. *Quaternary Research (Duluth)*, **37** (1), 101–16.

Hogg, I.D. & Williams, D.D. (1996) Response of stream invertebrates to a global-warming thermal regime: an ecosystem-level manipulation. *Ecology (Washington, DC)*, **77** (2), 395–407.

Hogg, J.C. & Hurd, H. (1997) The effects of natural *Plasmodium falciparum* infection on the fecundity and mortality of *Anopheles gambiae* s.l. in north east Tanzania. *Parasitology*, **114** (4), 325–31.

Hogmire, H.W. & Miller, S.S. (2005) Relative susceptibility of new apple cultivars to arthropod pests. *Hortscience*, **40** (7), 2071–5.

Hoke, R.A., Giesy, J.P., Zabik, M. & Unger, M. (1993) Toxicity of sediments and sediment pore waters from the Grand Calumet River–Indiana Harbor, Indiana, area of concern. *Ecotoxicology and Environmental Safety*, **26** (1), 86–112.

Holland, E.A. & Detling, J.K. (1990) Plant response to herbivory and below-ground nitrogen cycling. *Ecology*, **71**, 1040–9.

Holland, R.A., Wikelski M. & Wilcove D.S. (2006) How and why do insects migrate? *Science*, **313**, 794–6.

Hölldobler, B. & Wilson, E.O. (1990) *The Ants*. Springer Verlag, Berlin.

Holling, C.S. (1959) Some characteristics of simple types of predation and parasitism. *Canadian Entomologist*, **91**, 385–98.

Holling, C.S. (1965) The functional response of predators to prey density and its role in mimicry and population regulation. *Memoirs of the Entomological Society of Canada*, **45**, 1–60.

Holling, C.S. (1966) The functional response of invertebrate predators to prey density. *Memoirs of the Entomological Society of Canada*, **48**, 1–86.

Hollinger, D.Y. (1986) Herbivory and the cycling of nitrogen and phosphorus in isolated California oak trees. *Oecologia*, **70**, 291–7.

Holopainen, J.K., Rikala, R., Kainulainen, P. & Oksanen, J. (1995) Resource partitioning to growth, storage and defense in nitrogen-fertilized Scots pine and susceptibility of the seedlings to the tarnished plant bug *Lygus rugulipennus*. *New Phytologist*, **131**, 521–32.

Holt, J.A. (1987) Carbon mineralization in semiarid northeastern Australia – the role of termites. *Journal of Tropical Ecology*, **3**, 255–63.

Holt, R.D. (1977) Predation, apparent competition, and structure of prey communities. *Theoretical Population Biology*, **12**, 197–229.

Holt, R.D. (2000) Trophic cascades in terrestrial ecosystems. Reflections on Polis et al. *Trends in Ecology and Evolution*, **15**, 444–5.

Holt, R.D. & Polis, G.A. (1997) A theoretical framework for intraguild predation. *American Naturalist*, **149**, 745–64.

Holton, M.K., Lindroth, R.L. & Nordheim, E.V. (2003) Foliar quality influences tree–herbivore–parasitoid interactions: effects of elevated CO_2, O_3, and plant genotype. *Oecologia*, **137**, 233–44.

Holzinger, F. & Wink, M. (1996) Mediation of cardiac glycoside insensitivity in the monarch butterfly (*Danaus plexippus*): role of an amino acid substitution in the ouabain binding site of Na, K-ATPase. *Journal of Chemical Ecology*, **22**, 1921–37.

Honda, K. (1995) Chemical basis of different oviposition by lepidopterous insects. *Archives of Insect Biochemistry and Physiology*, **30**, 1–23.

Honek, A., Saska, P. & Martinkova, Z. (2006) Seasonal variation in seed predation by adult carabid beetles. *Entomologia Experimentalis et Applicata*, **118**, 157–62.

Hongoh, Y., Deevong, P., Inoue, T. et al. (2005) Intra- and interspecific comparisons of bacterial diversity and community structure support coevolution of gut microbiota and termite host. *Applied and Environmental Microbiology*, **71** (11), 6590–9.

Hooks, C.R.R., Pandey, R.R. & Johnson, M.W. (2006) Effects of spider presence on *Artogeia rapae* and host plant biomass. *Agriculture Ecosystems and Environment*, **112** (1), 73–7.

Hooper, D.U., Chapin, F.S., Ewel, J.J. et al. (2005) Effects of biodiversity on ecosystem functioning: a consensus of current knowledge. *Ecological Monographs*, **75**, 3–35.

Hoover, K., Wood, D.L., Storer, A.J., Fox, J.W. & Bros, W.E. (1996) Transmission of the pitch canker fungus, *Fusarium subglutinans* F. sp. *pini*, to Monterey pine, *Pinus radiata*, by cone- and twig-infesting beetles. *Canadian Entomologist*, **128** (6), 981–94.

Hopkins, G.W. & Freckleton, R.P. (2002) Declines in the numbers of amateur and professional taxonomists: implications for conservation. *Animal Conservation*, **5**, 245–9.

Hortal, J., Borges, P.A.V. & Gaspar, C. (2006) Evaluating the performance of species richness estimators: sensitivity to sample grain size. *Journal of Animal Ecology*, **75**, 274–87.

Horton, D.R., Chauvin, R.L., Hinojosa, T., Larson, D., Murphy, C. & Biever, K.D. (1997) Mechanisms of resistance to Colorado potato beetle in several potato lines and correlation with defoliation. *Entomologia Experimentalis et Applicata*, **82** (3), 239–46.

Horvitz, C.C. & Schemske, D.W. (1994) Effects of dispersers, gaps, and predators on dormancy and seedling emergence in a tropical herb. *Ecology*, **75**, 1949–58.

Hosking, G., Clearwater, J., Handiside, J., Kay, M., Ray, J. & Simmons, N. (2003) Tussock moth eradication – a success story from New Zealand. *International Journal of Pest Management*, **49** (1), 17–24.

Hougen-Eitzman, D. & Rausher, M.D. (1994) Interactions between herbivorous insects and plant–insect coevolution. *American Naturalist*, **143**, 677–97.

Houghton, J.T., Jenkins, G.J. & Ephraums, J.J. (1990) *Climate Change: the IPCC Scientific Assessment*. Cambridge University Press, New York.

Houghton, J.T., Meira Filko, L.G., Callander, B.A., Harris, M., Kattenberg, A. & Maskell, K. (1995) *Climate Changes 1995. Science of Climate Change*. Cambridge University Press, New York.

Houston, D.R. (1994) Major new tree disease epidemics: beech bark disease. *Annual Review of Phytopathology*, **32**, 75–87.

Howarth, F.G. & Ramsay, G.W. (1991) The conservation of island insects and their habitats. In *The Conservation of Insects and their Habitats* (eds N.M. Collins & J.A. Thomas), pp. 71–107. Academic Press, London.

Hoyle, M. & James, M. (2005) Global warming, human population pressure, and viability of the world's smallest butterfly. *Conservation Biology*, **19**, 1113–24.

Hsu, J.C. & Feng, H.T. (2006) Development of resistance to spinosad in oriental fruit fly (Diptera: Tephritidae) in laboratory selection and cross-resistance. *Journal of Economic Entomology*, **99** (3), 931–6.

Hu, X-P. & Appel, A-G. (2004). Seasonal variation of critical thermal limits and temperature tolerance in Formosan and eastern subterranean termites (Isoptera: Rhinotermitidae). *Environmental Entomology*, **33** (2), 197–205.

Huan, N.H., Thiet, L.V., Chien, H.V. & Heong, K.L. (2005) Farmers' participatory evaluation of reducing pesticides, fertilizers and seed rates in rice farming in the Mekong Delta, Vietnam. *Crop Protection*, **24** (5), 457–64.

Huang, F., Higgins, R.A. & Buschman, L.L. (1999) Heritability and stability of resistance to *Bacillus thuringiensis* in *Ostrinia nubilalis* (Lepidoptera: Pyralidae). *Bulletin of Entomological Research*, **89** (5), 449–54.

Huang, G., Vergne, E. & Gubler, D.J. (1992) Failure of dengue viruses to replicate in *Culex quinquefasciatus* (Diptera: Culicidae). *Journal of Medical Entomology*, **29** (6), 911–14.

Huang, H-C. (2003) Verticillium wilt of alfalfa: epidemiology and control strategies. *Canadian Journal of Plant Pathology*, **25** (4), 328–38.

Huang, J.K., Hu, R.F., Rozelle, S. & Pray, C. (2005) Insect-resistant GM rice in farmers' fields: assessing productivity and health effects in China. *Science*, **308** (5722), 688–90.

Huang, X., Renwick, J.A.A. & Sachdev-Gupta, K. (1994) Oviposition stimulants in *Barbarea vulgaris* for *Pieris rapae* and *P. napi oleracea*: isolation, identification and differential activity. *Journal of Chemical Ecology*, **20**, 423–38.

Huberty, A.F. & Denno, R.F. (2004) Plant water stress and its consequences for herbivorous insects: a new synthesis. *Ecology*, **85**, 1383–98.

Huberty, A.F. & Denno, R.F. (2006) Consequences of nitrogen and phosphorus limitation for the performance of two planthoppers with divergent life-history strategies. *Oecologia*, **149**, 444–55.

Huffaker, C.B. (1971) *Biological Control*, p. 511. Plenum, New York.

Hugentobler, U. & Renwick, J.A.A. (1995) Effects of plant nutrition on the balance of insect relevant cardenolides and glusinolates in *Erysimum cheiranthoides*. *Oecologia*, **102**, 95–101.

Huger, A.M. (2005) The *Oryctes* virus: its detection, identification, and implementation in biological control of the coconut palm rhinoceros beetle, *Oryctes rhinoceros* (Coleoptera: Scarabaeidae). *Journal of Invertebrate Pathology*, **89** (1), 78–84.

Hughes, L. & Westoby, M. (1992) Capitula on stick insect eggs and elaiosomes on seeds: convergent adaptaions for burial by ants. *Functional Ecology*, **6**, 642–8.

Hughes, P.R., Wood, H.A., Breen, J.P., Simpson, S.F., Duggan, A.J. & Dybas, J.A. (1997) Enhanced bioactivity of recombinant baculoviruses expressing insect-specific spider toxins in lepidopteran crop pests. *Journal of Invertebrate Pathology*, **69** (2), 112–18.

Hunter, A.F. (1993) Gypsy moth population sizes and the window of opportunity in spring. *Oikos*, **68** (3), 531–8.

Hunter, A.F. (1995) The ecology and evolution of reduced wings in forest macrolepidoptera. *Evolutionary Ecology*, **9** (3), 275–87.

Hunter, A.F. (2000) Gregariousness and repellent defences in the survival of phytophagous insects. *Oikos*, **91**, 213–24.

Hunter, D.M. (1989) The response of Mitchell grasses (*Astrebia* spp.) and button grass (*Dactylctenium radulus*) to rainfall and their importance to the survival of the Australian plague locust, *Chortoicetes terminifera* (Walker) in the arid zone. *Australian Journal of Ecology*, **14** (4), 467–72.

Hunter, D.M. (2004) Advances in the control of locusts (Orthoptera: Acrididae) in eastern Australia: from crop protection to preventive control. *Australian Journal of Entomology*, **43** (3), 293–303.

Hunter, D.M. & Cosenzo, E.L. (1990) The origin of plagues and recent outbreaks of the South American locust, *Schistocerca cancellata* (Orthoptera: Acrididae) in Argentina. *Bulletin of Entomological Research*, **80** (3), 295–300.

Hunter, M.D. (1987) Opposing effects of spring defoliation on late season oak caterpillars. *Ecological Entomology*, **12**, 373–82.

Hunter, M.D. (1990) Differential susceptibility to variable plant phenology and its role in competition between two insect herbivores on oak. *Ecological Entomology*, **15**, 401–8.

Hunter, M.D. (1992a) Interactions within herbivore communities mediated by the host plant: the keystone herbivore concept. In *The Effects of Resource Distribution on Animal–Plant Interactions* (eds M.D. Hunter, T. Ohgushi & P.W. Price), pp. 287–325. Academic Press, San Diego.

Hunter, M.D. (1992b) A variable insect plant interaction – the relationship between tree budburst phenology and population levels of insect herbivores among trees. *Ecological Entomology*, **17**, 91–5.

Hunter, M.D. (1997) Incorporating variation in plant chemistry into a spatially-explicit ecology of phytophagous insects. In *Forests and Insects* (eds A.D. Watt, N.E. Stork & M.D. Hunter), pp. 81–96. Chapman & Hall, London.

Hunter, M.D. (1998) Interactions between *Operophtera brumata* and *Tortrix viridana* on oak: new evidence from time-series analysis. *Ecological Entomology*, **23**, 168–73.

Hunter, M.D. (2001a) Multiple approaches to estimating the relative importance of top-down and bottom-up forces on insect populations: experiments, life tables, and time-series analysis. *Basic and Applied Ecology*, **2**, 295–309.

Hunter, M.D. (2001b) Out of sight, out of mind: the impacts of root-feeding insects in natural and managed systems. *Agricultural and Forest Entomology*, **3**, 3–10.

Hunter, M.D. (2001c) Effects of elevated atmospheric carbon dioxide on insect–plant interactions. *Agricultural and Forest Entomology*, **3**, 153–9.

Hunter, M.D. (2001d) Insect population dynamics meets ecosystem ecology: effects of herbivory on soil nutrient dynamics. *Agricultural and Forest Entomology* **3**, 77–84.

Hunter, M.D. (2002) A breath of fresh air: beyond laboratory studies of plant volatile–natural enemy interactions. *Agricultural and Forest Entomology*, **4**, 81–6.

Hunter, M.D. (2003a) Effects of plant quality on the population ecology of parasitoids. *Agricultural and Forest Entomology*, **5**, 1–8.

Hunter, M.D., Biddinger, D.J., Carlini, E.J., Mcpheron, B.A. & Hull, L.A. (1994) Effects of apple leaf allelochemistry on tufted apple bud moth (Lepidoptera: Tortricidae) resistance to azinphosmethyl. *Journal of Economic Entomology*, **87**, 1423–9.

Hunter, M.D. & Forkner R.E. (1999) Hurricane damage influences foliar polyphenolics and subsequent herbivory on surviving trees. *Ecology*, **80**, 2676–82.

Hunter, M.D., Forkner, R.E. & McNeil, J.N. (2000) Heterogeneity in plant quality and its impact on the population ecology of insect herbivores. In *The Ecological Consequences of Environmental Heterogeneity* (eds M.A. Hutchings, E.A. John & A.J.A. Stewart), pp. 155–79. Blackwell Publishing, Oxford.

Hunter, M.D., Malcolm, S.B. & Hartley, S.E. (1996) Population-level variation in plant secondary chemistry and the population biology of herbivores. *Chemoecology*, **7**, 45–56.

Hunter, M.D., Ohgushi, T. & Price, P.W. (1992) *The Effects of Resource Distribution on Animal–Plant Interactions*. Academic Press, San Diego.

Hunter, M.D. & Price, P.W. (1992) Playing chutes and ladders – heterogeneity and the relative roles of bottom-up and top-down forces in natural communities. *Ecology*, **73**, 724–32.

Hunter, M.D. & Price, P.W. (1998) Cycles in insect populations: delayed density dependence or exogenous driving variables? *Ecological Entomology*, **23** (2), 216–22.

Hunter, M.D. & P.W. Price. 2000. Detecting cycles and delayed density dependence: a reply to Turchin and Berryman. *Ecological Entomology*, **25**, 122–4.

Hunter, M.D. & Schultz, J.C. (1995) Fertilization mitigates chemical induction and herbivore responses within damaged oak trees. *Ecology*, **76**, 1226–32.

Hunter, M.D. & Willmer, P.G. (1989) The potential for interspecific competition between two abundant defoliators on oak: leaf damage and habitat quality. *Ecological Entomology*, **14**, 267–77.

Hunter, M.D., Varley, G.C. & Gradwell, G.R. (1997) Estimating the relative roles of top-down and bottom-up forces on insect herbivore populations: a classic study re-visited. *Proceedings of the National Academy of Sciences of the USA*, **94**, 9176–81.

Hunter, M.D., Watt, A.D. & Docherty, M. (1991) Outbreaks of the winter moth on Sitka spruce in Scotland are not influenced by nutrient deficiencies of trees, tree budburst, or pupal predation. *Oecologia*, **86**, 62–9.

Hunter, P.R. (2003b) Climate change and waterborne and vector-borne disease. *Society for Applied Microbiology Symposium Series*, **94**, 37S–46S.

Hurd, H. (2003) Manipulation of medically important insect vectors by their parasites. *Annual Review of Entomology*, **48**, 141–61.

Husband, B.C. & Barrett, S.C.H. (1996) A metapopulation perspective in plant population biology. *Journal of Ecology*, **84** (3), 461–9.

Hutchinson, G.E. (1957) Concluding remarks. *Cold Spring Harbor Symposium, Quantitative Biology*, **22**, 415–27.

Hutchinson, G.E. (1959) Homage to Santa Rosalia, or why are there so many kinds of animals? *American Naturalist*, **93**, 145–59.

Hutchinson, G.E. (1978) *An Introduction to Population Ecology*. Yale University Press, New Haven, CT.

Hwang, H.M., Green, P.G. & Young, T.M. (2006) Tidal salt marsh sediment in California, USA. Part 1: Occurrence and sources of organic contaminants. *Chemosphere*, **64** (8), 1383–92.

Ichimori, K., Graves, P.M., Crump, A. (2007) Lymphatic filariasis elimination in the Pacific: PacELF replicating Japanese success. *Trends in Parasitology*, **23** (1), 36–40.

Idris, A.B. & Grafius, E. (1995) Wildflowers as nectar sources for *Diadegma insulare* (Hymenoptera: Ichneumonidae), as a parasitoid of diamondback moth (Lepidoptera: Yponomeutidae). *Environmental Entomology*, **24** (6), 1726–35.

Inbar, M., Doostdar, H., Leibee, G.L. & Mayer, R.T. (1999) The role of plant rapidly induced responses in asymmetric interspecific interactions among insect herbivores. *Journal of Chemical Ecology*, **25**, 1961–79.

Inceoglu, A.B., Kamita, S.G. & Hammock, B.D. (2006) Genetically modified baculoviruses: a historical overview and future outlook. *Insect Viruses: Biotechnological Applications*, **68**, 323–60.

Inoue, T. & Kato, M. (1992) Inter and intra specific morphological variation in bumblebee species and competition in flower utilization In *Effects of Resource Distribution on Animal–Plant Interactions* (ed. M.D. Hunter, T. Ohgushi & P.W. Price), pp. 393–427. Academic Press, San Diego.

Inoue, T., Murashima, K., Azuma, J.I., Sugimoto, A. & Slaytor, M. (1997) Cellulose and xylan utilisation in the lower termite *Reticulitermes speratus*. *Journal of Insect Physiology*, **43** (3), 235–42.

IPCC (2001) *Climate Change 2001: The Scientific Basis. Contribution of Working Group I to the Third Assessment Report of the Intergovernmental Panel on Climate Change (IPCC)*. Cambridge University Press, Cambridge, UK.

IRRI (2004) *Rice Around the World*. International Rice Research Organisation, www.irri.org.

IRRI (2007) *Rice News*. International Rice Research Organisation, www.irri.org.

Irwin, M.E. & Thresh, J.M. (1990) Epidemiology of barley yellow dwarf: study in ecological complexity. *Annual Review of Phytopathology*, **28**, 393–424.

Ishihara, M. & Shimada, M. (1995) Photoperiodic induction of larval diapause and temperature-dependent termination in a wild multivoltine bruchid, *Kytorhinus sharpianus*. *Entomologia Experimentalis et Applicata*, **75** (2), 127–34.

Ishitani, M., Kotze, D.J. & Niemelä, J. (2003) Changes in carabid beetle assemblages across an urban–rural gradient in Japan. *Ecography*, **26**, 481–9.

Itioka, T. & Inoue, T. (1996) Density-dependent ant attendance and its effects on the parasitism of a honeydew-producing scale insect, *Ceroplastes rubens*. *Oecologia*, **106**, 448–54.

IUCN (1994) *IUCN Red List Categories*. IUCN – The World Conservation Union, Gland.

IUCN (2001) *IUCN Red List Categories and Criteria: Version 3.1 (IUCN Species Survival Commission)* IUCN – The World Conservation Union, Gland, Switzerland and Cambridge, UK.

Ivanovic, J., Dordevic, S., Ilijin, L., Jankovic, T.M. & Nenadovic, V. (2002). Metabolic response of cerambycid beetle (*Morimus funereus*) larvae to starvation and food quality. *Comparative Biochemistry and Physiology Part A, Molecular and Integrative Physiology*, **132A**, 555–66.

Ives, W.G.H. (1974) *Weather and Outbreaks of the Spruce Budworm*, Choristoneura rosaceana. Information Report NOR-X-118. Northern Forest Research Center Department of Environment, Canada.

Ivory, M. & Speight, M.R. (1994) Pests and diseases. In *Tropical Forestry Handbook*, Vol. II (ed. L. Pancel), Section 19. Springer Verlag, Berlin.

Iwao, K. & Rausher, M.D. (1997) Evolution of plant resistance to multiple herbivores: quantifying diffuse coevolution. *American Naturalist*, **149**, 316–35.

Izquierdo, J.I. (1996) *Helicoverpa armigera* (Hübner) (Lep., Noctuidae): relationship between captures in pheromone traps and egg counts in tomato and carnation crops. *Journal of Applied Entomology*, **120**, 281–90.

Izzo, T.J. & Vasconcelos, H.L. (2002) Cheating the cheater: domatia loss minimizes the effects of ant castration in an Amazonian ant-plant. *Oecologia*, **133**, 200–5.

Jackai, L.E.N. & Oghiakhe, S. (1989) Pod wall trichomes and resistance of two wild cowpea, *Vigna vexillata*, accessions to *Muruca testulalis* Geyer (Lepidoptera: Pyralidae) and *Clavigralla tomentosicollis* Stal (Hemiptera. *Coreidae*). *Entomological Bulletin of Research*, **79**, 595–605.

Jacobs, J.M., Spence, J.R. & Langor, D.W. (2007) Influence of boreal forest succession and dead wood qualities on saproxylic beetles. *Agricultural and Forest Entomology*, **9**, 3–16.

Jactel, H., Menassieu, P., Vetillard, F. et al. (2006) Population monitoring of the pine processionary moth (Lepidoptera: Thaumetopoeidae) with pheromone-baited traps. *Forest Ecology and Management*, **235** (1–3), 96–106.

Jakus, R., Schlyter, F., Barthelemy, B. et al. (2003) Overview of development of an anti-attractant based technology for spruce protection against *Ips typographus*: from past failures to future success. *Anzeiger fuer Schaedlingskunde*, **76** (4), 89–99.

James, A.A. (2002) Engineering mosquito resistance to malaria parasites: the avian malaria model. *Insect Biochemistry and Molecular Biology*, **32** (10), 1317–23.

Jamieson-Dixon, R.W. & Wrona, F.J. (1992) Life history and production of the predatory caddisfly *Rhyacophila vao* Milne in a spring-fed stream. *Freshwater Biology*, **27** (1), 1–11.

Jankowiak, R. (2005) Fungi associated with *Ips typographus* on *Picea abies* in southern Poland and their succession into the phloem and sapwood of beetle-infested trees and logs. *Forest Pathology*, **35** (1), 37–55.

Janmaat, A-F. & Myers, J. (2003) Rapid evolution and the cost of resistance to *Bacillus thuringiensis* in greenhouse populations of cabbage loopers, *Trichoplusia ni*. *Proceedings of the Royal Society of London Series B, Biological Sciences*, **270** (1530), 2263–70.

Jannin, J.G. (2005) Commentary: sleeping sickness – a growing problem? *British Medical Journal*, **331** (7527), 1242.

Janz, N., Nylin, S. & Wahlberg, N. (2006) Diversity begets diversity: host expansions and the diversification of plant-feeding insects. *Bmc Evolutionary Biology*, **6**, 10.

Janzen, D.H. (1966) Coevolution of mutualism between ants and acacias in Central America. *Evolution*, **20**, 249–75.

Janzen, D.H. (1967) Interaction of the bull's horn acacia (*Acacia cornigera* L.) with an ant inhabitant (*Pseudomyrmex ferruginea* F. Smith) in eastern Mexico. *Kansas University Science Bulletin*, **47**, 315–558.

Janzen, D.H. (1968) Host plants in evolutionary and contemporary time. *American Naturalist*, **102**, 592–5.

Janzen, D.H. (1973) Host plants as islands. II. Competition in evolutionary and contemporary time. *American Naturalist*, **107**, 786–90.

Janzen, D.H. (1980) When is it coevolution? *Evolution*, **34**, 611–12.

Janzen, D.H. (1997) Wildland biodiversity management in the tropics. In *Biodiversity II, Understanding and Protecting our Biological Resources* (eds M.L. Reaka-Kudla, D.E. Wilson & E.O. Wilson), pp. 411–31. Joseph Henry Press, Washington, DC.

Janzen, D.H. (2004) Setting up tropical biodiversity for conservation through non-damaging use: participation by parataxonomists. *Journal of Applied Ecology*, **41**, 181–7.

Janzen, D.H. & Pond, C.M. (1975) Comparison, by sweep sampling, of arthropod fauna of secondary vegetation in Michigan, England and Costa-Rica. *Transactions of the Royal Entomological Society of London*, **127**, 33–50.

Jayanthi, R., David, H. & Goud, Y.S. (1994) Physical characters of sugarcane plant in relation to infestation by *Melanaspis glomerata* (G.) and *Saccharicoccus sacchari* (Ckll.). *Journal of Entomological Research (New Delhi)*, **18**, 305–14.

Jeffree, C.E. (1986) Insects and the plant surface. In *The Cuticle, Epicuticular Waxes and Trichomes of Plants, with Reference to their Structure, Functions and Evolution* (eds B.E. Juniper & R. Southwood), pp. 23–64. Edward Arnold, London.

Jeffree, C.E. & Jeffree, E.P. (1996) Redistribution of the potential geographical ranges of mistletoe and Colorado beetle in Europe in response to the temperature component of climate change. *Functional Ecology*, **10** (5), 562–77.

Jeffries, M.J. & Lawton, J.H. (1984) Enemy-free space and the structure of ecological communities. *Biological Journal of the Linnean Society*, **23** (4), 269–86.

Jeger, M-J., Holt, J., van-den-Bosch, F. & Madden, L-V. (2004). Epidemiology of insect-transmitted plant viruses: modelling disease dynamics and control interventions. *Physiological Entomology*, **29** (3), 291–304.

Jennings, V.H. & Tallamy, D.W. (2006) Composition and abundance of ground-dwelling Coleoptera in a fragmented and continuous forest. *Environmental Entomology*, **35**, 1550–60.

Jepson, P. & Canney, S. (2003) Values-led conservation. *Global Ecology and Biogeography*, **12**, 271–4.

Jervis, M.A., Kidd, N.A.C. & Walton, M. (1992) A review of methods for determining dietary range in adult parasitoids. *Entomophaga*, **37**, 565–74.

Ji, R. & Brune, A. (2006) Nitrogen mineralization, ammonia accumulation, and emission of gaseous NH_3 by soil-feeding termites. *Biogeochemistry*, **78** (3), 267–83.

Ji, R., Li, D.M., Xie, B.Y., Li, Z. & Meng, D.L. (2006) Spatial distribution of oriental migratory locust (Orthoptera: Acrididae) egg pod populations: implications for site-specific pest management. *Environmental Entomology*, **35**, 1244–8.

Jiang, L. & Shao, N. (2003) Autocorrelated exogenous factors and the detection of delayed density dependence. *Ecology*, **84**, 2208–13.

Jiggins, F.M., Bentley, J.K., Majerus, M.E.N. & Hurst, G.D.D. (2002) Recent changes in phenotype and patterns of host specialization in *Wolbachia* bacteria. *Molecular Ecology*, **11** (8), 1275–83.

Jimenez-Martinez, E.S., Bosque-Perez, N.A., Berger, P.H. & Zemetra, R.S. (2004) Life history of the bird cherry-oat aphid, *Rhopalosiphum padi* (Homoptera: Aphididae), on transgenic and untransformed wheat challenged with barley yellow dwarf virus. *Journal of Economic Entomology*, **97** (2), 203–12.

Joern, A. & Mole, S. (2005) The plant stress hypothesis and variable responses by blue grama grass (*Bouteloua gracilis*) to water, mineral nitrogen, and insect herbivory. *Journal of Chemical Ecology*, **31**, 2069–90.

Johnson, D.M., Bjornstad, O.N. & Liebhold, A.M. (2004) Landscape geometry and travelling waves in the larch budmoth. *Ecology Letters*, **7** (10), 967–74.

Johnson, K.H., Vogt, K.A., Clark, H.J., Schmitz, O.J. & Vogt, D.J. (1996) Biodiversity and the productivity and stability of ecosystems. *Trends in Ecology and Evolution*, **11**, 372–7.

Johnson, M.T. & Gould, F. (1992) Interaction of genetically engineered host plant-resistance and natural enemies of *Heliothis-virescens* (Lepidoptera, Noctuidae) in tobacco. *Environmental Entomology*, **21**, 586–97.

Johnson, R.A. (2000) Water loss in desert ants: caste variation and the effect of cuticle abrasion. *Physiological Entomology*, **25** (1), 48–53.

Johnson, R.H. & Lincoln, D.E. (1990) Sagebrush and grasshopper responses to atmospheric carbon dioxide concentration. *Oecologia*, **84**, 103–10.

Johnson, S.N., Crawford, J.W., Gregory, P.J. et al. (2007) Non-invasive techniques for investigating and modelling root-feeding insects in managed and natural systems. *Agricultural and Forest Entomology*, **9** (1), 39–46.

Johnson, S.N., Mayhew, P.J., Douglas, A.E. & Hartley, S.E. (2002) Insects as leaf engineers: can leaf-miners alter leaf structure for birch aphids? *Functional Ecology*, **16**, 575–84.

Johst, K., Drechsler, M., Thomas, J. & Settele, J. (2006) Influence of mowing on the persistence of two endangered large blue butterfly species. *Journal of Applied Ecology*, **43**, 333–42.

Jones, C.G. & Lawton, J.H. (1991) Plant chemistry and insect species richness of British umbellifers. *Journal of Animal Ecology*, **60** (3), 767–78.

Jones, C.G. & Lawton, J.H. (1995) *Linking Species and Ecosystems*. Chapman & Hall, New York.

Jones, C.G., Lawton, J.H. & Shachak, M. (1994) Organisms as ecosystem engineers. *Oikos*, **69**, 1–14.

Jones, D.T., Susilo, F.X., Bignell, D.E., Hardiwinoto, S., Gillison, A.N. & Eggleton, P. (2003) Termite assemblage collapse along a land-use intensification gradient in lowland central Sumatra, Indonesia. *Journal of Applied Ecology*, **40**, 380–91.

Jouquet, P., Dauber, J., Lagerlof, J., Lavelle, P., Lepage, M. (2006) Soil invertebrates as ecosystem engineers: intended and accidental effects on soil and feedback loops. *Applied Soil Ecology*, **32** (2), 153–64.

Judd, G.J.R. & Gardiner, M.G.T. (2005) Towards eradication of codling moth in British Columbia by complimentary actions of mating disruption, tree banding and sterile insect technique: five-year study in organic orchards. *Crop Protection*, **24** (8), 718–33.

Judd, G.J.R., Gardiner, M.G.T. & Thomson, D.R. (1996) Commercial trials of pheromone-mediated mating disruption with isomate-C to control codling moth in British Columbia apple and pear orchards. *Journal of the Entomological Society of British Columbia*, **93**, 23–34.

Juniper, B. & Southwood, R. (1986) *Insects and the Plant Surface* Edward Arnold, London.

Jupp, P.G. & Lyons, S.F. (1987) Experimental assessment of bedbugs (*Cimex lectularius* and *Cimex hemipterus*) and mosquitoes (*Aedes aegypti formosus*) as vectors of human immunodeficiency virus. *AIDS*, **1** (3), 171–4.

Jupp, P.G. & Williamson, C. (1990) Detection of multiple blood meals in experimentally fed bedbugs (Hemiptera: Cimicidae). *Journal of the Entomological Society of Southern Africa*, **53** (2), 137–40.

Juutinen, A., Monkkonen, M. & Sippola, A.L. (2006) Cost-efficiency of decaying wood as a surrogate for overall species richness in boreal forests. *Conservation Biology*, **20**, 74–84.

Kabayo, J.P. (2002) Aiming to eliminate tsetse from Africa. *Trends in Parasitology*, **18** (11), 473–5.

Kagata, H. & Ohgushi, T. (2002) Clutch size adjustment of a leaf-mining moth (Lyonetiidae: Lepidoptera) in response to resource availability. *Annals of the Entomological Society of America*, **95** (2), 213–17.

Kagata, H. & Ohgushi, T. (2001) Preference and performance linkage of a leaf-mining moth on different Salicaceae species. *Population Ecology*, **43** (3), 141–7.

Kaiser, J. (1997) Biodiversity – unique, all-taxa survey in Costa Rica "self-destructs". *Science*, **276**, 893.

Kaitala, A. (1991) Phenotypic plasticity in reproductive behavior of waterstriders: trade-offs between reproduction and longevity during food stress. *Functional Ecology*, **5** (1), 12–18.

Kaller, M.D. & Kelso, W.E. (2006) Swine activity alters invertebrate and microbial communities in a coastal plain watershed. *American Midland Naturalist*, **156**, 163–77.

Kambhampati, S. (1995) A phylogeny of cockroaches and related insects based on DNA sequence of mitochondrial ribosomal RNA genes. *Proceedings of the National Academy of Sciences of the USA*, **92** (6), 2017–20.

Kaneko, S. (1995) Frequent successful multiparasitism by five parasitoids attacking the scale insect *Nipponaclerda biwakoensis*. *Researches on Population Ecology*, **37**, 225–8.

Kang, L.L.I.H.C. & Chen, Y.L. (1989) Analyses of numerical character variations of geographical populations of *Locusta migratoria* phase *solitaria* in China. *Acta Entomologica Sinica*, **32** (4), 418–26.

Kapari, L., Haukioja, E., Rantala, M.J. & Ruuhola, T. (2006) Defoliating insect immune defense interacts with induced plant defense during a population outbreak. *Ecology*, **87**, 291–6.

Kaplan, I. & Eubanks, M.D. (2005) Aphids alter the community-wide impact of fire ants. *Ecology*, **86** (6), 1640–9.

Karban, R. (1989) Fine-scale adaptation of herbivorous thrips to individual host plants. *Nature*, **340**, 60–1.

Karban, R. (1993) Costs and benefits of reduced resistance and plant density for a native shrub, *Gossypium thurberi*. *Ecology*, **74**, 9–19.

Karban, R. & Baldwin, I.T. (1997) *Induced Responses to Herbivory*. University of Chicago Press, Chicago.

Karban, R., Huntzinger, M. & McCall, A. C. (2004) The specificity of eavesdropping on sagebrush by other plants. *Ecology*, **85**, 1846–52.

Karban, R. & Strauss, S.Y. (1993) Effects of herbivores on growth and reproduction of their perennial host, *Erigeron glaucus*. *Ecology*, **74**, 39–46.

Karban, R. & Strauss, S.Y. (2004). Physiological tolerance, climate change, and a northward range shift in the spittlebug, *Philaenus spumarius*. *Ecological Entomology*, **29** (2), 251–4.

Karlsson, B. & Van Dyck, H. (2005) Does habitat fragmentation affect temperature-related life-history traits? A laboratory test with a woodland butterfly. *Proceedings of the Royal Society of London Series B, Biological Sciences*, **272**, 1257–63.

Karlsson, B. & Wiklund, C. (2005) Butterfly life history and temperature adaptations; dry open habitats select for increased fecundity and longevity. *Journal of Animal Ecology*, **74**, 99–104.

Kato, M. (1994) Structure, organization, and response of a species-rich parasitoid community to host leafminer population-dynamics. *Oecologia*, **97**, 17–25.

Katzourakis, A., Purvis, A., Azmeh, S., Rotheray, G. & Gilbert, F. (2001) Macroevolution of hoverflies (Diptera: Syrphidae): the effect of using higher-level taxa in studies of biodiversity, and correlates of species richness. *Journal of Evolutionary Biology*, **14**, 219–27.

Kawakita, A., Takimura, A., Terachi, T., Sota, T. & Kato, M. (2004) Cospeciation analysis of an obligate pollination mutualism: have *Glochidion* trees (Euphorbiaceae) and pollinating *Epicephala* moths (Gracillariidae) diversified in parallel? *Evolution*, **58**, 2201–14.

Kawazu, K., Kamimuro, T., Kamiwada, H. et al. (2004) Effective pheromone lures for monitoring the rice leaffolder moth, *Cnaphalocrocis medinalis* (Lepidoptera: Crambidae). *Crop Protection*, **23**, 589–93.

Kearsley, M.J.C. & Whitham, T.G. (1992) Guns and butter, a no cost defense against predation for *Chrysomela confluens*. *Oecologia (Heidelberg)*, **92**, 556–62.

Keasar, T., Segoli, M., Barak, R. et al. (2006) Costs and consequences of superparasitism in the polyembryonic parasitoid *Copidosoma koehleri* (Hymenoptera: Encyrtidae). *Ecological Entomology*, **31** (3), 277–83.

Keating, S.T., Hunter, M.D. & Schultz, J.C. (1990) Leaf phenolic inhibition of gypsy-moth nuclear polyhedrosis-virus – role of polyhedral inclusion body aggregation. *Journal of Chemical Ecology*, **16**, 1445–57.

Keeling, M.J. & Rand, D.A. (1995) A spatial mechanism for the evolution and maintenance of sexual reproduction. *Oikos*, **74** (3), 414–24.

Keenlyside, J. (1989) *Host choice in aphids*. DPhil Thesis, University of Oxford.

Keil, P. & Konvicka, M. (2005) Local species richness of Central European hoverflies (Diptera: Syrphidae): a lesson taught by local faunal lists. *Diversity and Distributions*, **11**, 417–26.

Keller, I., Fluri, P. & Imdorf, A. (2005) Pollen nutrition and colony development in honey bees: part I. *Bee World*, **86** (1), 3–10.

Kellert, S.R. & Wilson, E.O. (eds.) (1995) *The Biophilia Hypothesis*. Island Press, Washington, DC.

Kellou, R., Mahjoub, N., Benabdi, A. & Boulahya, M.S. (1990) Algerian case study and the need for permanent desert locust monitoring. *Philosophical Transactions of the Royal Society of London Series B, Biological Sciences*, **328** (1251), 573–84.

Kemp, D.J. & Rutowski, R.L. (2004) A survival cost to mating in a polyandrous butterfly, *Colias eurytheme. Oikos*, **105**, 65–70.

Kendall, D.A., George, S. & Smith, B.D. (1996) Occurrence of barley yellow dwarf viruses in some common grasses (Gramineae) in south west England. *Plant Pathology (Oxford)*, **45** (1), 29–37.

Kennedy, C.E.J. (1986) Attachment may be a basis for specialization in oak aphids. *Ecological Entomology*, **11**, 291–300.

Kennedy, C.E.J. & Southwood, T.R.E. (1984) The number of species of insects associated with British trees – a re-analysis. *Journal of Animal Ecology*, **53**, 455–78.

Kennedy, G.G., Farrar, R.R. & Kashyap, R.K. (1991) 2-Tridecanone glandular trichome-mediated insect resistance in tomato – effect on parasitoids and predators of *Heliothis zea*. *American Chemical Society Symposium Series*, **449**, 150–65.

Keogh, R.G. & Lawrence, T. (1987) Influence of *Acremonium lolii* presence on emergence and growth of ryegrass seedlings. *New Zealand Journal of Agricultural Research*, **30**, 507–10.

Kerslake, J.E., Kruuk, L.E.B., Hartley, S.E. & Woodin, S.J. (1996) Winter moth (*Operophtera brumata* (Lepidoptera,

Geometridae)) outbreaks on Scottish heather moorlands – effects of host-plant and parasitoids on larval survival and development. *Bulletin of Entomological Research*, **86**, 155–64.

Kery, M. & Plattner, M. (2007) Species richness estimation and determinants of species detectability in butterfly monitoring programmes. *Ecological Entomology*, **32**, 53–61.

Kessler, A. & Baldwin, I.T. (2001) Defensive function of herbivore-induced plant volatile emissions in nature. *Science*, **291**, 2141–4.

Kessler, A. & Baldwin, I.T. (2002) Plant responses to insect herbivory: the emerging molecular analysis. *Annual Review of Plant Biology*, **53**, 299–328.

Kessler, M. (2002) Environmental patterns and ecological correlates of range size among bromeliad communities of Andean forests in Bolivia. *Botanical Review*, **68**, 100–27.

Kessler, M., Herzog, S., Fjeldså, J. & Bach, K. (2001) Species richness and endemism of plant and bird communities along two gradients of elevation, humidity and land use in the Bolivian Andes. *Diversity and Distributions*, **7**, 61–77.

Kettle, D.S. (1995) *Medical and Veterinary Entomology*, 2nd edn. CAB International, Wallingford, UK.

Kidd, N.A.C. & Jervis, M.A. (eds) (1996) *Insect Natural Enemies: Practical approaches to their study and evaluation*. Chapman & Hall, London.

Kielland, K., Bryant, J.P. & Ruess, R.W. (1997) Moose herbivory and carbon turnover of early successional stands in interior Alaska. *Oikos*, **80**, 25–30.

Kier, G. & Barthlott, W. (2001) Measuring and mapping endemism and species richness: a new methodological approach and its application on the flora of Africa. *Biodiversity and Conservation*, **10**, 1513–29.

Kim, K.C. & Byrne, L.B. (2006) Biodiversity loss and the taxonomic bottleneck: emerging biodiversity science. *Ecological Research*, **21**, 794–810.

Kimberling, D.N., Scott, E.R. & Price, P.W. (1990) Testing a new hypothesis: plant vigor and *Phylloxera* distribution in wild grape in Arizona (USA). *Oecologia (Heidelberg)*, **84**, 1–8.

King, B.H. (2002) Offspring sex ratio and number in response to proportion of host sizes and ages in the parasitoid wasp *Spalangia cameroni*. *Environmental Entomology*, **31**, 505–8.

King, C.J., Fielding, N.J. & O'Keefe, T. (1991) Observations of the life cycle and behaviour of the *Rhizophagus grandis* (Coleoptera: Rhizophagidae) in Britain (UK). *Journal of Applied Entomology*, **11** (3), 286–96.

King, J.R. & Porter, S.D. (2005) Evaluation of sampling methods and species richness estimators for ants in upland ecosystems in Florida. *Environmental Entomology*, **34**, 1566–78.

Kingsland, S.E. (1985) *Modeling Nature: Episodes in the History of Population Ecology*. University of Chicago Press, Chicago.

Kingsolver, J.G. & Koehl, M.A.R. (1994) Selective factors in the evolution of insect wings. *Annual Review of Entomology*, **39**, 425–51.

Kinuura, H. & Kobayashi, M. (2006) Death of *Quercus crispula* by inoculation with adult *Platypus quercivorus* (Coleoptera: Platypodidae). *Applied Entomology and Zoology*, **41**, 123–8.

Kinzig, A.P. & Harte, J. (2000) Implications of endemics–area relationships for estimates of species extinctions. *Ecology*, **81**, 3305–11.

Kirby, P. (1992) *Habitat Management for Invertebrates: a Practical Handbook*. RSPB, Sandy, UK.

Kirby, W. & Spence, W. (1822) *An Introduction to Entomology: or Elements of the Natural History of Insects*. Printed for Longman, Hurst, Rees, Orme & Brown, London.

Kirk, A.A. & Wallace, M.M.H. (1990) Seasonal variations in numbers, biomass and breeding patterns of dung beetles (Coleoptera. *Scarabidae*) in Southern France. *Entomophaga*, **35**, 569–82.

Kirk, I.W., Hoffmann W.C & Carlton J.B (2001) Aerial electrostatic spray system performance. *Transactions of the Asabe*, **44** (5), 1089–92.

Kitching, R.L., Li, D.Q. & Stork, N.E. (2001) Assessing biodiversity 'sampling packages': how similar are arthropod assemblages in different tropical rainforests? *Biodiversity and Conservation*, **10**, 793–813.

Klaper, R.D. & Hunter, M.D. (1998) Genetic versus environmental effects on the phenolic chemistry of turkey oak, *Quercus laevis*. In *Diversity and Adaptation in Oak Species* (ed. K.C. Steiner), pp. 262–8. Pennsylvania State University Press, Pennsylvania.

Kleidon, A. & Mooney, H.A. (2000) A global distribution of biodiversity inferred from climatic constraints: results from a process-based modelling study. *Global Change Biology*, **6**, 507–23.

Kleijn, D., Baquero, R.A., Clough, Y. et al. (2006) Mixed biodiversity benefits of agri-environment schemes in five European countries. *Ecology Letters*, **9**, 243–54.

Kleijn, D., Berendse, F., Smit, R. & Gilissen, N. (2001) Agri-environment schemes do not effectively protect biodiversity in Dutch agricultural landscapes. *Nature*, **413**, 723–5.

Kleijn, D. & Sutherland, W.J. (2003) How effective are European agri-environment schemes in conserving and promoting biodiversity? *Journal of Applied Ecology*, **40**, 947–69.

Klein, B.C. (1989) Effects of forest fragmentation on dung and carrion beetle communities in central Amazonia. *Ecology*, **70**, 1715–25.

Klemola, T., Hanhimaki, S., Senn, J. et al. (2003) Performance of the cyclic autumnal moth, *Epirrita autumnata*, in relation to birch mast seeding. *Oecologia Berlin*, **135** (3), 354–61.

Klemola, T., Huitu, O. & Ruohomaki, K. (2006) Geographically partitioned spatial synchrony among cyclic moth populations. *Oikos*, **114**, 349–59.

Klug, R. & Bradler, S. (2006) The pregenital abdominal musculature in phasmids and its implications for the basal phylogeny of Phasmatodea (Insecta: Polyneoptera). *Organisms Diversity and Evolution* **6** (3), 171–84.

Kluth, S., Kruess, A. & Tscharntke, T. (2002) Insects as vectors of plant pathogens: mutualistic and antagonistic interactions. *Oecologia*, **133** (2), 193–9.

Knepp, R.G., Hamilton, J.G., Mohan, J.E., Zangerl, A.R., Berenbaum, M.R. & DeLucia, E.H. (2005) Elevated CO_2 reduces leaf damage by insect herbivores in a forest community. *New Phytologist*, **167**, 207–18.

Knight, A.L. & Light, D.M. (2005) Developing action thresholds for codling moth (Lepidoptera: Tortricidae) with pear ester- and codlemone-baited traps in apple orchards treated with sex pheromone mating disruption. *Canadian Entomologist*, **137** (6), 739–47.

Knox, O.G.G., Constable, G.A., Pyke, B. & Gupta, V. (2006) Environmental impact of conventional and *Bt* insecticidal cotton expressing one and two Cry genes in Australia. *Australian Journal of Agricultural Research*, **57**, 501–9.

Koch, T., Krumm, T., Jung, V., Engelberth, J. & Boland, W. (1999) Differential induction of plant volatile biosynthesis in the lima bean by early and late intermediates of the octadecanoid-signaling pathway. *Plant Physiology*, **121**, 153–62.

Kocourek, F., Havelka, J., Berankova, J. & Jarosik, V. (1994) Effect of temperature on development rate and intrinsic rate of increase of *Aphis gossypii* reared on greenhouse cucumbers. *Entomologia Experimentalis et Applicata*, **71** (1), 59–64.

Koh, L.P., Dunn, R.R., Sodhi, N.S., Colwell, R.K., Proctor, H.C. & Smith, V.S. (2004) Species coextinctions and the biodiversity crisis. *Science*, **305**, 1632–4.

Koh, L.P., Sodhi, N.S., Tan, H.T.W. & Peh, K.S.H. (2002) Factors affecting the distribution of vascular plants, springtails, butterflies and birds on small tropical islands. *Journal of Biogeography*, **29**, 93–108.

Koivisto, R.K.K. & Braig, H.R. (2003) Microorganisms and parthenogenesis. *Biological Journal of the Linnean Society*, **79** (1), 43–58.

Koivula, M. & Niemelä, J. (2003) Gap felling as a forest harvesting method in boreal forests: responses of carabid beetles (Coleoptera, Carabidae). *Ecography*, **26**, 179–87.

Koivula, M.J. & Vermeulen, H.J.W. (2005) Highways and forest fragmentation – effects on carabid beetles (Coleoptera, Carabidae). *Landscape Ecology*, **20**, 911–26.

Kolsch, G., Jakobi, K., Wegener, G. & Braune, H.J. (2002). Energy metabolism and metabolic rate of the alder leaf beetle *Agelastica alni* (L.) (Coleoptera, Chrysomelidae) under aerobic and anaerobic conditions: a microcalorimetric study. *Journal of Insect Physiology*, **48**, 143–51.

Komatsu, K., Okuda, S., Takahashi, M. & Matsunaga, R. (2004) Antibiotic effect of insect-resistant soybean on common cutworm (*Spodoptera litura*) and its inheritance. *Breeding Science*, **54** (1), 27–32.

Komonen, A., Ikavalko, J. & Wang, W.Y. (2003) Diversity patterns of fungivorous insects: comparison between glaciated vs. refugial boreal forests. *Journal of Biogeography*, **30**, 1873–81.

Konate, S., Le Roux, X., Verdier, B. & Lepage, M. (2003) Effect of underground fungus-growing termites on carbon dioxide emission at the point- and landscape-scales in an African savanna. *Functional Ecology*, **17**, 305–14.

Konig, H. (2006) *Bacillus* species in the intestine of termites and other soil invertebrates. *Journal of Applied Microbiology*, **101** (3), 620–7.

Koricheva, J., Larsson, S. & Haukioja, E. (1998) Insect performance on experimentally stressed woody plants: a meta-analysis. *Annual Review of Entomology*, **43**, 195–216.

Kost, C. & Heil, M. (2006) Herbivore-induced plant volatiles induce an indirect defence in neighbouring plants. *Journal of Ecology*, **94**, 619–28.

Kostal, V. & Finch, S. (1994) Influence of background on host-plant selection and subsequent oviposition by the cabbage root fly (*Delia radicum*). *Entomologia Experimentalis et Applicata*, **70** (2), 153–63.

Kotze, A.C., Sales, N. & Barchia, I.M. (1997) Diflubenzuron tolerance associated with monooxygenase activity in field strain larvae of the Australian sheep blowfly (Diptera: Calliphoridae). *Journal of Economic Entomology*, **90** (1), 15–20.

Kouki, J., Niemelä, P. & Viitasaari, M. (1994) Reversed latitudinal gradient in species richness of sawflies (Hymenoptera, Symphyta). *Annales Zoologici Fennici*, **31** (1), 83–8.

Kovats, R.S., Campbell-Lendrum, D. & Matthies, F. (2005) Climate change and human health: estimating avoidable deaths and disease. *Risk Analysis*, **25** (6), 1409–18.

Krafsur, E.S. (2003) Tsetse fly population genetics: an indirect approach to dispersal. *Trends in Parasitology*, **19** (4), 162–6.

Kramer, L.D., Styer, L.M. & Ebel, G.D. (2008) A global perspective on the epidemiology of West Nile virus. *Annual Review of Entomology*, **53**, 61–81.

Kranz, B.D., Schwarz, M.P., Morris, D.C. & Crespi, B.J. (2002) Life history of *Kladothrips ellobus* and *Oncothrips rodwayi*: insight into the origin and loss of soldiers in gall-inducing thrips. *Ecological Entomology*, **27** (1), 49–57.

Krasnov, B.R., Shenbrot, G.I., Khokhlova, I.S. & Poulin, R. (2007) Geographical variation in the 'bottom-up' control of diversity: fleas and their small mammalian hosts. *Global Ecology and Biogeography*, **16**, 179–86.

Krassilov, V.A. & Rasnitsyn, A.P. (1996) Pollen in the guts of Permian insects: first evidence of pollenivory and its evolutionary significance. *Lethaia*, **29** (4), 369–72.

Krauss, J., Harri, S.A., Bush, L. et al. (2007) Effects of fertilizer, fungal endophytes and plant cultivar on the performance of insect herbivores and their natural enemies. *Functional Ecology*, **21**, 107–16.

Krauss, J., Steffan-Dewenter, I. & Tscharntke, T. (2003) Local species immigration, extinction, and turnover of butterflies in relation to habitat area and habitat isolation. *Oecologia*, **137**, 591–602.

Krebs, C.J. (1999) *Ecological Methodology*, 2nd edn. Addison-Welsey Educational Publishers, Menlo Park, CA.

Krell, F.T. (2004) Parataxonomy vs. taxonomy in biodiversity studies – pitfalls and applicability of 'morphospecies' sorting. *Biodiversity and Conservation*, **13**, 795–812.

Kromp, B. (1999) Carabid beetles in sustainable agriculture: a review on pest control efficacy, cultivation impacts and enhancement. *Agriculture Ecosystems and Environment*, **74**, 187–228.

Kruess, A. & Tscharntke, T. (1994) Habitat fragmentation, species loss and biological control. *Science*, **264** (5165), 1581–54.

Kubatova-Hirsova, H. (2005) Temporal patterns in shore fly (Diptera, Ephydridae) community structure in a salt marsh habitat. *Ecological Entomology*, **30**, 234–40.

Kuefler, D. & Haddad, N.M. (2006) Local versus landscape determinants of butterfly movement behaviors. *Ecography*, **29**, 549–60.

Kuhn, K.G., Campbell-Lendrum, D.H., Armstrong, B. & Davies, C.R. (2003) Malaria in Britain: past, present, and future. *Proceedings of the National Academy of Sciences of the USA*, **100** (17), 9997–10,001.

Kumar, H. (1997) Inhibition of ovipositional responses of *Chilo partellus* (Lepidoptera: Pyralidae) by the trichomes on the lower leaf surafce of a maize cultivar. *Journal of Economic Entomology*, **85** (5), 1736–9.

Kumar, H. & Kumar, V. (2004) Tomato expressing Cry1A(b) insecticidal protein from *Bacillus thuringiensis* protected against tomato fruit borer, *Helicoverpa armigera* (Hubner) (Lepidoptera: Noctuidae) damage in the laboratory, greenhouse and field. *Crop Protection*, **23**, 135–9.

Kumar, N.S., Hewavitharanage, P. & Adikaram, N.K.B. (1995) Attack on tea by *Xyleborus fornicatus*: inhibition of the symbiote, *Monacrosporium ambrosium*, by caffeine. *Phytochemistry (Oxford)*, **40** (4), 1113–16.

Kunert, G. & Weisser, W.W. (2005) The importance of antennae for pea aphid wing induction in the presence of natural enemies. *Bulletin of Entomological Research*, **95** (2), 125–31.

Kusumawathie, P.H.D., Wickremasinghe, A.R., Karunaweera, N.D., Wijeyaratne, M.J.S. & Yapabandara, A.M.G.M. (2006) Anopheline breeding in river bed pools below major dams in Sri Lanka. *Acta Tropica*, **99** (1), 30–3.

Kwon, H.W., Lu, T., Rutzler, M. & Zwiebel, L.J. (2006) Olfactory responses in a gustatory organ of the malaria vector mosquito *Anopheles gambiae*. *Proceedings of the National Academy of Sciences of the USA*, **103** (36), 13526–31.

Kysela, E. (2002) Zikaden als Schmuck- und Trachtbestandteil in Romischer Kaiserzeit und Volkerwanderungszeit. *Denisia*, **4**, 21–8.

Kyto, M., Niemelä, P. & Larsson, S. (1996) Insects on trees: population and individual response to fertilization. *Oikos*, **75**, 148–59.

Labandeira, C.C., Beall, B.S. & Hueber, F.M. (1988) Early insect diversification: evidence from a lower Devonian bristletail from Quebec (Canada). *Science (Washington DC)*, **242** (4880), 913–16.

Labandeira, C.C., Dilcher, D.L., Davis, D.R. & Wagner, D.L. (1994) Ninety-seven million years of angiosperm–insect association; paleobiological insights into the meaning of co-evolution. *Proceedings of the National Academy of Sciences of the USA*, **91** (25), 12278–82.

Labandeira, C.C. & Phillips, T.L. (1996) A Carboniferous insect gall: insight into early ecologic history of the Holometabola. *Proceedings of the National Academy of Sciences of the USA*, **93** (16), 8470–4.

Labandeira, C.C. & Phillips, T.L. (2002) Stem borings and petiole galls from Pennsylvanian tree ferns of Illinois, USA: implications for the origin of the borer and galler functional-feeding-groups and holometabolous insects. *Palaeontographica Abteilung a Palaeozoologie Stratigraphie*, **264** (1–4), 1–84.

Labandeira, C.C. & Sepkoski, J.J.J.R. (1993) Insect diversity in the fossil record. *Science*, **261** (5119), 310–15.

Lacey, L.A. & Shapiro-Ilan, D.I. (2008) Microbial control of insect pests in temperate orchard systems: potential for incorporation into IPM. *Annual Review of Entomology*, **53**, 121–44.

Lack, D. (1969) The number of bird species on islands. *Bird Study*, **16**, 193–209.

Lack, D. (1976) *Island Birds*. Blackwell Scientific Publications, Oxford.

Lale, N.E.S (1998) Neem in the conventional Lake Chad basin area and the threat of oriental yellow scale insect (*Aonidiella orientalis* Newstead) (Homoptera: Diaspididae). *Journal of Arid Environments*, **40** (2), 191–7.

Lamb, C. & Dixon, R.A. (1997) The oxidative burst in plant disease resistance. *Plant Physiology and Plant Molecular Biology*, **48**, 251–75.

Lamb, R.J. & Mackay, P.A. (1987) *Acyrthosiphon kondoi* influences alata production by the pea aphid, *A. pisum*. *Entomologia Experimentalis et Applicata*, **45**, 195–8.

Lambert, A.L., Mcpherson, R.M. & Espelie, K.E. (1995) Soybean host plant resistance mechanisms that alter abundance of whiteflies (Homoptera: Aleyrodidae). *Environmental Entomology*, **24**, 1381–6.

Lamberti, G.A. & Resh, V.H. (1985) Distribution of benthic algae and macroinvertebrates along a thermal stream gradient. *Hydrobiologia*, **128** (1), 13–22.

Lamy, M. (1990) Contact dermatitis (erucism) produced by processionary caterpillars (genus *Thaumetopoea*). *Journal of Applied Entomology* **110** (5), 425–37.

Lancien, J. (1991) Campaign against sleeping sickness in south-west Uganda by trapping tsetse flies. *Annales de la Societé Belge de Médecine Tropicale*, **71** (Suppl 1), 35–47.

Landsberg, J. & Gillieson, D.S. (1995) Regional and local variation in insect herbivory, vegetation and soils of eucalypt associations in contrasted landscape positions along a climatic gradient. *Australian Journal of Ecology*, **20**, 299–315.

Langhof, M., Gathmann, A., Poehling, H.M. & Meyhofer, R. (2003) Impact of insecticide drift on aphids and their parasitoids: residual toxicity, persistence and recolonisation. *Agriculture Ecosystems and Environment*, **94**, 265–74.

Langley, P.A. (1994) Understanding tsetse flies. *Onderstepoort Journal of Veterinary Research*, **61** (4), 361–7.

Lanza, J., Schmitt, M.A. & Awad, A.B. (1992) Comparative chemistry of elaiosomes of three species of *Trillium*. *Journal of Chemical Ecology*, **18**, 209–21.

Larsen, T.B. (2005) *The Butterflies of West Africa*. Apollo Books, Denmark.

Larsson, S. (1989) Stressful times for the plant stress–insect performance hypothesis. *Oikos*, **56**, 277–83.

Larsson, S. & Tenow, O. (1984) Areal distribution of a *Neodiprion sertifer* (Hym., Diprionidae) outbreak on Scots pine as related to stand condition. *Holarctic Ecology*, **7**, 81–90.

Larsson, T-B. (2001) Biodiversity evaluation tools for european forests. *Ecological Bulletins*, **50**.

Laurance, W.F., Lovejoy, T.E., Vasconcelos, H.L. et al. (2002) Ecosystem decay of Amazonian forest fragments: a 22-year investigation. *Conservation Biology*, **16**, 605–18.

Lawrence, R., Potts, B.M. & Whitham, T.G. (2003) Relative importance of plant ontogeny, host genetic variation, and leaf age for a common herbivore. *Ecology*, **84**, 1171–8.

Lawson, G.L. (1994) Indigenous trees in West African forest plantations: the need for domestication by clonal techniques. In *Tropical Trees: the Potential for Domestication and the Rebuilding of Forest Resources* (eds R.R.B. Leakey & A.C. Newton), pp. 112–23. HMSO, London.

Lawson, S.A., Furuta, K. & Katagiri, K. (1996) The effect of host tree on the natural enemy complex of *Ips typographus japonicus* Niijima (Col, Scolytidae) in Hokkaido, Japan. *Zeitschrift fuer Angewandte Entomologie*, **120**, 77–86.

Lawton, J.H. (1982) Vacant niches and unsaturated communities: a comparison of bracken herbivores at sites on two continents. *Journal of Animal Ecology*, **51**, 573–95.

Lawton, J.H. (1984) Herbivore community organization: general models and specific tests with phytophagous insects. In *A New Ecology – Novel Approaches to Interactive Systems* (eds P.W. Price, C.N. Slobodchikoff & W.S. Gaud), pp. 329–52. John Wiley, New York.

Lawton, J.H. (1991) From physiology to population dynamics and communities. *Functional Ecology*, **5**, 155–61.

Lawton, J.H. (1992) There are not 10 million kinds of population-dynamics. *Oikos*, **63**, 337–8.

Lawton, J.H. (1994) What do species do in ecosystems? *Oikos*, **71**, 367–74.

Lawton, J.H. (1995) The response of insects to environmental change. In *Insects in a Changing Environment* (eds R. Harrington & N.E. Stork), pp. 3–26. Academic Press, London.

Lawton, J.H. (1999) Are there general laws in ecology? *Oikos*, **84**, 177–92.

Lawton, J.H., Bignell, D.E., Bolton, B. et al. (1998) Biodiversity inventories, indicator taxa and effects of habitat modification in tropical forest. *Nature*, **391**, 72–6.

Lawton, J.H. & Schroder, D. (1977) Effects of plant type, size of geographical range and taxonomic isolation on number of insect species associated with British plants. *Nature*, **265**, 137–40.

Lawton, J.H. & Strong, D.R. (1981) Community patterns and competition in folivorous insects. *American Naturalist*, **118**, 317–38.

Leather, S.R. (1986) Insect species richness of the British Rosaceae – the importance of host range, plant architecture, age of establishment, taxonomic isolation and species area relationships. *Journal of Animal Ecology*, **55**, 841–60.

Leather, S.R. (1989) Do alate aphids produce fitter offspring? The influence of maternal rearing history and morph on life-history parameters of *Rhopalosiphum padi* (L). *Functional Ecology*, **3** (2), 237–44.

Leather, S.R. (2004) *Insect Sampling in Forest Ecosystems*. Blackwell Publishing, Oxford.

Leather, S.R. & Owuor, A. (1996) The influence of natural enemies and migration on spring populations of the green spruce aphid, *Elatobium abietinum* Walker (Hom., Aphididae). *Journal of Applied Entomology*, **120** (9), 529–36.

Leather, S.R. & Walsh, P.J. (1993) Sublethal plant defenses – the paradox remains. *Oecologia*, **93**, 153–5.

Lecoq, M. (2001) Recent progress in desert and migratory locust management in Africa. Are preventative actions possible? *Journal of Orthoptera Research*, **10** (2), 277–91.

Lee, K.P., Behmer, S.T. & Simpson, S.J. (2006) Nutrient regulation in relation to diet breadth: a comparison of *Heliothis* sister species and a hybrid. *Journal of Experimental Biology*, **209**, 2076–84.

Lee, S.C. (1992) Towards integrated pest management of rice in Korea. *Korean Journal of Applied Entomology*, **31** (3), 205–40.

Legaspi, J.C. & Legaspi, B.C. (2005) Life table analysis for *Podisus maculiventris* immatures and female adults under four constant temperatures. *Environmental Entomology*, **34** (5), 990–8.

Legrand, P. (1993) Management of biological control against *Dendroctonus micans* infestation of spruce in Auvergne and Limousin (Coleoptera: Scolytidae). *Revue des Sciences Naturelles d'Auvergne*, **56**, 49–57.

Lehmann, F.O. & Heymann, N. (2006) Dynamics of in vivo power output and efficiency of *Nasonia* asynchronous flight muscle. *Journal of Biotechnology*, **124** (1), 93–107.

Lehtonen, P., Helander, M. & Saikkonen, K. (2005) Are endophyte-mediated effects on herbivores conditional on soil nutrients? *Oecologia*, **142**, 38–45.

Leibold, M.A., Chase, J., Shurin, J.B. & Dowding, L.A. (1997) Species turnover and regulation of trophic structure. *Annual Review of Ecology and Systematics*, **29**, 467–94.

Leimar, O. (1996) Life history plasticity: influence of photoperiod on growth and development in the common blue butterfly. *Oikos*, **76** (2), 228–34.

Lelis, A.T. (1992) The loss of intestinal flagellates in termites exposed to the juvenile hormone analogue (JHA) methoprene. *Material und Organismen (Berlin)*, **27** (3), 171–8.

Lesnik, M., Pintar, C., Lobnik, A. & Kolar, M. (2005) Comparison of the effectiveness of standard and drift-reducing nozzles for control of some pests of apple. *Crop Protection*, **24** (2), 93–100.

Letourneau, D.K. & Goldstein, B. (2001) Pest damage and arthropod community structure in organic vs. conventional tomato production in California. *Journal of Applied Ecology*, **38**, 557–70.

Leveque, C., Balian, E.V. & Martens, K. (2005) An assessment of animal species diversity in continental waters. *Hydrobiologia*, **542**, 39–67.

Levie, A., Legrand, M-A., Dogot, P., Pels, C., Baret, P.V. & Hance, T. (2005) Mass releases of *Aphidius rhopalosiphi* (Hymenoptera: Aphidiinae), and strip management to control of wheat aphids. *Agriculture Ecosystems and Environment*, **105**, 17–21.

Levins, R. (1970). Extinction. In *Some Mathematical Questions in Biology. Lectures on Mathematics in the Life Sciences*, Vol. 2. (ed. M. Gertenhaber), pp. 77–107. American Mathematical Society, Providence, RI.

Levinson, H. & Levinson, A. (1990) The Egyptian plagues and the beginnings of pest control in the ancient Orient. *Anzeiger fur Schadlingskunde, Pflanzenschutz, Umweltschutz*, **63** (5), 81–96.

Lewinsohn, T.M., Novotny, V. & Basset, Y. (2005) Insects on plants: diversity of herbivore assemblages revisited. *Annual Review of Ecology, Evolution and Systematics*, **36**, 597–620.

Lewinsohn, T.M. & Prado, P.I. (2005) How many species are there in Brazil? *Conservation Biology*, **19**, 619–24.

Li, X.C., Schuler, M.A. & Berenbaum, M.R. (2002) Jasmonate and salicylate induce expression of herbivore cytochrome P450 genes. *Nature*, **419**, 712–15.

Liadouze, I., Febvay, G., Guillaud, J. & Bonnot, G. (1995) Effect of diet on the free amino acid pools of symbiotic and aposymbiotic pea aphids, *Acyrthosiphon pisum*. *Journal of Insect Physiology*, **41** (1), 33–40.

Libert, M. (1994) Biodiversity: *Rhopalocera* fauna of two hills of Yaounde area, Cameroun (Lepidoptera). *Bulletin de la Société Entomologique de France*, **99** (4), 335–55.

Liechti, R. & Farmer, E.E. (2002) The jasmonate pathway. *Science*, **296**, 1649–50.

Lighton, J.R.B., Brownell, P.H., Joos, B. & Turner, R.J. (2001) Low metabolic rate in scorpions: implications for population biomass and cannibalism. *Journal of Experimental Biology*, **204**, 607–13.

Lilburn, T.C., Kim, K.S., Ostrom, N.E., Byzek, K.R., Leadbetter, J.R. & Breznak, J.A. (2001) Nitrogen fixation by symbiotic and free-living spirochetes. *Science*, **292**, 2495–8.

Lill, J.T. & Marquis, R.J. (2001) The effects of leaf quality on herbivore performance and attack from natural enemies. *Oecologia*, **126**, 418–28.

Lill, J.T. & Marquis, R.J. (2003) Ecosystem engineering by caterpillars increases insect herbivore diversity on white oak. Ecology, 84, 682–90.

Lincoln, D.E., Sionit, N. & Strain, B. R. (1984) Growth and feeding response of *Pseudoplusia includens* (Lepidoptera: Noctuidae) to host plants grown in controlled carbon dioxide atmospheres. *Environmental Entomology*, **13**, 1527–30.

Lincoln, P.E. & Couvet, P. (1989) The effects of carbon supply on allocation to allelochemicals and caterpillar consumption of peppermint. *Oecologia*, **78**, 112–14.

Lincoln, P.E., Fajer, E.D. & Johnson, R.H. (1993) Plant–insect herbivore interactions in elevated CO_2. *Trends in Ecology and Evolution*, **8**, 64–8.

Lindblade, K.A., Eisele, T.P., Gimnig, J.E. et al. (2004) Sustainability of reductions in malaria transmission and infant mortality in western Kenya with use of insecticide-treated bednets – 4 to 6 years of follow-up. *Journal of the American Medical Association*, **291**, 2571–80.

Lindroth, R.L., Arteel, G.E. & Kinney, K.K. (1995) Responses of three sturniid species to paper birch grown under enriched CO_2 atmospheres. *Functional Ecology*, **9**, 306–11.

Lindroth, R.L. & Hwang, S-Y. (1996) Clonal variation of foliar chemistry of quaking aspen (*Populus tremuloides* Michx.). *Biochemical Systematics and Ecology*, **24**, 357–64.

Lindroth, R.L., Kinney, K.K. & Platz, C.L. (1993) Responses of deciduous trees to elevated atmospheric CO_2: productivity, phytochemistry and insect performance. *Ecology*, **74**, 763–77.

Lindsay, S.W., Jawara, M., Paine, K., Pinder, M., Walraven, G.E.L. & Emerson, P.M. (2003) Changes in house design reduce exposure to malaria mosquitoes. *Tropical Medicine and International Health*, **8** (6), 512–17.

Linfield, M.C.J., Raubenheimer, D., Hambler, C. & Speight, M.R. (1993) Leaf miners on *Ochna ciliata* (Ochnaceae) growing on Aldabra Atoll. *Ecological Entomology*, **18**, 332–8.

Linnaeus, C. (1758) *Systema naturae per regna tria naturae, secundum classes, ordines, genera, species, cum characteribus, differentiis, synonymis, locis*, 10th edn.

Liu, C. (1990) Comparative studies on the role of *Anopheles anthropophagus* and *Anopheles sinensis* in malaria transmission in China. *Chung Hua Liu Hsing Ping Hsueh-Tsa-Chih*, **11** (6), 360–3.

Liu, H. & Beckenbach, A.T. (1992) Evolution of the mitochondrial cytochrome oxidase II gene among 10 orders of insects. *Molecular Phylogenetics and Evolution*, **1** (1), 41–52.

Liu, S., Yang, Z., Li, T., Wang, Y. & Wu, T. (1992) Studies on the mutagenicity of human body cells affected by insect viruses. *Sichuan Daxue Xuebo (Ziran Kexueban)*, **29** (4), 541–5.

Liu, W.H., Wang, Y.F. & Xu, R.M. (2006a) Habitat utilization by ovipositing females and larvae of the marsh fritillary (*Euphydryas aurinia*) in a mosaic of meadows and croplands. *Journal of Insect Conservation*, **10**, 351–60.

Liu, Z.D., Gong, P.Y., Wu, K.J., Sun, J.H. & Li, D.M. (2006b) A true summer diapause induced by high temperatures in the cotton bollworm, *Helicoverpa armigera* (Lepidoptera: Noctuidae). *Journal of Insect Physiology*, **52**, 1012–20.

Loader, C. & Damman, H. (1991) Nitrogen content of food plants and vulnerability of *Pieris rapae* to natural enemies. *Ecology*, **72**, 1586–90.

Lobl, I. & Leschen, R.A.B. (2005) Demography of coleopterists and their thoughts on DNA barcoding and the phylocode, with commentary. *Coleopterists Bulletin*, **59**, 284–92.

Lock, K., Verslycke, T. & Janssen, C.R. (2006) Energy allocation in brachypterous versus macropterous morphs of the pygmy grasshopper *Tetrix subulata* (Orthoptera: Tetrigidae). *Entomologia Generalis*, **28** (4), 269–74.

Lombardero, M.J., Ayres, M.P. & Ayres, B.D. (2006) Effects of fire and mechanical wounding on *Pinus resinosa* resin defenses, beetle attacks, and pathogens. *Forest Ecology and Management*, **225**, 349–58.

Lomolino, M.V. (2000) Ecology's most general, yet protean pattern: the species–area relationship. *Journal of Biogeography*, **27**, 17–26.

Longino, J.T., Coddington, J. & Colwell, R.K. (2002) The ant fauna of a tropical rain forest: estimating species richness three different ways. *Ecology*, **83**, 689–702.

Longo, S. (1994) The role of beekeeping within agrarian and natural ecosystems. *Ethology, Ecology and Evolution, Special Issue*, **3**, 5–9.

Loomis, W.E. (1932) Growth-differentiation balance vs. carbohydrate-nitrogen ratio. *American Society of Horticultural Science*, **29**, 240–5.

Loomis, W.E. (1953) Growth and differentiation – an introduction and summary. In *Growth and Differentiation in Plants* (ed. W.E. Loomis), pp. 1–17. Iowa State College Press, Ames, IA.

Lopez-Carretero, A., Cruz, M. & Eben, A. (2005) Phenotypic plasticity of the reproductive system of female *Leptinotarsa undecimlineata*. *Entomologia Experimentalis et Applicata*, **115** (1), 27–31.

Lopez-Vaamonde, C., Godfray, C.H. & Cook, J.M. (2003) Evolutionary dynamics of host-plant use in a genus of leaf-mining moths. *Evolution*, **57**, 1804–21.

Loreau, M., Naeem, S., Inchausti, P. et al. (2001) Biodiversity and ecosystem functioning: current knowledge and future challenges. *Science*, **294**, 804–8.

Lorio, P.L. Jr., Stephen, F.M. & Paine, T.D. (1995) Environment and ontogeny modify loblolly pine response to induced acute water deficits and bark beetle attack. *Forest Ecology and Management*, **73**, 97–110.

Lotka, A.J. (1925) *Elements of Physical Biology*. Williams & Wilkins, Baltimore, MD.

Lotka, A.J. (1932) The growth of mixed populations: two species competing for a common food supply. *Journal of Washington Academy of Science*, **22**, 461–9.

Louda, S.M. & Collinge, S.K. (1992) Plant resistance to insect herbivores: a field test on the environmental stress hypothesis. *Ecology*, **73**, 153–69.

Louda, S.M., Huntly, N. & Dixon, P.M. (1987b) Insect herbivory across a sun/shade gradient: response to experimentally-induced in situ plant stress. *Acta Oecologica Oecologia Generalis*, **8**, 357–64.

Louda, S.M., Kendall, D., Connor, J. & Simberloff, D. (1997) Ecological effects of an insect introduced for the biological control of weeds. *Science*, **277** (5329), 1088–90.

Louda, S.M., Rand, T.A., Russell, F.L. & Arnett, A.E. (2005) Assessment of ecological risks in weed biocontrol: input from retrospective ecological analyses. *Biological Control*, **35** (3), 253–264.

Louda, S.M. & Stiling, P. (2004) The double-edged sword of biological control in conservation and restoration. *Conservation Biology*, **18** (1), 50–3.

Loughrin, J.H., Manukian, A., Heath, R.R. & Tumlinson, J.H. (1995) Volatiles emitted by different cotton varieties damaged by feeding beet armyworm larvae. *Journal of Chemical Ecology*, **21**, 1217–27.

Loughrin, J.H., Manukian, A., Heath, R.R., Turlings, T.C.J. & Tumlinson, J.H. (1994) Diurnal cycle of emission of induced volatile terpenoids by herbivore-injured cotton plants. *Proceedings of the National Academy of Sciences of the USA*, **91**, 11836–40.

Lounibos, L.P. (2002) Invasions by insect vectors of human disease. *Annual Review of Entomology*, **47**, 233–66.

Lounibos, L.P., O'Meare, G.F., Nishimura, N. & Escher, R.L. (2003) Interactions with native mosquito larvae regulate the production of *Aedes albopictus* from bromeliads in Florida. *Ecological Entomology*, **28**, 551–8.

Lourenco de Oliveira, R., Castro, M.G., Braks, M.A.H. & Lounibos, L.P. (2004) The invasion of urban forest by dengue vectors in Rio de Janeiro. *Journal of Vector Ecology*, **29** (1), 94–100.

Louton, J., Gelhaus, J. & Bouchard, R. (1996) The aquatic macrofauna of water-filled bamboo (Poaceae: Bambusoideae: Guadua) internodes in a Peruvian lowland tropical forest. *Biotropica*, **28** (2), 228–42.

Lovei, G.L., Magura, T., Tothmeresz, B. & Kodobocz, V. (2006) The influence of matrix and edges on species richness patterns of ground beetles (Coleoptera: Carabidae) in habitat islands. *Global Ecology and Biogeography*, **15**, 283–9.

Lovei, G.L. & Sunderland, K.D. (1996) Ecology and behavior of ground beetles (Coleoptera: Carabidae). *Annual Review of Entomology*, **41**, 231–56.

Lovejoy, T.E., Erwin, N. & Boren, S. (1997) Insect conservation. In *Forests and Insects* (eds A.D. Watt, N.E. Stork & M.D. Hunter), pp. 395–400. Chapman & Hall, London.

Lovett, G.M., Christenson, L.M., Groffman, P.M., Jones, C.G., Hart, J.E. & Mitchell, M.J. (2002) Insect defoliation and nitrogen cycling in forests. *Bioscience*, **52**, 335–41.

Lowman, M.D. & Nadkarni, N.M. (1995) *Forest Canopies*. Academic Press, New York.

Lu, J.B. & Li, X. (2006) Review of rice-fish-farming systems in China – One of the Globally Important Ingenious Agricultural Heritage Systems (GIAHS). *Aquaculture*, **260** (1–4), 106–13.

Lu, J.H., Liu, S.S. & Shelton, A.M. (2004) Laboratory evaluations of a wild crucifer *Barbarea vulgaris* as a management tool for the diamondback moth *Plutella xylostella* (Lepidoptera: Plutellidae). *Bulletin of Entomological Research*, **94** (6), 509–16.

Lucky, A., Erwin, T.L. & Witman, J.D. (2002) Temporal and spatial diversity and distribution of arboreal Carabidae (Coleoptera) in a western Amazonian rain forest. *Biotropica*, **34**, 376–86.

Lugo, A.E. (1995) Management of tropical biodiversity. *Ecological Applications*, **5**, 956–61.

Lundberg, P. & Astrom, M. (1990) Low nutritive quality as a defense against optimally foraging herbivores. *American Naturalist*, **135**, 547–62.

Luoto, M., Poyry, J., Heikkinen, R.K. & Saarinen, K. (2005) Uncertainty of bioclimate envelope models based on the geographical distribution of species. *Global Ecology and Biogeography*, **14**, 575–84.

Luttrell, R.G., Fitt, G.P., Ramalho, F.S. & Sugonyaev, E.S. (1994) Cotton pest management, Part 1. A worldwide perspective. *Annual Review of Entomology*, **39**, 517–26.

Lykouressis, D., Perdikis, D., Samartzis, D., Fantinou, A. & Toutouzas, S. (2005) Management of the pink bollworm *Pectinophora gossypiella* (Saunders) (Lepidoptera: Gelechiidae) by mating disruption in cotton fields. *Crop Protection*, **24**, 177–83.

Lynch, A.M. (2004) Fate and characteristics of *Picea* damaged by *Elatobium abietinum* (Walker) (Homoptera: Aphididae) in the White Mountains of Arizona. *Western North American Naturalist*, **64** (1), 7–17.

Lyons, D.B. (1994) Development of the arboreal stages of the pine false webworm (Hymenoptera: Pamphiliidae). *Environmental Entomology*, **23** (4), 846–54.

Lytle, D.A. & Peckarsky, B.L. (2001) Spatial and temporal impacts of a diesel fuel spill on stream invertebrates. *Freshwater Biology*, **46**, 693–704.

Lyytikainen-Saarenmaa, P., Varama, M., Anderbrant, O. et al. (2006) Monitoring the European pine sawfly with pheromone traps in maturing Scots pine stands. *Agricultural and Forest Entomology*, **8** (1), 7–15.

Ma, J.W., Han, X.Z., Hasibagan, et al. (2005) Monitoring East Asian migratory locust plagues using remote sensing data and field investigations. *International Journal of Remote Sensing*, **26**, 629–34.

Ma, K.C. & Lee, S.C. (1996) Occurrence of major rice insect pests at different transplanting times and fertilizer levels in paddy field. *Korean Journal of Applied Entomology*, **35** (2), 132–6.

MacArthur, R.H. & Levins, R. (1967) The limiting similarity, convergence, and divergence of competing species. *American Naturalist*, **101**, 377–85.

MacArthur, R.H. & Wilson, E.O. (1967) *The Theory of Island Biogeography*, p. 203. Princeton University Press, Princeton, NJ.

MacDonald, R.E. & Bishop, C.J. (1952) Phloretin: an antibacterial substance obtained from apple leaves. *Canadian Journal of Botany*, **30**, 486–9.

Mace, G.M. & Stuart, S. (1994) Draft IUCN Red List categories, version 2.2. *Species*, **21/22**, 13–24.

Mace, J.M., Boussinesq, M., Ngoumou, P., Oye, J.E., Koeranga, A. & Godin, C. (1997) Country-wide rapid epidemiological mapping of onchocerciasis (REMO) in Cameroon. *Annals of Tropical Medicine and Parasitology*, **91** (4), 379–91.

MacLean, R.H., Litsinger, J.A., Moody, K., Watson, A.K. & Libetario, E.M. (2003) Impact of *Gliricidia sepium* and *Cassia spectabilis* hedgerows on weeds and insect pests of upland rice. *Agriculture Ecosystems and Environment*, **94** (3), 275–88.

Madden, D. & Young, T.P. (1992) Symbiotic ants as an alternative defense against giraffe herbivory in spinescent *Acacia drepanolobium*. *Oecologia (Heidelberg)*, **91**, 235–8.

Madders, M. & Whitfield, D.P. (2006) Upland raptors and the assessment of wind farm impacts. *Ibis*, **148**, 43–56.

Madibela, O.R., Seitiso, T.K., Thema, T.F. & Letso, M. (2007) Effect of traditional processing methods on chemical composition and in vitro true dry matter digestibility of the Mophane worm (*Imbrasia belina*). *Journal of Arid Environments*, **68**, 492–500.

Madritch, M.D. & Hunter, M.D. (2002) Phenotypic diversity influences ecosystem functioning in an oak sandhills community. *Ecology*, **83**, 2084–90.

Maehara, N. & Futai, K., (2002) Factors affecting the number of *Bursaphelenchus xylophilus* (Nematoda: Aphelenchoididae) carried by several species of beetles. *Nematology*, **4**, 653–8.

Maeta, Y., Sugiura, N. & Goubara, M. (1992) Patterns of offspring production and sex allocation in the small carpenter bee, *Ceratina flavipes* Smith (Hymenoptera, Xylocopinae). *Japanese Journal of Entomology*, **60** (1), 175–90.

Magura, T. & Kodobocz, V. (2007) Carabid assemblages in fragmented sandy grasslands. *Agriculture Ecosystems and Environment*, **119**, 396–400.

Magurran, A.E. (2004) *Measuring Biological Diversity* Blackwell Publishing, Oxford.

Maharaj, R. (2003) Life table characteristics of *Anopheles arabiensis* (Diptera: Culicidae) under simulated seasonal conditions. *Journal of Medical Entomology*, **40** (6), 737–42.

Majer, J., Recher, H. & Postle, A. (1994) Comparison of arthropod species richness in eastern and western australian canopies: a contribution to species number debate. *Memoirs of the Queensland Museum*, **36**, 121–31.

Makkouk, K.M. & Kumari, S.G. (2001) Reduction of incidence of three persistently transmitted aphid-borne viruses affecting legume crops by seed-treatment with the insecticide imidacloprid (Gaucho ®). *Crop Protection*, **20**, 433–7.

Malcolm, S.B., Cockrell, B.J. & Brower, L.P. (1989) Cardenolide fingerprint of monarch butterfly reared on common milkweed, *Asclepias syriaca* L. *Journal of Chemical Ecology*, **15** (3), 819–54.

Malcolm, S.B. & Zalucki, M.P. (eds) (1993) *Biology and Conservation of the Monarch Butterfly*. Science Series No. 38. Natural History Museum of Los Angeles County, Los Angeles.

Malcolm, S.B. & Zalucki, M.P. (1996) Milkweed latex and cardenolide induction may resolve the lethal plant defence paradox. *Entomologia Experimentalis et Applicata*, **80**, 193–6.

Malkin, E., Dubovsky, F. & Moree, M. (2006) Progress towards the development of malaria vaccines. *Trends in Parasitology*, **22** (7), 292–5.

Mallatt, J.M., Garey, J.R. & Shultz, J.W. (2004) Ecdysozoan phylogeny and Bayesian inference: first use of nearly complete 28S and 18S rRNA gene sequences to classify the arthropods and their kin. *Molecular Phylogenetics and Evolution*, **31**, 178–91.

Mallatt, J.M. & Giribet, G. (2006) Further use of nearly complete, 28S and 18S rRNA genes to classify Ecdysozoa: 37 more arthropods and a kinorhynch. *Molecular Phylogenetics and Evolution*, **40**, 772–94.

Malmqvist, B., Sjostrom, P. & Frick, K. (1991) The diet of two species of *Isoperla* (Plecoptera: Perlodidae) in relation to season, site, and sympatry. *Hydrobiologia*, **213** (3), 191–204.

Mamiya, Y. (1984) The pine wood nematode. In *Plant and Insect Nematodes* (ed. W.R. Nickle), pp. 589–626. Marcel Dekker Inc., New York.

Mann, J.A., Harrington, R., Carter, N. & Plumb, R.T. (1997) Control of aphids and barley yellow dwarf virus in spring-sown cereals. *Crop Protection*, **16** (1), 81–7.

Manojlovic, B., Zabel, A., Peric, P., Stankovic, S., Rajkovic, S. & Kostic, M. (2003) *Dendrosoter protuberans* (Hymenoptera: Braconidae), an important elm bark beetle parasitoid. *Biocontrol Science and Technology*, **13**, 429–39.

Mansour, S.A. (2004) Pesticide exposure – Egyptian scene. *Toxicology*, **198** (1–3), 91–115.

Marchand, D. & McNeil, J.N. (2004) Avoidance of intraspecific competition via host modification in a grazing, fruit-eating insect. *Animal Behaviour*, **67**, 397–402.

Margules, C., Higgs, A.J. & Rafe, R.W. (1982) Modern biogeographic theory: are there any lessons for nature reserve design? *Biological Conservation*, **24**, 115–28.

Maris, P.C., Joosten, N.N., Goldbach, R.W. & Peters, D. (2004) Tomato spotted wilt virus infection improves host suitability for its vector *Frankliniella occidentalis*. *Phytopathology*, **94** (7), 706–11.

Markiewicz, D.A. & O'Donnell, S. (2001) Social dominance, task performance and nutrition: implications for reproduction in eusocial wasps. *Journal of Comparative Physiology a, Neuroethology Sensory Neural and Behavioral Physiology*, **187**, 327–33.

Maron, J.L. & Harrison, S. (1997) Spatial pattern formation in an insect host–parasitoid system. *Science*, **278**, 1619–21.

Marques, M.I., Adis, J., da Cunha, C.N. & dos Santos, G.B. (2001) Arthropod biodiversity in the canopy of *Vochysia divergens* (Vochysiaceae), a forest dominant in the Brazilian Pantanal. *Studies on Neotropical Fauna and Environment*, **36**, 205–10.

Marris, G.C. & Caspard, J. (1996) The relationship between conspecific superparasitism and the outcome of *in vitro* contests staged between different larval instars of the solitary endoparasitoid *Venturia canescens*. *Behavioral Ecology Sociobiology*, **39**, 61–9.

Marron, M.T., Markow, T.A., Kain, K.J. & Gibbs, A.G. (2003) Effects of starvation and desiccation on energy metabolism in desert and mesic Drosophila. *Journal of Insect Physiology*, **49**, 261–70.

Marshall, E.J.P., West, T.M. & Kleijn, D. (2006) Impacts of an agri-environment field margin prescription on the flora and fauna of arable farmland in different landscapes. *Agriculture Ecosystems and Environment*, **113**, 36–44.

Martel, J.W. & Malcolm, S.B. (2004) Density-dependent reduction and induction of milkweed cardenolides by a sucking insect herbivore. *Journal of Chemical Ecology*, **30**, 545–61.

Martikainen, P. & Kouki, J. (2003) Sampling the rarest: threatened beetles in boreal forest biodiversity inventories. *Biodiversity and Conservation*, **12**, 1815–31.

Martin, G.J. (1995) *Ethnobotany*. Chapman & Hall, London.

Martin, M.M. (1991) The evolution of cellulose digestion in insects. *Philosophical Transactions of the Royal Society of London Series B, Biological Sciences*, **333** (1267), 281–8.

Martinez, A.S., Fernandez-Arhex, V. & Corley, J.C. (2006) Chemical information from the fungus *Amylostereum areolatum* and host-foraging behaviour in the parasitoid *Ibalia leucospoides*. *Physiological Entomology*, **31** (4), 336–40.

Martinez-Delclos, X. (1996) The fossil record of insects. *Boletin de la Asociacion Espanola de Entomologia*, **20** (1/2), 9–30.

Martinsen, G.D., Floate, K.D., Waltz, A.M., Wimp, G.M. & Whitham, T.G. (2000) Positive interactions between leafrollers and other arthropods enhance biodiversity on hybrid cottonwoods. *Oecologia*, **123**, 82–9.

Marucci, R.C., Lopes, J.R.S., Vendramim, J.D. & Corrente, J.E. (2005) Influence of *Xylella fastidiosa* infection of citrus on host selection by leafhopper vectors. *Entomologia Experimentalis et Applicata*, **117**, 95–103.

Maschinski, J. & Whitham, T.G. (1989) The continuum of plant responses to herbivory: the influence of plant association, nutrient availability, and timing. *American Naturalist*, **134**, 1–19.

Massafera, R., da Silva, A.M., de Carvalho, A.P., dos Santos, D.R., Galati, E.A.B. & Teodoro, U. (2005) Phlebotomine sandflies of southern Brazil. *Revista de Saude Publica*, **39** (4), 571–7.

Massey, F.P., Ennos, A.R. & Hartley, S.E. (2006) Silica in grasses as a defence against insect herbivores: contrasting effects on folivores and a phloem feeder. *Journal of Animal Ecology*, **75**, 595–603.

Masters, G.J. (1995a) The effect of herbivore density on host plant mediated interactions between two insects. *Ecological Research*, **10**, 125–33.

Masters, G.J. (1995b) The impact of root herbivory on aphid performance: field and laboratory evidence. *Acta Oecologia*, **16**, 135–42.

Masters, G.J. (1999) Upstairs-downstairs interactions: above- and below-ground insect herbivores. *Bulletin of the Royal Entomological Society*, **23**, 242–7.

Masters, G.J. & Brown, V.K. (1992) Plant-mediated interactions between two spatially separated insects. *Functional Ecology*, **6**, 175–9.

Masters, G.J. & Brown, V.K. (1997) Host-plant mediated interactions between spatially separated herbivores: effects on community structure. In *Multitrophic Interactions in Terrestrial Ecosystems. 36th Symposium of the British Ecological Society* (eds A.C. Gange & V.K. Brown), pp. 217–37. Blackwell Science, Oxford.

Masters, G.J., Jones, T.H. & Rogers, M. (2001) Host-plant mediated effects of root herbivory on insect seed predators and their parasitoids. *Oecologia*, **127**, 246–50.

Mathias, J.K., Ratcliffe, R.H. & Hellman, J.L. (1990) Association of an endophytic fungus in perennial ryegrass and resistance to the hairy chinch bug (Hemiptera. Lygaeidae). *Journal of Economic Entomology*, **83**, 1640–6.

Matsumoto, T., Itioka, T., Nishida, T. & Inoue, T. (2004) A test of temporal and spatial density dependence in the parasitism rates of introduced parasitoids on host, the arrowhead scale (*Unaspis yanonensis*) in stable host-parasitoids system. *Journal of Applied Entomology*, **128** (4), 267–72.

Matsumura, M., Trafelet-Smith, G.M., Gratton, C., Finke, D.L., Fagan, W.F. & Denno, R.F. (2004) Does intraguild predation enhance predator performance? A stoichiometric perspective. *Ecology*, **85** (9), 2601–15.

Matteson, P.C. (2000) Insect pest management in tropical Asian irrigated rice. *Annual Review of Entomology*, **45**, 549–74.

Matthews, G.A. (1992) *Pesticide Application Methods*, 2nd edn. Longman, Harlow, UK.

Matthews, G.A. & Friedrich, T. (2004) Sprayer quality in developing countries. *International Pest Control*, **46** (55), 254–8.

Matthews, G.A., Wiles, T. & Baleguel, P. (2003) A survey of pesticide application in Cameroon. *Crop Protection*, **22** (5), 707–14.

Matthews, R.W., Flage, L.R. & Matthews, J.R. (1997) Insects as teaching tools in primary and secondary education. *Annual Review of Entomology*, **42**, 269–90.

Mattson, W.J. Jr. (1980) Herbivory in relation to plant nitrogen content. *Annual Review Ecology and Systematics*, **11**, 119–61.

Mattson, W.J. & Addy, N.D. (1975) Phytophagous insects as regulators of forest primary production. *Science*, **190**, 515–22.

Maurer, P., Debieu, D., Malosse, C., Leroux, P. & Riba, G. (1992) Sterols and symbiosis in the leaf-cutting ant *Acromyrmex octospinosus* (Reich) (Hymenoptera, Formicidae: Attini). *Archives of Insect Biochemistry and Physiology*, **20** (1), 13–21.

Mavarez, J., Salazar, C.A., Bermingham, E., Salcedo, C., Jiggins, C.D. & Linares, M. (2006) Speciation by hybridization in *Heliconius* butterflies. *Nature*, **441** (7095), 868–71.

Mawdsley, N.A. & Stork, N.E. (1995) Species extinctions in insects: ecological and biogeographical considerations.

In *Insects in a Changing Environment* (eds R. Harrington & N.E. Stork), pp. 321–69. Academic Press, London.

Mawdsley, N.A. & Stork, N. (1997) Host-specificity and effective specialization of tropical canopy beetles. In *Canopy Arthropods* (eds N. Stork, J. Adis & R. Didham), pp. 104–130. Chapman & Hall, London.

Maxwell, M.R. (2000) Does a single meal affect female reproductive output in the sexually cannibalistic praying mantid *Iris oratoria*? *Ecological Entomology*, **25** (1), 54–62.

May, R.M. (1973) *Stability and Complexity in Model Ecosystems*. Princeton University Press, Princeton, NJ.

May, R.M. (1974) Biological populations with nonoverlapping generations: stable points, stable cycles and chaos. *Science*, **186**, 645–7.

May, R.M. (ed.) (1978) *Theoretical Ecology, Principles and Applications*, 2nd edn. Blackwell Scientific Publications, Oxford.

May, R.M. (1988) How many species are there on earth? *Science*, **241**, 1441–9.

May, R.M. (1990) How many species. *Philosophical Transactions of the Royal Society of London Series B, Biological Sciences*, **330**, 293–304.

May, R.M. (1992) How many species inhabit the earth? *Scientific American*, **267** (4), 18–24.

May, R.M., Lawton, J.H. & Stork, N.E. (1995) Assessing extinction rates. In *Extinction Rates* (eds J.H. Lawton & R.M. May), pp. 1–24. Oxford University Press, Oxford.

Mayhew, P.J. & Godfray, H.C.J. (1997) Mixed sex allocation strategies in a parasitoid wasp. *Oecologia*, **110**, 218–21.

Mayntz, D. & Toft, S.R. (2006) Nutritional value of cannibalism and the role of starvation and nutrient imbalance for cannibalistic tendencies in a generalist predator. *Journal of Animal Ecology*, **75** (1), 288–97.

Mbogo, C.M., Snow, R.W., Khamala, C.P.M. et al. (1995) Relationships between *Plasmodium falciparum* transmission by vector populations and the incidence of severe disease at nine sites on the Kenya coast. *American Journal of Tropical Medicine and Hygiene*, **52** (3), 201–6.

McAuslane, H.J., Johnson, F.A., Colvin, D.L. & Sojack, B. (1995) Influence of foliar pubescence on abundance and parasitism of *Bemisia agentifolia* (Homoptera: Aleyrodidae) on soybean and peanut. *Environmental Entomology*, **24** (5), 1135–43.

McCall, A.C. & Irwin, R. E. (2006) Florivory: the intersection of pollination and herbivory. *Ecology Letters*, **9**, 1351–65.

McClure, M.S. (1980) Competition between exotic species: scale insects on hemlock. *Ecology*, **61**, 1391–401.

McCornack, B.P., Carrillo, M.A., Venette, R.C. et al. (2005) Physiological constraints on the overwintering potential of the soybean aphid (Homoptera: Aphididae). *Environmental Entomology*, **34** (2), 235–40.

McGeoch, M.A. (1998) The selection, testing and application of terrestrial insects as bioindicators. *Biological Reviews*, **73**, 181–201.

McGill, N-G., Bade, G-S., Vitelli, R.A. & Allsopp, P.G. (2003) Imidacloprid can reduce the impact of the whitegrub *Antitrogus parvulus* on Australian sugarcane. *Crop Protection*, **22** (10), 1169–76.

McKenzie, C.L., Weathersbee, A.A., Hunter, W.B. et al. (2004) Sucrose octanoate toxicity to brown citrus aphid (Homoptera: Aphididae) and the parasitoid *Lysiphlebus testaceipes* (Hymenoptera: Aphidiidae). *Journal of Economic Entomology*, **97** (4), 1233–8.

McKirdy, S.J. & Jones, R.A.C. (1997) Effect of sowing time on barley yellow dwarf virus infection in wheat: virus incidence and grain yield losses. *Australian Journal of Agricultural Research*, **48** (2), 199–206.

McLain, D.K. (1991) The *r–K* continuum and the relative effectiveness of sexual selection. *Oikos*, **60** (2), 263–5.

McLaughlin, A. & Mineau, P. (1995) The impact of agricultural practices on biodiversity. *Agriculture, Ecosystems and Environment*, **55** (3), 201–12.

McManus, M.L. & Giese, R.L. (1968) The Columbia timber beetle, *Corthylus columbianus*. VII. The effect of climatic integrants on historic density fluctutations. *Forest Science*, **14**, 242–53.

McMichael, A.J. & Beers, M.Y. (1994) Climate change and human population health: global and South Australian perspectives. *Transactions of the Royal Society of South Australia*, **118** (1/2), 91–8.

McMillin, J.D. & Wagner, M.R. (1995) Season and intensity of water stress: host-plant effects on larval survival and fecundity of *Neodiprion gilletei* (Hymenoptera. Diprionidae). *Environmental Entomology*, **24**, 1251–7.

McNaughton, S.J., Ruess, R.W. & Seagle, S.W. (1988) Large mammals and process dynamics in African ecosystems: herbivorous mammals affect primary productivity and regulate recycling balances. *BioScience*, **38**, 794–800.

McNeil, J.N. & Quiring, D.T. (1983) Evidence of an oviposition-deterring pheromone in the alfalfa blotch leafminer, *Agromyza frontella* Rondani (Diptera: agromyzidae). *Environmental Entomology*, **12**, 990–2.

McNeill, S. & Southwood, T.R.E. (1978) The role of nitrogen in the development of insect–plant relationships. In *Biochemical Aspects of Plant and Animal Coevolution* (ed. J.B. Marborne), pp. 77–98. Academic Press, London.

McVean, R.I.K., Sait, S.M., Thompson, D.J. & Begon, M. (2002) Effects of resource quality on the population dynamics of the Indian meal moth *Plodia interpunctella* and its granulovirus. *Oecologia Berlin*, **131** (1), 71–8.

Medeiros, R.B., Resende, R.D. & De Avila, A.C. (2004) The plant virus tomato spotted wilt *Tospovirus* activates the immune system of its main insect vector, *Frankliniella occidentalis*. *Journal of Virology*, **78** (10), 4976–82.

Mehdiabadi, N.J., Hughes, B. & Muelle, U.G. (2006) Cooperation, conflict and coevolution in the attine ant–fungus symbiosis. *Behavioral Ecology*, **17** (2), 291–6.

Mehta, P.K. & Sandhu, G.S. (1992) Feeding behaviour of red pumpkin beetle, *Aulacophora fovecollis* (Lucas) on leaf extracts of different cucurbits. *Uttar Pradesh Journal of Zoology*, **12**, 87–94.

Mendis, C., Herath, P.R.J., Rajakaruna, J. et al. (1992) Method to estimate relative transmission efficiencies of *Anopheles* spp. (Diptera: Culicidae) in human malaria transmission. *Journal of Medical Entomology*, **29** (2), 188–96.

Mendoza-Aldana, J., Piechulek, H. & Maguire, J. (1997) Forest onchocerciasis in Cameroon: its distribution and implications for selection of communities for control programmes. *Annals of Tropical Medicine and Parasitology*, **91** (1), 79–86.

Merlin, J., Lemaitre, O. & Gregoire, J.C. (1996) Chemical cues produced by conspecific larvae deter oviposition by the coccidophagous ladybird beetle, *Cryptolaemus montrouzieri*. *Entomologia Experimentalis et Applicata*, **79**, 147–51.

Merritt, R.W. & Cummins, K.W. (eds) (1984) *An Introduction to the Aquatic Insects of North America*. Kendall/Hunt, Dubuque, IA.

Merritt, R.W., Eversham, B.C. & Moore, N.W. (1996) *Atlas of Dragonflies in Britain and Ireland*. HMSO, London.

Mesfin, T., Thottappilly, G. & Singh, S.R. (1992) Feeding behaviour of *Aphis craccivora* (Koch) on cowpea cultivars with different levels of aphid resistance. *Annals of Applied Biology*, **121** (3), 493–501.

Messing, R.H., Klungness, L.M. & Jang, E.B. (1997) Effects of wind on movement of *Diachasmimorpha longicaudata*, a parasitoid of tephritid fruit flies, in a laboratory flight tunnel. *Entomologia Experimentalis et Applicata*, **82** (2), 147–52.

Meyer, J.L. & O'Hop, J. (1983) Leaf-shredding insects as a source of dissolved organic carbon in headwater streams. *American Midland Naturalist*, **109**, 175–83.

Michelakis, S. (1973) *A study of the laboratory interactions between* Coccinella septempunctata *larvae and its prey* Myzus persicae. MSc Thesis, University of London.

Miliczky, E.R. & Horton, D.R. (2005) Densities of beneficial arthropods within pear and apple orchards affected by distance from adjacent native habitat and association of natural enemies with extra-orchard host plants. *Biological Control*, **33** (3), 249–59.

Millar, J.G., Paine, T.D., Joyce, A.L. & Hanks, L.M. (2003) The effects of eucalyptus pollen on longevity and fecundity of eucalyptus longhorned borers (Coleoptera: Cerambycidae). *Journal of Economic Entomology*, **96**, 370–6.

Millennium Ecosystem Assessment (2005) *Ecosystems and Human Well-being: Synthesis*. Island Press, Washington, DC.

Miller, K., Allegretti, M.H., Johnson, N. & Jonsson, B. (1995) Measures for conservation of biodiversity and sustainable use of its components. In *Global Biodiversity Assessment* (ed. V.H. Heywood), pp. 915–1061. Cambridge University Press, Cambridge, UK.

Miller, R.F. (1991) Chitin paleoecology. *Biochemical Systematics and Ecology*, **19** (5), 401–12.

Milli, R., Koch, U.T. & de Kramer, J.J. (1997) EAG measurement of pheromone distribution in apple orchards treated for mating disruption of *Cydia pomonella*. *Entomologia Experimentalis et Applicata*, **82** (3), 289–97.

Mills, A.P., Rutter, J.F. & Rosenberg, L.J. (1996) Weather associated with spring and summer migrations of rice pests and other insects in south-eastern and eastern Asia. *Bulletin of Entomological Research*, **86** (6), 683–94.

Mills, N.J. (1994) Parasitoid guilds: defining the structure of the parasitoid communities of endopterygoteinsect hosts. *Environmental Entomology*, **23** (5), 1066–83.

Mills, N.J. (2001) Factors influencing top-down control of insect pest populations in biological control systems. *Basic and Applied Ecology*, **2** (4), 323–32.

Milne, A. (1957a) The natural control of insect populations. *Canadian Entomologist*, **89**, 193–213.

Milne, A. (1957b) Theories of natural control of insect populations. *Cold Spring Harbor Symposia on Quantitative Biology*, **22**, 253–67.

Minckley, R.L., Wcislo, W.T., Yanega, D. & Buchmann, S.L. (1994) Behavior and phenology of a specialist bee (*Dieunomia*) and sunflower (*Helianthus*) pollen availability. *Ecology*, **75**, 1406–19.

Miner, A.L., Rosenberg, A.J. & Nijhout, H.F. (2000) Control of growth and differentiation of the wing imaginal disk of *Precis coenia* (Lepidoptera: Nymphalidae). *Journal of Insect Physiology*, **46**, 251–8.

Mira, A. (2000) Exuviae eating: a nitrogen meal? *Journal of Insect Physiology*, **46** (4), 605–10.

Mitchell, R.J., Runion, G.B., Kelley, W.D., Gjerstad, D.H. & Brewer, C.H. (1991) Factors associated with loblolly pine mortality on former agricultural sites in the Conservation Reserve Program. *Journal of Soil and Water Conservation*, **46**, 306–11.

Mitsuhashi, W., Fukuda, H., Nicho, K. & Murakami, R. (2004) Male-killing *Wolbachia* in the butterfly *Hypolimnas bolina*. *Entomologia Experimentalis et Applicata*, **112** (1), 57–64.

Mizuno, T. & Kajimura, H. (2002) Reproduction of the ambrosia beetle, *Xyleborus pfeili* (Ratzeburg) (Col., Scolytidae), on semi-artificial diet. *Journal of Applied Entomology, Zeitschrift Fur Angewandte Entomologie*, **126**, 455–62.

Moe, S.J., Stelzer, R.S., Forman, M.R., Harpole, W.S., Daufresne, T. & Yoshida, T. (2005) Recent advances in ecological stoichiometry: insights for population and community ecology. *Oikos*, **109**, 29–39.

Mogi, M. (1969) Predation response of the larvae of *Harmonia axyridis* Pallas (Coccinellidae) to the different prey density. *Japanese Journal of Applied Entomology and Zoology*, **13**, 9–16.

Mogi, M. & Sembel, D.T. (1996) Predator–prey system structure in patchy and ephemeral phytotelmata – aquatic communities in small aroid axils. *Researches on Population Ecology*, **38**, 95–103.

Mohandass, S.M., Arthur, F.H., Zhua, K.Y. & Throne, J.E. (2006) Hydroprene: mode of action, current status in stored-product pest management, insect resistance, and future prospect. *Crop Protection*, **25** (9), 902–9.

Molau, U. (2004) Mountain biodiversity patterns at low and high latitudes. *Ambio*, Suppl. 13, 24–8.

Moldovan, O-T., Jalzic, B. & Erichsen, E. (2004) Adaptation of the mouthparts in some subterranean Cholevinae (Coleoptera, Leiodidae). *Natura Croatica*, **13** (1), 1–18.

Molina-Montenegro, M.A., Avila, P., Hurtado, R., Valdivia, A.I. & Gianoli, E. (2006) Leaf trichome density may explain herbivory patterns of *Actinote* sp (Lepidoptera: Acraeidae) on *Liabum mandonii* (Asteraceae) in a montane humid forest (Nor Yungas, Bolivia). *Acta Oecologica, International Journal of Ecology*, **30**, 147–50.

Moloo, S.K., Kutuza, S.B. & Desai, J. (1988) Infection rates in sterile males of *morsitans, palpalis* and *fusca* groups *Glossina* for pathogenic *Trypanosoma* spp. from East and West Africa. *Acta Tropica*, **45** (2), 145–52.

Molyneux, D.H. (2005) Onchocerciasis control and elimination: coming of age in resource-constrained health systems. *Trends in Parasitology*, **21** (11), 525–9.

Mommaerts, V., Sterk, G. & Smagghe, G. (2006) Hazards and uptake of chitin synthesis inhibitors in bumblebees *Bombus terrestris*. *Pest Management Science*, **62** (8), 752–8.

Monge, J.P., Dupont, P., Idi, A. & Huignard, J. (1995) The consequences of interspecific competition between *Dinarmus basalis* (rond) (Hymenoptera, Pteromalidae) and *Eupelmusvuiletti* (crw) (Hymenoptera, Pteromalidae) on the development of their host population. *Acta Oecologica International Journal of Ecologica*, **16**, 19–30.

Montgomery, M.E. & Arn, H. (1974) Feeding response of *Aphis pomi, Myzus persicae*, and *Amphorophora agathonica* to phlorizin. *Journal of Insect Physiology*, **20**, 413–21.

Moon, A.M., Dyer, M.I., Brown, M.R. & Crossley, D.A. Jr. (1994) Epidermal growth factor interacts with indole-3-acetic acid and promotes coleoptile growth. *Plant and Cell Physiology*, **35**, 1173–7.

Moon, D.C., Rossi, A.M. & Stiling, P. (2000) The effects of abiotically induced changes in host plant quality (and morphology) on a salt marsh planthopper and its parasitoid. *Ecological Entomology*, **25**, 325–31.

Moon, D.C. & Stiling, P. (2000) Relative importance of abiotically induced direct and indirect effects on a salt-marsh herbivore. *Ecology*, **81**, 470–81.

Moorhead, D.L. & Reynolds, J.F. (1991) A general model of litter decomposition in the northern Chihuahuan Desert. *Ecological Modeling*, **56**, 197–220.

Mopper, S. (1996) Adaptive genetic structure in phytophagus insect populations. *Trends in Ecology and Evolution*, **11**, 235–8.

Mopper, S., Beck, M., Simberloff, D. & Stiling, P. (1995) Local adaptation and agents of selection in a mobile insect. *Evolution*, **49**, 810–15.

Mopper, S., Mitton, J.B., Whitham, T.G., Cobb, N.S. & Christiansen, K.M. (1991) Genetic differentiation and heterozygosity in pinyon pine associated with resistance to herbivory and environmental stress. *Evolution*, **45**, 989–99.

Mopper, S., Stiling, P., Landau, K., Simberloff, D. & VanZandt, P. (2000) Spatiotemporal variation in leafminer population structure and adaptation to individual oak trees. *Ecology*, **81**, 1577–87.

Mopper, S. & Strauss, S.Y. (eds) (1998) *Genetic Structure and Local Adaptation in Natural Insect Populations. Effects of Ecology, Life History, and Behavior*. Chapman & Hall, New York.

Mopper, S. & Whitham, T.G. (1992) The plant stress paradox: effects on pinyon sawfly sex ratios and fecundity. *Ecology*, **73**, 515–25.

Moran, M.D., Rooney, T.P. & Hurd, L.E. (1996) Top-down cascade from a bitrophic predator in an old-field community. *Ecology*, **77**, 2219–27.

Moran, N.A. & Whitham, T.G. (1990) Interspecific competition between root-feeding and leaf-galling aphids mediated by host-plant resistance. *Ecology*, **71**, 1050–8.

Moran, P.J. & Schultz, J.C. (1998) Ecological and chemical associations among late-season squash pests. *Environmental Entomology*, **27**, 39–44.

Moran, V.C. & Southwood, T.R.E. (1982) The guild composition of arthropod communities in trees. *Journal of Animal Ecology*, **51**, 289–306.

Moreau, G., Eveleigh, E.S., Lucarotti, C.J. & Quiring, D.T. (2006) Stage-specific responses to ecosystem alteration in an eruptive herbivorous insect. *Journal of Applied Ecology*, **43**, 28–34.

Moreau, J. & Rigaud, T. (2003) Variable male potential rate of reproduction: high male mating capacity as an adaptation to parasite-induced excess of females? *Proceedings of the Royal Society of London series B, Biological Sciences*, **270** (1523), 1535–40.

Morewood, P., Gries, G., Liska, J. et al. (2000) Towards pheromone-based monitoring of nun moth, *Lymantria monacha* (L.) (Lep., Lymantriidae) populations. *Journal of Applied Entomology, Zeitschrift Fur Angewandte Entomologie*, **124**, 77–85.

Mori, M.N., Oikawa, H., Sampa, M.H.O. & Duarte, C.L. (2006) Degradation of chlorpyrifos by ionizing radiation. *Journal of Radioanalytical and Nuclear Chemistry*, **270** (1), 99–102.

Morimoto, S., Imamura, T., Visarathanonth, P. & Miyanoshita, A. (2007) Effects of temperature on the development and reproduction of the predatory bug *Joppeicus paradoxus* Puton (Hemiptera: Joppeicidae) reared on *Tribolium confusum* eggs. *Biological Control*, **40**, 136–41.

Morin, P.J. (1999) *Community Ecology*. Blackwell Science, Malden, MA.

Morisawa, S., Kato, A., Yoneda, M. & Shimada, Y. (2002) The dynamic performances of DDTs in the environment and Japanese exposure to them: a historical perspective after the ban. *Risk Analysis*, **22** (2), 245–63.

Morris, W.F. (1992) The effects of natural enemies, competition and host plant water availability on an aphid population. *Oecologia*, **90**, 359–65.

Morse, G.E. & Farrell, B.D. (2005) Ecological and evolutionary diversification of the seed beetle genus *Stator* (Coleoptera: Chrysomelidae: Bruchinae). *Evolution*, **59**, 1315–33.

Morse, S.S. (1994) The viruses of the future? Emerging viruses and evolution. In *The Evolutionary Biology of Viruses* (ed. S.S. Morse), pp. 325–35. Raven Press, New York.

Moscardi, F. (1999) Assessment of the application of baculoviruses for control of Lepidoptera. *Annual Review of Entomology*, **44**, 257–89.

Mouquet, N., Belrose, V., Thomas, J.A., Elmes, G.W., Clarke, R.T. & Hochberg, M.E. (2005a) Conserving community modules: a case study of the endangered lycaenid butterfly *Maculinea alcon*. *Ecology*, **86**, 3160–73.

Mouquet, N., Thomas, J.A., Elmes, G.W., Clarke, R.T. & Hochberg, M.E. (2005b) Population dynamics and conservation of a specialized predator: a case study of *Maculinea arion*. *Ecological Monographs*, **75**, 525–42.

Moura, R., Garcia, P., Cabral, S. & Soares, A.O. (2006) Does pirimicarb affect the voracity of the euriphagous predator, *Coccinella undecimpunctata* L. (Coleoptera: Coccinellidae)? *Biological Control*, **38**, 363–8.

Mousson, L., Neve, G. & Baguette, M. (1999) Metapopulation structure and conservation of the cranberry fritillary *Boloria aquilonaris* (lepidoptera, nymphalidae) in Belgium. *Biological Conservation*, **87**, 285–93.

Mueller, U.G., Gerardo, N.M., Aanen, D.K. et al. (2005) The evolution of agriculture in insects. *Annual Review of Ecology, Evolution and Systematics*, **36**, 563–95.

Muenster-Swendsen, M. (1987) The effect of precipitation on radial increment in Norway spruce (*Picea abies* Karst.) and on the dynamics of a lepidopteran pest insect. *Journal of Applied Ecology*, **24**, 563–72.

Mulder, C.P.H. (1999) Vertebrate herbivores and plants in the Arctic and subarctic: effects on individuals, populations, communities and ecosystems. *Perspectives in Plant Ecology, Evolution and Systematics*, **2**, 29–55.

Mulder, P.G, Reid, W., Grantham, R.A. et al. (2003) Evaluations of trap designs and a pheromone formulation used for monitoring pecan weevil, *Curculio caryae*. *Southwestern Entomologist*, **27** (Suppl.), 85–99.

Mulla, M-S., Thavara, U., Tawatsin, A. & Chompoosri, J. (2004) Procedures for the evaluation of field efficacy of slow-release formulations of larvicides against *Aedes aegypti* in water-storage containers. *Journal of the American Mosquito Control Association*, **20** (1), 64–73.

Müller-Schärer, H. & Brown, V.K. (1995) Direct and indirect effects of aboveground and belowground insect herbivory on plant-density and performance of *Tripleurospermum perforatum* during early plant succession. *Oikos*, **72**, 36–41.

Mulrooney, J.E., Howard, K.D., Hanks, J.E. & Jones, R.J. (1997) Application of ultra-low-volume malathion by air assisted ground sprayer for boll weevil (Coleoptera: Curculionidae) control. *Journal of Economic Entomology*, **90** (2), 639–45.

Munck, I.A. & Manion, P.D. (2006) Landscape-level impact of beech bark disease in relation to slope and aspect in New York State. *Forest Science*, **52** (5), 503–10.

Munoz de Toro, M., Beldomenico, H.R., Garcia, S.R. et al. (2006) Organochlorine levels in adipose tissue of women from a littoral region of Argentina. *Environmental Research*, **102** (1), 107–12.

Muralidharan, K. & Pasalu, I.C. (2006) Assessments of crop losses in rice ecosystems due to stem borer damage (Lepidoptera: Pyralidae). *Crop Protection*, **25** (5), 409–17.

Murdoch, W.W., Briggs, C.J. & Nisbet, R.M. (1996) Competitive displacement and biological control in parasitoids: a model. *American Naturalist*, **148** (5), 807–26.

Murdoch, W.W., Briggs, C.J. & Swarbrick, S. (2005) Host suppression and stability in a parasitoid–host system: experimental demonstration. *Science*, **309** (5734), 610–13.

Murdoch, W.W., Luck, R.F., Swarbrick, S.L., Walde, S., Yu, D.S. & Reeve, J.D. (1995) Regulation of an insect population under biological control. *Ecology (Washington DC)*, **76** (1), 206–17.

Murdoch, W.W., Swarbrick, S.L. & Briggs, C.J. (2006) Biological control: lessons from a study of California red scale. *Population Ecology*, **48**, 297–305.

Murray, M.D. (1995) Influences of vector biology on transmission of arboviruses and outbreaks of disease: the *Culicoides brevitarsus* model. *Veterinary Microbiology*, **46** (1–3), 91–9.

Murray, P.J., Hatch, D.J. & Cliquet, J.B. (1996) Impact of insect herbivory on the growth and nitrogen and carbon contents of white clover (*Trifolium repens*) seedlings. *Canadian Journal of Botany*, **74**, 1591–5.

Musser, R.O., Hum-Musser, S.M., Eichenseer, H. et al. (2002) Herbivory: caterpillar saliva beats plant defences – a new weapon emerges in the evolutionary arms race between plants and herbivores. *Nature*, **416**, 599–600.

Mutinga, M.J., Kamau, C.C., Basimike, M., Mutero, C.M. & Kyai, F.M. (1992) Studies on the epidemiology of leishmaniasis in Kenya: flight range of phlebotomine sandflies in Marigat, Baringo District. *East African Medical Journal*, **69** (1), 9–13.

Muzika, R-M. & Pregitzer, K.S. (1992) Effect of nitrogen fertilization on leaf phenolic production of grand fir seedlings. *Trees*, **6**, 241–4.

Myers, C.T., Hull, L.A. & Krawczyk, G. (2007) Effects of orchard host plants (apple and peach) on development of oriental fruit moth (Lepidoptera: Tortricidae). *Journal of Economic Entomology*, **100** (2), 421–30.

Myers, J.H., Savoie, A. & van Randen, E. (1998) Eradication and pest management. *Annual Review of Entomology*, **43**, 471–91.

Myers, N. (1988) Threatened biotas: 'hot spots' in tropical forests. *Environmentalist*, **8**, 187–208.

Myers, N. (1990) The biodiversity challenge: expanded hotspots analysis. *Environmentalist*, **19**, 243–56.

Myers, N. (1997) The rich diversity of biodiversity issues. In *Biodiversity II, Understanding and Protecting our Biological Resources* (eds M.L. Reaka-Kudla, D.E. Wilson & E.O. Wilson), pp. 125–38. Joseph Henry Press, Washington, DC.

Myers, N. (2003) Biodiversity hotspots revisited. *Bioscience*, **53**, 916–17.

Myers, N. & Mittermeier, R.A. (2003) Impact and acceptance of the hotspots strategy: response to Ovadia and to Brummitt and Lughadha. *Conservation Biology*, **17**, 1449–50.

Myers, N., Mittermeier, R.A., Mittermeier, C.G., da Fonseca, G.A.B. & Kent, J. (2000) Biodiversity hotspots for conservation priorities. *Nature*, **403**, 853–8.

Nagadhara, D., Ramesh, S., Pasalu, I.C. et al. (2003) Transgenic indica rice resistant to sap-sucking insects. *Plant Biotechnology Journal*, **1** (3), 231–40.

Naidu, S.G. (2001) Water balance and osmoregulation in *Stenocara gracilipes*, a wax-blooming tenebrionid beetle from the Namib Desert. *Journal of Insect Physiology*, **47**, 1429–40.

Nair, K.S.S. & Sudheendrakumar, V.V. (1986) The teak defoliator, *Hyblaea puera*: defoliation dynamics and evidences of short range migration of moths. *Proceedings of the Indian Academy of Sciences, Animal Sciences*, **95** (1), 7–22.

Nair, K.S.S. & Varma, R.V. (1985) Some ecological aspects of the termite problem in young eucalypt plantations in Kerala, India. *Forest Ecology and Management*, **12**, 287–304.

Nakai, M. & Kunimi, Y. (1997) Granulosis-virus infection of the small tea tortrix (Lepidoptera, Tortricidae) – effect on the development of the endoparasitoid, *Ascogaster reticulatus* (Hymenoptera, braconidae). *Biological Control*, **8** (1), 74–80.

Nakajima, S., Kitamura, T., Baba, N., Iwasa, J. & Ichikawa, T. (1995) Oleuropein, a secoiridoid glucoside from olive, as a feeding stimulant to the olive weevil (*Dyscerus perforatus*). *Bioscience, Biotechnology and Biochemistry*, **59**, 769–70.

Nakamura, M. & Ohgushi, T. (2003) Positive and negative effects of leaf shelters on herbivorous insects: linking multiple herbivore species on a willow. *Oecologia*, **136**, 445–9.

Nakanishi, H. (1994) Myrmecochorous adaptations of *Corydalis* species (papaveraceae) in southern Japan. *Ecological Research*, **9**, 1–8.

Naranjo, S.E. & Ellsworth, P.C. (2005) Mortality dynamics and population regulation in *Bemisia tabaci*. *Entomologia Experimentalis et Applicata*, **116** (2), 93–108.

Narberhaus, I., Zintgraf, V. & Dobler, S. (2005) Pyrrolizidine alkaloids on three trophic levels – evidence for toxic and deterrent effects on phytophages and predators. *Chemoecology*, **15**, 121–5.

Nauen, R. & Elbert, A. (1997) Apparent tolerance of a field-collected strain of *Myzus nicotianae* to imidacloprid due to strong antifeedant responses. *Pesticide Science*, **49** (3), 252–8.

Nault, B.A. & Speese, J. III (2002) Major insect pests and economics of fresh-market tomato in eastern Virginia. *Crop Protection*, **21** (5), 359–66.

Naves, P.M., De Sousa, E.M. & Quartau, J.A. (2006) Feeding and oviposition preferences of *Monochamus galloprovincialis* for certain conifers under laboratory conditions. *Entomologia Experimentalis et Applicata*, **120**, (2), 99–104.

Neal, J.W. Jr., Buta, J.G., Pitarelli, G.W., Lusby, W.R. & Bentz, J.A. (1994) Novel sucrose esters from *Nicotiana gossei*.

Effective biorationals against selected horticultural insect pests. *Journal of Economic Entomology*, **87**, 1600–7.

Nealis, V.G., Oliver, D. & Tchir, D. (1996) The diapause response to photoperiod in Ontario populations of *Cotesia melanoscela* (Ratzeburg) (Hymenoptera: Braconidae). *Canadian Entomologist*, **128** (1), 41–6.

Nebeker, T.E., Schmitz, R.F., Tisdale, R.A. & Hobson, K.R. (1995) Chemical and nutritional status of dwarf mistletoe, *Amarilla roor* rot, and *Comandra* blister rust infected trees which may influence tree susceptability to bark beetle attack. *Canadian Journal of Botany*, **73**, 360–9.

Nelson, D.R. & Charlet, L.D. (2003) Cuticular hydrocarbons of the sunflower beetle, *Zygogramma exclamationis*. *Comparative Biochemistry and Physiology Part B, Biochemistry and Molecular Biology*, **135B** (2), 273–84.

Nespolo, R.F., Castaneda, L.E. & Roff, D.A. (2005) Dissecting the variance-covariance structure in insect physiology: the multivariate association between metabolism and morphology in the nymphs of the sand cricket (*Gryllus firmus*). *Journal of Insect Physiology*, **51** (8), 913–21.

Ness, J.H. (2003) *Catalpa bignonioides* alters extrafloral nectar production after herbivory and attracts ant bodyguards. *Oecologia*, **134**, 210–18.

Ness, J.H. & Bronstein I.L. (2004) The effects of invasive ants on prospective ant mutualists. *Biological Invasions*, **6**, 445–61.

Ness, J.H., Bronstein J.L., Andersen A.N. & Holland J.N. (2004) Ant body size predicts dispersal distance of ant-adapted seeds: implications of small-ant invasions. *Ecology*, **85**, 1244–50.

Netherer, S. & Nopp-Mayr, U. (2005) Predisposition assessment systems (PAS) as supportive tools in forest management – rating of site and stand-related hazards of bark beetle infestation in the High Tatra Mountains as an example for system application and verification. *Forest Ecology and Management*, **207** (1/2), 99–107.

Neuenschwander, P. (2001) Biological control of the cassava mealybug in Africa: a review. *Biological Control*, **21** (3), 214–29.

Neuenschwander, P., Hammond, W.N.O., Ajuonu, O. et al. (1990) Biological control of the cassava mealybug, *Phenacoccus manihoti* (Homoptera: Pseudococcidae) by *Epidinocarsis lopezi* (Hymenoptera: Encrytidae), in West Africa, as influenced by climate and soil. *Agriculture, Ecosystems and Environment*, **32** (1–2), 39–56.

Neuenschwander, P., Hammond, W.N.O., Gutierrez, A.P. et al. (1989) Impact assessment of the biological control of the cassava mealybug, *Phenacoccus manihoti* (Homoptera: Pseudococcidae) by the introduced parasitoid, *Epidinocarsis lopezi* (Hymenoptera: Encrytidae). *Bulletin of Entomological Research*, **79** (4), 579–94.

New, T.R. (1997) Are Lepidoptera an effective umbrella group for biodiversity conservation? *Journal of Insect Conservation*, **1**, 5–12.

Newman, D.A. & Thomson, J.D. (2005) Effects of nectar robbing on nectar dynamics and bumblebee foraging strategies in *Linaria vulgaris* (Scrophulariaceae). *Oikos*, **110** (2), 309–20.

Newton, C. & Dixon, A.F.G. (1990) Embryonic growth rate and birth weight of the offspring of apterous and alate aphids: a cost of dispersal. *Entomologia Experimentalis et Applicata*, **55** (3), 223–30.

Newton, S.F. & Newton, A.V. (1997) The effect of rainfall and habitat on abundance and diversity of birds in a fenced protected area in the central Saudi Arabian desert. *Journal of Arid Environments*, **35** (4), 715–35.

Ng, J.C.K. & Falk, B.W. (2006) Virus-vector interactions mediating nonpersistent and semipersistent transmission of plant viruses. *Annual Review of Phytopathology*, **44**, 183–212.

Nicholson, A.J. (1933) The balance of animal populations. *Journal of Animal Ecology*, **2**, 131–78.

Nicholson, A.J. (1957) The self-adjustment of populations to change. *Cold Spring Harbor Symposia on Quantitative Biology*, **22**, 153–72.

Nicholson, A.J. (1958) Dynamics of insect populations. *Annual Review of Entomology*, **3**, 107–36.

Nicholson, A.J. & Bailey, V.A. (1935) The balance of animal populations. *Proceedings of the Zoological Society of London, Part 1*, **3**, 551–98.

Nicholson, B.D., Walley, J.D. & Baguley, D.S. (2006) Leishmaniasis, Chagas disease and human African trypanosomiasis revisited: disease control priorities in developing countries. *Tropical Medicine and International Health*, **11** (9), 1339–40.

Nickel, H. & Hildebrandt, J. (2003) Auchenorrhyncha communities as indicators of disturbance in grasslands (Insecta, Hemiptera) – a case study from the Elbe flood plains (northern Germany). *Agriculture Ecosystems and Environment*, **98** (1–3), 183–99.

Nieh, J.C., Leon, A., Cameron, S. et al. (2006) Hot bumble bees at good food: thoracic temperature of feeding *Bombus wilmattae* foragers is tuned to sugar concentration. *Journal of Experimental Biology*, **209** (21), 4185–92.

Niemelä, J. (1996) From systematics to conservation – carabidologists do it all. Preface. *Annales Zoologici Fennici*, **33**, 1–4.

Niemelä, J. (2001) Carabid beetles (Coleoptera: Carabidae) and habitat fragmentation: a review. *European Journal of Entomology*, **98**, 127–32.

Niemelä, J., Kotze, D.J., Venn, S. et al. (2002) Carabid beetle assemblages (Coleoptera, Carabidae) across urban–rural gradients: an international comparison. *Landscape Ecology*, **17**, 387–401.

Niemelä, J., Langor, D. & Spence, J.R. (1993) Effects of clearcut harvesting on boreal ground-beetle assemblages (Coleoptera, Carabidae) in western Canada. *Conservation Biology*, **7**, 551–61.

Niemälä, P., Tuomi, J., Sorjonen, J., Hokkanen, T. & Neuvonen, S. (1982) The influence of host plant growth form

a phenology on the life strategies of Finnish macrolepidopterous larvae. *Oikos*, **39**, 164–70.

Niemelä, J., Young, J., Alard, D. et al. (2005) Identifying, managing and monitoring conflicts between forest biodiversity conservation and other human interests in Europe. *Forest Policy and Economics*, **7**, 877–90.

Nishida, H. & Hayashi, N. (1996) Cretaceous coleopteran larva fed on a female fructification of extinct gymnosperm. *Journal of Plant Research*, **109** (1095), 327–30.

Nishida, R. (1994) Oviposition stimulant of a Zeryntiine swallowtail butterfly, *Luehdorfia japonica*. *Phytochemistry (Oxford)*, **36**, 873–7.

Njiwa, J.R.K., Muller, P. Klein, R. (2004) Life cycle stages and length of zebrafish (*Danio rerio*) exposed to DDT. *Journal of Health Science*, **50** (3), 220–5.

Noda, H. & Kodama, K. (1996) Phylogenetic position of yeastlike endosymbionts of anobiid beetles. *Applied and Environmental Microbiology*, **62** (1), 162–7.

Nola, P., Morales, M., Motta, R. & Villalba, R. (2006) The role of larch budmoth (*Zeiraphera diniana* Gn.) on forest succession in a larch (*Larix decidua* Mill.) and Swiss stone pine (*Pinus cembra* L.) stand in the Susa Valley (Piedmont, Italy). *Trees, Structure and Function*, **20** (3), 371–82.

Nomura, M., Itioka, T. & Itino, T. (2000) Variations in abiotic defense within myrmecophytic and non-myrmecophytic species of *Macaranga* in a Bornean dipterocarp forest. *Ecological Research*, **15**, 1–11.

Norgaard, R.B. (1988) The biological-control of cassava mealybug in Africa. *American Journal of Agricultural Economics*, **70**, 366–71.

Norton, A.P. & Welter, S.C. (1996) Augmentation of the egg parasitoid *Anaphes iole* (Hymenoptera: Mymaridae) for *Lygus hesperus* (Heteroptera: Miridae) management in strawberries. *Environmental Entomology*, **25** (6), 1406–14.

Noss, R.F. (1990) Indicators for monitoring biodiversity: a hierarchical approach. *Conservation Biology*, **4**, 355–64.

Novotny, J. (2004) Community involvement in forest protection in Slovakia. *Unasylva*, **55** (217), 26–8.

Novotny, V. & Basset, Y. (2005) Review – host specificity of insect herbivores in tropical forests. *Proceedings of the Royal Society of London Series B, Biological Sciences*, **272**, 1083–90.

Novotny, V., Basset, Y., Miller, S.E. et al. (2004) Local species richness of leaf-chewing insects feeding on woody plants from one hectare of a lowland rainforest. *Conservation Biology*, **18**, 227–37.

Novotny, V., Basset, Y., Miller, S.E. et al. (2002) Low host specificity of herbivorous insects in a tropical forest. *Nature*, **416**, 841–4.

Novotny, V., Drozd, P., Miller, S.E. et al. (2006) Why are there so many species of herbivorous insects in tropical rainforests? *Science*, **313**, 1115–18.

Noyes, J.S. (1974) *The biology of the leek moth* Acrolepia assectella (Zeller). PhD Thesis, University of London.

Nozawa, A.K. & Ohgushi, T. (2002) Indirect effects mediated by compensatory shoot growth on subsequent generations of a willow spittlebug. *Population Ecology*, **44**, 235–9.

Nuckols, M.S. & Connor, E.F. (1995) Do trees in urban or ornamental plantings receive more damage by insects than trees in natural forests? *Ecological Entomology*, **20**, 253–60.

Nummelin, M., Lodenius, M., Tulisalo, E., Hirvonen, H. & Alanko, T. (2007) Predatory insects as bioindicators of heavy metal pollution. *Environmental Pollution*, **145** (1), 339–47.

Nyarango, P.M., Gebremeskel, T., Mebrahtu, G. et al. (2006) A steep decline of malaria morbidity and mortality trends in Eritrea between 2000 and 2004: the effect of combination of control methods. *Malaria Journal*, **5**, Article number 33.

Nylin, S. & Gotthard, K. (1998) Plasticity in life-history traits. *Annual Review of Entomology*, **43**, 63–83.

Nyman, T. & Julkunen-Tiitto, R. (2000) Manipulation of the phenolic chemistry of willows by gall-inducing sawflies. *Proceedings of the National Academy of Sciences of the USA*, **97**, 13184–7.

Oba, G. (1994) Responses of *Digofera spinosa* to simulated herbivory in a semidesert of north-west Kenya. *Acta Oecologia*, **15**, 105–17.

Oberrath, R. & Bohning-Gaese, K. (2002) Phenological adaptation of ant-dispersed plants to seasonal variation in ant activity. *Ecology*, **83**, 1412–20.

O'Callaghan, M., Glare, T.R., Burgess, E.P.J. & Malone, L.A. (2005) Effects of plants genetically modified for insect resistance on nontarget organisms. *Annual Review of Entomology*, **50**, 271–92.

Öckinger, E. & Smith, H.G. (2006) Landscape composition and habitat area affects butterfly species richness in semi-natural grasslands. *Oecologia*, **149**, 526–34.

Ockroy, M.L.B., Turlings, T.C.J., Edwards, P.J. et al. (2001) Response of natural populations of predators and parasitoids to artificially induced volatile emissions in maize plants (*Zea mays* L.). *Agricultural and Forest Entomology*, **3**, 201–9.

O'Connor, J.P. (1982) *Microchironomus deribae* (Freeman) (Dipt., Chironomidae): a fuel contaminant in an Irish helicopter. *Entomologist's Monthly Magazine*, **118**, 44.

Ødegaard, F. (2000a) How many species of arthropods? Erwin's estimate revised. *Biological Journal of the Linnean Society*, **71**, 583–97.

Ødegaard, F. (2000b) The relative importance of trees versus lianas as hosts for phytophagous beetles (Coleoptera) in tropical forests. *Journal of Biogeography*, **27**, 283–96.

O'Dowd, D.J., Green, P.T. & Lake, P.S. (2003) Invasional 'meltdown' on an oceanic island. *Ecology Letters*, **6**, 812–17.

Odum, E.P. (1969) The strategy of ecosystem development. *Science*, 262–70.

Oehlschlager, A.C., McDonald, R.S., Chinchilla, C.M. & Patschke, S.N. (1995) Influence of a pheromone-based mass-trapping system on the distribution of *Rhynchophorus palmarum* (Coleoptera: Curculionidae) in oil palm. *Environmental Entomology*, **24** (5), 1005–12.

Oerke, E.C. (2006) Crop losses to pests. *Journal of Agricultural Science*, **144**, 31–43.

Oertli, S., Muller, A., Steiner, D., Breitenstein, A. & Dorn, S. (2005) Cross-taxon congruence of species diversity and community similarity among three insect taxa in a mosaic landscape. *Biological Conservation*, **126**, 195–205.

Ogden, T.H. & Whiting, M.F. (2005) Phylogeny of Ephemeroptera (mayflies) based on molecular evidence. *Molecular Phylogenetics and Evolution*, **37**, 625–43.

Ogol, C.K.P.O., Spence, J.R. & Keddie, A. (1999) Maize stem borer colonization, establishment and crop damage levels in a maize–leucaena agroforestry system in Kenya. *Agriculture Ecosystems and Environment*, **76** (1), 1–15.

Ohba, N. (2003) Ecological notes and external morphology of four species of marine insects. *Science Report of the Yokosuka City Museum*, **50**.

Ohgushi, T. (1992) Resource limitation on insect herbivore populations. In *Effects of Resource Distribution on Animal–Plant Interactions* (eds M.D Hunter, T. Ohgushi, & P.W Price), pp. 199–241. Academic Press, San Diego.

Ohgushi, T. & Sawada, H. (1985) Population equilibrium with respect to available food resource and its behavioural basis in an herbivorous lady beetle *Henosepilachna niponica*. *Journal of Animal Ecology*, **54**, 781–96.

Ojeda-Avila, T., Woods, H.A. & Raguso, R.A. (2003) Effects of dietary variation on growth, composition, and maturation of *Manduca sexta* (Sphingidae: Lepidoptera). *Journal of Insect Physiology*, **49**, 293–306.

Okland, B. & Bjornstad, O.N. (2006) A resource-depletion model of forest insect outbreaks. *Ecology*, **87** (2), 283–90.

Okoth, J.O., Okethi, V. & Ogola, A. (1991) Control of tsetse trypanosomiasis transmission in Uganda by applications of lambda-cyhalothrin. *Medical and Veterinary Entomology*, **5** (1), 121–8.

Olafsson, E., Carlstrom, S. & Ndaro, S.G.M. (2000) Meiobenthos of hypersaline tropical mangrove sediment in relation to spring tide inundation. *Hydrobiologia*, **426** (1–3), 57–64.

Olaleye, O.D., Tomori, O. & Schmitz, H. (1996) Rift Valley fever in Nigeria: infections in domestic animals. *Revue Scientifique et Technique, Office International des Epizooties*, **15** (3), 937–46.

Oliveira, E.E., Guedes, R.N.C., Correa, A.S., Damasceno, B.L. & Santos, C.T. (2005) Pyrethroid resistance vs susceptibility in *Sitophilus zeamais* Motschulsky (Coleoptera: Curculionidae): is there a winner? *Neotropical Entomology*, **34** (6), 981–90.

Oliver, I. & Beattie, A.J. (1993) A possible method for the rapid assessment of biodiversity. *Conservation Biology*, **7**, 562–8.

Oliver, I. & Beattie, A.J. (1996a) Designing a cost-effective invertebrate survey – a test of methods for rapid assessment of biodiversity. *Ecological Applications*, **6**, 594–607.

Oliver, I. & Beattie, A.J. (1996b) Invertebrate morphospecies as surrogates for species – a case-study. *Conservation Biology*, **10**, 99–109.

Oliver, I., Beattie, A.J. & York, A. (1998) Spatial fidelity of plant, vertebrate, and invertebrate assemblages in multiple-use forest in eastern Australia. *Conservation Biology*, **12**, 822–35.

Omkar & Pervez, A. (2004) Temperature-dependent development and immature survival of an aphidophagous ladybeetle, *Propylea dissecta* (Mulsant). *Journal of Applied Entomology*, **128**, 510–14.

Opler, P.A. (1974) Oaks as evolutionary islands for leaf-mining insects. *American Scientist*, **62**, 67–73.

Opler, P.A., Pavulaan, H. & Stanford, R.E. (1995) *Butterflies of North America*. Northern Prairie Wildlife Research Center, Jamestown, ND.

Orme, C.D.L., Davies, R.G., Burgess, M. et al. (2005) Global hotspots of species richness are not congruent with endemism or threat. *Nature*, **436**, 1016–19.

Orshan, L., Szekely, D., Schnur, H. et al. (2006) Attempts to control sand flies by insecticide-sprayed strips along the periphery of a village. *Journal of Vector Ecology*, **31** (1), 113–17.

Ortega, G.M. & Oliver, C.M. (1990) Entomology of onchocerciasis in the Soconusgo area, Chiapas (Mexico): IV. Altitudinal distribution and seasonal variation of immature forms of the three simuliid species considered as vectors. *Folia Entomologica Mexicana*, **79**, 175–96.

Osier, T.L. & Lindroth, R.L. (2006) Genotype and environment determine allocation to and costs of resistance in quaking aspen. *Oecologia*, **148**, 293–303.

Otuka, A., Dudhia, J., Watanabe, T. & Furuno, A. (2005) A new trajectory analysis method for migratory planthoppers, *Sogatella furcifera* (Horvath) (Homoptera: Delphacidae) and *Nilaparvata lugens* (Stal), using an advanced weather forecast model. *Agricultural and Forest Entomology*, **7** (1), 1–9.

Otuka, A., Watanabe, T., Suzuki, Y., Matsumura, M., Furuno, A. & Chino, M. (2005) A migration analysis of the rice planthopper *Nilaparvata lugens* from the Philippines to East Asia with three-dimensional computer simulations. *Population Ecology*, **47** (2), 143–50.

Owen, D.F. (1980) *Camouflage and Mimicry*. Oxford University Press, Oxford.

Owen, D.F. & Wiegert, R.G. (1976) Do consumers maximize plant fitness? *Oikos*, **27**, 488–92.

Owensby, C.E., Ham, J.M. Knapp, A.K. & Allen, L.M. (1999) Biomass production and species composition change in a tallgrass prairie ecosystem after long-term exposure to elevated atmospheric CO_2. *Global Change Biology*, **5**, 497–506.

Oxbrough, A.G., Gittings, T., O'Halloran, J., Giller, P.S. & Smith, G.F. (2005) Structural indicators of spider communities across the forest plantation cycle. *Forest Ecology and Management*, **212**, 171–83.

Ozanne, C.M.P., Hambler, C., Foggo, A. & Speight, M.R. (1997) The significance of edge effects in the management of forests for invertebrate biodiversity. In *Canopy Arthropods* (eds N.E. Stork, J. Adis & R.K. Didham), pp. 534–50. Chapman & Hall, London.

Padovan, A.C. & Gibb, K.S. (2001) Epidemiology of phytoplasma diseases in papaya in northern Australia. *Journal of Phytopathology (Berlin)*, **149**, 649–58.

Paetzold, A., Schubert, C.J. & Tockner, K. (2005) Aquatic terrestrial linkages along a braided-river: riparian arthropods feeding on aquatic insects. *Ecosystems*, **8**, 748–59.

Paige, K.N. (1992) Overcompensation in response to mammalian herbivory: from mutalistic to antagonistic interactions. *Ecology*, **73**, 2076–85.

Paige, K.N. & Whitham, T.G. (1987) Overcompensation in response to mammalian herbivory: the advantage of being eaten. *American Naturalist*, **129**, 407–16.

Paine, T.D., Malinoski, M.K. & Scriven, G.T. (1990) Rating *Eucalyptus* vigor and the risk of insect infestation: leaf surface area and sapwood : heartwood ratio. *Canadian Journal of Forest Research*, **20**, 1485–9.

Palma, R.L. & Price, R.D. (2004) *Apterygon okarito*, a new species of chewing louse (Insecta: Phthiraptera: Menoponidae) from the Okarito brown kiwi (Aves: Apterygiformes: Apterygidae). *New Zealand Journal of Zoology*, **31**, 67–73.

Palmer, T.M. (2003) Spatial habitat heterogeneity influences competition and coexistence in an African acacia ant guild. *Ecology*, **84** (11), 2843–55.

Palumbo, J.C., Horowitz, A.R. & Prabhaker, N. (2001) Insecticidal control and resistance management for *Bemisia tabaci*. *Crop Protection*, **20** (9), 739–65.

Palumbo, M.J., Putz, F.E. & Talcott, S.T. (2007) Nitrogen fertilizer and gender effects on the secondary metabolism of yaupon, a caffeine-containing North American holly. *Oecologia*, **151**, 1–9.

Panagiotakopulu, E. (2004) Pharaonic Egypt and the origins of plague. *Journal of Biogeography*, **31** (2), 269–75.

Pandey, S. & Singh, R. (1999) Host size induced variation in progeny sex ratio of an aphid parasitoid *Lysiphlebia mirzai*. *Entomologia Experimentalis et Applicata*, **90**, 61–7.

Panzer, R. & Schwartz, M.W. (1998) Effectiveness of a vegetation-based approach to insect conservation. *Conservation Biology*, **12**, 693–702.

Paré, P.W., Alborn, H.T. & Tumlinson, J.H. (1998) Concerted biosynthesis of an insect elicitor of plant volatiles. *Proceedings of the National Academy of Sciences of the USA*, **95**, 13971–5.

Paré, P.W. & Tumlinson, J.H. (1996) Plant volatile signals in response to herbivore feeding. *Florida Entomologist*, **19**, 93–103.

Parker, A.R. & Lawrence, C.R. (2001) Water capture by a desert beetle. *Nature*, **414** (6859), 33–4.

Parmesan, C. (2006) Ecological and evolutionary responses to recent climate change. *Annual Review of Ecology Evolution and Systematics*, **37**, 637–69.

Parmesan, C. & Yohe, G. (2003) A globally coherent fingerprint of climate change impacts across natural systems. *Nature*, **421**, 37–42.

Parr, C.L., Robertson, H.G., Biggs, H.C. & Chown, S.L. (2004) Response of African savanna ants to long-term fire regimes. *Journal of Applied Ecology*, **41**, 630–42.

Parsons, M.S. (1996) *A Review of the Scarce and Threatened Ethmiine, Stathmopodine and Gelechiid Moths of Great Britain*. UK Joint Nature Conservation Committee, Peterborough, UK.

Partridge, L., Barrie, B., Fowler, K. & French, V. (1994) Thermal evolution of pre-adult life-history traits in *Drosophila melanogaster*. *Journal of Evolutionary Biology*, **7**, 645–63.

Passos, L. & Ferreira, S.O. (1996) Ant dispersal of *Croton priscus* (Euphobiaceae) seeds in a tropical semideciduous forest in Southeastern Brazil. *Biotropica*, **28**, 697–700.

Passos, L. & Oliveira, P.S. (2003) Interactions between ants, fruits and seeds in a restinga forest in south-eastern Brazil. *Journal of Tropical Ecology*, **19**, 261–70.

Pasteels, J.M., Dobler, S., Rowell-Rahier, M., Ehmke, A. & Hartman, T. (1995) Distribution of autogenous and host-derived chemical defenses in *Oreina* leaf beetles (Coleoptera. Chrysomelidae). *Journal of Chemical Ecology*, **21**, 1163–79.

Pasteels, J.M. & Rowell-Rahier, M. (1991) Proximate and ultimate causes for host plant influence on chemical defense of leaf beetles (Coleoptera Chrysomelidae). *Entomologia Generalis*, **15**, 227–35.

Pastor, J., Dewey, B., Naiman, R.J., McInnes, P.F. & Cohen, Y. (1993) Moose browsing and soil fertility in the boreal forests of Isle Royal National Park. *Ecology*, **74**, 467–80.

Pates, H. & Curtis, C. (2005) Mosquito behavior and vector control. *Annual Review of Entomology*, **50**, 53–70.

Pattison, J.E. (2003) Effect of the bubonic plague epidemic on inbreeding in 14th century Britain. *American Journal of Human Biology*, **15** (1), 101–11.

Patz, J.A. & Olson, S.H. (2006) Climate change and health: global to local influences on disease risk. *Annals of Tropical Medicine and Parasitology*, **100**, 535–49.

Paul, M.J., Meyer, J.L. & Couch, C.A. (2006) Leaf breakdown in streams differing in catchment land use. *Freshwater Biology*, **51** (9), 1684–95.

Paul, R.E.L., Ariey, F. & Robert, V. (2003) The evolutionary ecology of *Plasmodium*. *Ecology Letters*, **6** (9), 866–80.

Pauly, D. & Watson, R. (2005) Background and interpretation of the 'Marine Trophic Index' as a measure of biodiversity. *Philosophical Transactions of the Royal Society of London Series B, Biological Sciences*, **360**, 415–23.

Pauw, B. & Memelink, J. (2004) Jasmonate-responsive gene expression. *Journal of Plant Growth Regulation*, **23**, 200–10.

Payne, C.C. (1988) Pathogens for the control of insects: where next? *Philosophical Transactions of the Royal Society of London Series B, Biological Sciences*, **318**, 225–48.

Pearce, D.W. & Moran, D. (1994) *The Economic Value of Biodiversity*. Earthscan Publications, London.

Pearson, D.L. & Carroll, S.S. (1998) Global patterns of species richness: spatial models for conservation planning using bioindicator and precipitation data. *Conservation Biology*, **12**, 809–21.

Pearson, D.L. & Cassola, F. (1992) World-wide species richness patterns of tiger beetles (Coleoptera: Cicinelidae): indicator taxon for biodiversity and conservation studies. *Conservation Biology*, **6**, 376–91.

Pearson, D.L. & Cassola, F. (2005) A quantitative analysis of species descriptions of tiger beetles (Coleoptera: Cicindelidae), from 1758 to 2004, and notes about related developments in biodiversity studies. *Coleopterists Bulletin*, **59**, 184–93.

Peat, J., Darvill, B., Ellis, J. & Goulson, D. Effects of climate on intra- and interspecific size variation in bumble-bees. *Functional Ecology*, **19** (1), 145–51.

Peck, G.M., Andrews, P.K., Reganold, J.P. & Fellman, J.K. (2006) Apple orchard productivity and fruit quality under organic, conventional, and integrated management. *Hortscience*, **41** (1), 99–107.

Peck, S.B. & Thayer, M.K. (2003) The cave-inhabiting rove beetles of the United States (Coleoptera; Staphylinidae; excluding Aleocharinae and Pselaphinae): diversity and distributions. *Journal of Cave and Karst Studies*, **65** (1), 3–8.

Peck, S.B., Wigfull, P. & Nishida, G. (1999) Physical correlates of insular species diversity: the insects of the Hawaiian islands. *Annals of the Entomological Society of America*, **92**, 529–36.

Pedersen, B.S. & Mills, N.J. (2004) Single vs. multiple introduction in biological control: the roles of parasitoid efficiency, antagonism and niche overlap. *Journal of Applied Ecology*, **41** (5), 973–84.

Pellegrini, G., Levre, E., Valentini, P. & Cadoni, M. (1992) Cockroaches infestation and possible contribution in the spreading of some Enterobacteria. *Igiene Moderna*, **97** (1), 19–30.

Pemberton, R.W. (1988) Myrmecochory in the introduced range weed leafy spurge *Euphorbia esula* L. *American Midland Naturalist*, **119**, 431–5.

Pemberton, R.W. (1992) Fossil extrafloral nectaries, evidence for the ant-guard antiherbivore defense in an Oligocene *Populus*. *American Journal of Botany*, **79**, 1242–6.

Pemsl, D., Waibel, H. & Orphal, J. (2004) A methodology to assess the profitability of *Bt*-cotton: case study results from the state of Karnataka, India. *Crop Protection*, **23** (12), 1249–57.

Pena, C-J., Gonzalvez, G. & Chadee, D-D. (2003) Seasonal prevalence and container preferences of *Aedes albopictus* in Santo Domingo City, Dominican Republic. *Journal of Vector Ecology*, **28** (2), 208–12.

Percy, D.M., Page, R.D.M. & Cronk, Q.C.B. (2004) Plant–insect interactions: double-dating associated insect and plant lineages reveals asynchronous radiations. *Systematic Biology*, **53**, 120–7.

Percy, K.E., Awmack, C.S., Lindroth, R.L. et al. (2002) Altered performance of forest pests under atmospheres enriched by CO_2 and O_3. *Nature*, **420**, 403–7.

Perez, J.M. & Palma, R.L. (2001) A new species of Felicola (Phthiraptera: Trichodectidae) from the endangered

Iberian lynx: another reason to ensure its survival. *Biodiversity and Conservation*, **10**, 929–37.

Peterman, R.M., Clark, W.C. & Holling, C.S. (1979) The dynamics of resilience: shifting stability domains in fish and insect systems. In *Population Dynamics* (eds R.M. Anderson, B.D. Turner & L.R. Taylor), pp. 321–41. Blackwell Scientific Publications, Oxford.

Peters, M., Oberrath, R. & Bohning-Gaese, K. (2003) Seed dispersal by ants: are seed preferences influenced by foraging strategies or historical constraints? *Flora*, **198**, 413–20.

Petersen, M.J., Parker, C.R. & Bernard, E. (2005) The crane flies (Diptera: Tipuloidea) of Great Smoky Mountains National Park. *Zootaxa*, **1013**, 1–18.

Peterson, K.J., Lyons, J.B., Nowak, K.S., Takacs, C.M., Wargo, M.J. & McPeek, M.A. (2004) Estimating metazoan divergence times with a molecular clock. *Proceedings of the National Academy of Sciences of the USA*, **101** (17), 6536–41.

Petit, S., Firbank, R., Wyatt, B. & Howard, D. (2001) MIRABEL: models for integrated review and assessment of biodiversity in European landscapes. *Ambio*, **30**, 81–8.

Pexton, J.J. & Mayhew, P.J. (2005) Clutch size adjustment, information use and the evolution of gregarious development in parasitoid wasps. *Behavioral Ecology and Sociobiology*, **58** (1), 99–110.

Phelps, D.G. & Gregg, P.C. (1991) Effects of water stress on curly Mitchell grass, the common army worm and the Australian plague locust. *Australian Journal of Experimental Agriculture*, **31** (3), 325–32.

Philippe, R., Veyrunes, J.C., Mariau, D. & Bergoin, M. (1997) Biological control using entomopathogenic viruses. Application to oil palm and coconut pests. *Plantations Recherche Developpement*, **4** (1), 39–45.

Pichancourt, J.B., Burel, F. & Auger, P. (2006) Assessing the effect of habitat fragmentation on population dynamics: an implicit modelling approach. *Ecological Modelling*, **192**, 543–56.

Pielou, E.C. (1995) Biodiversity versus old-style diversity: measuring biodiversity for conservation. In *Measuring and Monitoring Biodiversity in Tropical and Temperate Forests* (eds T.J.B. Boyle & B. Boontawee), pp. 19–46. Centre for International Forestry Research (CIFOR), Bogor, Indonesia.

Pik, A.J., Oliver, I. & Beattie, A.J. (1999) Taxonomic sufficiency in ecological studies of terrestrial invertebrates. *Australian Journal of Ecology*, **24**, 555–62.

Pimentel, D., Wilson, C., McCullum, C. et al. (1997) Economic and environmental benefits of biodiversity. *Bioscience*, **47**, 747–57.

Pimm, S.L., Ayres, M., Balmford, A. et al. (2001) Environment – can we defy nature's end? *Science*, **293**, 2207–8.

Pitelka, F.A. (1964) The nutrient-recovery hypothesis for arctic microtine cycles. I. Introduction. In *Grazing in Terrestrial and Marine Environments* (ed. D.J. Crisp), pp. 55–6. Blackwell Science, Oxford.

Podoler, H. & Rogers, D. (1975) A new method for the identification of key factors from life-table data. *Journal of Animal Ecology*, **44**, 85–114.

Poitrineau, K., Brown, S.P. & Hochberg, M.E. (2003) Defence against multiple enemies. *Journal of Evolutionary Biology*, **16**, 1319–27.

Polis, G.A., Myers, C.A. & Holt, R.D. (1989) The ecology and evolution of intraguild predation – potential competitors that eat each other. *Annual Review of Ecology and Systematics*, **20**, 297–330.

Pollard, E., Rothery, P. & Yates, T.J. (1996) Annual growth rates in newly established populations of the butterfly *Pararge aegeria*. *Ecological Entomology*, **21** (4), 365–9.

Pollard, E. & Yates, T.J. (1993) *Monitoring Butterflies for Ecology and Conservation*. Chapman & Hall, London.

Pons, X., Nunez, E., Lumbierres, B. & Albajes, R. (2005) Epigeal aphidophagous predators and the role of alfalfa as a reservoir of aphid predators for arable crops. *European Journal of Entomology*, **102** (3), 519–25.

Poorter, L., de Plassche, M.V., Willems, S. & Boot, R.G.A. (2004) Leaf traits and herbivory rates of tropical tree species differing in successional status. *Plant Biology*, **6**, 746–54.

Porter, D.R., Burd, J.D., Shufran, K.A. & Webster, J.A. (2000) Efficacy of pyramiding greenbug (Homoptera: Aphididae) resistance genes in wheat. *Journal of Economic Entomology*, **93** (4), 1315–18.

Potts, G.R. & Vickerman, G.P. (1974) Studies on the cereal ecosystem. *Advances in Ecological Research*, **8**, 107–97.

Poulsen, M. & Boomsma, J-J. (2005). Mutualistic fungi control crop diversity in fungus-growing ants. *Science (Washington DC)*, **307** (5710), 741–4.

Poulsen, M., Bot, A.N.M., Currie, C.R. & Boomsma, J.J. Mutualistic bacteria and a possible trade-off between alternative defence mechanisms in *Acromyrmex* leaf-cutting ants. *Insectes Sociaux*, **49** (1), 15–19.

Powell, G.V.N. & Bjork, R. (1995) Implications of intratropical migration on reserve design – a case-study using *Pharomachrus mocinno*. *Conservation Biology*, **9**, 354–62.

Powell, S-J. & Bale, J-S. (2004) Cold shock injury and ecological costs of rapid cold hardening in the grain aphid *Sitobion avenae* (Hemiptera: Aphididae). *Journal of Insect Physiology*, **50** (4), 277–84.

Power, A.G. (1992) Host plant dispersion, leafhopper movement and disease transmission. *Ecological Entomology*, **17** (1), 63–8.

Prance, G.T. (1994) A comparison of the efficacy of higher taxa and species numbers in the assessment of biodiversity in the neotropics. *Philosophical Transactions of the Royal Society of London Series B, Biological Sciences*, **345**, 89–99.

Prasad, R.P. & Snyder, W.E. (2004) Predator interference limits fly egg biological control by a guild of ground-active beetles. *Biological Control*, **31** (3), 428–37.

Prendergast, J.R. (1997) Species richness covariance in higher taxa: empirical tests of the biodiversity indicator concept. *Ecography*, **20**, 210–16.

Prendergast, J.R., Quinn, R.M., Lawton, J.H., Eversham, B.C. & Gibbons, D.W. (1993) Rare species, the coincidence of diversity hotspots and conservation strategies. *Nature*, **365**, 335–7.

Prestidge, R.A. & Gallagher, R.T. (1988) Endophyte fungus confers resistance to ryegrass: argentine stem weevil larval studies. *Ecological Entomology*, **13**, 429–36.

Preston, C.A., Laue, G. & Baldwin, I.T. (2004) Plant-plant signaling: application of *trans*- or *cis*-methyl jasmonate equivalent to sagebrush releases does not elicit direct defenses in native tobacco. *Journal of Chemical Ecology*, **30**, 2193–214.

Preszler, R.W. & Boecklen, W.J. (1994) A three-trophic level analysis of the effects of plant hybridization on a leaf-mining moth. *Oecologia (Berlin)*, **100**, 66–73.

Preszler, R.W. & Boecklen, W.J. 1996. The influence of elevation on tri-trophic interactions: opposing gradients of top-down and bottom-up effects on a leaf-mining moth. *Ecoscience*, **3**, 75–80.

Preszler, R.W., Gaylord, E.S. & Boecklen, W.J. (1996) Reduced parasitism of a leaf-mining moth on trees with high infection frequencies of an endophytic fungus. *Oecologia*, **108**, 159–66.

Preszler, R.W. & Price, P.W. (1988) Host quality and sawfly populations: a new approach to life table analysis. *Ecology*, **69**, 2012–20.

Price, P.W. (1975) *Insect Ecology*. John Wiley, New York.

Price, P.W. (1989) Clonal development of coyote willow, *Salix exigua* (Salicaceae), and attack by the shoot-galling sawfly, *Euura exiguae* (Hymenoptera. Tenthredinidae). *Environmental Entomology*, **18**, 61–8.

Price, P.W. (1990) Evaluating the role of natural enemies in latent and eruptive species: new approaches in life table construction. In *Population Dynamics of Forest Insects* (eds A.D. Watt, S.R. Leather, M.D. Hunter & N.A.C. Kidd), pp. 221–32. Intercept, Andover, UK.

Price, P.W. (1991) The plant vigor hypothesis and herbivore attack. *Oikos*, **62**, 244–51.

Price, P.W. (1996) *Biological Evolution*. Saunders College Publishing, Philadelphia, PA.

Price, P.W., Andrade, I., Pires, C., Sujii, E. & Vieira, E.M. (1995) Gradient analysis using plant architecture and insect herbivore utilization. *Environmental Entomology*, **24**, 497–505.

Price, P.W., Bouton, P., Gross, B.A., McPheron, J.N., Thompson, J.N. & Weis, A.E. (1980) Interactions among three trophic levels: influence of plants on interactions between insect herbivores and natural enemies. *Annual Review of Ecology and Systematics*, **11**, 41–65.

Price, P.W., Craig, T.P. & Hunter, M.D. (1998) Population ecology of a gall-forming sawfly, *Euura lasiolepis*, and relatives. In *Insect Populations: in Theory and Practice* (eds J.P. Dempster & I.F.G. Mclean), pp. 323–40. Kluwer Academic Publishers, London.

Price, P.W. & Hunter, M. D. (2005) Long-term population dynamics of a sawfly show strong bottom-up effects. *Journal of Animal Ecology*, **74**, 917–25.

Priddel, D., Carlile, N., Humphrey, M., Fellenberg, S. & Hiscox, D. (2003) Rediscovery of the 'extinct' Lord Howe Island stick-insect (*Dryococelus australis* (Montrouzier)) (Phasmatodea) and recommendations for its conservation. *Biodiversity and Conservation*, **12**, 1391–403.

Prinzing, A., Klotz, S., Stadler, J. & Brandl, R. (2003) Woody plants in Kenya: expanding the higher-taxon approach. *Biological Conservation*, **110**, 307–14.

Proctor, H. & Grigg, A. (2006) Aquatic invertebrates in final void water bodies at an open-cut coal mine in central Queensland. *Australian Journal of Entomology*, **45**, 107–21.

Prokop, J., Smith, R., Jarzembowski, E.A. & Nel, A. (2006) New homoiopterids from the Late Carboniferous of England (Insecta: Palaeodictyoptera). *Comptes Rendus Palevol*, **5**, 867–73.

Pugachev, K.V., Guirakhoo, F., Trent, D.W. & Monath, T.P. (2003) Traditional and novel approached to flavivirus vaccines. *International Journal for Parasitology*, **33**, 567–82.

Pugalenthi, P. & Livingstone, D. (1995) Cardenolides (heart poisons) in the painted grasshopper *Poecilocerus pictus* F. (Orthoptera: Pyrgomorphidae) feeding on the milkweed *Calotropis gigantea* L. Asclepiadaceae). *Journal of the New York Entomological Society*, **103**, 191–6.

Pullin, A.S. (1996) Restoration of butterfly populations in Britain. *Restoration Ecology*, **4**, 71–80.

Pureswaran, D.S., Sullivan, B.T. & Ayres, M.P. (2006) Fitness consequences of pheromone production and host selection strategies in a tree-killing bark beetle (Coleoptera: Curculionidae: Scolytinae). *Oecologia*, **148**, 720–8.

Purrington, C.B. (2000) Costs of resistance. *Current Opinion in Plant Biology*, **3**, 305–8.

Purvis, A. & Hector, A. (2000) Getting the measure of biodiversity. *Nature*, **405**, 212–19.

Pustejovsky, D.E. & Smith, J.W. (2006) Partial ecological life table of immature *Helicoverpa zea* (Lepidoptera: Noctuidae) in an irrigated cotton cropping system in the trans-Pecos region of Texas, USA. *Biocontrol Science and Technology*, **16** (7), 727–42.

Putman, R.J. & Wratten, S.D. (1984) *Principles of Ecology*. Croom Helm, London.

Pyke, C.R. & Fischer, D.T. (2005) Selection of bioclimatically representative biological reserve systems under climate change. *Biological Conservation*, **121**, 429–41.

Pywell, R.F., Warman, E.A., Hulmes, L. et al. (2006) Effectiveness of new agri-environment schemes in providing foraging resources for bumblebees in intensively farmed landscapes. *Biological Conservation*, **129**, 192–206.

Quan, X., Zhao, X., Chen, S., Zhao, H.M., Chen, J.W. & Zhao, Y.Z. (2005) Enhancement of p,p'-DDT photodegradation on soil surfaces using TiO_2 induced by UV-light. *Chemosphere*, **60** (2), 266–73.

Quek, S.P., Davies, S.J., Itino, T. & Pierce, N.E. (2004) Codiversification in an ant–plant mutualism: stem texture and the evolution of host use in *Crematogaster* (Formicidae:

Myrmicinae) inhabitants of *Macaranga* (Euphorbiaceae). *Evolution*, **58**, 554–70.

Quiring, D.T. & McNeil, J.N. (1984) Intraspecific larval competition reduces efficacy of oviposition-deterring pheromone in the alfalfa blotch leafminer, *Agromyza frontella* (Diptera: agromyzidae). *Environmental Entomology*, **13**, 675–8.

Qureshi, J.A. & Michaud, J.P. (2005) Interactions among three species of cereal aphids simultaneously infesting wheat. *Journal of Insect Science* **5**, Article number 13.

Raa, J. (1968) Polyphenols and natural resistance of apple leaves against *Venturia inaequalis*. *Netherlands Journal of Plant Pathology*, **74**, 37–45.

Rafai, M.A., Boulaajaj, F.Z., Bourezgui, M. et al. (2007) Clinical and electrophysiological aspects of acute organophosphate intoxication. *Neurophysiologie Clinique, Clinical Neurophysiology*, **37** (1), 35–9.

Rainio, J. & Niemelä, J. (2003) Ground beetles (Coleoptera: Carabidae) as bioindicators. *Biodiversity and Conservation*, **12**, 487–506.

Rainio, J. & Niemelä, J. (2006) Comparison of carabid beetle (Coleoptera: Carabidae) occurrence in rain forest and human-modified sites in south-eastern Madagascar. *Journal of Insect Conservation*, **10**, 219–28.

Rajendran, R., Rajendran, N. & Venugopalan, V.K. (1990) Effect of organochlorine pesticides on the bacterial population of a tropical estuary. *Microbios Letters*, **44** (174), 57–64.

Rajska, P., Pechanova, O., Takac, P. et al. (2003) Vasodilatory activity in horsefly and deerfly salivary glands. *Medical and Veterinary Entomology*, **17** (4), 395–402.

Rajukkannu, K., Basha, A.A., Habeebullah, B., Duraisamy, P. & Balasubramanian, M. (1985) Degradation and persistence of DDT, BHC, carbaryl and malathion in soils. *Indian Journal of Environmental Health*, **27** (3), 237–43.

Raman, A. & Abrahamson, W.G. (1995) Morphometric relationships and energy allocation in the apical rosette galls of *Solidago altissima* (Asteraceae) induced by *Rhopalomyia solidaginis* (Diptera: Cecidomyiidae). *Environmental Entomology*, **24**, 635–9.

Ramesh, A., Tanabe, S., Kannan, K., Subramanian, A.N., Kumaran, P.L. & Tatsukawa, R. (1992) Characteristic trend of persistent organochlorine contamination in wildlife from a tropical agricultural watershed, South India. *Archives of Environmental Contamination and Toxicology*, **23** (1), 26–36.

Ramle, M., Wahid, M.B., Norman, K., Glare, T.R. & Jackson, T.A. (2005) The incidence and use of *Oryctes* virus for control of rhinoceros beetle in oil palm plantations in Malaysia. *Journal of Invertebrate Pathology*, **89** (1), 85–90.

Ramlov, H. & Lee, R.E. (2000) Extreme resistance to desiccation in overwintering larvae of the gall fly *Eurosta solidaginis* (Diptera, Tephritidae). *Journal of Experimental Biology*, **203**, 783–9.

Ramos-Elorduy, J. (1997) The importance of edible insects in the nutrition and economy of people of the rural areas of Mexico. *Ecology of Food and Nutrition*, **36**, 347–66.

Ramos-Elorduy, J., Moreno, J.M.P., Prado, E.E., Perez, M.A., Otero, J.L. & Larron de Guevara, O. (1997) Nutritional value of edible insects from the state of Oaxaca, Mexico. *Journal of Food Composition and Analysis*, **10** (2), 142–57.

Randall, M.G.M. (1982a) The dynamics of an insect population throughout its altitudinal distribution – *Coleophora alticolella* (Lepidoptera) in northern England. *Journal of Animal Ecology*, **51**, 993–1016.

Randall, M.G.M. (1982b) The ectoparasitization of *Coleophora alticolella* (Lepidoptera) in relation to its altitudinal distribution. *Ecological Entomology*, **7**, 177–85.

Ranger, C.M. & Hower, A.A. (2002) Glandular trichomes on perennial alfalfa affect host-selection behavior of *Empoasca fabae*. *Entomologia Experimentalis et Applicata*, **105**, 71–81.

Rantalainen, M.L., Fritze, H., Haimi, J., Pennanen, T. & Setala, H. (2005) Species richness and food web structure of soil decomposer community as affected by the size of habitat fragment and habitat corridors. *Global Change Biology*, **11**, 1614–27.

Rasnitsyn, A.P., Basibuyuk, H.H. & Quicke, D.L.J. (2004) A basal chalcidoid (Insecta: Hymenoptera) from the earliest Cretaceous or latest Jurassic of Mongolia. *Insect Systematics and Evolution*, **35** (2), 123–35.

Rathcke, B.J. (1976) Competition and coexistence within a guild of herbivorous insects. *Ecology*, **57**, 76–87.

Rathcke, B.J. (1992) Nectar distributions, pollinator behavior and plant reproductive success. In *Effects of Resource Distribution on Animal–Plant Interactions* (eds M.D. Hunter, T. Ohgushi & P.W. Price), pp. 113–138. Academic Press, San Diego, CA.

Rathor, H.R. (2000) *The Role of Vectors in Emerging and Re-emerging Diseases in the Eastern Mediterranean Region*. Dengue Bulletin No. 24. World Health Organization, Cairo, Egypt.

Raupp, M.J. (1985) Effects of leaf toughness on mandibular wear of the leaf beetle, *Plagiodera versicolora*. *Ecological Entomology*, **10**, 73–9.

Raven, P.H., Berg, L.R. & Johnson, G.B. (1993) *Environment*. Saunders College Publishing, Philadelphia, PA.

Rawn, D.F.K., Quade, S.C., Shields, J.B. et al. (2006) Organophosphate levels in apple composites and individual apples from a treated Canadian orchard. *Journal of Agricultural and Food Chemistry*, **54** (5), 1943–8.

Raworth, D.A. & Schade, D. (2006) Life-history parameters and population dynamics of *Ericaphis fimbriata* (Hemiptera: Aphididae) on blueberry, *Vaccinium corymbosum*. *Canadian Entomologist*, **138** (2), 205–17.

Raymond, B., Hartley, S.E., Cory, J.S. & Hails, R.S. (2005) The role of food plant and pathogen-induced behaviour in the persistence of a nucleopolyhedrovirus. *Journal of Invertebrate Pathology*, **88** (1), 49–57.

Rebe, M., Berg, J.V.D. & Donaldson, G. (2004) The status of leaf feeding resistance and oviposition preference of *Busseola fusca* (Fuller) (Lepidoptera: Noctuidae) and *Chilo partellus* (Swinhoe) (Lepidoptera: Crambidae) for sweet sorghum (*Sorghum bicolor*) landraces. *International Journal of Pest Management*, **50** (1), 49–53.

Redfern, M. & Hunter, M.D. (2005) Time tells: long-term patterns in the population dynamics of the yew gall midge, *Taxomyia taxi* (Cecidomyiidae), over 35 years. *Ecological Entomology*, **30**, 86–95.

Redmond, C.T. & Potter, D.A. (1995) Lack of efficacy of *in vivo* and putatively *in vitro* produced *Bacillus popilliae* against field populations of Japanese beetle (Coleoptera: Scarabaeidae) grubs in Kentucky. *Journal of Economic Entomology*, **88** (4), 846–54.

Reed, G.L., Jensen, A.S., Riebe, J., Head, G. & Duan, J.J. (2001) Transgenic *Bt* potato and conventional insecticides for Colorado potato beetle management: comparative efficacy and non-target impacts. *Entomologia Experimentalis et Applicata*, **100**, 89–100.

Reeve, R. (2002) *Policing International Trade in Endangered Species. The CITES Treaty and Compliance*. Earthscan Publications, London.

Reeves, W.K. & McCreadie, J.W. (2001) Population ecology of cavernicoles associated with carrion in caves of Georgia, USA. *Journal of Entomological Science*, **36** (3), 305–11.

Reichard, R.E. (2002) Area-wide biological control of disease vectors and agents affecting wildlife. *Revue Scientifique et Technique de l'office International des Epizooties*, **21** (1), 179–85.

Reid, W.V. (1992) How many species will there be? In *Tropical Deforestation and Species Extinction* (eds T.C. Whitmore & J.A. Sayer), pp. 55–73. Chapman & Hall, London.

Reid, W.V. & Miller, K.R. (1989) *Keeping Options Alive: the Scientific Basis for Conserving Biodiversity*. World Resources Institute, Washington, DC.

Reisen, W.K., Milby, M.M. & Meyer, R.P. (1992) Population dynamics of adult *Culex* mosquitoes (Diptera: Culicidae) along the Kern River, Kern County, California, in 1990. *Journal of Medical Entomology*, **29** (3), 531–43.

Reitz, S.R. (1996) Interspecific competition between two parasitoids of *Helicoverpa zea*, *Eucelatoria bryani* and *E. rubentis*. *Entomologia Experimentalis et Applicata*, **79**, 227–34.

Rengam, S.V. (1992) IPM: The role of governments and citizens' groups. In *Integrated Pest Management in the Asia–Pacific Region* (eds P.A.C. Ooi, G.S. Lim, T.H. Ho, P.L. Manalo & J. Waage), pp. 13–20. CAB International, Wallingford, UK.

Resh, V.H., Leveque, C. & Statzner, B. (2004) Long-term, large-scale biomonitoring of the unknown: assessing the effects of insecticides to control river blindness (Onchocerciasis) in West Africa. *Annual Review of Entomology*, **49**, 115–39.

Reynolds, B.C., Hunter, M.D. & Crossley, D.A. Jr. (2000) Effects of canopy herbivory on nutrient cycling in a northern hardwood forest in western North Carolina. *Selbyana*, **21**, 74–8.

Reznick, D., Bryant, M.J. & Bashey, F. (2002) *r*- and *K*-selection revisited: the role of population regulation in life-history evolution. *Ecology (Washington DC)*, **83** (6), 1509–20.

Rhainds, M. & Ho, C.T. (2002) Size-dependent reproductive output of female bagworms (Lepidoptera: Psychidae): implications for inter-generational variations of population density. *Applied Entomology and Zoology*, **37**, 357–64.

Rhoades, D.F. (1983) Responses of alder and willow to attack by tent caterpilars and webworms: evidence for pheromonal sensitivity of willows. In *Plant Resistance to Insects* (ed. P.A. Hedin), pp. 55–68. The American Chemical Society, Washington, DC.

Rhoades, D.F. (1985) Offensive–defensive interactions between herbivores and plants: their relevance to herbivore population dynamics and community theory. *American Naturalist*, **125**, 205–38.

Rhoades, D.F. & Cates, R.G. (1976) Toward a general theory of plant antiherbivore chemistry. *Record Advances in Phytochemistry*, **10**, 168–213.

Ribas-Fito, N., Torrent, M., Carrizo, D. et al. (2006) *In utero* exposure to background concentrations of DDT and cognitive functioning among preschoolers. *Epidemiology*, **17** (6), S103–S103.

Ribeiro, J.M.C. & Francischetti, I.M.B. (2003) Role of arthropod saliva in blood feeding: sialome and post-sialome perspectives. *Annual Review of Entomology*, **48**, 73–88.

Rice, W.R. (1983) Sexual reproduction: an adaptation reducing parent-offspring contagion. *Evolution*, **37**, 1317–20.

Richard, F.J., Mora, P., Errard, C. & Rouland, C. (2005) Digestive capacities of leaf-cutting ants and the contribution of their fungal cultivar to the degradation of plant material. *Journal of Comparative Physiology B, Biochemical Systemic and Environmental Physiology*, **175** (5), 297–303.

Richards, A., Matthews, M. & Christian, P. (1998) Ecological considerations for the environmental impact evaluation of recombinant baculovirus insecticides. *Annual Review of Entomology*, **43**, 493–517.

Richardson, M.D. & Bacon, C.W. (1993) Cyclic hydroxamic acid accumulation in corn seedlings exposed to reduced water potentials before, during, and after germination. *Journal of Chemical Ecology*, **19**, 1613–24.

Richter, S. (2002) The Tetraconata concept: hexapod–crustacean relationships and the phylogeny of Crustacea. *Organisms Diversity and Evolution* **2** (3), 217–37.

Ricketts, T.H. (2001) The matrix matters: effective isolation in fragmented landscapes. *American Naturalist*, **158**, 87–99.

Ricklefs, R.E. & Lovette, I.J. (1999) The roles of island area per se and habitat diversity in the species–area relationships of four Lesser Antillean faunal groups. *Journal of Animal Ecology*, **68**, 1142–60.

Riddick, E.W. (2006) Egg load and body size of lab-cultured *Cotesia marginiventris*. *Biocontrol*, **51** (5), 603–10.

Riedell, W.E., Kieckhefer, R.W., Langham, M.A.C. & Hesler, L.S. (2003) Root and shoot responses to bird cherry-oat aphids and barley yellow dwarf virus in spring wheat. *Crop Science*, **43**, 1380–6.

Risley, L.S. (1986) The influence of herbivores on seasonal leaf-fall: premature leaf abscission and petiole clipping. *Journal of Agricultural Entomology*, **3**, 152–62.

Ritchie, M.E., Tilman, D. & Knops, J.M.H. (1998) Herbivore effects on plant and nitrogen dynamics in oak savanna. *Ecology*, **79**, 165–77.

Ritchie, S.A., Fanning, I.D., Phillips, D.A., Standfast, H.A., McGinn, D. & Kay, B.H. (1997) Ross river virus in mosquitoes (Diptera: Culicidae) during the 1994 epidemic around Brisbane, Australia. *Journal of Medical Entomology*, **34** (2), 156–9.

Ritland, D.B. & Brower, L.P. (1993) A reassessment of the mimicry relationship among viceroys, queens and monarchs in Florida. *Natural History Museum of Los Angeles County Science Series*, **38**, 129–39.

Rivas, F., Diaz, L.A., Cardenas, V.M. et al. (1997) Epidemic Venezuelan equine encephalitis in La Guajira, Colombia, 1995. *Journal of Infectious Diseases*, **175** (4), 828–32.

Rivero, A. & Ferguson, H.M. (2003) The energetic budget of *Anopheles stephensi* infected with *Plasmodium chabaudi*: is energy depletion a mechanism for virulence? *Proceedings of the Royal Society of London Series B, Biological Sciences*, **270** (1522), 1365–71.

Robbins, R.K. & Opler, P.A. (1997) Butterfly diversity and a preliminary comparison with bird and mammal diversity. In *Biodiversity II, Understanding and Protecting our Biological Resources* (eds M.L. Reaka-Kudla, D.E. Wilson & E.O. Wilson), pp. 69–82. Joseph Henry Press, Washington, DC.

Roberge, J.M. & Angelstam, P. (2004) Usefulness of the umbrella species concept as a conservation tool. *Conservation Biology*, **18**, 76–85.

Robert, V., Le Goff, G., Essong, J., Tchuinkam, T., Faas, B. & Verhave, J.P. (1995) Detection of falciparum malarial forms in naturally infected anophelines in Cameroon using a fluorescent anti-25-kD monoclonal antibody. *American Journal of Tropical Medicine and Hygiene*, **52** (4), 366–9.

Roberts, L. (2002) Mosquitoes and disease. *Science*, **298** (5591), 82–3.

Robertson, L.N. (1993) Population dynamics of false wireworms (*Gonocephalum macleayi, Pterohelaeus alternatus, P. darlingensis*) and development of an integrated pest management program in central Queensland field crops: a review. *Australian Journal of Experimental Agriculture*, **33** (7), 953–62.

Robertson, M.P., Villet, M.H., Fairbanks, D.H.K. et al. (2003) A proposed prioritization system for the management of invasive alien plants in South Africa. *South African Journal of Science*, **99**, 37–43.

Robertson, S.P., Hull, L.A. & Calvin, D.D. (2005) Tufted apple bud moth (Lepidoptera: Tortricidae) management model for processing apples based on early season pheromone trap capture. *Journal of Economic Entomology*, **98** (4), 1229–35.

Roces, F. & Hoelldobler, B. (1994) Leaf density and a trade-off between load-size selection and recruitment behavior in the ant *Atta cephalotes*. *Oecologia*, **97**, 1–8.

Rodriguero, M.S. & Gorla, D.E. (2004) Latitudinal gradient in species richness of the New World Triatominae (Reduviidae). *Global Ecology and Biogeography*, **13**, 75–84.

Rodrigues, A.S.L., Pilgrim, J.D., Lamoreux, J.F., Hoffmann, M. & Brooks, T.M. (2006) The value of the IUCN Red List for conservation. *Trends in Ecology and Evolution*, **21**, 71–6.

Rodriguez-Saona, C., Chalmers, J.A., Raj, S. & Thaler, J.S. (2005) Induced plant responses to multiple damagers: differential effects on an herbivore and its parasitoid. *Oecologia*, **143**, 566–77.

Rodriguez-Saona, C., Crafts-Brandner, S.J., Paré, P.W. & Henneberry, T.J. (2001) Exogenous methyl jasmonate induces volatile emissions in cotton plants. *Journal of Chemical Ecology*, **27**, 679–96.

Roff, D.A. (1990) The evolution of flightlessness in insects. *Ecological Monographs*, **60** (4), 389–421.

Rogers, D.J. & Hassell, M.P. (1974) General models for insect parasite and predator searching behaviour: interference. *Journal of Animal Ecology*, **43**, 239–53.

Rogers, D.J. & Randolph, S.E. (2002). A response to the aim of eradicating tsetse from Africa. *Trends in Parasitology*, **18**, 534–6.

Rogowitz, G.L. & Chappell, M.A. (2000) Energy metabolism of eucalyptus-boring beetles at rest and during locomotion: gender makes a difference. *Journal of Experimental Biology*, **203**, 1131–9.

Rohde, K. (1992) Latitudinal gradients in species diversity: the search for the primary cause. *Oikos*, **65**, 514–27.

Roland, J. & Kaupp, W.J. (1995) Reduced transmission of forest tent caterpillar NPV at the forest edge. *Environmental Entomology*, **24**, 1175–8.

Roland, J., Keyghobadi, N. & Fownes, S. (2000) Alpine Parnassius butterfly dispersal: effects of landscape and population size. *Ecology*, **81**, 1642–53.

Roland, J. & Taylor, P.D. (1997) Insect parasitoid species respond to forest structure at different spatial scales. *Nature*, **386**, 710–13.

Roland, J., Taylor, P. & Cooke, B. (1997) Forest structure and the spatial pattern of parasitoid attack. In *Forests and Insects* (eds A.D. Watt, N.E. Stork & M.D. Hunter), pp. 97–106. Chapman & Hall, London.

Rolland, C. & Lemperiere, G. (2004) Effects of climate on radial growth of Norway spruce and interactions with attacks by the bark beetle *Dendroctonus micans* (Kug., Coleoptera: Scolytidae): a dendroecological study in the French Massif Central. *Forest Ecology and Management*, **201** (1), 89–104.

Romanow, L.R. & Ambrose, J.T. (1981) Effects of solid rocket fuel exhaust of honey bee colonies. *Environmental Ecology*, **10** (5), 812–16.

Romero, G.Q. & Izzo, T.J. (2004) Leaf damage induces ant recruitment in the Amazonian ant-plant *Hirtella myrmecophila*. *Journal of Tropical Ecology*, **20**, 675–82.

Roncin, E. & Deharveng, L. (2003) *Leptogenys khammouanensis* sp. nov. (Hymenoptera: Formicidae). A possible troglobitic species of Laos, with a discussion on cave ants. *Zoological Science Tokyo*, **20** (7), 919–24.

Root, R.B. (1973) Organization of a plant–arthropod association in simple and diverse habitats: the fauna of collards (*Brassica oleracea*). *Ecological Monography*, **43**, 95–124.

Root, T.L., Price, J.T., Hall, K.R., Schneider, S.H., Rosenzweig, C. & Pounds, J.A. (2003) Fingerprints of global warming on wild animals and plants. *Nature*, **421**, 57–60.

Röse, U.S.R., Lewis, W.J. & Tumlinson, J.H. (1998) Specificity of systemically released cotton volatiles as attractants for specialist and generalist parasitic wasps. *Journal of Chemical Ecology*, **24**, 303–19.

Rosemond, A.D., Reice, S.R., Elwood, J.W. & Mulholland, P.J. (1992) The effects of stream acidity on benthic invertebrate communities in the south-eastern United States. *Freshwater Biology*, **27**, 193–209.

Rosenheim, J.A. (1998) Higher-order predators and the regulation of insect herbivore populations. *Annual Review of Entomology*, **43**, 421–47.

Rosenheim, J.A. (2005) Intraguild predation of *Orius tristicolor* by *Geocoris* spp. and the paradox of irruptive spider mite dynamics in California cotton. *Biological Control*, **32**, 172–9.

Rosenheim, J.A., Wilhoit, L.R. & Armer, C.A. (1993) Influence of intraguild predation among generalist insect predators on the suppression of an herbivore population. *Oecologia*, **96**, 439–49.

Rosenthal, G.A. & Berenbaum, M. (eds) (1991) *Herbivores: their Interaction with Secondary Plant Metabolites*. Academic Press, New York.

Rosenthal, J.P. & Kotanen, P.M. (1994) Terrestrial plant tolerance to herbivory. *Trends in Ecology and Evolution*, **9**, 145–8.

Rossiter, M.C. (1994) Maternal effects hypothesis of herbivore outbreak. *Bioscience*, **44**, 752–63.

Rossiter, M.C. (1996) Incidence and consequences of inherited environmental effects. *Annual Review of Ecology and Systematics*, **27**, 451–76.

Rossiter, M.C., Schultz, J.C. & Baldwin, I.T. (1988) Relationships among defoliation, red oak phenolics, and gypsy-moth growth and reproduction. *Ecology*, **69**, 267–77.

Roth, S.K. & Lindroth, R.L. (1994) Effects of CO_2-mediated changes in paper birch and white pine chemistry on gypsy moth performance. *Oecologia*, **98**, 133–8.

Roth, S.K., McDonald, E.P. & Lindroth, R.L. (1997) Atmospheric CO_2 and soil water availability: consequences for tree-insect interactions. *Canadian Journal of Forest Research*, **27**, 1281–90.

Rothschild, M. (1973) Secondary plant substances and warning colouration in insects. *Symposium of the Royal Entomological Society London*, **6**, 59–79.

Rouault, G., Candau, J-N., Lieutier, F., Nageleisen, L.M., Martin, J-C. & Warzee, N. (2006) Effects of drought and heat on forest insect populations in relation to the 2003 drought in Western Europe. *Annals of Forest Science*, **63** (6), 613–24.

Roubik, D.W. (1992) Loose niches in tropical communities: why are there so few bees and so many trees? In *The Effects of Resource Distribution on Animal–Plant Interactions* (eds M.D. Hunter, T. Ohgushi & P.W. Price), pp. 326–54. Academic Press, San Diego.

Rouget, M., Cowling, R.M., Lombard, A.T., Knight, A.T. & Graham, I.H.K. (2006) Designing large-scale conservation corridors for pattern and process. *Conservation Biology*, **20**, 549–61.

Round, P.D. (1985) *The Status and Conservation of Resident Forest Birds in Thailand*. Association for the Conservation of Wildlife, Bangkok.

Roush, R.T. (1998) Two-toxin strategies for management of insecticidal transgenic crops: can pyramiding succeed where pesticide mixtures have not? *Philosophical Transactions of the Royal Society of London Series B, Biological Sciences*, **353**, 1777–86.

Rowell-Rahier, M. & Pasteels, J.N. (1992) Third trophic level influences of plant allelochemicals. *Ecological and Evolutionary Processes*, **2**, 243–77.

Roy, D.B., Rothery, P., Moss, D., Pollard, E. & Thomas, J.A. (2001) Butterfly numbers and weather: predicting historical trends in abundance and the future effects of climate change. *Journal of Animal Ecology*, **70**, 201–17.

Royama, T. (1970) Factors governing the hunting behaviour and selection of food by the great tit (*Parus major* L.). *Journal of Animal Ecology*, **39**, 619–68.

Rudgers, J.A., Koslow, J.M. & Clay, K. (2004) Endophytic fungi alter relationships between diversity and ecosystem properties. *Ecology Letters*, **7**, 42–51.

Rudgers, J.A. & Whitney, K.D. (2006) Interactions between insect herbivores and a plant architectural dimorphism. *Journal of Ecology*, **94** (6), 1249–60.

Ruess, R.W., Hendrick, R.L. & Bryant, J.P. (1998) Regulation of fine root dynamics by mammalian browsers in early successional Alaskan taiga forests. *Ecology*, **79** (8), 2706–20.

Rukazambuga, N.D.T.M., Gold, C.S., Gowen, S.R. & Ragama, P. (2002) The influence of crop management on banana weevil, *Cosmopolites sordidus* (Coleoptera: Curculionidae) populations and yield of highland cooking banana (cv. Atwalira) in Uganda. *Bulletin of Entomological Research*, **92** (5), 413–21.

Russell, R.C. (2002) Ross River virus: ecology and distribution. *Annual Review of Entomology*, **47**, 1–31.

Russell, T.L., Brown, M.L.D., Purdie, D.M., Ryan, P.A. & Kay, B.H. (2003) Efficacy of VectoBac (*Bacillus thuringiensis* variety *israelensis*) formulations for mosquito control in Australia. *Journal of Economic Entomology*, **96**, 1786–91.

Ryan, J.D., Gregory, P. & Tingey, W.M. (1982) Phenolic oxidase activities in the glandular trichomes of *Solanum berthaultii*(Hawkes). *Phytochemistry*, **21**, 1885–7.

Ryan, R.B. (1990) Evaluation of biological control, introduced parasites of larch casebearer (Lepidoptera, Coleophoridae) in Oregon. *Environmental Entomology*, **19**, 1873–81.

Saarinen, K., Valtonen, A., Jantunen, J. & Saarnio, S. (2005) Butterflies and diurnal moths along road verges: does road type affect diversity and abundance? *Biological Conservation*, **123**, 403–12.

Saayman, D. & Lambrechts, J.J.N. (1993) The possible cause of red leaf disease and its effect on Barlinka table grapes. *South African Journal for Enology and Viticulture*, **14** (2), 26–32.

Sabzalian, M.R., Hatami, B. & Mirlohi, A. (2004) Mealybug, *Phenococcus solani*, and barley aphid, *Sipha maydis*, response to endophyte-infected tall and meadow fescues. *Entomologia Experimentalis et Applicata*, **113**, 205–9.

Sachs, J.D. (2002) A new global effort to control malaria. *Science (Washington DC)*, **298**, 122–4.

Sadanandane, C., Sahu, S.S., Gunasekaran, K., Jambulingam, P. & Das, P.K. (1991) Pattern of rice cultivation and anopheline breeding in Koraput district of Orissa state (India). *Journal of Communicable Diseases*, **23** (1), 59–65.

Sagarra, L.A., Vincent, C. & Stewart, R.K. (2001) Body size as an indicator of parasitoid quality in male and female *Anagyrus kamali* (Hymenoptera: Encyrtidae). *Bulletin of Entomological Research*, **91**, 363–7.

Sagers, C.L. & Coley, P.D. (1995) Benefits and costs of defense in a neotropical shrub. *Ecology*, **76**, 1835–43.

Saha, P., Majumder, P., Dutta, I., Ray, T., Roy, S.C. & Das, S. (2006) Transgenic rice expressing *Allium sativum* leaf lectin with enhanced resistance against sap-sucking insect pests. *Planta*, **223** (6), 1329–43.

Sain, M. & Kalode, M.B. (1994) Greenhouse evaluation of rice cultivars for resistance to gall midge, *Orseolia oryzae* (Wood-Mason) and studies on the mechanism or resistance. *Insect Science and its Application*, **15** (1), 67–74.

Sait, S.M., Begon, M. & Thompson, D.J. (1994) Long-term population dynamics of the Indian meal moth *Plodia interpunctella* and its granulosis virus. *Journal of Animal Ecology*, **63** (4), 861–70.

Sala, O.E., Chapin, F.S., Armesto, J.J. et al. (2000) Biodiversity – global biodiversity scenarios for the year 2100. *Science*, **287**, 1770–4.

Sales, J. (2005) The endangered kiwi: a review. *Folia Zoologica*, **54**, 1–20.

Salt, D.E., Prince, R.C., Pickering, I.J. & Raskin, I. (1995) Mechanisms of cadmium mobility and accumulation in Indian mustard. *Plant Physiology*, **109**, 1427–33.

Salt, D.T., Fenwick, P. & Wjitaker, J.B. (1996) Interspecific herbivore interactions in a high CO_2 environment – root and shoot aphids feeding on a cardamine. *Oikos*, **77**, 326–30.

Samuel, T. & Pillai, K.K. (1989) The effect of temperature and solar radiations on volatilization, mineralization and degradation of carbon-14 DDT in soil. *Environmental Pollution*, **57** (1), 63–78.

Samways, M.J. (1996) Insects on the brink of a major discontinuity. *Biodiversity and Conservation*, **5**, 1047–58.

Samways, M.J. (2007) Insect conservation: a synthetic management approach. *Annual Review of Entomology*, **52**, 465–87.

Sanchez, L., Perez, D., Perez, T. et al. (2005) Intersectoral coordination in *Aedes aegypti* control. A pilot project in Havana City, Cuba. *Tropical Medicine and International Health*, **10**, 82–91.

Sanchis, V., Chaufaux, J. & Lereclus, D. (1995) The use of *Bacillus thuringiensis* in crop protection and the development of pest resistance. *Cahiers Agricultures*, **4** (6), 405–16.

Sanders, H.R., Evans, A.M., Ross, L.S. & Gill, S.S. (2003) Blood meal induces global changes in midgut gene expression in the disease vector, *Aedes aegypti*. *Insect Biochemistry and Molecular Biology*, **33** (11), 1105–12.

Sankar, T.V., Zynudheen, A.A., Anandan, R. & Nair, P.G.V. (2006) Distribution of organochlorine pesticides and heavy metal residues in fish and shellfish from Calicut region, Kerala, India. *Chemosphere*, **65** (4), 583–90.

Sastawa, B-M., Lawan, M. & Maina, Y.T. (2004) Management of insect pests of soybean: effects of sowing date and intercropping on damage and grain yield in the Nigerian Sudan savanna. *Crop Protection*, **23** (2), 155–61.

Sauberer, N., Zulka, K.P., Abensperg-Traun, M. et al. (2004) Surrogate taxa for biodiversity in agricultural landscapes of eastern Austria. *Biological Conservation*, **117**, 181–90.

Savage, H.M., Niebylski, M.L., Smith, G.C., Mitchell, C.J. & Craig, G.B. (1993) Host-feeding patterns of *Aedes albopictus* (Diptera: Culicidae) at a temperate North American site. *Journal of Medical Entomology*, **30** (1), 27–34.

Savolainen, V., Cowan, R.S., Vogler, A.P., Roderick, G.K. & Lane, R. (2005) Towards writing the encyclopaedia of life: an introduction to DNA barcoding. *Philosophical Transactions of the Royal Society of London Series B, Biological Sciences*, **360**, 1805–11.

Saxe, H., Ellsworth, D.S. & Heath, J. (1998) Tree and forest functioning in an enriched CO_2 atmosphere. *New Phytologist*, **139**, 395–436.

Sayer, J.A. & Stuart, S. (1988) Biological diversity and tropical forests. *Environmental Conservation*, **15**, 193–4.

Schädler, M., Jung, G., Brandl, R. & Auge, H. (2004) Secondary succession is influenced by belowground insect herbivory on a productive site. *Oecologia*, **138**, 242–52.

Schaefer, A., Konrad, R., Kuhnigk, T., Kaempfer, P., Hertel, H. & Koenig, H. (1996) Hemicellulose-degrading bacteria and yeasts from the termite gut. *Journal of Applied Bacteriology*, **80** (5), 471–8.

Schafer, M.L. & Lundstrom, J.O. (2006) Different responses of two floodwater mosquito species, *Aedes vexans* and *Ochlerotatus sticticus* (Diptera: Culicidae), to larval habitat drying. *Journal of Vector Ecology*, **31** (1), 123–8.

Schapira, A. (2006) DDT: a polluted debate in malaria control. *Lancet*, **368** (9553), 2111–13.

Schellhorn, N.A. & Sork, V.L. (1997) The impact of weed diversity on insect population dynamics and crop yield in collards, *Brassica oleracea* (Brassicaceae). *Oecologia*, **111**, 233–40.

Schenk, D., Bersier, L.F. & Bacher, S. (2005) An experimental test of the nature of predation: neither prey- nor ratio-dependent. *Journal of Animal Ecology*, **74** (1), 86–91.

Schlick-Steiner, B.C., Steiner, F.M., Moder, K. et al. (2006) A multidisciplinary approach reveals cryptic diversity in Western Palearctic Tetramorium ants (Hymenoptera: Formicidae). *Molecular Phylogenetics and Evolution*, **40**, 259–73.

Schlyter, P., Stjernquist, I., Barring, L., Jonsson, A.M. & Nilsson, C. (2006) Assessment of the impacts of climate change and weather extremes on boreal forests in northern Europe, focusing on Norway spruce. *Climate Research*, **31** (1), 75–84.

Schmickl, T. & Crailsheim, K. (2001) Cannibalism and early capping: strategy of honeybee colonies in times of experimental pollen shortages. *Journal of Comparative Physiology a, Sensory Neural and Behavioral Physiology*, **187**, 541–7.

Schmidt, B.C. & Roland, J. (2006) Moth diversity in a fragmented habitat: importance of functional groups and landscape scale in the boreal forest. *Annals of the Entomological Society of America*, **99** (6), 1110–20.

Schmidt, M.H., Lauer, A., Purtauf, T., Thies, C., Schaefer, M. & Tscharntke, T. (2003) Relative importance of predators and parasitoids for cereal aphid control. *Proceedings of the Royal Society of London Series B, Biological Sciences*, **270**, 1905–9.

Schmidt, M.H., Thewes, U., Thies, C. & Tscharntke, T. (2004) Aphid suppression by natural enemies in mulched cereals. *Entomologia Experimentalis et Applicata*, **113** (2), 87–93.

Schmitt, T.M., Hay, M.E. & Lindquist, N. (1995) Constraints on chemically mediated coevolution: multiple functions for seaweed secondary metabolites. *Ecology*, **76**, 107–23.

Schoener, T.W. (1983) Field experiments on interspecific competition. *American Science*, **70**, 586–95.

Schoener, T.W. (1988) On testing the MacArthur–Wilson model with data on rates. *American Naturalist*, **131** (6), 847–64.

Schofield, C.J., Jannin, J. & Salvatella, R. (2006) The future of Chagas disease control. *Trends in Parasitology*, **22**, 583–8.

Schonrogge, K., Barr, B., Wardlaw, J.C. et al. (2002) When rare species become endangered: cryptic speciation in myrmecophilous hoverflies. *Biological Journal of the Linnean Society*, **75**, 291–300.

Schoonhoven, L.M. (2005) Insect–plant relationships: the whole is more than the sum of its parts. *Entomologia Experimentalis et Applicata*, **115** (1), 5–6.

Schoonhoven, L.M., Beerling, E.A.M., Klijnstra, J.W. & Van-Vugt, Y. (1990) Two related butterfly species avoid oviposition near each other's eggs. *Experientia (Basel)*, **46** (5), 526–8.

Schops, K., Syrett, P. & Emberson, R.M. (1996) Summer diapause in *Chrysolina hyperici* and *C. quadrigemina* (Coleoptera: Chrysomelidae) in relation to biological control of St John's wort, *Hypericum perforatum* (Clusiaceae). *Bulletin of Entomological Research*, **86** (5), 591–7.

Schowalter, T.D., Hargrove, W.W. & Crossley, D.A. Jr. (1986) Herbivory in forested ecosystems. *Annual Review of Entomology*, **31**, 177–96.

Schreiber, S.J. & Vejdani, M. (2006) Handling time promotes the coevolution of aggregation in preclator–prey systems. *Proceedings of the Royal Society of London Series B, Biological Sciences*, **273** (1583), 185–91.

Schtickzelle, N., Choutt, J., Goffart, P., Fichefet, V. & Baguette, M. (2005) Metapopulation dynamics and conservation of the marsh fritillary butterfly: population viability analysis and management options for a critically endangered species in Western Europe. *Biological Conservation*, **126**, 569–81.

Schulthess, F., Baumgaertner, J.U., Delucchi, V. & Gutierrez, A.P. (1991) The influence of the cassava mealybug, *Phenacoccus manihoti* Mat. Ferr.) (Homoptera: Pseudococcidae) on yield formation of cassava, *Manihot esculenta* Crantz. *Journal of Applied Entomology*, **111** (2), 155–65.

Schulthess, F., Chabi-Olaye, A. & Gounou, S. (2004) Multitrophic level interactions in a cassava–maize mixed cropping system in the humid tropics of West Africa. *Bulletin of Entomological Research*, **94** (3), 261–72.

Schultz, C.B. (2001) Restoring resources for an endangered butterfly. *Journal of Applied Ecology*, **38**, 1007–19.

Schultz, C.B. & Crone, E.E. (2005) Patch size and connectivity thresholds for butterfly habitat restoration. *Conservation Biology*, **19**, 887–96.

Schultz, J.C. (1992) Factoring natural enemies into plant tissue availability to herbivores. In *Effects of Resource Distribution on Animal–Plant Interactions* (eds M.D. Hunter, T. Ohgushi & P.W. Price), pp. 175–97. Academic Press, San Diego, CA.

Schultz, J.C. (2002) How plants fight dirty – biochemical ecology. *Nature*, **416**, 267.

Schultz, J.C. & Baldwin, I. T. (1982) Oak leaf quality declines in response to defoliation by gypsy moth larvae. *Science*, **217**, 149–51.

Schultz, J.C., Foster, M.A. & Montgomery, M.E. (1990) Host plant-mediated impacts of a baculovirus on gypsy moth populations. In *Population Dynamics of Forest Insects* (eds A.D. Watt, S.R. Leather, M.D. Hunter & N.A.C. Kidd), pp. 303–13. Intercept, Andover, UK.

Schultz, J.C., Nothnagle, P.J. & Baldwin, I.T. (1982) Seasonal and individual variation in leaf quality of two northern hardwoods trees species. *American Journal of Botany*, **69** (5), 753–9.

Schultze-Lam, S., Ferris, F.G., Sherwood-Lollar, B. & Gerits, J.P. (1996) Ultrastructure and seasonal growth patterns of microbial mats in a temperate climate saline–alkaline lake: Goodenough Lake, British Columbia, Canada. *Canadian Journal of Microbiology*, **42** (2), 147–61.

Schulze, C.H., Linsenmair, K.E. & Fiedler, K. (2001) Understorey versus canopy: patterns of vertical stratification and diversity among Lepidoptera in a Bornean rain forest. *Plant Ecology*, **153**, 133–52.

Schumacher, M.J. & Egen, N.B. (1995) Significance of Africanised bees for public health. *Archives of Internal Medicine*, **155** (19), 2038–43.

Schuman, G.L. (1991) *Plant Diseases: their Biology and Social Impact*. APS Press, St Paul, MN.

Schwartz, E., Mendelson, E. & Sidi, Y. (1996) Dengue fever among travelers. *American Journal of Medicine*, **101** (5), 516–20.

Schweizer, P., Buchala, A., Dudler, R. & Metraux, J-P. (1998) Induced systemic resistance in wounded rice plants. *Plant Journal*, **14**, 475–81.

Schwenke, W. (1968) Neve Hinweise Auf einer Abhaengigkeit der Vermehrung blattund nadelfressender Forstinsekten vom Zuckergehalt ihrer Nahrung. *Zeitschrift fur Angewandte Entomologie*, **61**, 365–9.

Schwenke, W. (1994) On the fundamentals of forest insect outbreaks and of counter measures. *Anzeiger fuer Schaedlingskunde Pflanzenschulz Umweltschultz*, **67**, 120–4.

Scott, A.C., Stephenson, J. & Chaloner, W.G. (1992) Interaction and coevolution of plants and arthropods during the Palaeozoic and Mesozoic. *Philosophical Transactions of the Royal Society of London Series B, Biological Sciences*, **335** (1274), 129–65.

Scott, M.P. (2006) The role of juvenile hormone in competition. and cooperation by burying beetles. *Journal of Insect Physiology*, **52** (10), 1005–11.

Scott, S. & Duncan, C.J. (2001) *Biology of Plagues: Evidence from Historical Populations*. Cambridge University Press, Cambridge, UK.

Scott, T.W., Weaver, S.C. & Mallampalli, V.L. (1994) Evolution of mosquito-borne viruses. In *The Evolutionary Biology of Viruses* (ed. S.S. Morse), pp. 293–324. Raven Press, New York.

Scutareanu, P., Drukker, B., Bruin, J., Posthumus, M.A. & Sabelis, M.W. (1997) Volatiles from *Psylla*-infested pear trees and their possible involvement in attraction of anthocorid predators. *Journal of Chemical Ecology*, **23**, 2241–60.

Seastedt, T.R. & Crossley, D.A. Jr. (1984) The influence of arthropods on ecosystems. *Bioscience*, **34**, 157–61.

Seketeli, A. & Kuzoe, F.A.S. (1994) Diurnal resting sites of *Glossina palpalis palpalis* (Robineau-Desvoidy) in a forest edge habitat of Ivory Coast. *Insect Science and its Application*, **15** (1), 75–85.

Sequeira, A.S. & Farrell, B.D. (2001) Evolutionary origins of Gondwanan interactions: how old are Araucaria beetle herbivores? *Biological Journal of the Linnean Society*, **74** (4), 459–74.

Sergio, F., Newton, I., Marchesi, L. & Pedrini, P. (2006) Ecologically justified charisma: preservation of top predators delivers biodiversity conservation. *Journal of Applied Ecology*, **43**, 1049–55.

Service, M.W. (1991) Agricultural development and arthropod-borne disease: a review. *Revista de Saude Publica*, **25** (3), 165–78.

Service, M.W. (1996) *Medical Entomology for Students*. Chapman & Hall, London.

Service, M.W. (2004) *Medical Entomology for Students*, 3rd edn. Cambridge University Press, Cambridge, UK.

Shapiro, M., Robertson, J.L. & Webb, R.E. (1994) Effect of neem seed extract upon the gypsy moth (Lepidoptera: Lymantriidae) and its nuclear polyhedrosis virus. *Journal of Economic Entomology*, **87** (2), 356–60.

Sharkey, M.J. (2001) The all taxa biological inventory of the Great Smoky Mountains National Park. *Florida Entomologist*, **84**, 556–64.

Sharma, H.C., Franzmann, B.A. & Henzell, R.G. (2002) Mechanisms and diversity of resistance to sorghum midge, *Stenodiplosis sorghicola* in Sorghum bicolour. *Euphytica*, **124** (1), 1–12.

Sharma, S.K., Tyagi, P.K., Padhan, K. et al. (2006) Epidemiology of malaria transmission in forest and plain ecotype villages in Sundargarh District, Orissa, India. *Transactions of the Royal Society of Tropical Medicine and Hygiene*, **100** (10), 917–25.

Sharma, Y.D., Biswas, S., Pillai, C.R., Ansari, M.A., Adak, T. & Devi, C.U. (1996) High prevalence of chloroquine resistant *Plasmodium falciparum* infection in Rajasthan epidemic. *Acta Tropica*, **62** (3), 135–41.

Shelton, A.M. & Badenes-Perez, E. (2006) Concepts and applications of trap cropping in pest management. *Annual Review of Entomology*, **51**, 285–308.

Shelton, A.M. & Nault, B.A. (2004) Dead-end trap cropping: a technique to improve management of the diamondback moth, *Plutella xylostella* (Lepidoptera: Plutellidae). *Crop Protection*, **23** (6), 497–503.

Shen, B.Z., Zheng, Z.W. & Dooner, H.K. (2000) A maize sesquiterpene cyclase gene induced by insect herbivory and volicitin: characterization of wild-type and mutant alleles. *Proceedings of the National Academy of Sciences of the USA*, **97**, 14807–12.

Shintani, Y., Munyiri, F.N. & Ishikawa, Y. (2003) Change in significance of feeding during larval development in the yellow-spotted longicorn beetle, Psacothea hilaris. *Journal of Insect Physiology*, **49**, 975–81.

Shirt, B.D. (1987) *British Red Data Book: 2, Insects*. Nature Conservancy Council, Peterborough, UK.

Shonle, I. & Bergelson, J. (1995) Interplant communication revisited. *Ecology*, **76**, 2660–3.

Short, R.A. & Maslin, P.E. (1977) Processing of leaf litter by a stream detritivore: effect on nutrient availability to collectors. *Ecology*, **58**, 935–8.

Showler, A.T. (1995) Locust (Orthoptera: Acrididae) outbreak in Africa and Asia, 1992–4: an overview. *American Entomologist*, **41** (3), 179–85.

Showler, A.T. (2002) A summary of control strategies for the desert locust, Schistocerca gregaria (Forskal). *Agriculture Ecosystems and Environment*, **90** (1), 97–103.

Showler, A.T. (2003) The importance of armed conflict to desert locust control, 1986–2002. *Journal of Orthoptera Research*, **12**, 127–33.

Showler, A.T. & Potter, C.S. (1991) Synopsis of the 1986–9 desert locust (Orthoptera: Acrididae) plague and the concept of strategic control. *American Entomologist*, **37** (2), 106–10.

Shultz, J.W. & Regier, J.C. (2000) Phylogenetic analysis of arthropods using two nuclear protein-encoding genes supports a crustacean plus hexapod clade. *Proceedings of the Royal Society of London Series B, Biological Sciences*, **267** (1447), 1011–19.

Siepielski, A.M. & Benkman, C.W. (2004) Interactions among moths, crossbills, squirrels, and lodgepole pine in a geographic selection mosaic. *Evolution*, **58**, 95–101.

Siitonen, J. (2001) Forest management, coarse woody debris and saproxylic organisms: Fennoscandian boreal forests as an example. *Ecological Bulletins*, **49**, 11–41.

Silva, A., Bacci, M. Jr., Gomes, D.S.C., Correa, B.O., Pagnocca, F.C. & Aparecida, H.M.J. (2003) Survival of *Atta sexdens* workers on different food sources. *Journal of Insect Physiology*, **49** (4), 307–13.

Silva-Bohorquez, I. (1987) *Interspecific interactions between insects on oak trees, with special reference to defoliators and the oak aphid*. DPhil Thesis, University of Oxford.

Silverman, A.L., McCray, D.C., Gordon, S.C., Morgan, W.T. & Walker, E.D. (1996) Experimental evidence against replication or dissemination of hepatitis C virus in mosquitoes (Diptera: Culicidae) using detection by reverse transcriptase polymerase chain reaction. *Journal of Medical Entomology*, **33** (3), 398–401.

Simard, J.R. & Fryxell, J.M. (2003) Effects of selective logging on terrestrial small mammals and arthropods. *Canadian Journal of Zoology, Revue Canadienne de Zoologie*, **81**, 1318–26.

Simberloff, D. & Stiling, P. (1996) How risky is biological control? *Ecology*, **77** (7), 1965–74.

Simberloff, D. & Stiling, P. (1998) How risky is biological control? Reply. *Ecology*, **79** (5). 1834–6.

Simchuk, A.P. & Ivashov, A.V. (2006) Ecological-genetic aspects of trophic preference partitioning in a micro-assemblage of oak herbivores. *Zhurnal Obshchei Biologii*, **67** (1), 53–61.

Sime, K.R. & Brower, A.V.Z. (1998) Explaining the latitudinal gradient anomaly in ichneumonid species richness: evidence from butterflies. *Journal of Animal Ecology*, **67**, 387–99.

Simms, E.L. & Rausher, M.D. (1989) The evolution of resistance to herbivory in *Ipomoea purpurea*: II. Natural selection by insects and costs of resistance. *Evolution*, **43**, 573–85.

Simms, E.L. & Triplett, J. (1994) Costs and benefits of plant responses to diseases. *Resistance and Tolerance Evolution*, **48**, 1973–85.

Simon, J.C., Carre, S., Boutin, M. et al. (2003) Host-based divergence in populations of the pea aphid: insights from nuclear markers and the prevalence of facultative symbionts. *Proceedings of the Royal Society of London Series B, Biological Sciences*, **270** (1525), 1703–12.

Simon, J.C., Rispe, C. & Sunnucks, P. (2002) Ecology and evolution of sex in aphids. *Trends in Ecology and Evolution*, **17** (1), 34–9.

Sinclair, B.J., Vernon, P., Klok, C.J. & Chown, S.L. (2003). Insects at low temperatures: an ecological perspective. *Trends in Ecology and Evolution*, **18** (5), 257–62.

Singer, M.S., Carriere Y., Theuring C. & Hartmann T. (2004a) Disentangling food quality from resistance against parasitoids: diet choice by a generalist caterpillar. *American Naturalist*, **164**, 423–9.

Singer, M.S., Rodrigues, D., Stireman, J.O. & Carriere, Y. (2004b) Roles of food quality and enemy-free space in host use by a generalist insect herbivore. *Ecology*, **85**, 2747–53.

Singh, B.K. & Walker, A. (2006) Microbial degradation of organophosphorus compounds. *Fems Microbiology Reviews*, **30** (3), 428–71.

Sinsabaugh, R.L., Likens, A.E. & Benfield, E.F. (1985) Cellulose digestion and assimilation by three leaf-shredding insects. *Ecology*, **66**, 1464–71.

Siva-Jothy, M.T. (2006) Trauma, disease and collateral damage: conflict in cimicids. *Philosophical Transactions of the Royal Society of London Series B, Biological Sciences*, **361** (1466), 269–75.

Sjoeib, F., Anwar, E. & Tungguldihardjo, M.S. (1994) Behaviour of DDT and DDE in Indonesian tropical environments. *Journal of Environmental Science and Health, Part B – Pesticides, Food, Contaminants and Agricultural Wastes*, **29** (1), 17–24.

Skaf, R., Popov, G.B. & Roffer, J. (1990) The desert locust: an international challenge. *Philosophical Transactions of the Royal Society of London Series B, Biological Sciences*, **328** (1251), 525–38.

Skare, J.U., Stenersen, J., Kveseth, N. & Polder, A. (1985) Time trends of organochlorine residues in 7 sedentary marine fish species from a Norwegian fjord during the period 1972–82. *Archives of Environmental Contamination and Toxicology*, **14** (1), 33–42.

Sket, B. (2004) The cave hygropetric – a little known habitat and its inhabitants. *Archiv fuer Hydrobiologie*, **160** (2), 413–25.

Skovgard, H. & Pats, P. (1997) Reduction of stemborer damage by intercropping maize with cowpea. *Agriculture, Ecosystems and Environment*, **62** (1), 13–19.

Slosser, J.E., Bordovsky, D.G. & Bevers, S.J. (1994) Damage and costs associated with insect management options in irrigated cotton. *Journal of Economic Entomology*, **87** (2), 436–45.

Smiley, J.T. (1985) Are chemical barriers necessary for evolution of butterfly–plant associations? *Oecologia*, **65**, 580–3.

Smith, A.M. & Ward, S.A. (1995) Temperature effects on larval and pupal development, adult emergence, and survival of the pea weevil (Coleoptera: Chrysomelidae). *Environmental Entomology*, **24** (3), 623–34.

Smith, C.M. & Boyko, E. V. (2007) The molecular bases of plant resistance and defense responses to aphid feeding: current status. *Entomologia Experimentalis et Applicata*, **122**, 1–16.

Smith, D.A.S. & Owen, D.F. (1997) Colour genes as markers for migratory activity: the butterfly *Danaus chrysippus* in Africa. *Oikos*, **78** (1), 127–35.

Smith, G-R. & Candy, J-M. (2004). Improving Fiji disease resistance screening trials in sugarcane by considering virus transmission class and possible origin of Fiji disease virus. *Australian Journal of Agricultural Research*, **55** (6), 665–72.

Smith, J.W.J.R. & Johnson, S.J. (1989) Natural mortality of the lesser cornstalk borer (Lepidoptera: Pyralidae) in a peanut agroecosystem. *Environmental Entomology*, **18** (1), 69–77.

Smith, R.J., Muir, R.D.J., Walpole, M.J., Balmford, A. & Leader-Williams, N. (2003) Governance and the loss of biodiversity. *Nature*, **426**, 67–70.

Snow, R.S. & Gilles, H.M. (2002) The epidemiology of malaria. In *Essential Malariology*, 4th edn (eds D.A. Warrell & H.M. Gilles), pp. 85–106. Arnold, London.

Snyder, W.E., Ballard, S.N., Yang, S. et al. (2004) Complementary biocontrol of aphids by the ladybird beetle *Harmonia axyridis* and the parasitoid *Aphelinus asychis* on greenhouse roses. *Biological Control*, **30**, 229–35.

Socha, R. (2006) Endocrine control of wing morph-related differences in mating success and accessory gland size in male firebugs. *Animal Behaviour*, **71**, 1273–81.

Socha, R. & Zemek, R. (2003) Wing morph-related differences in the walking pattern and dispersal in a flightless bug, *Pyrrhocoris apterus* (Heteroptera). *Oikos*, **100**, 35–42.

Soe, A.R.B., Bartram, S., Gatto, N. & Boland, W. (2004) Are iridoids in leaf beetle larvae synthesized de novo or derived from plant precursors? A methodological approach. *Isotopes in Environmental and Health Studies*, **40**, 175–80.

Solla, A. & Gil, L. (2002) Influence of water stress on Dutch elm disease symptoms in *Ulmus minor*. *Canadian Journal of Botany*, **80**, 810–17.

Somta, P., Talekar, N.S. & Srinives, P. (2006) Characterization of *Callosobruchus chinensis* (L.) resistance in *Vigna umbellata* (Thunb.) Ohwi & Ohashi. *Journal of Stored Products Research*, **42** (3), 313–27.

Soto-Pinto, L., Perfecto, I. & Caballero-Nieto, J. (2002) Shade over coffee: its effects on berry borer, leaf rust and spontaneous herbs in Chiapas, Mexico. *Agroforestry Systems*, **55** (1), 37–45.

Southwood, T.R.E. (1961) The number of species of insect associated with various trees. *Journal of Animal Ecology*, **30**, 1–8.

Southwood, T.R.E. (1973) The insect/plant relationship – an evolutionary perspective. *Symposium of the Royal Entomological Society of London*, **6**, 3–30.

Southwood, T.R.E. (1977) Habitat, the templet for ecological strategies. *Journal of Animal Ecology*, **46**, 337–65.

Southwood, T.R.E. (2003) *The Story of Life*. Oxford University Press, Oxford.

Southwood, T.R.E. & Henderson, P.A. (2000) *Ecological Methods*, 3rd edn. Blackwell Publishing, Oxford.

Southwood, T.R.E., Moran, V.C. & Kennedy, C.E.J. (1982) The richness, abundance and biomass of the arthropod communities on trees. *Journal of Animal Ecology*, **51**, 635–49.

Sparks, T.H., Dennis, R.L.H., Croxton, P.J. & Cade, M. (2007) Increased migration of Lepidoptera linked to climate change. *European Journal of Entomology*, **104**, 139–43.

Specty, O., Febvay, G., Grenier, S. et al. (2003) Nutritional plasticity of the predatory ladybeetle *Harmonia axyridis* (Coleoptera: Coccinellidae): comparison between natural and substitution prey. *Archives of Insect Biochemistry and Physiology*, **52**, 81–91.

Speight, M.R. (1992) The impact of leaf-feeding by nymphs of the horse chestnut scale *Pulvinaria regalis* on young host trees. *Journal of Applied Entomology*, **112**, 389–99.

Speight, M.R. (1994) Reproductive capacity of the horse chestnut scale insect, *Pulvinaria regalis* Canard (Hom., Coccidae). *Journal of Applied Entomology*, **118** (1), 59–67.

Speight, M.R. (1997a) The relationship between host tree stresses and insect attack in tropical forest plantations, and its relevance to pest management. In *Impact of Diseases and Insect Pests in Tropical Forests, International Union of Forest Research Organizations (IUFRO) Symposium, Kerala Forest Research Institute, Peechi, India* (eds K.S.S. Nair, J.K. Sharma & R.C. Varma).

Speight, M.R. (1997b) Forest pests in the tropics: current status and future threats. In *Insects and Trees, Royal Entomological Society Symposium on Forest and Insects* (eds A.D. Watt, M.D. Hunter & N.E. Stork). Chapman & Hall, London.

Speight, M.R. & Cory, J.S. (2001) Integrated pest management of *Hysipyla* shoot borers. In *Proceedings of an International Workshop on Hypsipyla Shoot Borers of the Meliaceae at Kandy Sri Lanka, August 20–23, 1996* (eds R. Floyd & C. Hauxwell). Australian Centre for International Agricultural Research, Canberra, Australia.

Speight, M.R & Evans, H.F. (2004) Health and protection: integrated pest management principles. In *Encyclopedia of Forest Sciences*, Vol. 1: A–L (eds J. Burley, J. Evans & J.A. Youngquist), pp. 305–18. Elsevier, Amsterdam.

Speight, M.R., Hails, R.S., Gilbert, M. & Foggo, A. (1998) Horse chestnut scale (*Pulvinaria regalis*) (Homoptera: Coccidae) and urban host tree environment. *Ecology (Washington DC)*, **79** (5), 1503–13.

Speight, M.R., Intachat, J., Chey, V.K. & Chung, A.Y.C. (2003). Canopy arthropods and rainforest manipulation. In *Influences of Forest Management on Insects* (eds Y. Basset, R.L. Kitching, S.E. Miller & V. Novotny), pp. 380–9. Cambridge University Press, Cambridge, UK.

Speight, M.R., Kelly, P.M., Sterling, P.H. & Entwistle, P.F. (1992) Field application of a nuclear polyhedrosis virus against the brown-tail moth, *Euproctis chrysorrhoea* (L.) (Lepidoptera, Lymantriidae). *Journal of Applied Entomology*, **113** (3), 295–306.

Speight, M.R. & Lawton, J.H. (1976) The influence of weed-cover on the mortality imposed on artificial prey by predatory ground beetles in cereal fields. *Oecologia*, **23**, 211–23.

Speight, M.R. & Wainhouse, D. (1989) *Ecology and Management of Forest Insects*. Oxford Science Publications, Oxford.

Speight, M.R. & Wylie, F.R. (2001) *Insect Pests in Tropical Forestry*. Oxford University Press, Oxford.

Spencer, K.A. (1972) *Handbooks for the Identification of British Insects*, Vol. 10. Royal Entomological Society, London.

Spies, T.A., Hemstrom, M.A., Youngblood, A. & Hummel, S. (2006) Conserving old-growth forest diversity in disturbance-prone landscapes. *Conservation Biology*, **20**, 351–62.

Srivastava, D.S. & Vellend, M. (2005) Biodiversity-ecosystem function research: is it relevant to conservation? *Annual Review of Ecology Evolution and Systematics*, **36**, 267–94.

St George, R.A. (1930) Drought affected and injured trees attractive to bark beetles. *Journal of Economic Entomology*, **23**, 825–8.

Stabentheiner, A., Vollmann, J., Kovac, H. & Crailsheim, K. (2003) Oxygen consumption and body temperature of active and resting honeybees. *Journal of Insect Physiology*, **49**, 881–9.

Stadler, B. (2004) Wedged between bottom-up and top-down processes: aphids on tansy. *Ecological Entomology*, **29**, 106–16.

Stadler, B. & Mackauer, M. (1996) Influence of plant quality on interactions between the aphid parasitoid *Ephedrus californicus* Baker (Hymenoptera: Aphidiidae) and its host, *Acyrthosiphon pisum* (Harris) (Homoptera: Aphididae). *Canadian Entomologist*, **128** (1), 27–39.

Stadler, B., Muller, T. & Orwig, D. (2006) The ecology of energy and nutrient fluxes in hemlock forests invaded by hemlock woolly adelgid. *ECOLOGY*, **87** (7), 1792–804.

Staley, J.T., Mortimer, S.R., Masters, G.J., Morecroft, M.D., Brown, V.K. & Taylor, M.E. (2006) Drought stress differentially affects leaf-mining species. *Ecological Entomology*, **31**, 460–9.

Stamp, N. (2003) Out of the quagmire of plant defense hypotheses. *Quarterly Review of Biology*, **78**, 23–55.

Standley, L.J. & Sweeney, B.W. (1995) Organochlorine pesticides in stream mayflies and terrestrial vegetation of undisturbed tropical catchments exposed to long-range atmospheric transport. *Journal of the North American Benthological Society*, **14** (1), 38–49.

Stefanescu, C., Herrando, S. & Paramo, F. (2004) Butterfly species richness in the north-west Mediterranean basin: the role of natural and human-induced factors. *Journal of Biogeography*, **31**, 905–15.

Steffan-Dewenter, I. & Tscharntke, T. (2000) Butterfly community structure in fragmented habitats. *Ecology Letters*, **3**, 449–56.

Steinbauer, M-J., Kriticos, D-J., Lukacs, Z. & Clarke, A.R. (2004) Modelling a forest lepidopteran: phenological plasticity determines voltinism which influences population dynamics. *Forest Ecology and Management*, **198**, 117–31.

Steineke, S.B. & Jehle, J.A. (2004) Investigating the horizontal transmission of the *Cydia pomonella* granulovirus (CpGV) in a model system. *Biological Control*, **30** (3), 538–45.

Stelinski, L.L., Gut, L.J., Pierzchala, A.V. & Miller, J.R. (2004) Field observations quantifying attraction of four tortricid moths to high-dosage pheromone dispensers in untreated and pheromone-treated orchards. *Entomologia Experimentalis et Applicata*, **113** (3), 187–96.

Stenseth, N.C., Samia, N.I., Viljugrein, H. et al. (2006) Plague dynamics are driven by climate variation. *Proceedings of the National Academy of Sciences of the USA*, **103** (35), 13110–15.

Sterling, P.H. & Speight, M.R. (1989) Comparative mortalities of the brown-tail moth, *Euproctis chrysorrhoea* (L.) (Lepidoptera: Lymantriidae), in south-east England. *Botanical Journal of the Linnean Society*, **101**, 69–78.

Stern, G.A., Macdonald, C.R., Armstrong, D. et al. (2005) Spatial trends and factors affecting variation of organochlorine contaminants levels in Canadian Arctic beluga (*Delphinapterus leucas*). *Science of the Total Environment*, **351**, 344–68.

Sterner, R.W. & Elser, J.J. (2002) *Ecological Stoichiometry: the Biology of Elements from Molecules to the Biosphere*. Princeton University Press, Princeton, NJ.

Stevens, M., Smith, H.G. & Hallsworth, P.B. (1994) The host range of beet yellowing viruses along common arable weed species. *Plant Pathology (Oxford)*, **43** (3), 579–88.

Stevenson, P.C., Blaney, W.M., Simmonds, M.J.S. & Wightman, J.A. (1993) The identification and characterization of resistance in wild species of *Arachis* to *Spodoptera litura* (Lepidoptera: Noctuidae). *Bulletin of Entomological Research*, **83**, 421–9.

Steverding, D. & Troscianko, T. (2004) On the role of blue shadows in the visual behaviour of tsetse flies. *Proceedings of the Royal Society of London Series B, Biological Sciences*, **271**, S16–S17.

Stich, A., Barrett, M.P. & Krishna, S. (2003) Waking up to sleeping sickness. *Trends in Parasitology*, **19**, 195–7.

Stiling, P. (1987) The frequency of density dependence in insect host parasitoid systems. *Ecology*, **68**, 844–56.

Stiling, P. (1988) Density-dependent processes and key factors in insect populations. *Journal of Animal Ecology*, **57**, 581–93.

Stiling, P. (2002) *Ecology – Theories and Applications*, 4th edn. Prentice Hall, New Jersey.

Stiling, P. (2004) Biological control not on target. *Biological Invasions*, **6** (2), 151–9.

Stiling, P., Moon, D.C., Hunter, M.D. et al. (2003) Elevated CO_2 lowers relative and absolute herbivore density across all species of a scrub-oak forest. *Oecologia*, **134**, 82–7.

Stiling, P. & Rossi, A.M. (1996) Complex effects of genotype and environment on insect herbivores and their enemies. *Ecology*, **77**, 2212–18.

Stiling, P. & Rossi, A.M. (1998) Deme formation in a dispersive gall-forming midge. In *Genetic Structure and Local Adaptation in Natural Insect Populations. Effects of Ecology, Life History, and Behavior* (eds S. Mopper & S.Y. Strauss), pp. 22–36. Chapman & Hall, New York.

Stiling, P., Rossi, A.M., Hungate, B. et al. (1999) Decreased leaf-miner abundance in elevated CO_2: reduced leaf quality and increased parasitoid attack. *Ecological Applications*, **9**, 240–4.

Stoate, C., Henderson, I.G. & Parish, D.M.B. (2004) Development of an agri-environment scheme option: seed-bearing crops for farmland birds. *Ibis*, **146**, 203–9.

Stocks, K.I. & Grassle, J.F. (2001) Effects of microalgae and food limitation on the recolonization of benthic macrofauna into in situ saltmarsh-pond mesocosms. *Marine Ecology, Progress Series*, **221**, 93–104.

Stoll-Kleemann, S. (2001) Reconciling opposition to protected areas management in Europe: the German experience. *Environment*, **43**, 32–44.

Stoll-Kleemann, S. (2005) Voices for biodiversity management in the 21st century. *Environment*, **47**, 24–36.

Stone, C. & Bacon, P.E. (1994) Relationship among moisture stress, insect herbivory, foliar cineole content and the growth of river red gum *Eucalyptus camaldulensis*. *Journal of Applied Ecology*, **31**, 604–12.

Stork, N.E. (1987) Guild structure of arthropods from Bornean rain-forest trees. *Ecological Entomology*, **12**, 69–80.

Stork, N.E. (1988) Insect diversity – facts, fiction and speculation. *Biological Journal of the Linnean Society*, **35**, 321–37.

Stork, N.E. (1990) *The Role of Ground Beetles in Ecological and Environmental Studies*. Intercept, Andover, UK.

Stork, N.E. (1991) The composition of the arthropod fauna of Bornean lowland rain forest trees. *Journal of Tropical Ecology*, **7** (2), 161–88.

Stork, N.E. (1993) How many species are there. *Biodiversity and Conservation*, **2**, 215–32.

Stork, N.E. (1995) Measuring and monitoring arthropod diversity in temperate and tropical forests. In: *Measuring and Monitoring Biodiversity in Tropical and Temperate Forests* (eds T.J.B. Boyle & B. Boontawee), pp. 257–70. Centre for International Forestry Research (CIFOR), Bogor, Indonesia.

Stork, N.E. (1997) Measuring global biodiversity and its decline. In *Biodiversity II, Understanding and Protecting our Biological Resources* (eds M.L. Reaka-Kudla, D.E. Wilson & E.O. Wilson), pp. 41–68. Joseph Henry Press, Washington, DC.

Stork, N.E., Adis, J. & Didham, R.K. (1997) *Canopy Arthropods*. Chapman & Hall, London.

Stork, N.E., Balston, J., Farquhar, G.D., Franks, P.J., Holtum, J.A.M. & Liddell, M.J. (2007) Tropical rainforest canopies and climate change. *Austral Ecology*, **32**, 105–12.

Stork, N.E. & Brendell, M.J.D. (1993) Arthropod abundance in lowland rain forest of Seram. In *Natural History of Seram* (eds I.D. Edwards, A.A. MacDonald & J. Proctor), pp. 115–30. Intercept, Andover, UK.

Stork, N.E. & Gaston, K.J. (1990) Counting species one by one. *New Scientist*, **1729**, 43–7.

Stork, N.E. & Lyal, C.H.C. (1993) Extinction or co-extinction rates. *Nature* **366**, 307.

Stork, N.E., Srivastava, D.S., Watt, A.D. & Larsen, T.B. (2003) Butterfly diversity and silvicultural practice in lowland rainforests of Cameroon. *Biodiversity and Conservation*, **12**, 387–410.

Stowe, K.A. (1998) Experimental evolution of resistance in *Brassica rapa*: correlated response of tolerance in lines selected for glucosinolate content. *Evolution*, **52**, 703–12.

Strassmann, J.E., Solis, C.R., Hughes, C.R., Goodnight, K.F. & Queller, D.C. (1997) Colony life history and demography of a swarm-founding social wasp. *Behavioral Ecology and Sociobiology*, **40** (2), 71–7.

Strathdee, A.T. & Bale, J.S. (1998) Life on the edge: insect ecology in arctic environments. *Annual Review of Entomology*, **43**, 85–106.

Strathdee, A.T., Bale, J.S., Block, W.C., Webb, N.R., Hodkinson, I.D. & Coulson, S.J. (1993) Extreme adaptive life-cycle in a high arctic aphid, *Acyrthosiphon svalbardicum*. *Ecological Entomology*, **18** (3), 254–8.

Strausbaugh, C.A., Gillen, A.M., Gallian, J.J., Camp, S. & Stander, J.R. (2006) Influence of host resistance and insecticide seed treatments on curly top in sugar beets. *Plant Disease*, **90** (12), 1539–44.

Strauss, S.Y. & Murch, P. (2004) Towards an understanding of the mechanisms of tolerance: compensating for herbivore damage by enhancing a mutualism. *Ecological Entomology*, **29**, 234–9.

Strauss, S.Y., Rudgers, J.A., Lau, J.A. & Irwin, R.E. (2002) Direct and ecological costs of resistance to herbivory. *Trends in Ecology and Evolution*, **17**, 278–85.

Strauss, S.Y., Siemens, D.H., Decher, M.B. & Mitchell-Olds, T. (1999) Ecological costs of plant resistance to herbivores in the currency of pollination. *Evolution*, **53**, 1105–13.

Stride, B., Shah, A. & Sadeed, S.M. (2003) Recent history of Moroccan locust control and implementation of mechanical control methods in northern Afghanistan. *International Journal of Pest Management*, **49** (4), 265–70.

Strong, D.R. (1974) Rapid asymptotic species accumulation in phytophagous insect communities: the pests of cacao. *Science*, **185**, 1064–6.

Strong, D.R. (1982) Potential interspecific competition and host specificity: hispine beetles on *Heliconia*. *Ecological Entomology*, **7**, 217–20.

Strong, D.R., Lawton, J.H. & Southwood, T.R.E. (1984) *Insects on Plants: Community Patterns and Mechanisms*. Blackwell Scientific, Oxford.

Strong, D.R., Maron, J.L., Connors, P.G., Whipple, A., Harrison, S. & Jeffries, R.L. (1995) High mortality, fluctuation in numbers, and heavy subterranean insect herbivory in bush lupine, *Lupinus arboreus*. *Oecologia*, **104**, 85–92.

Strong, D.R., McCoy, E.D. & Rye, J.R. (1977) Time and number of herbivore species: the pests of sugarcane. *Ecology*, **58**, 167–75.

Stuening, D. (1988) Biological–ecological investigations on the Lepidoptera of the supralittoral zone of the North Sea coast. *Faunistisch-Oekologische Mitteilungen*, **7** (Suppl), 1–116.

Sturtevant, B.R., Gustafson, E.J. & He, H.S. (2004) Modeling disturbance and succession in forest landscapes using LANDIS: introduction. *Ecological Modelling*, **180** (1), 1–5.

Styer, L.M., Minnick, S.L., Sun, A.K. & Scott, T.W. (2007) Mortality and reproductive dynamics of *Aedes aegypti* (Diptera: Culicidae) fed human blood. *Vector-Borne and Zoonotic Diseases*, **7** (1), 86–98.

Styles, C.V. & Skinner, J.D. (1996) Possible factors contributing to the exclusion of saturniid caterpillars (mopane worms) from a protected area in Botswana. *African Journal of Ecology*, **34** (3), 276–83.

Subramanian, S., Santharam, G., Sathiah, N., Kennedy, J.S. & Rabindra, R.J. (2006) Influence of incubation temperature on productivity and quality of *Spodoptera litura* nucleopolyhedrovirus. *Biological Control*, **37** (3), 367–74.

Sulaiman, S., Pawanchee, Z.A., Othman, H.F. et al. (2002) Field evaluation of cypermethrin and cyfluthrin against dengue vectors in a housing estate in Malaysia. *Journal of Vector Ecology*, **27**, 230–4.

Sulaiman, S., Pawanchee, Z.A., Wahab, A., Jamal, J. & Sohadi, A.R. (1997) Field evaluation of Vectobac G, Vectobac 12AS and Bactimos WP against the dengue vector *Aedes albopictus* in tires. *Journal of Vector Ecology*, **22** (2), 122–4.

Summers, C.G., Newton, A.S. & Opgenorth, D.C. (2004) Overwintering of corn leafhopper, *Dalbulus maidis* (Homoptera: Cicadellidae), and *Spiroplasma kunkelii* (Mycoplasmatales: Spiroplasmataceae) in California's San Joaquin valley. *Environmental Entomology*, **33** (6), 1644–51.

Sunderland, K.D. (1988) Quantitative methods for detecting invertebrate predation occurring in the field. *Annals of Applied Biology*, **112**, 201–24.

Sunderland, K.D., Crook, N.E., Stacey, D.L. & Fuller, B.T. (1987) A study of feeding by polyphagous predators on cereal aphids using ELISA and gut dissection. *Journal of Applied Ecology*, **24**, 907–33.

Sunderland, K.D., Fraser, A.M. & Dixon, A.F.G. (1986) Field and laboratory studies on money spiders (Linyphiidae) as predators of cereal aphids. *Journal of Applied Ecology*, **23**, 433–47.

Sunderland, K.D. & Vickerman, G.P. (1980) Aphid feeding by some polyphagous predators in relation to aphid density in cereal fields. *Journal of Applied Ecology*, **17**, 389–96.

Suomi, D.A. & Akre, R.D. (1993) Biological studies of *Hemicoelus gibbicollis* (Leconte) (Coleoptera: Anobiidae), a serious structural pest along the Pacific coast: larval and pupal stages. *Pan-Pacific Entomologist*, **69** (3), 221–35.

Suominen, O., Niemelä, J., Martikainen, P., Niemelä, P. & Kojola, I. (2003) Impact of reindeer grazing on ground-dwelling Carabidae and Curculionidae assemblages in Lapland. *Ecography*, **26**, 503–13.

Sutton, S.L. & Collins, N.M. (1991) Insects and tropical forest conservation. In *The Conservation of Insects and their Habitats* (eds N.M. Collins & J.A. Thomas), pp. 405–24. Academic Press, London.

Swank, W.T., Waide, J.B., Crossley, D.A. Jr. & Todd, R.L. (1981) Insect defoliation enhances nitrate export from forest ecosystems. *Oecologia*, **51**, 297–9.

Swetnam, T.W. & Lynch, A.M. (1993) Multicentury, regional-scale patterns of western spruce budworm outbreaks. *Ecological Monographs*, **63** (4), 399–424.

Symmons, P.M. (1986) Locust (*Chortoicetes terminifera*) displacing winds in eastern Australia. *International Journal of Biometeorology*, **30** (1), 53–64.

Symmons, P.M. (1992) Strategies to combat the desert locust. *Crop Protection*, **11** (3), 206–12.

Symonds, M.R.E. & Elgar, M.A. (2004) Species overlap, speciation and the evolution of aggregation pheromones in bark beetles. *Ecology Letters*, **7** (3), 202–12.

Szewczyk, B., Hoyos-Carvajal, L., Paluszek, M., Skrzecz, W. & de Souza, M.L. (2006) Baculoviruses – re-emerging biopesticides. *Biotechnology Advances*, **24** (2), 143–60.

Tabashnik, B.E., Biggs, R.W., Fabrick, J.A. et al. (2006) High-level resistance to *Bacillus thuringiensis* toxin CrylAc and cadherin genotype in pink bollworm. *Journal of Economic Entomology*, **99** (6), 2125–31.

Takiya, D.M., Tran, P.L., Dietrich, C.H. et al. (2006) Co-cladogenesis spanning three phyla: leafhoppers (Insecta: Hemiptera: Cicadellidae) and their dual bacterial symbionts. *Molecular Ecology*, **15** (13), 4175–91.

Tanabe, S., Subramanian, A., Ramesh, A., Kumaran, P.L., Miyazaki, N. & Tatsuhawa, R. (1993) Persistent organochlorine residues in dolphins from the Bay of Bengal, South India. *Marine Pollution Bulletin*, **26** (6), 311–16.

Tanaka, K. & Matsumura, M. (2000) Development of virulence to resistant rice varieties in the brown planthopper, *Nilaparvata lugens* (Homoptera: Delphacidae), immigrating into Japan. *Applied Entomology and Zoology*, **35** (4), 529–33.

Tang, W., Chen, H., Xu, C.G., Li, X.H., Lin, Y.J. & Zhang, Q.F. (2006) Development of insect-resistant transgenic indica rice with a synthetic cry1C* gene. *Molecular Breeding*, **18** (1), 1–10.

Tanner, R.A. & Gange, A.C. (2005) Effects of golf courses on local biodiversity. *Landscape and Urban Planning*, **71**, 137–46.

Tansley, A.G. (1935) The use and abuse of vegetational concepts and terms. *Ecology*, **16**, 284–307.

Tartes, U., Vanatoa, A. & Kuusik, A. (2002) The insect abdomen – a heartbeat manager in insects? *Comparative Biochemistry and Physiology a, Molecular and Integrative Physiology*, **133**, 611–23.

Tauil, P.L. (2006) Perspectives of vector borne diseases control in Brazil. *Revista da Sociedade Brasileira de Medicina Tropical*, **39** (3), 275–7.

Taylor, P.S., Shields, L.J., Tauber, M.J. & Tauber, C.A. (1995) Induction of reproductive diapause in *Empoasca fabae* (Homoptera: Cicadellidae) and its implications regarding southward migration. *Environmental Entomology*, **24** (5), 1086–95.

Tcheslavskaia, K.S., Thorpe, K.W., Brewster, C.C. et al. (2005) Optimization of pheromone dosage for gypsy moth mating disruption. *Entomologia Experimentalis et Applicata*, **115**, 355–61.

Tebayashi, S.I., Matsuyama, S., Suzuki, T., Kuwahara, Y., Nemoto, T. & Fujii, K. (1995) Quercimeritrin: the third oviposition stimulant of the Azuki bean weevil from the host azuki bean. *Journal of Pesticide Science*, **20**, 299–305.

Teder, T., Tammaru, T. & Pedmanson, R. (1999) Patterns of host use in solitary parasitoids (Hymenoptera, Ichneumonidae): field evidence from a homogeneous habitat. *Ecography*, **22**, 79–86.

Teixeira, M.L.F., Coutinho, H.L.C. & Franco, A.A. (1996) Effects of *Cerotoma arcuata* (Coleoptera: Chrysomelidae) on predation of nodules and on N$_2$ fixation of *Phaseolus vulgaris*. *Journal of Economic Entomology*, **89**, 165–9.

Tekie, H., Seyoum, E. & Saxena, R.C. (2006) Potential of neem, *Azadirachta indica* A. Juss., in the management of *Chilo partellus* (Swinhoe) on maize in Kenya. *African Entomology*, **14** (2), 373–9.

Telang, A., Booton, V., Chapman, R.F. & Wheeler, D.E. (2001) How female caterpillars accumulate their nutrient reserves. *Journal of Insect Physiology*, **47**, 1055–64.

Telonis-Scott, M., Guthridge, K.M. & Hoffmann, A.A. (2006) A new set of laboratory-selected *Drosophila melanogaster* lines for the analysis of desiccation resistance: response to selection, physiology and correlated responses. *Journal of Experimental Biology*, **209**, 1837–47.

Temple, B., Pines, P.A. & Hintz, W.E. (2006) A nine-year genetic survey of the causal agent of Dutch elm disease, *Ophiostoma novo-ulmi* in Winnipeg, Canada. *Mycological Research*, **110**, 594–600.

Teng, P.S. (1994) Integrated pest management in rice. *Experimental Agriculture*, **30** (2), 115–37.

Terras, F.R.G., Penninckx, I.A.M.A., Goderis, I.J. & Broekaert, W.F. (1998) Evidence that the role of plant defensins in radish defense responses is independent of salicylic acid. *Planta*, **206**, 117–24.

Thacker, J.R.M., Bryan, W.J., McGinley, C., Heritage, S. & Strang, R.H.C. (2003) Field and laboratory studies on the effects of neem (*Azadirachta indica*) oil on the feeding activity of the large pine weevil (*Hylobius abietis* L.) and implications for pest control in commercial conifer plantations. *Crop Protection*, **22** (5), 753–60.

Thaler, J.S. (1999) Jasmonate-inducible plant defences cause increased parasitism of herbivores. *Nature*, **399**, 686–8.

Thaler, J.S., Fidantsef, A.L., Duffey, S.S. & Bostock, R.M. (1999) Trade-offs in plant defense against pathogens and herbivores: a field demonstration of chemical elicitors of induced resistance. *Journal of Chemical Ecology*, **25**, 1597–609.

Thebault, E. & Loreau, M. (2006) The relationship between biodiversity and ecosystem functioning in food webs. *Ecological Research*, **21**, 17–25.

Thelen, G.C., Vivanco, J.M., Newingham, B. et al. (2005) Insect herbivory stimulates allelopathic exudation by an invasive plant and the suppression of natives. *Ecology Letters*, **8**, 209–17.

Theunisse, J. & Schelling, G. (1996) Pest and disease management by intercropping: suppression of thrips and rust in leek. *International Journal of Pest Management*, **42** (4), 227–34.

Thiery, D., Gabel, D., Farkas, P. & Jarry, M. (1995) Egg dispersion in codling moth – influence of egg extract and of its fatty-acid constituents. *Journal of Chemical Ecology*, **21**, 2015–26.

Thirakhupt, V. & Araya, J.E. (1992) Survival and life table statistics of *Rhopalosiphum padi* (L.) and *Sitobion avenae* (F.) (Hom., Aphididae) in single or mixed colonies in laboratory wheat cultures. *Journal of Applied Entomology*, **113** (4), 368–75.

Thireau, J.C. & Regniere, J. (1995) Development, reproduction, voltinism and host synchrony of *Meteorus trachynotus* with its hosts *Choristoneura fumiferana* and *C. rosaceana*. *Entomologia Experimentalis et Applicata*, **76** (1), 67–82.

Thomas, A.T. & Hodkinson, I.D. (1991) Nitrogen, water and stress and the feeding efficiency of lepidopteran herbivores. *Journal of Applied Ecology*, **28**, 703–20.

Thomas, C.D. (1990) Fewer species. *Nature*, **347**, 237.

Thomas, C.D., Franco, A.M.A. & Hill, J.K. (2006) Range retractions and extinction in the face of climate warming. *Trends in Ecology and Evolution*, **21**, 415–16.

Thomas, C.D. & Lennon, J.J. (1999) Birds extend their ranges northwards. *Nature*, **399**, 213.

Thomas, C.D., Cameron, A., Green, R.E. et al. (2004a) Extinction risk from climate change. *Nature*, **427**, 145–8.

Thomas, C.D., Williams, S.E., Cameron, A. et al. (2004b) Biodiversity conservation – uncertainty in predictions of extinction risk – effects of changes in climate and land use – climate change and extinction risk – reply. *Nature*, **430**.

Thomas, D.J., Tracey, B., Marshall, H. & Norstrom, R.J. (1992) Arctic terrestrial ecosystem contamination. *Science of the Total Environment*, **122** (1/2), 135–64.

Thomas, J.A. (1995) The ecology and conservation of *Maculinen arion* and other European species of large blue butterfly. In *Ecology and Conservation of Butterflies* (ed. A.S. Pullin), pp. 180–97. Chapman & Hall, London.

Thomas, J.A. (2005) Monitoring change in the abundance and distribution of insects using butterflies and other indicator groups. *Philosophical Transactions of the Royal Society series B, Biological Sciences*, **360**, 339–57.

Thomas, J.A., Elmes, G.W., Wardlaw, J.C. & Woyciechowski, M. (1989) Host specificity among Maculinea butterflies in Myrmica ant nests. *Oecologia*, **79**, 452–7.

Thomas, J.A., Telfer, M.G., Roy, D.B. et al. (2004c) Comparative losses of British butterflies, birds, and plants and the global extinction crisis. *Science*, **303**, 1879–81.

Thompson, A.J. & Shrimpton, D.M. (1984) Weather associated with the start of mountain pine beetle outbreaks. *Canadian Journal of Forest Research*, **14**, 255–8.

Thompson, G.B. & Drake, B.G. (1994) Insects and fungi on a C3 and a C4 grass exposed to elevated atmospheric CO$_2$ concentrations in open-top chambers in the field. *Plant, Cell and Environment*, **17**, 1161–7.

Thompson, J.N. (1986) Patterns in coevolution. In *Coevolution and Systematics* (eds A.R. Stone & D.H. Hawksworth), pp. 119–43. Clarendon Press, Oxford.

Thompson, J.N. (1989) Concepts of coevolution. *Trends in Ecology and Evolution*, **4**, 179–83.

Thompson, J.N. (1994) *The Coevolutionary Process*. University of Chicago Press, Chicago.

Thompson, J.N. (1997) Evaluating the dynamics of coevolution among geographically structured populations. *Ecology*, **78**, 1619–23.

Thompson, J.N. & Cunningham B.M. (2002) Geographic structure and dynamics of coevolutionary selection. *Nature*, **417**, 735–8.

Thompson, R. (2003) *Close-Up on Insects: a Photographer's Guide*. Guild of Master Craftsman, Lewes, UK.

Thompson, R.A., Quisenberry, S.S., N'guessan, F.K., Heagler, A.M. & Giesler, G. (1994) Planting date as a potential method for managing the rice water weevil (Coleoptera: Curculionidae) in water-seeded rice in southwest Louisiana. *Journal of Economic Entomology*, **87** (5), 1318–24.

Thomson, M.C., Adiamah, J.H., Connor, S.J. et al. (1995) Entomological evaluation of the Gambia's National Impregnated Bednet Programme. *Annals of Tropical Medicine and Parasitology*, **89** (3), 229–41.

Thorne, A.D., Pexton, J.J., Dytham, C. & Mayhew, P.J. (2006) Small body size in an insect shifts development, prior to adult eclosion, towards early reproduction. *Proceedings of the Royal Society of London Series B, Biological Sciences*, **273** (1590), 1099–103.

Thorne, B.L. (1991) Ancestral transfer of symbionts between cockroaches and termites: an alternative hypothesis. *Proceedings of the Royal Society of London Series B, Biological Sciences*, **246** (1317), 191–6.

Thuy, N.N. & Thieu, D.V. (1992) Status of integrated pest management programme in Viet Nam. In *Integrated Pest Management in the Asia–Pacific Region* (eds P.A.C. Ooi, G.S. Lim, T.H. Ho, P.L. Manalo & J. Waage), pp. 237–50. CAB International, Wallingford, UK.

Tibbets, T.M. & Faeth, S.H. (1999) *Neotyphodium* endophytes in grasses: deterrents or promoters of herbivory by leaf-cutting ants? *Oecologia*, **118**, 297–305.

Tillman, P.G. (2006) Sorghum as a trap crop for *Nezara viridula* L. (Heteroptera: Pentatomidae) in cotton in the southern United States. *Environmental Entomology*, **35** (3), 771–83.

Tilman, D. (1987) The importance of the mechanisms of interspecific competition. *American Naturalist*, **129**, 769–74.

Tilman, D. & Downing, J.A. (1994) Biodiversity and stability in grasslands. *Nature*, **367**, 363–5.

Tilman D., Downing J.A. & Wedin, D.A. (1994) Does Diversity Beget Stability – Reply. *Nature*, **371** (6493), 114.

Toews, M.D. & Subramanyam, B. (2004) Survival of stored-product insect natural enemies in spinosad-treated wheat. *Journal of Economic Entomology*, **97** (3), 1174–80.

Togashi, K. (1991) Spatial pattern of pine wilt diseases caused by *Bursaphelencus xylophilus* (Nematoda: Aphelenchoididae) within a *Pinus thunbergii* stand. *Researches on Population Ecology (Kyoto)*, **33** (2), 245–56.

Togashi, K. & Shigesada, N. (2006) Spread of the pinewood nematode vectored by the Japanese pine sawyer: modeling and analytical approaches. *Population Ecology*, **48**, 271–83.

Tongren, J-E., Zavala, F., Roos, D-S. & Riley, E-M. (2004). Malaria vaccines: if at first you don't succeed. *Trends in Parasitology*, **20** (12), 604–10.

Torr, S.J. Hargrove, J.W. & Vale, G.A. (2005) Towards a rational policy for dealing with tsetse. *Trends in Parasitology*, **21** (11), 537–41.

Torre, A.D., Merzagora, L., Powell, J.R. & Coluzzi, M. (1997) Selective introgression of paracentric inversions between two sibling species of the *Anopheles gambiae* complex. *Genetics*, **146** (1), 239–44.

Torres, J.A. & Snelling, R.R. (1997) Biogeography of Puerto Rican ants: a non-equilibrium case? *Biodiversity and Conservation*, **6**, 1103–21.

Torrie, L-S., Radford, J-C., Southall, T.D. et al. (2004) Resolution of the insect ouabain paradox. *Proceedings of the National Academy of Sciences of the USA*, **101** (37), 13,689–93.

Tostowaryk, W. (1972) The effect of prey defense on the functional response of *Podisus modestus* (Hemiptera: Pentatomidae) to the densities of the sawflies *Neodiprion swainei* and *N. pratti banksianae* (Hymenoptera: Neodiprionidae). *Canadian Entomologist*, **104**, 61–9.

Touré, Y.T., Oduola, A.M.J. & Morel, C.M. (2004) The *Anopheles gambiae* genome: next steps for malaria vector control. *Trends in Parasitology*, **20** (3), 142–9.

Toussaint, J.F., Kerkhofs, P. & De Clercq, K. (2006) Influence of global climate changes on arboviruses spread. *Annales de Medecine Veterinaire*, **150** (1), 56–63.

Tovar-Sánchez, E. & Oyama, K. (2006) Effect of hybridization of the *Quercus crassifolia* × *Quercus crassipes* complex on the community structure of endophagous insects. *Oecologia*, **147**, 702–13.

Trager, M.D. & Bruna, E.M. (2006) Effects of plant age, experimental nutrient addition and ant occupancy on herbivory in a neotropical myrmecophyte. *Journal of Ecology*, **94**, 1156–63.

Traveset, A. (1990) Bruchid egg mortality on *Accacia farnesiana* caused by ants and abiotic factors. *Ecological Entomology*, **15**, 463–8.

Travis, J.M.J. (2003) Climate change and habitat destruction: a deadly anthropogenic cocktail. *Proceedings of the Royal Society of London Series B, Biological Sciences*, **270**, 467–73.

Traynor, R.E. & Mayhew, P.J. (2005) Host range in solitary versus gregarious parasitoids: a laboratory experiment. *Entomologia Experimentalis et Applicata*, **117** (1), 41–9.

Trezzi, G. (2003) Nuovi *Trechini troglobi* del Guatemala (Coleoptera, Carabidae). *Fragmenta Entomologica*.

Trimble, R.M. (1995) Mating disruption for controlling the codling moth, *Cydia Pomonella* (L.) (Lepidoptera: Tortricidae), in organic apple production in southwestern Ontario. *Canadian Entomologist*, **127** (4), 493–505.

Tripp, R., Wijeratne, M. & Piyadasa, V.H. (2005) What should we expect from farmer field schools? A Sri Lanka case study. *World Development*, **33** (10), 1705–20.

Trivedi, J.P., Srivastava, A.P., Narain, K. & Chatterjee, R.C. (1991) The digestion of wool fibres in the alimentary system of *Anthrenus flavipes* larvae. *International Biodeterioration*, **27** (4), 327–36.

Trumble, J.T., Carson, W.G. & Kund, G.S. (1997) Economics and environmental impact of a sustainable integrated pest management program in celery. *Journal of Economic Entomology*, **90** (1), 139–46.

Trumbo, S.T. & Robinson, G.E. (2004) Nutrition, hormones and life history in burying beetles. *Journal of Insect Physiology*, **50**, 383–91.

Trumbo, S.T. & Fernandez, A.G. (1995) Regulation of brood size by male parents and cues employed to assess resource size by burying beetles. *Ethology, Ecology and Evolution*, **7**, 313–22.

Trutmann, P. & Graf, W. (1993) The impact of pathogens and arthropod pests on common bean production in Rwanda. *International Journal of Pest Management*, **39** (3), 328–33.

Tscharntke, T., Steffan-Dewenter, I., Kruess, A. & Thies, C. (2002) Characteristics of insect populations on habitat fragments: a mini review. *Ecological Research*, **17**, 229–39.

Tscharntke, T., Thiessen, S., Dolch, R. & Boland, W. (2001) Herbivory, induced resistance, and interplant signal transfer in *Alnus glutinosa*. *Biochemical Systematics and Ecology*, **29**, 1025–47.

Tschen, J. & Fuchs, W.H. (1969) Wirkung von Phloridzin auf die infektion von *Phaseolus vulgaris* durch *Uromyces phaseoli*. *Naturwissenshaft*, **56**, 643–8.

Tsubaki, Y., Siva-Jothy, M.T. & Ono, T. (1994) Re-copulation and post-copulatory mate guarding increase immediate female reproductive output in the dragonfly *Nannophya pygmaea* Rambur. *Behavioral Ecology and Sociobiology*, **35**, 219–25.

Tsuchida, T., Koga, R., Sakurai, M. & Fukatsu, T. (2006) Facultative bacterial endosymbionts of three aphid species, *Aphis craccivora*, *Megoura crassicauda* and *Acyrthosiphon pisum*, sympatrically found on the same host plants. *Applied Entomology and Zoology*, **41** (1), 129–37.

Tukey, H.B. & Morgan, J.V. (1963) Injury to foliage and its effects upon the leaching of nutrients from above-ground plant parts. *Physiologia Planta*, **16**, 557–64.

Turchin, P. (1990) Rarity of density dependence or population regulation with lags? *Nature*, **344**, 660–3.

Turchin, P., Wood, S.N., Ellner, S.P. et al. (2003) Dynamical effects of plant quality and parasitism on population cycles of larch budmoth. *Ecology*, **84**, 1207–14.

Turlings, T.C.J., Alborn, H.T., Loughrin, J.H. & Tumlinson, J.H. (2000) Volicitin, an elicitor of maize volatiles in oral secretion of *Spodoptera exigua*: isolation and bioactivity. *Journal of Chemical Ecology*, **26**, 189–202.

Turlings, T.C.J. & Tumilinson, J.H. (1992) Systematic release of chemical signals by herbivore-injured corn. *Proceedings of the National Academy of Sciences of the USA*, **89**, 8399–402.

Turlings, T.C.J., Tumlinson, J.H. & Lewis, W.J. (1990) Exploitation of herbivore-induced plant odors by host-seeking parasitic wasps. *Science*, **250**, 1251–3.

Turner, C.L., Saestedt, T.R. & Dyer, M.I. (1993) Maximization of aboveground grassland production: the role of defoliation frequency, intensity, and history. *Ecological Applications*, **3**, 175–86.

Turner, R.K., Paavola, J., Cooper, P., Farber, S., Jessamy, V. & Georgiou, S. (2003) Valuing nature: lessons learned and future research directions. *Ecological Economics*, **46**, 493–510.

Turnipseed, S.G. (1977) Influence of trichome variations on populations of small phytophagous insects on soybean. *Environmental Entomology*, **6**, 815–17.

Uchman, A., Drygant, D., Paszkowski, M., Porebski, S.J. & Turnau, E. (2004) Early Devonian trace fossils in marine to non-marine redbeds in Podolia, Ukraine: palaeoenvironmental implications. *Palaeogeography, Palaeoclimatology, Palaeoecology*, **214**, 67–83.

Ugbogu, O.C., Nwachukwu, N.C. & Ogbuagu, U.N. (2006) Isolation of *Salmonella* and *Shigella* species from house flies (*Musca domestica* L.) in Uturu, Nigeria. *African Journal of Biotechnology*, **5** (11), 1090–1.

Ulmer, B.J. & Dosdall, L.M. (2006) Glucosinolate profile and oviposition behavior in relation to the susceptibilities of Brassicaceae to the cabbage seedpod weevil. *Entomologia Experimentalis et Applicata*, **121**, 203–213.

Umana, V. & Constenla, M.A. (1984) Organochlorinated pesticides in human milk. *Revista de Biologia Tropical*, **32** (2), 233–40.

Underwood, N. (1999) The influence of plant and herbivore characteristics on the interaction between induced resistance and herbivore population dynamics. *American Naturalist*, **153**, 282–94.

Underwood, N. (2000) Density dependence in induced plant resistance to herbivore damage: threshold, strength and genetic variation. *Oikos*, **89**, 295–300.

Underwood, N. (2004) Variance and skew of the distribution of plant quality influence herbivore population dynamics. *Ecology*, **85**, 686–93.

Underwood, N. & Rausher, M.D. (2000) The effects of host-plant genotype on herbivore population dynamics. *Ecology*, **81**, 1565–76.

Underwood, N. & Rausher, M. (2002) Comparing the consequences of induced and constitutive plant resistance for herbivore population dynamics. *American Naturalist*, **160**, 20–30.

UNEP (2004) United Nations Environment Programme, www.unep.org.

University of California (1998) UC pest management guidelines: citrus—California red scale. http://www.ipm.ucdavis.edu/PMG.

Unruh, T.R. & Luck, R.F. (1987) Deme formation in insects: a test with the pinyon needle scale and review of other evidence. *Ecological Entomology*, **12**, 439–49.

Uriarte, M. (2000) Interactions between goldenrod (*Solidago altissima* L.) and its insect herbivore (*Trirhabda virgata*) over the course of succession. *Oecologia*, **122**, 521–8.

US EPA (2004) US Environmental Protection Agency, www.epa.gov.

Usher, M.B. (1995) A world of change: land-use patterns and arthropod communities. In *Insects in a Changing Environment* (eds R. Harrington & N.E. Stork), pp. 372–97. Academic Press, London.

Uvah, I.I.L. & Coaker, T.H. (1984) Effects of mixed cropping on some insect pests of carrots and onions. *Entomologia Experimentalis et Applicata*, **36** (2), 159–68.

Vaisanen, R. & Heliovaara, K. (1994) Hot-spots of insect diversity in northern Europe. *Annales Zoologici Fennici*, **31** (1), 71–81.

Valand, G.B. & Muniyappa, V. (1992) Epidemiology of tobacco leaf curl virus in India. *Annals of Applied Biology*, **120** (2), 257–67.

Valkama, E., Koricheva, J. & Oksanen, E. (2007) Effects of elevated O-3, alone and in combination with elevated CO_2, on tree leaf chemistry and insect herbivore performance: a meta-analysis. *Global Change Biology*, **13**, 184–201.

Van Alphen, J.J.M. & Jervis, M.A. (1996) Foraging behaviour. In *Insect Natural Enemies: Practical Approaches to their Study and Evaluation* (eds M. Jervis & N. Kidd), pp. 1–62. Chapman & Hall, London.

van Borm, S., Buschinger, A., Boomsma, J.J. & Billen, J. (2002) Tetraponera ants have gut symbionts related to nitrogen-fixing root-nodule bacteria. *Proceedings of the Royal Society of London Series B, Biological Sciences*, **269** (1504), 2023–7.

Van Dam, N.M. & Baldwin, I.T. (1998) Costs of jasmonate-induced responses in plants competing for limited resources. *Ecology Letters*, **1**, 30–3.

Van Dam, N.M., De Jong, T.J., Iwasa, Y. & Kubo, T. (1996) Optimal distribution of defences: are plants smart investors? *Functional Ecology*, **10**, 128–36.

Van Dam, N.M., Harvey, J.A., Wackers, F.L., Bezemer, T.M., van der Putten, W.H. & Vet, L.E.M. (2003) Interactions between aboveground and belowground induced responses against phytophages. *Basic and Applied Ecology*, **4**, 63–77.

Van Dam, N.M., Raaijmakers, C.E. & van der Putten, W.H. (2005) Root herbivory reduces growth and survival of the shoot feeding specialist *Pieris rapae* on *Brassica nigra*. *Entomologia Experimentalis et Applicata*, **115**, 161–70.

Van Den Born, R.J.G., Lenders, R.H.J., De Groot, W. & Huijsman, E. (2001) The new biophilia: an exploration of visions of nature in Western countries. *Environmental Conservation*, **28**, 65–75.

Van den Bos, J. & Rabbinge, R. (1976) *Simulation of the Fluctuations of the Grey Larch Bud Moth*. Pudoc, Wageningen.

Van der Meijden, E., Vanwijk, C.A.M. & Kooi, R.E. (1991) Population-dynamics of the cinnabar moth (*Tyria jacobaeae*) – oscillations due to food limitation and local extinction risks. *Netherlands Journal of Zoology*, **41** (2/3), 158–73.

Van der Meijden, E., Wijn, M. & Verkaar, H.J. (1988) Defense and regrowth: alternative plant strategies in the struggle against herbivores. *Oikos*, **51**, 355–63.

van der Wal, R., van Wijnen, H., van Wieren, S., Beucher, O. & Bos, D. (2000) On facilitation between herbivores: how Brent geese profit from brown hares. *Ecology*, **81**, 969–80.

Van Driesche, R.G. (1983) Meaning of percent parasitism in studies of insect parasitoids. *Environmental Entomology*, **12**, 1611–22.

van Emden, H.F. & Service, M.W. (2004) *Pest and Vector Control*. Cambridge University Press, Cambridge, UK.

van Ham, R.C.H.J., Kamerbeek, J., Palacios, C. et al. (2003). Reductive genome evolution in *Buchnera aphidicola*. *Proceedings of the National Academy of Sciences of the USA*, **100** (2), 581–6.

van Huis, A. (2005) Insects eaten in Africa (Coleoptera, Hymenoptera, Diptera, Heteroptera, Homoptera). In *Ecological Implications of Minilivestock. Potential of Insects, Role of Rodents, Frogs, Snails and Insects for Sustainable Development* (ed. M.G. Paoletti), pp. 232–45. Enfield Science Publishers, Enfield, NH.

Van Lenteren, J.C., Hua, L.Z., Kamerman, J.W. & Rumei, X. (1995) The parasite–host relationship between *Encarsia formosa* (Hymenoptera: Aphelinidae) and *Trialeurodes vaporariorum* (Homoptera: Aleyrodidae): XXVI. Leaf hairs reduce the capacity of *Encarsia* to control greenhouse whitefly on cucumber. *Journal of Applied Entomology*, **119** (8), 553–9.

Van Roermund, H.J.W., Van Lenteren, J.C. & Rabbinge, R. (1996) The analysis of foraging behaviour of the whitefly parasitoid *Encarsia formosa* (Hymenoptera: Aphelinidae) in an experimental arena – a simulation study. *Journal of Insect Behaviour*, **9**, 771–97.

van Ruijven, J., De Deyn, G.B., Raaijmakers, C.E., Berendse, F. & van der Putten, W.H. (2005) Interactions between spatially separated herbivores indirectly alter plant diversity. *Ecology Letters*, **8**, 30–7.

Van Swaay, C.A.M. & Warren, M.S. (1999) *Red Data Book of European Butterflies (Rhopalocera)*. Council of Europe, Strasbourg.

Van Swaay, C.A.M., Warren, M. & Lois, G. (2006) Biotope use and trends of European butterflies. *Journal of Insect Conservation*, **10**, 189–209.

Van Teeffelen, A.J.A., Cabeza, M. & Moilanen, A. (2006) Connectivity, probabilities and persistence: comparing reserve selection strategies. *Biodiversity and Conservation*, **15**, 899–919.

Van Zandt, P.A. & Agrawal, A.A. (2004) Community-wide impacts of herbivore-induced plant responses in milkweed (*Asclepias syriaca*). *Ecology*, **85**, 2616–29.

Vanatoa, A., Kuusik, A., Tartes, U., Metspalu, L. & Hiiesaar, K. (2006) Respiration rhythms and heartbeats of diapausing Colorado potato beetles, *Leptinotarsa decemlineata*, at low temperatures. *Entomologia Experimentalis et Applicata*, **118**, 21–31.

Vanbergen, A.J., Woodcock, B.A., Watt, A.D. & Niemelä, J. (2005) Effect of land-use heterogeneity on carabid communities at the landscape scale. *Ecography*, **28**, 3–16.

Vanbuskirk, J. (1993) Population consequences of larval crowding in the dragonfly *Aeshna juncea*. *Ecology*, **74**, 1950–8.

Vandekerkhove, B., Van Baal, E., Bolckmans, K. & De Clercq, P. (2006) Effect of diet and mating status on ovarian development and oviposition in the polyphagous predator *Macrolophus caliginosus* (Heteroptera: Miridae). *Biological Control*, **39**, 532–8.

Vargas, R.I., Walsh, W.A., Kanehisa, D., Stark, J.D. & Nishida, T. (2000) Comparative demography of three Hawaiian fruit flies (Diptera: Tephritidae) at alternating temperatures. *Annals of the Entomological Society of America*, **93** (1), 75–81.

Varley, G.C. (1947) The natural control of population balance in the knapweed gall fly (*Urophora jaceana*). *Journal of Animal Ecology*, **16**, 139–87.

Varley, G.C. (1949) Population changes in German forest pests. *Journal of Animal Ecology*, **18**, 117–22.

Varley, G.C. & Gradwell, G.R. (1968) Population models for the winter moth. In *Insect Abundance* (ed. T.R.E. Southwood), pp. 132–42. Blackwell Scientific Publications, Oxford.

Varley, G.C. & Gradwell, G.R. (1971) The use of models and life tables in assessing the role of natural enemies. In *Biological Control*, (ed. C.B. Huffaker), pp. 93–112. Plenum Press, New York.

Varley, G.C., Gradwell, G.R. & Hassell, M.P. (1973) *Insect Population Ecology: an Analytical Approach*. Blackwell Scientific Publications, Oxford.

Vasconcelos, H.L. & Casimiro, A.B. (1997) Influence of *Azteca alfari* ants on the exploitation of *Cecropia* trees by a leaf-cutting ant. *Biotropica*, **29**, 84–92.

Vasconcelos, S.D., Hails, R.S., Speight, M.R. & Cory, J.S. (2005) Differential crop damage by healthy and nucleopolyhedrovirus-infected *Mamestra brassicae* L. (Lepidoptera: Noctuidae) larvae: a field examination. *Journal of Invertebrate Pathology*, **88** (2), 177–9.

Vega, F.E., Barbosa, P., Kuo-sell, H.L., Fisher, D.B. & Nelsen, T.C. (1995) Effects of feeding on healthy and diseased corn plants on a vector and on a non-vector insect. *Experimentia (Basel)*, **51** (3), 293–9.

Velisek, J., Wlasow, T., Gomulka, P. et al. (2006) Effects of cypermethrin on rainbow trout (*Oncorhynchus mykiss*). *Veterinarni Medicina*, **51** (10), 469–76.

Verhulst, P.F. (1838) Notice sur la loi que la population suit dans son accroissement. *Correspondances Mathématiques et Physiques*, **10**, 113–21.

Vessby, K., Soderstrom, B., Glimskar, A. & Svensson, B. (2002) Species-richness correlations of six different taxa in Swedish seminatural grasslands. *Conservation Biology*, **16**, 430–9.

Vezina, A. & Peterman, R.M. (1985) Tests of the role of a nuclear polyhedrosis virus in the population dynamics of its host, Douglas-fir tussock moth *Orgyia pseudotsugata* (Lepidoptera: Lymantriidae). *Oecologia (Heidelberg)*, **67** (2), 260–6.

Vezzani, D., Velazquez, S-M. & Schweigmann, N. (2004) Seasonal pattern of abundance of *Aedes aegypti* (Diptera: Culicidae) in Buenos Aires city, Argentina. *Memorias do Instituto Oswaldo Cruz*, **99** (4), 351–5.

Vialatte, A., Dedryver, C.A., Simon, J.C., Galman, M. & Plantegenest, M. (2005) Limited genetic exchanges between populations of an insect pest living on uncultivated and related cultivated host plants. *Proceedings of the Royal Society of London Series B, Biological Sciences*, **272**, 1075–82.

Vieira, N.K.M., Clements, W.H., Guevara, L.S. & Jacobs, B.F. (2004) Resistance and resilience of stream insect communities to repeated hydrologic disturbances after a wildfire. *Freshwater Biology*, **49**, 1243–59.

Villa-Castillo, J. & Wagner, M.R. (2002) Ground beetle (Coleoptera: Carabidae) species assemblage as an indicator of forest condition in northern Arizona ponderosa pine forests. *Environmental Entomology*, **31**, 242–52.

Villanueva, V.D. & Trochine, C. (2005) The role of microorganisms in the diet of *Verger* cf. *limnophilus* (Trichoptera: Limnephilidae) larvae in a Patagonian Andean temporary pond. *Wetlands*, **25** (2), 473–9.

Villasenor, J.L., Ibarra-Manriquez, G., Meave, J.A. & Ortiz, E. (2005) Higher taxa as surrogates of plant biodiversity in a megadiverse country. *Conservation Biology*, **19**, 232–8.

Vinhaes, M.C. & Schofield, C.J. (2003) Trypanosomiasis control: surmounting diminishing returns. *Trends in Parasitology*, **19**, 112–13.

Visser, M.E. & Driessen, G. (1991) Indirect mutual interference in parasitoids. *Netherlands Journal of Zoology*, **41**, 214–27.

Vite, J.P. (1961) The influence of water supply on oleoresin exudation pressure and resistance to bark beetle attack in *Pinus ponderosa*. *Contributions of the Boyce Thompson Institute*, **21**, 37–66.

Vogt, J.T., Appel, A.G. & West, M.S. (2000) Flight energetics and dispersal capability of the fire ant, *Solenopsis invicta* Buren. *Journal of Insect Physiology*, **46**, 697–707.

Voigt, C.C., Lehmann, G.U.C., Michener, R.H., Joachimski, M.M. (2006) Nuptial feeding is reflected in tissue nitrogen isotope ratios of female katydids. *Functional Ecology*, **20** (4), 656–61.

Volterra, V. (1931) Variations and fluctuations of the number of individuals in animal species living together. In *Animal Ecology* (ed. R.N. Chapman), pp. 409–48. McGraw-Hill, New York.

von Dahl, C.C. & Baldwin, I.T. (2004) Methyl jasmonate and cis-jasmone do not dispose of the herbivore-induced jasmonate burst in *Nicotiana attenuata*. *Physiologia Plantarum*, **120**, 474–81.

von Dungern, P. & Briegel, H. (2001) Enzymatic analysis of uricotelic protein catabolism in the mosquito *Aedes aegypti*. *Journal of Insect Physiology*, **47**, 73–82.

Voothuluru, P., Meng, J.Y., Khajuria, C. et al. (2006) Categories and inheritance of resistance to Russian wheat aphid (Homoptera: Aphididae) biotype 2 in a selection from wheat cereal introduction 2401. *Journal of Economic Entomology*, **99** (5), 1854–61.

Vorburger, C., Lancaster, M. & Sunnucks, P. (2003) Environmentally related patterns of reproductive modes in the aphid *Myzus persicae* and the predominance of two 'super-clones' in Victoria, Australia. *Molecular Ecology*, **12** (12), 3493–504.

Vos, J.G.M. & Nurtika, N. (1995) Transplant production techniques in integrated crop management of hot pepper (*Capsicum* spp.) under tropical lowland conditions. *Crop Protection*, **14** (6), 453–9.

Vrieling, K. (1991) Cost assessment of the production of pyrrolizidine alkaloids by *Senecio jacobaea* L. II. The generative phase. *Mededelingen Van de Faculteit Landbouwwetenschappen Rijksuniversiteit Gent*, **56**, 781–8.

Wagner, D.L. & Liebherr, J.K. (1992) Flightlessness in insects. *Trends in Ecology and Evolution*, **7** (7), 216–20.

Wagner, R. (1991) The influence of the diel activity pattern of the larvae of *Sericostoma personatum* (Kirby and Spence) (Trichoptera) on organic matter distribution in stream sediments: a laboratory study. *Hydrobiologia*, **224**, 65–70.

Wagner, T. (1997) The beetle fauna of different tree species in forests of Rwanda and East Zaire. In *Canopy Arthropods* (eds N. Stork, J. Adis & R. Didham), pp. 169–183. Chapman & Hall, London.

Wainhouse, D. & Howell, R.S. (1983) Intraspecific variation in beech scale populations and susceptibility of their host, *Fagus sylvatica*. *Ecological Entomology*, **8**, 351–9.

Wale, M., Schulthess, F., Kairu, E.W. & Omwega, C.O. (2006) Cereal yield losses caused by lepidopterous stemborers at different nitrogen fertilizer rates in Ethiopia. *Journal of Applied Entomology*, **130** (4), 220–9.

Walklate, P.J., Cross, J.V., Richardson, G.M. & Baker, D.E. (2006) Optimising the adjustment of label-recommended dose rate for orchard spraying. *Crop Protection*, **25** (10), 1080–6.

Wallace, J.B., Cuffney, T.F., Webster, J.R., Lughart, G.J., Chung, K. & Goldowitz, B.S. (1991) Export of fine organic particles from headwater streams: effects of season, extreme discharges, and invertebrate manipulation. *Limnology and Oceanography*, **36**, 670–82.

Wallace, J.B., Vogel, D.S. & Cuffney, T.F. (1986) Recovery of a headwater stream from an insecticide-induced community disturbance. *Journal of the North American Benthological Society*, **5**, 115–26.

Wallace, J.B. & Webster, J.R. (1996) The role of macroinvertebrates in stream ecosystem function. *Annual Review of Entomology*, **41**, 115–39.

Wallace, J.B., Webster, J.R. & Cuffney, T.F. (1982) Stream detritus dynamics: regulation by invertebrate consumers. *Oecologia*, **53**, 197–200.

Walters, D.S., Craig, R. & Mumma, R.O. (1989a) Glandular trichome exudate is the critical factor in geranium resistance to foxglove aphid. *Entomologia Experimentalis et Applicata*, **53**, 105–10.

Walters, D.S., Craig, R. & Mumma, R.O. (1990) Fatty acid incorporation in the biosynthesis of anacardic acids of geraniums. *Phytochemistry*, **29**, 1815–22.

Walters, D.S., Grossman, H., Craig, R. & Mumma, R.O. (1989b) Geranium defensive agents. IV. *Journal of Chemical Ecology*, **15**, 357–72.

Walther, G.R., Berger, S. & Sykes, M.T. (2005) An ecological 'footprint' of climate change. *Proceedings of the Royal Society of London Series B, Biological Sciences*, **272**, 1427–32.

Walther, G.R., Post, E., Convey, P. et al. (2002) Ecological responses to recent climate change. *Nature*, **416**, 389–95.

Walton, V.M., Daane, K.M. & Pringle, K.L. (2004) Monitoring *Planococcus ficus* in South African vineyards with sex pheromone-baited traps. *Crop Protection*, **23**, 1089–96.

Wang, W., Mo, J.C., Cheng, J.A. & Tang, Z.H. (2006) Selection and characterization of spinosad resistance in *Spodoptera exigua* (Hubner) (Lepidoptera: Noctuidae). *Pesticide Biochemistry and Physiology*, **84** (3), 180–7.

Wang, Y.M., Ge, F., Liu, X.H., Feng, F. & Wang, L.J. (2005) Evaluation of mass-trapping for control of tea tussock moth *Euproctis pseudoconspersa* (Strand) (Lepidoptera: Lymantriidae) with synthetic sex pheromone in south China. *International Journal of Pest Management*, **51** (4), 289–95.

Wantzen, K.M. (2006) Physical pollution: effects of gully erosion on benthic invertebrates in a tropical clear-water stream. *Aquatic Conservation – Marine and Freshwater Ecosystems*, **16**, 733–49.

Wantzen, K.M. & Wagner, R. (2006) Detritus processing by invertebrate shredders: a neotropical–temperate comparison. *Journal of the North American Benthological Society*, **25** (1), 216–32.

Ward, D.F. & Lariviere, M.C. (2004) Terrestrial invertebrate surveys and rapid biodiversity assessment in New Zealand: lessons from Australia. *New Zealand Journal of Ecology*, **28**, 151–9.

Waring, G.L. & Price, P.W. (1990) Plant water stress and gall formation (Cecidomyiidae: *Asphondylia* spp.) on creosote bush. *Ecological Entomology*, **15**, 87–96.

Warren, M.S., Hill, J.K., Thomas, J.A. et al. (2001) Rapid responses of British butterflies to opposing forces of climate and habitat change. *Nature*, **414**, 65–9.

Warren, M.S. & Key, R.S. (1991) Woodlands: past, present and potential for insects. In *The Conservation of Insects and their Habitats* (eds N.M. Collins & J.A. Thomas), pp. 155–211. Academic Press, London.

Warren, M.S., Thomas, C.D. & Thomas, J.A. (1984) The status of the heath fritillary butterfly, *Mellicta athalia* Rott., in Britain. *Biological Conservation*, **29**, 287–305.

Watanabe, H., Takase, A., Tokuda, G., Yamada, A. & Lo, N. (2006) Symbiotic 'archaezoa' of the primitive termite *Mastotermes darwiniensis* still play a role in cellulase production. *Eukaryotic Cell*, **5** (9), 1571–6.

Waterhouse, D.F. (1977) The biological control of dung. In *The Insects* (eds T. Eisner & E.O. Wilson), pp. 314–22. Scientific American W.H. Freeman, San Francisco, CA.

Waterman, P.G. & Mole, S. (1994) *Analysis of Phenolic Plant Metabolites.* Blackwell Scientific Publications, Oxford.

Watson, E.J. & Carlton, C.E. (2005) Insect succession and decomposition of wildlife carcasses during fall and winter in Louisiana. *Journal of Medical Entomology,* **42,** 193–203.

Watt, A.D. (1988) Effects of stress-induced changes in plant quality and host-plant species on the population dynamics of the pine beauty moth in Scotland: partial life tables of natural and manipulated populations. *Journal of Applied Ecology,* **25,** 209–21.

Watt, A.D. (1989) The growth and survival of *Panolis flammea* larvae in the absence of predators on Scots pine and lodgepole pine. *Ecological Entomology,* **14,** 225–34.

Watt, A.D. (1990) The consequences of natural, stress-induced and damage-induced differences in tree foliage on the population dynamics of the pine beauty moth. In *Population Dynamics of Forest Insects* (eds A.D. Watt, S.R. Leather, M.D. Hunter & N.A.C. Kidd), pp. 157–68. Intercept, Andover, UK.

Watt, A.D. (1998) Measuring disturbance in tropical forests: a critique of the use of species-abundance models and indicator measures in general. *Journal of Applied Ecology,* **35,** 467–9.

Watt, A.D., Barbour, D.A., McBeath, C., Worth, S. & Glimmerveen, I. (1998). The abundance, diversity and management of arthropods in spruce forests. In *Birch in Spruce Plantations Management for Biodiversity* (eds J. J. Humphrey, K. Holl & A. Broome), pp. 31–41. Forestry Commission, Edinburgh.

Watt, A.D., Carey, P.D. & Eversham, B.C. (1997a) Implications of climate change for biodiversity. In *Biodiversity in Scotland: Status, Trends and Initiatives* (eds L.V. Flemming, A.C. Newton, J.A. Vickery & M.B. Usher), pp. 147–59. The Stationery Office, Edinburgh.

Watt, A.D., Leather, S.R. & Evans, H.F. (1991) Outbreaks of the pine beauty moth on pine in Scotland – influence of host plant-species and site factors. *Forest Ecology and Management,* **39,** 211–21.

Watt, A.D., Stork, N.E. & Bolton, B. (2002) The diversity and abundance of ants in relation to forest disturbance and plantation establishment in southern Cameroon. *Journal of Applied Ecology,* **39,** 18–30.

Watt, A.D., Stork, N.E., Eggleton, P. et al. (1997b) Impact of forest loss and regeneration on insect abundance and diversity. In *Forests and Insects* (eds A.D. Watt, N.E. Stork & M.D. Hunter), pp. 274–86. Chapman & Hall, London.

Watt, A.D., Stork, N.E., McBeath, C. & Lawson, G.L. (1997c) Impact of forest management on insect abundance and damage in a lowland tropical forest in southern Cameroon. *Journal of Applied Ecology,* **34,** 985–98.

Way, M.J & Heong, K.L (1994) The role of biodiversity in the dynamics and management of insect pests of tropical irrigated rice – a review. *Bulletin of Entomological Research,* **84** (4), 567–87.

Wearing, C.H., Colhoun, K., Attfield, B., Marshall, R.R. & McLaren, G.F. (2003) Screening for resistance in apple cultivars to lightbrown apple moth, *Epiphyas postvittana,* and greenheaded leafroller, *Planotortrix octo,* and its relationship to field damage. *Entomologia Experimentalis et Applicata,* **109,** 39–53.

Weathersbee, A.A. & McKenzie, C.L. (2005) Effect of a neem biopesticide on repellency, mortality, oviposition, and development of *Diaphorina citri* (Homoptera: Psyllidae). *Florida Entomologist,* **88** (4), 401–7.

Weaver, S.C., Ferro, C., Barrera, R., Boshell, J. & Navarro, J.C. (2004) Venezuelan equine encephalitis. *Annual Review of Entomology,* **49,** 141–74.

Webb, B.A., Strand, M.R., Dickey, S.E. et al. (2006) Polydnavirus genomes reflect their dual roles as mutualists and pathogens. *Virology,* **347** (1), 160–74.

Webster, J.R., Wallace, J.B. & Benfield, E.F. (1995) Organic processess in streams of the eastern United States. In *River and Stream Ecosystems* (eds C.E. Cushing, K.W. Cummins & G.W. Marshall), pp. 117–87. Elsevier Science Publishers B.V., Amsterdam.

Wedmann, S., Bradler, S. & Rust, J. (2007) The first fossil leaf insect: 47 million years of specialized cryptic morphology and behavior. *Proceedings of the National Academy of Sciences of the USA,* **104** (2), 565–9.

Weidner, H. (1986) The migration routes of the European locust, *Locusta migratoria* in Europe in the year 1693 (Saltatoria, Acridiidae, Oedipodinae). *Anzeiger fuer Schaedlingskunde Pflanzenschutz Umweltschutz,* **59** (3), 41–51.

Weinig, C., Stinchcombe, J.R. & Schmitt, J. (2003) Evolutionary genetics of resistance and tolerance to natural herbivory in *Arabidopsis thaliana. Evolution,* **57,** 1270–80.

Weintraub, P.G. & Beanland, L. (2006) Insect vectors of phytoplasmas. *Annual Review of Entomology,* **51,** 91–111.

Weis, A.E. & Campbell, D.R. (1992) Plant genotype: a variable factor in insect–plant interactions. In *Effects of Resource Distribution on Animal–Plant Interactions* (eds M.D. Hunter, T. Ohgushi & P.W. Price), pp. 75–111. Academic Press, San Diego, CA.

Welburn, S.C., Coleman, P.G., Maudlin, I., Fevre, E.M., Odiit, M. & Eisler, M.C. (2006) Crisis, what crisis? Control of Rhodesian sleeping sickness. *Trends in Parasitology,* **22** (3), 123–8.

Wellington, W.G. (1950) Climate and spruce budworm outbreaks. *Canadian Journal of Forest Research,* **28,** 308–31.

Wermelinger, B. (2004). Ecology and management of the spruce bark beetle *Ips typographus* – a review of recent research. *Forest Ecology and Management,* **202,** 67–82.

Werren, J.H. (1983) Sex ratio evolution under local mate competition in a parasitic wasp. *Evolution,* **37,** 116–24.

Weseloh, R.M. (1976) Behaviour of forest insect parasitoids. In *Perspectives in Forest Entomology* (eds J.F. Anderson & H.K. Kaya), pp. 99–110. Academic Press, New York.

Wesolowski, T. & Rowinski, P. (2006) Tree defoliation by winter moth *Operophtera brumata* L. during an outbreak affected by structure of forest landscape. *Forest Ecology and Management*, **221** (1–3), 299–305.

West, C. (1985) Factors underlying the late seasonal appearance of the lepidopterous leaf-mining guild on oak. *Ecological Entomology*, **10**, 111–20.

Westerbergh, A. (2004) An interaction between a specialized seed predator moth and its dioecious host plant shifting from parasitism to mutualism. *Oikos*, **105**, 564–74.

Westerbergh, A. & Westerbergh, J. (2001) Interactions between seed predators/pollinators and their host plants: a first step towards mutualism? *Oikos*, **95**, 324–34.

Westneat, M.W., Betz, O., Blob, R.W., Fezzaa, K., Cooper, W.J. & Lee, W. K. (2003) Tracheal respiration in insects visualized with synchrotron X-ray imaging. *Science*, **299**, 558–60.

Wheeler, W.M. (1910) *Ants: their Structure, Development and Behaviour*. Columbia University Press, New York.

Whiles, M.R. & Wallace, J.B. (1995) Macroinvertebrate production in a headwater stream during recovery from anthropogenic disturbance and hydrologic extremes. *Canadian Journal of Fish Aquatic Science*, **52**, 2402–22.

Whitaker, P.M., Mahr, D.L. & Clayton, M. (2006) Verification and extension of a sampling plan for apple aphid, *Aphis pomi* DeGeer (Hemiptera: Aphididae). *Environmental Entomology*, **35** (2), 488–96.

White, A.J., Wratten, S.D., Berry, N.A. & Weigman, U. (1995) Habitat manipulation to enhance biological control of *Brassica* pests by hover flies (Diptera: Syrphidae). *Journal of Economic Entomology*, **88** (5), 1171–6.

White, M.J.D. (1974) Speciation in the Australian morabine grasshoppers: the cytogenetic evidence. In *Genetic Evidence of Speciation in Insects* (ed. M.J.D. White), pp. 57–68. Australia and New Zealand Book Company, Sydney.

White, T.C.R. (1969) An index to measure weather-induced stress of trees associated with outbreaks of psyllids in Australia. *Ecology*, **50**, 905–9.

White, T.R.C. (1976) Weather, food and plagues of locust. *Oecologia*, **22**, 119–34.

White, T.C.R. (1984) The abundance of vertebrate herbivores in relation to the availability of nitrogen in stressed food plants. *Oecologia*, **63**, 90–105.

White, T.C.R. (1993) *The Inadequate Environment: Nitrogen and the Abundance of Animals*. Springer Verlag, Berlin.

Whitehead, L.F. & Douglas, A.E. (1993) Populations of symbiotic bacteria in the parthenogenetic pea aphid (*Acyrthosiphon pisum*) symbiosis. *Proceedings of the Royal Society of London Series B, Biological Sciences*, **254** (1339), 29–32.

Whiteman, N.K. & Parker, P.G. (2005) Using parasites to infer host population history: a new rationale for parasite conservation. *Animal Conservation*, **8**, 175–81.

Whitfield, J.B. & Kjer, K.M. (2008) Ancient rapid radiations of insects: challenges for phylogenetic analysis. *Annual Review of Entomology*, **53**, 449–72.

Whitford, W.G., Stinnett, K. & Anderson, J. (1988) Decomposition of roots in a Chihuahuan Desert ecosystem. *Oecologia (Heidelberg)*, **75**, 8–11.

Whitham, T.G. (1978) Habitat selection by *Pemphigus* aphids in response to resource limitation and competition. *Ecology*, **59**, 1164–76.

Whitham, T.G. (1986) Costs and benefits of territoriality: behavioural and reproductive release by competing aphids. *Ecology*, **67**, 139–47.

Whitham, T.G. (1989) Plant hybrid zones as sinks for pests. *Science*, **23**, 1490–3.

Whitham, T.G., Martinsen, G.D., Floate, K.D., Dungey, H.S., Potts, B.M. & Keim, P. (1999) Plant hybrid zones affect biodiversity: tools for a genetic-based understanding of community structure. *Ecology*, **80**, 416–28.

Whitham, T.G., Morrow, P.A. & Potts, B.M. (1994) Plant hybrid zones as centers of biodiversity: the herbivore community of two endemic Tasmanian eucalypts. *Oecologia (Berlin)*, **97**, 481–90.

Whitham, T.G., Young, W.P., Martinsen, G.D. et al. (2003) Community and ecosystem genetics: a consequence of the extended phenotype. *Ecology*, **84**, 559–73.

Whiting, M.F., Bradler, S. & Maxwell, T. (2003) Loss and recovery of wings in stick insects. *Nature*, **421**, 264–7.

Whittaker, R.J., Araujo, M.B., Paul, J., Ladle, R.J., Watson, J.E.M. & Willis, K.J. (2005) Conservation biogeography: assessment and prospect. *Diversity and Distributions*, **11**, 3–23.

WHO (1997) African Programme for Onchocerciasis Control (APOC), http://www.who.org/programmes/ocp/apoc/index.html.

WHO (2003) http://www.who.int/.

WHO (2006) http://www.who.int/topics/.

WHO (2007) http://www.who.int/blindness/partnerships/onchocerciasis_OCP/en/.

Wichard, W., Kraemer, M.M.S. & Luer, C. (2006) First caddisfly species from Mexican amber (Insecta: Trichoptera). *Zootaxa*, **1378**, 37–48.

Wiedenmann, R.N., Smith, J.W.J.R. & Darnell, P.O. (1992) Laboratory rearing and biology of the parasite *Cotesia flavipes* (Hymenoptera: Braconidae) using *Diatraea saccharalis* (Lepidoptera: Pyralidae) as a host. *Environmental Entomology*, **21** (5), 1160–7.

Wiggins, G.J., Grant, J.F., Windham, M.T. et al. (2004) Associations between causal agents of the beech bark disease complex [*Cryptococcus fagisuga* (Homoptera: Crypitococcidae) and *Nectria* spp.] in the Great Smoky Mountains National Park. *Environmental Entomology*, **33** (5), 1274–81.

Wilby, A., Lan, L.P., Heong, K.L. et al. (2006) Arthropod diversity and community structure in relation to land use in the mekong delta, Vietnam. *Ecosystems*, **9** (4), 538–49.

Wiley, M.J. & Warren, G.L. (1992) Territory abandonment, theft, and recycling by a lotic grazer – a foraging strategy for hard times. *Oikos*, **63**, 495–505.

Wilkinson, T.L. & Douglas, A.E. (1995) Aphid feeding, as influenced by disruption of the symbiotic bacteria: an analysis of the pea aphid (*Acyrthosiphon pisum*). *Journal of Insect Physiology*, **41** (8), 635–40.

Wilkinson, T.L. & Douglas, A.E. (2003) Phloem amino acids and the host plant range of the polyphagous aphid, *Aphis fabae*. *Entomologia Experimentalis et Applicata*, **106** (2), 103–13.

Wilkinson, T.L., Fukatsu, T. & Ishikawa, H. (2003) Transmission of symbiotic bacteria *Buchnera* to parthenogenetic embryos in the aphid *Acyrthosiphon pisum* (Hemiptera: Aphidoidea). *Arthropod Structure and Development*, **32**, 241–5.

Williams, G. & Adam, P. (1994) A review of rainforest pollination and plant–pollinator interactions with particular reference to Australian subtropical rainforests. *Australian Zoologist*, **29** (3/4), 177–212.

Williams, M.R. (1995) An extreme-value function model of the species incidence and species: area relations. *Ecology (Washington DC)*, **76** (8), 2607–16.

Williams, P., Hannah, L., Andelman, S. et al. (2005) Planning for climate change: identifying minimum-dispersal corridors for the Cape proteaceae. *Conservation Biology*, **19**, 1063–74.

Williams, P.H. & Gaston, K.J. (1994) Measuring more of biodiversity – can higher-taxon richness predict wholesale species richness? *Biological Conservation*, **67**, 211–17.

Williams, P.H., Humphries, C.J. & Gaston, K.J. (1994) Centers of seed-plant diversity – the family way. *Proceedings of the Royal Society of London Series B, Biological Sciences*, **256**, 67–70.

Williams, R.S., Lincoln, D.E. & Thomas, R.B. (1997) Effects of elevated CO_2-grown loblolly pine needled on the growth, consumption, development, and pupal weight of red-headed pine sawfly larvae reared within open-topped chambers. *Global Change Biology*, **3**, 501–11.

Williamson, S.C., Detling, J.K., Dodd, J.L. & Dyer, M.I. (1989) Experimental evaluation of the grazing optimization hypothesis. *Journal of Range Management*, **42**, 149–52.

Willig, M.R., Kaufman, D.M. & Stevens, R.D. (2003) Latitudinal gradients of biodiversity: pattern, process, scale, and synthesis. *Annual Review of Ecology Evolution and Systematics*, **34**, 273–309.

Willocquet, L., Elazegui, F.A., Castilla, N. et al. (2004) Research priorities for rice pest management in tropical Asia: a simulation analysis of yield losses and management efficiencies. *Phytopathology*, **94** (7), 672–82.

Willott, E. (2004) Restoring nature, without mosquitoes? *Restoration Ecology*, **12** (2), 147–53.

Wilson, A.C.C. & Sunnucks, P. (2006) The genetic outcomes of sex and recombination in long-term functionally parthenogenetic lineages of Australian *Sitobion* aphids. *Genetical Research*, **87** (3), 175–85.

Wilson, A.C.C., Sunnucks, P. & Hales, D.F. (2003) Heritable genetic variation and potential for adaptive evolution in asexual aphids (Aphidoidea). *Biological Journal of the Linnean Society*, **79** (1), 115–35.

Wilson, C. & Tisdell, C. (2005) Knowledge of birds and willingness to support their conservation: an Australian case study. *Bird Conservation International*, **15**, 225–35.

Wilson, D. (1995) Fungal endophytes which invade insect galls: insect pathogens, benign saprophytes, or fungal inquilines? *Oecologia (Berlin)*, **103**, 255–60.

Wilson, D. & Faeth, S.H. (2001) Do fungal endophytes result in selection for leafminer ovipositional preference? *Ecology*, **82**, 1097–111.

Wilson, E.O. (1987) The little things that run the world (the importance and conservation of invertebrates). *Conservation Biology*, **1**, 344–6.

Wilson, E.O. (1992) *The Diversity of Life*. Harvard University Press, Cambridge, MA.

Wilson, M.D., Akpabey, F.J., Osei-Atweneboana, M.Y. et al. (2005) Field and laboratory studies on water conditions affecting the potency of VectoBac (R) (*Bacillus thuringiensis* serotype H-14) against larvae of the blackfly, *Simulium damnosum*. *Medical and Veterinary Entomology*, **19** (4), 404–12.

Wimp, G.M., Martinsen, G.D., Floate, K.D., Bangert, R.K. & Whitham, T.G. (2005) Plant genetic determinants of arthropod community structure and diversity. *Evolution*, **59**, 61–9.

Winchester, N.N. (1997) Arthropods of coastal old-growth Sitka spruce forests: conservation of biodiversity with special reference to the Staphylinidae. In *Forests and Insects* (eds A.D. Watt, N.E. Stork & M.D. Hunter), pp. 365–79. Chapman & Hall, London.

Wink, M. & Witte, L. (1991) Storage of quinolizidine alkaloids in *Macrosiphum albifrons* and *Aphis genistae* (Homoptera: aphididae). *Entomologia Generalis*, **15**, 237–54.

Wint, G.R.W. (1983) The role of alternative host plant species in the life of a polyphagous moth, *Operophatera brumata* (Lepidoptera. Geometridae). *Journal of Animal Ecology*, **52**, 439–50.

Wipking, W. (1995) Influences of daylength and temperature on the period of diapause and its ending process in dormant larvae of burnet moths (Lepidoptera, Zygaenidae). *Oecologia (Berlin)*, **102** (2), 202–10.

Wiseman, V., Hawley, W.A., Ter, K.F.O. et al. (2003) The cost-effectiveness of permethrin-treated bed nets in an area of intense malaria transmission in western Kenya. *American Journal of Tropical Medicine and Hygiene*, **68**, 161–7.

Wisnivesky-Colli, C., Schweigmann, N.J., Pietrokovsky, S., Bottazzi, V. & Rabinovich, J.E. (1997) Spatial distribution of *Triatoma guasayana* (Hemiptera: Reduviidae) in hardwood forest biotopes in Santiago del Estero, Argentina. *Journal of Medical Entomology*, **34** (2), 102–9.

Witanachchi, J.P. & Morgan, F.D. (1981) Behavior of the bark beetle, *Ips grandicollis*, during host selection. *Physiological Entomology*, **6**, 219–23.

Wittstock, U., Agerbirk, N., Stauber, E.J. et al. (2004) Successful herbivore attack due to metabolic diversion of a plant chemical defense. *Proceedings of the National Academy of Sciences of the USA*, **101**, 4859–64.

Witzgall, P., Lindblom, T., Bengtsson, M. & Tóth, M. (2004) The pherolist. www-pherolist.slu.se.

Witzgall, P., Stelinski, L., Gut, L. & Thomson, D. (2008) Codling moth management and chemical ecology. *Annual Review of Entomology*, **53**, 503–22.

Woldewahid, G., van der Werf, W., van Huis, A. & Stein, A. (2004) Spatial distribution of populations of solitarious adult desert locust (*Schistocerca gregaria* Forsk.) on the coastal plain of Sudan. *Agricultural and Forest Entomology*, **6**, 181–91.

Wood, T.G. & Sands, W.A. (1978) The role of termites in ecosystems. In *Production Ecology of Ants and Termites* (ed. M.V. Brian), pp. 245–92. Cambridge University Press, Cambridge, UK.

Wootton, R.J. (1992) Functional morphology of insect wings. *Annual Review of Entomology*, **37**, 113–40.

Wootton, R.J. (2002) Design, function and evolution in the wings of holometabolous insects. *Zoologica Scripta*, **31** (1), 31–40.

Worland, M.R., Wharton, D.A. & Byars, S.G. (2004) Intracellular freezing and survival in the freeze tolerant alpine cockroach *Celatoblatta quinquemaculata*. *Journal of Insect Physiology*, **50**, 225–32.

World Conservation Monitoring Centre (1992) *Global Biodiversity: Status of the Earth's Living Resources*. Chapman & Hall, London.

Wotton, R.S. (1994) *The Biology of Particles in Aquatic Systems*. Lewis, Boca Raton, FL.

Wraight, C.L., Zangerl, A.R., Carroll, M.J. & Berenbaum, M.R. (2000) Absence of toxicity of *Bacillus thuringiensis* pollen to black swallowtails under field conditions. *Proceedings of the National Academy of Sciences of the USA*, **97**, 7700–3.

Wright, D.E., Hunter, D.M. & Symmons, P.M. (1988) Use of pasture growth indices to predict survival and development of *Chortoicetes terminifera* (Walker) (Orthoptera: Acrididae). *Journal of the Australian Entomological Society*, **27** (3), 189–92.

Wright, P.A. (1995) Nitrogen-excretion – 3 end-products, many physiological roles. *Journal of Experimental Biology*, **198** (2), 273–81.

Wu, D., Daugherty, S.C., Van Aken, S.E. et al. (2006) Metabolic complementarity and genomics of the dual bacterial symbiosis of sharpshooters. *Plos Biology*, **4**, 1079–92.

Wyatt, T.D. (1986) How a subsocial intertidal beetle, *Bledius spectabilis*, prevents flooding and anoxia in its burrow. *Behavioral Ecology and Sociobiology*, **19** (5), 323–32.

Wyatt, T.D. (2003) *Pheromones and Animal Behaviour: Communication by Smell and Taste*. Cambridge University Press, Cambridge, UK.

Xu, F.Y., Ge, M.H., Zhu, Z.C. & Zhu, K.G. (1996) Studies on resistance of pine species and Masson pine provenances to *Bursaphelenchus xylophilus* and the epidemic law of the nematode in Nanjing. *Forest Research*, **9** (5), 521–4.

Yamazaki, K., Sugiura, S. & Kawamura, K. (2003) Ground beetles (Coleoptera: Carabidae) and other insect predators overwintering in arable and fallow fields in central Japan. *Applied Entomology and Zoology* **38** (4), 449–59.

Yanoviak, S.P., Dudley, R. & Kaspari, M. (2005) Directed aerial descent in canopy ants. *Nature*, **433**, 624–6.

Yasuoka, J., Levins, R., Mangione, T.W. & Spielman, A. (2006) Community-based rice ecosystem management for suppressing vector anophelines in Sri Lanka. *Transactions of the Royal Society of Tropical Medicine and Hygiene*, **100** (11), 995–1006.

Yencho, G.C. & Tingey, W.M. (1994) Glandular trichomes of *Solanum berthaultii* alter host preference of the Colorado potato beetle, *Leptintarsa decemlineata*. *Entomologia Experimentalis et Applicata*, **70**, 217–25.

Yoo, H.J.S. (2006) Local population size in a flightless insect: importance of patch structure-dependent mortality. *Ecology*, **87** (3), 634–47.

Yoshimura, T., Azumi, J.I., Tsunoda, K. & Takahashi, M. (1993) Changes of wood-attacking activity of the lower termite, *Coptotermes formosanus* Shiraki in defaunation–refaunation process of the intestinal Protozoa. *Material und Organismen (Berlin)*, **28** (2), 153–64.

Yoshinaga, N., Morigaki, N., Matsuda, F., Nishida, R. & Mori, N. (2005) In vitro biosynthesis of volicitin in *Spodotera litura*. *Insect Biochemistry and Molecular Biology*, **35**, 175–84.

Young, J., Watt, A., Nowicki, P. et al. (2005) Towards sustainable land use: identifying and managing the conflicts between human activities and biodiversity conservation in Europe. *Biodiversity and Conservation*, **14**, 1641–61.

Young, M.R. & Barbour, D.A. (2004) Conserving the New Forest burnet moth (*Zygaena viciae* [Denis and Schiffermueller]) in Scotland: responses to grazing reduction and consequent vegetation changes. *Journal of Insect Conservation*, **8**, 137–48.

Yu, X.D., Luo, T.H. & Zhou, H.Z. (2003) Species diversity of litter-layer beetles in the Fengtongzhai National Nature Reserve, Sichuan Province. *Acta Entomologica Sinica*, **46** (5), 609–16.

Yukawa, J. (2000) Synchronization of gallers with host plant phenology. *Population Ecology*, **42**, 105–13.

Yuval, B. (2006) Mating systems of blood-feeding flies. *Annual Review of Entomology*, **51**, 413–40.

Zahiri, N.S. & Mulla, M.S. (2003) Susceptibility profile of *Culex quinquefasciatus* (Diptera: Culicidae) to *Bacillus sphaericus* on selection with rotation and mixture of *B. sphaericus* and *B. thuringiensis* israelensis. *Journal of Medical Entomology*, **40** (5), 672–7.

Zalucki, M.P. & Clarke, A.R. (2004) Monarchs across the Pacific: the Columbus hypothesis revisited. *Biological Journal of the Linnean Society*, **82**, 111–21.

Zangerl, A.R. & Berenbaum, M.R. (2003) Phenotype matching in wild parsnip and parsnip webworms: causes and consequences. *Evolution*, **57**, 806–15.

Zchori-Fein, E. & Perlman, S-J. (2004). Distribution of the bacterial symbiont *Cardinium* in arthropods. *Molecular Ecology*, **13** (7), 2009–16.

Zeddam, J.L., Cruzado, J.A., Rodriguez, J.L. & Ravallec, M. (2003) A new nucleopolyhedrovirus from the oil-palm leaf-eater *Euprosterna elaeasa* (Lepidoptera: Limacodidae): preliminary characterization and field assessment in Peruvian plantation. *Agriculture Ecosystems and Environment*, **96**, 69–75.

Zehnder, G.W., Sikora, E.J. & Goodman, W.R. (1995) Treatment decisions based on egg scouting for tomato fruitworm, *Helicoverpa zea* (Boddie), reduce insecticide use on tomato. *Crop Protection*, **14** (8), 683–7.

Zettler, J.A., Spira, T.P. & Allen, C.R. (2001) Ant–seed mutualisms: can red imported fire ants sour the relationship? *Biological Conservation*, **101**, 249–53.

Zhai, L., Berg, M.C., Cebeci, F.C. et al. (2006) Patterned superhydrophobic surfaces: toward a synthetic mimic of the Namib Desert beetle. *Nano Letters*, **6** (6), 1213–17.

Zhang, L.P., Zhang, G.Y., Zhang, Y.J., Zhang, W.J. & Liu, Z. (2005) Interspecific interactions between *Bemisia tabaci* (Hem., Aleyrodidae) and *Liriomyza sativae* (Dipt., Agromyzidae). *Journal of Applied Entomology*, **129**, 443–6.

Zhang, Q.B. & Alfaro, R.I. (2003) Spatial synchrony of the two-year cycle budworm outbreaks in central British Columbia, Canada. *Oikos*, **102** (1), 146–54.

Zhang, Z.Q. & McEvoy, P.B. (1995) Responses of ragwort flea beetle *Longitarsus jacobaeae* (Coleoptera: Chrysomelidae) to signals from host plants. *Bulletin of Entomological Research*, **85** (3), 437–44.

Zhao, J.Z., Cao, J., Li, Y.X. et al. (2003a) Transgenic plants expressing two *Bacillus thuringiensis* toxins delay insect resistance evolution. *Nature Biotechnology*, **21**, 1493–7.

Zhao, X.F., Li, Z.M., Wang, J.X., Wang, J.B., Wang, S.L. & Blankespoor, H.D. (2003b) Efficacy of RH-2485, a new non-steroidal ecdysone agonist, against the cotton boll worm; *Helicoverpa armigera* (Lepidoptera: Noctuidae) in the laboratory and field. *Crop Protection*, **22** (7), 959–65.

Zhaohui, Z., Yongfan, P. & Juru, L. (1992) National integrated pest management of major crops in China. In *Integrated Pest Management in the Asia–Pacific Region* (eds P.A.C. Ooi, G.S. Lim, T.H. Ho, P.L. Manalo & J. Waage), pp. 255–66. CAB International, Wallingford, UK.

Zheng, Y.P., Retnakaran, A., Krell, P.J., Arif, B.M., Primavera, M. & Feng, Q.L. (2003) Temporal, spatial and induced expression of chitinase in the spruce budworm, *Choristoneura fumiferana*. *Journal of Insect Physiology*, **49**, 241–7.

Zhou, G., Minakawa N., Githeko, A-K. & Yan, G. (2005) Climate variability and malaria epidemics in the highlands of East Africa. *Trends in Parasitology*, **21** (2), 54–6.

Zhou, X.D., Burgess, T.I., Beer, Z.W. et al. (2007) High intercontinental migration rates and population admixture in the sapstain fungus *Ophiostoma ips*. *Molecular Ecology*, **16** (1), 89–99.

Zhou, X. & Carter, N. (1992) Effects of temperature, feeding position and crop growth stage on the population dynamics of the rose-grain aphid, *Metopolophium dirhodum* (Hemiptera: Aphidae). *Annals of Applied Biology*, **121** (1), 27–37.

Zobayed, S.M.A., Afreen, F. & Kozai, T. (2007) Phytochemical and physiological changes in the leaves of St. John's wort plants under a water stress condition. *Environmental and Experimental Botany*, **59**, 109–16.

Zucker, W.V. (1983) Tannins: does structure determine function? An ecological perspective. *American Naturalist*, **121**, 335–65.

TAXONOMIC INDEX

Page numbers in **bold** refer to tables and page numbers in *italic* refer to figures

SUBJECT INDEX

Page numbers in **bold** refer to tables and page numbers in *italic* refer to figures